ALGEBRA AND TRIGONOMETRY WITH APPLICATIONS

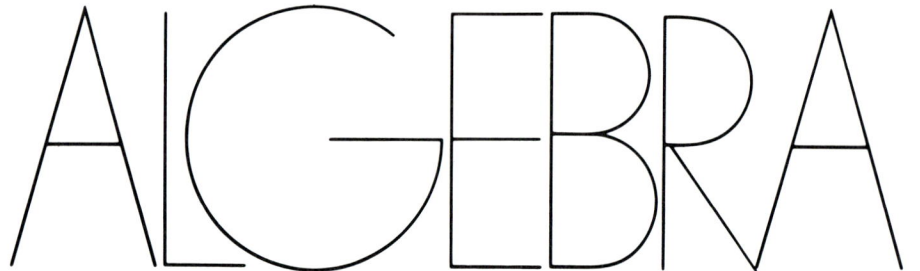

Jagdish C. Arya & Robin W. Lardner

Department of Mathematics, Simon Fraser University

Prentice-Hall, Inc., Englewood Cliffs, New Jersey 07632

& TRIGONOMETRY WITH APPLICATIONS

Library of Congress Cataloging in Publication Data

ARYA, JAGDISH C.
    Algebra and trigonometry with applications.

    Includes index.
    1. Algebra.   2. Trigonometry.   I. Lardner, Robin W.
II. Title.
QA154.2.A78   1983   512'.13   82-13179
ISBN 0-13-021675-5

To AMIT

Cover photo:
Golden Gate Bridge, San Francisco, California.
Photo by R. Krubner supplied by H. Armstrong Roberts.

ALGEBRA AND TRIGONOMETRY WITH APPLICATIONS
by Jagdish C. Arya and Robin W. Lardner

© 1983 by PRENTICE-HALL, INC., Englewood Cliffs, N.J. 07632

All rights reserved. No part of this book may be reproduced in any form or by any means without permission in writing from the publisher.

Printed in the United States of America

Production: Nicholas Romanelli
Design: Walter Behnke / Nicholas Romanelli
Manufacturing buyer: John Hall

ISBN 0-13-021675-5

Prentice-Hall International, Inc., *London*
Prentice-Hall of Australia Pty. Limited, *Sydney*
Editora Prentice-Hall do Brasil, Ltda., *Rio de Janeiro*
Prentice-Hall of Canada, Inc., *Toronto*
Prentice-Hall of India Private Limited, *New Delhi*
Prentice-Hall of Japan, Inc., *Tokyo*
Prentice-Hall of Southeast Asia Pte. Ltd., *Singapore*
Whitehall Books Limited, *Wellington, New Zealand*

CONTENTS

PREFACE xv

1 BASIC ALGEBRA OF REAL NUMBERS 1

A NOTE TO THE STUDENT 1
1.1 Number Systems 2
  THE REAL NUMBERS 3
  THE NUMBER LINE 3
  PRIME NUMBERS 5
1.2 Inequality Symbols and Absolute Values 6
  INEQUALITY SYMBOLS 6
  ABSOLUTE VALUES 8
1.3 Properties of Real Numbers 10
  COMMUTATIVE PROPERTY OF ADDITION 11
  COMMUTATIVE PROPERTY OF MULTIPLICATION 11
  ASSOCIATIVE PROPERTY OF ADDITION 11
  ASSOCIATIVE PROPERTY OF MULTIPLICATION 12
  DISTRIBUTIVE PROPERTIES 13
  IDENTITY ELEMENTS AND INVERSES 14

1.4 Fractions   17
MULTIPLICATION OF FRACTIONS   18
DIVISION OF FRACTIONS   18
CANCELLATION OF COMMON FACTORS   19
1.5 Addition and Subtraction of Fractions   22
1.6 Exponents   27
1.7 Radicals   33
RATIONALIZING THE DENOMINATOR   36
1.8 Fractional Exponents   38
CHAPTER REVIEW 1   43
REVIEW EXERCISES ON CHAPTER 1   44

2 ALGEBRAIC OPERATIONS   45

2.1 Algebraic Expressions   45
ADDITION AND SUBTRACTION OF EXPRESSIONS   46
MULTIPLICATION BY MONOMIALS   48
DIVISION BY MONOMIALS   49
NESTED PARENTHESES   50
2.2 Products of Multinomial Expressions   52
SPECIAL PRODUCTS   54
MORE ON RATIONALIZING DENOMINATORS   56
2.3 Factoring   59
MONOMIAL FACTORS   59
DIFFERENCE OF SQUARES   60
GROUPING   61
SUM AND DIFFERENCE OF TWO CUBES   62
SUMMARY OF FACTORING   63
2.4 Factoring Quadratic Expressions   65
2.5 Algebraic Fractions   70
I. SIMPLICATION OF FRACTIONS   70
II. ADDITION AND SUBTRACTION OF FRACTIONS   71
III. MULTIPLICATION OF FRACTIONS   74
IV. DIVISION OF FRACTIONS   74
CHAPTER REVIEW 2   77
REVIEW EXERCISES ON CHAPTER 2   77

3 EQUATIONS IN ONE VARIABLE   80

3.1 Solutions of Equations; Linear Equations   80
SOLUTIONS OF EQUATIONS   80
LINEAR EQUATIONS   84
IDENTITIES   87

3.2 *Applications of Linear Equations* 89 (REDO THIS SEC.)
    INTEREST PROBLEMS 93
    DISTANCE PROBLEMS 95
    WORK PROBLEMS 96
3.3 *Quadratic Equations (Solution by Factoring)* 99
3.4 *Quadratic Formula and Applications* 106
3.5 *Fractional and Irrational Equations* 116
    CHAPTER REVIEW 3 123
    REVIEW EXERCISES ON CHAPTER 3 124

4 INEQUALITIES 126

4.1 *Sets and Intervals* 126
    INTERVALS 129
4.2 *Linear Inequalities in One Variable* 132
4.3 *Quadratic Inequalities in One Variable* 139
4.4 *Absolute-value Equations and Inequalities* 146
    CHAPTER REVIEW 4 151
    REVIEW EXERCISES ON CHAPTER 4 151

5 STRAIGHT LINES 153

5.1 *Cartesian Coordinates* 153
5.2 *Graphs of Equations* 163
    TESTS FOR SYMMETRY 167
    ALTERNATIVE TEST FOR SYMMETRY ABOUT AN AXIS 168
5.3 *Straight Lines and Linear Equations* 171
    SLOPE 171
    POINT-SLOPE FORMULA 174
    GRAPHING BY SLOPE 176
    OTHER EQUATIONS OF STRAIGHT LINES 177
5.4 *More on Straight Lines* 181
    PERPENDICULAR LINES 183
    SOME APPLICATIONS 186
5.5 *Systems of Linear Equations in Two Unknowns* 190
    MARKET EQUILIBRIUM 198
    CHAPTER REVIEW 5 202
    REVIEW EXERCISES ON CHAPTER 5 202

6 FUNCTIONS AND GRAPHS 205

6.1 Functions 205
 GRAPHS OF FUNCTIONS 208
 VARIATION 213
6.2 Some Special Functions 218
 CONSTANT FUNCTIONS AND LINEAR FUNCTIONS 218
 POWER FUNCTIONS 219
 ABSOLUTE-VALUE FUNCTIONS 222
 SOME CLASSES OF FUNCTIONS 223
6.3 Combinations of Functions 226
 COMPOSITION OF TWO FUNCTIONS 228
6.4 Functions as Mappings; Inverse Functions 232
6.5 Horizontal and Vertical Asymptotes 239
 CHAPTER REVIEW 6 247
 REVIEW EXERCISES ON CHAPTER 6 248

7 CONICS 250

7.1 The Circle 251
7.2 The Parabola 257
 MAXIMUM AND MINIMUM VALUES 263
 HORIZONTAL PARABOLAS 266
7.3 The Ellipse and Hyperbola 269
 THE ELLIPSE 269
 THE HYPERBOLA 273
7.4 General Equation of Conics 280
 CHAPTER REVIEW 7 284
 REVIEW EXERCISES ON CHAPTER 7 284

8 EXPONENTIALS AND LOGARITHMS 287

8.1 Exponential Functions 287
 SOME APPLICATIONS 290
8.2 Logarithms 297
 LOGARITHMIC FUNCTIONS 299
 ELEMENTARY PROPERTIES 301
 LOGARITHMIC EQUATIONS 301
8.3 Properties of Logarithms 304
 COMMON LOGARITHMS 305
 NATURAL LOGARITHMS 308
 PROOFS OF THE BASIC PROPERTIES OF LOGARITHMS 309

    8.4 *The Use of Common Logarithms* *311*
        COMMON LOGARITHM TABLES 311
        MULTIPLICATION AND DIVISION USING LOGARITHMS 313
        POWERS, ROOTS, AND EXPONENTS USING LOGARITHMS 316
    8.5 *Applications of Logarithms* *319*
        EXPONENTIAL EQUATIONS 319
        RADIOACTIVE DECAY 321
        CONTINUOUS COMPOUNDING OF INTEREST 323
        LOGARITHMIC SCALES 325
    8.6 *Base-change Formulas* *330*
        CHAPTER REVIEW 8 335
        REVIEW EXERCISES ON CHAPTER 8 335

9 POLYNOMIALS 338

    9.1 *Division of Polynomials* *338*
        SYNTHETIC DIVISION 341
    9.2 *Complex Numbers* *346*
        ROOTS OF QUADRATIC EQUATIONS 346
        COMPLEX NUMBERS 349
        CONJUGATE COMPLEX NUMBERS 353
    9.3 *Factorization Theory* *356*
        COMPLETE FACTORIZATION 359
    9.4 *Polynomials with Real Coefficients* *364*
        POLYNOMIALS WITH INTEGER COEFFICIENTS 367
        DESCARTES' RULE OF SIGNS 370
        APPROXIMATE CALCULATION OF ZEROS 371
        CHAPTER REVIEW 9 375
        REVIEW EXERCISES ON CHAPTER 9 376

10 TOPICS IN ALGEBRA 378

    10.1 *Arithmetic Progressions* *378*
        THE nTH TERM OF AN A.P. 379
        SUM OF n TERMS OF AN A.P. 381
    10.2 *Geometric Progressions* *386*
        THE nTH TERM OF A G.P. 386
        SUM OF n TERMS OF A G.P. 388
        SUM OF AN INFINITE G.P. 390
    10.3 *Mathematical Induction* *393*
    10.4 *Permutations and Combinations* *398*
        FUNDAMENTAL PRINCIPLE OF COUNTING 399
    10.5 *The Binomial Theorem* *409*
        PROOF OF BINOMIAL THEOREM 413
        CHAPTER REVIEW 10 415
        REVIEW EXERCISES ON CHAPTER 10 416

11 MATRICES AND LINEAR EQUATIONS 419

11.1 Solution of Linear Systems by Row Reduction 419
11.2 Matrices 430
  SOME SPECIAL MATRICES 432
  EQUAL MATRICES 433
  SCALAR MULTIPLICATION OF A MATRIX 433
  ADDITION AND SUBTRACTION OF MATRICES 434
11.3 Products of Matrices 438
  SYSTEMS OF LINEAR EQUATIONS 443
11.4 Inverse of a Matrix 448
11.5 Determinants and Cramer's Rule 456
  PARENTHETICAL REMARK 459
  CRAMER'S RULE 460
  CHAPTER REVIEW 11 465
  REVIEW EXERCISES ON CHAPTER 11 465

12 TRIGONOMETRY 468

12.1 Angles 468
12.2 Right-triangle Trigonometry 476
  TRIGONOMETRIC FUNCTIONS OF SPECIAL ANGLES 479
  TABLES OF TRIGONOMETRIC FUNCTIONS 482
  CALCULATORS 484
12.3 Applications 487
12.4 Trigonometric Functions 492
  TRIGONOMETRIC FUNCTIONS OF AXIS ANGLES 496
  REFERENCE ANGLE 497
12.5 Graphs of Trigonometric Functions 503
  GRAPH OF $y = \cos x$ 504
  GRAPH OF $y = \tan x$ 508
  GRAPHS OF COTANGENT, SECANT, AND COSECANT FUNCTIONS 509
  CHAPTER REVIEW 12 512
  REVIEW EXERCISES ON CHAPTER 12 512

13 TRIGONOMETRIC IDENTITIES 514

    13.1 Fundamental Identities 514
        TRIGONOMETRIC FUNCTIONS OF $(\pi/2 - \theta)$ 520
    13.2 The Addition Formulas 522
        FORMULAS FOR ALLIED ANGLES 526
        COMBINATIONS OF SINE AND COSINE 528
    13.3 Multiple-angle Formulas 532
    13.4 The Inverse Trigonometric Functions 540
        CHAPTER REVIEW 13 549
        REVIEW EXERCISES ON CHAPTER 13 549

14 TRIGONOMETRIC APPLICATIONS 552

    14.1 The Law of Sines 552
    14.2 The Law of Cosines 560
        AREA OF A TRIANGLE 565
    14.3 Polar Coordinates and the Complex Plane 568
        THE COMPLEX PLANE 571
        MULTIPLICATION AND DIVISION 573
    14.4 De Moivre's Theorem 576
        CHAPTER REVIEW 14 580
        REVIEW EXERCISES ON CHAPTER 14 581

Appendix Tables:
    1 FOUR-PLACE COMMON LOGARITHMS 584
    2 NATURAL LOGARITHMS 586
    3 EXPONENTIAL FUNCTIONS 588
    4 TRIGONOMETRIC FUNCTIONS 590

Answers to Tests and Odd-numbered Exercises A-1

Index A-33

PREFACE

This book is designed for use in first-year college and university courses on algebra and pre-calculus mathematics. The orientation and level of approach reflect the needs and interests of most present-day college entrants. The emphasis is towards problem solving and applications rather than pure mathematics. While many theorems are stated, explained, and used, relatively few of them are proved. On the other hand, the development of mathematical skills is a primary goal and, in particular, the solution of "word problems" is stressed through both worked examples and the large number of exercises of this type.

The applications range from the physical and life sciences to the social sciences and business and economics. However, a larger proportion of them than usual is drawn from business and economics, reflecting the growing importance of mathematics in these fields.

The book has been written primarily for the student. Our aim has been to produce a book which an average student working on his own can read and understand. Thus, the discussions contain more explanatory points, and the solutions of examples are given in greater detail than in most books covering this area. Our thought has been that it is incomparably easier for the student (or instructor) who does not need it to skip over unnecessary detail than it is for the less prepared student to fill in detail which is missing.

In line with this general philosophy, we have also included in the first two chapters an extensive review of basic algebra. This material is definitely at a more basic level than the rest of the book, and in many cases an instructor will find that it can be covered very rapidly or perhaps even omitted. In most cases, however, we anticipated that a fairly complete coverage of these introductory chapters will be preferred.

In an attempt to escape from the stuffiness and formality of the old-style mathematics books, many current writers of college textbooks adopt a style which is so light and casual that mathematical clarity is often obscured. We firmly believe that a careful balance between informality and seriousness is essential, and that the pre-eminent requirement of a textbook must be clear presentation of the mathematics. For this reason you will find these pages unlittered by anecdotes or Charlie Brown cartoons or like gimmicks. We have tried to write all signal, no noise.

Calculators. The use of calculators in general is only to be encouraged. Nevertheless, from the point of view of learning basic algebraic techniques, they can play only a very limited role. Consequently, we have placed little emphasis on calculators in the first half-dozen chapters. However, in later chapters (for exponentials and logarithms, to some extent for roots of polynomials, and for the trigonometric functions) they become a useful aid indeed, and their use is referred to repeatedly in both the discussions and examples. (Ample tables are provided for those who wish to avoid calculators.)

Exercises. At the end of each section two exercise sets are given. The first, called a Diagnostic Test, contains a number of short questions, each addressing a single point and intended to be answered quickly in the student's head. The second is a set of longer exercises of the regular kind, usually consisting of a mixture of purely algebraic exercises and word problems. The more difficult exercises are indicated by asterisks. It hardly needs to be repeated that the only way to develop mathematical skills is to do lots of exercises.

Chapter Reviews. The Review at the end of each chapter consists of two parts. First, a summary of the topics covered in the chapter is given in a form which asks the student to provide an explanation or description of each topic. This forces the student into an active learning mode rather than the passive mode which would be induced by a simple list of topics. We strongly urge any student to use this learning tool seriously. Following this a set of exercises is given of the regular type covering the whole chapter.

Acknowledgements. It is a pleasure to acknowledge our indebtedness to those colleagues who have made suggestions for improving the book. Our appreciation and thanks go to Professors G. D. Allen, Texas A & M University; James E. Arnold, Jr., University of Wisconsin & Milwaukee; Thomas Bowman, University of Florida; Dennison R. Brown, University of Houston; William J. Gordon, State University of New York at Buffalo; Steven K. Ingram, Norwich University; John Kuisti, Michigan Technological University; William Ramaley, Fort Lewis College, and Carroll G. Wells, Western Kentucky University.

The solution of every worked example in the book has been checked by Dr. Gary Nicklason, and we are indebted to him for the thorough job he has done.

J.C.A., R.W.L.

ALGEBRA AND TRIGONOMETRY WITH APPLICATIONS

1 BASIC ALGEBRA OF REAL NUMBERS

A NOTE TO THE STUDENT

This book's readers bring to it a wide variety of mathematical backgrounds. For this reason it begins at a fairly basic level, and the first two chapters give a reasonably thorough coverage of introductory algebra.

If you find that you are already familiar with much of the material in Chapters 1 and 2, we suggest that you review this material rapidly as follows: First skim through each section, reading the *definitions* and the main results, which are *highlighted in boxes*. Study the *solved examples* in the section. Then answer the diagnostic test and work through selected exercises on that section. If you find you are making mistakes, a more thorough review is called for. Answering the Chapter Review will really test your knowledge of the material in the chapter. Further review exercises are also available at the end of the chapter if needed. Bon voyage.

1.1 Number Systems

Of the several types of numbers with which we deal, the simplest are the counting numbers—1, 2, 3, 4, and so on. These are called *natural numbers*.

If we add or multiply any two natural numbers, the result is always a natural number. For example, $3 + 5 = 8$ and $3 \cdot 5 = 15$;* both results, 8 and 15, are natural numbers. If, however, we subtract one natural number from another, the result is sometimes not a natural number. For example, $8 - 5 = 3$ is a natural number, but $5 - 8$ is not. Similarly if we divide one natural number by another, the result is not always a natural number.

To overcome the limitation on subtraction, we extend the natural-number system to the system of *integers*. This we do by including, together with all the natural numbers, all their negatives and the number 0 (zero). Thus we may represent the system of integers in the form

$$\ldots, -3, -2, -1, 0, 1, 2, 3, \ldots$$

Clearly, all the natural numbers are also integers.

If we now add, multiply, or subtract any two integers, the result will always turn out to be an integer. For example, we have the following:

$$8 + (-2) = 6, \qquad 2 - (-8) = 10$$
$$-5 - (-2) = -3, \qquad (5)(-3) = -15$$
$$(-2)(-4) = 8, \qquad (-3)(0) = 0.$$

In each case the answer is an integer. If, however, we divide one integer by another, the result is not always an integer. For example, $8 \div (-2)$ is an integer, namely -4, but $8 \div (-3)$ is not an integer.

To overcome this limitation on division, we extend the system of integers to the system of rational numbers. A number is defined to be a *rational number* if it can be expressed as the ratio of two integers with denominator nonzero. Thus, for example, the numbers $8/3$, $-5/7$, $-15/4$, and $\frac{9}{4}$ are all rational numbers. Note that any integer can also be regarded as a rational number with denominator equal to 1. (For example $6 = 6/1$ and $-3 = -3/1$.)

We can add, multiply, subtract, or divide any two rational numbers (except that division by zero is excluded), and the result is always a rational

---

*We shall use a dot to denote the product of two numbers. Alternatively the two numbers may be placed inside parentheses with no dot. Thus, for example, $6 \cdot 8$ and $(6)(8)$ both denote the product of 6 and 8.

number. Thus within the system of rational numbers all the four fundamental operations of arithmetic—addition, multiplication, subtraction, and division—are allowed.

Division by zero must be excluded for the following reason. The statement $a/b = c$ is true if and only if the inverse statement $a = bc$ is true. Now consider a fraction in which the denominator b is zero—for example, $\frac{3}{0}$. This cannot equal any number c, because the inverse statement $3 = (0)c$ cannot be true for any c. Therefore $\frac{3}{0}$ does not exist. Further, the fraction $\frac{0}{0}$ is not a number because the inverse statement $0 = (0)c$ is true for all numbers c. Thus we conclude that any fraction with a zero denominator is not a well-defined number or, equivalently, *division by zero is a meaningless operation*.

If you do not feel completely confident about calculating with negative numbers, it would be a good idea, before going any further, to work Exercises 1 through 24 at the end of this section.

THE REAL NUMBERS

When any rational number is expressed as a decimal, the decimal either terminates or else develops a pattern that repeats indefinitely. For example, $\frac{1}{4} = 0.25$ and $\frac{93}{80} = 1.1625$ both correspond to decimals that terminate, whereas $\frac{1}{60} = 0.016666\ldots$ and $\frac{3}{7} = 0.428571428571\ldots$ correspond to decimals with repeating patterns.

Many numbers in common use are not rational; that is, they cannot be expressed as the ratio of two integers. For example, $\sqrt{2}$, $\sqrt{3}$, and π are not rational numbers. Such numbers are called **irrational numbers**. We can see the essential difference between rational and irrational numbers in their decimal expressions. When an irrational number is represented by a decimal, the decimal never terminates, nor does it develop a repeating pattern. For example, $\sqrt{2} = 1.4142135623\ldots$ and $\pi = 3.1415926535\ldots$ to ten decimal places. The decimal expressions of these numbers never develop a repeating pattern, no matter to how many decimal places we expand them. Compare this with the form of decimals of rational numbers.

The term **real number** is used to mean a number that is either rational or irrational. The system of real numbers can be regarded as consisting of all possible decimals. Decimals that terminate or repeat correspond to the rational numbers; the rest correspond to the irrational numbers.

THE NUMBER LINE

We can represent the real numbers geometrically by the points on a horizontal straight line. We first select some point O on the line to represent the number 0 (zero) and also choose a unit of length along the line. Then the positive numbers are represented by the points that lie to the right of O and the negative numbers by the points to the left. If A_1 is a point to the right of O

such that OA_1 is of unit length,* then A_1 represents the number 1. The integers $2, 3, \ldots, n, \ldots$ are represented by the points $A_2, A_3, \ldots, A_n, \ldots$ on the right of O such that

$$OA_2 = 2OA_1, \quad OA_3 = 3OA_1, \quad \ldots, \quad OA_n = nOA_1, \quad \ldots.$$

```
       -n         -3 -2 -1  0  1  2  3         n
   ────┼──────────┼──┼──┼──┼──┼──┼──┼──────────┼────
    ···Bₙ ··· B₃ B₂ B₁  0  A₁ A₂ A₃···        Aₙ···
```

Similarly if $B_1, B_2, B_3, \ldots, B_n, \ldots$ are the points to the left of O such that the distances $OB_1, OB_2, OB_3, \ldots, OB_n, \ldots$ are respectively equal to the distances $OA_1, OA_2, OA_3, \ldots, OA_n, \ldots$, then these points represent the negative integers $-1, -2, -3, \ldots, -n, \ldots$. In this way all the integers can be represented by points on the line. The point O is called the **origin**.

Rational numbers can be represented by points on the number line that lie an appropriate fractional number of units from O. For example, the number $\frac{9}{2}$ is represented by the point that lies four and a half units to the right of O, and $-\frac{7}{3}$ is represented by the point that lies a distance of two and a third units to the left of O.

In general, to represent the rational number p/q, where p and q are positive integers, we proceed as follows. First we divide the unit interval OA_1 into q subintervals of equal length. Then we take the point that lies a distance equal to p times the length of one of these subintervals to the right of O. To represent the negative rational number $-p/q$ we take the point at the same distance to the left of O.

It turns out that every *irrational* number can also be represented by a point on the line. We do not intend to prove this fact, but we can easily demonstrate how the particular irrational number $\sqrt{2}$ can be so represented. As in the adjacent figure, draw a vertical line through A_1 and mark off a point P whose distance from A_1 is 1 unit. Then the triangle OPA_1 has a right angle at A_1 and $OA_1 = PA_1 = 1$. By Pythagoras's theorem, $OP^2 = OA_1^2 + PA_1^2 = 1^2 + 1^2 = 1 + 1 = 2$. Hence $OP = \sqrt{2}$. Now we draw a circle with center O and radius equal to OP. Let this cut the horizontal line at Q (lying between A_1 and A_2). Then $OQ = OP = \sqrt{2}$, so the point Q must represent the irrational number $\sqrt{2}$.

We have then that every real number, rational or irrational, can be represented by a point on the horizontal line. Furthermore, each point on the line corresponds to one and only one real number. A horizontal line used in this way to represent the real numbers is called the **real-number line**. Because of this correspondence it is quite common to use the word "point" to mean "real number."

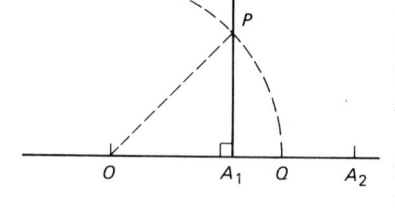

---

*The notation OA_1 is used to denote the distance between O and A_1. In general if P and Q are any two geometrical points, PQ denotes the distance between them.

PRIME NUMBERS

If the product of two natural numbers a and b is equal to c, that is, $c = ab$, then a and b are called *factors* of c. In other words, a natural number a is a factor of another natural number c if a divides c exactly with no remainder. For example, 2 and 3 are factors of 6; and 1, 2, 3, 4, and 6 are all factors of 12.

A natural number greater than 1 is said to be *prime* if its only factors are 1 and itself. For example, the only factors of 7 are 1 and 7, and so 7 is a prime number. On the other hand, 9 is not a prime number, because it has 3 as a factor as well as 1 and 9. The prime numbers less than 20 are 2, 3, 5, 7, 11, 13, 17, and 19.

Any natural number that is not itself prime can be expressed as the product of two or more factors that are prime numbers. For example, we can write

$$6 = (2)(3), \qquad 12 = (2)(2)(3), \qquad 70 = (2)(5)(7)$$

and so on. In each case the numbers in the product on the right are prime numbers. They are called the *prime factors* of the number on the left. (For example, the prime factors of 12 are 2, 2 and 3.)

Diagnostic Test 1.1

Fill in the blanks.

1. A number is said to be a _____ number, if it can be expressed in the form p/q ($q \neq 0$), where p and q are integers.
2. If a real number cannot be expressed in the form p/q ($q \neq 0$), where p and q are integers, then it is a (an) _____ number.
3. If a number has a terminating or recurring decimal representation, then it is a (an) _____ number.
4. If a number has a nonterminating and nonrecurring decimal representation, then it is a (an) _____ number.
5. A number that is either rational or irrational is a _____ number.
6. The sum of any two rational numbers is a _____ number.
7. The sum of a rational and an irrational number is always a (an) _____ number.
8. $a/a = $ _____ only if a _____.
9. The number $1/b$ is defined only if b _____.
10. The number $\sqrt{-x}$ is a real number only if x _____.

Sec. 1.1   Number Systems

11. The prime factors of 6 are __2, 3__ .

12. The prime factors of a 8 are __2, 2, 2__ .

Exercises 1.1

(1–24) Evaluate the following.

1. $0 - 3$.
2. $0(-3)$.
3. $2 - 5$.
4. $2 + (-5)$.
5. $-5 - 2$.
6. $-5 - (-2)$.
7. $4 - (-3)$.
8. $6 - (-7)$.
9. $-6 - (-7)$.
10. $5 - (-0)$.
11. $6(-0)$.
12. $-4 - 10$.
13. $-2(1 - 3)$.
14. $-4(4 - 1)$.
15. $(-4)(4) - 1$.
16. $-5(2 - 7)$.
17. $(2 - 7) - 5$.
18. $2 - (7 - 5)$.
19. $4 - 2(1 - 6)$.
20. $-3(-2 - 5) - 1$.
21. $-6 - 2(-3 - 2)$.
22. $-4(2 - 4) - (-7)$.
23. $-3(5 - 0) - (-1)$.
24. $-2(0 - 6) - (-2)$.

(25–36) State whether the given number is prime. If it is not, express it as a product of its prime factors.

25. 11. ✓
26. 14. $7 \cdot 2$
27. 20. $2 \cdot 10 = 2 \cdot 2 \cdot 5$
28. 17. ✓
29. 29. ✓
30. 51. $3 \cdot 17$
31. 62. $2 \cdot 31$
32. 41. ✓
33. 60. $2 \cdot 30 = 2 \cdot 15 \cdot 15$
34. 70. $2 \cdot 35$ $2 \cdot 7 \cdot 5$
35. 144. $12 \cdot 12$ $2 \cdot 6 \cdot 2 \cdot 6$
36. 108. $2 \cdot 54 = 2 \cdot 9 \cdot 6$ $2 \cdot 3 \cdot 32 \cdot 1$

1.2

Inequality Symbols and Absolute Values

INEQUALITY SYMBOLS

The symbol $a > 0$ is used to say that the number a is positive. It is read "a is greater than zero." Correspondingly, $b < 0$ is used to say that the number b is negative and it is read "b is less than zero."

If $a > 0$, then a is represented by a point to the right of the origin on the real-number line. If $b < 0$, then b is represented by a point to the left of the origin.

Examples 1 (a) The following statements are true:

$$4 > 0, \quad -4 < 0, \quad -\tfrac{3}{2} < 0, \quad \pi > 0.$$

(b) The following statements are false:

$$-\sqrt{2} > 0, \quad -(-8) < 0, \quad 3 + (-5) > 0.$$

For any two real numbers a and b we say that a is **greater than** b if the difference $a - b$ is positive. We write this $a > b$. If the difference $a - b$ is negative, we say that a is **less than** b, and we write $a < b$.

Examples 2 (a) $7 > 4$ because the difference $7 - 4 = 3$ is positive.
(b) $7 > -4$ because $7 - (-4) = 11$ is positive.
(c) $-2 > -6$ because $-2 - (-6) = 4$ is positive.
(d) $-1 < 4$ because $-1 - 4 = -5$ is negative.
(e) $-10 < -5$ because $-10 - (-5) = -5$ is negative.

To emphasize the definitions:

| $a > b$ | if and only if | $a - b > 0$ |
| $a < b$ | if and only if | $a - b < 0.$ |

If $a > b$, then the point representing a on the number line lies to the right of the point representing b. If $a < b$, then the point representing a lies to the left of the point representing b. These relationships are illustrated in the figure below.

Clearly the two statements $a > b$ and $b < a$ mean exactly the same thing.

The symbol $a \geq b$ is used to indicate that either $a > b$ or $a = b$. Correspondingly $a \leq b$ denotes that either $a < b$ or $a = b$.

Examples 3 The following statements are all true:

$$7 \geq 5, \quad -1 \geq -3, \quad 5 \geq 5, \quad 5 \leq 5.$$

(Note that it would not be correct to write $5 > 5$ or $5 < 5$.)

Examples 4 (a) The statement $x > 9$ means that the number x is greater than 9. The point representing x lies to the right of the point 9 on the number line.
(b) $x \geq 9$ means that x is greater than or equal to 9. The point x either is to the right of the point 9 or coincides with that point.
(c) $x \leq 9$ means that x is less than or equal to 9. The point x either is to the left of the point 9 or coincides with it.

The double inequality $a < b < c$ is used to denote that $a < b$ and $b < c$ both hold. Similarly $a > b > c$ denotes that both $a > b$ and $b > c$. In both cases the point b on the number line is between the points a and c. It may be remarked that we never use $>$ or $\geq$ together with $<$ or $\leq$ in the same double inequality. Thus, we shall never write $3 \leq 7 > 4$ or $7 > 3 < 5$. *In a double inequality, the inequality signs must open in the same direction.*

Sec. 1.2   Inequality Symbols and Absolute Values

Examples 5 (a) The following statements are true:
$$0 < 1 < 2, \quad -\tfrac{3}{2} < -1 < -\tfrac{1}{2}, \quad 3 > -1 > -3.$$
(b) The statement $1 < x < 2$ says that the number x lies between 1 and 2.
(c) The statement $-\tfrac{3}{4} < a \le \sqrt{2}$ states that the number a lies between $-\tfrac{3}{4}$ and $\sqrt{2}$ or is equal to $\sqrt{2}$.

ABSOLUTE VALUES

DEFINITION The ***absolute value*** of a real number x, denoted by $|x|$, is defined to be the distance (number of units) between the origin and the point x on the number line.

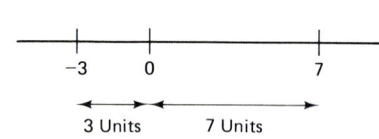

For example, the point 7 lies seven units to the right of O, so that $|7| = 7$. The point -3 lies 3 units to the left of O, so that $|-3| = 3$. The distance between O and O is 0, so that $|0| = 0$.

Clearly, if x is any positive number or zero, the absolute value of x is equal to x itself: if $x \ge 0$, $|x| = x$. If x is negative, then the absolute value of x is equal to the magnitude of x regardless of its negative sign. We obtain this magnitude by removing the negative sign from x. But the result of doing this is equal to $(-x)$. For example, take $x = -3$. Then $-x = -(-3) = 3$, which equals $|-3|$. In general, therefore, if $x < 0$, $|x| = -x$. To summarize:

$$\boxed{\begin{array}{ll} |x| = x & \text{if } x \ge 0 \\ |x| = -x & \text{if } x < 0. \end{array}}$$

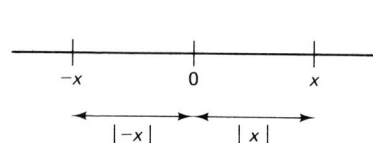

Note that the two real numbers x and $-x$ always lie at the same distance from the origin. This is illustrated in the adjacent figure for the case when x is positive. Thus we have

$$\boxed{|-x| = |x|}$$

for any real number x. For example, when $x = 7$, $|x| = |7| = 7$ and $|-x| = |-7| = 7$.

If x and y are any two real numbers, then $|x - y|$ has the geometrical interpretation that it is the distance (number of units) between the points x and y on the number line. For example, take $x = 2$ and $y = 5$. Then
$$|x - y| = |2 - 5| = |-3| = 3.$$

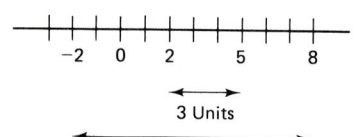

Clearly 2 and 5 are 3 units apart on the number line (see the adjacent figure). Or, taking $x = 8$ and $y = -2$, we have
$$|x - y| = |8 - (-2)| = |10| = 10.$$
Again it is clear from the figure that the distance between -2 and 8 is 10 units.

Note that it doesn't matter which number we call x and which we call y, because $|x - y|$ and $|y - x|$ are always equal to one another.

Examples 6 Express the following statements using absolute-value symbols.
- (a) x is 5 units distant from the origin.
- (b) The distance between x and 4 is equal to 3 units.
- (c) x is less than 3 units from the origin.
- (d) The distance between x and -1 is equal to twice the distance between x and -5.
- (e) x is more than 5 units from the point 3.
- (f) x is less than d units from the point c.
- (g) x is at least d units from the point c.

Solutions
- (a) $|x| = 5$.
- (b) $|x - 4| = 3$ (since the distance between x and 4 is $|x - 4|$).
- (c) $|x| < 3$.
- (d) The distance between x and -1 is $|x - (-1)| = |x + 1|$. The distance between x and -5 is $|x - (-5)| = |x + 5|$. Therefore
$$|x + 1| = 2|x + 5|.$$
- (e) The distance $|x - 3|$ is greater than 5: $|x - 3| > 5$.
- (f) The distance $|x - c|$ is less than d: $|x - c| < d$.
- (g) The distance $|x - c|$ is greater than or equal to d: $|x - c| \geq d$.

Diagnostic Test 1.2

Fill in the blanks.

1. $x < 0$ means that x is a _____ number.
2. If $x - 4 > 0$, then x _____ 4.
3. If $x \leq 0$, then either x _____ 0 or x _____ 0.
4. If $a \geq 4$ and $a \leq 4$, then a _____ 4.
5. $|x| = $ _-x ; 0 -(-x)=x_ when x is negative.
6. $|-x| = $ _-x>0_ when x is negative.
7. $|x| > x$ only if x is _____.
8. $|x| \geq -x$ for _____ x.
*9. $|x|/x = $ _____ if $x < 0$.
10. If the distance between x and -3 is greater than 5 units, then $|$_____$| > 5$.

Exercises 1.2

(1–18) Fill in the blank with either $>$, $<$, or $=$.

1. 10 ____ 4.
2. 3 ____ 13.
3. -1 ____ -3.
4. -2 ____ 1.
5. -5 ____ 4.
6. -6 ____ -4.
7. 3 ____ $|3|$.
8. 7 ____ $|4|$.
9. -3 ____ $|-3|$.

10. $|-4|$ _____ $|4|$. 11. $|-6|$ _____ $|-4|$. 12. $|-3|$ _____ $|-6|$.
13. $3 - |3|$ _____ 0. 14. $-6 - |-6|$ _____ 0. 15. $-5 + |-5|$ _____ 0.
16. $|-4| + |2|$ _____ $|-2|$. 17. $|3| + |2|$ _____ $|5|$. 18. $|-4| + |-2|$ _____ $|-6|$.

(19–26) State whether each inequality is true or false.

19. $-2 < -5$. 20. $-6 > -4$. 21. $|-3| \le 3$.
22. $|-4| \ge |4|$. 23. $|-5| + |-2| \ge |-5 - 2|$.
24. $|5| + |-3| \ge |5 - 3|$. 25. $|x| \ge x$ for any x.
26. $|x| \ge -x$ for any x.

(27–42) Express the following statements using inequality symbols and/or absolute values.

27. -1 is less than $+1$. 28. -3 is greater than -6.
29. x is greater than -5. 30. y is less than -2.
31. z is greater than 4 and less than 6. 32. x is greater than -1 and less than 3.
33. y is greater than -2 and less than -1. 34. z is greater than 6 and less than 15.
35. p is not greater than 2. 36. q is not less than -3.
37. x is less than 2.5 units from the origin. 38. x is more than 6 units from the origin.
39. The distance between x and -2 is less than or equal to 3 units. 40. The distance between x and 4 is less than or equal to 2 units.
41. The distance between x and -6 is half the distance between x and $+6$. 42. The distance between y and 5 is three times the distance between y and -3.

(43–44) Arrange the numbers in each list in order of increasing absolute value.

43. $5, -1, -6, 3, -\tfrac{3}{2}, -\sqrt{2}$. 44. $-\pi, 3.14, -3.142, -4, -3$.

(45–46) Arrange the numbers in each list in order of increasing distance from -3.

45. $-1, -2, 3, -4.5, -5.5$. 46. $-10, 10, -5, 5, -3$.

(47–52) Evaluate the following.

47. $|-3| - |-6|$. 48. $|-3 - (-6)|$. 49. $|-4| \div (-4)$.
50. $|-7| \div (-7)$. 51. $|-3|^2 - (-3)^2$. 52. $|-6|^2 - (-6)^2$.

1.3

Properties of Real Numbers

When two real numbers are added or multiplied together, the result is always a real number. These two operations of addition and multiplication are fundamental to the system of real numbers, and they possess certain properties, which we shall now discuss. These properties by themselves may appear rather obvious and elementary, but they are vital to our understanding the various algebraic manipulations with which we shall later be involved.

COMMUTATIVE PROPERTY OF ADDITION

When two real numbers are added together, the order in which they are added does not matter. For example, $3 + 7$ and $7 + 3$ give the same result: 10. Similarly $4 + (-9) = (-9) + 4$, since both sides are equal to -5. In general, if a and b are any two real numbers, then

$$a + b = b + a.$$

This is known as the **commutative property of addition**.

COMMUTATIVE PROPERTY OF MULTIPLICATION

In a similar way, when two real numbers are multiplied together, the order in which they are multiplied is immaterial. For example, $(3)(7) = (7)(3)$ (both equal to 21), and $4(-9) = (-9)4$ (both equal to -36). In general, if a and b are any two real numbers, then

$$ab = ba.$$

This is known as the **commutative property of multiplication**.

Examples 1 If x and y are any two real numbers, then on the basis of the two commutative properties the following equalities are true.
(a) $x + 2 = 2 + x$.
(b) $2x + 3y = 3y + 2x$.
(c) $(-2)y = y(-2)$.
(d) $(4x)y = y(4x)$.

ASSOCIATIVE PROPERTY OF ADDITION

When three real numbers are to be added together, they can only be added two at a time. It does not matter, however, which pair are added together first. For example,

$$(2 + 3) + 7 = 5 + 7 = 12 \quad \text{and} \quad 2 + (3 + 7) = 2 + 10 = 12$$

both resulting in the same answer. Thus $(2 + 3) + 7 = 2 + (3 + 7)$. This is true for any three real numbers. If a, b, and c stand for any three real numbers, then

$$(a + b) + c = a + (b + c).$$

This is called the **associative property of addition**.

Sec. 1.3   Properties of Real Numbers

ASSOCIATIVE PROPERTY OF MULTIPLICATION

A similar property holds for multiplication. When three numbers are multiplied together, it does not matter which pair are multiplied first. If a, b, and c are any three real numbers, then

$$(ab)c = a(bc).$$

For example, taking $a = 2$, $b = -3$, and $c = 7$, we get that

$$(ab)c = [2(-3)](7) = (-6)(7) = -42$$
$$a(bc) = 2[(-3)(7)] = 2(-21) = -42,$$

verifying the stated property. This is known as the **associative property of multiplication**.

Because of these associative properties it is not necessary to write parentheses in expressions such as $(a + b) + c$ and $(ab)c$. We can simply write $a + b + c$ for the sum of a, b, and c and abc for their product, and there will be no ambiguity.

Examples 2   If x is any real number, then the following are true.
(a) $(x + 3) + 4 = x + (3 + 4)$ (associative property of addition)
                 $= x + 7$.
(b) $2(3x) = (2 \cdot 3)x$ (associative property of multiplication)
               $= 6x$.

By combining the commutative and associative properties we can simplify many other algebraic expressions.

Examples 3   (a) $(3 + x) + 4 = (x + 3) + 4$ (commutative property)
                     $= x + 7$ [as in Example 2(a)]
(b) $(-2 + x) + (2y + 5)$
                $= [x + (-2)] + (5 + 2y)$ (commutative property)
                $= x + [(-2) + (5 + 2y)]$ (associative property)
                $= x + [(-2 + 5) + 2y]$ (associative property)
                $= x + (3 + 2y)$
                $= x + (2y + 3)$ (commutative property)
                $= x + 2y + 3$.

In expressions like the one above that involve the sum of a number of expressions inside parentheses, we can avoid going through all the steps above by simply removing the parentheses and reordering the terms. The next example illustrates this procedure.

(c) $(2a + 6) + (4b + 2) = 2a + 4b + 6 + 2$
                                $= 2a + 4b + 8$.
(d)   In a similar way, expressions involving products of several numbers and symbols can be rearranged. For example,

$$(2x)(3x) = [(2x) \cdot 3]x \quad \text{(associative property)}$$
$$= [3 \cdot (2x)]x \quad \text{(commutative property)}$$
$$= [(3 \cdot 2)x]x \quad \text{(associative property)}$$
$$= (6x)x$$
$$= 6(x \cdot x) \quad \text{(associative property)}$$
$$= 6x^2$$

where x^2 denotes $x \cdot x$. Note again that the final answer could be obtained by collecting together the similar terms in the original product: the numbers 2 and 3 multiplied together give the 6 and the two x's multiplied together give the x^2 in the final answer. The next example illustrates this procedure.

(e) $[5(3ab)]2a = (5 \cdot 3 \cdot 2)(a \cdot a)b = 30a^2b$. In order to justify this answer, we could follow through the appropriate sequence of associative and commutative properties as in the preceding example.

DISTRIBUTIVE PROPERTIES

If a, b, and c are any three real numbers, then

$$\boxed{a(b+c) = ab + ac} \quad \text{and} \quad (b+c)a = ba + ca$$

For example, take $a = 2$, $b = 3$, and $c = 7$ in the first of these properties:
$$a(b+c) = 2(3+7) = 2(10) = 20$$
$$ab + ac = (2)(3) + (2)(7) = 6 + 14 = 20.$$

The two sides are clearly equal to one another.

As a second example, take $a = -2$, $b = 3$, and $c = -7$. Then
$$a(b+c) = (-2)[3 + (-7)] = (-2)(-4) = 8$$
$$ab + ac = (-2)(3) + (-2)(-7) = -6 + 14 = 8$$

and again the equality is satisfied.

These two properties are called *distributive properties*. Observe that they involve *both* the operations addition and multiplication. (The earlier commutative and associative properties each involve only one of the operations.) The second of the distributive properties can actually be deduced from the first. (See Exercise 45.)

Examples 4
(a) $x(y + 2) = xy + x(2)$ (distributive property)
$ = xy + 2x$ (commutative property)
(b) $2x + 3x = (2 + 3)x = 5x$ (distributive property)
[*Note:* Compare this with Example 2(b).]
(c) $2x + 5(y + x) = 2x + (5y + 5x)$ (distributive property)
$ = 2x + (5x + 5y)$ (commutative property)
$ = (2x + 5x) + 5y$ (associative property)
$ = (2 + 5)x + 5y$ (distributive property)
$ = 7x + 5y.$

The distributive property extends to the case when more than two quantities are added together in the parentheses. That is,

$$a(b + c + d) = ab + ac + ad$$

and so on. (We leave it as an exercise for the more ambitious reader to prove this. Use only the associative and distributive properties.)

Examples 5 (a) $4(x + 3y + 4z) = 4x + 4(3y) + 4(4z)$ (distributive property)
$ = 4x + (4 \cdot 3)y + (4 \cdot 4)z$ (associative property)
$ = 4x + 12y + 16z.$

(b) $2(x + 2y) + 3(y + 2x + 3z) = 2x + 2(2y) + 3y + 3(2x) + 3(3z)$
(using distributive property twice)
$ = 2x + 4y + 3y + 6x + 9z$
$ = 2x + 6x + 4y + 3y + 9z$
(using the commutative property)
$ = (2 + 6)x + (4 + 3)y + 9z$
(distributive property)
$ = 8x + 7y + 9z.$

IDENTITY ELEMENTS AND INVERSES

If a is any real number, then

$$a + 0 = a \quad \text{and} \quad a \cdot 1 = a.$$

That is, if 0 is added to a, the result is still a; and if a is multiplied by 1, the result is again a. The numbers 0 and 1 are often called the **identity elements** for addition and multiplication, respectively, because they leave any number unchanged under these respective operations. They are the *only* numbers that have these respective properties.

If a is any real number, then there exists a unique real number called the **negative of a** and denoted by **$(-a)$** such that

$$a + (-a) = 0.$$

If a is nonzero, there also exists a unique real number called the **reciprocal of a** and denoted by $1/a$ or $\dfrac{1}{a}$ such that

$$a \cdot \frac{1}{a} = 1.$$

Observe the similarity between these two definitions: when $(-a)$ is added to a, the result is the additive identity element; when $1/a$ is multiplied by a, the result is the multiplicative identity element. We often refer to $(-a)$ as the **additive inverse of a** and to $1/a$ as the **multiplicative inverse of a**. (Sometimes $1/a$ is called just **the inverse of a**.)

Examples 6 (a) $x + 4x = 1x + 4x$
$= (1 + 4)x$ (distributive property)
$= 5x.$

(b) The additive inverse of 3 is -3, since $3 + (-3) = 0$. The additive inverse of -3 is 3, since $(-3) + 3 = 0$. But the additive inverse of -3 is denoted by $-(-3)$, so it follows that $-(-3) = 3$. A corresponding result in fact holds for any real number a:

$$\boxed{-(-a) = a.}$$

(c) The multiplicative inverse of 4 is $\frac{1}{4}$, since $4 \cdot \frac{1}{4} = 1$. The multiplicative inverse of $\frac{1}{4}$ would be denoted by $1/\frac{1}{4}$ and would be defined by the requirement that $\frac{1}{4}(1/\frac{1}{4}) = 1$. But since $\frac{1}{4} \cdot 4 = 1$, it follows that $1/\frac{1}{4}$ is equal to 4. Again this result generalizes to any nonzero real number a:

$$\boxed{\frac{1}{1/a} = a.}$$

(The inverse of the inverse of a is equal to a.)

We *define* the operations of *subtraction* and *divison* by making use of the additive and multiplicative inverses, respectively. We define $a - b$ to mean the number $a + (-b)$ (i.e., a plus the negative of b). And we define $a \div b$ to mean the number $a \cdot (1/b)$ (i.e., a multiplied by the reciprocal of b). This latter is defined only when $b \neq 0$. It is also denoted by the fraction a/b or $\frac{a}{b}$ and we have that

$$\boxed{\frac{a}{b} = a \cdot \frac{1}{b}.}$$ (definition)   (1)

Examples 7 $\frac{7}{(\frac{1}{3})} = 7 \cdot \frac{1}{\frac{1}{3}}$ using (1) with $a = 7$, $b = \frac{1}{3}$
$= 7(3)$
$= 21.$

This result extends to any pair of real numbers a and b ($b \neq 0$):

$$\boxed{\frac{a}{(1/b)} = ab.}$$

If 3 is multiplied by (-1), the result is -3, the additive inverse of 3. This result is true for any real number b—that multiplication by -1 produces the additive inverse:

$$\boxed{(-1)b = -b.}$$

Sec. 1.3   Properties of Real Numbers

(See Exercises 46 and 47.) Also, for any two real numbers a and b,

$$\boxed{a(-b) = -ab.} \qquad (2)$$

(See Exercise 48.) For example, $3(-7) = -(3 \cdot 7) = -21$.

Examples 8
$$\begin{aligned}
3(x - 2y) &= 3[x + (-2y)] && \text{(definition of subtraction)} \\
&= 3x + 3(-2y) && \text{(distributive property)} \\
&= 3x - [3(2y)] && \text{[from (2) above]} \\
&= 3x - [(3 \cdot 2)y] && \text{(associative property)} \\
&= 3x - 6y.
\end{aligned}$$

In general, the distributive property extends to parentheses that involve negative signs—for example,

$$\boxed{a(b - c) = ab - ac.}$$

Thus, we can write immediately

$$3(x - 2y) = 3x - 3(2y) = 3x - 6y.$$

Observe that when an expression inside parentheses is multiplied by a negative quantity, every term inside the parentheses must change sign:

$$\begin{aligned}
-(a + b) &= (-1)(a + b) = (-1)a + (-1)b \\
&= -a - b.
\end{aligned}$$

Examples 9 Simplify the expressions.
(a) $5x - 2(x - 3y)$.   (b) $2(2a - b) - 3(a - 3b)$.

Solutions (a) $\begin{aligned}[t] 5x - 2(x - 3y) &= 5x - 2x - 2(-3y) \\ &= (5 - 2)x - 2(-3)y \\ &= 3x + 6y. \end{aligned}$

(b) $\begin{aligned}[t] 2(2a - b) - 3(a - 3b) &= 2(2a) - 2b - 3a - 3(-3b) \\ &= 4a - 2b - 3a + 9b \\ &= (4 - 3)a + (-2 + 9)b \\ &= a + 7b. \end{aligned}$

Diagnostic Test 1.3

Fill in the blanks.

1. $2x + 5x = $ _____.
2. $8x - 3x = $ _____.
3. $3x - 7x = $ _____.
4. $3x - x/3 = $ _____.
5. $5(2x - y) = $ _____.
6. $-(x + 2y) = $ _____.
7. $-2(x - 3y) = $ _____.
8. $-3(2x + y) = $ _____.

9. $a - b = -($ _____ $)$.   10. $a \cdot \dfrac{1}{b} = a \div$ _____ .

11. $(-a)(-b) =$ _____ .   12. $2|-3| =$ _____ .

Exercises 1.3

(1–44) Simplify the expressions.

1. $3(x + 2y)$.
2. $4(2x + z)$.
3. $2(2x - y)$.
4. $3(4z - 2x)$.
5. $-(x - 6)$.
6. $-(-x - 3)$.
7. $3(x - 4)$.
8. $2(-x - 3)$.
9. $-2(-x - 2)$.
10. $-4(x - 6)$.
11. $-x(y - 6)$.
12. $-x(-y - 6)$.
13. $2(x - y) + 4x$.
14. $3y + 4(x + 2y)$.
15. $-2z - 3(x - 2z)$.
16. $-4x - 2(3z - 2x)$.
17. $(x + y) + 4(x - y)$.
18. $3(y - 2x) - 2(2x - 2y)$.
19. $5(7x - 2y) - 4(3y - 2x)$.
20. $4(8z - 2t) - 3(-t - 4z)$.
21. $x(-y)(-z)$.
22. $(-x)(-y)(-z)$.
23. $(-2)(-x)(x + 3)$.
24. $(-x)(-y)(2 - 3z)$.
25. $2(-a)(3 - a)$.
26. $(-3p)(2q)(q - p)$.
27. $x(-2)(-x - 4)$.
28. $(-2x)(-3)(-y - 4)$.
29. $-x(x - 2) + 2(x - 1)$.
30. $-2(-3x)(-2y + 1) - (-y)(4 - 5x)$.
31. $2x + 5 - 2(x + 2)$.
32. $3x - t - 2(x - t)$.
33. $2(x - y) - x$.
34. $4x(x + y) - x^2$.
35. $4[2(x + 1) - 3]$.
36. $x[3(x - 2) - 2x + 1]$.
37. $x[-3(-4 + 5) + 3]$.
38. $4[x(2 - 5) - 2(1 - 2x)]$.
39. $\dfrac{1}{x}(x + 2)$.
40. $\dfrac{1}{x}(2x - 1)$.
41. $\dfrac{1}{(-2x)}(3x - 1)$.
42. $\dfrac{1}{(-3x)}(6 + 2x)$.
43. $\dfrac{1}{xy}(x + y)$.
44. $\dfrac{1}{(-xy)}(2x - 3y)$.

*45. Prove that for any three real numbers a, b, and c, $(b + c)a = ba + ca$. [*Hint:* Use the first distributive property and the commutative property.]

*46. Prove that $0 \cdot b = 0$ for any real number b. [*Hint:* Show that $b + 0 \cdot b = b$, so $0 \cdot b$ leaves b unchanged when added to it.]

*47. Prove that $(-1)b = -b$ for any real number b. [*Hint:* Use Exercise 46 to show that $b + (-1)b = 0$.]

*48. Prove that for any two real numbers a and b, $a(-b) = -ab$. [*Hint:* Write $(-b) = (-1)b$ and use the associative property.]

1.4

Fractions

In the previous section we showed how the fraction a/b is defined as the product of a and the inverse of b:

$$\frac{a}{b} = a \cdot \frac{1}{b} \quad (b \neq 0).$$

From this definition it is possible to derive all the properties that are commonly used in calculating with fractions. In this section and the next we shall discuss such calculations. Proofs of the various rules are given in Exercise 49 at the end of the section.

MULTIPLICATION OF FRACTIONS

The product of two fractions is obtained by multiplying the two numerators together and the two denominators together:

$$\left(\frac{a}{b}\right)\left(\frac{c}{d}\right) = \frac{ac}{bd}.$$

This rule extends in a straightforward way to products of three or more fractions.

The following examples illustrate the use of this rule for arithmetical fractions.

Examples 1 (a) $\left(\dfrac{2}{3}\right)\left(\dfrac{5}{9}\right) = \dfrac{2 \cdot 5}{3 \cdot 9} = \dfrac{10}{27}.$

(b) $\left(\dfrac{4}{7}\right)\left(\dfrac{3}{5}\right) = \dfrac{4 \cdot 3}{7 \cdot 5} = \dfrac{12}{35}.$

(c) $\dfrac{2}{5} \cdot \dfrac{3}{5} \cdot \dfrac{7}{4} = \dfrac{2 \cdot 3 \cdot 7}{5 \cdot 5 \cdot 4} = \dfrac{42}{100}.$

(d) $5 \cdot \dfrac{3}{8} = \dfrac{5}{1} \cdot \dfrac{3}{8} = \dfrac{5 \cdot 3}{1 \cdot 8} = \dfrac{15}{8}.$

When the fractions involve algebraic symbols, there is essentially no change in the application of the rule for multiplication.

Examples 2 (a) $\left(\dfrac{2}{3}\right)\left(\dfrac{x}{y}\right) = \dfrac{2 \cdot x}{3 \cdot y} = \dfrac{2x}{3y}.$ In this example it is essential that $y \neq 0$, otherwise we would have zero denominators in some of the fractions.

(b) $\left(\dfrac{2x}{3}\right)\left(\dfrac{4}{y}\right) = \dfrac{(2x)4}{3 \cdot y} = \dfrac{8x}{3y}.$

(c) $3x\left(\dfrac{4}{5y}\right) = \left(\dfrac{3x}{1}\right)\left(\dfrac{4}{5y}\right) = \dfrac{(3x) \cdot 4}{1 \cdot (5y)} = \dfrac{12x}{5y}.$

(d) $\left(\dfrac{y}{2z}\right)\left(\dfrac{3}{2x}\right)\left(\dfrac{3y}{5}\right) = \dfrac{y \cdot 3 \cdot 3y}{2z \cdot 2x \cdot 5} = \dfrac{9y^2}{20xz}.$

(e) $\dfrac{y}{4} \cdot 5x \cdot \dfrac{3y}{2} = \dfrac{y}{4} \cdot \dfrac{5x}{1} \cdot \dfrac{3y}{2} = \dfrac{y \cdot 5x \cdot 3y}{4 \cdot 1 \cdot 2} = \dfrac{15xy^2}{8}.$

DIVISION OF FRACTIONS

In order to divide one fraction by another, we invert the second fraction and then multiply by the first. In other words,

$$\left(\frac{a}{b}\right) \div \left(\frac{c}{d}\right) = \left(\frac{a}{b}\right)\left(\frac{d}{c}\right) = \frac{ad}{bc}.$$

Ch. 1 Basic Algebra of Real Numbers

Examples 3 (a) $\left(\frac{3}{5}\right) \div \left(\frac{7}{9}\right) = \left(\frac{3}{5}\right)\left(\frac{9}{7}\right) = \frac{27}{35}$.

(b) $2 \div \left(\frac{3}{4}\right) = \left(\frac{2}{1}\right) \div \left(\frac{3}{4}\right) = \left(\frac{2}{1}\right)\left(\frac{4}{3}\right) = \frac{8}{3}$.

(c) $\frac{4}{5} \div 3 = \left(\frac{4}{5}\right) \div \left(\frac{3}{1}\right) = \left(\frac{4}{5}\right)\left(\frac{1}{3}\right) = \frac{4}{15}$.

The following examples illustrate this rule when algebraic symbols occur in the fractions.

Examples 4 (a) $\left(\frac{3}{4}\right) \div \left(\frac{2x}{y}\right) = \left(\frac{3}{4}\right) \cdot \left(\frac{y}{2x}\right) = \frac{3 \cdot y}{4 \cdot 2x} = \frac{3y}{8x}$. Again note that both x and y must be nonzero in this example if we are to avoid having a zero denominator.

(b) $\left(\frac{3x}{2}\right) \div \left(\frac{4}{y}\right) = \left(\frac{3x}{2}\right)\left(\frac{y}{4}\right) = \frac{3xy}{8}$.

(c) $5y \div \left(\frac{6}{5x}\right) = \left(\frac{5y}{1}\right)\left(\frac{5x}{6}\right) = \frac{25xy}{6}$.

(d) $\left(\frac{3}{2x}\right) \div (2y) = \left(\frac{3}{2x}\right) \div \left(\frac{2y}{1}\right) = \left(\frac{3}{2x}\right)\left(\frac{1}{2y}\right) = \frac{3}{4xy}$.

(e) $\frac{1}{a/b} = 1 \div \left(\frac{a}{b}\right) = 1 \cdot \frac{b}{a} = \frac{b}{a}$. (That is, the reciprocal of any fraction is obtained by turning the fraction upside down.)

CANCELLATION OF COMMON FACTORS

You will recall from arithmetic that any fraction can be "reduced to its lowest terms" if we divide out any factors that are common to the numerator and denominator. For example, the fraction $\frac{4}{6}$ is equal to $\frac{2}{3}$, since numerator and denominator can both be divided by 2. Similarly $\frac{36}{60} = \frac{3}{5}$ (dividing numerator and denominator by 12). We say that the common factor has been *canceled* from the numerator and denominator.

The following general principle underlies this cancellation of common factors: The numerator and denominator of any fraction can be multiplied or divided by any *nonzero* number, and the value of the fraction will not be changed:

$$\frac{a}{b} = \frac{ac}{bc} \quad (c \neq 0).$$

Examples 5 (a) $\frac{a}{b} = \frac{2a}{2b}$.

(b) $\frac{3}{5} = \frac{6}{10} = \frac{9}{15} = \frac{-12}{-20}$, and so on.

(c) $\frac{5x}{6} = \frac{10x^2}{12x}$ (provided that $x \neq 0$).

Sec. 1.4   Fractions

(d) $\dfrac{6x(x+y)}{3(x+y)} = \dfrac{2x}{1} = 2x$ (provided that $x+y \neq 0$). [Here we have divided numerator and denominator by $3(x+y)$.]

When we use this principle in order to reduce a fraction to its lowest terms, the procedure is to express the numerator and denominator in terms of their prime factors.* The common prime factors can then easily be divided out.

Examples 6 (a) Consider $\dfrac{70}{84}$. We can write
$$70 = 2 \cdot 5 \cdot 7, \qquad 84 = 2 \cdot 2 \cdot 3 \cdot 7$$
in terms of their prime factors. Therefore
$$\dfrac{70}{84} = \dfrac{2 \cdot 5 \cdot 7}{2 \cdot 2 \cdot 3 \cdot 7} = \dfrac{\cancel{2} \cdot 5 \cdot \cancel{7}}{\cancel{2} \cdot 2 \cdot 3 \cdot \cancel{7}} = \dfrac{5}{2 \cdot 3} = \dfrac{5}{6}.$$
Note that the common factors ($2 \cdot 7$ in this example) are indicated by cancellation marks as they are divided out.

(b) $\dfrac{264}{180} = \dfrac{2 \cdot 2 \cdot 2 \cdot 3 \cdot 11}{2 \cdot 2 \cdot 3 \cdot 3 \cdot 5} = \dfrac{\cancel{2} \cdot \cancel{2} \cdot 2 \cdot \cancel{3} \cdot 11}{\cancel{2} \cdot \cancel{2} \cdot \cancel{3} \cdot 3 \cdot 5} = \dfrac{2 \cdot 11}{3 \cdot 5} = \dfrac{22}{15}.$

A similar technique can be used for algebraic fractions.

Examples 7 (a) $\dfrac{2x^2}{6x} = \dfrac{2 \cdot x \cdot x}{2 \cdot 3 \cdot x} = \dfrac{\cancel{2} \cdot \cancel{x} \cdot x}{\cancel{2} \cdot 3 \cdot \cancel{x}} = \dfrac{x}{3}$ ($x \neq 0$). The numerator and denominator are first broken into their basic factors (which may be prime numbers or algebraic factors). Then the common factors ($2x$ in this example) are divided out.

(b) $\dfrac{6x^2y}{8xy^2} = \dfrac{2 \cdot 3 \cdot x \cdot x \cdot y}{2 \cdot 2 \cdot 2 \cdot x \cdot y \cdot y} = \dfrac{\cancel{2} \cdot 3 \cdot \cancel{x} \cdot x \cdot \cancel{y}}{\cancel{2} \cdot 2 \cdot 2 \cdot \cancel{x} \cdot y \cdot \cancel{y}} = \dfrac{3x}{4y}$ ($x \neq 0$). In this example, the numerator and denominator are divided by $(2xy)$ in the simplification.

(c) $\dfrac{2x(x+1)}{4y(x+1)} = \dfrac{x}{2y}$ ($x+1 \neq 0$). Here the common factor of $2(x+1)$ is divided from numerator and denominator.

It is often possible, after multiplying or dividing fractions, to simplify the result by dividing out common factors.

Examples 8 (a) $\left(\dfrac{8}{3}\right) \div \left(\dfrac{4}{9}\right) = \dfrac{8}{3} \cdot \dfrac{9}{4} = \dfrac{72}{12} = \dfrac{6}{1} = 6$ (common factor of 12 has been canceled)

(b) $\left(\dfrac{3x^2y}{2}\right)\left(\dfrac{8}{15x}\right) = \dfrac{24x^2y}{30x} = \dfrac{4xy}{5}$ (common factor $6x$)

---

*See Section 1.1.

(c) $\left(\dfrac{4x}{5y}\right)\left(\dfrac{3xy^2}{10}\right) \div 6x^2 = \dfrac{4x}{5y} \cdot \dfrac{3xy^2}{10} \div \dfrac{6x^2}{1}$
$= \dfrac{4x \cdot 3xy^2}{5y \cdot 10} \cdot \dfrac{1}{6x^2}$
$= \dfrac{12x^2y^2}{300x^2y} = \dfrac{y}{25}$ (common factor $12x^2y$)

Diagnostic Test 1.4

Fill in the blanks.

1. $\dfrac{a}{a+b} = $ _____.

2. $\dfrac{a+b}{c} = $ _____ + _____.

3. $\dfrac{a}{b} \cdot \dfrac{c}{d} = $ _____.

4. $\dfrac{a}{b} \div \dfrac{c}{d} = $ _____.

5. $a \cdot \dfrac{b}{c} = $ _____.

6. $a \div \dfrac{b}{c} = $ _____.

7. $\dfrac{a}{b} \div c = $ _____.

8. $1 \div a/b = $ _____.

9. $\dfrac{1}{(a/b)} = $ _____ provided a _____.

10. $\dfrac{a}{b} \div \dfrac{2}{b} = $ _____.

Exercises 1.4

(1–16) Evaluate the following.

1. $(\tfrac{2}{9})(\tfrac{5}{6})$.
2. $(\tfrac{3}{4})(\tfrac{5}{14})$.
3. $(\tfrac{2}{5}) \div (\tfrac{5}{2})$.
4. $(\tfrac{7}{3}) \div (\tfrac{4}{5})$.
5. $(\tfrac{1}{2})(\tfrac{3}{4})(\tfrac{3}{10})$.
6. $5(\tfrac{10}{3})(\tfrac{8}{7})$.
7. $\left(\dfrac{2x}{3}\right)\left(\dfrac{x}{5}\right)$.
8. $\left(\dfrac{3}{4y}\right)\left(\dfrac{5}{2y}\right)$.
9. $\left(\dfrac{3x}{2}\right)\left(\dfrac{xy}{10}\right)$.
10. $\left(\dfrac{7}{4x}\right)\left(\dfrac{3y^2}{4x}\right)$.
11. $\left(\dfrac{2x}{3}\right) \div \left(\dfrac{3y}{4}\right)$.
12. $\left(\dfrac{3xy}{11}\right) \div \left(\dfrac{2}{3y}\right)$.
13. $2xy \div \left(\dfrac{3}{2x}\right)$.
14. $4x^2y \div \left(\dfrac{1}{3y}\right)$.
15. $\left(\dfrac{2x}{3}\right) \div 15y$.
16. $\left(\dfrac{15}{8x}\right) \div 4xy$.

(17–28) Reduce the following fractions to their lowest terms.

17. $\tfrac{12}{16}$.
18. $\tfrac{18}{27}$.
19. $\tfrac{45}{60}$.
20. $\tfrac{70}{28}$.
21. $\tfrac{250}{1000}$.
22. $\tfrac{105}{75}$.
23. $\dfrac{14xy}{6x}$.
24. $\dfrac{2xy}{20y}$.
25. $\dfrac{6x^2y}{9xy^2}$.
26. $\dfrac{30x^2}{35xy}$.
27. $\dfrac{-2xy}{-8x^2}$.
28. $\dfrac{-15y^2}{-10x^2y}$.

(29–48) Evaluate the following and reduce the answer to the lowest terms.

29. $(\tfrac{2}{9})(\tfrac{6}{5})$.
30. $(\tfrac{8}{3})(\tfrac{15}{4})$.
31. $(\tfrac{3}{4})(\tfrac{8}{5})(\tfrac{4}{9})$.
32. $(\tfrac{2}{5})(\tfrac{3}{6})(\tfrac{10}{7})$.
33. $\left(\dfrac{3x}{25}\right)\left(\dfrac{25}{9x}\right)$.
34. $\left(\dfrac{14x}{15y}\right)\left(\dfrac{25y}{24}\right)$.

Sec. 1.4   Fractions

35. $7x^2 \left(\dfrac{6y}{21x}\right)$.

36. $\left(-\dfrac{2x}{3y}\right)(-5xy)$.

37. $\left(\dfrac{18}{11}\right) \div \left(\dfrac{8}{33}\right)$.

38. $\left(\dfrac{14}{3}\right) \div \left(\dfrac{6}{15}\right)$.

39. $\dfrac{4}{9} \div \left(\dfrac{2}{3} \cdot 8\right)$.

40. $\left(\dfrac{12}{25} \cdot \dfrac{15}{7}\right) \div \dfrac{20}{7}$.

41. $\left(\dfrac{7x}{10}\right) \div \left(\dfrac{21x}{5}\right)$.

42. $(2x) \div \left(\dfrac{3xy}{5}\right)$.

43. $4 \div \left(\dfrac{8}{9x}\right)$.

44. $\left(\dfrac{3}{8x}\right) \div \left(\dfrac{4x}{15}\right)$.

45. $\left(\dfrac{3x^2}{20} \cdot 4y\right) \div \left(\dfrac{6xy}{25}\right)$.

46. $\left(\dfrac{5x}{2} \cdot \dfrac{3y}{4}\right) \div \left(\dfrac{x^2y}{12}\right)$.

47. $8xy \div \left(\dfrac{2x}{3} \cdot \dfrac{2x}{5y}\right)$.

48. $6x^2 \div \left(\dfrac{4x}{y} \cdot \dfrac{3y^2}{2}\right)$.

REDO

*49. Using the definition of the multiplicative inverse, prove the following.

(a) $\left(\dfrac{1}{a}\right)\left(\dfrac{1}{b}\right) = \dfrac{1}{ab}$.

[*Hint:* show that $(ab)\left(\dfrac{1}{a}\right)\left(\dfrac{1}{b}\right) = 1.$]

(b) $\left(\dfrac{a}{b}\right)\left(\dfrac{c}{d}\right) = \dfrac{ac}{bd}$.

[*Hint:* use part (a).]

(c) $\dfrac{1}{a/b} = \dfrac{b}{a}$.

[*Hint:* show that $\left(\dfrac{b}{a}\right)\left(\dfrac{a}{b}\right) = 1.$]

(d) $\left(\dfrac{a}{b}\right) \div \left(\dfrac{c}{d}\right) = \left(\dfrac{a}{b}\right)\left(\dfrac{d}{c}\right)$.

[*Hint:* use part (c).]

(e) $\dfrac{a}{b} = \dfrac{ac}{bc}$ ($c \neq 0$).

[*Hint:* $\dfrac{ac}{bc} = \left(\dfrac{a}{b}\right)\left(\dfrac{c}{c}\right).$]

1.5

Addition and Subtraction of Fractions

When two fractions have the same denominator, we may easily add (or subtract) them simply by adding (or subtracting) the two numerators. For example,

$$\dfrac{5}{12} + \dfrac{11}{12} = \dfrac{5+11}{12} = \dfrac{16}{12} = \dfrac{4}{3}$$

and

$$\dfrac{5}{12} - \dfrac{11}{12} = \dfrac{5-11}{12} = \dfrac{-6}{12} = -\dfrac{1}{2}.$$

We can state this as a general rule:

$$\dfrac{a}{c} + \dfrac{b}{c} = \dfrac{a+b}{c}.$$

Note that this rule applies only when the two fractions have the same denominator. A similar rule applies to subtraction:

$$\frac{a}{c} - \frac{b}{c} = \frac{a-b}{c}.$$

Examples 1 (a) $\frac{1}{2x} + \frac{3}{2x} = \frac{1+3}{2x} = \frac{4}{2x} = \frac{2}{x}.$

(b) $\frac{2}{5y} - \frac{7}{5y} = \frac{2-7}{5y} = \frac{-5}{5y} = -\frac{1}{y}.$

These rules are an immediate consequence of the distributive property. (See Exercise 41.)

When two fractions with unequal denominators are to be added or subtracted, one or both of the fractions must first be changed so that they have the same denominator.

Example 2 Find $\frac{5}{6} + \frac{1}{2}$.

Solution We can write $\frac{1}{2} = \frac{3}{6}$ if we multiply numerator and denominator by 3. Then both fractions have the same denominator, and we have

$$\frac{5}{6} + \frac{1}{2} = \frac{5}{6} + \frac{3}{6} = \frac{5+3}{6} = \frac{8}{6} = \frac{4}{3}.$$

In this last example, only one of the fractions needed to be changed. Usually both fractions need to be changed to obtain a common denominator.

Example 3 Find $\frac{5}{6} - \frac{3}{4}$.

Solution We can change both fractions to have denominators of 12:

$$\frac{5}{6} = \frac{10}{12} \quad \text{and} \quad \frac{3}{4} = \frac{9}{12}.$$

Therefore,

$$\frac{5}{6} - \frac{3}{4} = \frac{10}{12} - \frac{9}{12} = \frac{10-9}{12} = \frac{1}{12}.$$

In general, when adding or subtracting fractions with different denominators we first replace each fraction by an equivalent fraction having some common denominator. To keep the numbers as small as possible we choose the smallest possible common denominator, called the **LCD (least common denominator)**. We would still get the right answer by using a larger common denominator, but it is preferable to use the smallest. For example, in Example 3 we could use 24 as common denominator:

$$\frac{5}{6} - \frac{3}{4} = \frac{20}{24} - \frac{18}{24} = \frac{20-18}{24} = \frac{2}{24} = \frac{1}{12}.$$

The answer is the same, but we had to calculate with bigger numbers.

Sec. 1.5   Addition and Subtraction of Fractions

In order to find the LCD of two or more fractions, we first break down the two denominators into their prime factors. Then we form the LCD by taking all the prime factors that occur in any of the denominators. Each such prime factor is included *as many times as it occurs in any one denominator.*

For example, to find the LCD of $\frac{5}{6}$ and $\frac{3}{4}$ we write the denominators as: $6 = 2 \cdot 3$ and $4 = 2 \cdot 2$. The prime factors that occur are 2 and 3, but 2 occurs twice in one denominator. So

$$\text{LCD} = 2 \cdot 2 \cdot 3 = 12.$$

Examples 4 (a) Find the LCD of $\frac{7}{18}$ and $\frac{11}{120}$.

(b) Find the LCD of $\frac{1}{12x}$ and $\frac{11}{10x^2y}$.

Solutions (a) Expressing the denominators in terms of their prime factors, we find

$$18 = 2 \cdot 3 \cdot 3 \quad \text{and} \quad 120 = 2 \cdot 2 \cdot 2 \cdot 3 \cdot 5.$$

The factors are 2 (which occurs three times in 120), 3 (which occurs twice in 18), and 5. Hence

$$\text{LCD} = 2 \cdot 2 \cdot 2 \cdot 3 \cdot 3 \cdot 5 = 360.$$

(b) The same method works with algebraic fractions. We just treat each algebraic factor as one of the prime factors. We have

$$12x = 2 \cdot 2 \cdot 3 \cdot x \quad \text{and} \quad 10x^2y = 2 \cdot 5 \cdot x \cdot x \cdot y.$$

The factors are therefore 2 (twice in $12x$), 3, 5, x (twice in $10x^2y$), and y. Therefore

$$\text{LCD} = 2 \cdot 2 \cdot 3 \cdot 5 \cdot x \cdot x \cdot y = 60x^2y.$$

The use of LCD's to add and subtract fractions is illustrated in the following examples.

Examples 5 Evaluate.

(a) $\frac{7}{18} - \frac{11}{120}$. (b) $\frac{5}{6} + \frac{2}{9} - \frac{3}{8}$.

Solutions (a) As in Example 4(a), the LCD is 360. We have

$$\frac{7}{18} = \frac{20 \cdot 7}{20 \cdot 18} = \frac{140}{360} \quad \text{and} \quad \frac{11}{120} = \frac{3 \cdot 11}{3 \cdot 120} = \frac{33}{360}.$$

Therefore

$$\frac{7}{18} - \frac{11}{120} = \frac{140}{360} - \frac{33}{360} = \frac{140 - 33}{360} = \frac{107}{360}.$$

(b) Here we have three denominators:

$$6 = 2 \cdot 3, \quad 9 = 3 \cdot 3 \quad \text{and} \quad 8 = 2 \cdot 2 \cdot 2.$$

The factors are 2 (three times in 8) and 3 (twice in 9). Thus

$$\text{LCD} = 2 \cdot 2 \cdot 2 \cdot 3 \cdot 3 = 72.$$

Then
$$\frac{5}{6} + \frac{2}{9} - \frac{3}{8} = \frac{60}{72} + \frac{16}{72} - \frac{27}{72} = \frac{60 + 16 - 27}{72} = \frac{49}{72}.$$

Examples 6 Evaluate.

(a) $\frac{x}{6} + \frac{3y}{4}.$ (b) $\frac{1}{6x} - \frac{2y}{15x^2}.$

Solutions (a) Here the LCD is 12 and we can write
$$\frac{x}{6} = \frac{2x}{12} \quad \text{and} \quad \frac{3y}{4} = \frac{9y}{12}.$$
Therefore,
$$\frac{x}{6} + \frac{3y}{4} = \frac{2x}{12} + \frac{9y}{12} = \frac{2x + 9y}{12}.$$

(b) We write for the two denominators:
$$6x = 2 \cdot 3 \cdot x \quad \text{and} \quad 15x^2 = 3 \cdot 5 \cdot x \cdot x.$$
Therefore,
$$\text{LCD} = 2 \cdot 3 \cdot 5 \cdot x \cdot x = 30x^2.$$
Then
$$\frac{1}{6x} = \frac{5x}{(5x)(6x)} = \frac{5x}{30x^2} \quad \text{and} \quad \frac{2y}{15x^2} = \frac{4y}{30x^2}.$$
Therefore we have
$$\frac{1}{6x} - \frac{2y}{15x^2} = \frac{5x}{30x^2} - \frac{4y}{30x^2} = \frac{5x - 4y}{30x^2}.$$

Example 7 Evaluate $\frac{2y}{5x^2} + \frac{1}{6xy} - \frac{3x^2}{12y^2}.$

Solution For the denominators we have
$$5x^2 = 5 \cdot x \cdot x, \quad 6xy = 2 \cdot 3 \cdot x \cdot y \quad \text{and} \quad 12y^2 = 2 \cdot 2 \cdot 3 \cdot y \cdot y.$$
Therefore,
$$\text{LCD} = 2 \cdot 2 \cdot 3 \cdot 5 \cdot x \cdot x \cdot y \cdot y = 60x^2y^2.$$
Then
$$\frac{2y}{5x^2} + \frac{1}{6xy} - \frac{3x^2}{12y^2} = \frac{(2y)(12y^2)}{(5x^2)(12y^2)} + \frac{1 \cdot (10xy)}{(6xy)(10xy)} - \frac{(3x^2)(5x^2)}{(12y^2)(5x^2)}$$
$$= \frac{24y^3}{60x^2y^2} + \frac{10xy}{60x^2y^2} - \frac{15x^4}{60x^2y^2}$$
$$= \frac{24y^3 + 10xy - 15x^4}{60x^2y^2}.$$

Example 8 Evalute $\dfrac{\frac{1}{6}+\frac{3}{8}}{\frac{3}{4}-\frac{1}{3}}$.

Solution We must first evaluate the numerator and denominator of the large fraction. We get

$$\frac{1}{6}+\frac{3}{8}=\frac{4+9}{24}=\frac{13}{24}$$

$$\frac{3}{4}-\frac{1}{3}=\frac{9-4}{12}=\frac{5}{12}.$$

Therefore,

$$\frac{\frac{1}{6}+\frac{3}{8}}{\frac{3}{4}-\frac{1}{3}}=\frac{\frac{13}{24}}{\frac{5}{12}}=\frac{13}{24}\div\frac{5}{12}=\left(\frac{13}{24}\right)\left(\frac{12}{5}\right)=\frac{13}{10}.$$

Diagnostic Test 1.5

Fill in the blanks.

1. $\dfrac{3}{x}+\dfrac{4}{x}=$ _____ .
2. $\dfrac{5}{x}-\dfrac{7}{x}=$ _____ .
3. $\dfrac{x}{3}+\dfrac{x}{4}=$ _____ .
4. $\dfrac{x}{7}-\dfrac{x}{5}=$ _____ .
5. $\dfrac{a}{c}+\dfrac{b}{c}=$ _____ .
6. $\dfrac{a}{b}+\dfrac{a}{c}=$ _____ .
7. $\dfrac{a}{b}+\dfrac{c}{d}=$ _____ .
8. $\dfrac{x}{a}-\dfrac{x}{b}=$ _____ .
9. LCD of $\dfrac{5}{8}+\dfrac{7}{12}$ is _____ .
10. LCD of $\dfrac{a}{b}+\dfrac{c}{2b}$ is _____ .

Exercises 1.5

(1–40) Evaluate the following, expressing the answer in the simplest terms.

1. $\frac{1}{6}-\frac{1}{2}$.
2. $\frac{2}{9}+\frac{4}{3}$.
3. $\frac{5}{8}+\frac{1}{2}$.
4. $\frac{13}{12}-\frac{5}{4}$.
5. $\frac{1}{10}+\frac{1}{15}$.
6. $\frac{7}{12}-\frac{5}{18}$.
7. $\frac{3}{8}+\frac{5}{12}$.
8. $\frac{5}{6}+\frac{8}{15}$.
9. $\dfrac{7x}{15}+\dfrac{3x}{10}$.
10. $\dfrac{3y}{8}-\dfrac{y}{6}$.
11. $\dfrac{5t}{12}+\dfrac{11x}{18}$.
12. $\dfrac{y}{9}-\dfrac{5z}{12}$.
13. $\dfrac{1}{x}+\dfrac{1}{2x}$.
14. $\dfrac{1}{2y}+\dfrac{1}{3y}$.
15. $\dfrac{3}{8x}-\dfrac{1}{12x}$.
16. $\dfrac{7}{6x}-\dfrac{4}{15x}$.
17. $\dfrac{a}{6b}-\dfrac{a}{2b}$.
18. $\dfrac{y}{2x}-\dfrac{y}{3x}$.
19. $\dfrac{7x}{6y}+\dfrac{3x}{8y}$.
20. $\dfrac{a}{6b}+\dfrac{a}{9b}$.
21. $\dfrac{7}{6x}+\dfrac{3}{4x^2}$.
22. $\dfrac{3y}{10x^2}-\dfrac{1}{6x}$.
23. $x+\dfrac{1}{x}$.
24. $\dfrac{y}{2}+\dfrac{2}{y}$.
25. $\dfrac{a}{b}+\dfrac{c}{d}$.
26. $\dfrac{p}{q}+\dfrac{r}{s}$.
27. $\dfrac{2}{x^2}+\dfrac{3}{xy}$.
28. $\dfrac{y}{6x^2}+\dfrac{3}{4y}$.
29. $\dfrac{x}{p^2}+\dfrac{y}{pq}$.
30. $\dfrac{x}{y}+\dfrac{y}{z}+\dfrac{z}{x}$.
31. $\dfrac{x}{y}-\dfrac{y}{x}$.
32. $\dfrac{x^2}{3y}+4y$.

33. $\dfrac{\frac{1}{2}+\frac{1}{3}}{\frac{1}{4}+\frac{1}{5}}.$ 34. $\dfrac{\frac{1}{3}-\frac{1}{4}}{\frac{1}{5}-\frac{1}{6}}.$ 35. $\dfrac{\frac{8}{5}+\frac{2}{3}}{2+\frac{4}{7}}.$ 36. $\dfrac{2-\frac{3}{4}}{3+\frac{1}{8}}.$

37. $\dfrac{7x-\frac{2x}{3}}{15y-\frac{y}{3}}.$ 38. $\dfrac{\frac{1}{2x}-\frac{1}{3x}}{\frac{1}{4y}-\frac{1}{5y}}.$ 39. $\dfrac{\left(\frac{2a}{3b}\right)\left(\frac{4b}{5}\right)+a}{2b+\frac{b}{15}}.$ 40. $\dfrac{\left(\frac{5p}{2q}\right)\left(\frac{p}{3}\right)+\frac{p^2}{8q}}{4p+\frac{p}{12}}.$

*41. Show that
$$\frac{a}{c}+\frac{b}{c}=\frac{a+b}{c}.$$

[*Hint:* Write $x/c = x(1/c)$ and use the distributive property.]

1.6

Exponents

We have already seen the notation a^2 used to denote the product of two a's: $a^2 = a \cdot a$. In a similar way, a^3 is used to denote the product of three a's, a^4 the product of four a's, and so on:

$$a^3 = a \cdot a \cdot a, \qquad a^4 = a \cdot a \cdot a \cdot a, \qquad a^5 = a \cdot a \cdot a \cdot a \cdot a, \qquad \ldots$$

In general, if m is any positive integer, then a^m is defined to be the product of m a's:

$$a^m = \underbrace{a \cdot a \cdot a \cdot \cdots \cdot a}_{m \text{ factors}}$$

The expression a^m is read "a to the power m" or "the mth power of a," except that when $m = 2$, the expression a^2 is read "a squared" and when $m = 3$, the expression a^3 is read "a cubed." For example,

$$2^4 = 2 \cdot 2 \cdot 2 \cdot 2 = 16 \qquad \text{(four factors of 2)}$$
$$3^5 = 3 \cdot 3 \cdot 3 \cdot 3 \cdot 3 = 243 \qquad \text{(five factors of 3)}$$

In the expression a^m, m is called the *power* or *exponent* and a the *base*. Thus in 2^4 (the fourth power of 2), 2 is the base and 4 the exponent. In the expression 3^5, 3 is the base and 5 the exponent. This definition of a^m when the exponent is a positive integer holds when a is any real number.

Now observe the pattern in the adjoining table in which several powers of 5 are given in decreasing order. Let us try to complete the table. We notice that every time the exponent is decreased by 1, the number in the right column is *divided* by 5. This suggests that we should complete the table by continuing to divide by 5 with each reduction in the exponent. Thus we are led to the following:

5^4	625
5^3	125
5^2	25
5^1	5
5^0	?
5^{-1}	?
5^{-2}	?
5^{-3}	?
5^{-4}	?

$$5^1 = 5, \qquad 5^0 = 5 \div 5 = 1, \qquad 5^{-1} = 1 \div 5 = \tfrac{1}{5}$$
$$5^{-2} = (\tfrac{1}{5}) \div 5 = \tfrac{1}{25}, \qquad 5^{-3} = (\tfrac{1}{25}) \div 5 = \tfrac{1}{125}$$

and so on.

Now observe the following relationships:

$$5^{-1} = \frac{1}{5^1}, \qquad 5^{-2} = \frac{1}{5^2}, \qquad 5^{-3} = \frac{1}{5^3}$$

and in general

$$5^{-n} = \frac{1}{5^n}.$$

This naturally leads to the following definition of a^m when the exponent m is zero or a negative integer.

DEFINITION If $a \neq 0$ and if m is any *positive* integer, then

$$a^0 = 1$$
$$a^{-m} = \frac{1}{a^m}.$$

Example 1

$$3^{-4} = \frac{1}{3^4} = \frac{1}{81}$$

$$2^{-5} = \frac{1}{2^5} = \frac{1}{32}$$

$$4^0 = 1, \qquad \left(\frac{3}{7}\right)^0 = 1, \qquad (-5)^0 = 1.$$

Note When the exponent m is zero or negative, a^m is defined only when $a \neq 0$. In particular 0^0 is not defined.

From these definitions it is possible to establish a number of properties that relate to the multiplication and division of expressions of the type a^m. These properties are often called **laws of exponents**.

PROPERTY I Consider the product $(3^2)(3^3)$. Writing out the two factors in full, we have

$$(3^2)(3^3) = (3 \cdot 3)(3 \cdot 3 \cdot 3) = 3 \cdot 3 \cdot 3 \cdot 3 \cdot 3 = 3^5.$$

Notice that we can obtain the exponent, 5, in the final answer by adding the two original exponents, $2 + 3$. This is an example of a general law that states that

$$a^m \cdot a^n = a^{m+n}.$$

In words: *when two powers of a common base are multiplied, the result is equal to the base raised to the sum of the two exponents.* This result holds for any real number a, except that if m or n is negative or zero, we require $a \neq 0$.

Examples 2 (a) $5^3 \cdot 5^4 = 5^{3+4} = 5^7$.

(b) $x^5 \cdot x^{-3} = x^{5+(-3)} = x^2$. Let us verify this by expanding the two powers:

$$x^5 \cdot x^{-3} = (x \cdot x \cdot x \cdot x \cdot x)\left(\frac{1}{x \cdot x \cdot x}\right) = x \cdot x = x^2$$

(c) $x^2 \cdot x^{-6} = x^{2+(-6)} = x^{-4}$. Again we can verify that this is correct:

$$x^2 \cdot x^{-6} = (x \cdot x)\left(\frac{1}{x \cdot x \cdot x \cdot x \cdot x \cdot x}\right) = \frac{1}{x \cdot x \cdot x \cdot x} = \frac{1}{x^4} = x^{-4}.$$

PROPERTY II Consider the ratio $(4^5) \div (4^3)$. Expanding the factors, we get

$$(4^5) \div (4^3) = \frac{4 \cdot 4 \cdot 4 \cdot 4 \cdot 4}{4 \cdot 4 \cdot 4} = 4 \cdot 4 = 4^2.$$

The final exponent, 2, is equal to the difference between the two original exponents, 5 − 3. Again there is a general law, as follows:

$$\boxed{\frac{a^m}{a^n} = a^{m-n} \quad (a \neq 0).}$$

That is, *when one power is divided by another with the same base, the result is equal to the base raised to an exponent that is the difference between the exponent in the numerator and the exponent in the denominator.*

Examples 3 (a) $\dfrac{5^7}{5^3} = 5^{7-3} = 5^4$.

(b) $\dfrac{4^3}{4^{-2}} = 4^{3-(-2)} = 4^{3+2} = 4^5$. Let us verify this:

$$\frac{4^3}{4^{-2}} = 4^3 \div (4^{-2}) = 4^3 \div \left(\frac{1}{4^2}\right)$$
$$= 4^3 \cdot 4^2 = 4^{3+2} \quad \text{(using Property I)}$$
$$= 4^5.$$

(c) $\dfrac{x^{-2}}{x} = \dfrac{x^{-2}}{x^1} = x^{-2-1} = x^{-3} \quad (x \neq 0)$.

PROPERTY III Consider the expression $(4^2)^3$. This means the product of three factors each equal to 4^2. Thus:

$$(4^2)^3 = (4^2)(4^2)(4^2)$$
$$= (4 \cdot 4)(4 \cdot 4)(4 \cdot 4) = 4^6.$$

This time the final exponent, 6, is equal to the product of the original exponents, 2 · 3. In general, we have

$$\boxed{(a^m)^n = a^{mn}} \quad (a \neq 0 \text{ if } m \text{ or } n \text{ is negative or zero}).$$

That is, *a power raised to a power is equal to the base raised to the product of the two exponents.*

Examples 4 (a) $(4^3)^{-2} = 4^{3 \cdot (-2)} = 4^{-6}$. We can verify that this is correct, since

$$(4^3)^{-2} = \frac{1}{(4^3)^2} = \frac{1}{(4^3)(4^3)} = \frac{1}{4^{3+3}} \quad \text{(using Property I)}$$
$$= \frac{1}{4^6} = 4^{-6}.$$

(b) $(x^3)^7 = x^{3 \cdot 7} = x^{21}$.
(c) $(y^{-2})^{-4} = y^{(-2)(-4)} = y^8$.

PROPERTY IV Consider the quantity $(2 \cdot 3)^3$. This is equal to 6^3, or $6 \cdot 6 \cdot 6$, which works out to 216. On the other hand, the product $(2^3)(3^3) = (8)(27)$ also equals 216, so we have the result that

$$(2 \cdot 3)^3 = 2^3 \cdot 3^3.$$

In general, in fact, we have that

$$\boxed{(ab)^m = a^m b^m} \qquad (ab \neq 0 \text{ if } m \leq 0).$$

That is, *the product of two numbers all raised to the mth power is equal to the product of the mth powers of the two numbers.*

Examples 5
(a) $10^4 = (2 \cdot 5)^4 = 2^4 \cdot 5^4$.
(b) $(x^2 y)^4 = (x^2)^4 \cdot y^4 = x^8 y^4$ (using Property III)
(c) $(3a^2 b^{-3})^2 = 3^2 \cdot (a^2)^2 (b^{-3})^2$
             $= 9a^4 b^{-6}$ (using Property III)

PROPERTY V A similar property relates to quotients. For example, $(\frac{10}{2})^4 = 5^4 = 625$ while $10^4/2^4 = 10{,}000/16$ also equals 625. Thus

$$\left(\frac{10}{2}\right)^4 = \frac{10^4}{2^4}.$$

In general, we have

$$\boxed{\left(\frac{a}{b}\right)^m = \frac{a^m}{b^m}} \qquad (b \neq 0 \text{ and } a \neq 0 \text{ if } m \leq 0).$$

That is, *the quotient of two numbers all raised to the mth power is equal to the quotient of the mth powers of the two numbers.*

Examples 6
(a) $\left(\dfrac{3}{2}\right)^4 = \dfrac{3^4}{2^4} = \dfrac{81}{16}$.
(b) $\left(\dfrac{x^2}{y}\right)^5 = \dfrac{(x^2)^5}{y^5} = \dfrac{x^{10}}{y^5} = x^{10} y^{-5}$.

Examples 7 Simplify, eliminating parentheses and negative exponents:
(a) $(ax)^5/x^{-7}$.   (b) $(x^{-2})^2/(x^2 z^3)^3$.
(c) $x^4(2x - 3x^{-2})$.   (d) $(x^{-1} + y^{-1})^{-1}$.

Solutions (a) $\dfrac{(ax)^5}{x^{-7}} = \dfrac{a^5 x^5}{x^{-7}}$ (Property IV)
           $= a^5 x^{5-(-7)}$ (Property II)
           $= a^5 x^{12}$.

(b) $\dfrac{(x^{-2})^2}{(x^2z^3)^3} = \dfrac{x^{(-2)(2)}}{(x^2)^3(z^3)^3}$ (Properties III and IV)

$= \dfrac{x^{-4}}{x^6 z^9}$ (Property III)

$= \dfrac{x^{-4-6}}{z^9}$ (Property II)

$= \dfrac{x^{-10}}{z^9} = \dfrac{1}{x^{10}z^9}.$

(c) $x^4(2x - 3x^{-2}) = x^4(2x) - x^4(3x^{-2})$ (distributive property)
$= 2x^{4+1} - 3x^{4+(-2)}$ (Property I, note $x = x^1$)
$= 2x^5 - 3x^2.$

(d) It would be *completely incorrect* in this example to write

$$(x^{-1} + y^{-1})^{-1} = (x^{-1})^{-1} + (y^{-1})^{-1} = x + y.$$

For example, suppose that $x = 2$ and $y = 4$. Then

$$(x^{-1} + y^{-1})^{-1} = (\tfrac{1}{2} + \tfrac{1}{4})^{-1} = (\tfrac{3}{4})^{-1} = \tfrac{4}{3}.$$

This is clearly not equal to $x + y$, which is 6. Instead we must first simplify the expression inside the parentheses:

$$x^{-1} + y^{-1} = \dfrac{1}{x} + \dfrac{1}{y} = \dfrac{y}{xy} + \dfrac{x}{xy} = \dfrac{y+x}{xy}.$$

Now recall that the reciprocal of a fraction is obtained by turning the fraction upside down:

$$(x^{-1} + y^{-1})^{-1} = \left(\dfrac{y+x}{xy}\right)^{-1} = \dfrac{xy}{y+x}.$$

Remark The above five properties are also true for fractional exponents, as we shall see in Section 1.8.

Diagnostic Test 1.6

Fill in the blanks.

1. $a^m \cdot a^n =$ _____.
2. $a^m \div a^n =$ _____.
3. $(x^a)^b =$ _____.
4. $(3^{281})^0 =$ _____.
5. $a^{p+q} =$ _____.
6. $x^{a-b} =$ _____.
7. $a^m - a^n = \underline{a^m - a^n}$.
8. $p^x + p^y = \underline{p^x + p^y}$.
9. $a^p \cdot b^p =$ _____.
10. $x^a \div y^a =$ _____.
11. $(2a)^4 =$ _____.
12. $a^{-m} =$ _____.
13. $1/2a^{-3} =$ _____.
14. $(x^{-1}y^2)^{-3} =$ _____.
15. $(-1)^{257} =$ _____.
16. $(-1)^{586} =$ _____.
17. $a^p \cdot b^q =$ _____.
18. $x^m/y^n =$ _____.

19. If $2^m = 2^n$, then _____.
20. $a^0 = $ _____ for all a _____.
*21. $(a^{-1} + b^{-1})^{-1} = $ _____.
22. $(x^{-1}y^{-1})^{-1} = $ _____.
23. $a^n + b^n = $ _____.
24. $(\frac{5}{7})^0 = (\frac{3}{57})$ _____.
25. $(-1)^n = $ _____ if n is odd.

Exercises 1.6

(1–56) Simplify the following, avoiding all parentheses and negative exponents in the final answer.

1. $(2^5)^2$.
2. $(3^4)^3$.
3. $(a^3)^7$.
4. $(x^4)^5$.
5. $(-x^2)^5$.
6. $(-x^5)^2$.
7. $y^2 \cdot y^5$.
8. $x^7 \cdot x^4$.
9. $a^3 \cdot a^{-5}$.
10. $b^{-2} \cdot b^6$.
11. $(3x)^2 x^{-7}$.
12. $(4x)^{-2} x^4$.
13. $(2x)^2 (2x^{-1})^3$.
14. $\frac{x^3}{2}(4x^{-1})^2$.
15. $(x^2yz)^3(xy)^4$.
16. $(3yz^2)^2(y^3z)^3$.
17. $(x^{-2}y)^{-2}$.
18. $(ab^{-3})^{-1}$.
19. $(xy^2z^3)^{-1}(xyz)^3$.
20. $(x^2pq^2)^2(xp^2)^{-1}$.
21. $\frac{(2^4)^2}{4^2}$.
22. $\frac{(3^3)^2}{3^5}$.
23. $\left(\frac{1}{3}\right)^{-2} \div 3^{-4}$.
24. $\left(\frac{1}{5}\right)^3 \div 5^{-2}$.
25. $\frac{x^5}{x^{-2}}$.
26. $\frac{y^{-3}}{y^{-7}}$.
27. $\frac{(x^2)^3}{x^4}$.
28. $\frac{z^{-8}}{(z^2)^4}$.
29. $\frac{(a^{-2})^6}{(a^4)^{-3}}$.
30. $\frac{(b^{-7})^2}{(b^3)^3}$.
31. $\frac{(-x^3)^2}{(-x)^{-3}}$.
32. $\frac{(-y^{-1})^{-3}}{(-y^2)^{-2}}$.
33. $\frac{(x^2y)^{-3}}{(xy)^2}$.
34. $\frac{(ab^{-2})^{-1}}{a^{-2}b^{-1}}$.
35. $\frac{(-2xy)^3}{x^3y}$.
36. $\frac{(-ab^2c)^{-1}}{a^{-2}bc^{-1}}$.
37. $\frac{(-3x)^2}{-3x^2}$.
38. $\frac{(2x^2y)^{-1}}{(-2x^2y^3)^2}$.
39. $\frac{(2a^{-1}b^2)^2}{(a^3b)^3}$.
40. $\frac{(x^{-3}y^4)^3}{(-3x^2y^{-2})^2}$.
41. $x^2(x^4 - 2x)$.
42. $x^3(x^{-1} - x)$.
43. $2x(x^5 + 3x^{-1})$.
44. $3x^2(x^4 + 2x^{-3})$.
45. $x^4(2x^2 - x - 3x^{-2})$.
46. $2x^{-3}(x^5 - 3x^4 + x)$.
47. $(2^{-1} + x^{-1})^{-1}$.
48. $[(2x)^{-1} + (2y)^{-1}]^{-1}$.
49. $(xy)^{-1}(x^{-1} + y^{-1})^{-1}$.
50. $(a^{-2} + b^{-2})^{-1}$.
51. $\left(\frac{7}{x}\right)\left(\frac{3}{14x}\right) + \left(\frac{3}{2x}\right)^2$.
52. $x^{-3}\left(\frac{6}{5x}\right)^{-1} - \left(-\frac{1}{2x}\right)^2$.
53. $\frac{3y}{10x^3} + \frac{2}{15xy}$.
54. $\frac{5}{12x^{-3}} - \frac{2}{15x^{-2}}$.
55. $\frac{1}{2x^{-2}} + \frac{1}{3x^{-2}}$.
56. $\frac{1}{4y^{-4}} - \frac{1}{3y^{-4}}$.

1.7

Radicals

The **square root** of a real number a, denoted by $\sqrt{a}$, is the nonnegative real number b for which $b^2 = a$. Thus, for example,

$$\sqrt{9} = 3 \quad \text{because } 3^2 = 9$$
$$\sqrt{25} = 5 \quad \text{because } 5^2 = 25.$$

It is equally true that $(-5)^2 = 25$. However, we do not write $\sqrt{25} = -5$, because by definition $\sqrt{25}$ means the *positive* number whose square is 25. We have

$$\sqrt{a} = b \quad \text{if and only if} \quad b^2 = a \quad \text{and} \quad b \geq 0.$$

Note that b^2 can never be negative. Thus, if a is negative, it has no real number as its square root. In other words, $\sqrt{a}$ is defined only for $a \geq 0$.

In a similar way we define the **cube root** of a real number a, denoted by $\sqrt[3]{a}$, as that real number whose cube is equal to a. For example,

$$\sqrt[3]{8} = 2 \quad \text{because} \quad 2^3 = 8$$
$$\sqrt[3]{-27} = -3 \quad \text{because } (-3)^3 = -27.$$

Observe that negative numbers do have cube roots and that the cube roots themselves can be negative. Thus, we have

$$\sqrt[3]{a} = b \quad \text{if and only if} \quad b^3 = a.$$

In general, we define the **nth root** of a to be the number whose nth power is equal to a. It is denoted by $\sqrt[n]{a}$, and we have

$$\sqrt[n]{a} = b \quad \text{if and only if} \quad b^n = a.$$
$$\text{If } n \text{ is even,} \quad a \geq 0 \text{ and } b \geq 0.$$

If $a < 0$, the even roots are not defined, and if $a > 0$ and n even we must restrict $b \geq 0$ because otherwise there would be two possible choices for the nth root. For example, take $\sqrt[4]{16}$. We have that $2^4 = 16$ and $(-2)^4 = 16$, so $\sqrt[4]{16}$ could mean either 2 or -2 unless we specify which one is to be taken. Since we stipulate that any even root must be nonnegative the ambiguity is removed, and $\sqrt[4]{16} = 2$.

Examples 1 (a) $\sqrt[5]{32} = 2$ since $2^5 = 32$.
(b) $\sqrt[5]{-32} = -2$ since $(-2)^5 = -32$.

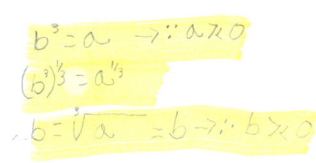

(c) $\sqrt[4]{625} = 5$ since $5^4 = 625$. (We take the positive fourth root, not -5.)
(d) $\sqrt[4]{-16}$ does not exist. (Negative numbers do not have even roots.)
(e) $\sqrt[3]{\frac{27}{125}} = \frac{3}{5}$ since $(\frac{3}{5})^3 = \frac{27}{125}$.

If we substitute $b = \sqrt[n]{a}$ into the equation $b^n = a$, we obtain the condition that

$$(\sqrt[n]{a})^n = a. \tag{1}$$

Similarly, substituting $a = b^n$ into the equation $b = \sqrt[n]{a}$, we obtain that

$$\sqrt[n]{b^n} = b. \tag{2}$$

Observe particularly that in the case of even roots, these two identities hold only when a and b are nonnegative. For example, when $n = 2$,

$$(\sqrt{a})^2 = a \quad \text{and} \quad \sqrt{b^2} = b \quad \text{provided } a \text{ and } b \text{ are nonnegative.}$$

Examples 2 (a) $(\sqrt{25})^2 = 5^2 = 25$ $(a = 25)$.
(b) $(\sqrt[3]{8})^3 = 2^3 = 8$ $(a = 8)$.
(c) $\sqrt[3]{4^3} = \sqrt[3]{64} = 4$ $(b = 4)$.
(d) $\sqrt{(-2)^2} = \sqrt{4} = 2 \neq -2$ $(b = -2)$.

In this case Equation (2) is not satisfied, because we have an even root with b negative.

The expression $\sqrt[n]{a}$ is called a **radical**. The symbol $\sqrt{}$ is called the **radical sign** and n is called the **index** of the radical.

Radicals have a number of properties just like exponents. The most useful of the properties of radicals are:

$$\sqrt[n]{ab} = \sqrt[n]{a} \cdot \sqrt[n]{b} \tag{3}$$

$$\sqrt[n]{\frac{a}{b}} = \frac{\sqrt[n]{a}}{\sqrt[n]{b}} \quad (b \neq 0) \tag{4}$$

provided $\sqrt[n]{a}$ and $\sqrt[n]{b}$ both exist. That is, *the nth root of the product (or quotient) of two numbers is equal to the product (or quotient) of the nth roots of the numbers*, provided they both exist.

To illustrate these properties, take $a = 4$ and $b = 9$ with square roots ($n = 2$). Then

$$\sqrt{ab} = \sqrt{4 \cdot 9} = \sqrt{36} = 6$$

and

$$\sqrt{a}\sqrt{b} = \sqrt{4} \cdot \sqrt{9} = 2 \cdot 3 = 6$$

so Equation (3) is verified. Equation (4) says that

$$\sqrt{\frac{a}{b}} = \frac{\sqrt{a}}{\sqrt{b}}$$

or

$$\sqrt{\frac{4}{9}} = \frac{\sqrt{4}}{\sqrt{9}} = \frac{2}{3}.$$

This is obviously correct, since $(\frac{2}{3})^2 = (\frac{4}{9})$.

Equations (3) and (4) are often of great help in simplifying radicals. For example, we can simplify $\sqrt{8}$:

$$\sqrt{8} = \sqrt{4 \cdot 2} = \sqrt{4} \cdot \sqrt{2} = 2\sqrt{2}.$$

Similarly

$$\sqrt{108} = \sqrt{36 \cdot 3} = \sqrt{36} \cdot \sqrt{3} = 6\sqrt{3}.$$

Observe what we did in these examples. In both cases we took out of the quantity under the radical a factor that was a perfect square. In general, with an nth root, we try to find factors that are perfect nth powers.

Examples 3 (a) $\sqrt[3]{48} = \sqrt[3]{8 \cdot 6}$ (we factor out 8 because it is a perfect cube)
$= \sqrt[3]{8} \cdot \sqrt[3]{6} = 2\sqrt[3]{6}$

(b) $\sqrt[3]{-81} = \sqrt[3]{(-27)(3)}$ (-27 is a perfect cube)
$= \sqrt[3]{-27} \cdot \sqrt[3]{3} = -3\sqrt[3]{3}.$

(c) $\sqrt{16x^5} = \sqrt{16x^4 \cdot x}$ $[16x^4 = (4x^2)^2$ is a perfect square]
$= \sqrt{16x^4} \cdot \sqrt{x} = 4x^2\sqrt{x}.$

(d) $\sqrt[3]{16x^7y^5} = \sqrt[3]{8x^6y^3 \cdot 2xy^2}$
$\qquad\qquad\qquad\qquad [8x^6y^3 = (2x^2y)^3$ is a perfect cube]
$= \sqrt[3]{8x^6y^3} \cdot \sqrt[3]{2xy^2}$
$= 2x^2y \sqrt[3]{2xy^2}.$

Example 4 Simplify $\dfrac{\sqrt{27} + \sqrt{75}}{2\sqrt{12}}$.

Solution We observe that we can simplify the three radicals in this expression by factoring out a perfect square from each of the numbers beneath them:

$$\sqrt{27} = \sqrt{9 \cdot 3} = \sqrt{9} \cdot \sqrt{3} = 3\sqrt{3}$$
$$\sqrt{75} = \sqrt{25 \cdot 3} = \sqrt{25} \cdot \sqrt{3} = 5\sqrt{3}$$
$$\sqrt{12} = \sqrt{4 \cdot 3} = \sqrt{4} \cdot \sqrt{3} = 2\sqrt{3}.$$

Therefore,

$$\frac{\sqrt{27} + \sqrt{75}}{2\sqrt{12}} = \frac{3\sqrt{3} + 5\sqrt{3}}{2(2\sqrt{3})}$$
$$= \frac{8\sqrt{3}}{4\sqrt{3}} = \frac{8}{4} = 2.$$

Sec. 1.7 Radicals

RATIONALIZING THE DENOMINATOR*

Given a fraction in which one or more radicals appear in the denominator, it is sometimes convenient to change it to an equivalent fraction in which all the radicals appear in the numerator. For example, suppose we had to evaluate $1/\sqrt{3}$ (without the help of a calculator). We could write $\sqrt{3} \approx 1.732$ to four figures and then do the long division:

```
              0.577
       1.732)1.0000
             0.8660
             ──────
              13400
              12124
              ─────
              12760
              12124
              ─────
                536
```

The result is $1/\sqrt{3} \approx 0.577$ to three figures.

This calculation is clearly pretty tedious. It is much easier if we multiply numerator and denominator of the given fraction by $\sqrt{3}$:

$$\frac{1}{\sqrt{3}} = \frac{\sqrt{3}}{\sqrt{3} \cdot \sqrt{3}} = \frac{\sqrt{3}}{(\sqrt{3})^2} = \frac{\sqrt{3}}{3}$$

$$\approx \frac{1.732}{3} \approx 0.577.$$

You can now do the division in your head.

For this reason it is quite common to express fractions in a form in which no radicals appear in the denominator. This is called **rationalizing the denominator**.

Examples 5 Rationalize the denominators.

(a) $\dfrac{2\sqrt{2}}{3\sqrt{7}}$.

(b) $\dfrac{1}{2\sqrt[3]{5}}$.

Solutions (a) Multiply numerator and denominator by $\sqrt{7}$:

$$\frac{2\sqrt{2}}{3\sqrt{7}} = \frac{2\sqrt{2} \cdot \sqrt{7}}{3(\sqrt{7})^2} = \frac{2\sqrt{14}}{3(7)} = \frac{2\sqrt{14}}{21}.$$

(b) Multiply numerator and denominator by $(\sqrt[3]{5})^2$:

$$\frac{1}{2\sqrt[3]{5}} = \frac{(\sqrt[3]{5})^2}{2(\sqrt[3]{5})(\sqrt[3]{5})^2} = \frac{\sqrt[3]{5}\sqrt[3]{5}}{2(\sqrt[3]{5})^3} = \frac{\sqrt[3]{5 \cdot 5}}{2 \cdot 5} = \frac{\sqrt[3]{25}}{10}.$$

---

*The advent of pocket calculators has rendered this topic less significant than it once was, and it may be omitted without loss of continuity.

Denominators can also be rationalized for algebraic fractions. Keep in mind that the aim is to get rid of all radicals from the denominator.

Examples 6   Rationalize the denominators.

(a) $\sqrt{\dfrac{x}{y}}.$ (b) $\dfrac{y}{2\sqrt[4]{x}}.$ (c) $\dfrac{3}{\sqrt{x}\sqrt[3]{y}}.$

Solutions   (a)   Multiply numerator and denominator by $\sqrt{y}$:

$$\sqrt{\dfrac{x}{y}} = \dfrac{\sqrt{x}}{\sqrt{y}} = \dfrac{\sqrt{x}\sqrt{y}}{(\sqrt{y})^2} = \dfrac{\sqrt{xy}}{y}.$$

(b)   Multiply numerator and denominator by $(\sqrt[4]{x})^3$:

$$\dfrac{y}{2\sqrt[4]{x}} = \dfrac{y(\sqrt[4]{x})^3}{2(\sqrt[4]{x})^4} = \dfrac{y\sqrt[4]{x^3}}{2x}.$$

Here we have used the following

$$(\sqrt[4]{x})^3 = \sqrt[4]{x} \cdot \sqrt[4]{x} \cdot \sqrt[4]{x} = \sqrt[4]{x \cdot x \cdot x} = \sqrt[4]{x^3}.$$

(See also Theorem 1 in Section 1.8.)

(c)   Multiply numerator and denominator by $\sqrt{x}(\sqrt[3]{y})^2$:

$$\dfrac{3}{\sqrt{x}\sqrt[3]{y}} = \dfrac{3\sqrt{x}(\sqrt[3]{y})^2}{(\sqrt{x})^2(\sqrt[3]{y})^3} = \dfrac{3\sqrt{x}\sqrt[3]{y^2}}{xy}.$$

Diagnostic Test 1.7

Fill in the blanks.

1. $\sqrt{a} = b$ only if _____ $= a$ and b _____.
2. If n is a positive odd integer, then $\sqrt[n]{a} = b$ only if _____ $= a$.
3. If n is a positive even integer, then $\sqrt[n]{a} = b$ only if _____ $= a$ and b _____.
4. $\sqrt[n]{a}$ is defined only if _____.
5. $\sqrt{a^2} =$ _____ if $a < 0$.
6. $\sqrt[3]{27} =$ _____ and $\sqrt[4]{16} =$ _____.
7. $\sqrt[3]{-8} =$ _____ and $\sqrt{-9} =$ _____.

Exercises 1.7

(1–12) Evaluate the following.

1. $\sqrt{81}$.
2. $\sqrt[3]{27}$.
3. $\sqrt{1\tfrac{9}{16}}$.
4. $\sqrt[3]{3\tfrac{3}{8}}$.
5. $\sqrt[5]{-32}$.
6. $\sqrt[3]{-0.125}$.
7. $\sqrt{(-3)^2}$.
8. $\sqrt{(-\tfrac{2}{5})^2}$.
9. $\sqrt[4]{-81}$.
10. $\sqrt[3]{\tfrac{125}{8}}$.
11. $\sqrt[4]{\tfrac{625}{1296}}$.
12. $\sqrt[4]{162(-3)^3}$.

(13–50) Simplify the following expressions.

13. $\sqrt{64x^4}$.
14. $\sqrt[3]{8y^6}$.
15. $\sqrt[3]{\dfrac{8a^3}{27b^3}}$.
16. $\sqrt[4]{\dfrac{625}{a^8}}$.

17. $\sqrt{8x}$.
18. $\sqrt{20a}$.
19. $\sqrt{27x^5}$.
20. $\sqrt{72b^3}$.

21. $\sqrt[3]{-8x^2}$.
22. $\sqrt[3]{-16x^4}$.
23. $\sqrt[5]{\frac{-512}{x^4}}$.
24. $\sqrt[4]{\frac{-16}{a^5}}$.

25. $\sqrt{8x^4y^3}$.
26. $\sqrt[3]{54a^4y^6}$.
27. $\sqrt{4xy^{-3}}$.
28. $\sqrt[3]{45x^{-2}y^{-4}}$.

29. $\sqrt[3]{\left(\frac{27x^3}{64}\right)^2}$.
30. $\sqrt[5]{32x^5y^{-10}}$.
31. $\sqrt{(4xy^2)^3}$.
32. $\sqrt[3]{(4xy^2)^2}$.

33. $(\sqrt[3]{-16ab^2})^2$.
34. $(\sqrt[3]{9a^4b^5})^2$.
35. $\sqrt{2xy}\sqrt{4x^2y^3}$.
36. $\sqrt{5x^3}\sqrt{6xy^3}$.

37. $\sqrt[3]{x^2y^{-1}}\sqrt[3]{2xy^2}$.
38. $\sqrt[3]{-5x^{-1}}\sqrt[3]{10x^3y^4}$.
39. $\sqrt[3]{\frac{4x^2}{y}}\sqrt[3]{\frac{4x}{y^2}}$.
40. $\sqrt{\frac{x}{2y^3}}\sqrt{\frac{y}{4x^3}}$.

41. $\sqrt[3]{\left(\frac{3p^{-1}}{q^2}\right)^2}\sqrt[3]{\left(\frac{p}{6q}\right)^5}$.
42. $\sqrt{\frac{x^{-2}y}{2z}}\sqrt{\frac{4z^{-1}}{xy}}$.
43. $3\sqrt{45}+\sqrt{20}$.
44. $2\sqrt{24}-\sqrt{54}$.

45. $2\sqrt{18}-\sqrt{32}$.
46. $\frac{8\sqrt{2}-4\sqrt{8}}{\sqrt{32}}$.
47. $\sqrt{63}-\sqrt{175}+4\sqrt{112}$.

48. $\sqrt{112}-\sqrt{63}+\frac{224}{\sqrt{28}}$.
49. $\frac{20}{\sqrt{5}}-2\sqrt{20}+\frac{50}{\sqrt{125}}$.
50. $2\sqrt[3]{-16}-\sqrt[3]{-54}$.

(51–62) Rationalize the denominators in the following fractions.

51. $\frac{6}{\sqrt{2}}$.
52. $\frac{2}{3\sqrt{3}}$.
53. $\frac{2\sqrt{7}}{3\sqrt{5}}$.
54. $\frac{4\sqrt{2}}{5\sqrt{3}}$.

55. $\frac{1}{\sqrt[3]{3}}$.
56. $\frac{2}{3\sqrt[3]{7}}$.
57. $\frac{x}{2\sqrt[3]{y}}$.
58. $\frac{2p}{\sqrt[3]{2q^2}}$.

59. $\frac{\sqrt{p}}{\sqrt{2q^3}}$.
60. $\frac{\sqrt{3x}}{2\sqrt{3y^3}}$.
61. $\frac{\sqrt[3]{x}}{\sqrt[3]{4yz^{-1}}}$.
62. $\frac{2\sqrt[3]{x}}{3\sqrt[3]{y}\sqrt[3]{z^2}}$.

*63. Show that the following are all true:

$$\sqrt{4^2}=|4|, \quad \sqrt{(-4)^2}=|-4|,$$
$$\sqrt{(-7)^2}=|-7|.$$

Show that for any real number x,

$$\sqrt{x^2}=|x|.$$

(This property is sometimes used as an alternative definition of $|x|$.)

*64. Prove that $\sqrt[n]{a}\sqrt[n]{b}=\sqrt[n]{ab}$. [*Hint:* Use property (1) to show that $(\sqrt[n]{a}\sqrt[n]{b})^n=ab$.]

1.8

Fractional Exponents

Having defined a^m when m is any integer, let us now attempt to extend the definition to the case when m is any rational number. We should like to make this extension in such a way that the Properties I–V in Section 1.6 continue to hold when m and n are no longer integers.

We consider first the definition of $a^{1/n}$, where n is a positive integer. If Property III is to remain true when $m = 1/n$, we must have that

$$(a^{1/n})^n = a^{(1/n)n} = a^1 = a.$$

So if we set $b = a^{1/n}$, then it is necessary that $b^n = a$.

What this means is that b must be the nth root of a. Thus we have the following definition.

DEFINITION   If n is any positive integer,

$$a^{1/n} = \sqrt[n]{a}.$$

Recall that there are certain restrictions when n is an even integer. If n is even, $a^{1/n}$ is not defined when a is negative; and when a is positive, $a^{1/n}$ denotes the *positive* nth root of a.

Examples 1   (a) $9^{1/2} = \sqrt{9} = 3$.
(b) $8^{1/3} = \sqrt[3]{8} = 2$.
(c) $(-243)^{1/5} = \sqrt[5]{-243} = -3$. Note that $(-3)^5 = -243$.
(d) $(-216)^{1/3} = -6$. Note that $(-6)^3 = -216$.
(e) $(-81)^{1/2}$ does not exist, because a negative number does not have a square root.
(f) $(-64)^{1/6}$ does not exist, because a negative number has no even roots.
(g) $1^{1/n} = 1$ for all positive integers n.
(h) $(-1)^{1/n} = -1$ if n is any odd integer.

Now we are in a position to define $a^{m/n}$ with a rational exponent, m/n.

DEFINITION   Let n be a positive integer and m any integer and let a be a real number. Then

$$a^{m/n} = (a^{1/n})^m.$$

That is, *the (m/n)th power of a is the mth power of the nth root of a*. If n is even, we must restrict $a > 0$. If m is negative or zero, we must restrict a to be nonzero.

Examples 2   (a) $9^{3/2} = (9^{1/2})^3 = 3^3 = 27$.
(b) $4^{-1/2} = (4^{1/2})^{-1} = 2^{-1} = \frac{1}{2}$.
(c) $16^{-3/4} = (16^{1/4})^{-3} = 2^{-3} = \frac{1}{8}$.

From the second of the examples we can generalize to the following result:

$$a^{-1/n} = \frac{1}{\sqrt[n]{a}}.$$

Sec. 1.8   *Fractional Exponents*

This follows since

$$a^{-1/n} = (a^{1/n})^{-1} = \frac{1}{a^{1/n}}.$$

It is possible to prove the following theorem:

THEOREM 1 $(a^{1/n})^m = (a^m)^{1/n}$ provided the left side is well defined.

That is, *the mth power of the nth root of a number is equal to the nth root of the mth power of the number.*

The condition that the left side be well defined is necessary to avoid such equations as $[(-1)^{1/2}]^2 = [(-1)^2]^{1/2}$, which is not true.

Observe that this theorem offers us an alternative method of calculating $a^{m/n}$:

$$a^{m/n} = (a^{1/n})^m = (a^m)^{1/n}.$$

Examples 3 (a) $16^{3/4} = (16^{1/4})^3 = 2^3 = 8$
or $16^{3/4} = (16^3)^{1/4} = (4096)^{1/4} = 8.$
(b) $36^{3/2} = (36^{1/2})^3 = 6^3 = 216$
or $36^{3/2} = (36^3)^{1/2} = (46{,}656)^{1/2} = 216.$

From these two examples it is clear that when evaluating $a^{m/n}$ it is generally easier to take the *n*th root first and then raise to the *m*th power, because then we shall be working with smaller numbers. In other words, in practice, to evaluate $a^{m/n}$ we use the definition $(a^{1/n})^m$ rather than $(a^m)^{1/n}$.

With these definitions we can show that all five laws of exponents stated in Section 1.6 remain valid for fractional exponents. Let us restate these laws and illustrate them with some examples.

I $a^m \cdot a^n = a^{m+n}.$ IV $(ab)^m = a^m b^m.$
II $\dfrac{a^m}{a^n} = a^{m-n}.$ V $\left(\dfrac{a}{b}\right)^m = \dfrac{a^m}{b^m}.$
III $(a^m)^n = a^{mn}.$

In writing these laws it must be borne in mind that certain restrictions apply: in any power if the exponent is negative or zero the base must not be zero, and if the exponent involves an even root the base must not be negative.

Examples 4 (a) $5^3 \cdot 5^{7/2} = 5^{3+7/2} = 5^{13/2}$ (by Law I)
(b) $4^{-2} \cdot 4^{7/3} = 4^{-2+7/3} = 4^{1/3}$ (by Law I)
(c) $\dfrac{(4)^{7/2}}{(4)^{3/2}} = 4^{7/2-3/2} = 4^2 = 16$ (by Law II)
(d) $\dfrac{9^{1/2}}{9^{-2}} = 9^{1/2-(-2)} = 9^{5/2} = (9^{1/2})^5 = 3^5 = 243$ (by Law II)

(e) $\dfrac{x^{9/4}}{x^4} = x^{(9/4)-4} = x^{-7/4}$ (by Law II)

(f) $(5^3)^{7/6} = 5^{3(7/6)} = 5^{7/2}$ (by Law III)

(g) $(3^{-4/3})^{-6/5} = 3^{(-4/3)(-6/5)} = 3^{8/5}$ (by Law III)

(h) $a^{-m} = (a^m)^{-1} = \dfrac{1}{a^m}$ for any rational number m (by Law III)

(i) $(36)^{1/2} = (4 \cdot 9)^{1/2} = 4^{1/2} \cdot 9^{1/2} = 2 \cdot 3 = 6$ (by Law IV)

(j) $(x^2 y)^{1/2} = (x^2)^{1/2} y^{1/2} = x^{2(1/2)} y^{1/2} = xy^{1/2}$ (by Law IV)

(k) $(3a^{2/5} b^{-4})^{-1/2} = 3^{-1/2}(a^{2/5})^{-1/2}(b^{-4})^{-1/2}$
$= 3^{-1/2} a^{-1/5} b^2$ (by Law IV)

(l) $\left(\dfrac{8}{27}\right)^{-2/3} = \dfrac{8^{-2/3}}{27^{-2/3}} = \dfrac{(8^{1/3})^{-2}}{(27^{1/3})^{-2}} = \dfrac{2^{-2}}{3^{-2}}$
$= (\tfrac{1}{4})/(\tfrac{1}{9}) = \tfrac{9}{4}$ (by Law V)

Examples 5 Evaluate.

(a) $(1\tfrac{64}{225})^{1/2}$.

(b) $\left(\dfrac{64x^3}{27}\right)^{-2/3}$.

Solutions (a) $(1\tfrac{64}{225})^{1/2} = \left(\dfrac{289}{225}\right)^{1/2} = \left(\dfrac{17^2}{15^2}\right)^{1/2}$
$= \left[\left(\dfrac{17}{15}\right)^2\right]^{1/2}$ (by Law V)
$= \left(\dfrac{17}{15}\right)^{2(1/2)}$ (by Law III)
$= \left(\dfrac{17}{15}\right)^1 = 1\tfrac{2}{15}$

(b) $\left(\dfrac{64x^3}{27}\right)^{-2/3} = \left(\dfrac{4^3 x^3}{3^3}\right)^{-2/3} = \left[\left(\dfrac{4x}{3}\right)^3\right]^{-2/3}$ (by Law V)
$= \left(\dfrac{4x}{3}\right)^{-2} = \dfrac{1}{\left(\dfrac{4x}{3}\right)^2}$ (by Law III)
$= \dfrac{1}{\dfrac{16x^2}{9}} = \dfrac{9}{16x^2}.$

Example 6 Simplify
$$\dfrac{(4^p)(27^{p/3})(125^p)(6^{2p})}{(8^{p/3})(9^{3p/2})(10^{3p})}.$$

Solution In expressions such as this, it is usually a good idea to express all the bases in terms of their prime factors. Thus we write:

$$\dfrac{(4^p)(27^{p/3})(125^p)(6^{2p})}{(8^{p/3})(9^{3p/2})(10^{3p})} = \dfrac{(2^2)^p (3^3)^{p/3} (5^3)^p (2 \cdot 3)^{2p}}{(2^3)^{p/3} (3^2)^{3p/2} (2 \cdot 5)^{3p}}$$

$$= \dfrac{2^{2p} \cdot 3^{3p/3} \cdot 5^{3p} \cdot 2^{2p} \cdot 3^{2p}}{2^{3 \cdot p/3} \cdot 3^{2 \cdot 3p/2} \cdot 2^{3p} \cdot 5^{3p}}$$

where we have used Laws III and IV,

$$= \dfrac{(2^{2p} \cdot 2^{2p})(3^p \cdot 3^{2p})(5^{3p})}{(2^p \cdot 2^{3p})(3^{3p})(5^{3p})}$$

Sec. 1.8 *Fractional Exponents*

where we have combined terms with like bases,

$$= \frac{2^{4p} \cdot 3^{3p} \cdot 5^{3p}}{2^{4p} \cdot 3^{3p} \cdot 5^{3p}} = 1.$$

Example 7 Which is greater, $2^{1/2}$ or $3^{1/3}$?

Solution We express the two exponents as equivalent fractions with the common denominator 6:

$$\tfrac{1}{2} = \tfrac{3}{6} \quad \text{and} \quad \tfrac{1}{3} = \tfrac{2}{6}.$$

Then

$$2^{1/2} = 2^{3/6} = (2^3)^{1/6} = 8^{1/6}$$

and

$$3^{1/3} = 3^{2/6} = (3^2)^{1/6} = 9^{1/6}.$$

Since $8 < 9$, it must follow that $8^{1/6} < 9^{1/6}$ and therefore $2^{1/2} < 3^{1/3}$.

Diagnostic Test 1.8

Fill in the blanks:

1. $\sqrt{a} = a$ _____.
2. $\sqrt[n]{a} = a$ _____.
3. $\sqrt[n]{a^m} = a$ _____.
4. $\dfrac{1}{\sqrt[3]{a^7}} = a$ _____.
5. If $a \geq 0$, m any integer, and n any positive even integer, then $(a^m)^{1/n} = ($ _____$)^m$.
6. If $a^n = b^n$ where n is an odd integer, then _____.
7. If n is an even integer and $a^n = b^n$, then _____.

Exercises 1.8

(1–22) Evaluate the following.

1. $81^{1/2}$.
2. $27^{1/3}$.
3. $(1\tfrac{9}{16})^{1/2}$.
4. $(3\tfrac{3}{8})^{1/3}$.
5. $(-32)^{1/5}$.
6. $(-0.125)^{1/3}$.
7. $9^{3/2}$.
8. $25^{-5/2}$.
9. $(-27)^{2/3}$.
10. $(-8)^{5/3}$.
11. $0.125^{-2/3}$.
12. $0.0016^{3/4}$.
13. $81^{-3/4}$.
14. $(8/27)^{-4/3}$.
15. $0.16^{-1/2}$.
16. $(-0.04)^{-1/2}$.
17. $(27)^{-2/3} \div (16)^{1/4}$.
18. $(9^{-3} \cdot 16^{3/2})^{1/6}$.
19. $9^{3/4} \cdot 3^{-1/2}$.
20. $16^{4/5} \cdot 8^{-2/5}$.
21. $25^{1/3}(\tfrac{1}{5})^{-2/3}$.
22. $-(\tfrac{1}{36})^{3/4} \div (6)^{-5/4}$.

(23–54) Simplify the following expressions.

23. $(16x^4)^{3/4}$.
24. $(3x^{5/2})(5x^{3/2})$.
25. $x(2x^{-1/4})$.
26. $\left(\dfrac{27x^3}{64}\right)^{2/3}$.
27. $\left(\dfrac{1}{8x}\right)^{1/3}$.
28. $\left(\dfrac{1}{16x^2}\right)^{-1/4}$.
29. $x^{1/2} \cdot x^{1/3} \cdot x^{1/4}$.
30. $z^{1/2} \cdot z^{1/3} \cdot z^{1/6}$.
31. $(32x^5 y^{-10})^{1/5}$.

32. $\left(\dfrac{8a^3}{27b^3}\right)^{1/3}$.

33. $(4x^3)^{1/2} \div (8x^2)^{1/3}$.

34. $(16x^{-4})^{-1/2} \div (8x^6)^{1/3}$.

35. $(16x^{1/2})^0$.

36. $(16x^0)^{1/2}$.

37. $\dfrac{x^{3/7}y^{2/5}}{x^{-1/7}y^{1/5}}$.

38. $\dfrac{a^{4/9}b^{-3/4}}{a^{2/9}b^{-1/2}}$.

39. $\left(\dfrac{p^{-1/5}q^{2/5}}{p^{-3/5}q^{-2/5}}\right)^{10}$.

40. $\dfrac{(x^2y)^{-1/3}(xy)^{1/4}}{(xy^{-2})^{1/12}}$.

41. $a^{2/3} \cdot a^{-3/4} \cdot (a^2)^{-1/6} \cdot \dfrac{1}{(a^{1/12})^5}$.

42. $a^{2/3} \cdot b^{-5/7} \cdot \left(\dfrac{a}{b}\right)^{7/8} \cdot \dfrac{a^{11/24}}{b^{23/56}}$.

43. $\dfrac{2^{3m} \cdot 3^{2m} \cdot 5^m \cdot 6^m}{8^m \cdot 9^{3m/2} \cdot 10^m}$.

44. $\dfrac{28^m \cdot 35^m \cdot 10^{3m}}{8^{5m/3} \cdot 49^m \cdot 25^{2m}}$.

45. $\dfrac{12^m \cdot 15^{2m} \cdot 6^m}{4^m \cdot 75^m \cdot 54^m}$.

46. $\dfrac{(x^{a+b})^2(y^{a-b})^2}{(xy)^{2a-b}}$.

47. $\left(\dfrac{x^a}{x^b}\right)^c \cdot \left(\dfrac{x^b}{x^c}\right)^a \cdot \left(\dfrac{x^c}{x^a}\right)^b$.

48. $\left(\dfrac{x^{a+b}}{x^{2b}}\right)\left(\dfrac{x^{b+c}}{x^{2c}}\right)\left(\dfrac{x^{c+a}}{x^{2a}}\right)$.

49. $\dfrac{(27)^{2n/3} \cdot (8)^{-n/6}}{(18)^{-n/2}}$.

50. $\dfrac{(4x^2)^{2/3}(125y^2)^{4/9}}{(10xy)^{1/3}}$.

51. $x^{1/2}(x^{1/2} + x^{-1/2})$.

52. $x^{-3/4}(2x^{3/2} + x^{3/4})$.

53. $\dfrac{x^{1/2} - 2x^{-1/2}}{x^{1/2}}$.

54. $\dfrac{2x^{-1/4} + 3x^{-3/4}}{(4x^{1/2})^{1/2}}$.

*(55–58) Find which is greater of each pair of numbers.

55. $3^{1/3}$ or $4^{1/4}$.

56. $2^{1/2}$ or $5^{1/5}$.

57. $3^{1/2}$ or $5^{1/3}$.

58. $3^{1/2}$ or $15^{1/5}$.

Chapter Review 1

Define and/or explain the following terms.

(a) Rational number, Irrational number, Real number.

(b) The real-number line.

(c) The four inequality symbols.

(d) The absolute value of a real number and its geometrical interpretation.

(e) Commutative properties of addition and multiplication.

(f) Associative properties of addition and multiplication.

(g) Distributive properties.

(h) Identity elements and the additive and multiplicative inverse of a real number.

(i) Subtraction.

(j) Division.

(k) Fraction.

(l) The procedures for multiplication and division of fractions.

(m) Reduction to lowest terms of a fraction.

(n) Least common denominator.

(o) The procedure for addition or subtraction of fractions.

(p) Positive integer exponents.

(q) Negative integer exponents.

(r) The five laws of exponents.

(s) Radical; nth root.

(t) Fractional exponents.

Review Exercises on Chapter 1

(1–2) Express each statement in terms of symbols.

1. x is greater than $-\frac{1}{2}$ and less than $\frac{3}{2}$.
2. The distance between y and $-\frac{3}{4}$ is less than $\frac{5}{2}$.

(3–4) Evaluate the following.

3. $|-6| - |-8|$.
4. $|-6 - (-8)|$.

(5–38) Simplify the following expressions.

5. $5(2a - 5b)$.
6. $7(-3x + 2y) - y$.
7. $-2a(1 - a) + 2a^2$.
8. $2p(q - 3p) + q(q - 2p)$.
9. $(2/x)(xy + 2) - (1/x)$.
10. $(pq)^{-1}(4p - 2q)$.
11. $\left(\dfrac{2p}{15q}\right)\left(\dfrac{3pq}{8}\right)$.
12. $\left(\dfrac{4}{9p^2}\right)\left(\dfrac{12pq}{5}\right)$.
13. $2xy \div \left(\dfrac{8x}{3y}\right)$.
14. $\left(-\dfrac{3a}{5b}\right) \div \left(\dfrac{6a^2b}{25}\right)$.
15. $\dfrac{3}{4x} - \dfrac{1}{6x}$.
16. $\dfrac{2a}{15b} - \dfrac{a}{6b}$.
17. $\dfrac{5x - \dfrac{3x}{4}}{\dfrac{21x}{8} - 2x}$.
18. $\dfrac{\left(\dfrac{2x}{3}\right)\left(\dfrac{6}{5x^2}\right) - x^{-1}}{4x - \dfrac{7x}{5}}$.
19. $\left[4x \div \left(\dfrac{24}{5x}\right)\right] + \dfrac{3x^2}{4}$.
20. $\dfrac{a}{2b} + \dfrac{c}{3d}$.
21. $(2x^5)^2 \div (2x^3)^3$.
22. $\dfrac{(3a^2b^{-1})^4}{(6a^{-1}b)^3}$.
23. $\dfrac{(-5p^{-2}q^{-3})^{-2}}{(-10p^2q)^{-3}}$.
24. $\dfrac{(-6x^{-1}y^{-2})^3}{(-2x^{-4}y^{-3})^4}$.
25. $x^{-5/2}(2x^6 - x^3)$.
26. $\dfrac{3x}{20y^{-2}} - \dfrac{2}{15x^{-3}}$.
27. $\sqrt[3]{\dfrac{-64x^4}{y^6}}$.
28. $\sqrt[4]{\dfrac{4x^3}{y^{-2}}} \sqrt{2xy}$.
29. $x\sqrt{18x^3} - 2\sqrt{8x^5}$.
30. $2\sqrt{\dfrac{3p^4}{q} - \dfrac{p^2}{\sqrt{3q}}}$.
31. $(2x^3)^{-1/2}(2x)^{1/2}$.
32. $(-2x^2y)^{1/5}(4^{-1}xy^{-2})^{-2/5}$.
33. $(r^{-2/5})^2(r^{3/10})^3(r^{-2/15})$.
34. $\dfrac{2x^{5/2}}{y^{3/4}} \div \dfrac{x^{2/3}}{3y^{2/5}}$.
35. $z^{-2/5}(z^{1/2} - 2z^{1/3})$.
36. $\dfrac{2p^{-1/4} - 3p^{-1/6}}{6p^{-1/3}}$.
37. $\dfrac{27^{3n/4} \cdot 32^{2n/5}}{24^{-n/2}}$.
38. $\dfrac{45^p \cdot 75^{2p/3}}{3^{4p/3} \cdot 15^{p/3}}$.

(39–40) Rationalize the denominators.

39. $2\sqrt{\dfrac{3}{x^3y}}$.
40. $\dfrac{\sqrt{x}}{\sqrt[3]{x^2}\sqrt{2y}}$.

2 ALGEBRAIC OPERATIONS

2.1

Algebraic Expressions

Quantities of the type $2x^2 - 3x + 7$, $5y^3 - y^2 + 6y + 2$, $2x - 3/y + 4$, and so on are known as *algebraic expressions*. The building blocks of an algebraic expression are called its **terms**. For example, the expression $2x^2 - 3x + 7$ has three terms: $2x^2$, $-3x$, and 7. The expression $x^2y/3 - y/x$ has two terms: $x^2y/3$ and $-y/x$.

In the term $2x^2$ the factor 2 is called the **numerical coefficient** (or simply the **coefficient**), and the factor x^2 is called the **literal part**. In the term $-3x$ the coefficient is -3 and the literal part is x. In the term $x^2y/3$ the coefficient is $1/3$ and the literal part is x^2y. The term 7 has no literal part and is called a **constant term**. Its coefficient is said to be 7.

An algebraic expression containing only one term is called a **monomial**. An expression that contains exactly two terms is a **binomial** and one that contains exactly three terms is a **trinomial**. Here are a few examples:

MONOMIALS: $2x^3, -5y^2, 7/t, 3, 2xy/z$.

BINOMIALS: $2x + 3, 3x^2 - 5/y, 6x^2y - 5zt$.

TRINOMIALS: $5x^2 + 7x - 1, 2x^3 + 4x - 3/x, 6y^2 - 5x + t$.

45

In general an algebraic expression containing more than one term is sometimes called a **multinomial**.

The letters appearing in an algebraic expression are called **variables** because we may wish to give them different values. The number of variables in an expression is simply the number of different letters that appear in it. For example, $5x^2 - 3x$ is an expression in the single variable x, while $3x^2y - \sqrt{xy}$ is an expression in two variables, x and y. We can **evaluate** an expression at a particular value of the variable or variables that appear in it simply by substituting the given value or values into the expression.

Example 1 Evaluate the expression $3x^2 + 4/x$ (a) when $x = 2$, (b) when $x = -1$.

Solution (a) We simply replace x by 2 wherever it appears in the given expression. We obtain
$$3(2)^2 + 4/2 = 3(4) + 2 = 12 + 2 = 14.$$

(b) For this part we replace x by -1, obtaining
$$3(-1)^2 + 4/(-1) = 3(1) - 4 = 3 - 4 = -1.$$

Example 2 When any object is released from rest and allowed to fall under gravity, the distance in feet that it falls in t seconds is given by the expression $16t^2$.* Find the distance fallen in (a) the first 3 seconds, (b) the first $4\frac{1}{2}$ seconds.

Solution (a) We simply need to replace t by 3 in the given expression. The distance fallen is then
$$16(3)^2 = 16(9) = 144 \text{ feet}.$$

(b) When $t = \frac{9}{2}$, the distance fallen is
$$16(\tfrac{9}{2})^2 = 16(\tfrac{81}{4}) = 324 \text{ feet}.$$

ADDITION AND SUBTRACTION OF EXPRESSIONS

When 4 apples are added to 3 apples we get 7 apples. In the same way, $4x + 3x = 7x$. This is simply a consequence of the distributive property:
$$4x + 3x = (4 + 3)x = 7x.$$

This distributive property enables us to add any two expressions whose literal parts are precisely the same. We simply add together the two numerical coefficients.

Examples 3 (a) $2x + 9x = (2 + 9)x = 11x.$
(b) $4ab + 3ab = (4 + 3)ab = 7ab.$
(c) $\dfrac{2x}{y} + \dfrac{x}{2y} = 2 \cdot \dfrac{x}{y} + \dfrac{1}{2} \cdot \dfrac{x}{y} = \left(2 + \dfrac{1}{2}\right)\dfrac{x}{y}$
$= \dfrac{5}{2} \cdot \dfrac{x}{y} = \dfrac{5x}{2y}.$

---

*This formula ignores air resistance.

Two or more terms in an algebraic expression are said to be *like terms* if they have identical literal parts. For example, the two terms $2x^2y$ and $5yx^2$ are like terms, since their literal parts, x^2y and yx^2, are identical. Similarly the three terms $3x^2yz^3$, $-7x^2z^3y$, and $yz^3x^2/2$ are all like terms. In general, two like terms can differ only in their numerical coefficients or in the order in which the variables appear.

We can add or subtract two or more like terms by making use of the distributive property just as in Examples 3 above.

Examples 4 (a) $2x^3 - 7x^3 = (2-7)x^3 = -5x^3$.
(b) $5x^2y - 3x^2y + 2yx^2 = (5-3+2)x^2y = 4x^2y$.

Terms that are not like cannot be combined in this manner. Thus the terms in the expression $2x^2 + 5xy$ cannot be combined into a single term.

When adding or subtracting two or more algebraic expressions, we rearrange the terms in the two expressions so that like terms are grouped together.

Example 5 Add $5x^2y^3 - 7xy^2 + 3x - 1$ to $6 - 2x + 4xy^2 + 3y^3x^2$.

Solution The required sum is

$$5x^2y^3 - 7xy^2 + 3x - 1 + (6 - 2x + 4xy^2 + 3y^3x^2)$$
$$= 5x^2y^3 - 7xy^2 + 3x - 1 + 6 - 2x + 4xy^2 + 3x^2y^3.$$

Rearranging the terms, so that like terms are grouped together, we obtain the sum in the form:

$$\underbrace{5x^2y^3 + 3x^2y^3}_{} - \underbrace{7xy^2 + 4xy^2}_{} + \underbrace{3x - 2x}_{} - \underbrace{1 + 6}_{}$$
$$= (5+3)x^2y^3 + (-7+4)xy^2 + (3-2)x + (-1+6)$$
$$= 8x^2y^3 + \quad (-3)xy^2 + \quad 1x + \quad 5$$
$$= 8x^2y^3 - 3xy^2 + x + 5.$$

Example 6 Subtract $3x^2 - 5xy + 7y^2$ from $7x^2 - 2xy + 4y^2 + 6$.

Solution In this case we want

$$7x^2 - 2xy + 4y^2 + 6 - (3x^2 - 5xy + 7y^2). \tag{i}$$

Upon removal of the parentheses, each term inside changes sign. Thus expression (i) is equivalent to

$$7x^2 - 2xy + 4y^2 + 6 - 3x^2 + 5xy - 7y^2$$
$$= 7x^2 - 3x^2 - 2xy + 5xy + 4y^2 - 7y^2 + 6$$
$$= (7-3)x^2 + (-2+5)xy + (4-7)y^2 + 6$$
$$= \quad 4x^2 + \quad 3xy + \quad (-3)y^2 + 6$$
$$= 4x^2 + 3xy - 3y^2 + 6.$$

Sec. 2.1 *Algebraic Expressions*

MULTIPLICATION BY MONOMIALS

The expression $a(x + y)$ denotes the product of a and $x + y$. To simplify this expression by removing the parentheses, we multiply each term inside by the number outside (in this case a)—that is,

$$a(x + y) = ax + ay.$$

We are simply using the distributive property. Similarly, for example,

$$-2(x - 3y + 7t^2) = (-2)x - (-2)(3y) + (-2)(7t^2)$$
$$= -2x + 6y - 14t^2.$$

This method works whenever a multinomial is multiplied by any monomial.

Examples 7 Simplify the following expressions by removing the parentheses.
(a) $2x(4x^{1/2} - 3x^{-1/2})$.
(b) $-x^2y(x^2 + 3x - 5y^3)$.

Solutions (a) Multiply each term inside the parentheses by $2x$:

$$2x(4x^{1/2} - 3x^{-1/2}) = 2x \cdot 4x^{1/2} - 2x \cdot 3x^{-1/2}$$
$$= 8x^{1+1/2} - 6x^{1-1/2} \qquad \text{(using the laws of exponents)}$$
$$= 8x^{3/2} - 6x^{1/2}.$$

(b) Multiply each term inside the parentheses by $-x^2y$:

$$-x^2y(x^2 + 3x - 5y^3) = (-x^2y)(x^2) + (-x^2y)(3x) - (-x^2y)(5y^3)$$
$$= -x^4y - 3x^3y + 5x^2y^4.$$

Example 8 Simplify the expression

$$3(x^3 + 2x - 4x^{-1}) - 2(x - 2x^3 - 3x^{-1}).$$

Solution We first use the distributive property to remove the parentheses, then we combine like terms. We obtain

$$3(x^3 + 2x - 4x^{-1}) - 2(x - 2x^3 - 3x^{-1})$$
$$= 3x^3 + 6x - 12x^{-1} - 2x + 4x^3 + 6x^{-1}$$
$$= 3x^3 + 4x^3 + 6x - 2x - 12x^{-1} + 6x^{-1}$$
$$= (3 + 4)x^3 + (6 - 2)x + (-12 + 6)x^{-1}$$
$$= 7x^3 + 4x - 6x^{-1}.$$

Example 9 Simplify the expression

$$xy(x - 3y) - 2x(2x^2 - xy - 4y^2) + 3x^2(y - 3x).$$

Solution Proceeding as in the last example, we get

$$xy(x - 3y) - 2x(2x^2 - xy - 4y^2) + 3x^2(y - 3x)$$
$$= xy(x) - xy(3y) + (-2x)(2x^2) - (-2x)(xy)$$
$$\qquad\qquad\qquad\qquad - (-2x)(4y^2) + 3x^2y - 3x^2(3x)$$
$$= x^2y - 3xy^2 - 4x^3 + 2x^2y + 8xy^2 + 3x^2y - 9x^3$$
$$= x^2y + 2x^2y + 3x^2y - 3xy^2 + 8xy^2 - 4x^3 - 9x^3$$
$$= (1 + 2 + 3)x^2y + (-3 + 8)xy^2 + (-4 - 9)x^3$$
$$= 6x^2y + 5xy^2 - 13x^3.$$

DIVISION BY MONOMIALS

We saw in Section 1.5 that the distributive law extends to division, and we have in general:

$$\frac{a + b}{c} = \frac{a}{c} + \frac{b}{c}.$$

This property is useful when we are dividing an algebraic expression by a monomial, since it allows us to divide each term separately by the monomial. For example, we can write

$$\frac{6x + 12}{3} = \frac{6x}{3} + \frac{12}{3} = 2x + 4.$$

Observe that the denominator 3 must be divided into *both* terms in the numerator.

Note It is *not true* in general that

$$\frac{c}{a + b} = \frac{c}{a} + \frac{c}{b}.$$

For example,

$$\frac{3}{6x + 12} \neq \frac{3}{6x} + \frac{3}{12} = \frac{1}{2x} + \frac{1}{4}.$$

(Just try putting $x = 0$ if you don't believe it.)

Examples 10 Simplify.

(a) $\dfrac{2x^2 + 4x}{2x}.$ (b) $\dfrac{25t^3 + 12t^2 + 15t - 6}{3t}.$

Solutions (a) $\dfrac{2x^2 + 4x}{2x} = \dfrac{2x^2}{2x} + \dfrac{4x}{2x} = x + 2$ after dividing out common factors.

(b) $\dfrac{25t^3 + 12t^2 + 15t - 6}{3t} = \dfrac{25t^3}{3t} + \dfrac{12t^2}{3t} + \dfrac{15t}{3t} - \dfrac{6}{3t}$

$$= \dfrac{25t^2}{3} + 4t + 5 - \dfrac{2}{t}.$$

Sec. 2.1 *Algebraic Expressions*

NESTED PARENTHESES

Sometimes we need to use expressions that involve parentheses inside other parentheses. For example, the expression $2[x + 3(x - 1)]$ is of this type. The expression $x + 3(x - 1)$ involves parentheses and is itself contained within another set of parentheses. Note the way in which parentheses of different shapes are used to make such expressions easier to read.

When only one set of parentheses is needed, we usually use round brackets (called simply parentheses). When two sets are needed, we usually use parentheses for the inner ones and square brackets, [. . .], for the outer ones. If three sets are needed, we use braces, {. . .}, for the outermost ones.

When simplifying expressions that involve parentheses nested inside other parentheses, we seek to minimize mistakes by always removing the innermost parentheses first.

Example 11 Simplify
$$3\{5x(2 - 3x) + 7[3 - 2(x - 4)]\}.$$

Solution The innermost parentheses are those containing $(x - 4)$, and we remove these first. We get
$$3\{5x(2 - 3x) + 7[3 - 2(x - 4)]\}$$
$$= 3\{5x(2 - 3x) + 7[3 - 2x + 8]\}$$
$$= 3\{10x - 15x^2 + 21 - 14x + 56\}$$
$$= 3\{-15x^2 + 10x - 14x + 21 + 56\}$$
$$= 3\{-15x^2 - 4x + 77\}$$
$$= -45x^2 - 12x + 231.$$

Diagnostic Test 2.1

Fill in the blanks.

1. $2x^2y - 3y^{-2}$ is a _____ nomial expression.
2. In the expression $3x^2 - 5xy^2$ the literal part in the second term is _____ and the coefficient is _____.
3. The value of $5x^2 - 2x$ when $x = -1$ is _____.
4. Two terms are like if they have equal _____.
5. $6zxy^2 - 3xy^2z + 2y^2xz =$ _____.
6. $5(x^2 - 2) - x(3x - 1) =$ _____.
7. $(3x^3y - 2xy^3)/2xy =$ _____.
8. $(15x^4 - 35x^2)/5x^2 =$ _____.

Exercises 2.1

(1–6) State whether each of the following expression is a monomial, binomial, or trinomial and give the number of variables.

1. $-2x^2 t$.
2. $\sqrt{x} - 2\sqrt{y}$.
3. $\sqrt{x - 2y}$.
4. $x^4 - 2\sqrt{x} + 1$.
5. $x^3 y + \dfrac{1}{3y} - \dfrac{y}{x}$.
6. $x^5 yz - y^3 t$.

(7–16) Evaluate the following expressions at the given value of the variables.

7. $3\sqrt{x} - 1;\ x = 4$.
8. $2x^4 - 3x^3;\ x = -1$.
9. $xy + y^3/2x;\ x = 2,\ y = 4$.
10. $\sqrt{x/y} + (y/x)^2;\ x = -1,\ y = -4$.
11. $3xy^2 - 4x^2 y + y^3;\ x = 1,\ y = -1$.
12. $2x^3 - 3xy + 5y^2;\ x = 2,\ y = -1$.
13. $2x^2/y - 3y^2/x + 1;\ x = -3,\ y = 4$.
14. $x^5 - 2x^3 y + 3y^3;\ x = -2,\ y = 3$.
15. $-x\sqrt{x^2 + y^2};\ x = 3,\ y = 4$.
16. $\sqrt{x^2 + 2xy + y^2};\ x = 5,\ y = 9$.

17. If an object travels a distance d (in kilometers) in a time t (in hours), then its average velocity is dt^{-1}. Calculate the average velocity when $d = 200$ kilometers and $t = 3$ hours.

18. The acceleration of an object is $(v - u)t^{-1}$, where u is the initial velocity, v is the final velocity and t is the time during which the velocity changes. Calculate the acceleration if (a) $u = 10$ feet per second, $v = 30$ feet per second, and $t = 4$ seconds and (b) $u = 50$ feet per second, $v = 20$ feet per second, and $t = 1\frac{1}{2}$ seconds.

(19–54) Simplify the following expressions.

19. $(x + 4y) + (3x - 2y)$.
20. $(a - 3b) - (2a - 5b)$.
21. $(5a + 7b - 3) + (3b - 2a + 9)$.
22. $(3x^2 - 5x + 7) + (-2 + 6x - 7x^2 + x^3)$.
23. $(2\sqrt{a} + 5\sqrt{b}) + (3\sqrt{a} - 2\sqrt{b})$.
24. $(4xy + 5x^2 y - 6x^3) + (3y^3 - 6xy^2 + 7xy + x^3 - 2x^2 y)$.
25. $(7t^2 + 6t - 1) - (3t - 5t^2 + 4 - t^3)$.
26. $(x^2 + 3xy + 4y^2) - (2x^2 - xy + 3y^2 - 5)$.
27. $(2\sqrt{x} + \sqrt{2y}) - (\sqrt{x} - 2\sqrt{2y})$.
28. $(5\sqrt{xy} - 3) - (2 - 4\sqrt{xy})$.
29. $2(x - 4y) + 3(2x + 3y)$.
30. $4(2x + 3y) + 2(5y + 3x)$.
31. $-(x - 7y) - 2(2y - 5x)$.
32. $5(2y - 4x) - 4(x - 2y)$.
33. $2(x^2 + 4xy - 2y^2) + 3(2x^2 - 2xy + 5y^2)$.
34. $3(x^2 - 2xy + y^2) - (2xy - x^2 + 2y^2)$.
35. $x(2x^2 + 3xy + y^2) - y(5x^2 - 2xy + y^2)$.
36. $a^2 b(a^3 + 5ab - b^3) + 2ab(a^4 - 2a^2 b + b^3 a)$.
37. $3xy(-x^3 + 2x^2 y - xy^2 + 3y^3) - xy^2(2x^2 - 3xy - y^2)$.
38. $2(x^4 - 3x^2 y^2 + 2xy^3) - 3x(2x^3 - x^2 y + 3xy^2) - 2x^2(3y^2 - 2x^2)$.
39. $\dfrac{4x^3 - 3x^2}{2x}$.
40. $\dfrac{15x^5 - 25x^3}{5x^2}$.
41. $\dfrac{4x^2 y + 8xy^2}{4xy}$.
42. $\dfrac{y^3 z^2 - 6yz^4}{3yz^2}$.
43. $\dfrac{x^3 + 7x^2 - 5x + 4}{x^2}$.
44. $\dfrac{y^4 + 6y^3 - 7y^2 + 9y - 3}{3y^2}$.
45. $\dfrac{t^2 - 2t + 7}{\sqrt{t}}$.
46. $\dfrac{t^3 + 2t^2 - 3t + 1}{t\sqrt{t}}$.
47. $\dfrac{6x^2 y - 8xy^2}{2xy} + \dfrac{x^3 y^2 + 2x^2 y^3}{x^2 y^2}$.
48. $\dfrac{3x^4 - 9x^2 y^2}{3x^3 y} - \dfrac{4x^3 - 8xy^2}{2x^2 y}$.

49. $\dfrac{p^3 - 2p^2q - pq^2 + 3q^3}{2pq} - \dfrac{4p^2 - 3pq - q^2}{2p}$.

50. $\dfrac{2t^4 - 4t^2 + 6}{4t^2} - \dfrac{t - 3t^3 + 2t^5}{3t^3}$.

51. $3\{x^2 - 5[x + 2(3 - 5x)]\}$.

52. $2\{a^2 - 2a[3a - 5(a^2 - 2)]\} + 7a^2 - 3a + 6$.

53. $2a[a + 3(a - 5)] - 4[a(a + 2) - 2(3 - 2a)]$.

54. $3\{2a[a + 3a(2 - a)] + 4\} - 5[2(a^2 - 4) - 3a(1 - a)]$.

2.2

Products of Multinomial Expressions

An expression such as $(x + 2)(y + 3)$ involves the product of two binomial expressions, $(x + 2)$ and $(y + 3)$. We can simplify expressions such as this by using the distributive property more than once. First we remove the first set of parentheses,

$$(x + 2)(y + 3) = x(y + 3) + 2(y + 3).$$

In this use of the distributive property, we treat $(y + 3)$ as a single real number that multiplies each of the numbers, x and 2, in the first parentheses. We now use the distributive property twice more and write

$$x(y + 3) = xy + x \cdot 3 = xy + 3x$$
$$2(y + 3) = 2y + 2 \cdot 3 = 2y + 6.$$

Combining these, we have

$$(x + 2)(y + 3) = xy + 3x + 2y + 6.$$

We say that the product on the left has been *expanded*.

Example 1 Expand the product $(x - 3)(2x + 5)$.

Solution Using the distributive property repeatedly, we get

$$(x - 3)(2x + 5) = x(2x + 5) - 3(2x + 5)$$
$$= x \cdot 2x + x \cdot 5 - 3 \cdot 2x - 3 \cdot 5$$
$$= 2x^2 + 5x - 6x - 15$$
$$= 2x^2 - x - 15.$$

This calculation can be set out like an ordinary arithmetical multiplication.

$$\begin{array}{rl} & 2x + 5 \\ & x - 3 \\ \hline 2x^2 + 5x & \leftarrow \quad x(2x + 5) \\ -6x - 15 & \leftarrow \quad -3(2x + 5) \\ \hline 2x^2 - x - 15 & \leftarrow \quad \text{Add} \end{array}$$

Such products as this arise very often, and we can expand them much more rapidly by using the *method of arcs* as explained below.

Example 2 Use the method of arcs to repeat Example 1.

Solution We write down the required product $(x - 3)(2x + 5)$ and draw arcs connecting each term in the first parentheses to each term in the second parentheses. We write down the product for each arc, then add all such products.

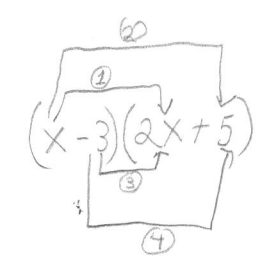

$$x \cdot 2x \qquad x \cdot 5 \qquad x \cdot 2x + x \cdot 5 = 2x^2 + 5x$$
$$(x - 3)(2x + 5) \qquad\qquad\qquad\qquad +$$
$$-3 \cdot 2x \qquad -3 \cdot 5 \qquad -3 \cdot 2x - 3 \cdot 5 = \underline{\quad -6x - 15\quad}$$
$$2x^2 - x - 15$$

Example 3 Use the method of arcs to expand the product

$$(2x - 3y)(3x + 2y).$$

Solution The diagram is as follows:

$$2x \cdot 3x \qquad 2x \cdot 2y \qquad 2x \cdot 3x + 2x \cdot 2y = 6x^2 + 4xy$$
$$(2x - 3y)(3x + 2y) \qquad\qquad\qquad\qquad +$$
$$-3y \cdot 3x \qquad -3y \cdot 2y \qquad -3y \cdot 3x - 3y \cdot 2y = \underline{\quad -9xy - 6y^2\quad}$$
$$6x^2 - 5xy - 6y^2$$

The result is that

$$(2x - 3y)(3x + 2y) = 6x^2 - 5xy - 6y^2.$$

Example 4 Expand the product

$$(3x - 4)(6x^2 - 5x + 2).$$

Solution Using the distributive property, we obtain

$$(3x - 4)(6x^2 - 5x + 2) = 3x(6x^2 - 5x + 2) - 4(6x^2 - 5x + 2)$$
$$= (3x)(6x^2) - (3x)(5x) + (3x)(2)$$
$$\quad + (-4)(6x^2) - (-4)(5x) + (-4)(2)$$
$$= 18x^3 - 15x^2 + 6x - 24x^2 + 20x - 8$$
$$= 18x^3 - 15x^2 - 24x^2 + 6x + 20x - 8$$

(after grouping like terms together)

$$= 18x^3 - (15 + 24)x^2 + (6 + 20)x - 8$$
$$= 18x^3 - 39x^2 + 26x - 8.$$

Alternatively, we can obtain the answer by drawing arcs connecting each term in the first parentheses to each term in the second. In this case there are six such arcs, giving six products in the expansion on the right.

Sec. 2.2   Products of Multinomial Expressions

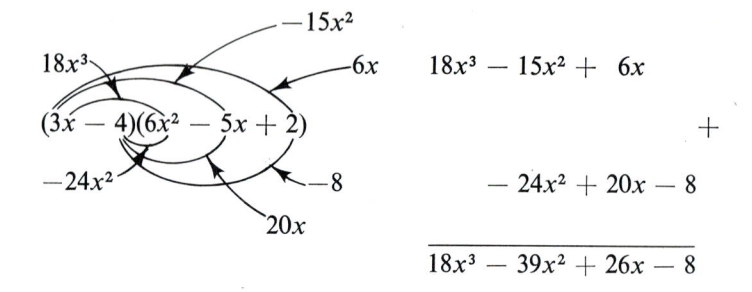

$$
\begin{array}{c}
\overset{\displaystyle 18x^3}{\underset{\displaystyle -24x^2}{(3x-4)(6x^2}} \overset{\displaystyle -15x^2}{\underset{\displaystyle 20x}{-5x}} \overset{\displaystyle -6x}{\underset{\displaystyle -8}{+2)}}
\end{array}
\qquad
\begin{array}{r}
18x^3 - 15x^2 + 6x \\
+ \\
-24x^2 + 20x - 8 \\
\hline
18x^3 - 39x^2 + 26x - 8
\end{array}
$$

SPECIAL PRODUCTS

Certain special products that are often encountered may be treated as standard formulas. First consider the product $(x + a)(x + b)$:

$$(x + a)(x + b) = x(x + b) + a(x + b)$$
$$= x^2 + bx + ax + ab$$
$$= x^2 + (b + a)x + ab.$$

Therefore,

$$(x + a)(x + b) = x^2 + (a + b)x + ab. \qquad (1)$$

Examples 5 Expand the products.

    (a) $(x + 2)(x + 7)$.   (b) $(x + 3)(x - 2)$.

Solutions (a) Taking $a = 2$ and $b = 7$ in Equation (1), we get

$$(x + 2)(x + 7) = x^2 + (2 + 7)x + 2 \cdot 7$$
$$= x^2 + 9x + 14.$$

(b) Taking $a = 3$ and $b = -2$, we get

$$(x + 3)(x - 2) = (x + 3)[x + (-2)]$$
$$= x^2 + [3 + (-2)]x + 3(-2)$$
$$= x^2 + x - 6.$$

We obtain another general formula called the ***binomial-square formula*** by taking b equal to a in Equation (1). We get

$$(x + a)(x + a) = x^2 + (a + a)x + a \cdot a$$

or

$$(x + a)^2 = x^2 + 2ax + a^2. \qquad (2)$$

This result gives an expansion for the square of a binomial. *The square of the sum of two terms is equal to the sum of the squares of the two terms plus twice their product.*

Examples 6 Expand.

(a) $(t + 2)^2$. (b) $(2t + 7)^2$. (c) $\left(3x + \dfrac{4}{y}\right)^2$.

Solutions (a) $(t + 2)^2 = t^2 + 2 \cdot 2 \cdot t + 2^2$
$= t^2 + 4t + 4.$

(b) $(2t + 7)^2 = (2t)^2 + 2 \cdot 7 \cdot 2t + 7^2$

[Note that $x = 2t$ and $a = 7$ in Equation (2)]

$= 4t^2 + 28t + 49.$

(c) $\left(3x + \dfrac{4}{y}\right)^2 = (3x)^2 + 2 \cdot 3x \cdot \dfrac{4}{y} + \left(\dfrac{4}{y}\right)^2$

$= 9x^2 + \dfrac{24x}{y} + \dfrac{16}{y^2}.$

If we set $a = -b$ in Equation (2), we get

$$(x - b)^2 = x^2 + 2(-b)x + (-b)^2 = x^2 - 2bx + b^2.$$

Replacing b by a to have the same notation as in (2), we have

$$\boxed{(x - a)^2 = x^2 - 2ax + a^2.} \qquad (3)$$

This expresses the *square of the difference of two terms as equal to the sum of the squares of the two terms minus twice their product*.

Example 7 Expand $(2y - 5)^2$.

Solution From Equation (3), with $x = 2y$ and $a = 5$,

$(2y - 5)^2 = (2y)^2 - 2 \cdot 5 \cdot 2y + 5^2$
$= 4y^2 - 20y + 25.$

A fourth general formula is as follows:

$$\boxed{(x + a)(x - a) = x^2 - a^2.} \qquad (4)$$

This result states that the *product of the sum and difference of two terms is the difference of their squares*. Equation (4) is commonly called the *difference-of-two-squares formula*.

We can deduce this formula by putting $b = -a$ in Equation (1). We get

$(x + a)[x + (-a)] = x^2 + [a + (-a)]x + a(-a)$
$= x^2 + 0x - a^2$

—that is,

$(x + a)(x - a) = x^2 - a^2$

as required. [Alternatively, we can prove Equation (4) directly by multiplying out the left-hand side.]

Sec. 2.2 Products of Multinomial Expressions

Examples 8 Expand the products.

(a) $(2x + 3)(2x - 3)$. (b) $(\sqrt{3} + \sqrt{2})(\sqrt{3} - \sqrt{2})$.
(c) $(3x - 4y)(3x + 4y)$.

Solutions (a) $(2x + 3)(2x - 3) = (2x)^2 - 3^2 = 4x^2 - 9$.
(b) $(\sqrt{3} + \sqrt{2})(\sqrt{3} - \sqrt{2}) = (\sqrt{3})^2 - (\sqrt{2})^2 = 3 - 2 = 1$.
(c) $(3x - 4y)(3x + 4y) = (3x)^2 - (4y)^2 = 9x^2 - 16y^2$.

Example 9 Simplify
$$(t + 3)[(2t + 1)(t - 1) - 2(t + 1)^2].$$

Solution We first simplify the expression inside the square brackets by removing the inner parentheses. Using the method of arcs, we find that
$$(2t + 1)(t - 1) = 2t^2 - t - 1 \quad \text{(check this)}.$$
From the binomial-square formula,
$$2(t + 1)^2 = 2(t^2 + 2t + 1) = 2t^2 + 4t + 2.$$
Therefore the given expression is
$$(t + 3)[(2t + 1)(t - 1) - 2(t + 1)^2]$$
$$= (t + 3)[2t^2 - t - 1 - (2t^2 + 4t + 2)]$$
$$= (t + 3)[2t^2 - t - 1 - 2t^2 - 4t - 2]$$
$$= (t + 3)(-5t - 3)$$
$$= -5t^2 - 15t - 3t - 9 \quad \text{(method of arcs again)}$$
$$= -5t^2 - 18t - 9.$$

MORE ON RATIONALIZING DENOMINATORS*

As we saw in Section 1.7, it is sometimes useful to be able to remove radicals from the denominator of a fraction. The formula for the difference of two squares can help us do this when the denominator contains two terms. For example, we can rationalize the denominators of fractions such as
$$\frac{2}{3 - \sqrt{2}} \quad \text{or} \quad \frac{x}{\sqrt{x + 2} - \sqrt{2}}$$
by making use of this formula.

Consider the first of these two fractions. What we do is to multiply numerator and denominator by $(3 + \sqrt{2})$:
$$\frac{2}{3 - \sqrt{2}} = \frac{2(3 + \sqrt{2})}{(3 - \sqrt{2})(3 + \sqrt{2})}.$$
The reason this works is that we can simplify the denominator in the new fraction by using the formula for the difference of two squares:
$$(3 - \sqrt{2})(3 + \sqrt{2}) = 3^2 - (\sqrt{2})^2 = 9 - 2 = 7.$$

---

*May be omitted without loss of continuity.

Therefore,
$$\frac{2}{3-\sqrt{2}} = \frac{2(3+\sqrt{2})}{7}$$
and the denominator has been rationalized.

In general, to rationalize a fraction involving an expression of the form $A + \sqrt{B}$ in the denominator, we multiply top and bottom by $A - \sqrt{B}$. If $A - \sqrt{B}$ occurs, we multiply top and bottom by $A + \sqrt{B}$. More generally, if a factor of the type $(P\sqrt{A} \pm Q\sqrt{B})$ occurs in the denominator, we multiply top and bottom by $(P\sqrt{A} \mp Q\sqrt{B})$ (note the change in sign of the second term). This is illustrated in the following example.

Examples 10 Rationalize the denominators.

(a) $\dfrac{2}{3+\sqrt{5}}$.

(b) $\dfrac{1}{2\sqrt{5} - 3\sqrt{3}}$.

Solution (a) Since the factor $(3 + \sqrt{5})$ occurs, we multiply top and bottom by $(3 - \sqrt{5})$:

$$\frac{2}{3+\sqrt{5}} = \frac{2(3-\sqrt{5})}{(3+\sqrt{5})(3-\sqrt{5})} = \frac{2(3-\sqrt{5})}{3^2 - (\sqrt{5})^2}$$

$$= \frac{2(3-\sqrt{5})}{9-5} = \frac{2(3-\sqrt{5})}{4} = \frac{1}{2}(3-\sqrt{5}).$$

(b) The factor $(2\sqrt{5} - 3\sqrt{3})$ occurs in the denominator, so we multiply by $(2\sqrt{5} + 3\sqrt{3})$:

$$\frac{1}{2\sqrt{5} - 3\sqrt{3}} = \frac{1(2\sqrt{5} + 3\sqrt{3})}{(2\sqrt{5} - 3\sqrt{3})(2\sqrt{5} + 3\sqrt{3})}$$

$$= \frac{2\sqrt{5} + 3\sqrt{3}}{(2\sqrt{5})^2 - (3\sqrt{3})^2}$$

$$= \frac{2\sqrt{5} + 3\sqrt{3}}{4(5) - 9(3)}$$

$$= \frac{2\sqrt{5} + 3\sqrt{3}}{20 - 27}$$

$$= -\frac{1}{7}(2\sqrt{5} + 3\sqrt{3}).$$

Diagnostic Test 2.2

Fill in the blanks.

1. $(a + b)^2 = $ _____.
2. $(x - a)^2 = $ _____.
3. $(2x + 3y)^2 = $ _____.
4. $(3x - 5y)^2 = $ _____.
5. $(2x - 3)(3x - 4) = $ _____.
6. $(x + 2)(4x - 7) = $ _____.
7. $(a + 4)(a - 4) = $ _____.
8. $(3x + 7)(3x - 7) = $ _____.
9. $2x(x - 3)(y + 4) = $ _____.
10. $3(x + 4)(2y - 5) = $ _____.
11. $2[3x - 5(2 - 4x)] = $ _____.
*12. $1/(\sqrt{3} - \sqrt{2}) = $ _____.

Exercises 2.2

(1–10) Use the distributive law to expand the following products:

1. $(x + 1)(y + 3)$.
2. $(x + 4)(y + 3)$.
3. $(a - 2)(b + 4)$.
4. $(a + 2)(b - 6)$.
5. $(3x - 2)(y - 5)$.
6. $(2x - 3)(3y - 4)$.
7. $(2x - 3)^2$.
8. $(3y + 2)^2$.
9. $(x + 2)(x^2 + 2x + 3)$.
10. $(y - 1)(y^2 + 3y - 2)$.

(11–24) Use the method of arcs to expand the following products:

11. $(x - 3)(y + 2)$.
12. $(x + 4)(y - 5)$.
13. $(2x + 1)(3y - 4)$.
14. $(5x - 2)(2y - 5)$.
15. $(a + 2)(3a - 4)$.
16. $(x + 3y)(2x + y)$.
17. $(x + 3)(2x^2 - 5x + 7)$.
18. $(a - 2b)(a^2 - 2ab + b^2)$.
19. $(\sqrt{a} - \sqrt{b})(\sqrt{a} + \sqrt{b})$.
20. $(\sqrt{x} + 3\sqrt{y})(\sqrt{x} - 3\sqrt{y})$.
21. $(x + y - z)(x + y + z)$.
22. $(x - 2y + z)(x + 2y + z)$.
23. $\left(x^2 - \dfrac{1}{x}\right)(x^3 + 2x)$.
24. $(y^2 + 2y)(y^3 - 2y^2 + 1)$.

(25–50) Expand and simplify the following expressions.

25. $(x + 6)(x + 8)$.
26. $(y + 2)(y + 10)$.
27. $\left(y - \dfrac{3}{y}\right)\left(y - \dfrac{8}{y}\right)$.
28. $(z - 4)(z - 2)$.
29. $(x + 10)^2$.
30. $(y + 6)^2$.
31. $(x - 5)^2$.
32. $(x - 8)^2$.
33. $(x + 4)(x - 4)$.
34. $(y^2 - 2)(y^2 + 2)$.
35. $(3t + 4x)^2$.
36. $(4x - 5y)^2$.
37. $(2t + 5x)(2t - 5x)$.
38. $(5\sqrt{x} + 2y)(5\sqrt{x} - 2y)$.
39. $(2x + 3y)^2 + (2x - 3y)^2$.
40. $(3a - b)^2 + 3(a + b)^2$.
41. $(\sqrt{2}x - \sqrt{3y})^2$.
42. $(\sqrt{x} + 2\sqrt{y})^2$.
43. $(x^2y + y/x)^2$.
44. $(2xy - x/y)(xy^2 + 2y/x)$.
45. $3[(x + y)^2 - (x - y)^2]$.
46. $xy[(x + y)^2 + (x - y)^2]$.
47. $(x^a + 1)(x^a + 2)$.
48. $(p^q - 3)^2$.
49. $3p\{[4(p + q)(p - q) + q^2] - (2p - q)^2\}$.
50. $(x - 1)[2(2x - 1)(x + 1) - (2x + 1)^2]$.
51. $2a\{(a + 2)(3a - 1) - [a + 2(a - 1)(a + 3)]\}$.
52. $(a + 3b)(a^2 - 3ab + b^2) - (a + b)^2(a + 2b)$.

(53–64) Rationalize the denominators of the following expressions.

53. $\dfrac{1}{3 + \sqrt{7}}$.
54. $\dfrac{3 + \sqrt{2}}{2 - \sqrt{3}}$.
55. $\dfrac{1 + \sqrt{2}}{\sqrt{5} + \sqrt{3}}$.
56. $\dfrac{6\sqrt{2}}{\sqrt{3} + \sqrt{6}}$.
57. $\dfrac{3}{3 + \sqrt{3}}$.
58. $\dfrac{1}{2\sqrt{3} - \sqrt{6}}$.
59. $\dfrac{1}{\sqrt{x} - \sqrt{y}}$.
60. $\dfrac{\sqrt{x} - \sqrt{y}}{\sqrt{x} + \sqrt{y}}$.
61. $\dfrac{x}{\sqrt{x + 2} - \sqrt{2}}$.
62. $\dfrac{x}{\sqrt{x + 1} - \sqrt{x - 1}}$.
63. $\dfrac{2x - 2}{\sqrt{x + 3} - 2\sqrt{x}}$.
64. $\dfrac{4 - x}{2\sqrt{x + 5} - 3\sqrt{x}}$.

65. Show that
$$\left(\dfrac{x^a}{x^b}\right)^{a+b}\left(\dfrac{x^b}{x^c}\right)^{b+c}\left(\dfrac{x^c}{x^a}\right)^{c+a} = 1.$$

66. Given that $\sqrt{5} \approx 2.236$, evaluate $1/(\sqrt{5} - 2)$ without using tables or calculator or long division.

2.3

Factoring

The idea of factors is used also for algebraic expressions. If two (or more) algebraic expressions are multiplied together, these expressions are said to be *factors* of the expression obtained as their product. For example, the expression $2xy$ is obtained by multiplying 2, x, and y, so these three are factors of $2xy$. Furthermore, for example, $2y$ is a factor of $2xy$, since we can obtain $2xy$ by multiplying $2y$ by x.

Similarly x is a factor of the expression $2x^2 + 3x$, since we can write $(2x^2 + 3x) = x(2x + 3)$; and x^2 is a factor of $6x^2 + 9x^3$, since we can write $(6x^2 + 9x^3) = x^2(6 + 9x)$.

The process of writing a given expression as the product of its factors is called *factoring*. In this section we shall discuss certain methods by which multinomial expressions can be factored.

MONOMIAL FACTORS

The first step in factoring a multinomial expression is to extract all the monomial factors that are common to all the terms. The following examples illustrate this step.

Examples 1 Factor all the common monomial factors from the following expressions.

(a) $x^2 + 2xy^2$. (b) $2x^2y + 6xy^2$.
(c) $6ab^2c^3 + 6a^2b^2c^2 + 18a^3bc^2$.

Solutions (a) Let us break each term in the given expression into its basic factors. We write

$$x^2 = x \cdot x, \qquad 2xy^2 = 2 \cdot x \cdot y \cdot y.$$

Looking at the two lists of basic factors, we see that only the factor x is common to both terms. So we write

$$x^2 + 2xy^2 = x \cdot x + x \cdot 2y^2$$
$$= x(x + 2y^2).$$

Observe how we use the distributive property in order to extract the common factor x.

(b) Expressing each term in terms of basic factors, we get

$$2x^2y = 2 \cdot x \cdot x \cdot y, \qquad 6xy^2 = 2 \cdot 3 \cdot x \cdot y \cdot y.$$

Sec. 2.3 Factoring   59

We notice that this time the factors 2, x, and y occur in both lists, so the common factor to be extracted will be $2xy$. So we write

$$2x^2y + 6xy^2 = 2xy \cdot x + 2xy \cdot 3y$$
$$= 2xy(x + 3y)$$

again using the distributive property.

(c) We write

$$6ab^2c^3 = 2 \cdot 3 \cdot a \cdot b \cdot b \cdot c \cdot c \cdot c$$
$$6a^2b^2c^2 = 2 \cdot 3 \cdot a \cdot a \cdot b \cdot b \cdot c \cdot c$$
$$18a^3bc^2 = 2 \cdot 3 \cdot 3 \cdot a \cdot a \cdot a \cdot b \cdot c \cdot c.$$

The common factors of these three terms are $2 \cdot 3 \cdot a \cdot b \cdot c \cdot c = 6abc^2$. So we write

$$6ab^2c^3 + 6a^2b^2c^2 + 18a^3bc^2 = 6abc^2 \cdot bc + 6abc^2 \cdot ab + 6abc^2 \cdot 3a^2$$
$$= 6abc^2(bc + ab + 3a^2).$$

DIFFERENCE OF SQUARES

Now let us turn to the question of extracting factors that are themselves binomial expressions. Some of the formulas established in Section 2.2 are useful for factoring, in particular the formula:

$$a^2 - b^2 = (a - b)(a + b). \qquad (I)$$

This formula can be used to factor any expression that is reducible to a *difference of two squares*.

Example 2 Factor completely.

(a) $x^2 - 9$. (b) $5x^4 - 80y^4$.

Solution (a) The given expression can be written as

$$x^2 - 3^2$$

which is the difference of two squares. Using Equation (I) with $a = x$ and $b = 3$, we have

$$x^2 - 9 = (x)^2 - 3^2 = (x - 3)(x + 3).$$

(b) $5x^4 - 80y^4$. First we check whether we can take out any common monomial factor. In this case, we take 5 as a common factor. Thus,

$$5x^4 - 80y^4 = 5(x^4 - 16y^4).$$

The term in parentheses on the right is the difference between two squares:

$$5x^4 - 80y^4 = 5[(x^2)^2 - (4y^2)^2]$$
$$= 5[(x^2 - 4y^2)(x^2 + 4y^2)]$$
$$= 5(x^2 - 4y^2)(x^2 + 4y^2). \qquad (*)$$

The factoring is not complete, because $x^2 - 4y^2 = x^2 - (2y)^2$ can be further factored into $(x - 2y)(x + 2y)$. Thus from (*) we have

$$5x^4 - 80y^4 = 5(x^2 - 4y^2)(x^2 + 4y^2)$$
$$= 5(x - 2y)(x + 2y)(x^2 + 4y^2).$$

Notes Equation (I) allows us to factor any expression that takes the form of the *difference* between two squares. There is no corresponding formula expressing the sum $a^2 + b^2$ as the product of two or more factors. An expression such as $a^2 + b^2$ or $4x^2 + 9y^2$ that involves the *sum* of two squares cannot be factored (except possibly for a monomial factor).

GROUPING

A technique useful in factoring multinomial expressions that involve an even number of terms is the ***grouping method***. The terms are grouped in pairs and the common monomial factors extracted from each pair. This often reveals a binomial factor that is common to all the pairs. This method is particularly useful for expressions containing four terms.

Example 3 Factor $ax^2 + by^2 + bx^2 + ay^2$.

Solution We can group the terms in the given expression into those that have x^2 as a factor and those that have y^2:

$$(ax^2 + bx^2) + (ay^2 + by^2). \qquad \text{(i)}$$

Each term in the first parentheses is divisible by x^2, whereas each term in the second parentheses is divisible by y^2, and we can write the expression (i) as

$$x^2(a + b) + y^2(a + b). \qquad \text{(ii)}$$

We notice that $(a + b)$ is a common factor of the two terms. Thus (ii) is equivalent to the expression

$$(a + b)(x^2 + y^2).$$

Hence the given expression has as its factors $(a + b)$ and $(x^2 + y^2)$.

Example 4 Factor the expression
$$2x^3y - 4x^2y^2 + 8xy - 16y^2.$$

Solution We first observe that the terms in this expression have a common monomial factor of $2y$, and we write

$$2x^3y - 4x^2y^2 + 8xy - 16y^2 = 2y(x^3 - 2x^2y + 4x - 8y). \qquad \text{(i)}$$

Inside the parentheses we group the first two terms together and take out their common factor x^2, and we group the last two terms together and take out their common factor 4:

$$\underbrace{x^3 - 2x^2y}_{x^2 \text{ common}} + \underbrace{4x - 8y}_{4 \text{ common}} = x^2(x - 2y) + 4(x - 2y)$$
$$= (x^2 + 4)(x - 2y).$$

Observe that we could obtain this same result by grouping together the first and third terms and the second and fourth terms:

$$\underbrace{x^3 + 4x}_{x \text{ common}} - \underbrace{2x^2y - 8y}_{-2y \text{ common}} = x(x^2 + 4) - 2y(x^2 + 4)$$
$$= (x - 2y)(x^2 + 4).$$

Returning to the original expression (i) therefore, we get

$$2x^3y - 4x^2y^2 + 8xy - 16y^2 = 2y(x - 2y)(x^2 + 4).$$

It is not possible to factor these expressions on the right any further, so the factoring is complete.

Example 5 Factor the expression

$$4x^4y + 12x^3y^2 - 9x^2y - 27xy^2.$$

Solution We first observe that there is a common monomial factor of xy, so we write

$$4x^4y + 12x^3y^2 - 9x^2y - 27xy^2 = xy(4x^3 + 12x^2y - 9x - 27y)$$
$$= xy[4x^2(x + 3y) - 9(x + 3y)]$$

(grouping the terms inside the parentheses)

$$= xy(4x^2 - 9)(x + 3y).$$

We still haven't finished because, using the difference-of-squares formula [Equation (I)], we can write

$$(4x^2 - 9) = (2x + 3)(2x - 3).$$

Therefore,

$$4x^4y + 12x^3y^2 - 9x^2y - 27xy^2 = xy(2x + 3)(2x - 3)(x + 3y).$$

SUM AND DIFFERENCE OF TWO CUBES

In factoring an expression that can be expressed as either the sum or the difference of two cubes, the following two formulas are helpful:

$$\boxed{\begin{array}{l} a^3 + b^3 = (a + b)(a^2 - ab + b^2) \\ a^3 - b^3 = (a - b)(a^2 + ab + b^2). \end{array}} \quad \text{(II)}$$

We can verify these formulas by expanding the products on the right, for example, by the method of arcs. The first formula is proved as follows:

$$(a + b)(a^2 - ab + b^2) = a(a^2 - ab + b^2) + b(a^2 - ab + b^2)$$
$$= a^3 - a^2b + ab^2 + ba^2 - ab^2 + b^3$$
$$= a^3 + b^3$$

as required.

Examples 6 Factor.

(a) $2y^3 - 2$. (b) $8x^3 + 27y^3$.

Solutions (a) First we extract the monomial factor 2:
$$2y^3 - 2 = 2(y^3 - 1) = 2[y^3 - (1)^3].$$

Now, using Equation (II) for the difference of two cubes with $a = y$ and $b = 1$, we get
$$y^3 - (1)^3 = (y - 1)(y^2 + y \cdot 1 + 1^2)$$
$$= (y - 1)(y^2 + y + 1).$$

Thus, finally,
$$2y^3 - 2 = 2(y - 1)(y^2 + y + 1).$$

(b) We have
$$8x^3 + 27y^3 = (2x)^3 + (3y)^3$$
$$= (2x + 3y)[(2x)^2 - (2x)(3y) + (3y)^2]$$
$$= (2x + 3y)(4x^2 - 6xy + 9y^2).$$

The expression $4x^2 - 6xy + 9^2y$ cannot be factored further. In fact, if all common monomial factors have been removed earlier, this second factor can never be factored further in such examples.

Note According to Equation (II) above, the sum and difference of two cubes can always be factored. In fact, every expression of the type $a^n + b^n$ or $a^n - b^n$ can be factored for all integers $n \geq 2$ with the single exception of the sum of two squares, $a^2 + b^2$. For example,

$$a^4 - b^4 = (a^2 - b^2)(a^2 + b^2) = (a - b)(a + b)(a^2 + b^2)$$
$$a^5 + b^5 = (a + b)(a^4 - a^3b + a^2b^2 - ab^3 + b^4)$$
$$a^4 + b^4 = (a^2 + \sqrt{2}\,ab + b^2)(a^2 - \sqrt{2}\,ab + b^2)$$

and so on.

SUMMARY OF FACTORING

We conclude this section with a summary of factoring for algebraic expressions.

1. Remove all common monomial factors.
2. If the remaining factor is any of the following types:
   (a) Difference of two squares
   (b) Difference of two cubes
   (c) Sum of two cubes
   then use the standard formula (I) or (II) in order to factor further.
3. In order to factor a multinomial expression consisting of four terms, use grouping.

4. A trinomial expression of the type $mx^2 + px + q$ can often be factored into the product of two factors of the type $(ax + b)(cx + d)$. This type of factoring will be the subject of the next section.

Diagnostic Test 2.3

Fill in the blanks (factor completely).

1. $x^2 - 5x =$ (____)(____).
2. $5x^2y - 20xy^2 =$ (____)(____).
3. $6x^3 - 4x^2y + 8xy^2 =$ (____)(____).
4. $a^2b - ab^2 =$ (____)(____).
5. $x^2 - y^2 =$ (____)(____).
6. $4a^2 - 9b^2 =$ (____)(____).
7. $25u^2 - 36v^2 =$ (____)(____).
8. $49a^2b^2 - 4(u^2/v^2) =$ (____)(____).
9. $2x^2 - 3y^2 =$ (____)(____).
10. $4a^2 - 5b^2 =$ (____)(____).
11. $y^3 + a^3 =$ (____)(____).
12. $8y^3 - 27x^3 =$ (____)(____).
13. $x^3 + x^2 + x + 1 =$ (____)(____).
14. $x + y + ax + ay =$ (____)(____).

Exercises 2.3

(1–8) Extract all the monomial factors from each expression.

1. $3a + 6b$.
2. $15x - 12y$.
3. $4xy - 6yz$.
4. $6x^2 + 9xy$.
5. $10x^3yz + 15xy^2z^2 - 5xy^4$.
6. $12p^2q^2r^2 - 16p^2qr^3 - 24pqr^4$.
7. $8x^5y^3z + 4x^3y^2z^4 - 12x^2y^2z^5$.
8. $42a^6b^2c^2 - 105a^4bc^5 - 63a^3b^3c^4$.

(9–54) Factor each expression completely.

9. $x^2 - 16$.
10. $4y^2 - 25$.
11. $3t^2 - 108a^2$.
12. $5x^2 - 20y^2$.
13. $x^3y - 25xy^3$.
14. $x^5 - 4x^3y^2$.
15. $x^5 - x$.
16. $y^7 - 4y^3$.
17. $x^3 - 27$.
18. $8t^3 + 125$.
19. $27u^3 + 8v^3$.
20. $128x^3 - 54$.
21. $x^5 - x^2$.
22. $y^7 + 8y^4$.
23. $xy + 4x - 2y - 8$.
24. $pq - 6q - 3p + 18$.
25. $2u + av - 2v - au$.
26. $px - qy + py - qx$.
27. $6xz - 16y - 24x + 4yz$.
28. $15ac - 9ad - 30bc + 18bd$.
29. $(x^3 - 9x) + (45 - 5x^2)$.
30. $x^2y^2 - a^2y^2 - b^2x^2 + a^2b^2$.
31. $x^2y^2 - 9y^2 - 4x^2 + 36$.
32. $5u^2v^2 - 20v^2 + 15u^2 - 60$.
33. $x^2z^2 - 4z^2 + x^4 - 4x^2$.
34. $ax^3 + by^3 + bx^3 + ay^3$.
35. $a^3 + (b + 2)^3$.
*36. $x^6 + y^6$.
*37. $x^6 - 64y^6$.
38. $x^3 + 1$.
39. $xa + a + x + a^2$.
40. $x^3 + y^3 + x^2y + xy^2$.
*41. $x^4 + 4y^4$.
*42. $16a^4 + b^4$.
43. $x^4 - 16y^4$.
*44. $x^5 + y^5$.
45. $x^3y - 8 + 8y - x^3$.
46. $2x^2y - 8y + 2x^2 - 8$.
47. $x^5 + 3x^4 - x - 3$.

48. $x^5 - 32 - 16x + 2x^4$.

*50. $8xy - 4ty + 2xt - t^2 - 6x + 3t$.

52. $4(x + 2)^2 - (2x - 3y)^2$.

53. $x^{2n} - 25$.

*49. $2x^2 + 6xy - xz - 3yz + 2x + 6y$.

51. $(x + y)^2 - (x - y)^2$.

54. $x^{3p} - 8y^{3q}$.

2.4

Factoring Quadratic Expressions

An expression involving a variable x of the type $ax + b$, where a and b are two constants, is called a *linear* expression. For example, the expressions $2x - 3$, $5x + \sqrt{2}$, and $4 - x$ are all linear expressions.

An expression of the type $mx^2 + px + q$, where m, p, and q are three constants with $m \neq 0$ is called a *quadratic* expression. For example, the expressions $2x^2 - 3x + 1$, $4 - x^2$, and $x^2 + 4x + 4$ are all quadratic expressions. Note that in a quadratic expression there are three terms in general—one term with literal part x^2, one term with literal part x, and a constant term. The x term can be missing if its coefficient p happens to be zero (for example, the expression $4 - x^2$ has no x term). The constant term can be missing if q happens to be zero (for example, $x^2 - 3x$ is a quadratic expression with a zero constant term).

Often a quadratic expression can be factored into the product of two linear factors. For example,

$$(2x - 3)(x + 4) = 2x^2 + 5x - 12.$$

Thus the quadratic expression $2x^2 + 5x - 12$ can be factored into the product of the two linear factors $(2x - 3)$ and $(x + 4)$. Finding the two linear factors of a given quadratic expression is a very important type of factoring that arises in a variety of different applications, as we shall see.

We shall begin by considering the case when the coefficient of the x^2 term is 1 (i.e., $m = 1$). The quadratic expression then has the form $x^2 + px + q$. Often such an expression can be written as the product of two linear factors of the type $(x + a)(x + b)$. For example, the expression $x^2 + 3x + 2$ (in which $p = 3$, $q = 2$) is equal to the product of $(x + 1)$ and $(x + 2)$:

$$x^2 + 3x + 2 = (x + 1)(x + 2).$$

So in this case $a = 1$ and $b = 2$.

In general, with p and q given, we wish to find a and b such that

$$x^2 + px + q = (x + a)(x + b).$$

But we saw in Section 2.2 that

$$(x + a)(x + b) = x^2 + (a + b)x + ab$$

and so

$$x^2 + px + q = x^2 + (a + b)x + ab.$$

Sec. 2.4 Factoring Quadratic Expressions

These two expressions must be identical for all values of the variable x. The x^2 terms are identical on the two sides. Making the coefficients of x and the constant term equal, we obtain the following conditions:

$$a + b = p \quad \text{and} \quad ab = q.$$

So, in order to determine a and b, we must find two numbers whose sum is equal to p and whose product is equal to q. In terms of the original expression $x^2 + px + q$, the sum $a + b$ is equal to the coefficient of x and the product ab is equal to the constant term.

The procedure in finding a and b is to examine all possible pairs of integers whose product is equal to q. We select the pair (if one exists) whose sum is the given coefficient of x.

Example 1 Factor $x^2 + 7x + 12$.

Solution Here $p = 7$ and $q = 12$. We must find two numbers a and b whose product is 12 and whose sum is 7 (i.e., $ab = 12$ and $a + b = 7$). Let us consider all the possible pairs of integer factors of 12:

$$a = 1, \quad b = 12: \quad a + b = 13$$
$$a = -1, \quad b = -12: \quad a + b = -13$$
$$a = 2, \quad b = 6: \quad a + b = 8$$
$$a = -2, \quad b = -6: \quad a + b = -8$$
$$a = 3, \quad b = 4: \quad a + b = 7$$
$$a = -3, \quad b = -4: \quad a + b = -7.$$

Clearly the appropriate choice is $a = 3$, $b = 4$. Therefore,

$$x^2 + 7x + 12 = (x + 3)(x + 4).$$

(*Note*: The choice $a = 4$, $b = 3$ gives exactly the same pair of factors.)

Example 2 Factor $x^2 - 5x + 6$.

Solution In order to factor $x^2 - 5x + 6$, we have to find two numbers whose product is $+6$ (the constant term) and whose sum is -5 (the coefficient of x). The possible integer factors of 6 are

$$(1)(6), \quad (-1)(-6), \quad (2)(3), \quad (-2)(-3).$$

The two factors of 6 that have the sum -5 are -2 and -3. Thus, we take $a = -2$ and $b = -3$, so that

$$x^2 - 5x + 6 = (x + a)(x + b)$$
$$= [x + (-2)][x + (-3)]$$
$$= (x - 2)(x - 3).$$

Example 3 Factor $3x^2 - 3x - 6$.

Solution First we observe that there is a common monomial factor of 3:
$$3x^2 - 3x - 6 = 3(x^2 - x - 2). \quad (*)$$
To factor $x^2 - x - 2$, we have to find two numbers whose product is -2 (the constant term) and whose sum is -1 (the coefficient of x). The possible integer factors of -2 are
$$(1)(-2) \quad \text{and} \quad (-1)(2).$$
Only the pair 1 and -2 have the sum -1: Thus we must have $a = 1$ and $b = -2$, and so
$$x^2 - x - 2 = (x + 1)[x + (-2)] = (x + 1)(x - 2).$$
Therefore from (*) above,
$$3x^2 - 3x - 6 = 3(x^2 - x - 2)$$
$$= 3(x + 1)(x - 2).$$

Example 4 Factor $x^2 + 6x + 9$.

Solution We have $p = 6, q = 9$. Clearly the two factors of 9 whose sum is 6 are 3 and 3. Thus the given expression factors to $x + 3$ and $x + 3$; that is,
$$x^2 + 6x + 9 = (x + 3)(x + 3) = (x + 3)^2.$$

Now let us turn to the problem of factoring an expression of the form
$$mx^2 + px + q \quad (m \neq 0)$$
where m, p, q are certain constants with $m \neq 1$ or -1. In this case, the first step consists of forming the product (mq) of the coefficient of x^2 and the constant term. We then seek two numbers whose product is equal to mq and whose sum is equal to p, the coefficient of x. Then we split p into the sum of these two numbers. This changes the given expression into the sum of four terms. These four terms can be combined two by two and factored by the method of grouping. The method is illustrated by the following solved examples.

Example 5 Factor $3x^2 + 11x + 6$.

Solution In this expression the coefficients are $m = 3, p = 11$, and $q = 6$. The product of the coefficient of x^2 and the constant term is $mq = 3(6) = 18$. We must find two factors of this product 18, such that their sum is equal to 11, the coefficient of x. Making the usual list of pairs of factors, we get

$$(1)(18), \ \text{sum} = 19, \quad (-1)(-18), \ \text{sum} = -19$$
$$(2)(9), \ \text{sum} = 11, \quad (-2)(-9), \ \text{sum} = -11$$
$$(3)(6), \ \text{sum} = 9, \quad (-3)(-6), \ \text{sum} = -9.$$

Clearly the correct factors are 2 and 9. Thus in the given expression we split 11, the coefficient of x, into $9 + 2$ and write

$$3x^2 + 11x + 6 = 3x^2 + (9 + 2)x + 6$$
$$= 3x^2 + 9x + 2x + 6.$$

We can take $3x$ as a common factor from the first two terms and 2 as a common factor from the last two terms. Thus,

$$3x^2 + 11x + 6 = 3x(x + 3) + 2(x + 3)$$
$$= (x + 3)(3x + 2)$$

where in the last step we have taken $(x + 3)$ as common factor from the two terms.

Example 6 Factor $6x^2 - 5x - 4$.

Solution The product of the coefficient of x^2 and the constant term is $6(-4) = -24$. We must find two factors of -24 that add up to -5, the coefficient of x. The pairs of factors are

$$(-1)(24), \quad (1)(-24), \quad (-2)(12), \quad (2)(-12),$$
$$(-3)(8), \quad (3)(-8), \quad (-4)(6), \quad (4)(-6).$$

Clearly the two factors of -24 that have the sum -5 are 3 and -8; therefore, we write -5 as $-8 + 3$ in the given expression. This gives

$$6x^2 - 5x - 4 = 6x^2 + (-8 + 3)x - 4$$
$$= 6x^2 - 8x + 3x - 4$$
$$= 2x(3x - 4) + 1(3x - 4)$$
$$= (3x - 4)(2x + 1).$$

Example 7 Factor $8x^2 - 6xy - 27y^2$.

Solution The given expression is not of the form $mx^2 + px + q$ but rather of the form $mx^2 + pxy + qy^2$. Expressions of this type can be factored using the same methods described above, except simply that the two factors are of the form $(ax + by)$. We proceed exactly as in Example 6. The coefficients in this example are $m = 8$, $p = -6$, and $q = -27$. Therefore

$$mq = 8(-27) = -216.$$

Therefore we must seek a pair of numbers whose product is -216 and whose sum is p (i.e., -6). The various pairs of factors of -216 are

$$(-216)(1), \quad (-108)(2), \quad (-72)(3), \quad (-54)(4),$$
$$(-36)(6), \quad (-27)(8), \quad (-24)(9), \quad (-18)(12)$$

and the same pairs with the signs reversed. The pair whose sum is -6 is

clearly -18 and 12. Thus we write
$$8x^2 - 6xy - 27y^2 = 8x^2 - 18xy + 12xy - 27y^2$$
$$= 2x(4x - 9y) + 3y(4x - 9y)$$
$$= (2x + 3y)(4x - 9y)$$

which is the required factoring.

Diagnostic Test 2.4

Fill in the blanks (Factor completely):

1. $x^2 - 7x + 12 = (\_\_\_\_)(\_\_\_\_)$.
2. $y^2 + 5y + 6 = (\_\_\_\_)(\_\_\_\_)$.
3. $a^2 + 6a + 8 = (\_\_\_\_)(\_\_\_\_)$.
4. $t^2 - 5t - 6 = (\_\_\_\_)(\_\_\_\_)$.
5. $x^2 - 4x - 21 = (\_\_\_\_)(\_\_\_\_)$.
6. $t^2 + 4t - 21 = (\_\_\_\_)(\_\_\_\_)$.
7. $x^2 + 2x + 1 = (\_\_\_\_)(\_\_\_\_)$.
8. $x^2 - 6x + 9 = (\_\_\_\_)(\_\_\_\_)$.
9. $t^2 + 4t + 4 = (\_\_\_\_)(\_\_\_\_)$.
10. $y^2 - 2y + 1 = (\_\_\_\_)(\_\_\_\_)$.

Exercises 2.4

Factor the following expressions.

1. $x^2 + 3x + 2$.
2. $x^2 + 5x + 6$.
3. $x^2 - x - 2$.
4. $x^2 + x - 2$.
5. $x^2 - 7x + 12$.
6. $x^2 - 8x + 12$.
7. $x^2 - 9$.
8. $x^2 - 25$.
9. $x^2 - 4x$.
10. $x^2 + 3x$.
11. $x^2 - 12x + 11$.
12. $x^2 - 9x + 20$.
13. $x^2 - 15x + 54$.
14. $x^2 - 14x + 48$.
15. $x^2 - 2x - 80$.
16. $x^2 - x - 90$.
17. $2x^2 + 2x - 12$.
18. $3x^2 - 6x + 3$.
19. $5y^4 + 25y^3 - 70y^2$.
20. $12x - 7x^2 + x^3$.
21. $2x^2 + 5x + 3$.
22. $6x^2 + 10x - 4$.
23. $6t^3 - 7t^2 - 20t$.
24. $6t^4 + 15t^3 - 9t^2$.
25. $9t^2 - 12t + 4$.
26. $9 + 12x + 4x^2$.
27. $10x^2 - 11x - 6$.
28. $5x^2 - 17x + 6$.
29. $2t^2 - 3t - 14$.
30. $2t^2 - 7t + 6$.
31. $6x^3y + 4x^2y - 10xy$.
32. $6p^3 + p^2 - 12p$.
33. $3q^2 + 20q + 32$.
34. $10p^2 + 3p - 18$.
35. $4p^3q + 10p^2q - 36pq$.
36. $18p^2q - 48pq - 18q$.
37. $x^2 + 6xy + 5y^2$.
38. $x^2 - 4xy - 5y^2$.
39. $p^2 - pq - 20q^2$.
40. $s^2 + 7st - 30t^2$.
41. $2t^2 + tu - 6u^2$.
42. $2x^2 - 9xy + 10y^2$.
43. $6a^2 + ab - 15b^2$.
44. $18u^2 + 15uv - 12v^2$.
45. $(x + y)^2 + 3(x + y) + 2$.
46. $2(p - q)^2 - (p - q) - 1$.
47. $3x^{2n} + 7x^n + 2$.
48. $2x^{2p} - 5x^p y^q - 12y^{2q}$.
49. $(x + 3)(x - 2) + (x + 2)(x - 1) - x + 5$.
50. $(3x + 1)(x - 2) + (x + 4)(x + 3) - 4(3x + 4)$.
51. $(2x + 1)^2 - (x + 3)(x + 1)$.
52. $5 + (2x + 3)^2 - (3x + 2)(x + 1)$.

*(53–56) Use the difference of squares formula to factor the following.

53. $x^2 + 4ax + 4a^2 - 4$.
54. $4x^2 + 4x + 1 - a^2$.
55. $x^2 + y^2 - z^2 + 2xy$.
56. $2x + y^2 - x^2 - 1$.

2.5

Algebraic Fractions

The term *algebraic fraction* is used generally for the ratio of two expressions containing one or more variables—for example:

$$\text{(a)} \quad \frac{x^2 - 7x + 5}{2x + 3}, \qquad \text{(b)} \quad \frac{x^2 y + xy^2}{x - y}.$$

Example (a) involves one variable, x, while example (b) involves two variables, x and y.

To make an algebraic fraction meaningful, it is understood that the variable or variables do not take values that make the denominator of the fraction zero. Thus, in example (a) above, $x \neq -\frac{3}{2}$, because otherwise $2x + 3 = 2(-\frac{3}{2}) + 3 = -3 + 3 = 0$, and the denominator is zero. Similarly in (b) $y \neq x$.

In this section we shall study methods of simplifying algebraic fractions and of adding, subtracting, multiplying, and dividing two or more such fractions. Factoring plays an important role in such operations, as the examples will make clear. *The basic principles involved will be the same as those described for simpler fractions in Sections 1.4 and 1.5.*

I. SIMPLIFICATION OF FRACTIONS

Often we may simplify algebraic fractions by dividing the numerator and denominator by any common factors they may have. Discovery of these common factors usually requires that the numerator and denominator be factored completely.

Example 1 Simplify the algebraic fraction

$$\frac{2x^2 + 4x}{3x^3 + 6x^2}.$$

Solution Completely factoring the numerator and denominator, we get

$$2x^2 + 4x = 2x(x + 2)$$
$$3x^3 + 6x^2 = 3x^2(x + 2).$$

Therefore,
$$\frac{2x^2 + 4x}{3x^3 + 6x^2} = \frac{2x(x+2)}{3x^2(x+2)}$$
$$= \frac{2}{3x}.$$

In the final step the common factors x and $(x + 2)$ have been divided out of the numerator and denominator.

Notes (a) It is perfectly acceptable to cancel binomial expressions such as $(x + 2)$ in this example as long as they are factors of both the numerator and denominator.

(b) The common factors that are canceled from numerator and denominator must always be nonzero. In this example x and $(x + 2)$ are canceled, and this is valid only if $x \neq 0$ and $x + 2 \neq 0$ (i.e., $x \neq -2$). Often the simplification of an algebraic fraction requires such restrictions on the variable or variables that occur.

Example 2 Simplify $\dfrac{4x^2 - 20x + 24}{6 + 10x - 4x^2}$.

Solution First we factor completely the expressions that appear in the numerator and denominator. We have
$$4x^2 - 20x + 24 = 4(x^2 - 5x + 6)$$
$$= 2 \cdot 2(x - 2)(x - 3)$$
and
$$6 + 10x - 4x^2 = -2(2x^2 - 5x - 3)$$
$$= -2(2x + 1)(x - 3).$$

Note that in factoring the denominator, we first made the coefficient of x^2 positive. This makes factoring easier and also means that the x terms in the factors are positive both in the numerator and denominator. Thus,
$$\frac{4x^2 - 20x + 24}{6 + 10x - 4x^2} = \frac{2 \cdot 2(x - 2)(x - 3)}{-2(2x + 1)(x - 3)}$$
$$= \frac{2(x - 2)}{-(2x + 1)}$$
$$= \frac{-2(x - 2)}{2x + 1} \quad (x \neq 3).$$

Observe that we have divided the numerator and denominator by the factors 2 and $(x - 3)$ that appear both in numerator and denominator.

II. ADDITION AND SUBTRACTION OF FRACTIONS

We can add or subtract two or more fractions that have a common denominator simply by adding or subtracting their numerator while keeping the denominator unchanged. Compare the following examples with Examples 1 in Section 1.5.

Sec. 2.5 Algebraic Fractions

Examples 3 (a) $\dfrac{2x+3}{x+1} + \dfrac{x-1}{x+1} = \dfrac{(2x+3)+(x-1)}{x+1}$

$= \dfrac{2x+x+3-1}{x+1}$

$= \dfrac{3x+2}{x+1}.$

(b) $\dfrac{2x+5}{x-1} - \dfrac{7}{x-1} = \dfrac{(2x+5)-7}{x-1}$

$= \dfrac{2x-2}{x-1}$

$= \dfrac{2(x-1)}{x-1} = 2 \quad (x \neq 1).$

In this second example we can simplify the resulting fraction by dividing out the common factor of $(x-1)$.

When the fractions to be added or subtracted do not have the same denominator, we may first find their *least common denominator (LCD)* and then replace each of the given fractions by an equivalent fraction having this LCD as its denominator. The method is in principle no different from that described in Section 1.5.

> To find the LCD of two or more fractions, we factor each denominator completely. We then obtain the LCD by multiplying all the distinct factors that appear in various denominators and raising each factor to the highest power to which it occurs in any one denominator.

Examples 4 (a) In the sum

$$\dfrac{2x+3}{x-3} + \dfrac{3x-1}{2x+7}$$

the LCD of the two fractions is $(x-3)(2x+7)$.

(b) The LCD of the three fractions

$$\dfrac{x+1}{(x-1)^2}, \quad \dfrac{5}{(x-1)(x+2)}, \quad \dfrac{7}{(x+2)^3(x+3)}$$

is $(x-1)^2(x+2)^3(x+3)$. [The factor $(x-1)$ appears with power 2 in the first denominator; $(x+2)$ appears with power 3 in the third denominator.]

Example 5 Simplify $\dfrac{2x+1}{x+2} + \dfrac{x-1}{3x-2}.$

Solution Here the denominators are already completely factored. The LCD in this case is

$$(x+2)(3x-2).$$

To replace the first fraction $(2x+1)/(x+2)$ by an equivalent fraction having the LCD $(x + 2)(3x - 2)$ as its denominator, we multiply the top and bottom of the fraction by $(3x - 2)$. Thus

$$\frac{2x + 1}{x + 2} = \frac{(2x + 1)(3x - 2)}{(x + 2)(3x - 2)}.$$

Similarly,

$$\frac{x - 1}{3x - 2} = \frac{(x - 1)(x + 2)}{(x + 2)(3x - 2)}.$$

Therefore,

$$\frac{2x + 1}{x + 2} + \frac{x - 1}{3x - 2} = \frac{(2x + 1)(3x - 2)}{(x + 2)(3x - 2)} + \frac{(x - 1)(x + 2)}{(x + 2)(3x - 2)}$$

$$= \frac{(2x + 1)(3x - 2) + (x - 1)(x + 2)}{(x + 2)(3x - 2)}$$

$$= \frac{(6x^2 - x - 2) + (x^2 + x - 2)}{(x + 2)(3x - 2)}$$

$$= \frac{7x^2 - 4}{(x + 2)(3x - 2)}.$$

Example 6 Simplify $\dfrac{5}{x^2 - 3x + 2} - \dfrac{1}{x + 2} + \dfrac{3}{x^2 - 4x + 4}$.

Solution The given expression, after we factor the denominators, is

$$\frac{5}{(x - 1)(x - 2)} - \frac{1}{x + 2} + \frac{3}{(x - 2)^2}$$

[Here the LCD is $(x - 1)(x - 2)^2(x + 2)$.]

$$= \frac{5(x - 2)(x + 2)}{(x - 1)(x - 2)^2(x + 2)} - \frac{(x - 1)(x - 2)^2}{(x + 2)(x - 1)(x - 2)^2}$$

$$+ \frac{3(x - 1)(x + 2)}{(x - 2)^2(x - 1)(x + 2)}$$

$$= \frac{5(x - 2)(x + 2) - (x - 1)(x - 2)^2 + 3(x - 1)(x + 2)}{(x - 1)(x + 2)(x - 2)^2}$$

$$= \frac{5(x^2 - 4) - (x - 1)(x^2 - 4x + 4) + 3(x^2 + x - 2)}{(x - 1)(x + 2)(x - 2)^2}$$

$$= \frac{5x^2 - 20 - (x^3 - 5x^2 + 8x - 4) + 3x^2 + 3x - 6}{(x - 1)(x + 2)(x - 2)^2}$$

$$= \frac{-x^3 + 13x^2 - 5x - 22}{(x - 1)(x + 2)(x - 2)^2}.$$

Example 7 Simplify $\sqrt{1 - x^2} + \dfrac{1 + x^2}{\sqrt{1 - x^2}}$.

Solution In this case we write both terms as fractions with LCD $= \sqrt{1 - x^2}$. We have

$$\sqrt{1 - x^2} = \frac{\sqrt{1 - x^2}\sqrt{1 - x^2}}{\sqrt{1 - x^2}}.$$

$$= \frac{1 - x^2}{\sqrt{1 - x^2}}.$$

Sec. 2.5 Algebraic Fractions

Therefore

$$\sqrt{1-x^2} + \frac{1+x^2}{\sqrt{1-x^2}} = \frac{1-x^2}{\sqrt{1-x^2}} + \frac{1+x^2}{\sqrt{1-x^2}}$$

$$= \frac{1-x^2+1+x^2}{\sqrt{1-x^2}} = \frac{2}{\sqrt{1-x^2}}.$$

III. MULTIPLICATION OF FRACTIONS

We can multiply two or more fractions together simply by multiplying their numerators and denominators, as illustrated below.

Examples 8 Evaluate the following products and simplify the resulting fractions if possible.

(a) $\left(\dfrac{2x+1}{2x-4}\right)\left(\dfrac{6-2x}{x+1}\right).$ (b) $\left(\dfrac{x^2-5x+6}{6x^2+18x+12}\right)\left(\dfrac{4x^2-16}{2x^2-5x-3}\right).$

Solutions (a) $\left(\dfrac{2x+1}{2x-4}\right)\left(\dfrac{6-2x}{x+1}\right) = \dfrac{(2x+1)(6-2x)}{(2x-4)(x+1)}$

$$= \frac{(2x+1)\cdot 2(3-x)}{2(x-2)(x+1)}$$

$$= \frac{(2x+1)(3-x)}{(x-2)(x+1)}.$$

In the last step the common factor 2 has been divided from numerator and denominator.

(b) $\left(\dfrac{x^2-5x+6}{6x^2+18x+12}\right)\left(\dfrac{4x^2-16}{2x^2-5x-3}\right)$

$$= \frac{(x^2-5x+6)(4x^2-16)}{(6x^2+18x+12)(2x^2-5x-3)}.$$

We can simplify the fraction on the right by factoring the numerator and denominator and dividing numerator and denominator by their common factors. We obtain

$$\frac{(x-2)(x-3)2\cdot 2(x-2)(x+2)}{2\cdot 3(x+1)(x+2)(x-3)(2x+1)} = \frac{2(x-2)(x-2)}{3(x+1)(2x+1)}$$

$$= \frac{2(x-2)^2}{3(x+1)(2x+1)} \quad (x \neq 3, -2).$$

IV. DIVISION OF FRACTIONS

To divide one fraction a/b by another fraction c/d, we invert c/d and multiply (see Section 1.4, p. 17),

$$\frac{a}{b} \div \frac{c}{d} = \frac{a/b}{c/d} = \frac{a}{b} \cdot \frac{d}{c}.$$

The method is illustrated for algebraic fractions by the following examples.

Examples 9 Simplify.

(a) $\left(\dfrac{2x+3}{x-1}\right) \div \left(\dfrac{x+3}{2x^2-2}\right).$ (b) $\dfrac{\dfrac{3x-1}{x-2}}{x+1}.$

Solutions (a) $\dfrac{2x+3}{x-1} \div \dfrac{x+3}{2x^2-2} = \dfrac{2x+3}{x-1} \cdot \dfrac{2x^2-2}{x+3}$

$= \dfrac{(2x+3) \cdot 2(x-1)(x+1)}{(x-1)(x+3)}$

$= \dfrac{2(x+1)(2x+3)}{x+3} \quad (x \neq 1).$

(b) $\dfrac{\dfrac{3x-1}{x-2}}{x+1} = \dfrac{\dfrac{3x-1}{x-2}}{\dfrac{x+1}{1}}$ (where the denominator is expressed also as a fraction)

$= \dfrac{3x-1}{x-2} \cdot \dfrac{1}{x+1}$

$= \dfrac{3x-1}{(x-2)(x+1)}.$

Example 10 Simplify $\dfrac{\left(x+2-\dfrac{4}{x-1}\right)}{\dfrac{x^2-5x+6}{x^2-1}}.$

Solution First of all we simplify the numerator. We have:

$x + 2 - \dfrac{4}{x-1} = \dfrac{x+2}{1} - \dfrac{4}{x-1}$

$= \dfrac{(x+2)(x-1)}{x-1} - \dfrac{4}{x-1}$

$= \dfrac{(x+2)(x-1) - 4}{x-1}$

$= \dfrac{x^2+x-6}{x-1}.$

Using this value for the numerator, the given expression reduces to

$\dfrac{\dfrac{x^2+x-6}{x-1}}{\dfrac{x^2-5x+6}{x^2-1}} = \left(\dfrac{x^2+x-6}{x-1}\right)\left(\dfrac{x^2-1}{x^2-5x+6}\right)$

$= \dfrac{(x^2+x-6)(x^2-1)}{(x-1)(x^2-5x+6)}$

$= \dfrac{(x-2)(x+3)(x-1)(x+1)}{(x-1)(x-2)(x-3)}$

$= \dfrac{(x+3)(x+1)}{x-3} \quad (x \neq 1, 2).$

Sec. 2.5 Algebraic Fractions

Diagnostic Test 2.5

Fill in the blanks (simplify).

1. $\dfrac{6x^2}{2x^2 + 4x} = $ _____.

2. $\dfrac{x^2 - 4}{2x - 4} = $ _____.

3. $\dfrac{x^2 + 5x + 6}{3 + 4x + x^2} = $ _____.

4. $\dfrac{x^2 + 2x + 1}{x^2 - 1} = $ _____.

5. $\dfrac{2}{x} - \dfrac{1}{x + 1} = $ _____.

6. $\dfrac{4(x - 1)}{2x + 1} - \dfrac{2x - 1}{2x + 1} = $ _____.

7. $\left(\dfrac{2x - 3}{x^2 + x}\right)\left(\dfrac{x + 1}{4x - 6}\right) = $ _____.

8. $\dfrac{x^2 - 2x}{2x + 4} \div \dfrac{x - 2}{x + 2} = $ _____.

Exercises 2.5

(1–12) Simplify each fraction by dividing out all the common factors.

1. $\dfrac{3x^2y + 6xy^2}{6x^2y + 9xy^2}$.

2. $\dfrac{5x^2 - 15x^3}{5x^3 - 15x^2}$.

3. $\dfrac{2x^2 + 10x^4}{2x^3 + 10x^6}$.

4. $\dfrac{4t + 6t^2 + 8t^3}{2t^2 + 4t^3}$.

5. $\dfrac{t^2 + t}{2t^2 - 2}$.

6. $\dfrac{t^3 - t}{t^2 - t}$.

7. $\dfrac{2t^2 - 8}{t^2 + 3t + 2}$.

8. $\dfrac{4t^3 - 4t}{2t^2 - 2t - 4}$.

9. $\dfrac{4x^3 - 4}{2x^3 + 2x^2 + 2}$.

10. $\dfrac{6x^3 + 21x^2 + 18x}{12x^2 + 66x + 72}$.

11. $\dfrac{x^2 - 5x + 6}{x^2 - x - 2}$.

12. $\dfrac{2x^3 - 7x^2 + 3x}{4x^2 + 6x - 4}$.

(13–52) Perform the indicated operations and simplify as far as possible.

13. $\dfrac{4x}{2x + 3} + \dfrac{6}{2x + 3}$.

14. $\dfrac{2x}{x - 2} - \dfrac{4}{x - 2}$.

15. $\dfrac{x^2}{x - 3} - \dfrac{5x - 6}{x - 3}$.

16. $\dfrac{2 - 3x}{x - 1} + \dfrac{x^2}{x - 1}$.

17. $\dfrac{2x + 1}{x + 2} + 3$.

18. $\dfrac{3x - 2}{x + 1} - 2$.

19. $\dfrac{x}{x + 2} + \dfrac{3}{2x - 1}$.

20. $\dfrac{x}{2x - 6} + \dfrac{x - 2}{x + 2}$.

21. $\dfrac{2}{x - 1} - \dfrac{3x + 1}{x + 1}$.

22. $\dfrac{x}{2x + 3} - \dfrac{2x - 3}{4x + 1}$.

23. $\dfrac{2x}{2x - 1} - \dfrac{x + 2}{x + 1}$.

24. $\dfrac{2}{5x - 6} - \dfrac{4}{10x - 2}$.

25. $\dfrac{1}{x^2 - 5x + 6} - \dfrac{1}{x^2 - 3x + 2}$.

26. $\dfrac{x}{x^2 + 2x - 3} + \dfrac{1}{x^2 + x - 2}$.

27. $\dfrac{x}{x^2 + 2x - 3} - \dfrac{1}{1 - 2x + x^2}$.

28. $\dfrac{2}{9x^2 - 6x + 1} - \dfrac{3}{x + 1} + \dfrac{1}{3x^2 + 2x - 1}$.

29. $\dfrac{1}{x^2 + 4x + 3} + \dfrac{3}{x^2 - 1} - \dfrac{2}{x + 3}$.

30. $\dfrac{x}{2x^2 - x - 1} - \dfrac{3}{1 - 2x + x^2} + 2$.

31. $\left(\dfrac{x^2 - 1}{x}\right)\left(\dfrac{x^2 + 2x}{x + 1}\right)$.

32. $\left(\dfrac{x^2 + 4x}{2x + 6}\right)\left(\dfrac{2x + 4}{x + 4}\right)$.

33. $\left(\dfrac{x^2 + 5x + 6}{x^2 - 6x + 8}\right)\left(\dfrac{2x^2 - 9x + 4}{2x^2 + 7x + 3}\right)$.

34. $\left(\dfrac{2x^4 - 2x}{2x^2 - 5x - 3}\right)\left(\dfrac{2x^2 - 3x - 2}{x^3 + x^2 + x}\right)$.

35. $\dfrac{2x + 4}{1 - x} \cdot \dfrac{x^2 - 1}{3x + 6}$.

36. $\dfrac{x^2 - 7x + 12}{x^2 - x - 2} \cdot \dfrac{x^2 + 4x + 3}{2x^2 - 5x - 3}$.

37. $\left(3 + \dfrac{1}{x-1}\right)\left(1 - \dfrac{1}{3x-2}\right).$

38. $\left(x - \dfrac{3}{x-2}\right)\left(\dfrac{5}{x^2-9} + 1\right).$

39. $\left(\dfrac{x^2+x}{2x+1}\right) \div \left(\dfrac{x^3-x}{4x+2}\right).$

40. $\left(\dfrac{3x-6}{2x^2+4x+2}\right) \div \left(\dfrac{x^2-4}{x^2+3x+2}\right).$

41. $\dfrac{3x^2-x-2}{x^2-x-2} \div \dfrac{3x^2+5x+2}{2x^2-5x+2}.$

42. $\dfrac{2x^2+x-1}{2x^2+10x+12} \div \dfrac{1-4x^2}{4x^2+8x-12}.$

43. $\dfrac{x^2+x-2}{2x+3} \bigg/ \dfrac{x^2-4}{2x^2+5x+3}.$

44. $\dfrac{1 - 1/t^2}{t + 1 - 2/t}.$

45. $\dfrac{x + 2 + \dfrac{3}{x-2}}{x - 6 + \dfrac{7}{x+2}}.$

46. $\left(p - \dfrac{2}{p+1}\right) \bigg/ \left(1 - \dfrac{4p+7}{p^2+4p+3}\right).$

47. $\dfrac{(x^{-1} + y^{-1})}{(x+y)^{-1}}.$

48. $\dfrac{(x-y)^{-1}}{(x^{-2} - y^{-2})^{-1}}.$

49. $\dfrac{x^{-2} + y^{-2}}{x^{-2} - y^{-2}} \cdot \dfrac{x-y}{x+y}.$

50. $\dfrac{y^{-2} - x^{-2}}{xy^{-1} - yx^{-1}}.$

51. $\dfrac{1}{h}\left(\dfrac{1}{x+h} - \dfrac{1}{x}\right).$

52. $\dfrac{1}{h}\left[\dfrac{1}{(x+h)^2} - \dfrac{1}{x^2}\right].$

Chapter Review 2

1. Define and/or explain the terms.
   (a) Literal part and coefficient.
   (b) Monomial, binomial, trinomial, multinomial.
   (c) Like terms.
   (d) Method of arcs.
   (e) Binomial-square formula.
   (f) Difference-of-two-squares formula.
   (g) Monomial factor.
   (h) Grouping method.
   (i) Sum-and-difference-of-two-cubes formula.
   (j) Linear and quadratic expressions.
   (k) Least common denominator.
   (l) Cancellation of common factors.

2. Describe the procedure you would use to perform the operations.
   (a) Evaluate an algebraic expression for given values of the variables.
   (b) Add or subtract two multinomial expressions.
   (c) Multiply or divide an expression by a monomial.
   (d) Multiply two multinomial expressions.
   (e) Simplify an expression involving parentheses nested inside other parentheses.
   (f) Rationalize the denominator of a fraction.
   (g) Factor a multinomial expression.
   (h) Factor a quadratic expression.
   (i) Simplify an algebraic fraction.
   (j) Add or subtract two algebraic fractions.
   (k) Multiply or divide two algebraic fractions.

Review Exercises on Chapter 2

(1–4) Evaluate the following expressions at the given value of the variables.

1. $x^3 - 2x + 1;\ x = -1.$

2. $\sqrt{x} - \dfrac{2}{\sqrt{x}};\ x = 4.$

3. $\dfrac{2x^2}{y} - \dfrac{y^2}{2x};\ x = 2,\ y = -4.$

4. $x^2yz + xy^2z + xyz^2;\ x = 1,\ y = -1,\ z = 2.$

(5–18) Simplify the following expressions.

5. $3(2x - y) - 3(2y - x)$.
6. $4t^2 - 3t - 2t(t - 2)$.
7. $3(x^4 + 2x^2 + 4) - 2x^2(2x^2 + 3)$.
8. $xy(3x - 4y) - x(x + y)^2$.
9. $\dfrac{2t^3 - 10t}{4t^2}$ (Divide).
10. $\dfrac{2x^2 - 3x + 1}{\sqrt{x}}$ (Divide).
11. $\dfrac{2(x^2 + 1)}{3x} - \dfrac{3x^2 + 2}{4x}$.
12. $\dfrac{4(p - q)^2}{3pq} - 2pq\left(\dfrac{1}{p} + \dfrac{1}{q}\right)^2$.
13. $(x - 4)(y - 4)$.
14. $(2x - 3)(5x + 2)$.
15. $(2\sqrt{x} - 1)(3\sqrt{x} + 2)$.
16. $(x + 2y)^2 - (x - 2y)^2$.
17. $2[x(x + 1) - 2(x - 1)] - 3(x - 2)$.
18. $p\{(p + 1)^2 + 2[p(p - 2) - (p - 1)^2]\} - p^2(2p - 3)$.

(19–20) Rationalize the denominators.

19. $\dfrac{\sqrt{5}}{2\sqrt{5} - 3}$.
20. $\dfrac{2x}{\sqrt{x + 1} + \sqrt{x - 1}}$.

21. Prove that $\dfrac{1}{\sqrt{2} - 1} + \dfrac{2}{\sqrt{3} + 1} = \sqrt{2} + \sqrt{3}$.

22. Prove that $\dfrac{1}{\sqrt{3} - \sqrt{2}} + \dfrac{1}{2 + \sqrt{3}} = \sqrt{2}(\sqrt{2} + 1)$.

23. Prove that $\dfrac{1}{\sqrt{2} + 1} + \dfrac{1}{\sqrt{3} + \sqrt{2}} + \dfrac{1}{2 + \sqrt{3}} = 1$.

24. Given $\sqrt{7} \approx 2.6458$ and $\sqrt{6} \approx 2.4495$, evaluate $1/(\sqrt{7} - \sqrt{6})$ without using tables, calculators, or long division.

25. Given $\sqrt{2} \approx 1.414$ and $\sqrt{3} \approx 1.732$, evaluate $1/(\sqrt{3} - \sqrt{2})$ without using tables or long division.

(26–50) Factor completely.

26. $3x^2 - 75y^2$.
27. $x^2 + 7x + 10$.
28. $6x^2 - x - 15$.
29. $2p^2 + p - 28$.
30. $(a + 4)(a - 3) + (2a + 3)(a + 1)$.
31. $(x + 2)(x^2 + x - 1) + (2x - 1)(x^2 - 3x - 2)$.
32. $4(x + 1)^2 - (2x + 5)^2$.
33. $54x^3 - 16y^3$.
34. $x^3 + \dfrac{8}{x^3}$.
35. $(x - 1)(x^2 + 1) + (x + 1)(x^2 - 1)$.
36. $3(x + 1)^3(2x - 1)^2 + 2(x + 1)^2(2x - 1)^3$.
37. $2x^2 - 10xy + 4x^3$.
38. $5x^2y + 10xy^2$.
39. $x^2 + 4x - 5$.
40. $x^2 - 3x - 4$.
41. $x^2 + x - 2$.
42. $u^2 - 2u - 3$.
43. $y^2 - 3y - 10$.
44. $k^2 + k - 20$.
45. $10t^2 + 3tu - u^2$.
46. $12x^2 + 7xy - 12y^2$.
47. $8x^2 - 18x + 9$.
48. $12x^2 + 20x - 25$.
49. $(p + q)^2 + 3(p + q) - 4$.
50. $x^2 - a^2 - 6x + 9$.

(51–61) Simplify.

51. $\dfrac{1}{x + 2} + \dfrac{2}{x - 3}$.
52. $\dfrac{1}{x^2 + 1} + 2$.
53. $\dfrac{1}{x - 1} + \dfrac{1}{x^2 - 3x + 2} - \dfrac{1}{x^2 - 2x + 1}$.
54. $\dfrac{2}{x^2 + 2x + 1} - \dfrac{1}{x^2 + 4x + 3} + \dfrac{3}{x^2 - x - 2}$.

55. $\dfrac{x+y}{p^2-q^2} \div \dfrac{x^2-y^2}{p+q}$.

56. $\dfrac{x^2+4x+4}{y^2-9} \div \dfrac{x^2+5x+6}{y^2-y-6}$.

57. $\dfrac{a^2-b^2}{2a+4} \div \dfrac{a^2-3ab+2b^2}{a^2-4}$.

58. $\dfrac{a^2+2ab+b^2}{x^2+5x+6} \cdot \dfrac{x^2-x-6}{a^2-b^2}$.

59. $\left(x - \dfrac{2}{x+1}\right)\left(x + \dfrac{1}{x+2}\right)$.

60. $\left(a + \dfrac{2}{a+3}\right)\left(a - \dfrac{9}{a}\right)$.

61. $\dfrac{x+1}{x-2} + \dfrac{3x^2-27}{x+3} - \dfrac{2x+1}{2x-1}$.

62. Show that $\left(\dfrac{x^p}{x^q}\right)^{p+q-r} \left(\dfrac{x^q}{x^r}\right)^{q+r-p} \left(\dfrac{x^r}{x^p}\right)^{r+p-q} = 1$.

3 EQUATIONS IN ONE VARIABLE

3.1

Solutions of Equations; Linear Equations

An *equation* is a statement that expresses the equality of two mathematical expressions. It generally involves one or more variables and the equality symbol "=". Examples of equations are:

(i) $2x - 3 = 9 - x$. (ii) $y^2 - 5y = 6 - 4y$. (iii) $2x + y = 7$.

In Equation (i) the variable is the letter x, and in Equation (ii) it is y. In Equation (iii) we have two variables, x and y.

The two expressions separated by the equality symbol are called the two *sides* of the equation, the *left side* and the *right side*.

SOLUTIONS OF EQUATIONS

Equations involving only constants and no variables are either true or false statements. For example,

$$3 + 2 = 5 \quad \text{and} \quad \tfrac{3}{15} = \tfrac{4}{20}$$

are true statements, while
$$2 + 5 = 6 \quad \text{and} \quad \tfrac{3}{2} = \tfrac{2}{3}$$
are false statements.

An equation containing a variable can become a true statement for certain values of the variable and a false statement for other values. For example, consider the equation
$$2x - 3 = x + 2.$$
If x takes the value 5, this equation becomes
$$2(5) - 3 = 5 + 2 \quad \text{—that is,} \quad 10 - 3 = 5 + 2$$
which is a true statement. On the other hand, if x takes the value 4, we get
$$2(4) - 3 = 4 + 2 \quad \text{or} \quad 5 = 6$$
which is a false statement.

A value of the variable that, when substituted in the given equation, makes the equation a true statement is called a *root* or *solution* of the given equation. We say that the equation is *satisfied* by such a value of the variable.

Thus, for example, the number 5 is a root of the equation $2x - 3 = x + 2$. Similarly -2 is a root of the equation
$$y^2 + 3y = 6 + 4y$$
because when we substitute -2 for y in the equation we obtain
$$(-2)^2 + 3(-2) = 6 + 4(-2)$$
—that is, $\quad\quad\quad\quad\quad\quad 4 - 6 = 6 - 8$

which is a true statement.

In the same way 5 is *not* a root of the equation
$$t^2 + 2t = 6 + 3t$$
because when t is replaced by 5, we have
$$(5)^2 + 2(5) = 6 + 3(5)$$
—that is, $\quad\quad\quad\quad\quad\quad 25 + 10 = 6 + 15$

which is *not* a true statement.

We are often interested in finding the roots of some given equation—that is, in determining all the values of the variable that make the equation a true statement. The process of finding the roots is called *solving the equation*. In carrying out this process we usually perform certain operations that transform the equation into a new one that is simpler to solve. Such simplifications must be made whenever possible in such a way that the new equation has the same roots as the original equation.

The following two operations satisfy this requirement of leaving the roots of the equation unchanged:

    I. Addition or subtraction of any real number to both sides of the equation.   (ADDITION PRINCIPLE)

II. Multiplication or division of both sides of an equation by any nonzero real number.   (MULTIPLICATION PRINCIPLE)

Consider, for example, the equation
$$x - 3 = 2. \qquad \text{(i)}$$
Let us add 3 to both sides; by the addition principle this operation will not change the roots of the equation. We get
$$x - 3 + 3 = 2 + 3.$$
After simplification, this becomes
$$x = 5.$$
We conclude, therefore, that if x satisfies Equation (i), then $x = 5$: the number 5 is the one and only solution of Equation (i).

As a second example, consider the equation
$$5x = 15. \qquad \text{(ii)}$$
Let us divide both sides of this equation by 5. By the multiplication principle this operation will not change the roots of the equation, since the number by which we are dividing is nonzero. We get
$$\frac{5x}{5} = \frac{15}{5}$$
or
$$x = 3.$$
Thus, the one and only solution of Equation (ii) is the value $x = 3$.

Two equations that have exactly the same solutions are said to be *equivalent* to one another. The operations I and II therefore transform a given equation into a new equation that is equivalent to the old one. In solving a given equation we may have to use these operations several times in succession.

Example 1 Solve the equation
$$5x - 3 = 2x + 9. \qquad \text{(i)}$$

Solution First let us subtract the real number $2x$ from both sides:
$$5x - 3 - 2x = 2x + 9 - 2x$$
—that is,
$$5x - 2x - 3 = 9$$
$$3x - 3 = 9. \qquad \text{(ii)}$$
Next let us add 3 to both sides:
$$3x - 3 + 3 = 9 + 3$$
or
$$3x = 12. \qquad \text{(iii)}$$

Finally let us divide both sides by 3 (which is nonzero):
$$\frac{3x}{3} = \frac{12}{3}$$
$$x = 4.$$

Thus, the solution of Equation (i) is the value $x = 4$.

We observe that in the last example we can obtain Equation (ii) from Equation (i) simply by moving the $2x$ term from the right side to the left and changing its sign. We get
$$5x - 3 - 2x = 9$$
or
$$3x - 3 = 9$$

which agrees with Equation (ii). Again, we can obtain Equation (iii) from Equation (ii) by simply moving the -3 term from the left side to the right and changing its sign. We get
$$3x = 9 + 3 \quad \text{—that is,} \quad 3x = 12.$$

Thus we can see that the addition principle stated above is equivalent to the following: *We can move any term from one side of an equation to the other side while changing its sign, and this will not affect the roots of the equation.*
According to this, for example, the equation $5x + 3 = 2x$ is equivalent to $5x - 2x + 3 = 0$ and also to $3 = 2x - 5x$.

Example 2 Solve the equation
$$5x + 7 = 21 - 2x. \tag{i}$$

Solution Moving the term containing $2x$ from the right side to the left side and changing its sign, we get
$$5x + 7 + 2x = 21$$
or
$$7x + 7 = 21.$$

Moving the constant term, 7, from the left side to the right side and changing its sign, we obtain
$$7x = 21 - 7$$
$$7x = 14.$$

Dividing both sides by 7, we get
$$\frac{7x}{7} = \frac{14}{7}$$
$$x = 2.$$

According to the multiplication principle, any real number by which we multiply or divide must be nonzero, and we must take care not to multiply or divide the equation by an expression that can be equal to zero. For example, consider the equation

$$x^2 = 5x.$$

Clearly $x = 0$ is a root of this equation. If we divide both sides by x, we obtain

$$x = 5.$$

Clearly $x = 0$ is *not* a root of this resulting equation, although 0 was a root of the original equation. The reason is that we have divided both sides by x, which can be zero, and this violates the multiplication principle. In dividing by x we have lost one root of the equation. To avoid such pitfalls *we must be careful not to multiply or divide by an expression that contains the variable unless we are sure that this expression is nonzero.*

LINEAR EQUATIONS

DEFINITION A *linear equation* in the variable x is an equation that can be expressed in the form

$$ax + b = 0 \qquad (a \neq 0)$$

where a and b are constants.

Example 3 (a) $x - 4 = 0$ is a linear equation. Moving the 4 to the right side and changing its sign, we get $x = 4$. (*Note:* This is equivalent to adding 4 to both sides.) Thus the number 4 is the only solution of this equation.

(b) $2x + 3 = 0$ is a linear equation. Moving 3 to the right side, we get $2x = -3$; and now, dividing by 2, we find $x = -\frac{3}{2}$. Thus the number $(-\frac{3}{2})$ is the only solution of the given equation.

(c) In the general case,

$$ax + b = 0$$

we can move the constant b to the right side:

$$ax = -b.$$

Now if we divide by a, we get

$$x = -\frac{b}{a}.$$

Thus the linear equation $ax + b = 0$ has one and only one solution, namely $x = -b/a$.

Observe that in solving these equations we kept the terms involving x on the left side and moved the constant terms to the right side. This is the general strategy for solving all linear equations. (We used it in solving Example 2 above.)

Often equations arise that at first glance do not appear to be linear but that may be reduced to linear equations by appropriate simplification. In carrying out such a reduction the following step-by-step procedure is often helpful.

Step 1. Remove any numerical fractions that occur in the equation by multiplying both sides by the least common denominator of the fractions involved.

Step 2. Expand any parentheses that occur. (Steps 1 and 2 may be interchanged.)

Step 3. Move all the terms containing the variable to the left side and all other terms to the right side, and simplify, if possible, by combining like terms.

This procedure is amplified in the following solved examples.

Example 4 Solve the equation
$$3x - 4(6 - x) = 15 - 6x.$$

Solution

Step 1. Since there are no fractions in the equation, we do not need Step 1.

Step 2. On expanding the parentheses, we obtain
$$3x - 24 + 4x = 15 - 6x.$$

Step 3. Moving all the terms containing the variable to the left and the constant terms to the right, remembering to change their signs, we get
$$3x + 4x + 6x = 15 + 24$$
$$13x = 39.$$

We now obtain the solution by dividing both sides by 13 (the coefficient of x):
$$x = \tfrac{39}{13} = 3.$$

Example 5 Solve the equation
$$\frac{5x}{3} - \frac{x-2}{4} = \frac{9}{4} - \frac{1}{2}\left(x - \frac{2x-1}{3}\right).$$

Solution After removing the last parentheses, we can write the given equation as
$$\frac{5x}{3} - \frac{x-2}{4} = \frac{9}{4} - \frac{x}{2} + \frac{2x-1}{6}.$$

In order to remove the fractions, we multiply both sides by 12, the common denominator. We get
$$12\left(\frac{5x}{3}\right) - 12\left(\frac{x-2}{4}\right) = 12\left(\frac{9}{4}\right) - 12\left(\frac{x}{2}\right) + 12\left(\frac{2x-1}{6}\right)$$
$$4(5x) - 3(x - 2) = 3(9) - 6x + 2(2x - 1)$$
$$20x - 3x + 6 = 27 - 6x + 4x - 2.$$

Moving the x terms all to the left and the constant terms to the right, we get
$$20x - 3x + 6x - 4x = 27 - 2 - 6$$
$$19x = 19.$$

Sec. 3.1 Solutions of Equations; Linear Equations

Finally, dividing both sides by 19, we obtain $x = 1$, which is the required solution.

Sometimes we come across an equation which involves more than one algebraic symbol. Such equations arise often, for example, as formulas in physics. In solving such an equation, we simply treat the other letters (except the variable for which the equation is being solved) as constants. The method is illustrated in the following example.

Example 6 Solve the equation $\dfrac{x - 2t}{a} = 3x$

(a) For x. (b) For t.

Solution Multiplying both sides by a to get rid of fractions, we obtain

$$x - 2t = 3ax. \qquad \text{(i)}$$

(Note that a cannot be zero, otherwise the given equation would have a fraction with zero denominator. Hence we are allowed to multiply by a.)

(a) Since we are solving for x, we move all the terms containing the variable x to the left and all the terms without x to the right:

$$x - 3ax = 2t$$
$$x(1 - 3a) = 2t.$$

Dividing both sides of the equation by $(1 - 3a)$, assuming this factor to be nonzero, we get

$$x = \frac{2t}{1 - 3a}.$$

(b) Since we are solving for t, we keep only those terms that contain the variable t on the left, obtaining

$$-2t = -x + 3ax.$$

Dividing both sides by -2, the coefficient of t, we have

$$t = \frac{-x + 3ax}{-2}$$
$$= \tfrac{1}{2}x(1 - 3a)$$

which is the required solution for the variable t.

Example 7 Solve the equation

$$(2x + 1)^2 = 4(x^2 - 1) + x - 1.$$

Solution At first glance this equation does not appear to be linear because of the x^2 terms. It does, however, reduce to a linear equation, as we shall see. Let us remove the parentheses and move all the terms involving x to the left side of

the equation. We obtain
$$4x^2 + 4x + 1 = 4x^2 - 4 + x - 1$$
$$4x^2 + 4x - 4x^2 - x = -4 - 1 - 1.$$

The two terms $4x^2$ cancel one another out, and we are left with
$$3x = -6.$$
Hence the solution is $x = -2$.

Not every equation involving a variable does have a solution. For example, the equation
$$2x + 1 = 2x + 7$$
has no solution. For if we subtract $2x$ from both sides, we obtain $1 = 7$ which is a false statement. In fact, if we substitute any number for x in the given equation, we shall always obtain a false statement.

Thus, *if on simplifying an equation by use of the addition principle and/or the multiplication principle, we get a false statement, the original equation has no root.*

Example 8 Solve the equation
$$3(x - 1) = 3(x + 2) - 1.$$

Solution We have
$$3(x - 1) = 3(x + 2) - 1$$
$$3x - 3 = 3x + 6 - 1$$
$$3x - 3x = 6 - 1 + 3$$
$$0 = 8$$
which is a false statement. Thus the given equation has no root.

IDENTITIES

Equations exist that are satisfied for all values of the variables involved. Such equations are known as *identities*. For example, the equation
$$x^2 - 9 = (x - 3)(x + 3)$$
is an identity, because on expanding the right side we obtain the equivalent equation
$$x^2 - 9 = x^2 - 9$$
which is always true, no matter what value x takes.

To verify that an equation is an identity, we simplify it to obtain an equivalent equation that has identical terms on the two sides.

Example 9 Show that the following equation is an identity:
$$(2x + 3)(x - 1) - (x + 2)(x + 1) = (x - 1)^2 - 6.$$

Solution Multiplying out the expressions on the two sides, we have

$$\text{left side} = (2x + 3)(x - 1) - (x + 2)(x + 1)$$
$$= 2x^2 + x - 3 - (x^2 + 3x + 2)$$
$$= 2x^2 + x - 3 - x^2 - 3x - 2$$
$$= x^2 - 2x - 5.$$
$$\text{right side} = (x - 1)^2 - 6$$
$$= x^2 - 2x + 1 - 6$$
$$= x^2 - 2x - 5.$$

This is exactly the same as the left side. Thus the given equation is an identity.

Diagnostic Test 3.1

Fill in the blanks.

1. If $2x = 7$, then $x =$ _____.
2. If $3x = -5$, then $x =$ _____.
3. If $x/3 = 4$, then $x =$ _____.
4. If $2x/5 = 8$, then $x =$ _____.
5. If $x + 2 = 7$, then $x =$ _____.
6. If $7 = 5 - x$, then $x =$ _____.
7. If $ax + b = 0$, then $x =$ _____.
8. If $x/2 + 3 = 0$, then $x =$ _____.
9. If $x/p + 1/q = 0$, then $x =$ _____.
10. If $x/2 + x/3 = 0$, then $x =$ _____.

Exercises 3.1

(1–6) Check whether the given numbers are solutions of the corresponding equations.

1. $3x + 7 = 12 - 2x$; 1.
2. $5t - 3 = 18 + 3(1 - t)$; 3.
3. $\dfrac{u+2}{3u-1} + 1 = \dfrac{6-u}{u+1}$; 2.
4. $y^2 + 12 = 7y$; 4, -3.
5. $\dfrac{7}{x+1} + \dfrac{15}{3x-1} = 8$; $-\tfrac{1}{2}, \tfrac{1}{3}$.
6. $\dfrac{5}{x} - \dfrac{3}{2x} = \dfrac{x}{2}$; 3.

(7–36) Solve the following equations.

7. $4(x - 3) = 8 - x$.
8. $2(3x - 1) = 3(x + 4)$.
9. $3 - 2(1 - x) = 5 + 7(x - 3)$.
10. $2x - 5(1 - 3x) = 1 - 3(1 - 2x)$.
11. $6y - 5(1 + 2y) = 3 + 4(1 - y)$.
12. $3z - 2 + 4(1 - z) = \tfrac{1}{2}(1 - 2z) - 12$.
13. $\dfrac{3x+7}{2} = \dfrac{1+x}{3}$.
14. $\dfrac{2x-7}{3} = 5 - \dfrac{3x-2}{4}$.
15. $1 - \dfrac{2u-3}{3} = \dfrac{2-5u}{3} - 3u$.
16. $\dfrac{5y-6}{2} = y - \dfrac{2-y}{3}$.
17. $5[1 - 2(2z - 1)] = -3(3z - 1) + 1$.
18. $5t - [3 + (2 - 3t)] = 6 - [3 - 2(t + 1)]$.
19. $\tfrac{1}{3}(2y + 1) + \tfrac{1}{2}y = \tfrac{2}{5}(1 - 2y) - 4$.
20. $\tfrac{1}{2}[1 + \tfrac{1}{4}(3z - 1)] = 2z/3 - \tfrac{1}{2}$.
21. $y^2 + 7 = (y - 1)^2 + 3y$.
22. $(u - 1)^2 = (u + 1)(u + 3) + 5$.
23. $(x - 4)^2 = (x - 2)^2$.
24. $(x - 1)(x + 3) = (x + 2)(x - 3) + 1$.

25. $x^2 + (x + 1)^2 = (2x - 1)(x + 3)$.
26. $(3x - 1)(x + 2) + 5x = (2x + 1)(x - 3) + x^2$.
27. $(2x + 1)(x - 1) + x^2 = 3(x - 1)(x + 2) - 3$.
28. $(3x + 1)(2x - 1) - 2x^2 = (2x - 3)^2 + 6x + 5$.
29. $(x - 1)(x + 2) = (x - 2)(x + 3)$.
30. $(x + 1)(x + 3) = (x + 2)^2$.
*31. $3^{x+1} = 9^x$.
*32. $2^{x-3} = 8^{1+x}$.
*33. $4^{3x-2} = 16 \cdot 8^x$.
*34. $25^x = 5(125^{x+3})$.
*35. $4^x = \dfrac{8^x}{2^{x-1}}$.
*36. $27^x = \dfrac{9^{2x+1}}{3^{x-2}}$.

(37–40) Solve the following equations for the indicated variables.

37. $ax + by = cz$: (i) for x, (ii) for b.
38. $\dfrac{1}{x} + \dfrac{1}{y} = \dfrac{1}{t}$: (i) for x, (ii) for t.
39. $S = \dfrac{a - rl}{1 - r}$: (i) for r, (ii) for l.
40. $\dfrac{2}{x} + \dfrac{3}{xy} = 1$: (i) for x, (ii) for y.

(41–44) Prove each equation is an identity.

41. $x^3 - 8 = (x - 2)(x^2 + 2x + 4)$.
42. $8 + (3x + 1)(x + 2) + (x - 2)(x + 3) = 4(x + 1)^2$.
43. $(x + 1)^2 = (x - 1)^2 + 4x$.
*44. $16(4^x) = \dfrac{8^{x+1}}{2^{x-1}}$.

45. For what value of k is 3 a root of $2x + 3 = 3k - x + 4$?
46. For what value of m is 2 a root of $x^2 + mx - 2 = 0$?

(47–50) In each of the following, determine whether the two given equations are equivalent.

47. $x^2 = 9$; $x = 3$.
48. $x = -\sqrt{4}$; $x = 2$.
49. $\dfrac{x^2 - 4}{x - 2} = 3x - 1$; $x + 2 = 3x - 1$.
50. $(x - 3)(x + 1) = (x - 3)(2x + 7)$; $x + 1 = 2x + 7$.

3.2

Applications of Linear Equations

Algebraic methods are very useful in solving applied problems in many different fields. Such problems are generally stated in verbal form, which we need to translate into corresponding algebraic statements before we can use our algebraic tools. The following step-by-step procedure is often helpful in carrying out this process.

Step 1. Represent the unknown quantity—that is, the quantity that is to be determined—by an algebraic symbol, say x. In some problems two or even more quantities must be determined, and we just choose one of them to be x.

Step 2. Express any other nonconstant quantities involved in the problem in terms of x.

Step 3. Translate verbal statements occurring in the problem into algebraic statements involving *x*. Words such as "is" or "was" are translated into the algebraic symbol "=".

Step 4. Solve the algebraic statements according to the familiar methods of algebra.

Step 5. Translate the algebraic solution back into verbal form.

Verbal statements of a problem often include phrases such as so many "more than" or "less than" a certain value or multiples such as "twice" or "half". The following examples illustrate the translation of such expressions into algebraic terms.

Example 1 (a) If Jack has x dollars and Jill has five more than Jack, then she has $(x + 5)$ dollars. If Sam has three less than Jack, then he has $(x - 3)$.
(b) If Chuck is x years old and his father is four years more than twice Chuck's age, then his father is $(2x + 4)$ years old.
(c) If a certain store sells x refrigerators per month and a second store sells five less than one-third as many, then this second store sells $(\frac{1}{3}x - 5)$ refrigerators.

We look now at some examples that illustrate the translation between verbal and algebraic forms.

Example 2 (Integer Problem) Determine two consecutive integers whose sum is 19.

Solution *Step 1.* Since there are two integers to be found, we must decide which of them to call x. Let us denote the smaller integer by x.

Step 2. The second integer is then $(x + 1)$, since the two consecutive integers differ by one.

Step 3. We are given that

sum of two consecutive integers is 19

$$x + (x + 1) = 19$$

Step 4. Therefore,

$$2x + 1 = 19$$
$$2x = 19 - 1 = 18$$
$$x = 9.$$

Step 5. The two integers are thus x and $x + 1$ or 9 and 10.

Example 3 (Age Problem) A man is seven years older than his wife. Ten years ago he was twice her age. How old is he now?

Solution *Step 1.* Let x denote the present age (in years) of the man.

Step 2. Since his wife is 7 years younger than he is, her present age is therefore $(x - 7)$ years. Ten years ago, their ages were 10 less than their present ages. Thus the man's age was $(x - 10)$ and his wife's age was $(x - 7) - 10$ or $x - 17$.

Step 3. We are told that 10 years ago

$$\underbrace{\text{man's age}}_{x - 10} \quad \underbrace{\text{was}}_{=} \quad \underbrace{\text{twice}}_{2} \quad \underbrace{\text{his wife's age}}_{(x - 17)}$$

Step 4. Therefore,
$$x - 10 = 2x - 34$$
and so
$$x - 2x = -34 + 10$$
$$-x = -24$$
$$x = 24.$$

Step 5. Thus the present age of the man is 24 years.

Example 3 (Coin Problem) Sue has three more nickels than dimes, five more dimes than quarters. In all she has $2.10. How many of each coin does she have?

Solution Let x be the number of quarters. Because she has five more dimes than quarters, the number of dimes she has is $(x + 5)$. Moreover, she has three more nickels than dimes and therefore she has $(x + 5) + 3$ or $x + 8$ nickels.

In coin problems of this type when we want to calculate the total value of various coins, it is helpful to set this out in a table as follows.

	Cents per coin	Number of coins	Total amount in cents
Nickels	5	$x + 8$	$5(x + 8)$
Dimes	10	$x + 5$	$10(x + 5)$
Quarters	25	x	$25x$

We are given that the

$$\underbrace{\text{total value of all coins (in cents)}}_{5(x + 8) + 10(x + 5) + 25x} \text{ is } \underbrace{210}_{= 210}$$

Therefore,
$$5x + 40 + 10x + 50 + 25x = 210$$
$$40x = 210 - 40 - 50$$
$$= 120$$
$$x = 3.$$

Thus Sue has $x = 3$ quarters, $x + 5 = 8$ dimes, and $x + 8 = 11$ nickels.

Example 4 **(Mixture Problem)** Peanuts at 70¢ per pound are mixed with walnuts at $1.20 per pound to obtain a mixture of 50 pounds. How much of each should be mixed if the mixture is to sell for 90¢ per pound?

Solution Let x be the number of pounds of peanuts in the mixture. Since the total mixture is 50 pounds, we must use $(50 - x)$ pounds of walnuts.

We compute the values of the nuts in the mixture:

value (in cents) of x lb peanuts at 70¢/lb $= 70x$

value (in cents) of $(50 - x)$ lb walnuts at 120¢/lb $= 120(50 - x)$

value (in cents) of 50 pounds of mixture at 90¢/lb $= 90(50)$.

Now we have

$$\underbrace{\text{value of peanuts}}_{70x} + \underbrace{\text{value of walnuts}}_{120(50-x)} = \underbrace{\text{value of mixture}}_{90(50)}$$

or

$$70x + 6000 - 120x = 4500$$
$$-50x = 4500 - 6000 = -1500$$

Therefore

$$x = 30$$

and

$$50 - x = 20.$$

We must use 30 pounds of peanuts and 20 pounds of walnuts in the mixture.

Example 5 **(Mixture Problem)** A vintner wishes to make 10,000 gallons of sherry by fortifying white wine, which has an alcohol content of 10%, with brandy, which has an alcohol content of 35% by volume. The sherry is to have an alcohol content of 15%. Determine the quantities of wine and brandy that should be mixed together to produce the desired result.

Solution Let x gallons of brandy be used in making the 10,000 gallons of sherry. Then the volume of white wine used will be $(10,000 - x)$ gallons. Since brandy contains 35% alcohol, the amount of alcohol in x gallons of brandy is $\frac{35}{100}x$. Similarly the wine contains 10% alcohol, so $(10,000 - x)$ gallons of it contain $(10/100)(10,000 - x)$ gallons of alcohol. Therefore the total amount of alcohol in the mixture will be

$$\tfrac{35}{100}x + \tfrac{10}{100}(10,000 - x) \text{ gallons.}$$

Now the mixture is to contain 15% alcohol, so in the 10,000 gallons there should be $\frac{15}{100}(10,000) = 1500$ gallons of alcohol. Therefore, we have the equation

$$\tfrac{35}{100}x + \tfrac{1}{10}(10,000 - x) = 1500.$$

That is,

$$\tfrac{35}{100}x + 1000 - \tfrac{1}{10}x = 1500$$

or
$$\tfrac{35}{100}x - \tfrac{1}{10}x = 1500 - 1000$$
$$= 500.$$

Multiplying through by 100, we get
$$35x - 10x = 50{,}000$$
$$25x = 50{,}000$$
$$x = \frac{50{,}000}{25}$$
$$= 2000.$$

So 2000 gallons of brandy (and therefore 8000 gallons of wine) must be mixed together.

Example 6 (**Profit-and-Loss Problem**) A cattle dealer bought 1000 steers for $150 each. He sold 400 of them at a profit of 25%. At what price must he sell the remaining 600 if his average profit on the whole lot is to be 30%?

Solution His profit on each steer in the 400 already sold is 25% of the cost price—that is, 25% of $150 or $37.50. On the total 400 his profit is therefore $(37.50)(400) = $15,000. Let his selling price on the remaining 600 steers be x dollars. Then his profit per steer is $(x - 150)$ and his profit on the whole 600 is $600(x - 150)$ dollars. Therefore his total profit on the whole purchase is
$$15{,}000 + 600(x - 150) \text{ dollars.}$$

It is required that this profit be 30% of the price he paid for the 1000 steers—that is, 30% of $150,000. This is equal to $\tfrac{30}{100}(150{,}000)$, or $45,000. Thus we arrive at the equation
$$15{,}000 + 600(x - 150) = 45{,}000$$
—that is,
$$15{,}000 + 600x - 90{,}000 = 45{,}000$$
$$600x = 45{,}000 - 15{,}000 + 90{,}000$$
$$= 120{,}000.$$

Therefore
$$x = \frac{120{,}000}{600} = 200.$$

The dealer must sell the remaining steers at $200 each if his average profit is to be 30%.

INTEREST PROBLEMS

Suppose you invest a sum of $500 at 6% simple interest per annum for three years. At the end of one year, $100 will earn an interest of $6 at a 6% rate and $500 will earn an interest of $5($6) = $30. Similarly, $30 will be the interest

Sec. 3.2 Applications of Linear Equations

earned for the second and for the third years. The total interest after three years will be 3($30) = $90.

In general, if a sum of money *P* (called the ***principal***) is invested at a simple interest rate of *R*% per annum for *n* years, then the interest *I* earned is given by

$$I = Prn$$

where $r = R/100$.

In the previous example, $500 is invested at a simple interest rate of 6% per annum for three years. Then

$$P = 500, \quad r = \frac{R}{100} = \frac{6}{100}, \quad n = 3 \text{ years.}$$

The interest earned will be

$$\begin{aligned} I &= Prn \\ &= (500)(\tfrac{6}{100})(3) \\ &= \$90. \end{aligned}$$

Example 7 An investor with $70,000 wants to receive an annual income of $5000. He can invest his funds in 6% government bonds and with greater risk in 8.5% mortgage bonds. How should he invest his money in order to earn exactly $5000?

Solution Let the amount invested in government bonds be *x* dollars. Then the amount invested in mortgage bonds is (70,000 − *x*) dollars.

$$\text{Income from government bonds at } 6\% = \frac{6}{100}x.$$

$$\begin{aligned} \text{Income from mortgage bonds at } 8.5\% &= \frac{8.5}{100}(70{,}000 - x) \\ &= \frac{85}{1000}(70{,}000 - x). \end{aligned}$$

Since the total income received from the two types of bonds must be $5000,

$$\tfrac{6}{100}x + \tfrac{85}{1000}(70{,}000 - x) = 5000.$$

Multiplying both sides by 1000, we get

$$60x + 85(70{,}000 - x) = 5{,}000{,}000$$
$$60x + 5{,}950{,}000 - 85x = 5{,}000{,}000$$
$$-25x = 5{,}000{,}000 - 5{,}950{,}000$$
$$= -950{,}000.$$

Therefore,

$$x = \frac{-950{,}000}{-25} = 38{,}000.$$

Thus he should invest $38,000 in government bonds and the remaining $32,000 in mortgage bonds.

DISTANCE PROBLEMS

Many distance and speed problems can be handled by algebraic methods. The basic relationship for such problems is that the distance d traveled by an object moving at a constant speed r for time t is given by

$$d = rt.$$

For example, consider a car traveling at a constant speed of 30 miles per hour for 6 hours. Then $r = 30$ and $t = 6$, and the car will travel a distance $d = rt = (30)(6) = 180$ miles.

Example 8 A man swims across a lake at the rate of 20 yards per minute and returns in a boat at the rate of 75 yards per minute. If the round trip takes him an hour and 16 minutes, how far is it from shore to shore?

Solution Let t be the time (in minutes) taken when swimming across. Then the time taken on the return journey by boat is $(76 - t)$ minutes, because the total round trip time is 76 minutes.

Since it takes t minutes to swim across at a speed of 20 yards per minute, the distance covered in t minutes is $20t$ yards. Similarly the distance covered by boat in $(76 - t)$ minutes at the rate of 75 yards per minute is $75(76 - t)$ yards.

Since these two distances are equal, we have   *2ND PIECE OF INFO DEDUCED*

$$20t = 75(76 - t)$$
$$20t = 5700 - 75t$$
$$95t = 5700$$
$$t = 60.$$

Therefore the distance from shore to shore is

$$20t = (20)(60) = 1200 \text{ yards.}$$

Example 9 A passenger train leaves Boston at 1 P.M. for New York, 190 miles away, and travels at a speed that averages 50 miles per hour. A freight train leaves New York at 4 P.M. for Boston at a speed averaging 30 miles per hour, traveling on a parallel track. When and where do the two trains meet?

Solution Let the freight train travel for t hours (after 4 P.M.) before it meets the oncoming train from Boston. Then the distance d_1 it has traveled is

$$d_1 = 30t. \quad \text{OR} \quad t = \frac{d}{30}$$

Since the passenger train leaves Boston 3 hours earlier than the freight train, it will have traveled for $(t + 3)$ hours before meeting the freight train. If d_2

Sec. 3.2   Applications of Linear Equations

denotes the distance covered by the passenger train in this time, then
$$d_2 = 50(t + 3).$$

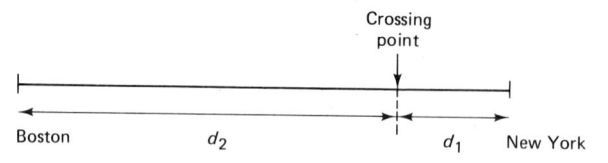

Now the sum of the distances traveled by the two trains must be the distance 190 miles between Boston and New York. Thus

$$d_1 + d_2 = 190$$
$$30t + 50(t + 3) = 190$$
$$80t + 150 = 190$$
$$80t = 40$$
$$t = \tfrac{1}{2} \text{ hour.}$$

Hence the two trains meet $\tfrac{1}{2}$ hour after 4 P.M. or at 4: 30 P.M., and they meet at a point $d_1 = 30t = 30(\tfrac{1}{2}) = 15$ miles from New York.

WORK PROBLEMS

Suppose a job is being done by several people, each working at his or her own rate. To determine how much of the job is done after a time t, we have to calculate the fraction of the job done by each person during the given time and then add up these fractions.

For example, if Jane can mow a lawn alone in 5 hours, then in 1 hour she will mow one-fifth of the lawn and in 3 hours three-fifths. If Dick can mow the same lawn in 8 hours, then in 1 hour he will mow one-eighth of it and in 3 hours three-eighths. If Dick and Jane mow together, then in one hour the fraction of the lawn they mow will be $\tfrac{1}{5} + \tfrac{1}{8} = \tfrac{13}{40}$ and in three hours $\tfrac{3}{5} + \tfrac{3}{8} = \tfrac{39}{40}$.

Example 10 It takes 3 hours for Sam to shovel the snow from the driveway; Joe can do it in 4 hours. How long will the job take, if they both shovel together?

Solution Let x be the number of hours it takes to shovel the snow together. Then in 1 hour Sam shovels one-third of the driveway and Joe shovels one-fourth. In x hours the fractions of the job done by Sam and Joe are $x/3$ and $x/4$. Now

$$\underbrace{\text{fraction of job done by Sam}}_{\tfrac{x}{3}} + \underbrace{\text{fraction of job done by Joe}}_{\tfrac{x}{4}} = \underbrace{1 \text{ (total job)}}_{1}.$$

Multiply both sides by 12:

$$4x + 3x = 12$$
$$7x = 12$$
$$x = \tfrac{12}{7}.$$

It takes $\tfrac{12}{7}$ hours for them to shovel the snow together.

Exercise 3.2

1. If Joe has x dollars, how many dollars does Judy have if she has
   (a) $4 more than Joe?
   (b) $3 less than twice as much as Joe?
   (c) $2 more than half as much as Joe?

2. If Joe is x years old and Judy is 4 years younger, how old is Fred if
   (a) He is 3 years older than Judy?
   (b) He is 1 year more than the average age of Joe and Judy?
   (c) He is 10 years less than the sum of Joe's and Judy's ages?

3. Find two consecutive integers whose sum is 43.

4. Find two consecutive odd integers whose sum is 308.

5. Find two consecutive even integers whose sum is 354.

6. Bruce and Jack together have $75. If Jack has $5 more than Bruce, how much does Jack have?

7. In a mathematics class there are 52 students. If the number of boys is seven more than twice the number of girls, determine the number of girls in the class.

8. A father is three times as old as his son. In 12 years he will be twice as old as his son. How old are the father and the son now?

9. Five years ago Steve was twice as old as his brother. Find Steve's present age if the sum of their ages today is 40 years.

10. Jack is five years older than Joe. Fifteen years ago Jack was twice as old as Joe. How old is Joe now?

11. A man has a son and a daughter. The man is four times as old as his daughter, who is two years younger than her brother. In five years the man will be three times as old as his son. How old is the son?

12. Peter is three times as old as Sue, who is four years older than Joe. In 20 years Peter will be twice as old as Joe. How old is Peter now?

13. Susan has two more nickels than dimes and three more dimes than quarters. In all she has $1.35. How many of each coin does she have?

14. I have twice as many dimes in my pocket as quarters. If I had four fewer dimes and three more quarters, I would have $2.60. How many dimes and quarters do I have?

15. Sam has 31 coins in dimes and quarters, which are worth $5.20. How many dimes does he have?

16. Celia has six more nickels than dimes and two more dimes than quarters. In all she has $2.60. How many nickels does she have?

17. Coffee worth $1.25 a pound is blended with coffee worth $1.60 a pound to produce 70 pounds of mixture worth $1.40 per pound. How many pounds of each type must be mixed?

18. How many pounds of hazel nuts at 90¢ per pound must be mixed with 20 pounds of peanuts at 75¢ per pound to obtain a mixture that is worth 80¢ per pound?

19. Ten pounds of peanuts worth 75¢ per pound and 12 pounds of walnuts worth 80¢ per pound are mixed with pecans worth $1.10 per pound to produce a mixture worth 90¢ per pound. How many pounds of pecans should be used?

20. Twenty pounds of peanuts at 75¢ per pound and five pounds of walnuts at 90¢ per pound are mixed with almonds at $1.90 per pound to produce a mixture worth $1.20 per pound. How many pounds of almonds should be used in the mixture?

21. How much of a 10% acid solution must be mixed with 10 ounces of a 15% acid solution to obtain a 12% acid solution?

22. Fifteen ounces of a 12% acid solution are mixed with a certain quantity of 25% acid solution to obtain a 20% acid solution. How much of the 25% acid solution should be used?

23. How much water should be added to 15 ounces of 20% acid solution to obtain a 12% acid solution?

24. An alloy that is 16% silver is melted with another alloy that is 25% silver to obtain 45 pounds of an alloy that is 20% silver. How much of each alloy should be used?

25. A sample of sea water has 20% salt content. Fresh water is added to yield 75 ounces of 8% salt solution. How much sea water was in the sample?

26. How much pure alcohol should be added to 48 gallons of a solution that is 30% alcohol to obtain a solution that is 36% alcohol?

27. How much water must be evaporated from 300 ounces of 12% salt solution to obtain a 15% salt solution?

28. A bandit robs a grocery store and drives away at an average speed of 40 miles per hour. After 20 minutes the police begin to chase the bandit at 50 miles per hour. After how long will the police car overtake the bandit's car?

29. Two cars start at the same time and in the same direction, traveling at speeds averaging 30 and 36 miles an hour. After how long will the two cars be 20 miles apart?

30. Two bicyclists are 21 miles apart and are traveling toward each other. One cycles at the rate of 6 miles per hour and the other at 8 miles per hour. After how long will they meet?

31. A train leaves Vancouver at 7 A.M. for Edmonton, a 490-mile route, at a speed averaging 55 miles per hour. At 9 A.M. another train leaves Edmonton for Vancouver at a speed averaging 40 miles per hour. When will the two trains cross each other?

32. A hitchhiker begins to walk to the next town, 8 miles away, at a rate of 4 miles per hour. After 20 minutes he is picked up by a car. If his total travel time to the town is 30 minutes, at what speed was the car traveling?

33. A farmer starts walking at the rate of 3 miles per hour to a town 15 miles away. After a while he is picked up by a car, which travels at an average speed of 33 miles per hour. If the total time of his journey is 50 minutes, after how long was the farmer picked up?

34. A man can paint a house in 5 hours and his son can paint it in 7 hours. If the father and son paint together, in how much time will the house be painted?

35. Three persons start mowing a lawn. One can do the job alone in 2 hours, one in $2\frac{1}{2}$ hours, and the last one in 4 hours. How long does it take them together to mow the lawn?

36. One pipe can fill a tank in 25 minutes and another pipe can fill it in 35 minutes. How long will it take to fill the tank if both the pipes are opened together?

37. A pipe can fill a tank in 20 minutes and another pipe can empty it in 24 minutes. If the tank is half full when both pipes are opened, how much time will it take to fill the tank?

38. During a sale an item is marked down 20%. If its sale price is $2, what was its original price?

Ch. 3 Equations in One Variable

39. An item sells for $12. If the price markup is 50% of the wholesale price, what is the wholesale price?

40. The cost of publishing each copy of a weekly magazine is 28¢. The revenue from dealer sales is 24¢ per copy and from advertising is 20% of the revenue obtained from sales in excess of 3000 copies. How many copies must be published and sold each week to earn a weekly profit of $1000?

41. A used-car dealer bought two cars for $2900. He sold one at a gain of 10% and the other at a loss of 5%, realizing a gain of $185 on the whole transaction. Find the cost of each car.

42. A trader offers a 30% discount on the marked price of an article and yet makes a profit of 10%. If the cost to the trader is $35, what must be the marked price?

43. A salesman earns a basic salary of $600 per month and a commission of 10% on the sales he makes. He finds that on average he takes $1\frac{1}{2}$ hours to make $100 worth of sales. How many hours must he work on average each month if his monthly earnings are to be $2000?

44. A man invests twice as much at 8% as at 5%. His total annual income from the two investments is $840. How much is invested at each rate?

45. A man invested $2000 more at 8% than at 10% and received a total interest income of $700 for one year. How much did he invest at each rate?

46. A college has $60,000 to invest in an endowment fund in order to yield an annual income of $5000 for a scholarship. A part of this $60,000 will be invested in government bonds at 8% and the remainder in long-term fixed deposits at 10.5%. How much should be invested in each so as to provide the required income?

47. The trustees of an endowment fund want to invest $18,000 in two kinds of securities paying 9% and 6% annual dividends, respectively. How much must be invested at each rate if the annual income is to be equivalent to a yield of 8% on the total investment?

48. A company invests $15,000 at 8%, $22,000 at 9%. At what rate should the remaining sum of $12,000 be invested so that the combined annual interest income from the three investments in $4500?

3.3

Quadratic Equations (Solution by Factoring)

An equation of the type
$$ax^2 + bx + c = 0 \quad (a \neq 0) \quad (1)$$
where a, b, c are all constants, is called a *quadratic equation* in the variable x.

We shall study three methods of solving such an equation: factoring, using the quadratic formula, and completing the square. Whichever method is used, the first step consists of arranging the equation in the standard form (1): the right side is zero, and on the left side the x^2 terms, x terms, and constant terms are collected together. In arriving at this standard form we first remove all fractions by multiplying through by their common denominator, then remove all parentheses, next transfer all terms to the left side of the equation, and finally group all like terms together.

Example 1 Reduce the quadratic equation
$$(x + 1)(x - 2) = \tfrac{1}{3}x^2 - 5$$
to the standard form of Equation (1).

Solution First we multiply each term by 3 to get rid of fractions. We obtain
$$3(x + 1)(x - 2) = 3(\tfrac{1}{3}x^2) - 3(5)$$
$$3(x + 1)(x - 2) = x^2 - 15.$$

Now we multiply the expressions in parentheses:
$$3(x^2 - x - 2) = x^2 - 15$$
$$3x^2 - 3x - 6 = x^2 - 15.$$

Next we move all the terms to the left.
$$3x^2 - 3x - 6 - x^2 + 15 = 0$$
$$2x^2 - 3x + 9 = 0.$$

This is the required quadratic equation in standard form.

In this section we illustrate the method of factoring for the solution of quadratic equations. This method is based on the following principle.

THEOREM If p and q are two real numbers such that their product pq is zero, then either p or q is zero.

Proof We leave the proof as an exercise. [*Hint:* Assume that $q \neq 0$ and then multiply both sides of the equation $pq = 0$ by q^{-1}. This shows that p must be zero.]

To solve a given quadratic equation by the method of factoring, we first express the equation in the standard form, remove any constant monomial factors and then express the left side as the product of two linear factors. By the above theorem one of these linear factors must be zero. Therefore we equate each linear factor to zero and solve the resulting linear equation. The method is illustrated by the following examples.

Example 2 Solve the equation
$$3(x^2 + 1) = 5(1 - x)$$
by factoring.

Solution There are no fractions in this equation. Removing the parentheses, we find
$$3x^2 + 3 = 5 - 5x.$$

Moving all the terms on the right to the left, we get
$$3x^2 + 3 - 5 + 5x = 0$$
$$3x^2 + 5x - 2 = 0. \tag{i}$$

This is now a quadratic equation in standard form.
Factoring the left side in (i), we have
$$(3x - 1)(x + 2) = 0.$$
Thus, the product of the two real numbers $(3x - 1)$ and $(x + 2)$ is zero. By the preceding theorem, therefore, one of these two numbers must be zero:

either $3x - 1 = 0$ | or $x + 2 = 0$
$3x = 1$ | $x = -2.$
$x = \frac{1}{3}$

Therefore there are two roots of the given equation: $\frac{1}{3}$ and -2.

Example 3 Solve the following equation by the method of factoring:
$$(2x + 3)(3x - 1) = -4.$$

Solution Rewriting the given equation with right side zero, we have
$$(2x + 3)(3x - 1) + 4 = 0$$
$$6x^2 + 7x - 3 + 4 = 0$$
$$6x^2 + 7x + 1 = 0.$$

Factoring the left side, we obtain
$$(6x + 1)(x + 1) = 0.$$

Therefore

either $6x + 1 = 0$ | or $x + 1 = 0$
$6x = -1$ | $x = -1.$
$x = -\frac{1}{6}$

Hence the required roots are $-\frac{1}{6}$ and -1.

Example 4 Solve the equation $x\left(\frac{x}{4} - 1\right) = -1.$

Solution The given equation is
$$\frac{x^2}{4} - x = -1.$$

Multiplying both sides by 4 to get rid of fractions, we obtain
$$x^2 - 4x = -4$$
$$x^2 - 4x + 4 = 0$$
$$(x - 2)(x - 2) = 0.$$

Our standard procedure of setting the two linear factors equal to zero therefore leads to

either $x - 2 = 0$ | or $x - 2 = 0$
$x = 2$ | $x = 2.$

Therefore, the given equation has only one root: $x = 2$.

Since the factor $(x - 2)$ appears twice in the solution, 2 is called a **double root** or a **root of multiplicity two**.

Consider the equation
$$x^2 = 4.$$
Since $2^2 = 4$ and $(-2)^2 = 4$ also, x can be equal to either 2 or -2. That is,
$$x = 2 \quad \text{or} \quad x = -2.$$
These two solutions are abbreviated together as $x = \pm 2$. (This is read as "x equals plus or minus 2"). Thus, if $x^2 = 4$, then $x = \pm 2$.

This result generalizes to the equation $x^2 = k$, where k is any positive real number. We can write the given equation
$$x^2 = k \quad (k > 0)$$
as
$$x^2 - k = 0.$$
Using the difference-of-two-squares formula, we can factor this:
$$(x - \sqrt{k})(x + \sqrt{k}) = 0.$$
Therefore

either $x - \sqrt{k} = 0$ $\quad$ or $\quad x + \sqrt{k} = 0$
$\qquad x = \sqrt{k}$ $\qquad\qquad\qquad x = -\sqrt{k}.$

Thus either $x = \sqrt{k}$ or $x = -\sqrt{k}$, and we condense this into the statement $x = \pm\sqrt{k}$. To summarize, therefore, for any $k > 0$,

$$\boxed{\text{If } x^2 = k \text{ then } x = \pm\sqrt{k}.}$$

We can say that in order to solve the equation $x^2 = k$, we "take the square root of both sides." In doing so, however, we must remember to include the $\pm$ sign: $x = \pm\sqrt{k}$.

Example 5 Solve the equation $(x - 3)^2 = 4$.

Solution The given equation can be written as
$$(x - 3)^2 - 4 = 0$$
or
$$x^2 - 6x + 5 = 0.$$
Factoring, we have
$$(x - 1)(x - 5) = 0.$$
Therefore

either $x - 1 = 0$ $\quad$ or $\quad x - 5 = 0$
$\qquad x = 1$ $\qquad\qquad\qquad x = 5.$

Hence the required roots are 1 and 5.

Alternative Solution We can solve the given equation in this case by taking the square root of both sides. We get
$$x - 3 = \pm 2.$$
(Note the $\pm$ sign on the right side.) Thus

either $x - 3 = +2$ which gives $x = 5$

or $x - 3 = -2$ which gives $x = 1$.

Hence the roots are 5 and 1.

Example 6 Solve the equation $x^4 - 13x^2 + 36 = 0$.

Solution The given equation is of fourth degree, but we can reduce it to a quadratic equation by substituting $x^2 = y$. On putting $x^2 = y$ in the given equation, we have
$$y^2 - 13y + 36 = 0$$
which factors to
$$(y - 9)(y - 4) = 0.$$
Thus

either $y - 9 = 0$ which gives $y = 9$

or $y - 4 = 0$ which gives $y = 4$.

But $y = x^2$. Thus $y = 9$ gives
$$x^2 = 9 \quad \text{or} \quad x = \pm 3.$$
Similarly $y = 4$ gives
$$x^2 = 4 \quad \text{or} \quad x = \pm 2.$$
Thus there are four roots of the given equation: $x = \pm 2$ or ± 3.

Example 7 Solve the equation $x^{2/3} - x^{1/3} - 6 = 0$.

Solution Substituting $x^{1/3} = y$, we have
$$y^2 - y - 6 = 0$$
—that is,
$$(y - 3)(y + 2) = 0.$$
Therefore, either $y = 3$ or $y = -2$. When $y = 3$, we have
$$x^{1/3} = 3 \quad \text{or} \quad x = 3^3 = 27.$$
When $y = -2$, we have
$$x^{1/3} = -2 \quad \text{or} \quad x = (-2)^3 = -8.$$
Thus the roots of the given equation are 27 and -8.

Example 8 A piece of wire 32 inches long is cut into two pieces and then each piece is bent into a square. Find the length of each piece of wire, if the sum of the areas enclosed in the two squares is 40 square inches.

Solution Let one piece be of length x inches. Since the total length is 32 inches, the other piece therefore will be of length $(32 - x)$ inches.

Now each piece is bent into a square. Thus the lengths x and $(32 - x)$ are the perimeters of the two squares. Therefore the lengths of sides of the two squares will be $x/4$ and $(32 - x)/4$ inches, respectively.

We are given that

$$\underbrace{\left(\frac{x}{4}\right)^2 + \left(\frac{32-x}{4}\right)^2}_{\text{sum of the areas of two squares}} = \underbrace{40}_{\text{is 40}}$$

or

$$\frac{x^2}{16} + \frac{1}{16}(32 - x)^2 = 40.$$

Multiplying throughout by 16, we get

$$x^2 + (32 - x)^2 = 640$$
$$x^2 + x^2 - 64x + 1024 = 640$$
$$2x^2 - 64x + 384 = 0.$$

Dividing throughout by 2, we have

$$x^2 - 32x + 192 = 0$$
$$(x - 8)(x - 24) = 0$$

which gives $x = 8$ or 24.

If $x = 8$, then $32 - x = 32 - 8 = 24$.
If $x = 24$, then $32 - x = 32 - 24 = 8$.

Thus in either case the lengths of the two pieces are 8 inches and 24 inches.

Example 9 A ball is thrown upward with an initial velocity of 64 feet per second. After t seconds the ball is at a height H above the ground, where

$$H = 64t - 16t^2.$$

(a) At what time will the ball be at a height of 48 feet?
(b) How long will it be before the ball returns to the ground?

Solution The height H attained by the ball at time t seconds (after being thrown) is given by

$$H = 64t - 16t^2. \qquad (i)$$

(a) The ball will be at a height of 48 feet when $H = 48$. Thus,

$$48 = 64t - 16t^2$$

or

$$16t^2 - 64t + 48 = 0$$

Dividing throughout by 16, we have

$$t^2 - 4t + 3 = 0$$
$$(t - 1)(t - 3) = 0$$

Thus $t = 1$ or 3; the ball will be at a height of 48 feet after 1 second and after 3 seconds. (The first of these situations occurs while the ball is traveling upward, the second while it is traveling back downward.)

(b) The ball will be on the ground when $H = 0$. Thus, from (i),
$$0 = 64t - 16t^2 \quad \text{or} \quad 16t(t - 4) = 0.$$
The roots are either $t = 0$ or $t = 4$. The value $t = 0$ corresponds to the time when the ball is thrown up and $t = 4$ seconds to the time when it returns to the ground.

Diagnostic Test 3.3

Fill in the blanks.

1. If $x^2 = 9$, then $x = $ _____.
2. If $x^2 = 4x$, then $x = $ _____.
3. If $x^2 = a^2$, then $x = $ _____.
4. If $x^4 = 25x^2$, then $x^2 = $ _____.
5. If $(x - 2)(x - 3) = 0$, then $x = $ _____.
6. If $(x + 5)(x + 3) = 0$, then $x = $ _____.
7. If $(x - 2)(x - 3) = 6$, then $x = $ _____.
8. If $(x - 1)(x - 2) = 2$, then $x = $ _____.

Exercises 3.3

(1–40) Solve the following equations by the method of factoring.

1. $x^2 + 5x + 6 = 0$.
2. $x^2 + 3x + 2 = 0$.
3. $x^2 - 9x + 14 = 0$.
4. $x^2 - 5x + 6 = 0$.
5. $x^2 - 6x + 9 = 0$.
6. $x^2 + 10x + 25 = 0$.
7. $x^2 = 7x - 12$.
8. $x^2 = -2x + 35$.
9. $9x^2 - 25 = 0$.
10. $20x^2 - 45 = 0$.
11. $x^2 - 8x = 0$.
12. $3x^2 + 5x = 0$.
13. $4x^2 - 12x + 9 = 0$.
14. $9x^2 - 6x + 1 = 0$.
15. $x^2 = 7x$.
16. $2x^2 = 3x$.
17. $x^2 = x + 2$.
18. $x^2 = x + 6$.
19. $2(x^2 + 1) = 5x$.
20. $3(x^2 + 1) = 10x$.
21. $(x + 2)(x - 1) = 2x$.
22. $(x + 3)(x - 3) = x - 9$.
23. $15x^2 = 40(x + 2)$.
24. $2x(x + 1) = x^2 - 1$.
25. $7x + 3(x^2 - 4) = x - 3$.
26. $2x(4x - 1) = 4 + 2x$.
27. $\frac{1}{2}x^2 + \frac{10}{3}x + 2 = 0$.
28. $6x^2 + \frac{5}{2}x + \frac{1}{4} = 0$.
29. $\frac{1}{3}x^2 = \frac{11}{6}x + 1$.
30. $5x^2 - \frac{7}{2}x = \frac{1}{2}x + 1$.
31. $x^4 - 5x^2 + 4 = 0$.
32. $x^4 - 3x^2 + 2 = 0$.
33. $x^4 - 4x^2 + 4 = 0$.
34. $81x^4 - 18x^2 + 1 = 0$.
35. $x^{2/3} - 3x^{1/3} + 2 = 0$.
36. $x^{2/3} + x^{1/3} - 2 = 0$.
37. $x^{2/5} + x^{1/5} - 2 = 0$.
38. $x^{2/5} - 4x^{1/5} + 3 = 0$.
*39. $2^{x^2} = 8/4^x$.
*40. $3^{x^2-2} = 27^x/9^{2x}$.

41. Find two numbers whose sum is 15 and the sum of whose squares is 137.
42. Find two consecutive integers whose product is 132.
43. Find two consecutive odd integers whose product is 143.
44. Find two consecutive even integers the sum of whose squares is 100.

45. The length of the hypotenuse of a right triangle is 13 centimeters. Find the other two sides of the triangle if their sum is 17 centimeters.

46. The diameter of a circle is 8 centimeters. By how much should the radius increase so that the area increases by 33π square centimeters?

47. The perimeter of a rectangle is 20 inches and its area 24 square inches. Find the lengths of its sides.

48. The perimeter of a rectangle is 24 centimeters and its area is 32 square centimeters. Find the lengths of its sides.

49. A ball is thrown upward with an initial velocity of 80 feet per second. The height h (in feet) traveled in t seconds is given by the formula
$$h = 80t - 16t^2.$$
   (a) After how many seconds will the ball reach a height of 64 feet?
   (b) How long will it be before the ball returns to the ground?
   (c) Find the maximum height that the ball will reach. [*Hint:* The time of upward travel equals half the time to return to the ground.]

50. A rocket is shot vertically upward from the ground with an initial velocity of 128 feet per second. The rocket is at a height h after t seconds of launching, where $h = 128t - 16t^2$.
   (a) After what time will the rocket be at a height of 192 feet above the ground?
   (b) When will the rocket return to the ground?
   (c) Find the maximum height the rocket will reach.

*51. The number of items of his product that a manufacturer can sell each week depends on the price he charges for them. Suppose that at a price of p dollars per item he can sell x items per week, where $x = 300(6 - p)$. What price p per item should he charge so as to generate a weekly revenue of $2400? [*Hint:* Revenue = (price per item)(number of units sold.)]

*52. A manufacturer can sell each week x units of his product at a price of $6 per unit, where
$$x = 160(10 - p).$$
Find the price p that he should charge in order to generate a revenue of $3840.

*53. In Exercise 51, suppose it costs $3 to manufacture each item. How much should the manufacturer charge per item so that his *weekly profit* is $600? [*Hint:* Profit = revenue − cost of manufacture of x items.]

*54. In Exercise 52, suppose it costs $2 to produce each item. At what price p should he sell the product in order to obtain a weekly profit of $2400?

3.4

Quadratic Formula and Applications

In this section we describe the method of completing the square for solving a quadratic equation, and then we go on to the so-called quadratic formula for the roots. The simplest way to explain the method is by solving a particular equation.

Example 1 Solve the equation
$$4x^2 + 24x + 11 = 0. \tag{i}$$

Solution First we divide both sides by the coefficient of x^2, in this case 4:
$$x^2 + 6x + \tfrac{11}{4} = 0. \tag{ii}$$
Then we move the constant term to the right side:
$$x^2 + 6x = -\tfrac{11}{4}. \tag{iii}$$
We observe from the binomial-square identity that
$$(x+3)^2 = x^2 + 6x + 9. \tag{iv}$$
Comparing the right side of Equation (iv) with the left side of Equation (iii), we see that they differ only by the constant 9. Therefore, we add 9 to both sides of Equation (iii), obtaining
$$x^2 + 6x + 9 = -\tfrac{11}{4} + 9 = \tfrac{25}{4}$$
or, in other words,
$$(x+3)^2 = \tfrac{25}{4}.$$
We can easily solve this equation by taking the square root:
$$x + 3 = \tfrac{5}{2} \quad \text{or} \quad -\tfrac{5}{2}$$
and so
$$\text{either } x = -3 + \tfrac{5}{2} = -\tfrac{1}{2} \quad \text{or} \quad x = -3 - \tfrac{5}{2} = -\tfrac{11}{2}.$$
The two solutions are therefore $x = -\tfrac{1}{2}$ and $x = -\tfrac{11}{2}$.

In equation (iv) we considered $(x+3)^2$ because when expanded it agrees with the left side of Equation (iii) as far as the x^2 and x terms are concerned. For example, if we took $(x-3)^2$ instead, we would get that $(x-3)^2 = x^2 - 6x + 9$. Although the x^2 term here is the same as on the left of Equation (iii), the x-term is different. In order to get the same coefficient of x in Equation (iii) we must consider $(x+k)^2$, where k is *half the coefficient of x in Equation* (iii)—that is, k equals half of 6, or 3.

The procedure for solving a quadratic equation in standard form by completing the square is as follows:

Step 1. Divide throughout by the coefficient of x^2.

Step 2. Move the constant term to the right side.

Step 3. Add k^2 to both sides, where k is half the coefficient of x on the left side.

Step 4. The left side of the equation will now be the perfect square, $(x+k)^2$, and we find the solution by taking the square root of both sides.

Example 2 Solve the equation
$$2x^2 - x - 2 = 0$$
by completing the square.

Solution   *Step 1.* Dividing through by 2, we get
$$x^2 - \tfrac{1}{2}x - 1 = 0.$$

Step 2. MOVE CONSTANT
$$x^2 - \tfrac{1}{2}x = 1.$$

Step 3. The coefficient of x is $(-\tfrac{1}{2})$. We must take k equal to half of this, namely $(-\tfrac{1}{4})$. So we must add $k^2 = (-\tfrac{1}{4})^2 = \tfrac{1}{16}$ to both sides, obtaining
$$x^2 - \tfrac{1}{2}x + \tfrac{1}{16} = 1 + \tfrac{1}{16} = \tfrac{17}{16}.$$

Step 4. The left side of this equation is now $(x + k)^2$—that is,
$$[x + (-\tfrac{1}{4})]^2 = (x - \tfrac{1}{4})^2.$$
So
$$(x - \tfrac{1}{4})^2 = \tfrac{17}{16}.$$

Taking the square root of both sides, we find that
$$x - \frac{1}{4} = \pm\sqrt{\frac{17}{16}} = \pm\frac{\sqrt{17}}{4}$$
and therefore $x = \tfrac{1}{4} \pm \sqrt{17}/4$.

Let us now solve the general quadratic equation $ax^2 + bx + c = 0$ ($a \neq 0$) by the method of completing the square. We begin by moving the constant term to the right:
$$ax^2 + bx = -c.$$
Dividing both sides by a (this is OK because $a \neq 0$), we get
$$x^2 + \frac{b}{a}x = -\frac{c}{a}.$$

According to the method of completing the square, we must divide the coefficient of x (which is b/a in this case) by 2 (giving $b/2a$), square the result, and add this to both sides. We get
$$x^2 + \frac{b}{a} + \left(\frac{b}{2a}\right)^2 = -\frac{c}{a} + \left(\frac{b}{2a}\right)^2$$
$$= -\frac{c}{a} + \frac{b^2}{4a^2}$$
$$= \frac{-4ac + b^2}{4a^2}.$$

But the left side here is $(x + b/2a)^2$, as can be seen from the binomial-square formula. Therefore, we obtain
$$\left(x + \frac{b}{2a}\right)^2 = \frac{b^2 - 4ac}{4a^2}.$$

After taking the square root of both sides, therefore, we get
$$x + \frac{b}{2a} = \pm\sqrt{\frac{b^2 - 4ac}{4a^2}} = \pm\frac{\sqrt{b^2 - 4ac}}{2a}.$$
Therefore
$$x = \frac{-b \pm \sqrt{b^2 - 4ac}}{2a}.$$

Thus we have established a very important result: The roots of the quadratic equation
$$ax^2 + bx + c = 0 \quad (a \neq 0)$$
are given by the formula

$$x = \frac{-b \pm \sqrt{b^2 - 4ac}}{2a}. \tag{1}$$

In other words, if p and q denote the two roots of the equation $ax^2 + bx + c = 0$ $(a \neq 0)$, then

$$p = \frac{-b + \sqrt{b^2 - 4ac}}{2a}, \quad q = \frac{-b - \sqrt{b^2 - 4ac}}{2a}. \tag{2}$$

Equation (1) above is known as the **quadratic formula**.

In order to use the quadratic formula to solve a quadratic equation the procedure is as follows. First we reduce the given equation to standard form and remove any constant monomial factors. Then we identify a, b, c, the three coefficients that appear in the equation, and finally we substitute these values of a, b, c into the quadratic formula.

A pocket calculator will often be useful to evaluate the square root in the solution.

Example 3 Solve the equation
$$3(x^2 + 1) = 5(1 - x)$$
by using the quadratic formula.

Solution The given equation is
$$3x^2 + 3 = 5 - 5x$$
or
$$3x^2 + 5x - 2 = 0.$$
This equation is now in standard form. Comparing this with the general equation
$$ax^2 + bx + c = 0$$
we have $a = 3$, $b = 5$, and $c = -2$. Therefore
$$x = \frac{-b \pm \sqrt{b^2 - 4ac}}{2a}$$
$$= \frac{-5 \pm \sqrt{25 - 4(3)(-2)}}{2(3)}$$
$$= \frac{-5 \pm \sqrt{25 + 24}}{6} = \frac{-5 \pm 7}{6}$$
$$= \frac{-5 + 7}{6} \quad \text{or} \quad \frac{-5 - 7}{6}$$
$$= \frac{1}{3} \quad \text{or} \quad -2.$$

Hence the roots are $\frac{1}{3}$ and -2. (Compare with Example 1 of the last section.)

Sec. 3.4 Quadratic Formula and Applications

Remark *Every* quadratic equation can be solved by using the quadratic formula. The method of factors is often quicker, but on many occasions it is difficult to spot the two factors of the quadratic expression. Furthermore, many quadratic expressions cannot be resolved into rational factors, and in such cases it is virtually impossible to factor by inspection.

Example 4 Solve the equation $2x^2 - x - 2 = 0$.

Solution Comparing the given equation with the standard equation

$$ax^2 + bx + c = 0$$

we see that the coefficients are

$$a = 2, \quad b = -1, \quad c = -2.$$

Therefore

$$\begin{aligned} x &= \frac{-b \pm \sqrt{b^2 - 4ac}}{2a} \\ &= \frac{-(-1) \pm \sqrt{(-1)^2 - 4(2)(-2)}}{2 \cdot 2} \\ &= \frac{1 \pm \sqrt{1 + 16}}{4} \\ &= \frac{1 \pm \sqrt{17}}{4}. \end{aligned}$$

Hence the two roots are $\frac{1}{4}(1 + \sqrt{17}) \approx 1.281$ and $\frac{1}{4}(1 - \sqrt{17}) \approx -0.781$, rounded off to three decimal places.

The quantity $b^2 - 4ac$ that appears under the radical sign in the quadratic formula is called the ***discriminant*** of the equation $ax^2 + bx + c = 0$. Since the square root of a negative number cannot be equal to any real number, the quadratic formula gives real solutions only when the discriminant $b^2 - 4ac$ is nonnegative. If $b^2 - 4ac$ is negative, then we do not have real solutions.* If $b^2 - 4ac = 0$, then the quadratic formula shows that there are two equal roots, each equal to $-b/2a$ [see (2) above].

To summarize:

> (i) If $b^2 - 4ac > 0$, the equation $ax^2 + bx + c = 0$ has *two real and unequal roots*.
> (ii) If $b^2 - 4ac = 0$, the equation $ax^2 + bx + c = 0$ *has two equal real roots*, each equal to $-b/2a$. This value is called a ***double root***.
> (iii) If $b^2 - 4ac < 0$, the equation $ax^2 + bx + c = 0$ has *no real roots*.

---

*See Section 9.2. If you wish to learn a little about complex numbers, you can study this section right away.

Example 5 Solve the equation $x^2 - 2x + 5 = 0$.

Solution Comparing the given equation with

$$ax^2 + bx + c = 0$$

we have $a = 1, b = -2, c = 5$. Therefore, by the quadratic formula, the roots of the given equation are given by

$$x = \frac{-b \pm \sqrt{b^2 - 4ac}}{2a}$$

$$= \frac{-(-2) \pm \sqrt{(-2)^2 - 4(1)(5)}}{2(1)}$$

$$= \frac{2 \pm \sqrt{-16}}{2}.$$

Since a negative number, -16, appears under the radical sign, the given equation has no real solutions.

Example 6 A box with no top is to be formed from a rectangular sheet of tin by cutting out 4-inch squares from each corner and folding up the sides. If the width of the box is 3 inches less than the length and the box is to hold 280 cubic inches, find the dimensions of the sheet of tin.

Solution Let x inches denote the width of the box; then its length is $(x + 3)$ inches and its height 4 inches. (See the figure below.) The volume of the box is given by

$$(\text{length})(\text{width})(\text{height}) = (x + 3)(x)(4)$$

$$= 4x(x + 3).$$

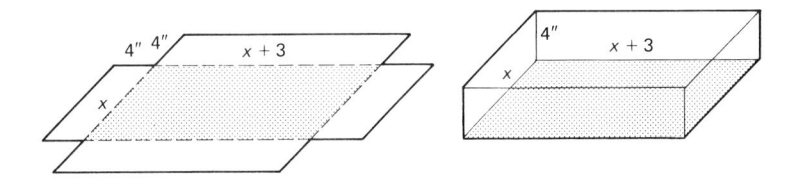

But the box is to hold 280 cubic inches, so

$$4x(x + 3) = 280.$$

Dividing both sides by 4, we have

$$x(x + 3) = 70$$

$$x^2 + 3x - 70 = 0. \qquad \text{(i)}$$

Comparing this with the general quadratic equation $ax^2 + bx + c = 0$ we have $a = 1, b = 3, c = -70$. Then by the quadratic formula the roots of (i)

are given by
$$x = \frac{-b \pm \sqrt{b^2 - 4ac}}{2a}$$
$$= \frac{-3 \pm \sqrt{9 - 4(1)(-70)}}{2(1)}$$
$$= \frac{-3 \pm \sqrt{9 + 280}}{2}$$
$$= \frac{-3 \pm 17}{2}$$
$$= \frac{-3 + 17}{2} \quad \text{or} \quad \frac{-3 - 17}{2}$$
$$= 7 \quad \text{or} \quad -10.$$

But $x = -10$ is unacceptable, because x represents the width of the box, and the width cannot be a negative number. Thus $x = 7$.

The dimensions of the tin sheet before we cut the corners are $x + 8$ and $(x + 3) + 8$. Since $x = 7$, the dimensions are 15 inches and 18 inches.

Example 7 Steve owns an apartment building that has 60 suites. He can rent all the suites if he charges $180 per month. At a higher rent some of the suites will remain empty: on average, for each increase of $5 in rent, one suite becomes vacant with no possibility of renting it. Find the rent he should charge per suite in order to obtain a total income of $11,475.

Solution Let n denote the number of $5 increases. Then the increase in rent per suite is $5n$, which means that the rent per suite is $(180 + 5n)$. The number of units unrented will then be n, so that the number rented will be $(60 - n)$. The total rent he will receive equals (rent per suite)(number of suites rented). Therefore,
$$11{,}475 = (180 + 5n)(60 - n)$$
$$= 5(36 + n)(60 - n).$$

Dividing both sides by 5, we get
$$2295 = (36 + n)(60 - n)$$
$$= 2160 + 24n - n^2.$$

Therefore,
$$n^2 - 24n + 135 = 0$$
$$(n - 9)(n - 15) = 0.$$

Thus $n = 9$ or 15. Hence the rent charged should be $180 + 5n = 180 + 45$ or $180 + 75$—that is, $225 or $255. In the first case, 9 of the suites will be vacant and the 51 rented suites will produce an income of $225 each. In the second case, when the rent is $255, 15 will be vacant and only 45 rented, but the total revenue will be the same.

Example 8 The egg marketing board of British Columbia knows from past experience that if it charges p dollars per dozen eggs the number sold per week will be x million dozen, where $p = 2 - x$. Its total weekly revenue would then be $R = xp = x(2 - x)$ million dollars. The cost to the industry of producing x million dozen eggs per week is given by $C = 0.25 + 0.5x$ million dollars. The weekly profit of the industry is $P = R - C$. What price should the marketing board set for eggs to ensure a weekly profit of 0.25 million dollars?

Solution The profit is given by
$$P = R - C$$
$$= x(2 - x) - (0.25 + 0.5x)$$
$$= -x^2 + 1.5x - 0.25.$$

Setting this equal to 0.25, we obtain the equation
$$-x^2 + 1.5x - 0.25 = 0.25$$
$$x^2 - 1.5x + 0.5 = 0.$$

Using the quadratic formula, we find the roots for x:
$$x = \frac{-b \pm \sqrt{b^2 - 4ac}}{2a}$$
$$x = \frac{-(-1.5) \pm \sqrt{(-1.5)^2 - 4(1)(0.5)}}{(2)(1)}$$
$$= \tfrac{1}{2}(1.5 \pm 0.5)$$
$$= 1 \quad \text{or} \quad 0.5.$$

Now $p = 2 - x$. So when $x = 1$, we have $p = 1$, and when $x = 0.5$, $p = 1.5$. So the marketing board has a choice of two policies. It can charge $1 per dozen, in which case the sales will be 1 million dozen, or it can charge $1.50 per dozen, when the sales will be 0.5 million dozen per week. In either case the profits to the industry will be $0.25 million per week.

Example 9 A corporation wishes to set aside a sum of $1 million to be invested at interest and used at a later date to repay two bond issues that will become due. One year after the sum is first invested, $250,000 will be required for the first issue and one year later a further $900,000 for the second. Determine the rate of interest necessary in order that the investment will cover both repayments.

Solution Let the rate of interest be $R\%$ per annum. When invested at this rate, a sum S gains interest during one year equal to $(R/100)S$. [For example, at 5% rate, the interest per annum is $(5/100)S$ or $0.05S$.] So after the first year the $1 million earns interest of $(R/100)(1)$ million. Adding this to the original sum, we see that the value of the investment after one year is

$$1 + \left(\frac{R}{100}\right)(1) = \left(1 + \frac{R}{100}\right) \text{ million dollars}$$

At this time 0.25 million is withdrawn, so at the beginning of the second year the amount remaining invested is (in millions)

$$S' = \left(1 + \frac{R}{100}\right) - 0.25 = 0.75 + \frac{R}{100}$$

After a second year at interest the value of the investment is

$$S' + \left(\frac{R}{100}\right)S' = S'\left(1 + \frac{R}{100}\right) = \left(0.75 + \frac{R}{100}\right)\left(1 + \frac{R}{100}\right).$$

It is required that this be the amount (0.9 million) necessary to pay off the second bond issue. Therefore, we arrive at the equation

$$\left(0.75 + \frac{R}{100}\right)\left(1 + \frac{R}{100}\right) = 0.9.$$

Thus,

$$0.75 + 1.75\left(\frac{R}{100}\right) + \left(\frac{R}{100}\right)^2 = 0.9.$$

Multiplying both sides by 100^2 to remove the fractions, we get the equation

$$7500 + 175R + R^2 = 9000$$
$$R^2 + 175R - 1500 = 0.$$

From the quadratic formula (with $a = 1, b = 175, c = -1500$) we find that

$$R = \frac{-175 \pm \sqrt{175^2 - 4(1)(-1500)}}{2(1)}$$
$$= \tfrac{1}{2}(-175 \pm \sqrt{30{,}625 + 6000})$$
$$= \tfrac{1}{2}(-175 \pm \sqrt{36{,}625})$$
$$\simeq \tfrac{1}{2}(-175 \pm 191.4) \quad \text{(to one decimal place)}$$
$$= 8.2 \quad \text{or} \quad -183.2$$

Clearly, the second solution makes no practical sense—a rate of interest would hardly be negative. The meaningful solution is $R = 8.2$. So the investment must earn 8.2% per annum in order to provide sufficient funds to pay off the bond issue.

Diagnostic Test 3.4

Fill in the blanks.
1. If $ax^2 + bx + c = 0$, then by the quadratic formula, $x = $ _____.
2. If $px^2 - 2qx + 3r = 0$, then by the quadratic formula, $x = $ _____.
3. The discriminant of the equation $ax^2 + bx + c = 0$ is given by _____.
4. If the discriminant of $ax^2 + bx + c = 0$ is positive, then the roots of the equation are _____ and different.

5. The two roots of $ax^2 + bx + c = 0$ are equal if _____.

6. If $ax^2 + bx + c = 0$ has no real roots, then _____.

Exercises 3.4

(1–12) Solve the following equations by completing the square.

1. $x^2 + 6x + 8 = 0$.
2. $x^2 + 2x - 3 = 0$.
3. $x^2 - 6x + 1 = 0$.
4. $x^2 + 2x - 4 = 0$.
5. $x^2 = 4(x - 1)$.
6. $x^2 = 3(2x - 3)$.
7. $x^2 = 3x + 1$.
8. $x^2 = 5(x - 1)$.
9. $4x^2 - 8x - 3 = 0$.
10. $2x^2 = 14x - 1$.
11. $7x + 3(x^2 - 5) = x - 3$.
12. $2x(4x - 1) = 4 + 2x$.

(13–24) Solve the following equations by using the quadratic formula.

13. $6x^2 - 5x - 6 = 0$.
14. $3x^2 + 5x - 1 = 0$.
15. $2x^2 + 3\sqrt{3}\, x - 6 = 0$.
16. $x^2 + \sqrt{2}\, x - 12 = 0$.
17. $4x^2 - 12x + 9 = 0$.
18. $4x^2 + 20x + 25 = 0$.
19. $5x(x + 2) + 6 = 3$.
20. $(4x - 1)(2x + 3) = 18x - 4$.
21. $(3x + 2)(x - 1) = (2x + 1)(x + 3)$.
22. $(2x + 1)(x - 3) = (x + 1)(x + 2)$.
23. $x^2 = 2(x - 1)(x + 2)$.
24. $(3x + 5)(2x - 3) = -8$.

(25–40) Solve the following equations by any method.

25. $6x^2 = 11$.
26. $5x^2 + 7 = 0$.
27. $3x^2 = 7x$.
28. $3x^2 = x + 2$.
29. $2x^2 - 3x - 1 = 0$.
30. $x^2 + 3x - 2 = 0$.
31. $3x^2 = 5x - 3$.
32. $2x^2 = 5x - 2$.
33. $(2x + 3)(x + 1) = (x + 2)(x - 1) + 2$.
34. $(3x - 1)(x + 2) = (2x - 1)(x + 3) + 5$.
35. $(4x - 1)(2x + 3) = 10x - 4$.
36. $x^2 = 2(x + 1)(x - 2)$.
37. $x^4 - 3x^2 - 4 = 0$.
38. $2x^4 - x^2 - 1 = 0$.
39. $2x^{2/3} + x^{1/3} - 1 = 0$.
40. $x^{2/5} - 3x^{1/5} + 2 = 0$.
41. Solve $s = ut + \frac{1}{2}gt^2$ for t.
42. Solve $A = 2\pi R(R + H)$ for R.
43. Solve $s = 2a/(1 + a^2)$ for a.
44. Solve $A = 2x^2 + 4xy$ for x.
45. If 2 is one root of $x^2 - kx + 2 = 0$, find the other root.
46. If -1 is one root of $2x^2 + 5x + k = 0$, find the other root.
47. Find the value of k so that the roots of $3x^2 + kx - 12 = 0$ are equal in magnitude but differ in sign.
48. Find the value of p so that the roots of $x^2 - px - 7 = 0$ are the negatives of each other.
49. Use the quadratic formula to solve the equation

$$x^2 - 2xy + 1 - 3y^2 = 0$$

(a) For x in terms of y.
(b) For y in terms of x.

50. Use the quadratic formula to solve the equation

$$3x^2 - 2y^2 = xy + 1$$

(a) For x in terms of y.
(b) For y in terms of x.

*51. If α and β are the roots of the equation $ax^2 + bx + c = 0$, show that
   (a) $\alpha + \beta = -b/a$ and $\alpha\beta = c/a$.
   (b) $ax^2 + bx + c = a(x - \alpha)(x - \beta)$.
   [*Hint:* For (a) use the two roots given by the quadratic formula, and use the results of (a) to prove (b)].

53. Equal squares are removed from each corner of a rectangular metal sheet with dimensions 20 by 16 inches. The sides are then folded up to form a rectangular box. If the base of the box has area 140 square inches, find the side of the square that is removed from each corner.

55. Royal Realty has built a new rental unit of 60 apartments. Past experience shows that if they charge a monthly rent of $150 per apartment, all the units will be occupied, but for each $3 increase in rent, one apartment unit is likely to remain vacant. What rent should be charged in order to generate the same $9000 total revenue as is obtained with $150 rent and at the same time to leave some suites vacant?

57. If a publisher prices one of his books at $20, he will sell 20,000 copies. For every dollar by which he increases the price his sales will fall by 500 copies. What should he charge in order to generate a total revenue from sales of $425,000?

52. A box with square base and no top is to be made from a square piece of metal by cutting 2-inch squares from each corner and folding up the sides. Find the dimensions of the metal sheet if the volume of the box is to be 50 cubic inches.

54. In two years the *XYZ* company will require $1,102,500 to retire some of its bonds. At what rate of interest compounded annually should $1 million be invested over the two-year period in order to earn the amount required to retire the bonds?

56. In Exercise 55 the maintenance service and other costs on the building amount to $5000 per month plus $50 per occupied suite and $20 per vacant suite. What rental should be charged if the profit is to be $1225 per month? (Profit = rental revenue minus all costs.)

3.5

Fractional and Irrational Equations

The term *fractional equation* refers to an equation in which algebraic fractions occur having the unknown variable in one or more of the denominators. When cleared of fractions, such an equation often reduces to either a linear or a quadratic equation.

Example 1 Find the solutions of the fractional equation

$$2x + 3 = \frac{5}{x}.$$

Solution In order to remove the fraction we must multiply both sides of the equation by x. However, we must be certain that x is nonzero, because, according to the multiplication principle, we are allowed to multiply or divide both sides of an equation only by a nonzero quantity; otherwise, we may change the roots. But x must be nonzero, otherwise the given equation itself would contain an undefined term. Multiplying by x, then, we obtain

$$2x^2 + 3x = 5$$
$$2x^2 + 3x - 5 = 0.$$

Factoring, we get

$$(2x + 5)(x - 1) = 0$$

and so either $2x + 5 = 0$ or $x - 1 = 0$. So there are two roots of the given fractional equation: $x = 1$ and $x = -\frac{5}{2}$.

When removing the fractions from a fractional equation we must multiply both sides by the least common denominator of all the fractions that occur. This common denominator will involve the unknown variable, and *we must always make sure that the common denominator is nonzero.* This places some restrictions on the allowed values of the variable; for instance, in Example 1 we had to restrict x to be nonzero.

Example 2 Solve the equation

$$\frac{3}{y-6} + \frac{7}{y-2} = \frac{10}{y-4}.$$

Solution The given equation requires that $y \neq 6, 2,$ or 4.

The common denominator of the fractions involved is

$$(y-6)(y-2)(y-4)$$

which is different from zero since y cannot equal 2, 4, or 6. Multiplying both sides by $(y-6)(y-2)(y-4)$ to get rid of fractions, we have

$$3(y-2)(y-4) + 7(y-6)(y-4) = 10(y-6)(y-2)$$
$$3(y^2 - 6y + 8) + 7(y^2 - 10y + 24) = 10(y^2 - 8y + 12)$$
$$3y^2 - 18y + 24 + 7y^2 - 70y + 168 = 10y^2 - 80y + 120.$$

Moving all the terms to the left and collecting like terms, we get

$$3y^2 + 7y^2 - 10y^2 - 18y - 70y + 80y + 24 + 168 - 120 = 0$$
$$-8y + 72 = 0$$

because all the y^2 terms cancel. The equation thus reduces to a linear equation and the solution is

$$y = \frac{-72}{-8} = 9.$$

Thus 9 is the root of the given equation.

Example 3 Solve the equation

$$\frac{x+3}{x-2} + \frac{x+1}{x+2} = \frac{8+6x}{x^2-4}.$$

Sec. 3.5   *Fractional and Irrational Equations*

Solution The given equation can be written as

$$\frac{x+3}{x-2} + \frac{x+1}{x+2} = \frac{8+6x}{(x-2)(x+2)}.$$

Multiplying both sides by $(x-2)(x+2)$, the common denominator, we obtain

$$(x+3)(x+2) + (x+1)(x-2) = 8 + 6x$$
$$(x^2 + 5x + 6) + (x^2 - x - 2) = 8 + 6x$$
$$2x^2 + 4x + 4 = 8 + 6x.$$

Therefore,
$$2x^2 - 2x - 4 = 0$$
$$2(x^2 - x - 2) = 0$$
$$2(x+1)(x-2) = 0.$$

Thus,

either $x + 1 = 0$ or $x - 2 = 0$
$x = -1$ $x = 2.$

So, there *appear* to be two roots, 2 and -1. But *the original equation is undefined when* $x = 2$, so 2 cannot be a root. Thus, the given equation has only one root, $x = -1$. (We can check that $x = -1$ is a genuine root by substituting in the original equation.)

This example illustrates the care that we must exercise when multiplying both sides of an equation by an expression that is not a constant. Such an operation can sometimes introduce a root that was not a root of the original equation.* When working with fractional equations, it is essential to make sure that the original equation remains defined for each value of the variable that is determined—just to make sure that these values are genuine roots.

Example 4 Fred drives on an average 5 miles per hour slower than Susan and takes one hour more than Susan to cover a distance of 280 miles. How fast does each drive?

Solution Let Fred's average speed in miles per hour be denoted by x. Then the speed with which Susan drives is $(x + 5)$ miles per hour. Since the time taken by a journey is equal to distance divided by speed, we have

$$\text{time taken by Fred to cover 280 miles} = \frac{280}{x} \text{ hours}$$

$$\text{time taken by Susan to cover 280 miles} = \frac{280}{x+5} \text{ hours.}$$

---

*A phoney solution of this kind is often called an extraneous root.

We are given that

$$\underbrace{\text{Fred's time}}_{\frac{280}{x}} \underbrace{\text{is}}_{=} \underbrace{\text{one hour more than}}_{1 +} \underbrace{\text{Susan's}}_{\frac{280}{x+5}}$$

or

$$\frac{280}{x} = 1 + \frac{280}{x+5}.$$

Multiplying both sides by the common denominator $x(x+5)$, we have

$$280(x+5) = x(x+5) + 280x$$
$$280x + 1400 = x^2 + 5x + 280x$$

—that is,

$$x^2 + 5x - 1400 = 0$$

or

$$(x-35)(x+40) = 0.$$

Therefore, $x = 35$ or -40. Since x denotes a speed, the solution $x = -40$ is rejected. Therefore, $x = 35$, and we conclude that

$$\text{Fred's speed} = x = 35 \text{ miles per hour}$$
$$\text{Susan's speed} = x + 5 = 40 \text{ miles per hour}.$$

Next we briefly consider what are called irrational equations. An *irrational equation* is one that involves one or more radical signs with the unknown variable occurring beneath them. By appropriate operations such equations can often be reduced to linear or quadratic equations. First, however, let us recall that the symbol $\sqrt{a}$ stands for the *positive* square root of a. Thus $\sqrt{4} = +2$ and not -2, even though the square of (-2) is also 4. Similarly $-\sqrt{9} = -3$.

> In an equation involving a square root, we can get rid of the square-root symbol by a process of squaring. Since squaring removes the difference between the positive and negative square roots, this process can often introduce new roots. We must always check the solution obtained by substituting into the original equation. An irrational equation may or may not have a root.

Example 5 Solve the following irrational equations.

(a) $\sqrt{3x+4} = 5.$ (b) $\sqrt{2x+3} = -1.$

Solution (a) The given equation is

$$\sqrt{3x+4} = 5. \tag{i}$$

Sec. 3.5 *Fractional and Irrational Equations*

Squaring both sides, we get

$$3x + 4 = 25$$
$$3x = 25 - 4 = 21$$
$$x = \tfrac{21}{3} = 7.$$

We must now verify that this is a root of the original equation. Substituting $x = 7$ in the given equation (i), we get

$$\sqrt{3(7) + 4} = 5$$
$$\sqrt{21 + 4} = 5$$

which is true. Thus $x = 7$ *is* a solution of the given equation.

(b) The given equation is

$$\sqrt{2x + 3} = -1. \qquad \text{(ii)}$$

It is obvious that this equation can have no solution, since the radical on the left means the *positive* square root of $2x + 3$, hence it *can never equal a negative number*. Let us see what happens if we square both sides.

$$2x + 3 = (-1)^2 = 1$$
$$2x = 1 - 3 = -2$$
$$x = -1.$$

Thus we appear to have found a solution. However, if we substitute $x = -1$ into the original equation (ii), we find

$$\sqrt{2(-1) + 3} = -1 \quad \text{—that is,} \quad \sqrt{1} = -1.$$

But since $\sqrt{1} = +1$, we see that the given equation is not satisfied when $x = -1$. (This is an extraneous root.)

Example 6 Solve the irrational equation

$$\sqrt{2x - 2} + \sqrt{x - 2} = 3.$$

Solution First we write the equation in the form

$$\sqrt{2x - 2} = 3 - \sqrt{x - 2}.$$

If we now square both sides, we will remove the radical on the left:

$$(\sqrt{2x - 2})^2 = (3 - \sqrt{x - 2})^2$$
$$2x - 2 = 9 - 2(3)(\sqrt{x - 2}) + (\sqrt{x - 2})^2$$
$$= 9 - 6\sqrt{x - 2} + x - 2.$$

Therefore, leaving only the radical term on the right, we get

$$2x - 2 - 9 - x + 2 = -6\sqrt{x - 2}$$

or

$$x - 9 = -6\sqrt{x - 2}.$$

Squaring both sides again, we remove the remaining radical:
$$(x - 9)^2 = (-6\sqrt{x - 2})^2$$
$$x^2 - 18x + 81 = 36(x - 2)$$
$$= 36x - 72.$$

Thus,
$$x^2 - 54x + 153 = 0.$$

The quadratic expression factors as
$$(x - 3)(x - 51) = 0$$

so that either $x = 3$ or $x = 51$. Because our squaring operations can introduce new roots, we must check each of these possible solutions by substituting it into the original equation:

(a) $x = 3$:
$$\sqrt{2x - 2} + \sqrt{x - 2} = \sqrt{2(3) - 2} + \sqrt{3 - 2}$$
$$= \sqrt{4} + \sqrt{1}$$
$$= 2 + 1 = 3$$

as required. So $x = 3$ *is* a solution.

(b) $x = 51$:
$$\sqrt{2x - 2} + \sqrt{x - 2} = \sqrt{2(51) - 2} + \sqrt{51 - 2}$$
$$= \sqrt{100} + \sqrt{49}$$
$$= 10 + 7 \neq 3.$$

So $x = 51$ is *not* a solution of the original equation.

Diagnostic Test 3.5

Fill in the blanks.

1. If $\sqrt{x} + 3 = 2$, then $x = $ _____.
2. The equation $\sqrt{x + 5} = 3$ has _____ solution.
3. The equation $\sqrt{x + 2} = -3$ has _____ solution.
4. The equation $1/(x - 2) = 0$ has _____ solution.
5. The equation $1/(x - 2) = 3$ has _____ solution.
6. The equation $\sqrt{x + 2} + \sqrt{2x - 3} = 0$ has _____ solution.
7. The equation $\sqrt{x - 3} + 3\sqrt{x + 1} = 0$ has _____ solution.
8. The equation $1/(x - 3) = 2/(x - 3)$ has _____ solution.

Exercises 3.5

(1–48) Solve the following equations.

1. $2x - 5 = \dfrac{3}{x}$.

2. $8x - 2 = \dfrac{3}{x}$.

3. $x + \dfrac{1}{x} = 3\tfrac{1}{3}$.

4. $x + \dfrac{1}{x} = 2\tfrac{1}{12}$.

Sec. 3.5 Fractional and Irrational Equations

5. $6x + \dfrac{15}{x} = 19$.

6. $6x - \dfrac{12}{x} = 1$.

7. $\dfrac{x}{4} + \dfrac{4}{x} = \dfrac{x}{9} + \dfrac{9}{x}$.

8. $\dfrac{x}{2} + \dfrac{2}{x} = \dfrac{x}{3} + \dfrac{3}{x}$.

9. $\dfrac{3x+4}{19} = \dfrac{x^2}{4x+3}$.

10. $\dfrac{3x-7}{2x+5} = \dfrac{5(x-2)}{x+4}$.

11. $\dfrac{x}{x+1} + \dfrac{x}{x-2} = 2$.

12. $\dfrac{3x+4}{2x+5} - \dfrac{1}{2x+1} = \dfrac{3}{4}$.

13. $\dfrac{1}{x-1} + \dfrac{2}{x-2} = \dfrac{3}{x^2-3x+2}$.

14. $\dfrac{x}{x+1} + \dfrac{2x-1}{x-1} = 3 + \dfrac{1}{x^2-1}$.

15. $\dfrac{3x+2}{x-2} - \dfrac{3x-1}{x+2} = \dfrac{32}{x^2-4}$.

16. $\dfrac{2x+1}{x-3} - \dfrac{2x+5}{x+1} = \dfrac{12}{x^2-2x-3}$.

17. $\dfrac{4}{x-1} - \dfrac{5}{x+2} = \dfrac{3}{x}$.

18. $\dfrac{2}{x-1} + \dfrac{3}{x+2} = \dfrac{5}{x+3}$.

19. $\dfrac{x+4}{x-4} + \dfrac{x-2}{x-3} = 6\tfrac{1}{3}$.

20. $\dfrac{x-1}{x+1} + \dfrac{x+2}{x-2} = 2$.

21. $\dfrac{x+1}{x+3} + \dfrac{x+5}{x-2} = \dfrac{14x+7}{x^2+x-6}$.

22. $\dfrac{2x+1}{x-2} + \dfrac{3x-1}{x+1} = \dfrac{23-4x}{x^2-x-2}$.

23. $\dfrac{2x+1}{x-1} + \dfrac{3x-1}{x-2} = \dfrac{4x-3}{x^2-3x+2}$.

24. $\dfrac{x+2}{2x-1} + \dfrac{3x-1}{x+1} = \dfrac{4(1-2x)}{2x^2+x-1}$.

25. $\sqrt{2x-7} = 1$.

26. $\sqrt{7-2x} = 3$.

27. $\sqrt{3x+4} = -2$.

28. $\sqrt{5-3x} + 7 = 0$.

29. $\sqrt{3x+1} = x-1$.

30. $\sqrt{x+3} = x+1$.

31. $\sqrt{x-2} = 2-x$.

32. $\sqrt{2x-1} + 2x = 1$.

33. $\sqrt{x+1} + \sqrt{x} = 1$.

34. $\sqrt{2x-1} + \sqrt{x} = 2$.

35. $\sqrt{x+4} + \sqrt{x-1} = 1$.

36. $\sqrt{x+4} + \sqrt{x-1} = 5$.

*37. $\sqrt{2x-1} + \sqrt{x-4} = 2$.

38. $\sqrt{3x-2} - \sqrt{3x-1} = 3$.

39. $\sqrt{1-5x} + \sqrt{1-3x} = 2$.

40. $\sqrt{2x+8} + \sqrt{x+5} = 7$.

41. $\sqrt{2x+6} - \sqrt{x-1} = 2$.

42. $\sqrt{4x+1} - \sqrt{x-1} = 2$.

43. $\sqrt{4x+1} + \sqrt{x-1} = 4$.

44. $\sqrt{2x-4} + \sqrt{x-3} = 3$.

45. $\sqrt{x^2-5x+31} = 2x+1$.

46. $\sqrt{x^2+3x-2} = 1+x$.

47. $x + \sqrt{2x-1} = 0$.

48. $4x - \sqrt{13-3x} = 0$.

49. A rectangular plot has an area of 600 square feet and is surrounded by a uniform strip 2 feet wide. If the area of the strip is 216 square feet, find the dimensions of the rectangular plot.

50. A rectangular plot has an area of 18,000 square feet and is surrounded by a uniform strip of 5-foot width. If the total area of the plot and the strip is 20,800 square feet, find the dimensions of the rectangular plot.

51. A car travels an average 10 miles per hour faster than a heavy truck. The car takes 1 hour less than the truck to cover 120 miles. Find the speeds of the car and the truck.

52. Sue drives on the average 5 miles per hour faster than Pat and takes 1 hour less than Pat to drive 360 miles. How fast does each drive?

53. A dealer sold a watch for $75. His percentage profit was equal to the cost price in dollars. Find the cost price of the watch.

54. The numerator of a fraction is 1 less than the denominator. If the numerator is increased by 2 and the denominator is decreased by 1, the new fraction is three times the original fraction. Find the original fraction.

55. The denominator of a fraction is 1 less than the numerator. If the numerator is increased by 10 and the denominator is decreased by 2, the new fraction is six times the original fraction. Find the original fraction.

56. A boat travels 12 miles up a river and then returns. If the speed of the current in the river is 4 miles per hour, how fast must the boat travel relative to the water if it is to cover the round trip in 4 hours?

57. A boat whose average speed is 15 miles per hour travels 20 miles up a river, then returns. How fast is the current in the river if the total travel time is 3 hours?

58. The smaller leg of a right triangle is 5 feet and the hypotenuse is 1 less than twice the difference of the two legs. Find the length of the hypotenuse.

59. One side of a rectangle is 12 feet. If the diagonal of the rectangle is five times the difference between two sides, find the other side of the rectangle.

*60. A person wishes to walk from A to B as shown on the figure below. He does this by walking a distance x along a highway to the point C and then walking in a straight line from C to B through light bush. His speed along the highway is 4 miles per hour and through the bush is 2 miles per hour. Find x if his total time is $2\frac{1}{2}$ hours.

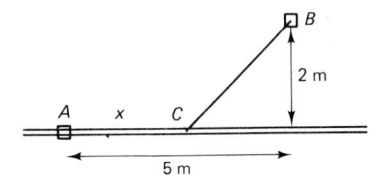

*61. In the preceding exercise find the value of x that makes the travel time (a) $2\frac{1}{8}$ hours, (b) 2 hours, (c) 3 hours. In cases (b) and (c) explain your results.

Chapter Review 3

1. Define or describe the following terms.
   (a) Linear equation.
   (b) Quadratic equation.
   (c) Fractional equation.
   (d) Irrational equation.
   (e) Addition principle for equations.
   (f) Multiplication principle for equations.
   (g) Roots of an equation.
   (h) Extraneous roots of an equation.
   (i) Equivalent equations.

2. Outline the procedure you would use to do the following.
   (a) Solve a linear equation.
   (b) Solve a quadratic equation:
       (1) by factoring.
       (2) by completing the square.
       (3) by the quadratic formula.
   (c) Solve a fractional equation.
   (d) Solve an irrational equation.

Review Exercises on Chapter 3

(1–46) Solve the following equations for x.

1. $3(2 - x) + x = 5(2x - 1) + 2$.
2. $2(1 - 4x) - 1 = x - 2(2 - 3x)$.
3. $4(3x - 1) - 3(2x + 1) = 1 - 7x$.
4. $5x - 2(3x - 1) = x - 2(1 - 5x)$.
5. $3(2x - 3) - 2(x + 7) = 4(x + 1) - 3$.
6. $(3x + 1)^2 - (3x - 1)^2 = 12x + 7$.
7. $(2x - 1)^2 = 3x^2 + (x - 1)(x - 2)$.
8. $(2x + 1)(x - 3) = (2x + 5)(x - 1)$.
9. $(x + 2)(x - 3) = 2 + (x - 1)(x - 2)$.
10. $(x + 1)(2x - 5) = 2(x + 2)(x - 3)$.
11. $\dfrac{1}{x} + \dfrac{1}{a} = \dfrac{1}{b}$.
12. $\dfrac{x}{bc} + \dfrac{x}{ca} + \dfrac{x}{ab} = a + b + c$.
13. $\dfrac{x}{p} + \dfrac{x}{q} + \dfrac{x}{r} = pq + qr + rp$.
14. $\dfrac{1}{x + 2} - \dfrac{1}{3} = \dfrac{1}{5}$.
15. $x^2 + 13x + 40 = 0$.
16. $3x^2 - 11x + 10 = 0$.
17. $6x^2 = x + 1$.
18. $5x^2 = 13x + 6$.
19. $(x + 1)(2x - 5) = (x + 2)(x - 3)$.
20. $1 + (3x + 4)(x - 2) = (2x + 1)(x - 3)$.
21. $(x + 2)(2x - 1) = 1 + (x + 3)(x + 1)$.
22. $28 + (x - 5)(x + 7) = (3x - 1)(x - 2)$.
23. $\dfrac{3}{x - 2} + \dfrac{2}{x - 3} = \dfrac{x + 1}{x^2 - 5x + 6}$.
24. $\dfrac{1}{x - 1} + \dfrac{2}{x - 2} = \dfrac{3}{x - 3}$.
25. $\dfrac{x}{x + 1} + \dfrac{x + 2}{x + 3} = 1$.
26. $\dfrac{x + 1}{x - 2} + \dfrac{x - 1}{x + 2} = \dfrac{28 - 2x}{x^2 - 4}$.
27. $\dfrac{x + 2}{x + 3} + \dfrac{x - 1}{x - 3} = \dfrac{9x - 15}{x^2 - 9}$.
28. $\dfrac{13 - 2x}{x^2 + x - 6} - \dfrac{x + 1}{x - 2} = \dfrac{x - 2}{x + 3}$.
29. $\dfrac{2x + 1}{x - 3} + \dfrac{1 - 2x}{x + 4} = \dfrac{49}{x^2 + x - 12}$.
30. $\dfrac{x}{x + 1} - \dfrac{10}{3x^2 + 2x - 1} = \dfrac{2x}{3x - 1}$.
31. $\sqrt{2x + 5} = x + 1$.
32. $\sqrt{x + 5} = x - 1$.
33. $x + 3 = \sqrt{5x + 11}$.
34. $\sqrt{3x - 1} = 1 - 3x$.
35. $\sqrt{3x + 1} + \sqrt{x - 1} = 2$.
36. $\sqrt{2x - 1} + \sqrt{x - 1} = 5$.
37. $\sqrt{5x - 6} - \sqrt{x + 1} = 5$.
38. $\sqrt{x^2 + 2x - 7} = x + 1$.

*39. $4^x = 8^{3-x}$.
*40. $2^{3x} = \dfrac{8^{x+1}}{4^{1-x}}$.
*41. $8^x = \dfrac{16}{\sqrt{4^{x^2}}}$.
*42. $3^{x^2} = \dfrac{9^x}{27}$.
43. $x^4 + x^2 - 20 = 0$.
44. $3^{2x} - 10(3^x) + 9 = 0$.
45. $2^{2x} - 7 \cdot 2^x - 8 = 0$.
46. $9^x - 26 \cdot 3^x - 27 = 0$.

47. The equation $R = R_1 R_2 / (R_1 + R_2)$ appears in electrical theory, where R, R_1, R_2 are positive. Solve for R_2 in terms of R and R_1.

48. Solve $a = (ne - ri)/ni$ for n.

49. Solve $\frac{1}{a} = \frac{1}{b} + \frac{1}{c} + \frac{1}{d}$ for c.

50. Solve $\frac{a^2}{x^2} - \frac{b^2}{y^2} = 1$ for y.

51. The winner of a Western Express Lottery wants to invest his prize money of $100,000 in two investments at 8% and 10%. How much should he invest in each if he wants to obtain an annual income of $8500?

52. A radiator contains 8 gallons of fluid that is 25% antifreeze and 75% water. How much of the fluid should be drained off and replaced by pure antifreeze so that the new mixture contains 40% antifreeze?

53. Two runners in a 30-mile cross-country race travel at 8 miles and 5 miles per hour, respectively. Suppose the slower runner starts 45 minutes earlier than the faster one. After what time will the two runners be 1 mile apart?

54. A ladder 13 feet long is placed against a wall so that its top is 12 feet above the ground. If the ladder slips so that its base moves 1 foot further away from the wall, how far does the top of the ladder fall?

55. The average speed of a commuter on a 25-mile journey to downtown Vancouver is 6 miles per hour faster at noon than during the morning rush hour and it takes $12\frac{1}{2}$ minutes less. What is the average speed of the commuter during the rush hour?

56. A rectangular sheet of metal has dimensions of 16 by 12 feet. Squares of equal area are cut from its four corners and then the edges are folded up to form a rectangular box. If the area of the base is 60 square feet, what is its height?

57. A rectangular window of height 5π feet is surmounted by a semicircular glass shade. If the total area of the rectangular window and the semicircular glass shade is 48π square feet, find the width of the window.

58. For every $100 invested for two years in secured commercial loans a bank receives $116.64, which represents capital and the interest compounded annually. What is the rate of interest?

59. A sum of $100 is invested at interest for one year; then, together with the interest earned, it is invested for a second year at twice the first rate of interest. If the total sum realized is $112.32, what are the two rates of interest?

60. In the preceding exercise $25 is withdrawn after the first year and the remainder is invested at twice the rate of interest. If the value of the investment at the end of the second year is $88, what are the two rates of interest?

61. The numerator of a fraction is 3 less than the denominator. If the numerator is increased by 1 and the denominator by 4, the fraction is unchanged. Find the fraction.

62. The numerator of a fraction is 1 less than the denominator. If the numerator is increased by 1 and the denominator is increased by 6, the new fraction is half of the original one. Find the original fraction.

63. One side of a rectangle is 4 feet. If the diagonal of the rectangle is 9 less than the perimeter, find the other side of the rectangle.

64. One side of a right triangle is 6 feet. If the hypotenuse is 3 more than the average of the other two sides, find all three sides of the triangle.

4 INEQUALITES

4.1

Sets and Intervals

Knowledge of sets and operations on sets is basic to all modern mathematics. Many lengthy statements in mathematics can be written concisely in terms of sets and set operations. We shall find them particularly useful in this chapter and in Chapter 6 on functions.

DEFINITION Any well-defined collection of objects is referred to as a *set*. The objects constituting the set are called its *members* or *elements*.

By *well-defined* collection we mean that, given any object, we should be able to decide unambiguously whether or not it belongs to that collection.

We may specify a set in two ways, either by making a list of all its members or else by stating a rule for membership in it. Let us examine these two methods in turn.

LISTING METHOD

If it is possible to specify all the elements of a set, we can describe the set by listing all the elements and enclosing the list inside braces. For example {1, 2, 3} denotes the set consisting of the three numbers 1, 2, and 3; {p, q} denotes the set whose only members are the two letters p and q. The order in which the elements are listed does not matter. Thus {1, 2, 3} and {3, 1, 2} represent the same set.

Often when the set contains a large number of elements we can employ a ***partial listing***. For example, {2, 4, 6, . . . , 100} denotes the set of all the even integers from 2 up to 100. The ellipsis, . . . , indicates that the sequence of elements continues in a manner made clear by the first few members listed. The sequence terminates at 100.

The ellipsis allows us to use the listing method in some cases when the set in question contains infinitely many members. For example {1, 3, 5, . . .} denotes the set of *all* the odd natural numbers. The absence of any number after the ellipsis indicates that the sequence continues on indefinitely.

RULE METHOD*

In many cases it is impossible or inconvenient to list all the members of a particular set. In such a case we can specify the set by stating a rule for membership. For example, consider the set of all people living in Canada at the present moment. To specify this set by listing all the members by name would be a prodigious task. Instead we can denote it as follows:

$\{x \mid x$ is a person currently living in Canada$\}$.

The symbol | stands for "such that," and this expression should be read: "the set of all x such that x is a person currently living in Canada." The statement that follows | inside the braces is the rule that specifies the membership in the set.

As a second example, consider the set $\{x \mid x$ is a real number greater than 0$\}$, which denotes the set of all positive real numbers. This is an example of a set that we cannot specify by the listing method even if we wanted to.

Many sets can be specified either by listing or by stating a rule, and we can choose whichever method we like. We shall give several examples of sets, specifying some of them in both ways.

Examples 1 (a) If N denotes the set of all natural numbers, then we can write

$$N = \{1, 2, 3, \ldots\}$$
$$= \{k \mid k \text{ is a natural number}\}.$$

---

*Sometimes called the **set-builder** notation.

(b) If P denotes the set of integers between -2 and $+3$ inclusive, then
$$P = \{-2, -1, 0, 1, 2, 3\}$$
$$= \{x \mid x \text{ is an integer}, -2 \leq x \leq 3\}.$$

Notice that the membership rule here consists of two conditions separated by a comma. Both conditions must be satisfied by any member of the set.

(c) $Q = \{1, 4, 7, \ldots, 37\}$
$= \{x \mid x = 3k + 1, \text{ where } k \text{ is an integer}, 0 \leq k \leq 12\}.$

(d) The set of all students currently enrolling at ABC Business College can be represented formally as
$$S = \{x \mid x \text{ is a student currently enrolled at ABC Business College}\}.$$

We could also specify this set by listing the names of all the students involved.

(e) The set of all real numbers that are greater than 1 and less than 2 can be specified by the rule method as
$$T = \{x \mid x \text{ is a real number}, 1 < x < 2\}.$$

DEFINITION A set is said to be *finite* if the elements belonging to it can be counted. A set which is not finite is called *infinite*.

In Example 1 the sets in parts (b), (c), and (d) are all finite, while those in parts (a) and (e) are infinite.

It is usual to use capital letters to denote sets and small letters to denote the elements in the sets. We have followed this convention in the preceding examples. If A is any set and x any object, the notation $x \in A$ is used to denote the fact that x is a member of A. The statement "$x \in A$" is read "x belongs to A" or "x is an element of A." We denote the negative statement "x is not an element of A" by writing $x \notin A$.

In part (b) of Example 1 we have that $2 \in P$, but $6 \notin P$. For the set in part (e), $\sqrt{2} \in T$ and $\frac{3}{2} \in T$, but $2 \notin T$ and $\pi \notin T$.

DEFINITION A set that contains no elements is called an *empty set*. (The terms *null* set and *void* set are also used.) The symbol $\emptyset$ is used to denote a set that is empty, and the statement $A = \emptyset$ means that the set A contains no members.

Here are some examples of empty sets:
$$\{x \mid x \text{ is an integer and } 3x = 2\} = \emptyset.$$
$$\{x \mid x \text{ is a real number and } x^2 + 1 = 0\} = \emptyset.$$
$$\text{The set of all living dragons} = \emptyset$$
$$\text{The set of all magnets having only one pole} = \emptyset.$$

DEFINITION A set A is said to be a *subset* of another set B if every element of A is also an element of B. In such a case we write $A \subseteq B$.

The set A is said to be a ***proper subset*** of the set B if every element of A is in B but at least one element in B is not in A. In this case we write $A \subset B$.

Examples 2 (a) Let $A = \{2, 4, 6\}$ and $B = \{1, 2, 3, 4, 5, 6, 7, 8\}$, then
$$A \subset B.$$

(b) If N is the set of all natural numbers, I is the set of all integers, Q is the set of all rational numbers, and R is the set of all real numbers, then
$$N \subset I \subset Q \subset R.$$

(c) The set of all women students at XYZ university is a subset of the set of all students at that university.

(d) Every set is a subset of itself; that is,
$$A \subseteq A \qquad \text{for any set } A.$$
However, the statement $A \subset A$ is not true.

(e) An empty set $\varnothing$ is a subset of any set A:
$$\varnothing \subseteq A \qquad \text{for any set } A.$$

(There exists no object that belongs to $\varnothing$ and does not belong to A for the simple reason that there exists no object that belongs to $\varnothing$ at all. Hence $\varnothing \subseteq A$.)

Two sets are equal to one another if they contain identical elements. More formally, we have the following definition.

DEFINITION Two sets A and B are said to be ***equal*** if $A \subseteq B$ and $B \subseteq A$. In such a case we write $A = B$. Thus $A = B$ if there is no object that belongs to A and does not belong to B and no object that belongs to B and does not belong to A. If two sets A and B are not equal, then we write $A \neq B$.

Examples 3 (a) If $A = \{x \mid x^2 = 1\}$ and $B = \{-1, +1\}$, then $A = B$.
(b) If $A = \{y \mid y^2 - 3y + 2 = 0\}$ and $B = \{1, 2\}$, then $A = B$.
(c) If $A = \{x \mid x^2 = 4\}$ and $B = \{2\}$, then $A \neq B$. In this case we have $B \subset A$.
(d) $\{x \mid x > 2\} \neq \varnothing$.

INTERVALS

Let a and b be two real numbers with $a < b$. Then the ***open interval*** from a to b denoted by (a, b) is the set of all real numbers x that lie between a and b. Thus,
$$(a, b) = \{x \mid x \text{ is a real number and } a < x < b\}.$$

Similarly, the ***closed interval*** from a to b, denoted by $[a, b]$, is the set of all real numbers that lie between a and b, together with a and b themselves. Thus
$$[a, b] = \{x \mid x \text{ is a real number and } a \leq x \leq b\}.$$

Semiclosed or ***semiopen*** intervals are defined as below:
$$(a, b] = \{x \mid a < x \leq b\}$$
$$[a, b) = \{x \mid a \leq x < b\}.$$

(Note that the statement that x is a real number has been omitted from the rules defining these sets. This is commonly done to avoid repetition when we know that we are dealing with sets of real numbers.)

For any of the intervals (a, b), $[a, b]$, (a, b), or $[a, b]$, a and b are called the **endpoints**. An open interval does not contain its endpoints, whereas a closed interval contains both its endpoints. A semiclosed interval contains only one of its endpoints.

Geometrically, these intervals are represented as shown below:

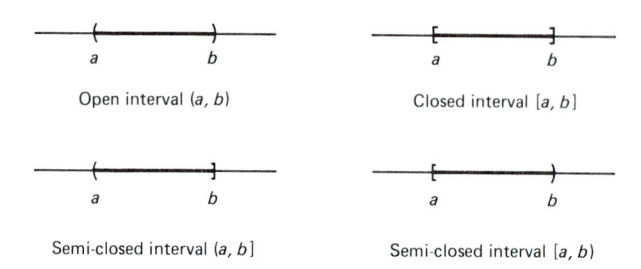

Open interval (a, b) Closed interval $[a, b]$

Semi-closed interval $(a, b]$ Semi-closed interval $[a, b)$

we have
$$(a, b) \subset [a, b) \subset [a, b] \quad \text{and} \quad (a, b) \subset (a, b] \subset [a, b].$$

The four intervals (a, b), $(a, b]$, $[a, b)$, and $[a, b]$ are **bounded intervals**. We use the symbols ∞ (infinity) and $-\infty$ (negative infinity) to describe **unbounded** intervals as follows:

$$(a, \infty) = \{x \mid x > a\}$$

$$[a, \infty) = \{x \mid x \geq a\}$$

$$(-\infty, a) = \{x \mid x < a\}$$

$$(-\infty, a] = \{x \mid x \leq a\}$$

$$(-\infty, \infty) = \text{set of all real numbers}$$

Note that ∞ and $-\infty$ are not real numbers.

Diagnostic Test 4.1

Fill in the blanks.

1. 2 _____ $\{1, 2, 3\}$.
2. 3 _____ $\{1, 2, 4\}$.
3. $\{1, 2\}$ _____ $\{1, 2, 3\}$.
4. $\{2, 4\}$ _____ $\{1, 2, 3, 4\}$.

5. 3 _____ (2, 4).
7. (a, b) _____ [a, b].
9. ∅ _____ {0}.
11. {2, 3} _____ [2, 3].
13. ∅ _____ A for all sets A.
14. If A is a subset of B and B is a subset of C, then A _____ C.
15. If $A = \left\{x \mid \frac{(x-2)^2}{x-2} = 0\right\}$, $B = \{x \mid x - 2 = 0\}$, then A _____ B.
16. If $A = \{x \mid x^2 - 5x + 6 = 0\}$, $B = \{2, 3\}$, then A _____ B.
17. If $A = \{3\}$ and $B = \{x \mid (x-4)^2 = 1\}$, then A _____ B.

6. −3 _____ (−1, 2).
8. (2, 3) _____ (1, 4).
10. 0 _____ ∅.
12. 2 _____ (2, 3).

Exercises 4.1

(1–10) Use the listing method to describe the following sets.

1. The set of all integers less than 5 and greater than −2.
2. The set of all natural numbers less than 50.
3. The set of all negative integers greater than −20.
4. The set of all even integers greater than −4 and less than 7.
5. The set of all prime numbers less than 20.
6. $\{x \mid x = 1/(n + 2)$, n is a natural number$\}$.
7. $\{x \mid x = (n + 1)/n$, n is a natural number$\}$.
8. $\{x \mid x$ is a prime factor of 36$\}$.
9. $\{x \mid x = 1/(p - 1)$, p is a prime number less than 20$\}$.
10. $\{x \mid x = (2n - 1)^2$, n is a natural number$\}$.

(11–20) Use the rule method to describe the following sets.

11. The set of all even numbers less than 100.
12. The set of all positive odd integers less than 15.
13. The set of all prime numbers less than 30.
14. $\{1, 3, 5, 7, 9, \ldots, 19\}$.
15. $\{3, 6, 9, \ldots\}$.
16. $\{1, \frac{1}{2}, \frac{1}{3}, \frac{1}{4}, \ldots\}$.
17. $\{1, 4, 9, 16, \ldots\}$.
18. $\{1, 8, 27, 64, \ldots\}$.
19. $\{\frac{1}{2}, \frac{2}{3}, \frac{3}{4}, \frac{4}{5}, \ldots\}$.
20. $\{\ldots, -4, -2, 0, 2, 4, \ldots\}$.

(21–30) Write the following sets of numbers in interval form.

21. $3 \leq x \leq 8$.
22. $-2 \leq y \leq 7$.
23. $5 < x \leq 8$.
24. $8 > x \geq -3$.
25. $-3 > t > -7$.
26. $0 > t > -5$.
27. $x \geq 5$.
28. $t < -6$.
29. $x < -3$.
30. $x \geq 10$.

(31–36) Write the following intervals as inequalities.

31. [2, 5].
32. (2, 7].
33. (−3, 2).
34. [−7, −2).
35. (−∞, 3].
36. (2, ∞).

37. Indicate whether the following statements are true or false.
    (a) 3 is in [−4, 7].
    (b) 0 is in (0, 3).
    (c) $-\frac{5}{2}$ is in [−2, 0].
    (d) −3 is in (−∞, −1).
    (e) −3 is in (−1, ∞).
    (f) −7 is in (−∞, −9].
    (g) $2 \in \{1, 2, 3\}$.
    (h) $3 \subseteq \{1, 2, 3, 4\}$.
    (i) $4 \in \{1, 2, 5, 7\}$.
    (j) $\{1, 2\} \subseteq \{1, 2, 3\}$.

(k) $\emptyset = \{0\}$.
(m) $(a, b) \in [a, b]$.
(o) $\{1, 2, 3, 4\} = \{4, 2, 1, 3\}$.
(q) $\{x \mid x^2 - 1 = 0\} \subset \{y \mid -1 \leq y \leq 1\}$.

(l) $0 \in \emptyset$.
(n) $(a, b) \subset (a, b]$.
(p) $\{2, 5\} \subseteq [2, 5]$.

38. If $A = (1, 3)$, $B = [-1, 3)$, $C = (1, 3]$, and $D = [1, 3]$, then which of these sets is a subset of one (or more than one) of the others?

39. If A = set of all squares in a plane,
B = set of all rectangles in a plane,
C = set of all isosceles triangles in a plane, and
D = set of all equilateral triangles in a plane,
then which of these sets is a subset of one (or more) of the others?

4.2

Linear Inequalities in One Variable

In this section we consider certain inequalities that involve a single variable. The following example gives a simple business problem that results in such an inequality.

Suppose the total cost of production of x units of a certain commodity consists of an overhead cost of $3100 plus $25 manufacturing cost per unit, and each unit sells for $37. The manufacturer wants to know how many units he should produce and sell so as to gain a profit of at least $2000. Suppose x units are produced and sold. The revenue R obtained by selling x units at $37 each is $R = 37x$ dollars. The cost C of producing x units is $C = 3100 + 25x$. The profit P (in dollars) obtained by producing and selling x units is then given by

$$\text{profit} = \text{revenue} - \text{cost}$$
$$P = 37x - (3100 + 25x)$$
$$= 12x - 3100.$$

Since the profit is required to be at least $2000—that is, $2000 or more—we must have

$$P \geq 2000$$

—that is,

$$12x - 3100 \geq 2000. \qquad (1)$$

This is an inequality involving the single variable x. We observe that the terms that occur in it are of two types, either constant terms or terms that are constant multiples of the variable x. Any inequality that has terms only of these two types is called a *linear inequality*. If the inequality symbol that occurs is either $>$ or $<$, the inequality is called a *strict* inequality. For example,

$$3 - x \leq 2x + 4$$

is a linear inequality in the variable x, and
$$\tfrac{1}{4}z + 3 > 5 - \tfrac{1}{3}z$$
is a strict linear inequality in the variable z.

DEFINITION The **solution** of an inequality in one variable is the set of all values of the variable for which the inequality is a true statement.

Consider, for example, the linear inequality
$$3x + 7 > 5x - 1 \tag{i}$$
in the variable x. If $x = 3$, then the inequality becomes
$$3(3) + 7 > 5(3) - 1 \quad \text{or} \quad 16 > 14$$
which is a true statement. Thus $x = 3$ is a solution of the inequality. On the other hand, $x = 5$ is not a solution, because when we substitute $x = 5$ in (i), we have
$$3(5) + 7 > 5(5) - 1 \quad \text{or} \quad 22 > 24$$
which is *not* a true statement. In fact, it turns out that any number less than 4 is a solution of the inequality (i). (See Example 4 below.)

Solving a given inequality means finding all its solutions. In solving an inequality, we reduce it to an equivalent inequality (that is, an inequality having the same solutions) that is easier to solve. There are two basic operations by means of which we can transform a given inequality into equivalent inequalities. The rules that govern these operations are as follows:

RULE 1 *When the same real number is added to or subtracted from both sides of an inequality, the direction of inequality is preserved.*

In symbols, if $a > b$ and c is any real number, then
$$a + c > b + c \quad \text{and} \quad a - c > b - c.$$

Examples 1 (a) $8 > 5$ is a true statement. If we add 4 to both sides, we get $8 + 4 > 5 + 4$, or $12 > 9$, which is still true. If we subtract 7 from both sides, we get $8 - 7 > 5 - 7$, or $1 > -2$, which is again true.

(b) Let $x - 1 > 3$. Adding 1 to both sides, we get
$$x - 1 + 1 > 3 + 1$$
or
$$x > 4.$$
The set of values of x for which $x - 1 > 3$ is the same as that for which $x > 4$.

In the last example we see that we can obtain the inequality $x > 4$ from the given inequality $x - 1 > 3$ by moving the term (-1) from left side to right side and changing its sign. In general the above rule allows us to do this kind of operation:

Sec. 4.2 *Linear Inequalities in One Variable* 133

Any term can be moved from one side of the inequality to the other while changing its sign and this does not affect the solutions of the inequality.

In symbols, if $a > b + c$, then $a - b > c$, and $a - c > b$.

Examples 2 (a) $8 > 5 + 2$, then $8 - 2 > 5$.

(b) Let $2x - 1 < x + 4$. Then, moving the x over to the left and (-1) over to the right, we get

$$2x - x < 4 + 1$$
$$x < 5.$$

RULE 2 *The direction of the inequality is preserved if both sides are multiplied (or divided) by the same positive number and is reversed if they are multiplied (or divided) by the same negative number.*

In symbols, if $a > b$ and c is any positive number, then

$$ac > bc \quad \text{and} \quad \frac{a}{c} > \frac{b}{c}$$

while if c is any negative number, then

$$ac < bc \quad \text{and} \quad \frac{a}{c} < \frac{b}{c}.$$

Examples 3 (a) $4 > -1$ is a true statement. Multiplying both sides by 2, we obtain $8 > -2$, which is still true. If, however, we multiply by (-2), we must reverse the direction of inequality. We get

$$(-2)(4) < (-2)(-1) \quad \text{or} \quad -8 < 2$$

which is again true.

(b) If $2x < 4$, then we can divide both sides by 2 and obtain the equivalent inequality

$$\frac{2x}{2} < \frac{4}{2} \quad \text{or} \quad x < 2.$$

(c) If $-3x < 12$, then we divide by (-3), which is negative, so we must reverse the inequality:

$$\frac{-3x}{-3} > \frac{12}{-3} \quad \text{or} \quad x > -4.$$

Example 4 Find all real numbers that satisfy the inequality

$$3x + 7 > 5x - 1.$$

Express the solution in interval form.

Solution We move all x terms to one side of the inequality and all constant terms to the other side. Thus, moving $5x$ to the left and 7 to the right side and changing their signs, we get

$$3x - 5x > -1 - 7 \qquad \text{(RULE 1)}$$
$$-2x > -8.$$

Dividing both sides by -2 and reversing the direction of inequality (because -2 is negative), we get

$$\frac{-2x}{-2} < \frac{-8}{-2} \qquad \text{(RULE 2)}$$

$$x < 4.$$

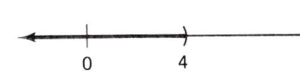

Therefore the solution consists of the set of real numbers in the interval $(-\infty, 4)$. This is illustrated on the adjacent figure.

Example 5 Solve the inequality

$$y + \frac{3}{4} \leq \frac{5y - 2}{3} + 1$$

and express the solution in interval form.

Solution First of all we must get rid of fractions. Here the common denominator is 12; so, multiplying both sides by 12, we get

$$12\left(y + \frac{3}{4}\right) \leq 12\left(\frac{5y - 2}{3} + 1\right)$$

$$12y + 9 \leq 4(5y - 2) + 12$$

$$12y + 9 \leq 20y - 8 + 12$$

$$12y + 9 \leq 20y + 4.$$

Moving y terms to the left and constant terms to the right, we get

$$12y - 20y \leq 4 - 9$$

$$-8y \leq -5$$

Dividing both sides by -8 and reversing the direction of inequality (because -8 is negative), we get

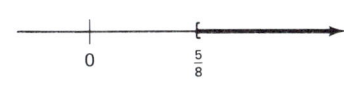

$$y \geq \frac{-5}{-8} \quad \text{or} \quad y \geq \frac{5}{8}.$$

Hence the solution consists of the set of all real numbers greater than or equal to $\frac{5}{8}$—that is, the numbers in the interval $[\frac{5}{8}, \infty)$. This set is illustrated in the adjacent figure.

Example 6 Solve the inequality

$$(2x + 3)(3x - 1) > (6x + 1)(x + 1).$$

Solution On multiplying out the parentheses, we have

$$6x^2 + 7x - 3 > 6x^2 + 7x + 1.$$

Moving the terms containing x from left side to the right, we get

$$6x^2 + 7x - 3 - 6x^2 - 7x > 1$$

$$-3 > 1.$$

This is not a true statement. Thus the given inequality has no solution. In other words, there is no real number x that satisfies the given inequality.

Sec. 4.2 Linear Inequalities in One Variable

Example 7 Solve the double inequality for x:
$$8 - 3x \leq 2x - 7 < 5x - 19.$$

Solution The given double inequality is equivalent to the following two inequalities:
$$8 - 3x \leq 2x - 7 \quad \text{and} \quad 2x - 7 < 5x - 19.$$

Solving these two inequalities separately by the methods described above, we obtain, respectively,
$$x \geq 3 \quad \text{and} \quad x > 4.$$

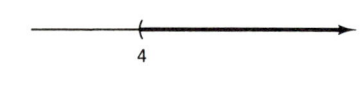

Both these inequalities must be satisfied by x. But whenever $x > 4$, the other inequality $x \geq 3$ is automatically satisfied. Thus the solution is given by $x > 4$ alone. The corresponding interval $(4, \infty)$ is shown in the adjacent figure.

Example 8 Solve the double inequality
$$2x - 7 < 4x - 5 \leq x + 1.$$

Solution We have that
$$2x - 7 < 4x - 5 \quad \text{and} \quad 4x - 5 \leq x + 1.$$

Solving each of these separately, we obtain that
$$x > -1 \quad \text{and} \quad x \leq 2.$$

Since both of these must be satisfied, the solution is
$$-1 < x \leq 2$$

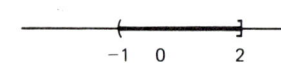

as shown in the adjacent figure.

Example 9 The manufacturer of a certain item can sell all he can produce at the selling price of $60 each. It costs him $(40x + 3000)$ dollars to produce x items per week. Find the number of units he should produce and sell so as to make a profit of at least $1000 per week.

Solution Let x be the number of items produced and sold each week. The total cost of producing x units is given by $\$(40x + 3000)$. The revenue obtained by selling x units at $60 each will be $60x$. Therefore,
$$\text{profit} = \text{revenue} - \text{cost}$$
$$= 60x - (40x + 3000) = 20x - 3000.$$

Since we want a profit of at least $1000 each week,
$$\text{profit} \geq 1000$$

—that is,
$$20x - 3000 \geq 1000$$
$$20x \geq 4000$$
$$x \geq 200.$$

Thus, the manufacturer should produce and sell at least 200 units each week.

To conclude, let us summarize the procedure for solving inequalities that are linear or reducible to linear.

Step 1. First we get rid of fractions on both sides by multiplying throughout by the least common denominator. (Here we assume that the fractions do not contain the variable in the denominator.)

Step 2. Remove all parentheses by multiplying them out.

Step 3. Move all the terms containing the variable to the left and the terms not containing the variable to the right side of the inequality, and simplify by combining like terms.

Step 4. Divide both sides by the coefficient of the variable. Do not forget to change the direction of the inequality if you are dividing by a negative number.

Step 5. If the simplified inequality in step 3 does not contain the variable, then the inequality will be either a true or a false statement. If it is a true statement, then the solution of the given inequality is the set of all real numbers. If this inequality is a false statement, then the original inequality has no solution.

Diagnostic Test 4.2

Fill in the blanks.

1. If $a > b > 0$, then $1/a$ _____<_____ $1/b$.
2. If $a < b$, then $-3a$ _____>_____ $-3b$.
3. If $a > b$, then $ac > bc$ only if c _____>0_____.
4. If $a \geq b$ and $b \geq a$, then a _____=_____ b.
5. If $a > b$ and $b > c$, then a _____>_____ c.
6. If $a < b$ and $a/c > b/c$, then c _____<0_____.
7. If $a > b$ then $a - c$ _____>_____ $b - c$.
8. If $a > b$, and $a/c > b/c$, then c _____>0_____.
9. If $2x < 6$, then x _____<_____ 3.
10. If $-3x > -12$, then x _____<_____ 4.
11. If $x/3 \leq 4$, then x _____≤_____ 12.

Exercises 4.2

(1–30) Solve the following inequalities and express the answer in interval form where possible.

1. $5 + 3x < 11$.
2. $3 - 2y \geq 7$.
3. $2u - 9 \leq 5u + 6$.
4. $5x + 7 < 31 - 3x$.
5. $3t + 10 < 2 + 7t$.
6. $2 - 5t \leq 3t + 18$.

7. $\dfrac{3x-1}{4} \geq \dfrac{2x+1}{3}$.

8. $\dfrac{5-2u}{3} > \dfrac{u+1}{2}$.

9. $x + \dfrac{4}{3} > \dfrac{2x-3}{4} + 1$.

10. $\tfrac{3}{2}(x+4) \geq 2 - \tfrac{1}{3}(1-4x)$.

11. $\tfrac{1}{4}(2x-1) - x < \dfrac{x}{6} - \dfrac{1}{3}$.

12. $\dfrac{x+2}{3} - \dfrac{2-3x}{4} < \dfrac{x+1}{2} + \dfrac{1-2x}{3}$.

13. $\dfrac{x-2}{3} - \dfrac{x+1}{2} \geq \dfrac{3x+2}{4} - \dfrac{x+1}{3}$.

14. $\dfrac{2x+1}{5} - \dfrac{x-3}{2} \leq \dfrac{x+3}{4} - \dfrac{x+1}{5}$.

15. $(x+3)^2 > (x-2)^2$.

16. $(x+2)(x-1) > (x+1)^2$.

17. $(2x+3)(3x-1) \leq (6x+1)(x-2)$.

18. $(2x+3)(5x-1) > (10x+3)(x-2)$.

19. $(3x-1)(2x+3) > (2x+1)(3x+2)$.

20. $(2x+1)(x-3) < (2x+3)(x-4)$.

21. $(3x+1)(x-2) > (x-3)(3x+4)$.

22. $(2x+5)(3x-1) < (6x+1)(x+2)$.

23. $3x - 4 < x - 2 < 2x + 1$.

24. $2x + 1 < 3 - x \leq 2x + 5$.

25. $3x + 7 > 5 - 2x > 13 - 6x$.

26. $2x - 3 \leq 1 + x \leq 3x - 1$.

27. $3x - 5 < 1 + x < 2x - 3$.

28. $x + 13 < 2x + 11 < 1 - 3x$.

29. $1 - 3x \geq 2x + 11 \geq x + 9$.

30. $5x - 7 \geq 3x + 1 \geq 6x - 11$.

31. A man has $7000 to invest. He wants to invest some of it at 8% and the rest at 10%. What is the maximum amount he should invest at 8% if he wants an annual interest income of at least $600 per year?

32. Mrs. K. has $5000 which she wants to invest, some at 6% and the rest at 8%. If she wants an annual interest income of at least $370, what is the minimum amount she should invest at 8%?

33. A manufacturer can sell all he produces at $30 per unit. The cost of producing x units is given by $(22x + 12{,}000)$ dollars. How many units must be produced and sold each month by the company so as to realize a profit?

34. A stereo manufacturer can sell all the units he produces at a price of $150 each. His weekly cost of producing x units is given by $(100x + 15{,}000)$ dollars. Find the number of stereos he must manufacture and sell each week so as to obtain a weekly profit of at least $1000?

35. The cost of publishing each copy of a weekly "Buy and Sell" magazine is 35¢. The revenue from the dealer sales is 30¢ per copy and from advertising is 20% of the revenue obtained from sales in excess of 2000 copies. How many copies must be published and sold each week so as to earn a weekly profit of at least $1000?

36. The publisher of a monthly magazine has a publishing cost of 60.5¢ per copy. The revenue from dealer sales is 70¢ per copy and from advertising is 15% of the revenue obtained from sales in excess of 20,000 copies. How many copies must be published and sold each month to earn a monthly profit in excess of $4000?

37. A developing country has 100 square kilometers of land to cultivate with potatoes and corn. A crop of potatoes yields an average of 16 metric tons of protein per square kilometer of planted area while corn yields 24 metric tons per square kilometer. What is the maximum area that should be used for the potato crop if the country needs total protein production of at least 2040 tons?

38. A hospital dietician plans to meet a patient's daily requirement of at least 72 milligrams of vitamin C by use of a combination of prune and orange juice. Prune juice contains 9 milligrams of vitamin C per ounce and orange juice 44 milligrams. What is the maximum amount of prune juice that can be given to the patient to meet his daily requirement of vitamin C, if the dietician decides to give a total of 8 ounces of the two juices?

4.3

Quadratic Inequalities in One Variable RE DO THIS SECTION

A *quadratic inequality in standard form* in one variable x is an inequality of the form

$$ax^2 + bx + c > 0 \quad \text{or} \quad < 0$$

or

$$ax^2 + bx + c \geq 0 \quad \text{or} \quad \leq 0$$

where a, b, and c are certain constants with $a \neq 0$. For example, the inequalities $x^2 + 3x - 2 < 0$ and $-2x^2 + 5x + 12 \geq 0$ are both quadratic inequalities in standard form.

Again we are interested in "solving" a given inequality—that is, in finding the set of values of the variable for which the inequality becomes a true statement. The method of solution will be explained through the following examples.

Example 1 Solve the inequality $x^2 + 3x < 4$.

Solution We first express the given inequality in standard form by moving the constant term 4 to the left:

$$x^2 + 3x - 4 < 0.$$

Factoring the left side, we have

$$(x - 1)(x + 4) < 0.$$

The expression $(x - 1)(x + 4)$ is the product of two numbers, $(x - 1)$ and $(x + 4)$. The product is required to be negative, so one of these factors is positive while the other is negative. Now

(a) $(x - 1)$ is positive when $x > 1$ and is negative when $x < 1$.
(b) $(x + 4)$ is positive when $x > -4$ and is negative when $x < -4$.

The two factors $(x - 1)$ and $(x + 4)$ change sign at $x = 1$ and $x = -4$, respectively. When plotted on the number line, as in the figure below, they can be seen to divide the line into three separate intervals, $x < -4$, $-4 < x < 1$, and $x > 1$.

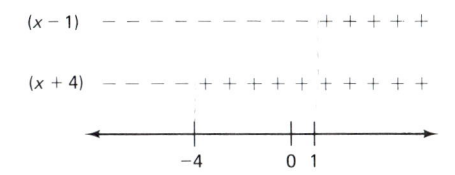

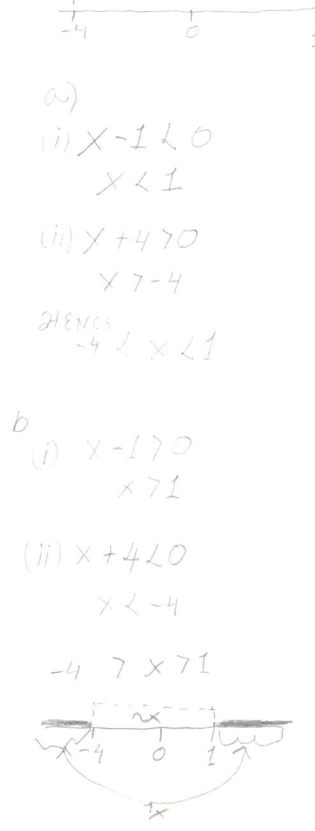

$(x - 1)$ — — — — — — — — — — + + + + +

$(x + 4)$ — — — — + + + + + + + + + + +

$\quad\quad\quad\quad\quad\quad\quad\quad -4 \quad\quad\quad 0\ 1$

Sec. 4.3   Quadratic Inequalities in One Variable   139

It can be seen from the figure that the factors $(x - 1)$ and $(x + 4)$ have opposite signs only in the middle interval, $-4 < x < 1$. We conclude therefore that the inequality

$$(x - 1)(x + 4) < 0$$

is satisfied when $-4 < x < 1$ and for no other values of x. The solution is illustrated on the adjacent figure. [We note that the endpoints $x = 1$ and $x = -4$ are not included in the interval, since at these points $(x - 1)(x + 4)$ is zero, not negative.]

Example 2 Solve the inequality

$$18 + (2x + 1)(x - 2) \leq 4(x - 1)^2.$$

Solution Multiplying out the parentheses, we get

$$18 + (2x^2 - 3x - 2) \leq 4(x^2 - 2x + 1)$$

—that is,

$$2x^2 - 3x + 16 \leq 4x^2 - 8x + 4.$$

Moving all the terms to the left, we get

$$-2x^2 + 5x + 12 \leq 0.$$

It is always convenient to have the square term positive, so let us multiply both sides by -1 (and reverse the direction of inequality):

$$2x^2 - 5x - 12 \geq 0. \qquad \text{(i)}$$

Factoring, we then have

$$(2x + 3)(x - 4) \geq 0.$$

The product $(2x + 3)(x - 4)$ will be positive when the two factors $(2x + 3)$ and $(x - 4)$ are either both positive or both negative. The signs of these two factors are shown on the figure below. Clearly both factors are negative

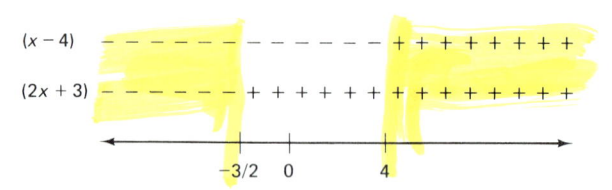

when $x < -\frac{3}{2}$, and both are positive when $x > 4$. The solution of the given inequality is therefore

$$x \leq -\tfrac{3}{2} \quad \text{or} \quad x \geq 4.$$

The endpoints are included because the inequality is not a strict one.

Example 3 Solve the inequality $x^2 < 1 + 2x$.

Solution In standard form the given inequality is

$$x^2 - 2x - 1 < 0.$$

The left side does not factor directly, and therefore we use the method of completing the square. We write

$$x^2 - 2x - 1 = (x^2 - 2x + 1) - 1 - 1$$
$$= (x - 1)^2 - 2$$
$$= (x - 1)^2 - (\sqrt{2})^2$$
$$= (x - 1 + \sqrt{2})(x - 1 - \sqrt{2}). \quad \text{(difference of squares)}.$$

```
(x - 1 - √2)   - - - - - - - - - - - - - - - - - - + + + + + + +
(x - 1 + √2)   - - - - - - - - - - + + + + + + + + + + + + + + +
              ←——————————————+———————————————+——————————————→
                        1 - √2 ≈ -0.41    1 + √2 ≈ 2.41
```

Therefore $(x^2 - 2x - 1)$ is negative when the two factors $(x - 1 + \sqrt{2})$ and $(x - 1 - \sqrt{2})$ have opposite signs. These signs are indicated on the figure above. The factors have opposite signs when

$$1 - \sqrt{2} < x < 1 + \sqrt{2},$$

which is therefore the solution of the given inequality.

Example 4 Solve the inequality

$$x^2 + 7 > 4x.$$

Solution In standard form the inequality is

$$x^2 - 4x + 7 > 0.$$

We use completing the square because again elementary factoring is not possible.

$$x^2 - 4x + 7 = (x^2 - 4x + 4) + 3$$
$$= (x - 2)^2 + 3.$$

Now $(x - 2)^2$ is a perfect square, hence is never negative. Thus $x^2 - 4x + 7$ is never less than 3. The given inequality is therefore satisfied for all real values of x.

Example 5 The monthly sales x of a certain commodity when its price is p dollars per unit is given by $p = 200 - 3x$. The cost of producing x units of the same commodity is $C = 650 + 5x$ dollars. How many units of this commodity should be produced and sold so as to yield a monthly profit of at least $2500?

Solution The revenue R (in dollars) obtained by selling x units at a price of p per unit is

$$R = (x)(\text{price per unit})$$
$$= xp$$
$$= x(200 - 3x)$$
$$= 200x - 3x^2.$$

Sec. 4.3 *Quadratic Inequalities in One Variable*

The cost C (in dollars) of manufacturing x units is
$$C = 650 + 5x.$$
The profit P (in dollars) obtained by producing and selling x units is given by
$$\begin{aligned} P &= \text{revenue} - \text{cost} \\ &= 200x - 3x^2 - (650 + 5x) \\ &= 195x - 3x^2 - 650. \end{aligned}$$
Since the profit P has to be at least $2500, we have
$$P \geq 2500$$
—that is,
$$195x - 3x^2 - 650 \geq 2500$$
$$-3x^2 + 195x - 3150 \geq 0.$$
Dividing both sides by -3 and reversing the direction of inequality, we have
$$x^2 - 65x + 1050 \leq 0$$
$$(x - 30)(x - 35) \leq 0.$$
Solving this inequality as in Example 2, we get
$$30 \leq x \leq 35.$$
Thus, to obtain a profit of at least $2500 per month the manufacturer must produce and sell any number of units between 30 and 35 (both inclusive).

Example 6 A hairdresser gets on the average 100 customers per week at his present charge of $6 per haircut. For each increase of $1 in the price he will lose 10 customers. What price should he charge so that his weekly earnings will not be less than they are at the $6 price?

Solution Let there be x increases of $1 in price beyond $6. Then the price per haircut is $(6 + x)$ dollars and the number of customers at this price will be $(100 - 10x)$ per week.
$$\begin{aligned} \text{total weekly earnings} &= (\text{number of customers})(\text{price per haircut}) \\ &= (100 - 10x)(6 + x) \text{ dollars.} \end{aligned}$$
Earnings from the present 100 customers are $100 \cdot 6 = 600$ dollars. Therefore, the new weekly earnings should be at least $600. Thus
$$(100 - 10x)(6 + x) \geq 600$$
—that is,
$$600 + 40x - 10x^2 \geq 600$$
$$40x - 10x^2 \geq 0$$
and so
$$10x(4 - x) \geq 0.$$

Solving this in the usual way we obtain $0 \leq x \leq 4$. Thus there should be, at most, four increases of $1—that is, the rate should be increased, at most, by $4 to a maximum of $10 per haircut.

Example 7 Solve the fractional inequality

$$\frac{3x+5}{x-2} \leq 2.$$

Solution First we combine the terms as a single fraction:

$$\frac{3x+5}{x-2} - 2 \leq 0$$

$$\frac{3x+5-2(x-2)}{x-2} \leq 0$$

$$\frac{x+9}{x-2} \leq 0. \qquad \text{(i)}$$

In this form, we can solve the fractional inequality by an argument similar to the one used for quadratic inequalities. The ratio of $(x+9)$ and $(x-2)$ is negative if and only if $(x+9)$ and $(x-2)$ have opposite signs. From the figure below we see that this occurs in the interval $-9 < x < 2$. Since the

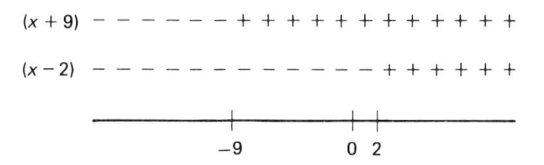

```
(x + 9)   - - - - - - + + + + + + + + + + + +

(x - 2)   - - - - - - - - - - - - - + + + + + +
         ─────────────┼──────────────┼┼─────────
                     -9              0 2
```

inequality (i) is not strict, it is also satisfied at the endpoint $x = -9$ (but not at $x = 2$). The solution of the given inequality is therefore

$$-9 \leq x < 2.$$

Diagnostic Test 4.3

Fill in the blanks:

1. If $(x-1)(x-3) < 0$, then _____ $< x <$ _____.
2. If $(x-2)(x+3) < 0$, then _____ $< x <$ _____.
3. The solution of $(x-1)(x-3) > 0$ is given by _____.
4. The solution of $(x+1)(x-5) > 0$ is given by _____.
*5. If $a < b$, then the solution of $(x-a)(x-b) < 0$ is given by _____.
*6. If $a < b$, then the solution of $(x-a)(x-b) \geq 0$ is given by _____.
7. The solution of $(x-2)^2 > -1$ is given by _____.

8. The solution of $(x - 3)^2 + 4 > 0$ is given by _____.

9. The solution of $x^2 < 9$ is given by _____.

Exercises 4.3

(1–32) Solve the following inequalities.

1. $(x - 2)(x - 5) < 0$.
2. $(x + 1)(x - 3) \leq 0$.
3. $(5 - 2x)(x + 3) \geq 0$.
4. $(1 - 3x)(x + 2) > 0$.
5. $(3x + 1)(2 - x) < 0$.
6. $(x + 5)(2 - x) \leq 0$.
7. $x^2 < 9$.
8. $25 \geq 4x^2$.
9. $7x > x^2 + 12$.
10. $9x > x^2 + 14$.
11. $y^2 > 3y + 10$.
12. $15 < t^2 + 2t$.
13. $y(2y + 1) > 6$.
14. $4 - 11y \leq 3y^2$.
15. $x^2 + 9 < 0$.
16. $x^2 + 1 < 0$.
17. $(2x + 1)(x + 3) < (x + 10)(x - 3)$.
18. $(3x + 1)(x - 2) < (2x + 1)(x - 3)$.
19. $x^2 + 3 > 0$.
20. $(2x + 1)(x + 2) > (x + 6)(x - 1)$.
21. $x^2 - 6x + 9 < 0$.
22. $x^2 + 4x + 4 \leq 0$.
23. $(x - 2)^2 + 5 \geq 0$.
24. $(x - 2)^2 + 4 \leq 0$.
25. $x^2 + 2x + 4 < 0$.
26. $x^2 + 2x + 1 > 0$.
27. $(2x + 1)(x - 2) > (x + 2)(x - 3)$.
28. $(2x + 3)(x - 3) > (x - 1)(3x + 2)$.
*29. $(1 - 2x)(x + 3) \geq (x + 4)(x - 1)$.
30. $(1 - 3x)(x + 2) > (3 - 2x)(x + 3)$.
*31. $9x < x^3$.
*32. $(x - 2)(x + 1)(x - 3) \geq 0$.

33. Find all values of k for which the equation $x^2 + kx + 4 = 0$ has (a) two real roots, (b) no real roots.

34. Find all values of k for which the equation $2x^2 - 3kx + 8 = 0$ has (a) no real roots, (b) two real roots.

35. Find the values of p for which the equation $px^2 - px + 1 = 0$ has (a) no real roots, (b) two real roots.

36. Find the values of p for which the equation $x^2 + 2px - 3p = 0$ has (a) two real roots, (b) no real roots.

37. A farmer wishes to enclose a rectangular field and has available 200 yards of fencing. What should be the maximum length of the field if its area is at least 2100 square yards?

38. One side of the rectangular field is bounded by a river. A farmer has 100 yards of fencing and wants to cover the field's other three sides. If he wants to enclose an area of at least 800 square yards, what are the possible values for the length of the field along the river?

39. A rectangular sheet of cardboard is 16 by 10 inches. Squares of equal sizes are cut from each corner and the edges of the cardboard are folded up to form an open box. What is the maximum height of this box if the base has an area of at least 72 square inches?

40. An open box is made from a rectangular sheet of metal 16 by 14 feet by cutting equal squares from each corner and folding up the edges. If the area of the base of the box is to be at least 80 square feet, what is the maximum possible height of the box?

41. A ball is thrown upward with an initial velocity of 80 feet per second. The height h (in feet) traveled in t seconds is given by the formula
$$h = 80t - 16t^2.$$
For what values of t will the ball be at a height of more than 64 feet?

42. A rocket is shot vertically upward from the ground with an initial velocity of 128 feet per second. The rocket is at a height h after t seconds of launching, where
$$h = 128t - 16t^2$$
During what time intervals will the rocket be at a height of less than 192 feet above the ground?

43. The sum S of first p natural number $S = 1 + 2 + 3 + \ldots, + p$ is given by the formula $S = \frac{1}{2}p(p+1)$.
   (a) For what values of p will the sum S be less than 66?
   (b) How many natural numbers must be added to obtain a sum S of more than 190?

44. A manufacturer can sell each week x units of his product at a price of \$$p$ per unit, where $x = 160(10 - p)$. Find the price p that he should charge so as to generate a weekly revenue of at least \$3840.

45. In Exercise 44, suppose it costs \$2 to produce each item. In what price range should he sell each unit so as to obtain a weekly profit of at least \$2400?

46. A manufacturer can sell x units per week at a price \$$p$ per unit, where $x = 300(6 - p)$.
   (a) What price per item should he charge so as to generate a weekly revenue of at least \$2400?
   (b) If it costs the manufacturer \$3 to produce each unit, how much should he charge per item so as to obtain a weekly profit of at least \$600?

47. Royal Trust owns an apartment building that has 60 suites. If the rent per suite is \$180 per month, all the suites are rented. At a higher rent, some of the suites will remain vacant. On average, for each increase of \$5 in monthly rent one suite becomes vacant with no possibility of renting it. Find the rent that should be charged in order to obtain a total income of at least \$11,475.

48. For what values of k is $x = 3$ a solution of $(x - k)(x - 2k) < 1$?

(49–64) Solve the following inequalities.

49. $\dfrac{3}{x} > -2$.

50. $\dfrac{2}{x - 3} < 1$.

51. $\dfrac{3}{x - 1} > 1$.

52. $\dfrac{5}{3 - x} \leq -1$.

53. $\dfrac{5}{x - 2} > 0$.

54. $(2 - 3x)^{-1} \leq 0$.

55. $\dfrac{x - 5}{2x - 3} < -1$.

56. $\dfrac{x + 1}{5 - 2x} \geq 1$.

57. $\dfrac{3}{x} + 1 \leq 2 - \dfrac{5}{x}$.

58. $\dfrac{2x + 3}{x - 1} < 3$.

59. $\dfrac{2x - 3}{x + 1} > 0$.

60. $\dfrac{3 - x}{2 - 3x} < 0$.

61. $\dfrac{x - 1}{x + 2} > 1$.

62. $\dfrac{2x + 1}{x - 7} > 2$.

63. $\dfrac{3x + 5}{x - 2} < 3$.

64. $\dfrac{2x + 3}{x + 1} > 2$.

65. $\dfrac{1}{x^2 + 4} < 0$.

66. $\dfrac{5}{x^2 + 1} < 0$.

67. $\dfrac{3}{(x - 2)^2} > 0$.

68. $\dfrac{4}{(x + 1)^2} < 0$.

69. The numerator of a fraction is 3 more than the denominator. If the value of the fraction is greater than 2, what are the possible values of the denominator?

70. The numerator of a fraction is 5 more than the denominator. What values of the denominator will make the value of the fraction less than 3?

71. For what values of k is $x = 3$ a solution of $(x - k)/(x + k) > 2$?

72. Find the values of p for which $x = -1$ is a solution of the inequality
$$\dfrac{2p - x}{3x - p} < 3.$$

Sec. 4.3 Quadratic Inequalities in One Variable

4.4

Absolute-Value Equations and Inequalities

We recall from Section 1.2 that if x is a real number, then the **absolute value** of x, denoted by $|x|$, is given by

$$|x| = x \quad \text{if } x \geq 0$$
$$ = -x \quad \text{if } x < 0.$$

For example: $|5| = 5, |-3| = -(-3) = 3, |0| = 0.$

From the definition, it is clear that the *absolute value of a number is always a nonnegative real number*; that is,

$$|x| \geq 0.$$

The absolute value of x is a measure of the "size" of x without regard as to whether it is positive or negative.

We use this basic definition to solve absolute-value equations and inequalities. Consider the equation

$$|x| = 3. \tag{i}$$

If $x \geq 0$, then $|x| = x$ and (i) becomes $x = 3$, which is a solution. If $x < 0$, then $|x| = -x$ and (i) becomes $-x = 3$ or $x = -3$. Thus, we have two solutions of (i), namely $x = 3$ and $x = -3$.

Geometrically this result is obvious. The equation $|x| = 3$ means that the point x on the number line is 3 units distant from the origin. There are two points that are 3 units from 0: 3 and -3.

The above result can be generalized as follows:

$$\text{If } a \geq 0, \text{ then } |x| = a \text{ implies } x = \pm a. \tag{1}$$

Again the geometrical interpretation is obvious.

Similarly, if $|x| = |y|$, then either $x = y$ or $x = -y$. In other words,

$$|x| = |y| \text{ implies } x = \pm y. \tag{2}$$

Example 1 Solve for x:

$$|2x - 3| = 5.$$

Solution Using the result (1) above, the given equation implies that either

$$2x - 3 = 5 \quad \text{or} \quad 2x - 3 = -5.$$

146   Ch. 4   Inequalities

Solving these two equations separately, we obtain $x = 4$ and $x = -1$, respectively. Hence the solutions are $x = 4$ and $x = -1$.

Example 2 Solve for x:
$$|3x - 2| = |2x + 7|.$$

Solution Using (2) above, this equation will be satisfied if either
$$3x - 2 = 2x + 7 \quad \text{or} \quad 3x - 2 = -(2x + 7).$$

Solving these two equations separately, we obtain respectively $x = 9$ and $x = -1$.

Example 3 Solve the equation
$$|7 - 3x| = -5.$$

Solution By the definition of absolute value, $|7 - 3x|$ is always nonnegative for any value of x and therefore cannot be equal to a negative number. Thus the given equation has no solution.

Let us now consider some inequalities involving absolute values. Consider, for example, the inequality $|x| < 5$. We can solve this inequality by using the geometric interpretation of $|x|$. From Section 1.2, we know that $|x|$ represents the distance between the point x and the origin on the number line, without regard to direction.

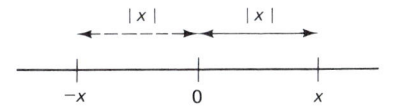

Thus the inequality $|x| < 5$ implies that the distance of x from the origin is less than 5 units. Since x can be on either side of 0, x lies between -5 and 5 or $-5 < x < 5$.

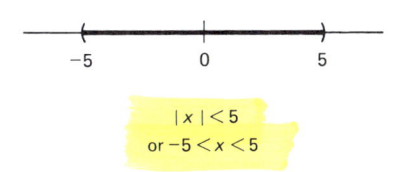

Arguing in the same manner, we can see that $|x| > 5$ implies that x is more than 5 units from the origin on either side. That is, $|x| > 5$ implies that either $x < -5$ or $x > 5$.

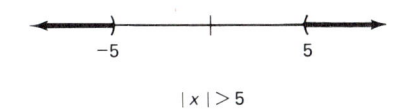

This result may be generalized as follows:

THEOREM 1 If $a > 0$, then

(i) $|x| < a$ if and only if $-a < x < a$.

(ii) $|x| > a$ if and only if either $x > a$ or $x < -a$. (3)

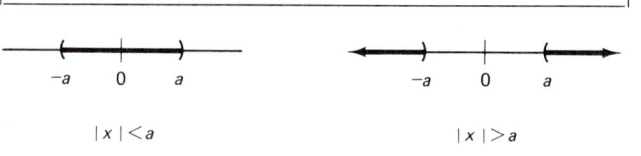

$|x| < a$ $|x| > a$

Example 4 Solve for x:
$$|2x - 3| < 5$$
and express the result in terms of intervals.

Solution Using Theorem 1, part (i) the given inequality implies that
$$-5 < 2x - 3 < 5.$$
Adding 3 to each member of the double inequality, we obtain
$$-5 + 3 < 2x - 3 + 3 < 5 + 3$$
$$-2 < 2x < 8.$$
Dividing throughout by 2, we get
$$-1 < x < 4.$$
Thus the solution consists of the real numbers x that lie in the open interval $(-1, 4)$ shown in the adjacent figure.

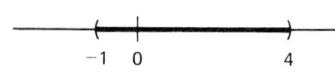

Example 5 Solve for x:
$$|2x - 3| + 5 \leq 0.$$

Solution The given inequality can be rewritten as
$$|2x - 3| \leq -5.$$
But $|2x - 3|$ can never be negative, so there are no values of x for which the given inequality is true: No solution exists.

Example 6 Solve for x:
$$|2 - 3x| > 7$$
and express the result in interval notation.

Solution Using Theorem 1, part (ii), the given inequality implies that either
$$2 - 3x > 7 \quad \text{or} \quad 2 - 3x < -7.$$
Taking the first inequality, we have
$$2 - 3x > 7.$$ 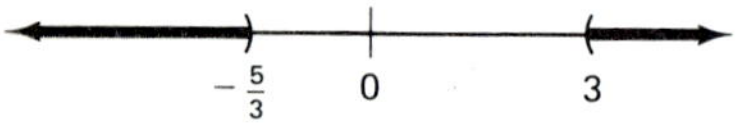
Subtracting 2 from both sides and dividing by -3, we have
$$x < -\tfrac{5}{3}.$$

Similarly, solving the second inequality, we obtain
$$x > 3.$$
Thus $|2 - 3x| < 7$ is equivalent to
$$x < -\tfrac{5}{3} \quad \text{or} \quad x > 3.$$
Thus the solution consists of all real numbers *not* in the closed interval $[-\tfrac{5}{3}, 3]$.

Example 7 Solve the inequality
$$|x^2 - 5x| < 6.$$

Solution Using part (i) of Theorem 1, the given inequality implies
$$-6 < x^2 - 5x < 6$$
—that is,
$$-6 < x^2 - 5x \quad \text{and} \quad x^2 - 5x < 6.$$
In standard form, these inequalities are
$$x^2 - 5x + 6 > 0 \quad \text{and} \quad x^2 - 5x - 6 < 0 \qquad \text{(i)}$$
Solving $x^2 - 5x + 6 > 0$ or $(x - 2)(x - 3) > 0$, as described in the last section, we obtain the solution
$$x < 2 \quad \text{or} \quad x > 3.$$
Similarly $x^2 - 5x - 6 < 0$ can be written $(x - 6)(x + 1) < 0$, and its solution is given by $-1 < x < 6$. These solutions are illustrated in the adjacent figure.

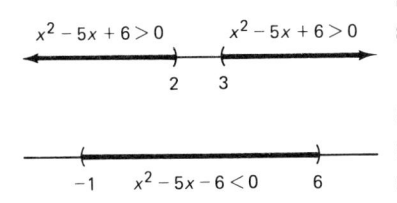

We are interested in those values of x that satisfy *both* the inequalities
$$x^2 - 5x + 6 > 0 \quad \text{and} \quad x^2 - 5x - 6 < 0.$$
That is, x must lie in *both* of the sets shown in the above figure. The required solution thus consists of the common part of the two sets, as shown in the figure below. Clearly the solution is
$$-1 < x < 2 \quad \text{or} \quad 3 < x < 6.$$

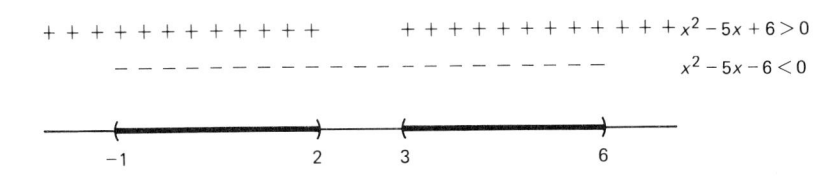

It may be remarked that, if $a > 0$, then the two inequalities $x^2 < a^2$ and $|x| < a$ have the same solution, which is $-a < x < a$. Similarly the two inequalities $x^2 > a^2$ and $|x| > a$ have identical solutions. In other words, if $a > 0$,
$$x^2 < a^2 \quad \text{is equivalent to} \quad |x| < a$$
and
$$x^2 > a^2 \quad \text{is equivalent to} \quad |x| > a.$$

Thus if we "take the square root" on both sides of the inequality $x^2 < a^2$, we should write $|x| < |a|$ and not $x < a$. Similarly, if we take the square root on both sides of $x^2 > a^2$, we should write $|x| > |a|$ and not $x > a$.

Diagnostic Test 4.4

Fill in the blanks.

1. If $x^2 \geq 9$, then _____ ≥ 3.
2. If $x^2 < a^2$, then $|x|$ _____ $|a|$.
3. If $x < -2$, then x^2 _____ 4.
4. If $x > 3$, then x^2 _____ 9.
5. If $a > 0$, then the solution of $|x| < a$ is _____.
6. If $a > 0$, then the solution of $|x| > a$ is _____.
7. The solution of $|x| < -3$ is given by _____.
8. The solution of $|x - 2| > -3$ is given by _____.
9. The solution of $x^2 < 4$ is given by _____.
10. The solution of $x^2 > 9$ is given by _____.
11. The equation $|x| = 3$ has _____ as solution.
12. The equation $|x - 2| = 3$ has _____ as solution.
13. The equation $|x - 2| + 3 = 0$ has _____ solution.
14. The equation $|x - 2| + |x - 3| = 0$ has _____ solution.
15. Taking square roots on both sides of the inequality $x^2 < 9$, we obtain _____.
16. Taking square roots on both sides of the inequality $x^2 > 25$, we obtain _____.

Exercises 4.4

(1–12) Solve the following absolute-value equations.

1. $|3 - 7x| = 4$.
2. $|2x + 5| = 7$.
3. $|3x - 2| = 4 - x$.
4. $|3x - 2| = x - 4$.
5. $|x + 2| = 3 - x$.
6. $\left|\dfrac{2x + 1}{3}\right| = |3x - 7|$.
7. $|x - 3| + 7 = 0$.
8. $|2x + 1| + 3 = 0$.
9. $|3x - 7| + |2x - 5| = 0$.
10. $|5 - 2x| + |3 - x| = 0$.
11. $\left|\dfrac{x - 3}{3x - 5}\right| = 6$.
12. $\left|\dfrac{-5x - 2}{x + 3}\right| = 3$.

(13–40) Solve the following absolute-value inequalities and express the solution in the interval form if possible.

13. $|3x + 7| < 4$.
14. $|2x - 6| \leq 3$.
15. $|2 - 5x| \geq 3$.
16. $|3 - 4x| > \tfrac{1}{2}$.
17. $\left|\dfrac{4x - 7}{3}\right| \leq 2$.
18. $\left|\dfrac{2 - 5x}{4}\right| \geq 3$.
19. $|2x - 7| > 0$.
20. $|5 - 3x| < 0$.
21. $|3x + 4| \leq 0$.
22. $|2x + 7| \geq 0$.
23. $7 + |3x - 5| \leq 5$.
24. $|3x - 13| + 6 \geq 0$.

25. $|2x - 3| + 4 \geq 1$.
26. $\left|\dfrac{5-x}{3}\right| + 4 \leq 2$.
27. $|x + 2| + |2x - 1| \geq 0$.
28. $|3x - 2| + |2x - 7| < 0$.
29. $|2x + 3| + |5 - x| \leq 0$.
30. $|2x - 3| + |x - 7| > 0$.
31. $|x^2 - 5| < 4$.
32. $|x^2 - 7| < 9$.
33. $|x^2 + 10| < 3$.
34. $|x^2 - 13| < 3$.
35. $|x^2 - 7| > 3$.
36. $|x^2 - 5| \geq 4$.
37. $|x^2 + x| < 2$.
38. $|x^2 + 2x| > 3$.
39. $|x^2 - 5x| \geq 6$.
40. $|x^2 - 2x| \leq 3$.

Chapter Review 4

1. Define and/or explain the following terms.
   (a) Set; finite, infinite, and empty set; subset of a set; proper subset; equality of two sets.
   (b) Interval; open, closed and semiclosed or semi-open interval.
   (c) Inequality; double inequality; strict inequalities; linear and quadratic inequalities; solution of an inequality in one variable.
   (d) Absolute-value equation; absolute-value inequality.

2. Explain the method of solving the following:
   (a) A linear inequality in one variable.
   (b) A quadratic inequality in one variable.
   (c) An absolute-value equation.
   (d) An absolute-value inequality.
   (e) A fractional inequality.

Review Exercises on Chapter 4

(1–34) Solve the following inequalities.

1. $(2x + 1)(x + 2) > 2(x + 3)(x - 1)$.
2. $x^2 + 3(x - 2) < (x + 3)(x + 2)$.
3. $\dfrac{2x - 5}{4} - \dfrac{1 - 2x}{3} > \dfrac{x + 3}{2}$.
4. $\dfrac{x + 1}{3} - \dfrac{2x + 1}{6} < \dfrac{1 - 3x}{2}$.
5. $(3x - \tfrac{1}{4})^2 < 9(x + \tfrac{1}{2})^2$.
6. $(2x + \tfrac{1}{3})^2 > 4(x - \tfrac{1}{2})^2$.
7. $(3x - 1)(x + 2) > (3x + 2)(x + 1)$.
8. $(x + 5)(x + 7) < (x + 9)(x + 3)$.
9. $5x < 2(x^2 + 1)$.
10. $3x^2 > 7x - 2$.
11. $3(x^2 + 1) \leq 10x$.
12. $9x + 5 \leq 2x^2$.
13. $3 - x > 2x^2$.
14. $15 - 2x^2 > x$.
15. $x^2 + 9 > 4x$.
16. $x^2 + 12 \geq 6x$.
17. $(2x + 1)(x - 2) < (x + 2)(x - 3)$.
18. $(3x + 2)(x - 1) \leq (2x - 3)(x + 2)$.
*19. $x^3 + 12x > 7x^2$.
*20. $x^3 \leq 2x^2 + 15x$.
21. $\dfrac{3x + 2}{x - 1} > 1$.
22. $\dfrac{5 - 2x}{2x - 3} < 3$.
23. $\dfrac{5x + 3}{x - 2} < 5$.
24. $\dfrac{2x - 3}{x + 1} > 2$.

25. $|3 - 5x| < 2$.
26. $|2x - 3| \leq 7$.
27. $|4x - 7| \geq 3$.
28. $|2 - 3x| > 7$.
29. $7 - |x - 3| < 0$.
30. $5 + |2x - 5| \geq 0$.
31. $9 + |2x - 7| \leq 0$.
32. $|x^2 + 9| < 7x - 3$.
33. $|2x - 3| + |7 + 3x| < 0$.
34. $|3x - 5| + |x - 2| \geq 0$.

35. A ball is thrown upward with an initial velocity of 48 feet per second. The height h in feet after t seconds is given by the formula
$$h = 48t - 16t^2.$$
When will the ball be at a height of more than 32 feet?

36. The numerator of a fraction is 2 less than the denominator. If the value of the fraction is less than 3, what are the possible values for the denominator?

37. The numerator of a fraction is 2 more than three times the denominator. If the value of the fraction is at least 2, what are the possible values for the denominator of the fraction?

38. A newspaper costs 25 cents per copy to produce and sell. The publisher receives 20 cents per copy from the sales, and in addition he receives revenue from advertising that is equal to 30% of the revenue from sales in excess of 20,000 copies. How many copies must the publisher sell if (a) he is at least to break even and (b) he is to make a profit of at least $1000 per issue of the newspaper?

39. The owner of an apartment building can rent all its 50 suites if he charges $150 per month for each. For each increase of $5 in the monthly rent he will have one suite vacant with no possibility of renting it. What maximum rent should he charge for each suite so as to obtain a monthly revenue of at least $8000?

40. A state liquor board buys whiskey for $2 a bottle and sells it for p per bottle. The volume of sales x (in hundreds of thousands of bottles per week) is given by $x = 24 - 2p$, when the price is p. What value of p gives a total revenue of $7 million per week? What values of p gives a profit to the liquor board of at least $4.8 million per week?

5 STRAIGHT LINES

5.1

Cartesian Coordinates

A relationship between two variables is commonly expressed by means of an algebraic equation that involves the two variables. For example, if x is the length of one side of a square (in inches, say) and if y is the area of the square (in square inches), then the relation between x and y is expressed by the equation $y = x^2$.

An algebraic equation of this kind can be represented pictorially by means of a graph. Significant features of the given relationship often are much more apparent from the graph than they are from the algebraic relation between the variables. Particularly in areas of mathematical applications, it is useful to develop the habit, when presented with an algebraic relation, of asking oneself what its graph looks like.

Graphs are constructed using what are called cartesian coordinates. We draw two perpendicular lines called *coordinate axes*, one horizontal and the other vertical, intersecting each other at a point O. The horizontal line is called the *x axis*, the vertical line the *y axis*, and O the *origin*. The plane with such coordinate axes is called the *cartesian plane* or simply the *xy plane*.

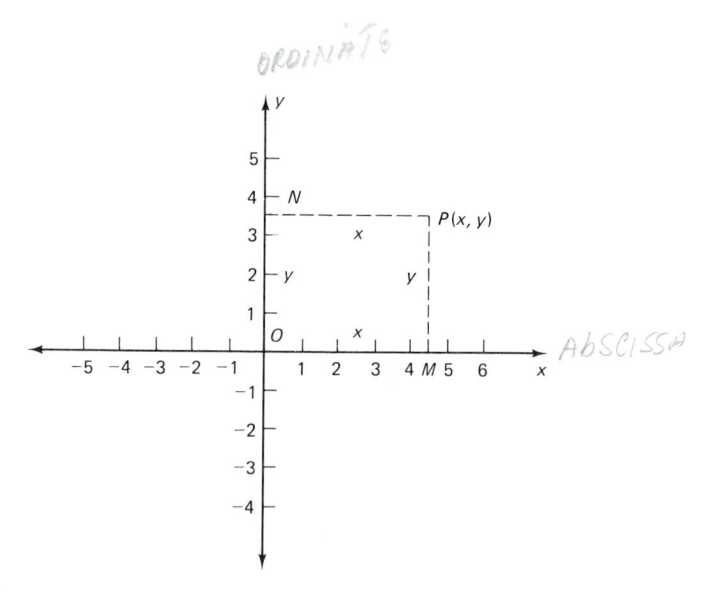

We select a unit of length on the two axes (usually the units of length on both axes are the same, but they need not be). Starting from the origin as zero, we then mark off number scales on each axis, as shown in the above figure. Positive numbers are marked to the right of O on the x axis and above O on the y axis.

Consider a point P in the plane and from it draw PM perpendicular to the x axis and PN perpendicular to the y axis, as shown in the figure above. If the point M represents the number x on the x axis and the point N the number y on the y axis, then x and y are called the **cartesian coordinates** of the point P. We write these two coordinates enclosed in parentheses in the order (x, y).

In this way, corresponding to each point P in the plane there is a unique ordered pair of real numbers (x, y), which are the point's coordinates. Conversely, we can see that corresponding to each ordered pair (x, y) of real numbers there is a unique point in the plane. This representation of points in the plane by ordered pairs of real numbers is called the **cartesian coordinate system**.

If the ordered pair (x, y) represents a point P in the plane, then x (the first member) is called the **abscissa** or **x coordinate** or **x value** of the point P and y (the second member) is called the **ordinate** or **y coordinate** or **y value** of P. The x and y coordinates of P together are the cartesian coordinates of the point P. The notation $P(x, y)$ is used to denote a point P whose coordinates are (x, y).

The coordinates of the origin are $(0, 0)$. *For each point on the x axis, the y coordinate is zero, and each point on the y axis has a zero x coordinate.* The following figure shows several ordered pairs of real numbers and the corresponding points.

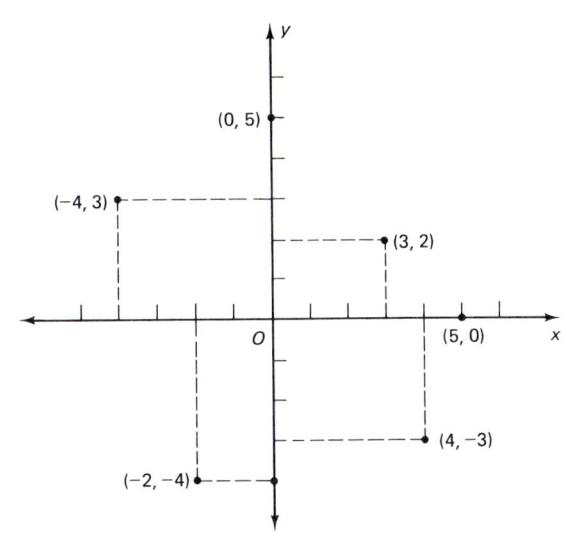

The coordinate axes divide the xy plane into four parts called **quadrants**. The quadrants are numbered first, second, third, and fourth, as shown in the figure below. We have:

(x, y) is in the first quadrant if $x > 0$ and $y > 0$,
(x, y) is in the second quadrant if $x < 0$ and $y > 0$,
(x, y) is in the third quadrant if $x < 0$ and $y < 0$,
(x, y) is in the fourth quadrant if $x > 0$ and $y < 0$.

The points on the coordinate axes do not belong to any quadrant.

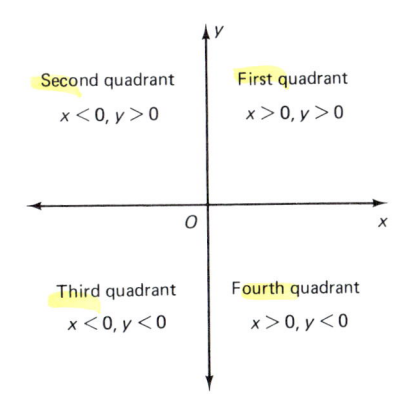

THEOREM 1 (Distance Formula) If $P(x_1, y_1)$ and $Q(x_2, y_2)$ are any two points in the plane, then the distance d between P and Q is given by

$$d = \sqrt{(x_2 - x_1)^2 + (y_2 - y_1)^2}.$$

Sec. 5.1 Cartesian Coordinates

Proof Let PM and QN be perpendiculars from the two points $P(x_1, y_1)$ and $Q(x_2, y_2)$ onto the x axis, and PA and QB perpendiculars onto the y axis, as shown in the adjacent figure. Then the coordinates of the points M, N, A, and B are as shown in the figure. Let the line PA produced meet the line QN at the point R, so that PQR is a right triangle with right angle at R.

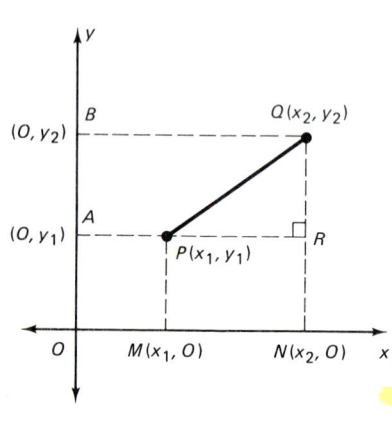

From the figure,* we have

$$PR = MN = ON - OM = x_2 - x_1$$

and

$$RQ = AB = OB - OA = y_2 - y_1.$$

Now recall Pythagoras's theorem, which states that in any right triangle the square of the hypotenuse is equal to the sum of the squares of the other two sides. The triangle PQR is a right triangle whose hypotenuse is PQ so we have

$$PQ^2 = PR^2 + RQ^2$$

that is,

$$d^2 = (x_2 - x_1)^2 + (y_2 - y_1)^2.$$

Taking the square root (the nonnegative square root because the distance is always nonnegative), we have

$$d = \sqrt{(x_2 - x_1)^2 + (y_2 - y_1)^2} \qquad (1)$$

which proves the result. The formula (1) is known as the **distance formula in the plane**.

Example 1 Find the distance between the two points $A(-1, 3)$ and $B(4, 15)$.

Solution Identifying the two given points as

$$(-1, 3) = (x_1, y_1), \qquad (4, 15) = (x_2, y_2)$$

and using the distance formula, we have

$$AB = \sqrt{(x_2 - x_1)^2 + (y_2 - y_1)^2}$$
$$= \sqrt{(4 - (-1))^2 + (15 - 3)^2}$$
$$= \sqrt{5^2 + 12^2} = \sqrt{169} = 13.$$

---

*The figure has been drawn with P and Q both in the first quadrant and with Q above and to the right of P, so that $x_2 > x_1$ and $y_2 > y_1$. In the more general case when these conditions are not satisfied, the horizontal and vertical distances between P and Q are given by $PR = |x_2 - x_1|$ and $RQ = |y_2 - y_1|$. It can be seen that formula (1) for d continues to apply in this general case.

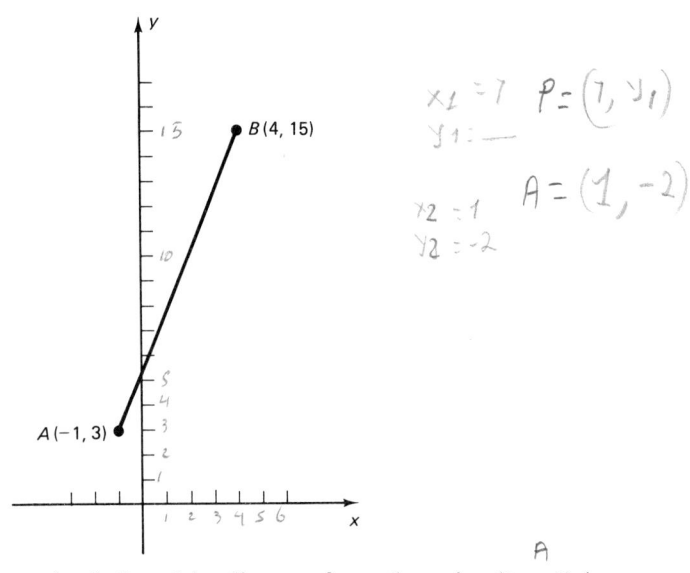

Example 2 The x coordinate of a point is 7 and its distance from the point $(1, -2)$ is 10. Find the y coordinate of the point.

Solution Let P be the point whose y coordinate, y, is required, and A be the point $(1, -2)$. Then the coordinates of P are $(7, y)$, because its x coordinate is given to be 7. From the statement of problem, we are given that

$$PA = 10. \tag{i}$$

Identifying the two points P and A as

$$(7, y) = (x_1, y_1) \quad \text{and} \quad (1, -2) = (x_2, y_2)$$

and using the distance formula, we get

$$PA = \sqrt{(x_2 - x_1)^2 + (y_2 - y_1)^2}$$
$$= \sqrt{(1 - 7)^2 + (-2 - y)^2}.$$

Using (i) above, we then have

$$10 = \sqrt{36 + (2 + y)^2}.$$

Squaring both sides, we get

$$100 = 36 + (2 + y)^2$$
$$= 36 + 4 + 4y + y^2.$$

Therefore,

$$y^2 + 4y - 60 = 0$$
$$(y + 10)(y - 6) = 0.$$

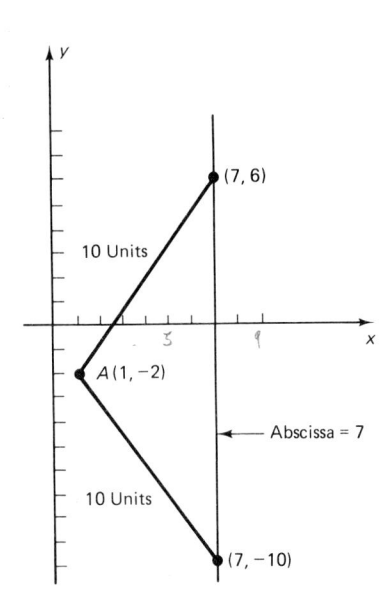

Sec. 5.1 Cartesian Coordinates

Thus,

$$\text{either } y + 10 = 0 \quad | \quad \text{or } y - 6 = 0$$
$$y = -10 \quad | \quad y = 6.$$

Thus the ordinate of the required point P is either 6 or -10.

THEOREM 2 **(Midpoint Formula)** If $P(x_1, y_1)$ and $Q(x_2, y_2)$ are any two points in the plane, then the coordinates of the midpoint of the line segment PQ are given by

$$x = \frac{x_1 + x_2}{2}, \quad y = \frac{y_1 + y_2}{2}.$$

Proof Let $R(x, y)$ denote the midpoint of the line segment PQ. Draw PA, RC, and QB perpendiculars onto the x axis and let the horizontal line through P meet RC at L and QB at M (see the adjacent figure). The triangles PRL and PQM are similar; that is, their angles are equal. Therefore, the ratios of corresponding sides in the two triangles are equal:

$$\frac{PL}{PM} = \frac{RL}{QM} = \frac{PR}{PQ}. \quad (i)$$

Now

$$PL = AC = OC - OA = x - x_1$$
$$PM = AB = OB - OA = x_2 - x_1$$
$$RL = RC - LC = RC - PA = y - y_1$$
$$QM = QB - MB = QB - PA = y_2 - y_1.$$

Also, since R is the midpoint of PQ,

$$\frac{PR}{PQ} = \frac{1}{2}.$$

From (i), taking the first and last members, we have

$$\frac{PL}{PM} = \frac{PR}{PQ}$$

—that is,

$$\frac{x - x_1}{x_2 - x_1} = \frac{1}{2}$$
$$2x - 2x_1 = x_2 - x_1$$
$$2x = x_2 - x_1 + 2x_1 = x_1 + x_2$$
$$x = \frac{x_1 + x_2}{2}.$$

Similarly in (i), by taking the second and third members, we can show that

$$y = \frac{y_1 + y_2}{2}.$$

This proves the required midpoint formula. (*Note:* This proof does not work when PQ is horizontal or vertical. You can prove the midpoint formula yourself in these two cases.)

Example 3 If A is the point $(2, -3)$ and the midpoint of the line AB is $M(-2, 1)$, find the coordinates of the point B.

Solution Let (x, y) denote the coordinates of B. Then, by the midpoint formula, the coordinates of the midpoint of AB are

$$\left(\frac{x+2}{2}, \frac{y-3}{2}\right).$$

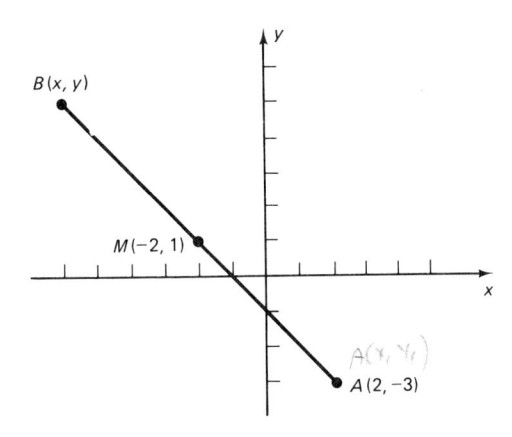

But the coordinates of the midpoint are given to be $(-2, 1)$. Thus, we must have

$$\frac{x+2}{2} = -2, \quad \frac{y-3}{2} = 1.$$

Solving these equations, we get $x = -6$, $y = 5$. Thus B is the point $(-6, 5)$.

Example 4 Prove that the points $A(-1, 2)$, $B(3, 4)$, and $C(1, -2)$ form the vertices of an isosceles triangle. If M is the midpoint of BC, show that AM is perpendicular to BC.

Solution By the distance formula,

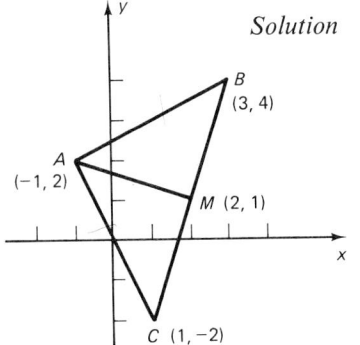

$$AB = \sqrt{(-1-3)^2 + (2-4)^2} = \sqrt{16+4} = \sqrt{20}$$
$$AC = \sqrt{(1+1)^2 + (-2-2)^2} = \sqrt{4+16} = \sqrt{20}$$
$$BC = \sqrt{(1-3)^2 + (-2-4)^2} = \sqrt{4+36} = \sqrt{40}.$$

Since $AB = AC$, the points A, B, C as given form the vertices of an isosceles triangle. The midpoint M of the side BC has the coordinates

$$\left(\frac{3+1}{2}, \frac{4+(-2)}{2}\right) \quad \text{—that is,} \quad (2, 1).$$

Sec. 5.1   Cartesian Coordinates

Now by the distance formula we have

$$AM^2 = (-1-2)^2 + (2-1)^2 = 9 + 1 = 10$$
$$BM^2 = (2-3)^2 + (1-4)^2 = 1 + 9 = 10$$
$$AB^2 = (-1-3)^2 + (2-4)^2 = 16 + 4 = 20.$$

Since $AM^2 + BM^2 = AB^2$, by Pythagoras's theorem the triangle AMB is a right triangle with hypotenuse AB and right angle at M. Thus AM is perpendicular to BC.

The distance formula and the midpoint formula can be used to prove a number of properties of certain geometric figures such as triangles and quadrilaterals. Many properties of these geometric figures remain the same, no matter how we select the axes of our coordinate system. Thus, in proving such properties, a proper choice of coordinate axes will usually simplify the computational work. This is illustrated in the example that follows.

Example 5 Show that in any right triangle the midpoint of the hypotenuse is equidistant from the three vertices.

Solution Let ABC be any right triangle with the right angle at the vertex A. Without loss of generality we can select the vertex A as the origin and take the x axis along the side AB and y axis along the side AC. Let $AB = a$ and $AC = b$.
Then the coordinates of A, B, and C are

$$A(0, 0), \quad B(a, 0) \quad \text{and} \quad C(0, b).$$

If M denotes the midpoint of BC, then the coordinates of M are

$$\left(\frac{a+0}{2}, \frac{0+b}{2}\right) \quad \text{—that is,} \quad \left(\frac{a}{2}, \frac{b}{2}\right).$$

To prove that M is equidistant from the three vertices, we must show that

$$AM = BM = CM$$

or

$$AM^2 = BM^2 = CM^2.$$

Using the distance formula, we have

$$AM^2 = \left(\frac{a}{2} - 0\right)^2 + \left(\frac{b}{2} - 0\right)^2 = \frac{a^2}{4} + \frac{b^2}{4}$$
$$BM^2 = \left(\frac{a}{2} - a\right)^2 + \left(\frac{b}{2} - 0\right)^2 = \frac{a^2}{4} + \frac{b^2}{4}$$
$$CM^2 = \left(0 - \frac{a}{2}\right)^2 + \left(b - \frac{b}{2}\right)^2 = \frac{a^2}{4} + \frac{b^2}{4}.$$

Clearly $AM^2 = BM^2 = CM^2$. Thus M is equidistant from the three vertices.

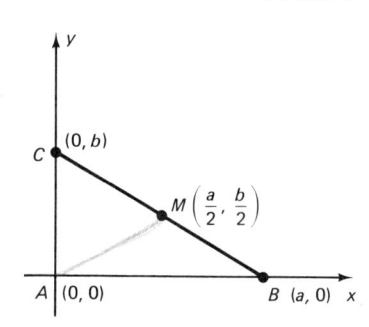

Diagnostic Test 5.1

Fill in the blanks.

1. The point $P(2, 5)$ is at a distance of _____ units from the x axis and _____ units from the y axis.
2. The point $P(x, y)$ lies above the x axis if _____.
3. The point $P(x, y)$ lies below the x axis if _____.
4. The point $P(a, b)$ lies to the right of the y axis if _____.
5. The point $P(x, y)$ lies to the left to the y axis if _____.
6. The point $P(x, y)$ lies on the x axis if _____.
7. If the point $P(x, y)$ lies on the y axis, then _____.
8. The point that lies on both the x axis *and* the y axis is called the _____.
9. If the point $P(x, y)$ lies in the first quadrant, then x _____ and y _____.
10. If $x < 0$ and $y > 0$, then the point $P(x, y)$ lies in the _____ quadrant.
11. If the point $P(x, y)$ lies in the fourth quadrant, then x _____ and y _____.
12. If x and y are of the same sign, then the point $P(x, y)$ lies in the _____ quadrant.
13. If x and y are of opposite sign, then the point $P(x, y)$ lies in _____ quadrant.
14. If the point $P(x, y)$ lies in the second quadrant, then the ratio x/y is _____ (positive or negative).
15. If the ratio x/y is positive, then the point $P(x, y)$ lies in the _____ quadrant.
16. If the ratio x/y is zero, then the point $P(x, y)$ lies on the _____.
17. If the point $P(a, b)$ lies on the x axis, then the ratio a/b is _____.
18. The distance between the two points (x_1, y_1) and (x_2, y_2) is given by _____.
19. The distance between the points (x_1, y_1) and the origin is _____.
20. The distance between the two points $(x_1, 0)$ and $(x_2, 0)$ is given by _____.
21. The distance between the two points $(0, y_1)$ and $(0, y_2)$ is given by _____.
22. The midpoint of the line joining the two points (a, b) and (c, d) has the coordinates _____.

Exercises 5.1

1. Plot the following points whose coordinates are given. Determine the quadrant in which each point lies.
   $A(2, -5); \quad B(-1, 4); \quad C(0, 2);$
   $D(-3, -2); \quad E(5, 0).$

2. Repeat Exercise 1 for the following points.
   $A(3, 5); \quad B(-2, -1); \quad C(3, 0);$
   $D(1, -2); \quad E(-3, 0).$

Sec. 5.1 Cartesian Coordinates

(3–8) Find the distances between the following pairs of points. Determine also the midpoint of the line joining each pair of points.

3. (4, −1) and (2, 0).
4. (−3, 1) and (−2, −3).
5. ($\frac{1}{2}$, 2) and (−2, 1).
6. (a, 2) and (b, 2).
7. (a, b) and (0, 0).
8. (3, p) and (3, q).

(9–10) Prove that the three given points are the vertices of a right triangle. At which vertex is the right angle?

9. A(2, −1), B(3, 2), C(4, 1).
10. A(−1, 6), B(5, 2), C(3, −1).

11. Prove that the points A(1, 1), B(2, −1), and C(3, 2) form the vertices of an isosceles triangle.
12. Prove that the points A(0, 2), B(2$\sqrt{3}$, 4), and C(2$\sqrt{3}$, 0) form the vertices of an equilateral triangle. What is the length of each side of the triangle?

13. Prove that the points A(2, 3), B(5, −1), C(1, −4), and D(−2, 0) taken in order form the vertices of a square.
14. Prove that the points (1, 2), (3, −4), (4, −2), and (2, 4) taken in order form the vertices of a parallelogram.

15. The y-coordinate of a point is 6 and its distance from the point (−3, 2) is 5. Find the x-coordinate of the point.
16. The x-coordinate of a point is 2 and its distance from the point (3, −7) is $\sqrt{5}$. Find the y-coordinate of the point.

17. P is the point (1, a) and its distance from the point (6, 7) is 13. Find the value of a.
18. P is the point (x, 3). The distance of P from the point (4, 5) is $\sqrt{8}$. Find the value of x.

19. P is the point (−1, y) and its distance from the origin is half its distance from the point A(1, 3). Determine the value of y.
20. P is the point (x, 2). The distance of P from the point A(9, −6) is twice its distance from the point B(−1, 5). Find the value of x.

21. Find all the points on the x axis that are 5 units from the point (1, 4).
22. Find the points on the y axis that are 3$\sqrt{2}$ units from the point (−3, −2).

23. Given P(3, −4), find the coordinates of the point Q such that M(−1, −1) is the midpoint of PQ.
24. Given A(2, 7) and B(6, 3), find the coordinates of the point that is one-fourth of the way from A to B.

(25–30) Find the equation that the coordinates of the point P(x, y) must satisfy in order that the following conditions are met.

25. P is at a distance of 5 units from the point (2, −3).
26. P is at a distance of 3 units from the point (−1, 3).

27. P is at a distance of 4 units from the x axis and is above it.
28. P is at a distance of 3 units from the y axis and is to the left of it.

29. The distance of P from the point A(2, 1) is twice the distance from the point B(−1, 3).
30. The sum of squares of the distances of the points A(0, 1) and B(−1, 0) from P is 3.

31. Find all the points (x, y) with x = y that are 2 units from (5, 3).
32. Find all the points (x, y) with x = 2y that are 5 units from (3, −1).

*(33–34) Prove that each set of three points lie on a straight line.

33. (0, −2), (−3, −4), and (6, 2).
34. (1, 1,), (−1, 2), and (−3, 3).

(35–38) Use the distance formula and the midpoint formula, where necessary, to prove the following statements.

35. The diagonals of a rectangle are equal in length and bisect each other.
36. The diagonals of a square bisect each other at right angles.

37. The line joining the midpoints of any two sides of a triangle is half the length of the third side.

38. The lines joining the midpoints of opposite sides of any quadrilateral bisect each other.

39. Show that the points $A(0, 0)$, $B(a, 0)$, and $C(a/2, \sqrt{3}a/2)$ form the vertices of an equilateral triangle. Show that for this triangle the line joining C to the midpoint of AB is perpendicular to AB.

40. Prove that the points $A(1, -2)$, $B(2, 3)$, and $C(6, -1)$ form the vertices of an isosceles triangle. If M denotes the midpoint of BC, show that AM is perpendicular to BC.

5.2

Graphs of Equations

DEFINITION The *graph* of an equation in two variables, say x and y, is the set of all those points whose coordinates (x, y) satisfy the equation.

Consider, for example, the equation $2x - y - 3 = 0$. One of the points whose coordinates satisfy this equation is $(1, -1)$, since upon substituting $x = 1$ and $y = -1$ into the equation we get $2x - y - 3 = 2(1) - (-1) - 3 = 2 + 1 - 3 = 0$, and so the equation is satisfied. Other such points are $(0, -3)$ and $(2, 1)$, as you can easily verify. The graph of this equation is obtained by plotting these points in the xy plane as well as all the other points that satisfy this equation.

Drawing the *exact* graph of an equation in two variables is usually an impossible task, because it would involve plotting infinitely many points. In general practice, enough points satisfying the given equation are chosen so that the nature of the graph is clearly evident. These points are plotted and joined by a smooth curve.

When finding the points satisfying the given equation, it is useful to solve the equation for one variable in terms of the other. For example, if we solve the above equation $2x - y - 3 = 0$ for y in terms of x, we have

$$y = 2x - 3.$$

Now if we give values to x, we can easily calculate the corresponding values for y. For example, if $x = 1$, $y = 2(1) - 3 = -1$; if $x = 5$, $y = 10 - 3 = 7$; and so on.

Example 1 Sketch the graph of the equation $2x - y - 3 = 0$.

Solution Solving the given equation for y, we have

$$y = 2x - 3.$$

For different values of x, the corresponding values of y are given in the following table.

x	−1	0	1	2	3	4
y	−5	−3	−1	1	3	5

Plotting these points, we observe that they lie on a straight line (see the following figure). This line provides the graph of the given equation.

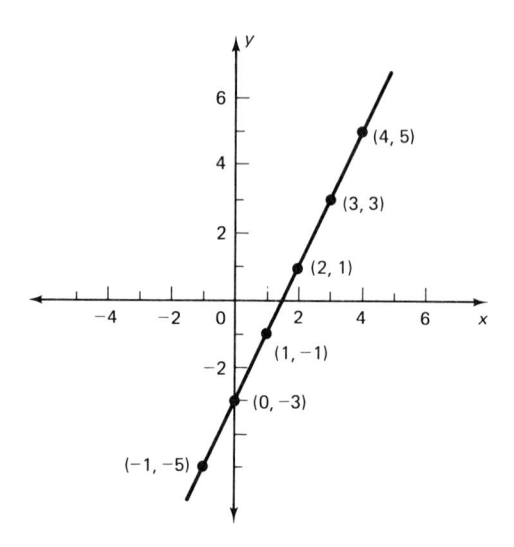

As we shall see in Section 3, this example is typical of a class of equations whose graphs are straight lines. (See Exercises 17–28 of this section for more examples.)

Example 2 Sketch the graph of the equation $y = 5 - x^2$.

Solution For different values of x, the corresponding values of y are given in the following table.

x	−3	−2	−1	0	1	2	3
y	−4	1	4	5	4	1	−4

We plot these points and join them by a smooth curve, obtaining the graph of the equation $y = 5 - x^2$, as shown in the following figure.

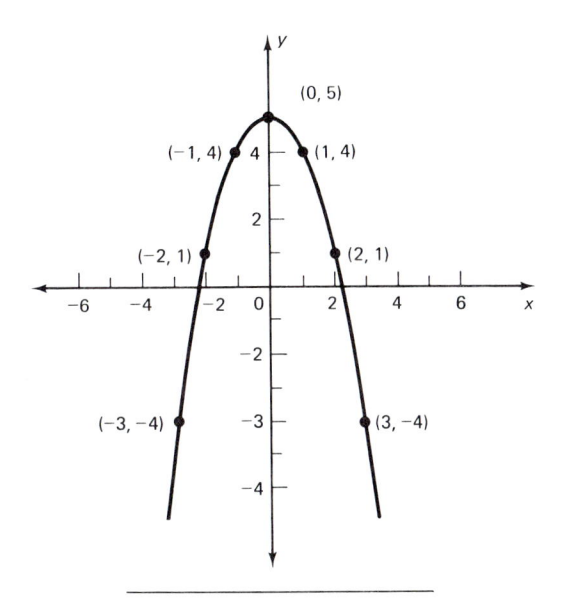

The graph in Example 2 is called a *parabola*. (See section 7.2.)

In the figure of Example 2, the part of the graph on the left of the y axis is the mirror image of the part on the right of the y axis. In this case we say that the graph is symmetric about the y axis.

DEFINITION A graph is *symmetric about the y axis* if the point $(-x, y)$ lies on the graph whenever the point (x, y) lies on it.

In Example 2, for instance, $(1, 4)$ lies on the graph of $y = 5 - x^2$ and so does $(-1, 4)$; $(3, -4)$ lies on the graph and so does $(-3, -4)$. If we change the sign of x, we still get a point on the graph in each case. The basic reason is that the equation $y = 5 - x^2$ involves only the square of x, and so it remains unchanged when x is changed to $(-x)$. In Example 1, however, the graph is not symmetric about the y axis. In this case, for example, $(2, 1)$ lies on the graph but $(-2, 1)$ does not. The equation $y = 2x - 3$ in this example does not remain unchanged when x is replaced by $(-x)$.

The following figures illustrate two graphs that are symmetric about the y axis.

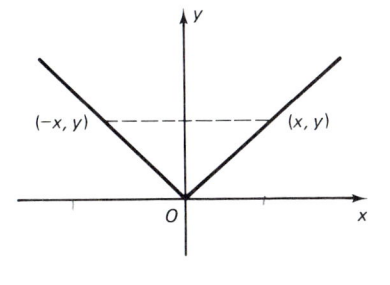

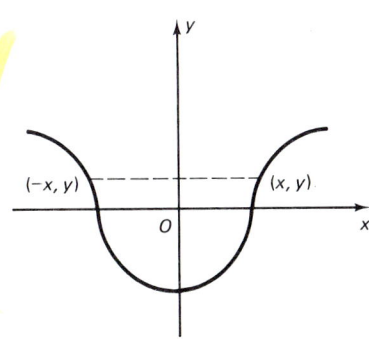

Sec. 5.2   Graphs of Equations

DEFINITION   A graph is said to be **symmetric about the x axis** if the point $(x, -y)$ lies on the graph whenever the point (x, y) lies on it.

This symmetry is illustrated in the following figures. In these cases, the upper part of the graph is the mirror image of the lower part.

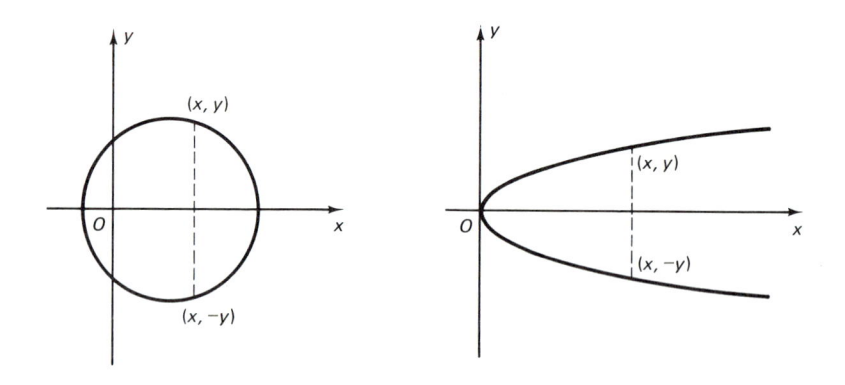

Another type of symmetry that the graphs of certain equations possess is symmetry about the origin.

DEFINITION   We say that a graph is **symmetric about the origin** if the point $(-x, -y)$ lies on the graph whenever the point (x, y) lies on it.

This symmetry is illustrated in the following figures. Note in the figures that the origin is the midpoint on the line joining (x, y) and $(-x, -y)$.

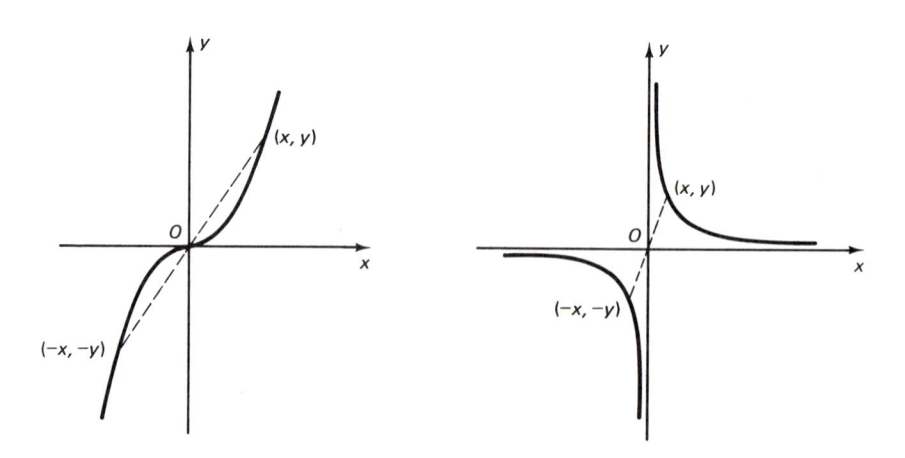

The definitions given above lead to the following tests.

166   Ch. 5   Straight Lines

TESTS FOR SYMMETRY

(a) The graph of an equation is *symmetric about the y axis* if, when x is replaced by $-x$, an equivalent equation is obtained.
(b) The graph of an equation is *symmetric about the x axis* if, when y is replaced by $-y$, an equivalent equation is obtained.
(c) The graph of the equation is *symmetric about the origin*, if an equivalent equation is obtained when x is replaced by $-x$ and simultaneously y is replaced by $-y$.

Example 3 Test the symmetry about the coordinate axes of the graphs of the following equations.
(a) $2x^2 - 3y + 7 = 0$. (b) $x^6 + y^6 = 3$.

Solution (a) The given equation is

$$2x^2 - 3y + 7 = 0. \quad (i)$$

Changing x to $-x$ in this equation, we have

$$2(-x)^2 - 3y + 7 = 0$$
$$2x^2 - 3y + 7 = 0.$$

This is exactly the same as the original equation. Thus the graph of Equation (i) must be symmetric about the y axis.

If we change y to $-y$ in Equation (i), we have

$$2x^2 - 3(-y) + 7 = 0$$
$$2x^2 + 3y + 7 = 0.$$

This equation is not equivalent to the original equation. Therefore the graph is *not* symmetric about the x axis.

(b) The given equation is

$$x^6 + y^6 = 3. \quad (ii)$$

Changing x to $-x$, we obtain

$$(-x)^6 + y^6 = 3$$
$$x^6 + y^6 = 3$$

which is the same equation as (ii). Thus the graph is symmetric about the y axis.

Again, changing y to $-y$ in (ii), we obtain

$$x^6 + (-y)^6 = 3$$
$$x^6 + y^6 = 3$$

which is again the same as equation (ii). Therefore the curve is symmetric about the x axis also. In this case the curve is symmetric about both the coordinate axes.

Example 4 Show that the graph of the equation $2y + x^5 = 0$ is symmetric about the origin but not about either coordinate axis.

Solution To test for symmetry about the origin we must simultaneously change x to $-x$ and y to $-y$. The given equation becomes

$$2(-y) + (-x)^5 = 0$$
$$-2y - x^5 = 0$$
$$2y + x^5 = 0.$$

Since this is exactly the same as the original equation, we conclude that the graph is symmetric about the origin.

On the other hand, if we change x to $-x$ but leave y unchanged, we get

$$2y + (-x)^5 = 0$$
$$2y - x^5 = 0.$$

Since this is not equivalent to the given equation, the graph is not symmetric about the y axis.

Similarly, the graph is not symmetric about the x axis, since the equation does not remain unchanged if we change y to $-y$, with x unchanged.

Suppose that the given equation involves only powers of x and y with integer exponents. When n is any even integer, $(-x)^n = x^n$. Therefore if only *even* powers of x appear in the equation, the equation will not change when x is replaced by $(-x)$. A similar conclusion holds for y. This leads to the following alternative tests.

ALTERNATIVE TEST FOR SYMMETRY ABOUT AN AXIS (INTEGER POWERS)

(a) The graph of an equation will be symmetric about the y axis if the powers of x that occur in the equation are all even.
(b) The graph of an equation will be symmetric about the x axis if the powers of y that occur in the equation are all even.

As illustration, the graph of $x^2 + 2x^4 = y$ will be symmetric about the y axis because only even powers of x occur in the equation. It will not be symmetric about the x axis because y appears with the first power, which is not even. The graph of $y = 3x^2 + 7x$ is symmetric neither about the x axis nor about the y axis, since odd powers of both x and y appear.

If we know that a graph is symmetric about an axis, then we need to plot the graph only in half the plane. We can draw the other half by making use of the symmetry.

Example 5 Sketch the graph of the equation $y^2 = x + 1$.

Solution Since the equation contains only even powers of y, the graph is symmetric about the x axis. Therefore, if we sketch the part of the graph that lies above the x axis, we can draw the graph below the x axis by symmetry. Solving the given equation for y and taking the positive square root (because $y \geq 0$ for the graph above the x axis), we have

$$y = \sqrt{x+1}.$$

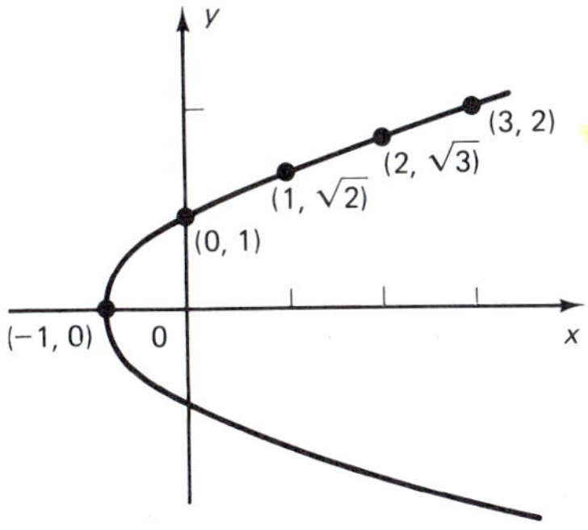

Since y is undefined if $x + 1$ is negative, we must take only those values of x that make $x + 1$ nonnegative—that is, $x \geq -1$. Thus, we obtain the table of values as follows.

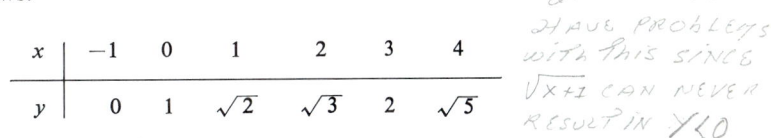

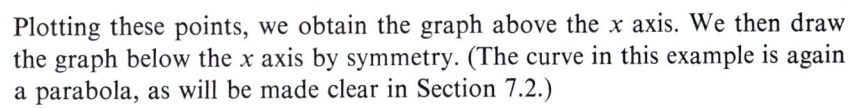

x	-1	0	1	2	3	4
y	0	1	$\sqrt{2}$	$\sqrt{3}$	2	$\sqrt{5}$

Plotting these points, we obtain the graph above the x axis. We then draw the graph below the x axis by symmetry. (The curve in this example is again a parabola, as will be made clear in Section 7.2.)

Since the points (x, y) and $(-x, -y)$ lie in opposite quadrants, the graph of an equation that is symmetric about the origin can be drawn in one quadrant once it has been constructed in the opposite quadrant.

Example 6 Sketch the graph of the equation $xy = -1$.

Solution The given equation is

$$xy = -1. \qquad \text{(i)}$$

The graph of this equation is symmetric neither about the x axis nor about the y axis. However, if we change both x to $-x$ and y to $-y$ in (i), we get

$$(-x)(-y) = -1$$
$$xy = -1$$

which is the same equation as (i). Thus the graph must be symmetric about the origin. Therefore we need only construct the graph for positive values of x and the symmetry will then allow the rest of the graph to be drawn.

Solving the given equation for y, we obtain

$$y = -\frac{1}{x}. \qquad \text{(ii)}$$

Clearly x cannot take the value zero, because then y is undefined. The following table gives some points that lie on the graph for $x > 0$.

x	$\frac{1}{5}$	$\frac{1}{2}$	1	2	3	4
y	-5	-2	-1	$-\frac{1}{2}$	$-\frac{1}{3}$	$-\frac{1}{4}$

Plotting these points, we obtain the graph sketched in the first figure below.

Sec. 5.2 *Graphs of Equations*

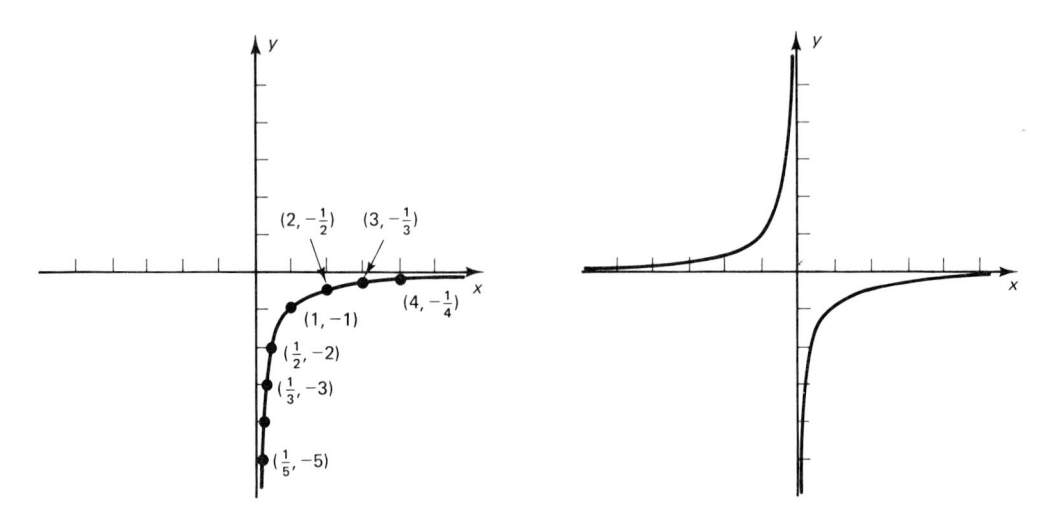

This graph is in the fourth quadrant. By symmetry, we can sketch the graph in the opposite (second) quadrant as shown in the second figure.

Diagnostic Test 5.2

Fill in the blanks.

1. The graph of an equation is symmetric about _____ y axis _____ if the equation contains only even powers of x.
2. The graph of an equation is symmetric about the y axis if the equation contains only even powers of _____ x _____.
3. If the graph of an equation is symmetric about the origin, then the equation remains unchanged when x is replaced by _____ $-x$ _____ and y is replaced by _____ $-y$ _____.
4. The graph of $x^2 y^3 = 1$ is symmetric about the _____ y _____.
5. The graph of $x^3 y^5 + 2 = 0$ is symmetric about the _____ origin _____.
6. The graph $y = |x|$ is symmetric about the _____ y axis _____.
7. The graph of $|x| = |y|$ is symmetric about the _____ origin _____.
8. The graph of the equation $x^6 + 2x^2 y^2 + y^4 = 1$ is symmetric about _____ both / origin _____.
9. The graph of the equation $y^2 + 2x + x^3 = 0$ is symmetric about _____ x axis _____.

Exercises 5.2

(1–16) State whether the graphs of the following equations are symmetric about the x axis, the y axis, the origin, or none of these.

1. $2y + 3x^2 = 0$. y
2. $x + 2y^2 = 0$.
3. $y^2 = x^3 - 1$. x
4. $y = 1 - x^4$.
5. $y = 5x^3$. none
6. $y = -2x^5$.
7. $y = |x|$. y
8. $y = 2 - |x|$.
9. $y = 1/x$. orig
10. $xy = -3$.
11. $x^2 + y^2 = 4$. or, both
12. $x^4 + y^4 = 1$.

13. $x^3y = 1$. *both*   **14.** $x^3y^2 = 1$.   **15.** $y = \sqrt{|x|}$.   **16.** $y = \sqrt{-x}$.

(17–40) Sketch the graphs of the following equations. Use symmetry whenever possible.

17. $y = x + 1$.   **18.** $x = y + 1$.   **19.** $y = 2x - 3$.   **20.** $2x - 3y = 6$.
21. $x + y + 2 = 0$.   **22.** $x - 2y = 4$.   **23.** $\frac{x}{2} + \frac{y}{3} = 1$.   **24.** $\frac{x}{4} + \frac{y}{3} = 1$.
25. $y = 4$.   **26.** $2y - 3 = 0$.   **27.** $3x - 2 = 0$.   **28.** $2x - 5 = 0$.
29. $2y = x^3$.   **30.** $3y + x^3 = 0$.   **31.** $y = x^3 - 1$.   **32.** $y = x^3 + 2$.
33. $y = \sqrt{x}$.   **34.** $y = \sqrt{-x}$.   **35.** $y = \sqrt{x} + 1$.   **36.** $y = \sqrt{x+1}$.
37. $y = \sqrt{2-x}$.   **38.** $y = 2 - \sqrt{x}$.   **39.** $xy = -1$.   **40.** $xy - 2 = 0$.

5.3

Straight Lines and Linear Equations

SLOPE

In this section we examine a number of properties of straight lines. Our first aim is to investigate the algebraic equation that has a given line as its graph.

One of the most important properties of a straight line is how steeply it rises or falls, and we shall first introduce a quantity that will measure this steepness of a given line. Let us begin by considering an example. The equation $y = 2x - 4$ has as its graph the straight line shown in the figure below. Let us choose two points on this line, say the points (3, 2) and (5, 6), which are denoted respectively by P and Q in the figure. The difference between the

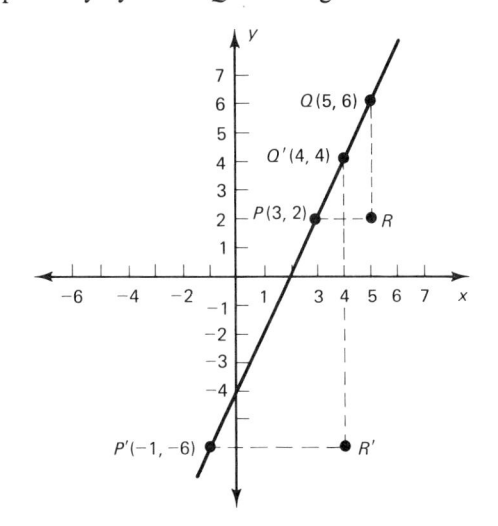

x coordinates of these two points, denoted by PR in the figure, is called the **run** from P to Q:

$$\text{run} = PR = 5 - 3 = 2.$$

The difference between the y coordinates of P and Q, equal to the distance QR, is called the **rise** from P to Q:

$$\text{rise} = QR = 6 - 2 = 4.$$

We note that the rise is equal to twice the run. This would have turned out to be the case no matter which pair of points we had chosen on the given graph. For example, let us take the two points $P'(-1, -6)$ and $Q'(4, 4)$ (see the above figure). Then

$$\text{run} = P'R' = 4 - (-1) = 5; \quad \text{rise} = Q'R' = 4 - (-6) = 10.$$

Again we see that the ratio rise/run is equal to 2.

The reason why the same ratio of rise to run is obtained in the two cases is that the two triangles PQR and $P'Q'R'$ are similar to one another. Therefore the ratios of corresponding sides are equal: $QR/PR = Q'R'/P'R'$. *This ratio is called the **slope** of the given straight line.* The line in the preceding figure has a slope equal to 2.

The slope of a general straight line is defined similarly. Let P and Q be any two points on the given line (see the adjacent figure). Let them have coordinates (x_1, y_1) and (x_2, y_2), respectively. Let R be the intersection of the horizontal line through P and the vertical line through Q. Then the horizontal displacement PR is called the **run** from P to Q and the vertical displacement RQ is called the **rise** from P to Q.

In terms of the coordinates,

$$\text{rise} = RQ = y_2 - y_1$$
$$\text{run} = PR = x_2 - x_1.$$

(Note that if Q turns out to lie below R, which happens when the line slopes downward to the right, then the rise is negative. We could also choose Q to lie to the left of P, in which case $x_2 < x_1$ and the run would be negative.)

The **slope** of the line is defined to be the ratio of rise to run. It is usually denoted by the letter m. Hence

$$\text{Slope:} \quad m = \frac{\text{rise}}{\text{run}} = \frac{y_2 - y_1}{x_2 - x_1}. \tag{1}$$

The expression (1) for slope is meaningful as long as $x_2 - x_1 \neq 0$, that is, provided that the line is nonvertical. Slope is *not* defined for vertical lines.

Note, too, that the slope of a line remains the same, no matter how we choose the positions of the two points P and Q on the line.

If the slope m of a line is positive, the line ascends to the right. The larger the value of m, the more steeply the line is inclined to the horizontal.

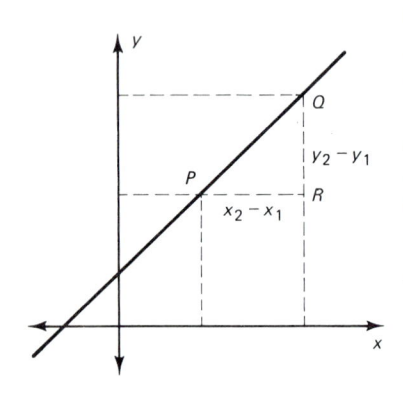

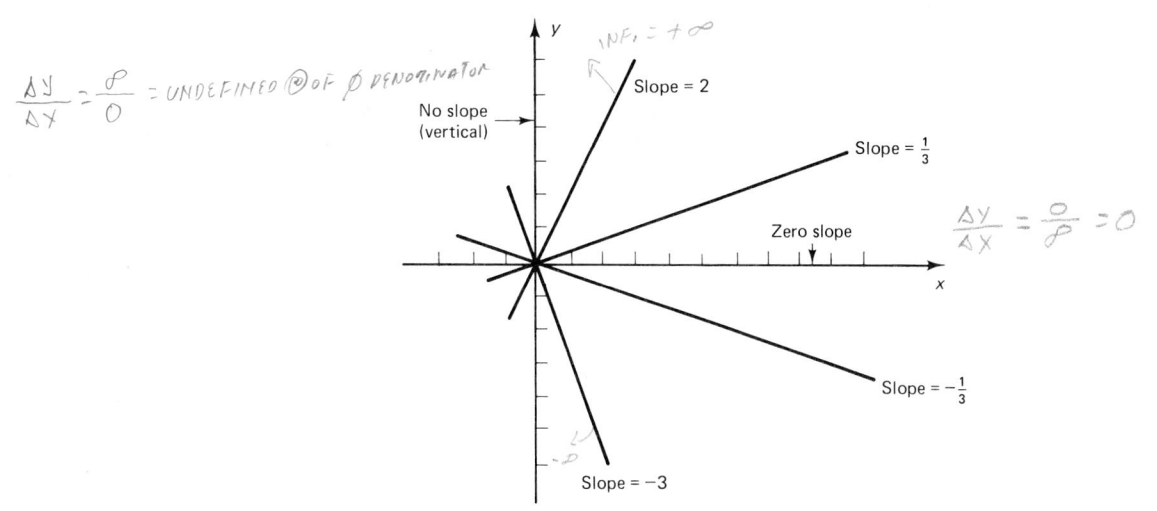

$\frac{\Delta y}{\Delta x} = \frac{\rho}{0}$ = UNDEFINED ② OF ∅ DENOMINATOR

INF. = +∞

$\frac{\Delta y}{\Delta x} = \frac{0}{\rho} = 0$

If m is negative, then the line descends to the right. If $m = 0$, the line is a horizontal one. These properties are illustrated in the above figure.

Example 1   Find the slope of the line joining the two points $(1, -3)$ and $(3, 7)$.

Solution   Using the formula (1), we obtain the slope as

$$m = \frac{7 - (-3)}{3 - 1} = \frac{10}{2} = 5.$$

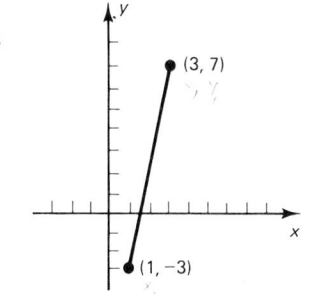

Example 2   The slope of the line joining the two points $(3, 2)$ and $(5, 2)$ is

$$m = \frac{2 - 2}{5 - 3} = 0.$$

Thus the line joining these two points is a horizontal one.

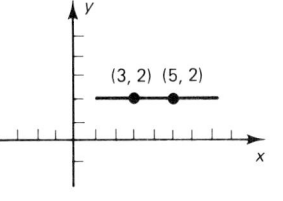

Example 3   The slope of the line joining $P(2, 3)$ and $Q(2, 6)$ is given by

$$m = \frac{6 - 3}{2 - 2} = \frac{3}{0}$$

which is undefined. Thus the line joining P and Q has *no* slope. Note that the line PQ is vertical.

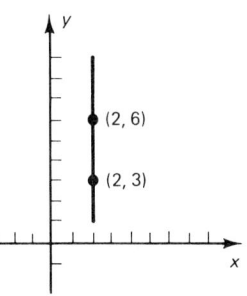

Sec. 5.3   *Straight Lines and Linear Equations*

POINT-SLOPE FORMULA

What information do we need in order to draw a particular straight line? One way in which we can specify a line is by giving two points that lie on it. Once two points are given, the whole line is determined, since through any two points there is only one straight line.

Through any *one* point there are, of course, many different straight lines with slopes ranging from large to small, positive or negative. If the slope is given, however, then there is only one line through the point in question. Thus a second way in which we can specify a straight line is by giving one point on it and the value of its slope.

Our immediate task will be to determine the equation of a nonvertical straight line that has a given slope (equal to m) and that passes through a given point (x_1, y_1).

Let (x, y) be a point on the line different from the given point (x_1, y_1). Then the slope m of the line joining the two points (x_1, y_1) and (x, y) is given by

$$m = \frac{y - y_1}{x - x_1}.$$

It follows therefore that

$$y - y_1 = m(x - x_1). \tag{2}$$

This is called the ***point-slope formula*** for the line. Every point (x, y) on the line satisfies this equation.

Example 4 Find the equation of the line through the point $(5, -3)$ whose slope is -2.

Solution Using (2) with $m = -2$ and $(x_1, y_1) = (5, -3)$, we find that the required equation is

$$y - (-3) = -2(x - 5)$$
$$y + 3 = -2x + 10$$
$$y = -2x + 7.$$

This equation is satisfied by every point (x, y) on the line.

Example 5 Find the equation of the straight line passing through the two points $(1, -2)$ and $(5, 6)$.

Solution The slope of the line joining $(1, -2)$ and $(5, 6)$ is

$$m = \frac{6 - (-2)}{5 - 1} = \frac{8}{4} = 2.$$

Thus, from the point-slope formula the equation of the straight line through

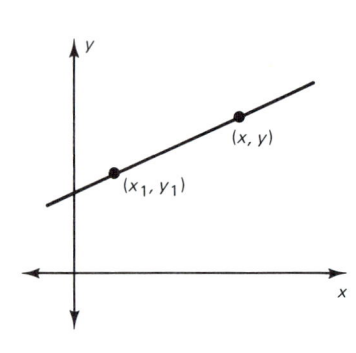

$(1, -2)$ with slope $m = 2$ is
$$y - (-2) = 2(x - 1)$$
$$y = 2x - 4.$$

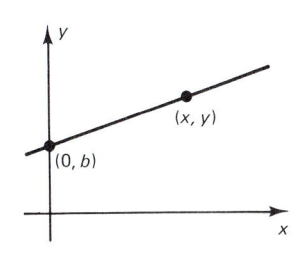

In the point-slope formula let us take (x_1, y_1) to be $(0, b)$. Then Equation (2) becomes
$$y - b = m(x - 0)$$
$$\boxed{y = mx + b.} \qquad (3)$$

The quantity b, which gives the displacement along the y axis that is cut off by the straight line, is called the ***y intercept*** of the line. Equation (3) is called the ***slope-intercept*** formula of the line.

In Examples 4 and 5 the final answers were given in slope-intercept form. In Example 4, $y = -2x + 7$ has slope -2 and y intercept 7. In Example 5, $y = 2x - 4$ has slope 2 and y intercept -4.

Example 6 Show that the three points $A(1, 3)$, $B(-1, 7)$, and $C(3, -1)$ are collinear.

Solution One way to prove that three given points are collinear is first to find the equation of the line through two of the points and then show that the coordinates of the third point satisfy this equation.

Let us find the equation of the line through $A(1, 3)$ and $B(-1, 7)$. The slope of this line is
$$m = \frac{7 - 3}{-1 - 1} = \frac{4}{-2} = -2.$$

Using the point-slope formula, the equation of the line through $A(1, 3)$ with slope -2 is given by
$$y - 3 = -2(x - 1)$$
$$y = -2x + 5. \qquad (i)$$

The point $C(3, -1)$ lies on this line if its coordinates satisfy Equation (i). Substituting $x = 3$, $y = -1$, we obtain
$$-1 = -2(3) + 5$$
which is true. Thus C lies on the line joining A and B.

Alternative Solution We can use the idea of slope alone to prove that the three given points are collinear. If three points A, B, C lie on the same straight line, then the slopes of the line segments AB and AC must be the same. Now the slope of the line joining $A(1, 3)$ and $B(-1, 7)$ is
$$m_1 = \frac{7 - 3}{-1 - 1} = -2.$$

And the slope of the line joining $C(3, -1)$ and $A(1, 3)$ is

$$m_2 = \frac{3 - (-1)}{1 - 3} = \frac{4}{-2} = -2.$$

Since $m_1 = m_2$ the points A, B, C lie on the same line.

GRAPHING BY SLOPE

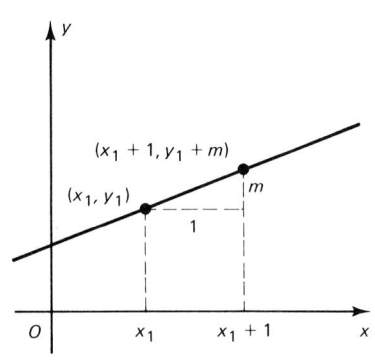

We can use the idea of slope to sketch the graph of a straight line. Suppose the given line passes through (x_1, y_1) and has a slope m. Then by the point-slope formula its equation is

$$y - y_1 = m(x - x_1).$$

When $x = x_1 + 1$, we have

$$y - y_1 = m(x_1 + 1 - x_1) = m$$
$$y = y_1 + m.$$

Thus a run of 1 unit (from x_1 to $x_1 + 1$) causes a rise of m units (from y_1 to $y_1 + m$). Therefore the point $(x_1 + 1, y_1 + m)$ also lies on the given line. Thus we know two points (x_1, y_1) and $(x_1 + 1, y_1 + m)$ on the line, and we can plot these points and join them by a straight line.

Example 7 Sketch the graph of the line that passes through the point $(1, 3)$ and has a slope of -2.

Solution We can sketch the line once we know two different points on it. We are given that one point on the line is $(1, 3)$. To determine the second point on the line we use the given slope. If we start at the given point $(1, 3)$ and *increase* the value of x by 1, the value of y must *decrease* by 2, since $m = -2$. That is, the second point is $(1 + 1, 3 - 2)$ or $(2, 1)$.

Plotting the two points $(1, 3)$ and $(2, 1)$ and joining them by a straight line, we have the required graph, as shown in the figure below.

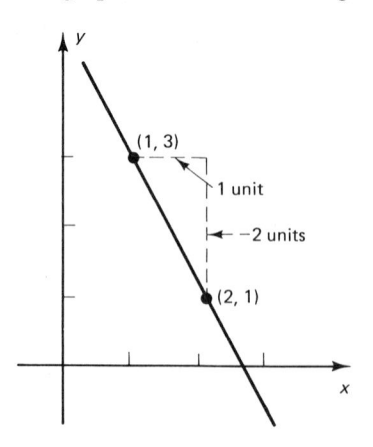

OTHER EQUATIONS OF STRAIGHT LINES

If the given line is horizontal, then its slope is $m = 0$ and Equation (3) above reduces to

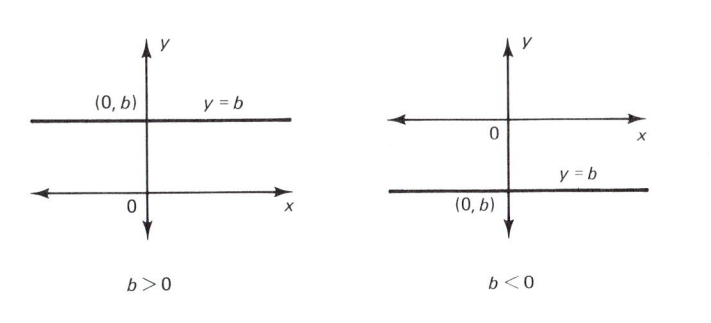

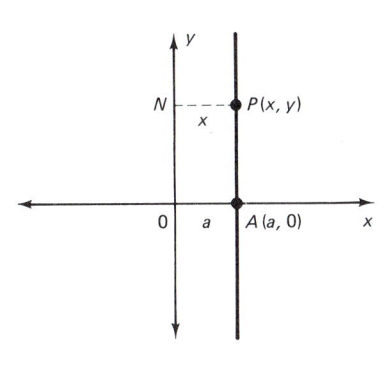

$$\boxed{y = b.} \quad\longleftrightarrow\quad \tag{4}$$

b > 0   *b < 0*

This is the equation of a horizontal line at a distance $|b|$ above or below the x axis, as shown in the above figure. In particular, if we take $b = 0$ in (4), we get $y = 0$, which is the equation of the x axis itself.

Next, suppose that the line in question is a *vertical* one, and let it intersect the x axis at the point $A(a, 0)$, as shown in the adjacent figure. If $P(x, y)$ is a general point on the line, then the two points P and A have the same abscissa—that is, $x = a$. Every point on the line satisfies this condition, so we can say that

$$\boxed{x = a.} \tag{5}$$

is the equation of the vertical line.

For example, taking $a = 0$, we obtain the equation $x = 0$, which is the equation of the y axis. Similarly, $x = 2$ is the equation of the vertical line lying 2 units to the right of the y axis, and $x = -4$ is the equation of the vertical line lying 4 units to the left of the y axis.

A **general linear equation** (or a first-degree equation) in two variables x and y has the form

$$Ax + By + C = 0 \tag{6}$$

where A, B, C are constants and A, B are *not both zero*. In the light of the above discussion, we are in a position to describe the graph of the general linear equation (6) for different values of A and B.

1. $B \neq 0$, $A \neq 0$. Equation (6) in this case when solved for y takes the form

$$y = -\frac{A}{B}x - \frac{C}{B}.$$

Sec. 5.3   Straight Lines and Linear Equations

In view of Equation (3), this is the equation of a straight line whose slope is $-A/B$ and y intercept $-C/B$.

2. $B \neq 0$, $A = 0$. The equation in this case when solved for y becomes

$$y = -\frac{C}{B}.$$

In view of Equation (4), this is the equation of a horizontal line whose y intercept is $-C/B$.

3. $A \neq 0$, $B = 0$. The general equation (6) when $B = 0$ can be written in the form

$$x = -\frac{C}{A}.$$

In view of Equation (5), this is the equation of a vertical line that intersects the x axis at the point $(-C/A, 0)$.

Thus the graph of the general linear equation (6) is in each case a straight line. A linear equation of the form (6) is often referred to as the **general equation** of the straight line.

We summarize in the table below the various forms taken by the equation of a straight line.

Name	Equation
1. Point-slope formula	$y - y_1 = m(x - x_1)$
2. Slope-intercept formula	$y = mx + b$
3. General formula	$Ax + By + C = 0$; A, B not both zero
4. Horizontal line	$y = b$
5. Vertical line	$x = a$

Example 8 Given the linear equation $2x + 3y = 6$, find the slope and y intercept of its graph.

Solution The graph of the given linear equation is a straight line. To find the slope and y intercept of the line, we must express the given equation in the form

$$y = mx + b$$

(that is, we must solve the equation for y in terms of x). We have

$$2x + 3y = 6$$
$$3y = -2x + 6$$
$$y = -\tfrac{2}{3}x + 2.$$

Comparing with the general form $y = mx + b$, we have

$$m = -\tfrac{2}{3} \quad \text{and} \quad b = 2.$$

Thus the slope is equal to $-\tfrac{2}{3}$ and the y intercept is equal to 2.

Example 9 Sketch the graph of the linear equation

$$3x - 4y = 12.$$

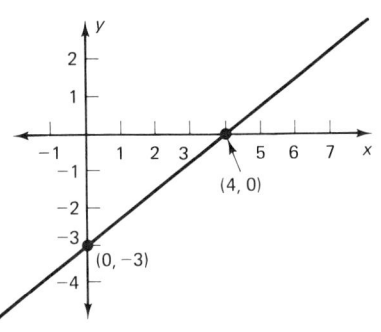

Solution We know that the graph of a linear equation in two variables is always a straight line, and a straight line is completely determined by two points on it. Thus to sketch the graph of the given linear equation, we find two *different* points (x, y) satisfying the given equation, plot them, and then join them by a straight line. Putting $x = 0$ in the given equation, we get

$$-4y = 12 \quad \text{or} \quad y = -3.$$

Thus one point on the line is $(0, -3)$. Putting $y = 0$ in the given equation, we get

$$3x = 12 \quad \text{or} \quad x = 4.$$

Hence $(4, 0)$ is a second point on the line. Plotting these two points $(0, -3)$ and $(4, 0)$ that lie on the two coordinate axes, and joining them by a straight line, we obtain the graph of the given equation as shown in the adjacent figure.

When we are plotting the graph of a linear relation, usually the simplest procedure is to find the two points where the graph crosses the coordinate axes, as we did in Example 9. Occasionally, however, this is inconvenient. For example, one of these points of intersection may be off the graph paper we are using. Or, if the graph happens to pass through the origin, it is impossible to use this technique. In such cases we can use any pair of points on the graph in order to draw it, and we simply choose two convenient values of x at which to calculate y.

Alternatively we can use the idea of slope in order to sketch the graph, as discussed earlier in this section.

Diagnostic Test 5.3

Fill in the blanks.

1. The steepness of a line is measured by its _____.
2. A horizontal line has _____ slope.
3. A vertical line has _____ slope.
4. The slope m of a nonvertical line joining two points (x_1, y_1) and (x_2, y_2) is given by $m = $ _____.
5. If a line rises to the right, then its slope m is _____.
6. If a line falls to the right, then its slope m is _____.
7. If a line with slope m_1 rises more steeply than a line with slope m_2, then m_1 _____ m_2.
8. If the slope m of a line is negative, then the line _____ to the right.
9. If the slope m of a line is positive, then the line _____ to the right.
10. If two points (x_1, y_1) and (x_2, y_2) lie on a horizontal line, then _____.
11. If two points (a, b) and (c, d) lie on a vertical line, then _____.
12. The two points $(3, 5)$ and $(3, -5)$ lie on a _____ line.

Sec. 5.3 *Straight Lines and Linear Equations*

13. The two points $(-3, 5)$ and $(3, 5)$ lie on a _____ line.
14. The equation of the x axis is _____.
15. The equation of a line parallel to the x axis and passing through (p, q) is _____.
16. The equation of a line perpendicular to the x axis and passing through (a, b) is _____.
17. The equation of a line through (x_1, y_1) and with slope m is given by _____.
18. The equation of a line with slope m and y intercept b is given by _____.
19. The equation of a horizontal line through $(2, -3)$ is given by _____.
20. The equation of a vertical line through $(-2, 5)$ is given by _____.
21. The line given by $x + 3 = 0$ is parallel to the _____ axis.
22. The line given by $2y = 3$ is perpendicular to the _____ axis.
23. The line given by $2x + y = 3$ has slope _____ and y intercept _____.
24. The slope of a line given by $x = my + c$ is _____.
25. The graph of the linear equation $ax + by + c = 0$ (a, b not both zero) is _____.

Exercises 5.3

(1–6) Find the slopes of the lines joining the pairs of points given below.

1. $(2, 1)$ and $(5, 7)$.
2. $(5, -2)$ and $(1, -6)$.
3. $(2, -1)$ and $(4, -1)$.
4. $(3, 5)$ and $(-1, 5)$.
5. $(-3, 2)$ and $(-3, 4)$.
6. $(1, 2)$ and $(1, 5)$.

(7–24) Find the equation of the straight lines satisfying the conditions in each of the following exercises.

7. Passing through $(2, 1)$ and with slope 5.
8. Passing through $(1, -2)$ with slope -3.
9. Passing through $(3, 4)$ with zero slope.
10. Passing through $(2, -3)$ with no slope.
11. Passing through $(3, -1)$ and $(4, 5)$.
12. Passing through $(2, 1)$ and $(3, 4)$.
13. Passing through $(3, -2)$ and $(3, 7)$. [*Hint:* A vertical line has equation $x = a$.]
14. Passing through $(2, -1)$ and $(-3, -1)$.
15. With slope -2 and y intercept 5.
16. With slope $1/3$ and y intercept -4.
17. With slope 3 and x intercept 3.
18. With slope -2 and x intercept -5.
19. With x intercept -3 and y intercept 4.
20. With x intercept 0 and passing through $(2, -3)$.
21. Passing through $(2, 1)$ and parallel to the x axis.
22. Passing through $(-1, 3)$ and parallel to the y axis.
23. Bisecting the first and third quadrants.
24. Bisecting the second and fourth quadrants.

(25–30) Find the slope and y intercept for each of the following linear relations.

25. $3x - 2y = 6$.
26. $4x + 5y = 20$.
27. $y - 2x + 3 = 0$.
28. $\dfrac{x}{3} + \dfrac{y}{4} = 1$.
29. $2y - 3 = 0$.
30. $3x + 4 = 0$.

(31–32) Use slopes to show that the following sets of three points lie on a straight line.

31. $(1, -1), (2, 1)$, and $(-1, -5)$.

32. $(-1, 1), (2, 2)$, and $(-7, -1)$.

(33–34) Show that the following sets of three points are collinear by first finding the equation of a line through two given points and then showing that the third point lies on this line.

33. $(1, 1), (-1, 5)$, and $(2, -1)$.

34. $(4, 1), (-2, 3)$, and $(7, 0)$.

35. Find all values of k such that the slope of the line joining the two points $(k, 1)$ and $(3, 2 - 3k)$ is less than 2.

36. Find all values of p such that the slope of the line joining the two points $(-p, 3)$ and $(2, 5 - p)$ is greater than 5.

*37. If a line makes intercepts of a and b on the x axis and y axis, respectively, show that the equation of the line is given by

$$\frac{x}{a} + \frac{y}{b} = 1.$$

(This is called the *intercept form* of the equation of the line.)

*38. Prove that the equation of a line passing through two points (x_1, y_1) and (x_2, y_2) is given by

$$\frac{y - y_1}{y_2 - y_1} = \frac{x - x_1}{x_2 - x_1}.$$

(This is called the *two-point form* for the equation of a line.)

39. Find the equation of a line that passes through $(2, -3)$ and that has a slope of 5. Find the intercepts made by this line on the two coordinate axes.

40. A given line passes through two points $(1, 2)$ and $(2, -1)$. What intercepts does this line make on the two axes?

(41–42) In a triangle, a *median* is a line joining a vertex to the middle point of opposite side. Given a triangle with vertices $A(2, -1), B(4, 3)$, and $C(-6, 5)$:

41. Find the equation of the median through the vertex A.

42. Find the equation of the median through the vertex C.

(43–48) Sketch the graphs of the following linear equations.

43. $3x - 5y = 15$.

44. $x = 2y - 7$.

45. $2x - 7 = 0$.

46. $5 - 2y = 0$.

47. $\frac{x}{3} - \frac{y}{4} = 1$.

48. $\frac{2x}{3} + \frac{3y}{2} = 1$.

5.4

More on Straight Lines

In the last section we described the various forms of equations of a line. In this section we discuss the conditions for two given lines to be parallel or perpendicular, and describe some applications of straight lines.

THEOREM 1 If two nonvertical lines with slopes m_1 and m_2 are parallel, then $m_1 = m_2$.

Proof Let l_1 and l_2 be two given straight lines with slopes m_1 and m_2. Take P and R any two points on l_1 and draw a horizontal line through P and a vertical line through R to meet at Q. Then in the right triangle PQR,

$$\frac{QR}{PQ} = \frac{\text{rise}}{\text{run}} = m_1.$$

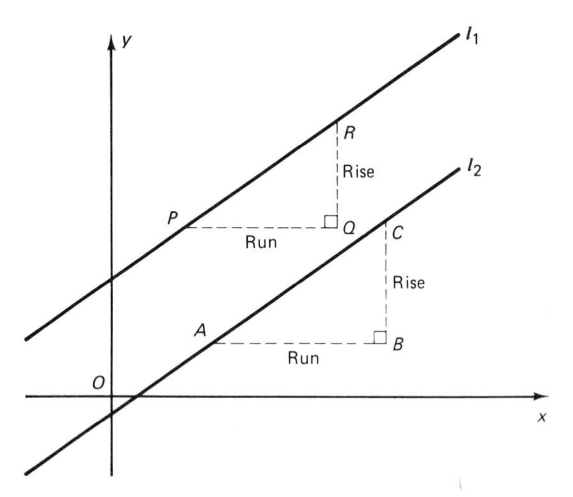

Similarly, if we take any two points A, C on l_2 and form a right triangle ABC, as shown, then

$$\frac{BC}{AB} = \frac{\text{rise}}{\text{run}} = m_2.$$

But clearly the two triangles PQR and ABC are similar, since their corresponding angles are equal. Therefore the ratios of corresponding sides are equal:

$$\frac{QR}{PQ} = \frac{BC}{AB}.$$

That is, $m_1 = m_2$.

Thus, *two parallel nonvertical lines have equal slopes*. If both lines are vertical (and therefore parallel), then both lines have no slope.

Example 1 Find the equation of the straight line passing through (2, 1) and parallel to $3x - 4y + 5 = 0$.

Solution The given line has the equation

$$3x - 4y + 5 = 0. \tag{i}$$

When solved for y, this takes the form

$$4y = 3x + 5$$
$$y = \tfrac{3}{4}x + \tfrac{5}{4}.$$

Thus this line has a slope of $\tfrac{3}{4}$. Since the required line is parallel to this given line, its slope must also equal $\tfrac{3}{4}$. We can therefore use the point-slope formula since we need the equation of the line passing through the point $(x_1, y_1) = (2, 1)$ and having slope $m = \frac{3}{4}$:

$$y - y_1 = m(x - x_1)$$
$$y - 1 = \tfrac{3}{4}(x - 2)$$
$$4y - 4 = 3x - 6$$
$$3x - 4y + 2 = 0. \tag{ii}$$

Observe that Equation (ii) in Example 1 differs from Equation (i) only by a constant. The x and y terms in the two equations are exactly the same. This is true for any two parallel straight lines. In other words, the equation of any line parallel to $ax + by + c = 0$ has an equation of the form $ax + by + k = 0$ for some constant k. This new constant k must be determined from some additional condition.

Example 2 Find the equation of the line parallel to $2x + 3y - 7 = 0$ and passing through $(-2, 1)$.

Solution The equation of any line parallel to $2x + 3y - 7 = 0$ is

$$2x + 3y + k = 0. \tag{i}$$

Since this line passes through $(-2, 1)$, we have

$$2(-2) + 3(1) + k = 0.$$

This gives $k = 1$. Substituting this value of k in (i), we get

$$2x + 3y + 1 = 0$$

which is the required equation.

PERPENDICULAR LINES

THEOREM 2 If two nonvertical lines with slopes m_1 and m_2 are perpendicular, then $m_1 m_2 = -1$.

Proof Let l_1 and l_2 be the two given lines with slopes m_1 and m_2. Then let l_1' and l_2' be two parallel lines through the origin as in the figure on page 184. By Theorem 1, the slopes of l_1' and l_2' are also m_1 and m_2, respectively; furthermore, l_1' and l_2' are perpendicular. Let $P(x, y_1)$ be a point on l_1' and $Q(x, y_2)$ be a point on l_2' with the same abscissa x. Then the slopes are given by

$$m_1 = \frac{y_1}{x}, \quad m_2 = \frac{y_2}{x}.$$

The triangle OPQ has a right angle at O, so by Pythagoras's theorem,

$$PQ^2 = OP^2 + OQ^2. \tag{i}$$

But from the distance formula,

$$OP^2 = x^2 + y_1^2, \quad OQ^2 = x^2 + y_2^2, \quad \text{and} \quad PQ^2 = (y_1 - y_2)^2.$$

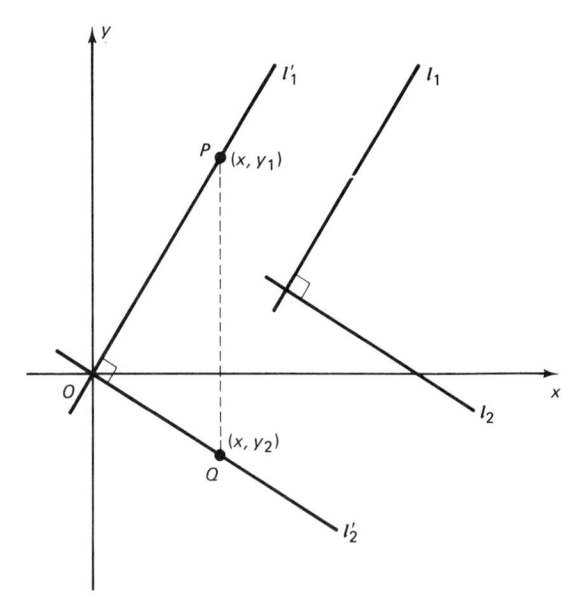

Substituting these values of (i), we obtain
$$(y_1 - y_2)^2 = (x^2 + y_1^2) + (x^2 + y_2^2)$$
—that is,
$$y_1^2 - 2y_1y_2 + y_2^2 = 2x^2 + y_1^2 + y_2^2$$
$$-2y_1y_2 = 2x^2$$
$$\frac{y_1y_2}{x^2} = -1.$$

Therefore
$$m_1m_2 = \left(\frac{y_1}{x}\frac{y_2}{x}\right) = \frac{y_1y_2}{x^2} = -1.$$

Thus we have shown that *if two nonvertical lines are perpendicular then the product of their slopes is* -1.

Since we can write the relation $m_1m_2 = -1$ as $m_2 = -1/m_1$, it follows that *the slope of a line perpendicular to a given line is the negative reciprocal of the slope of the given line* (provided both slopes are defined).

Example 3 Find the equation of the line passing through $(3, 4)$ and perpendicular to $2x - 3y + 4 = 0$.

Solution The given line has the equation
$$2x - 3y + 4 = 0.$$
Solving this for y, we have
$$3y = 2x + 4$$
$$y = \tfrac{2}{3}x + \tfrac{4}{3}.$$

Thus the slope of this line is $\frac{2}{3}$. Since the required line is perpendicular to this given line, its slope is given by the negative reciprocal of $\frac{2}{3}$,

$$m = \frac{-1}{\frac{2}{3}} = -\frac{3}{2}.$$

Thus the required line has a slope of $m = -\frac{3}{2}$ and passes through the point (3, 4), and so from the point-slope formula its equation is

$$y - y_1 = m(x - x_1)$$
$$y - 4 = -\tfrac{3}{2}(x - 3)$$
$$2y - 8 = -3(x - 3)$$
$$3x + 2y - 17 = 0.$$

Example 4 Find the equation of the line passing through the point (2, −1) and perpendicular to $3x - 7 = 0$.

Solution The given line has equation

$$x = \tfrac{7}{3}$$

and so is a vertical line. Its slope is undefined. A line perpendicular to a vertical line is a horizontal line and therefore has a slope of zero. Thus, the required line passes through (2, −1) and has a slope $m = 0$, and so its equation is given by

$$y = -1.$$

Example 5 Use slopes to prove that the points (1, 2), (3, 4), (2, 5), and (0, 3) taken in order form the vertices of a rectangle.

Proof Let the given four points be denoted by A, B, C, D. To prove that $ABCD$ is a rectangle, it will be sufficient to show that $AB \parallel CD$, $AD \parallel BC$, and also $AD \perp AB$.

If m_{AB} denotes the slope of line joining A and B, then

$$m_{AB} = \frac{4 - 2}{3 - 1} = \frac{2}{2} = 1.$$

Similarly,

$$m_{CD} = \frac{5 - 3}{2 - 0} = \frac{2}{2} = 1$$

$$m_{AD} = \frac{3 - 2}{0 - 1} = -1$$

$$m_{BC} = \frac{5 - 4}{2 - 3} = -1.$$

Since $m_{AB} = m_{CD}$ and $m_{AD} = m_{BC}$, we have $AB \parallel CD$ and $AD \parallel BC$. Since $m_{AD} \cdot m_{AB} = (-1)(1) = -1$, it follows that $AD \perp AB$. Thus A, B, C, D are the vertices of a rectangle.

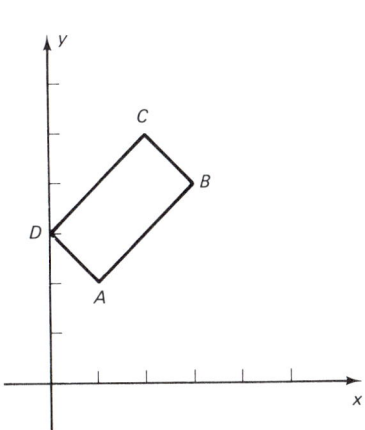

SOME APPLICATIONS

Example 6 Suppose that only dried lentils and dried soybeans are available to satisfy a person's daily requirement for protein, which is 75 grams. One gram of lentils contains 0.26 gram of protein and one gram of soybeans contains 0.35 gram of protein. Let his daily consumption be x grams of lentils and y grams of soybeans. What is the relationship between x and y that exactly satisfies his protein requirement?

Solution The amount of protein in grams obtained from x grams of lentils is $0.26x$ and from y grams of soybeans is $0.35y$. The total protein is therefore

$$(0.26x + 0.35y) \text{ grams.}$$

This must be equal to his daily need, which is 75 grams. Thus

$$0.26x + 0.35y = 75.$$

Multiplying both sides by 100, we have

$$26x + 35y = 7500$$

$$y = -\tfrac{26}{35}x + \tfrac{7500}{35}.$$

We observe that x and y satisfy a linear relation.

Example 7 The cost of manufacturing 10 units of a certain commodity per day is \$350, while it costs \$600 to produce 20 units. Determine the relation representing the total cost y of producing x units per day, assuming the relation to be linear. What will be the total cost if 30 units are produced per day?

Solution Let y represent the total cost of producing x units per day. We are given that $y = 350$ when $x = 10$; and $y = 600$ when $x = 20$. Thus we are given two points, $(10, 350)$ and $(20, 600)$, that lie on the graph of the linear relation between x and y. The slope of the line joining these two points is

$$m = \frac{600 - 350}{20 - 10} = \frac{250}{10} = 25.$$

Using the point-slope formula, we obtain the equation of the straight line with slope $m = 25$ and passing through the point $(10, 350)$ as

$$y - y_1 = m(x - x_1)$$
$$y - 350 = 25(x - 10)$$
$$y = 25x + 100. \tag{i}$$

This is the required linear relation between x and y.

The cost of producing 30 units per day is obtained by substituting $x = 30$ in (i):

$$y = 25(30) + 100 = 850.$$

Therefore the cost of producing 30 units per day would be \$850.

Equation (i) in the foregoing example is typical of what is called the *linear cost model*. Note that when $x = 0$, we get $y = 100$, so in this example it costs $100 per day even if no items are produced at all. This cost is called the *fixed cost* or *overhead cost*. In addition to this fixed cost, it costs a further $25x$ dollars to produce x items per day—that is, $25 per item. This part of the total production cost is called the *variable cost*.

In a general linear cost equation $y = mx + b$, the y intercept b is equal to the fixed cost while the slope m is equal to the variable cost per item produced.

Example 8 The quantity x of any commodity that will be purchased by consumers depends on the price p per unit at which that commodity is made available. The relationship between p and x in this case is known as a *demand relation*.

A dealer can sell 20 electric shavers per day at $25 each, whereas he can sell 30 shavers if he charges $20 each. Determine the demand relation, assuming it to be linear.

Solution Taking the quantity x demanded as abscissa and the price p per unit as ordinate, the two points on the graph of the demand relation are

$$x = 20, p = 25, \text{ and } x = 30, p = 20$$

—that is, (20, 25) and (30, 20). Since the demand equation is linear, it is given by the equation of the straight line passing through the two given points (20, 25) and (30, 20). The slope of the line joining these two points is

$$m = \frac{20 - 25}{30 - 20} = -\frac{5}{10} = -0.5.$$

From the point-slope formula, the equation of the line through (20, 25) with slope $m = -0.5$ is (note that $y = p$)

$$y - y_1 = m(x - x_1)$$
$$p - 25 = -0.5(x - 20)$$
$$p = -0.5x + 35$$

which is the required demand equation.

Note that the graph has been drawn only in the first quadrant, because neither p nor x can be negative.

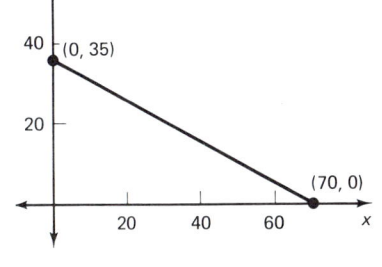

Diagnostic Test 5.4

Fill in the blanks.

1. If two nonvertical lines with slopes m_1 and m_2 are parallel, then _____.

2. If two nonvertical lines with slopes m_1 and m_2 are perpendicular, then $m_2 = $ _____.

3. The two lines given by $2x + 3y + 7 = 0$ and $2x + 3y - 5 = 0$ are _____ (parallel or perpendicular).

4. The two lines given by $x = 3$ and $2x - 5 = 0$ are _____.

5. The two lines given by $y = 7$ and $2x + 3 = 0$ are _____.

6. The two lines given by $3y - 7 = 0$ and $8 - 3y = 0$ are _____.

Exercises 5.4

(1–14) Are the lines determined by the following pairs of linear equations parallel, perpendicular, or neither?

1. $2x - 3y = 7$; $6y - 4x + 5 = 0$.
2. $x + 2y - 3 = 0$; $y = 3 - \dfrac{x}{2}$.
3. $5x - y + 4 = 0$; $x + 5y - 3 = 0$.
4. $2x + 7y - 1 = 0$; $7x = 5 + 2y$.
5. $\dfrac{x}{2} + \dfrac{y}{3} = 1$; $\dfrac{x}{3} - \dfrac{y}{2} = 1$.
6. $y = 2x + 3$; $x = 5 - 2y$.
7. $2x + 3 = 0$; $x - 5 = 0$.
8. $3y + 7 = 0$; $5 - 2y = 0$.
9. $2x + 7 = 0$; $3y + 4 = 0$.
10. $3x - 4 = 0$; $2y + 3 = 0$.
11. $2x + 3y - 7 = 0$; $3x + 2y - 5 = 0$.
12. $x + y - 3 = 0$; $y = 2x + 7$.
13. $5x - y + 4 = 0$; $2x - 3 = 0$.
14. $y - 7 = 0$; $2x + 3y - 4 = 0$.

(15–24) Find the equations of the straight lines satisfying the conditions in each of the following exercises.

15. Passing through (4, 1) and parallel to $3x - 4y + 4 = 0$.
16. Passing through (1, −3) and parallel to $x = y + 3$.
17. Passing through (4, −1) and parallel to $2x - 3 = 0$.
18. Passing through (2, 5) and parallel to $4 - 3y = 0$.
19. Passing through (1, −1) and perpendicular to $2x + 3y - 4 = 0$.
20. Passing through (2, 3) and perpendicular to $3x - y + 7 = 0$.
21. Passing through (3, −2) and perpendicular to $5 - 2x = 0$.
22. Passing through (1, 4) and perpendicular to $3y + 7 = 0$.
23. Passing through (2, 4) and parallel to the line joining the points (1, 2) and (2, −1).
24. Passing through (3, −2) and perpendicular to the line joning the points (2, 5) and (−1, 3).

(25–32) Use slopes to prove the following exercises.

25. The four points (1, 2), (3, −4), (4, −2), and (2, 4) taken in order form the vertices of a parallelogram.
26. The four points (1, 1), (−1, −3), (−2, −1), and (0, 3) taken in order form the vertices of a parallelogram.
27. The four points (−1, 2), (2, 5), (1, 6), and (−2, 3) taken in order form the vertices of a rectangle.
28. The three points (2, −1), (3, 2), and (4, 1) form the vertices of a right triangle.
29. The line joining the midpoints of two sides of a triangle is parallel to the third side.
30. The diagonals of any square are perpendicular to one another.
31. The lines joining the midpoints of the sides of any quadrilateral form a parallelogram.
32. The lines joining the midpoints of the sides of any square form a rectangle. (In this case the rectangle is actually a square.)

(33–36) In a triangle, a line through a vertex and perpendicular to the opposite side is called an *altitude* of the triangle. Given a triangle with vertices $P(2, 1)$, $Q(-3, 4)$, and $R(1, 2)$:

33. Find the equation of the altitude through the vertex P.

34. Find the equation of the altitude through the vertex Q.

35. Find the equation of the right bisector of the side QR.

36. Find the equation of the right bisector of the side PQ.

37. Show that the two lines given by equations $a_1x + b_1y + c_1 = 0$ and $a_2x + b_2y + c_2 = 0$ are
    (a) Parallel if $a_1b_2 - b_1a_2 = 0$.
    (b) Perpendicular if $a_1a_2 + b_1b_2 = 0$.

38. Find the value of m if the two lines given by the equations $y = 2x - 3$ and $x = my + c$ are
    (a) Parallel.
    (b) Perpendicular.

39. A company manufactures two types of a certain product. Each unit of the first type requires 2 machine-hours and each unit of the second type requires 5 machine-hours. There are 280 machine-hours available each week.
    (a) If x units of first type and y units of second type are made each week, find the relation between x and y if all the machine-hours are used.
    (b) How many units of the first type can be manufactured if 40 units of the second type are manufactured in a particular week?

40. The Boss-Toss Company manufactures two products X and Y. Each unit of X requires 3 man-hours and each unit of Y requires 4 man-hours. There are 120 man-hours available each day.
    (a) If x units of X and y units of Y are manufactured each day and all the available man-hours are used up, find the relation between x and y.
    (b) How many units of X can be made in one day if 15 units of Y are made the same day?
    (c) How many units of Y can be made in one day if 16 units of X are made the same day?

41. A patient in the hospital who is on a liquid diet has the choice of prune juice and orange juice to satisfy his daily requirement of thiamine, which is 1 milligram. One ounce of prune juice contains 0.05 milligram of thiamine, and 1 ounce of orange juice contains 0.08 milligram of thiamine. Let his daily consumption be x ounces of prune juice and y ounces of orange juice. What relationship between x and y exactly satisfies his thiamine requirement?

42. An individual on a strict diet plans to breakfast on cornflakes, milk, and a boiled egg. After allowing for the egg, his diet allows a further 300 calories for this meal. One ounce of milk contains 20 calories and 1 ounce (about one cupful) of cornflakes (plus sugar) contains 160 calories. What is the relation between the number of ounces of milk and of cornflakes that can be consumed?

43. A detergent manufacturer finds that his weekly sales are 10,000 packets when the price is $1.20 per packet but that the sales increase to 12,000 when he reduces the price to $1.10 per packet. Determine the demand relation, assuming that it is linear.

44. A television manufacturer finds that at $500 per set he can sell 2000 sets per month, whereas at $450 per set he can sell 2400. Determine the demand equation, assuming it to be linear.

45. The amount x of a particular commodity that its suppliers are willing to make available depends on the price p per unit at which they can sell it. In general, suppliers will make a greater quantity available if they can get a higher price. The relation between x and p in this case is known as a *supply relation.*

    At a price of $2.50 per unit a firm will supply 8000 shirts a month, and at $4 per unit it will supply 14,000. Determine the supply relation, assuming it to be linear.

46. At a price of $10 per unit a firm will supply 1200 units of its product, and at $15 per unit, 4200 units. Determine the supply relation, assuming it to be linear.

5.5

Systems of Linear Equations in Two Unknowns

Many problems lead to what are called systems of linear equations. Consider, for example, the following situation:

The owner of a television store wants to expand his business by buying and displaying two new types of TV sets. Each set of the first type costs him $300 and of the second type $400. Each set of the first type occupies 4 square feet of floor space and each set of the second type 5 square feet. If he has only $2000 available for this expansion and 26 square feet of floor space, how many sets of each type should he buy and display so as to make full use of the available capital and space?

Suppose he buys x TV sets of the first type and y sets of the second type. Then it costs him $300x$ to buy sets of the first type and $400y$ to buy those of the second type. Since the total amount to be spent is $2000, it is necessary that

$$300x + 400y = 2000. \tag{1}$$

Also the amounts of space occupied by the two types of sets are $4x$ square feet and $5y$ square feet, respectively. The total space available for these two types of sets is 26 square feet. Therefore,

$$4x + 5y = 26. \tag{2}$$

To find the number of sets of each type that can be bought and displayed, we must solve the above two equations for x and y. That is, we must find the values of x and y that satisfy both the equations (1) and (2) simultaneously. Observe that each of these equations is a linear one in x and y.

DEFINITION A *system of two linear equations* in two variables, x and y, consists of equations of the type

$$a_1 x + b_1 y = c_1 \tag{3}$$
$$a_2 x + b_2 y = c_2 \tag{4}$$

where $a_1, b_1, c_1, a_2, b_2, c_2$ are six given constants. The **solution** of the system (3), (4) is the set of ordered pairs (x, y) that satisfy both equations.

Equations (1) and (2) form such a system of linear equations. If we identify (1) with (3) and (2) with (4), the six coefficients have the values $a_1 = 300, b_1 = 400, c_1 = 2000$ and $a_2 = 4, b_2 = 5, c_2 = 26$.

Our main concern in this section is to solve systems of linear equations algebraically. The solution by the use of algebraic methods involves the elimination of one of the variables x or y from the two equations, thus

190   Ch. 5   Straight Lines

allowing determination of the value of the other variable. Elimination of one of the variables can be achieved by either of two methods, which we shall explain separately. First we illustrate the *method of addition*. We make the coefficients of one of the variables in the two equations exactly the same in magnitude and opposite in sign by multiplying the equations by appropriate numbers. Adding the two resulting equations then eliminates that variable.

Example 1 Solve the following system of equations by the method of addition:

$$3x + 4y = 5 \qquad (i)$$
$$2x - 3y = -8 \qquad (ii)$$

Solution Suppose we eliminate y from the two equations. To do so, we must make the coefficient of y in the two equations exactly the same in magnitude and opposite in sign. We observe that if we multiply the first equation by 3 and the second by 4, the y terms in the two equations will become $12y$ and $-12y$, respectively:

$$9x + 12y = 15 \qquad \text{(three times first equation)}$$
$$8x - 12y = -32 \qquad \text{(four times second equation)}$$

The coefficients of y in the two equations are now equal in magnitude but opposite in sign. If we add these two equations, the y terms cancel. Adding vertically, we obtain

$$9x + 12y = 15$$
$$\underline{8x - 12y = -32}$$
$$17x = -17$$

and so $x = -1$. Substituting $x = -1$ in one of the given equations, say (i), we get

$$3(-1) + 4y = 5$$
$$4y = 5 + 3 = 8$$

which gives $y = 2$. Thus the solution of the given system is $x = -1$, $y = 2$.

Example 2 Solve the following system of equations:

$$\frac{x-y}{3} = \frac{y-1}{4} \qquad (i)$$

$$\frac{4x - 5y}{7} = x - 7. \qquad (ii)$$

Solution The first step is to get rid of fractions in the given equations. Multiplying both sides of the first equation by 12, the common denominator, we have

$$4(x - y) = 3(y - 1)$$
$$4x - 4y = 3y - 3$$

or

$$4x - 7y = -3.$$

Sec. 5.5 Systems of Linear Equations in Two Unknowns

Multiplying both sides of the second equation by 7, we obtain
$$4x - 5y = 7(x - 7) = 7x - 49$$
and so
$$-3x - 5y = -49.$$
Multiplying throughout by -1, we get
$$3x + 5y = 49.$$
Thus the given system of Equations (i) and (ii) is equivalent to
$$4x - 7y = -3 \qquad \text{(iii)}$$
$$3x + 5y = 49. \qquad \text{(iv)}$$
Let us eliminate x from these two equations. We multiply (iii) by -3 and (iv) by 4, obtaining
$$-12x + 21y = 9$$
$$\underline{12x + 20y = 196}$$
$$41y = 205$$
In the third line the two equations have been added vertically. Thus
$$y = \tfrac{205}{41} = 5.$$
Substituting $y = 5$ into (iv), we get
$$3x + 5(5) = 49$$
$$3x = 49 - 25 = 24$$
$$x = 8.$$
Therefore the solution of the given system is $x = 8$, $y = 5$.

A system of linear equations and its solution has an important geometrical interpretation. Consider, for example, the system
$$x + y = 3 \qquad \text{(5)}$$
$$3x - y = 1. \qquad \text{(6)}$$
By simply adding these equations, we can easily see that the solution is $x = 1$, $y = 2$.

Now each of Equations (5) and (6) is a linear equation in x and y and so has as its graph a straight line in the xy plane. In the case of (5) we find the point where the line meets the x axis by setting $y = 0$, which gives $x = 3$, so the line passes through the point $(3, 0)$. Similarly, setting $x = 0$, we find $y = 3$, so that the line meets the y axis at the point $(0, 3)$. These two points are shown on the adjoining figure, and the graph of Equation (5) is drawn as the straight line joining them.

Proceeding in a similar way for Equation (6), we find the points $(0, -1)$ and $(\tfrac{1}{3}, 0)$ on the coordinate axes.

Any pair of values of x and y that satisfies Equation (5) corresponds to a point (x, y) on the first straight line. Any pair of values that satisfies Equation (6) corresponds to a point (x, y) on the second straight line. Thus if x

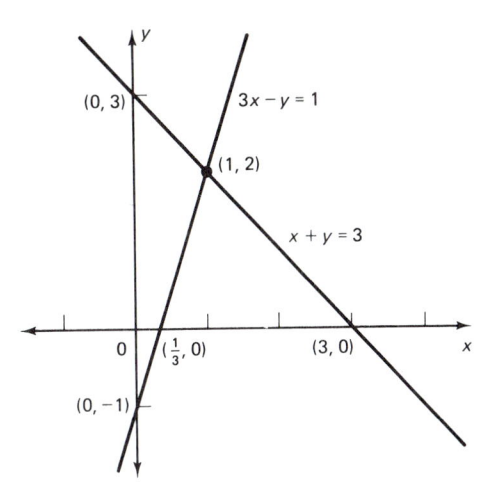

and y satisfy *both* equations, then the point (x, y) must lie on *both* lines. In other words, (x, y) must be the point at which the lines intersect. From the figure we see that this point is $(1, 2)$, so the solution in this case is $x = 1$, $y = 2$, as stated earlier.

Now let us turn to the general system of linear equations:

$$a_1 x + b_1 y = c_1 \tag{3}$$

$$a_2 x + b_2 y = c_2. \tag{4}$$

The graphs of these two equations consist of two straight lines in the xy plane, since any linear equation always represents a straight line. Any pair of values of x and y that satisfies both (3) and (4) must correspond to a point (x, y) that lies on both of the lines—that is, to their point of intersection.

Let us denote the two straight lines by L and M, respectively. Three possible kinds of behavior can be found.

1. The lines L and M are distinct and intersect each other. Since the point of intersection lies on both the lines, its coordinates (x_0, y_0) satisfy the equations of both lines and hence they provide a solution of the given system. This solution is unique, because if two distinct straight lines intersect, they intersect each other only at one point. (See the left-hand figure below.)

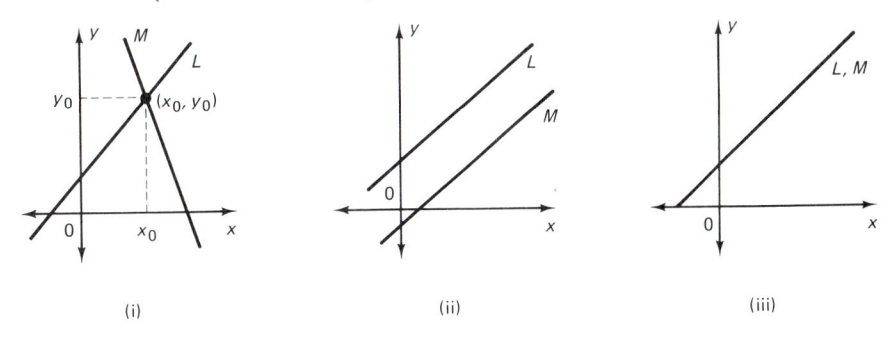

Sec. 5.5 *Systems of Linear Equations in Two Unknowns*

2. The lines *L* and *M* are parallel. They do not meet each other and there is no point that lies on both lines. Thus, there are no values of *x* and *y* that will satisfy both equations. In other words, the equations have *no solution*. (See the middle figure above.)
3. The lines *L* and *M* coincide. Every point on the line *L* is also on the line *M*. In this case the system has an *infinite number of solutions*—namely, every (*x*, *y*) that lies on *L* (= *M*) (See the right-hand figure above).

Example 3 Solve the equations

$$x + 2y = 4 \qquad \text{(i)}$$
$$3x + 6y - 8 = 0. \qquad \text{(ii)}$$

Solution If we multiply (i) by -3, the coefficient of *x* will be the negative of that in (ii). The two equations are then

$$-3x - 6y = -12$$
$$3x + 6y - 8 = 0.$$

Adding, we find that both the *x* and *y* terms cancel, and we are left with the equation

$$-8 = -12.$$

This is impossible. Thus the given equations have *no solution*. This is illustrated graphically in the following figure. The two straight lines are parallel and do not intersect. We can see this easily by writing the given equations in the slope-intercept form:

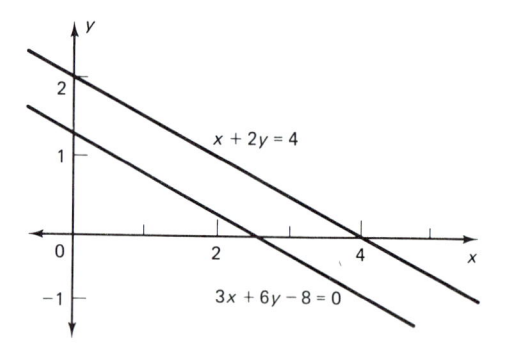

$$y = -\tfrac{1}{2}x + 2$$

and

$$y = -\tfrac{1}{2}x + \tfrac{4}{3}.$$

The two lines have the *same slope* (namely $-\tfrac{1}{2}$) but different *y* intercepts. Thus the two lines are parallel with no common point.

194   Ch. 5   Straight Lines

Example 4 Solve the equations
$$2x - 3y = 6 \qquad \text{(i)}$$
$$\frac{x}{3} - \frac{y}{2} = 1. \qquad \text{(ii)}$$

Solution Multiplying both sides of (ii) by -6, we obtain
$$-2x + 3y = -6. \qquad \text{(iii)}$$

Adding both sides of (i) and (iii), we get
$$0 = 0$$

an equation that is always true. Writing the two equations in the slope-intercept form, we find that they both reduce to
$$y = \tfrac{2}{3}x - 2.$$

Since the two equations are identical, the two lines coincide in this case and the given equations are equivalent to one another. We have an infinite number of solutions in this case: any pair of values (x, y) that satisfies the one independent equation provides a solution. One such pair is $(6, 2)$, for example; another is $(0, -2)$.

For any system of two linear equations in two variables we have one of the following possibilities.

(A) *The system has no solution*. In such a case the system is said to be ***inconsistent***.
(B) *The system has one or more solutions*. In such a case the system is said to be ***consistent***. There are two subcases:
   (B1) *The system has exactly one solution*. We say that the system is ***independent***.
   (B2) *The system has an infinite number of solutions*. We say that the system is ***dependent***.

Examples 1 and 2 above involved systems that were consistent and independent; Example 3 was an inconsistent system; Example 4 was a consistent but dependent system.

A second method of solving systems of linear equations is the ***method of substitution***. We shall explain it by means of an example.

Example 5 Solve the two equations that resulted from the problem posed in the beginning of this section—that is,
$$300x + 400y = 2000 \qquad \text{(i)}$$
$$4x + 5y = 26. \qquad \text{(ii)}$$

Sec. 5.5 *Systems of Linear Equations in Two Unknowns*

Solution Solving Equation (ii) for x, we have

$$4x = 26 - 5y$$

$$x = \frac{26 - 5y}{4}. \qquad \text{(iii)}$$

Substituting this value of x in Equation (i), we obtain

$$300\left(\frac{26 - 5y}{4}\right) + 400y = 2000$$

$$75(26 - 5y) + 400y = 2000$$

$$1950 - 375y + 400y = 2000.$$

Therefore,

$$25y = 2000 - 1950 = 50$$

$$y = 2.$$

Substituting $y = 2$ in (iii), we have

$$x = \tfrac{1}{4}(26 - 10) = 4.$$

Thus the solution of the system of Equations (i) and (ii) above is $x = 4$, $y = 2$. In other words, the dealer should buy and display four sets of the first type and two of the second type to make use of all the available space and capital.

We can summarize the method of substitution in the following steps. (Compare with the solution of Example 5 above.) Suppose the system consists of two equations in two unknowns, x and y.

Step 1. Solve one of the given equations for y in terms of x or for x in terms of y, whichever is simpler.

Step 2. Substitute the expression for y (or x) obtained in step 1 into the other equation, thus obtaining an equation in the one variable x (or y) alone.

Step 3. Solve the equation obtained in step 2 for x (or y).

Step 4. Substitute the values of x (or y) obtained in step 3 into the equation that was obtained at the end of step 1. This allows us to find the value of the second variable.

Example 6 Solve the following system by using the substitution method:

$$3x + y = 5 \qquad \text{(i)}$$

$$2x - 3y = 7. \qquad \text{(ii)}$$

Solution   *Step 1.* Here it is easy to solve Equation (i) for y in terms of x:

$$y = 5 - 3x. \qquad \text{(iii)}$$

Step 2. Substituting this value of y into Equation (ii), we obtain
$$2x - 3(5 - 3x) = 7$$
$$2x - 15 + 9x = 7$$
$$11x = 22.$$

Step 3. Solving this equation for x, we get
$$x = 2.$$

Step 4. Substituting this value of $x = 2$ into (iii), we obtain
$$y = 5 - 3(2) = -1.$$

Thus the solution of the system is $x = 2, y = -1$.

Example 7 The "Britannia Stores," which specializes in selling all kinds of nuts, sells peanuts at $0.70 and cashews at $1.60 per pound. At the end of the month the store owner finds that peanuts are not selling well. He decides to mix peanuts and cashews to make a mixture of 45 pounds that could sell for $1.00 per pound. How many pounds of peanuts and cashews should he mix so as to keep the same revenue?

Solution Let the mixture contain x pounds of peanuts and y pounds of cashews. Then, since the total mixture is 45 pounds,
$$x + y = 45.$$

The revenue from x pounds of peanuts at $0.70 per pound is $0.7x$, and the revenue from y pounds of cashews at $1.60 per pound is $1.6y$. The revenue obtained from the mixture of 45 pounds at $1.00 per pound will be $45. Since the revenue from the mixture must be the same as that from the separate nuts,

revenue from peanuts + revenue from cashews = revenue from mixture
$$0.7x + 1.6y = 45$$
$$7x + 16y = 450.$$

Thus we arrive at a system of linear equations:

$$x + y = 45 \qquad \text{(i)}$$
$$7x + 16y = 450. \qquad \text{(ii)}$$

Multiplying (i) by -7 and adding the result to (ii), we get

$$-7x - 7y = -315$$
$$\underline{7x + 16y = 450}$$
$$9y = 135$$

Therefore $y = \frac{135}{9} = 15$. Substituting $y = 15$ into (i), we then see that $x = 30$. Thus 30 pounds of peanuts should be mixed with 15 pounds of cashews.

MARKET EQUILIBRIUM

If the price of a certain commodity is too high consumers will not purchase it, whereas if the price is too low suppliers will not sell it. In a competitive market when the price per unit depends only on the quantity demanded and the supply available, there is always a tendency for the price to adjust itself so that the quantity demanded by purchasers matches the quantity that suppliers are willing to supply at the given price. *Market equilibrium* is said to occur at a price when the quantity demanded is equal to quantity supplied. This corresponds to the point of intersection of the demand and supply curves.

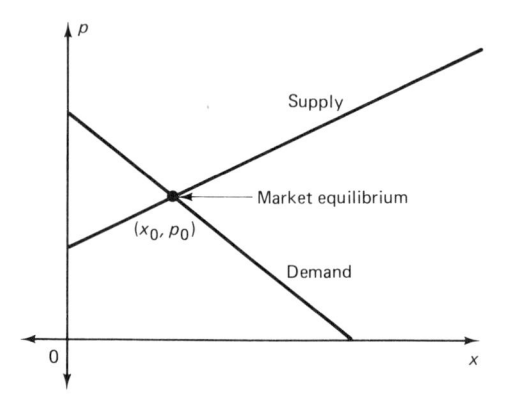

Algebraically, the market equilibrium price p_0 and quantity x_0 are determined by solving the demand and supply equations simultaneously for p and x. Note that equilibrium price and quantity are meaningful only if they are not negative.

Example 8 Determine the equilibrium price and quantity for the following demand and supply laws:

$$D: \quad p = 25 - 2x \tag{i}$$
$$S: \quad p = 3x + 5. \tag{ii}$$

Solution Substituting the value of p from Equation (ii) into Equation (i), we have

$$3x + 5 = 25 - 2x.$$

The solution is readily found to be $x = 4$. Substituting $x = 4$ in (i), we get

$$p = 25 - 8 = 17.$$

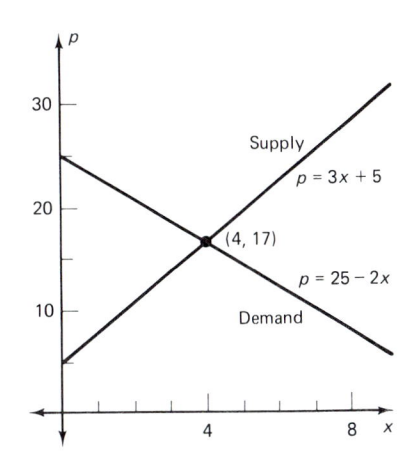

Thus the equilibrium price is 17 and the quantity is 4 units. The supply and demand curves are graphed in the above figure.

Systems of equations quite often arise in which one or more of the equations are not linear. The method of substitution can sometimes be used to solve such systems. In the case of a system of two equations in two variables when *one* of the equations is linear, we can always use this method to eliminate one of the variables.

Example 9 Solve the following equations for x, y:

$$2x + y = 4 \tag{i}$$
$$x^2 + y^2 = 5. \tag{ii}$$

Solution We observe that the first equation is linear and can readily be solved for either x or y. Solving this equation for y, we have

$$y = 4 - 2x. \tag{iii}$$

Substituting this value of y in (ii), we obtain

$$x^2 + (4 - 2x)^2 = 5$$
$$x^2 + 16 - 16x + 4x^2 = 5$$
$$5x^2 - 16x + 11 = 0.$$

This quadratic equation can be solved by factoring:

$$(5x - 11)(x - 1) = 0.$$

Therefore:

$$\text{either} \quad 5x - 11 = 0, \quad \text{which gives} \quad x = \tfrac{11}{5}$$
$$\text{or} \quad x - 1 = 0, \quad \text{which gives} \quad x = 1.$$

When $x = 1$, Equation (iii) gives

$$y = 4 - 2(1) = 2.$$

Thus $x = 1$, $y = 2$ is one solution. Again, substituting $x = \frac{11}{5}$ in (iii), we get

$$y = 4 - 2(\tfrac{11}{5}) = -\tfrac{2}{5}.$$

Thus $x = \frac{11}{5}$, $y = -\frac{2}{5}$ is another solution. In this case we have two solutions of the system:

$$x = 1, y = 2 \quad \text{and} \quad x = \tfrac{11}{5}, y = -\tfrac{2}{5}.$$

We also encounter systems in which there are three (or more) equations involving three (or more) unknown quantities. The two elimination methods can also be used for these larger systems, and a few problems of this type have been included at the end of the following exercises. However there is a more systematic approach called the **method of row reduction**, which is preferable for such systems. It will be described in Section 11.1, but you can proceed there immediately if you wish.

Exercises 5.5

(1–10) Use the method of addition to solve the following systems of linear equations.

1. $x + y = 3$; $2x - 3y = 1$.
2. $x - y = 1$; $2x + 3y + 8 = 0$.
3. $4x - y = -2$; $3x + 4y = 27$.
4. $3u + 2v = 9$; $u + 3v = 10$.
5. $2p - q = 3$; $p = 5 - 3q$.
6. $7x - 8y = 4$; $\tfrac{1}{2}x + \tfrac{1}{3}y = 3$.
7. $2x - 3y = 12$; $6x - 9y = 7$.
8. $2u + 5v = 7$; $v = 4 - \tfrac{2}{3}u$.
9. $2x + 3y = 6$; $\tfrac{1}{3}x + \tfrac{1}{2}y = 1$.
10. $3u - 4v = 12$; $\tfrac{1}{4}u = 1 + \tfrac{1}{3}v$.

(11–20) Use the method of substitution to solve the following systems.

11. $2x - 5y = 7$; $3x + y = 2$.
12. $x + 2y = 4$; $2x + 3y = 7$.
13. $3x - 4y = 1$; $2x + 5y + 7 = 0$.
14. $5u + 2v = 1$; $3u - v = 5$.
15. $\dfrac{x}{3} - \dfrac{y}{2} = -1$; $\dfrac{x}{2} + \dfrac{y}{3} = 5$.
16. $\dfrac{x}{3} + \dfrac{y}{2} = 7$; $\dfrac{x}{4} - \dfrac{y}{3} = 1$.
17. $2u + 3v = 12$; $\dfrac{u}{3} + \dfrac{v}{2} = 3$.
18. $\dfrac{u}{3} + \dfrac{v}{2} = 7$; $4u + 6v = 5$.
19. $3x - 5y = 15$; $\dfrac{x}{5} - \dfrac{y}{3} = 1$.
20. $\dfrac{x}{4} + \dfrac{y}{3} = 1$; $6x + 8y = 24$.

(21–30) Solve the following systems of linear equations by any appropriate method.

21. $\dfrac{x}{3} + \dfrac{y}{2} = \dfrac{x}{2} + \dfrac{y}{3} = 5$.
22. $\dfrac{x}{2} - \dfrac{y}{3} + 6 = \dfrac{x}{3} + \dfrac{y}{4} = 5$.
23. $\dfrac{x+y}{2} - \dfrac{x-y}{3} = 8$; $\dfrac{x+y}{3} + \dfrac{x-y}{4} = 11$.
24. $\dfrac{x - 2y}{3} = 2 + \dfrac{2x + 3y}{4}$; $\dfrac{3x - 2y}{2} = \dfrac{-y + 5x + 11}{4}$.
25. $\dfrac{x+y}{2} - \dfrac{x-y}{3} = 2$; $\dfrac{2x+y}{3} + \dfrac{y-x}{2} = 1$.
26. $\dfrac{x + 2y}{4} - \dfrac{2x - y}{3} = 1$; $2y - x = 3$.

27. $\frac{4}{x} + \frac{3}{y} = 3; \frac{1}{x} - \frac{1}{y} = \frac{1}{6}.$

[*Hint:* Set $\frac{1}{x} = u, \frac{1}{y} = v.$]

28. $\frac{3}{x} - \frac{2}{y} = 5; \frac{1}{3x} + \frac{1}{2y} = 2.$

29. $\frac{x+y}{3} + \frac{x-y}{2} = 1; \frac{2x-y}{3} - \frac{x+3y-2}{4} = 1 - y.$

30. $\frac{2x+y}{3} - \frac{x-2y}{2} = 1; \frac{3x+4y}{4} - \frac{2x+y}{3} = \frac{1}{2}.$

(31–34) Find the equilibrium price and quantity for the following demand and supply relations.

31. D: $2p + 3x = 100$; S: $p = \frac{1}{10}x + 2$.

32. D: $3p + 5x = 200$; S: $7p - 3x = 56$.

33. D: $4p + x = 50$; S: $6p - 5x = 10$.

34. D: $5p + 8x = 80$; S: $3x = 2p - 1$.

35. Find the values of a and b for which the line whose equation is $ax + by + 5 = 0$ passes through the points $(2, 3)$ and $(-1, 1)$.

36. Find the values of p and q such that the line with equation $px + qy = 6$ passes through $(3, 1)$ and has a y intercept of 2.

37. Determine the value(s) of the constant k such that the system
$$4x - 6y = 12, \quad 6x = k + 9y$$
(a) Has no solution.
(b) Has only one solution.
(c) Has an infinite number of solutions.

38. Determine the constant k such that the system
$$\frac{x+y}{3} - \frac{x-y}{2} = 1, \quad \frac{3x-y}{2} - \frac{4x+y}{3} = k$$
(a) Has no solution.
(b) Has only one solution.
(c) Has an infinite number of solutions.

39. How many ounces of 10-karat gold and how many ounces of 24-karat (pure) gold should be melted together to obtain 35 ounces of 18-karat gold?

40. A chemical store has two types of acid solutions, one containing 25% acid and the other 15% acid. How many gallons of each type should be mixed to obtain 200 gallons of mixture containing 18% acid?

41. A store sells two types of coffee at $2 per pound and $1.50 per pound. The store owner makes 50 pounds of a new blend of coffee by mixing these two types and sells it at $1.60 per pound. How many pounds of coffee of each type should be mixed so as to keep the revenue unchanged?

42. Two metals, X and Y, can be extracted from two types of ore, I and II. One hundred pounds of ore I yields 3 ounces of X and 5 ounces of Y, and 100 pounds of ore II yields 4 ounces of X and 2.5 ounces of Y. How many pounds of ores I and II will be required to produce 72 ounces of X and 95 ounces of Y?

43. Substance A contains 5 milligrams of niacin per ounce and substance B 2 milligrams. In what proportions should A and B be mixed so that the resulting mixture contains 4 milligrams of niacin per ounce?

44. A crop of potatoes yields an average 16 metric tons of protein per square kilometer of planted area, while corn yields 24 metric tons per square kilometer. In what proportions must potatoes and corn be planted in order to yield 21 tons of protein per square kilometer from the combined crop?

(45–54) Solve the following nonlinear systems.

45. $x + y = 3$
    $x^2 + y^2 = 29.$

46. $2x + y = 5$
    $x^2 + y^2 = 5.$

47. $x + 2y = 1$
    $x^2 + 2y^2 = 11.$

48. $x - y = 4$
    $2x^2 + y^2 = 19.$

49. $2x + y = 5$
    $xy = 2.$

50. $x - 2y = 1$
    $xy = 3.$

51. $x + y = 2$
    $x^2 + xy + y^2 = 3.$

52. $x + y = 3$
    $x^2 - xy + y^2 = 3.$

53. $x + y = 3$
    $\frac{x}{y} + \frac{y}{x} = 2\frac{1}{2}.$

54. $x + 2y = 1$
    $\frac{y}{x} - \frac{x}{y} = 2\frac{2}{3}.$

(55–58) Find the equilibrium price and quantity for the following demand and supply relations.

55. D: $p^2 + x^2 = 25$
S: $p = x + 1$.

56. D: $p^2 + 2x^2 = 114$
S: $p = x + 3$.

57. D: $p^2 + x^2 = 169$
S: $p = x + 7$.

58. D: $2p^2 + 3x^2 = 77$
S: $x = p - 2$.

(59–64) Solve the following systems of equations.

*59. $x + y + z = 5$
$x - y + z = 3$
$-x + y + z = 1$.

*60. $x + 2y + 3z = 6$
$2x - y + z = 2$
$3x + y + 4z = 8$.

*61. $x + 3y + 4z = 1$
$2x + 7y + 3z = -5$
$3x + 10y + 8z = -3$.

*62. $3x - 2y + 4z = 3$
$4x + 3y = 9$
$2x + 4y + z = 0$.

*63. $x + y - 2z = -1$
$2x - 3y + z = 13$
$-3x + 2y + 5z = -8$.

*64. $u + 2v + 3w = 17$
$2u - v + 4w = 11$
$3u + 7v - 8w = -14$.

Chapter Review 5

Define and or explain the following terms:

(a) Distance formula.
(b) Midpoint formula.
(c) Graph of an equation.
(d) Symmetry of the graph of an equation about the x axis, the y axis, and the origin.
(e) Slope of a nonvertical line.
(f) Slope of a vertical line.
(g) Point-slope formula of a line.
(h) Slope-intercept formula of a line.
(i) Linear equation in two variables x and y.
(j) Conditions for lines to be parallel or perpendicular.
(k) System of linear equations and the methods of solution of such systems.
(l) Consistent, inconsistent, and dependent systems of linear equations.

Review Exercises on Chapter 5

1. Find the distance of the point $A(2, -3)$ from the midpoint of the line joining $B(3, 7)$ and $C(-1, 5)$.

2. Given the points $A(3, 1)$, $B(2, -1)$, $C(4, 3)$, $D(-3, 1)$, find the distance between the midpoints of AB and CD.

(3–8) Sketch the graphs of the following equations.

3. $2x - 3y = 12$.
4. $x^2 = y + 2$.
5. $x = y^2 - 1$.
6. $4y = x^4$.
7. $y = -x^4$.
8. $y^2 + x^2 + 9 = 0$.

(9–20) Find the equation of the straight line satisfying the condition given in each of the following exercises.

9. Passing through $(-1, -3)$ with slope 2.
10. Passing through $(0, 2)$ with zero slope.
11. Passing through $(2, 1)$ with no slope.
12. Passing through $(3, 1)$ and $(2, -1)$.
13. Passing through $(2, 5)$ and $(2, -3)$.
14. Passing through $(1, 2)$ and parallel to $3x - y + 2 = 0$.
15. Passing through $(2, -1)$ and parallel to $3x + 7 = 0$.
16. Passing through $(1, -1)$ and perpendicular to $2x - y + 4 = 0$.

17. Passing through $(1, -3)$ and perpendicular to $x/2 + y/3 = 1$.

18. Passing through the point of intersection of the lines $2x - 3y = 1$ and $x + 2y = 4$ and parallel to $3x + 4y = 7$.

19. Passing through the point of intersection of the lines $2x - 3 = 0$ and $x + y = 2$ and parallel to $2x - 3y + 7 = 0$.

20. Perpendicular to $2x - 5y + 4 = 0$ and passing through the point of intersection of the lines $x - y = 3$ and $2x + 3y = 1$.

21. Find the interecepts made by the line $2x - 3y = 6$ on the coordinate axes.

22. A line passes through the points $(1, 2)$ and $(2, 1)$. Find the coordinates of the points where this line meets the coordinate axes.

23. A line is parallel to $3x - 4y = 7$ and passes through the point $(1, 1)$. Find the intercepts made by this line on the coordinate axes.

24. A line passes through the point $(1, 1)$ and is perpendicular to the line joining $(1, 2)$ and $(1, 4)$. Find the y intercept of this line.

(25–26) Use slopes to prove the following exercises.

25. The points $(4, -1)$, $(-2, 3)$ and $(1, 1)$ lie on the same line.

26. The points $(1, 2)$, $(2, 3)$, $(4, 7)$, and $(3, 6)$ taken in order form the vertices of a parallelogram.

(27–30) A triangle has vertices $P(1, 2)$, $Q(3, 5)$, and $R(1, -3)$.

27. Find the equation of the median that passes through the vertex P.

28. Find the equation of the altitude through Q.

29. Find the equation of the right bisector of the side QR.

30. Show that the line joining the midpoints of the sides PQ and PR is parallel to QR and half its size.

(31–54) Solve the following systems of equations.

31. $x + 2y = 3$; $2x - 3y = -1$.

32. $3u - 5v = 8$; $2u + 3v + 1 = 0$.

33. $\frac{x}{2} - \frac{y}{3} = -1$; $\frac{2x}{3} + \frac{y}{2} = 10$.

34. $2p - q = \frac{3}{4}$; $\frac{1}{p} + \frac{1}{q} = 6$.

35. $\frac{1}{x} + \frac{1}{y} = 12$; $\frac{2}{3x} + \frac{3}{2y} = 13$.

36. $\frac{2}{u} - \frac{3}{v} = -15$; $\frac{1}{2u} + \frac{1}{3v} = 6$.

37. $2x + 3y = 5$; $\frac{x}{3} + \frac{y}{2} = 1$.

38. $3x - 4y = 12$; $\frac{x}{4} = 1 + \frac{y}{3}$.

39. $2x - y = 1$; $xy = 1$.

40. $x + y = 3$; $x^2 - xy + y^2 = 3$.

41. $x + y = \frac{1}{x} + \frac{1}{y} = 2\frac{1}{2}$.

42. $\frac{x}{2} + \frac{y}{3} = \frac{13}{36}$; $\frac{2}{x} + \frac{3}{y} = 13$.

43. $x + y = 1$; $x^2 + xy + y^2 = x - 2$.

44. $x + y = 2$; $x^2 + xy + y^2 = 3$.

45. $2^{x+1} + 3^{y+2} = 11$; $2^{x-1} + 3^{y+1} = 3$. [Hint: Put $2^x = u$, $3^y = v$.]

46. $3^{x+1} - 2^{y-2} = 8$; $3^{x-1} + 2^{y+2} = 17$.

47. $2^{x+1} - 3^{y+2} = 16$; $2^{x+2} + 3^{y+1} + 10 = 0$.

48. $|2x| + |3y| = 13$; $\left|\frac{x}{2}\right| + \left|\frac{y}{3}\right| = 2$.

*49. $2^x = \frac{8}{4^y}$; $9^{x^2} = \frac{27}{3^{y^2}}$.

*50. $x + y - 2z = 5$; $2x - y + 3z = -3$; $-x + 2y - z = 4$.

*51. $\frac{1}{x} + \frac{1}{y} + \frac{1}{z} = 6$; $\frac{2}{x} - \frac{1}{y} + \frac{3}{z} = 4$; $\frac{3}{x} + \frac{2}{y} - \frac{1}{z} = 11$.

*52. $\frac{2}{u} - \frac{3}{v} + \frac{4}{w} = 20$; $\frac{1}{u} + \frac{2}{v} + \frac{3}{w} = 6$; $\frac{3}{u} - \frac{1}{v} + \frac{5}{w} = 20$.

*53. $x + y - z = 1$; $2x + 3y + 5z = 10$; $x + 2y + 6z = 4$.

*54. $3x - y + 2z = 4$; $2x + y - 4z = -1$; $x - 2y + 6z = 5$.

*55. Find the values of the constants a, b, c so that the curve $y = ax^2 + bx + c$ passes through the points $(1, 2)$, $(-2, 7)$, and $(0, 3)$.

*56. Find the values of the constants a, b, c such that the graph of the equation
$$x^2 + y^2 + ax + by + c = 0$$
contains the points $(1, 2)$, $(2, -1)$, and $(3, 4)$.

57. The total cost of manufacturing 100 cameras per week is $700 and of 120 cameras per week is $800.
   (a) Determine the cost equation, assuming it to be linear.
   (b) What are the fixed cost and the variable cost per unit?

58. It costs a company $75 to produce 10 units of a certain item per day and $120 to produce 25 units per day.
   (a) Determine the cost equation, assuming it to be linear.
   (b) What are the variable cost per item and the fixed cost?

59. A firm manufactures two products, A and B. Each product has to be processed by two machines, I and II. Each unit of type A requires 1 hour of processing by machine I and 1.5 hours by machine II, and each unit of type B requires 3 hours by machine I and 2 hours by machine II. If machine I is available for 300 hours each month and machine II for 250 hours, how many units of each type can be manufactured in a month if the total available time on the two machines is to be utilized?

60. A company is trying to purchase and store two types of items X and Y. Each item X costs $3 and each item Y costs $2.50. Each item X occupies 2 square feet of floor space and each item Y 1 square foot. How many units of each type can be purchased and stored if $400 is available for purchasing and 240 square feet of floor space is available for storing these items?

61. A dealer can sell 200 units of a certain commodity per day at $30 per unit and 250 units at $27 per unit. The supply equation for that commodity is $6p = x + 48$.
   (a) Find the demand equation for the commodity, assuming it to be linear.
   (b) Find the equilibrium price and quantity.

FUNCTIONS AND GRAPHS

6.1

Functions

$x \to y$

The concept of function is one of the basic ideas in all mathematics. It expresses the idea of one quantity depending on or being determined by another.

Let the two quantities be denoted by y and x, respectively. Then, roughly speaking, we say that y is a function of x if to each value of x there corresponds one and only one value of y. Before giving a more precise definition, let us consider a few examples:

1. The area of a circle is a function of its radius, because to each value of the radius there corresponds one and only one value for the area. Once the radius is specified, the area is determined.
2. The population of New York City is a function of time, because to each instant of time there corresponds one and only value for the population.
3. A wage-earner's weekly pay is a function of the number of hours he has worked during the week.
4. On the other hand, among human beings, eye color is *not* a function

of body weight. In other words, if you were told how much a person weighs, you would not be able to determine the color of his or her eyes.

DEFINITION Let X and Y be two nonempty sets. Then a **function** from X to Y is a *rule* that assigns to each element x belonging to X a single y in Y.

If a function assigns y to a particular x belonging to X, then we speak of y as being the **value** of the function at x. A function is generally denoted by a single letter, say f or g or F or G.

Let f denote a given function. The set X for which f assigns a unique y in Y is called the **domain** of the function f. It is often denoted by D_f. The set of all the values of the function is called the **range** of the function and is denoted by R_f. The range is a subset of Y.

The following are examples of functions.

1. Let X be the set of students in a class. Let f be the rule that assigns to each student his/her final grade. Since to each student there corresponds exactly one final grade, this rule does define a function. In this case the domain is the set of all students in the class and the range is the set of all the different grades awarded (for example, R_f might be the set $\{A, B, C, D, F\}$).
2. The Dow-Jones industrial average index is a function of time. Here the domain is a certain set of values of time and the range of the function is the set of values of the index.

If a function f assigns a value y in the range to a certain x in the domain, we write

$$y = f(x).$$

$f(x)$ is read as "f of x". Note that $f(x)$ is not the product of f and x but is the value of f at x.

If a function f is expressed by a relation of the type $y = f(x)$, then x is called the **independent variable** and y is called the **dependent variable**. In the expression $f(x)$, x is called the **argument** of the function f.

Generally, we shall come across functions that are expressed by stating the value of the function by means of an algebraic expression in terms of the independent variable involved. For example, if $f(x)$ is the function that gives the area of a circle when x is the radius, then

$$f(x) = \pi x^2.$$

Other examples of functions are

$$g(x) = 5x^2 - 7x + 2$$

$$h(x) = 2x^3 + \frac{7}{x+1}.$$

When a function is given by an algebraic expression, we can calculate the value of the function by substituting any given value of the independent

206 Ch. 6 Functions and Graphs

variable into the expression. For example, for $f(x) = \pi x^2$, we find the area of the circle of radius 2 by substituting $x = 2$:

$$\text{area} = f(2) = \pi(2)^2 = 4\pi \approx 12.57 \text{ square units.}$$

Example 1 Given $f(x) = 2x^2 - 5x + 1$, find the value of f when $x = a$, $x = 3$, $x = -2$ and $x = -\frac{1}{4}$; that is, find $f(a), f(3), f(-2),$ and $f(-\frac{1}{4})$.

Solution We have

$$f(x) = 2x^2 - 5x + 1. \quad \text{(i)}$$

To find $f(a)$, we replace x by a in the given expression:

$$f(a) = 2a^2 - 5a + 1.$$

To evaluate $f(3)$, we substitute 3 for x on both sides of (i):

$$f(3) = 2(3)^2 - 5(3) + 1$$
$$= 18 - 15 + 1 = 4.$$

Similarly,

$$f(-2) = 2(-2)^2 - 5(-2) + 1 = 19$$

and

$$f(-\tfrac{1}{4}) = 2(-\tfrac{1}{4})^2 - 5(-\tfrac{1}{4}) + 1 = \tfrac{19}{8}.$$

Example 2 Given $g(x) = 3x^2 - 2x + 5$, evaluate (a) $g(1 + h)$, (b) $g(1) + g(h)$, (c) $g(x + h) - g(x)$.

Solution We have

$$g(x) = 3x^2 - 2x + 5. \quad \text{(i)}$$

(a) To evaluate $g(1 + h)$, we must replace the argument x in (i) by $1 + h$:

$$g(1 + h) = 3(1 + h)^2 - 2(1 + h) + 5$$
$$= 3(1 + 2h + h^2) - 2 - 2h + 5$$
$$= 3h^2 + 4h + 6.$$

(b) Replacing x by 1 and h, respectively, in (i), we get

$$g(1) = 3(1^2) - 2(1) + 5$$
$$= 3 - 2 + 5 = 6$$

$$g(h) = 3h^2 - 2h + 5.$$

Therefore

$$g(1) + g(h) = 6 + 3h^2 - 2h + 5$$
$$= 3h^2 - 2h + 11.$$

Observe that $g(1) + g(h) \neq g(1 + h)$.

(c) Replacing the argument x in (i) by $x + h$, we have

$$g(x + h) = 3(x + h)^2 - 2(x + h) + 5$$
$$= 3(x^2 + 2xh + h^2) - 2x - 2h + 5$$
$$= 3x^2 + 6xh + 3h^2 - 2x - 2h + 5.$$

Therefore

$$g(x+h) - g(x)$$
$$= 3x^2 + 6xh + 3h^2 - 2x - 2h + 5 - (3x^2 - 2x + 5)$$
$$= 6xh + 3h^2 - 2h.$$

Example 3 Given the function $g(x) = 3$, compute = CONSTANT

$$g(-3), \quad g(a^2), \quad \text{and} \quad g(a+h).$$

Solution The function $g(x)$ has the value 3 no matter what the value of its argument. Thus,

$$g(-3) = 3$$
$$g(a^2) = 3$$
$$g(a+h) = 3.$$

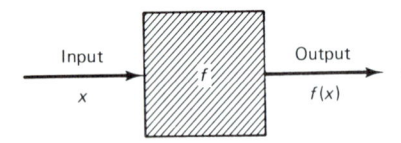

The adjoining figure provides a useful picture of a function. The function f is viewed as a "black box." The value of the argument, x, is the *input* into this black box, and the *output* is the corresponding value of the function, $f(x)$.

For example, suppose f is the function that squares real numbers, $f(x) = x^2$. Then if the input into the black box is 2, the output will be 4; if the input is $-\frac{3}{2}$, the output will be $\frac{9}{4}$, and so on.

GRAPHS OF FUNCTIONS

In most cases of interest the domains and ranges of the function we are concerned with are subsets of the real numbers. In such cases the function is commonly represented by its **graph**. The graph of a function f is obtained by plotting all the points (x, y), where x belongs to the domain of f and $y = f(x)$.

Example 4 Draw the graph of the function

$$f(x) = 2 + 0.5x^2.$$

Solution The domain of f is the set of all real numbers, since we can evaluate $f(x)$ for any real value of x. Some of the values of this function are set out in the following table; certain values of x are listed in the top row and the corresponding value of $y = f(x)$ is given beneath each value of x:

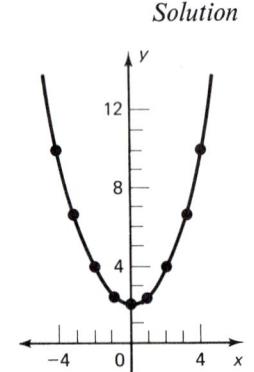

x	1	2	3	4	-1	-2	-3	-4
$y = f(x)$	2.5	4	6.5	10	2.5	4	6.5	10

The points corresponding to the values of x and y are plotted as heavy dots in the adjoining figure. The graph of the function $f(x) = 2 + 0.5x^2$ is shown as the U-shaped curve passing through the heavy dots.

Note that the range of this function—that is, the set of y values—is the set $\{y \mid y \geq 2\}$.

Any vertical line corresponds to some particular value, say $x = x_0$, of the independent variable, and the point where this vertical line meets the graph of a function f determines the value of y that corresponds to x_0. Thus the graph itself provides the rule that relates each value of x to some value of y. If the vertical line $x = x_0$ does not meet the graph at all, then x_0 simply does not belong to the domain of f.

The following graphs represent functions. [Note that in the last example the domain of the function is the set of integers {1, 2, 3, 4, 5}, so that the graph consists of simply five points rather than a curve.]

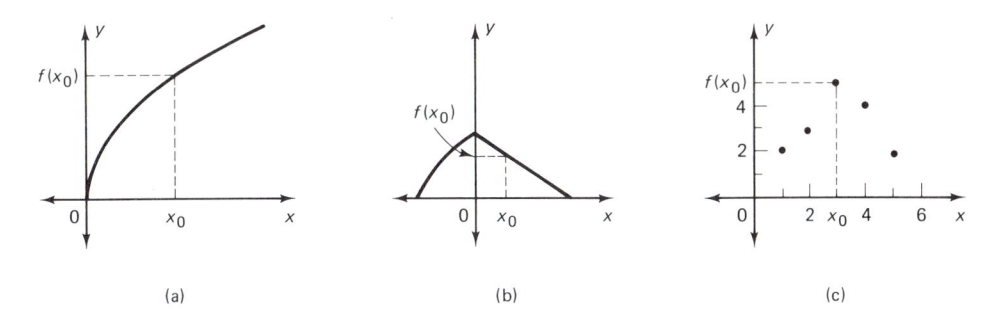

(a) (b) (c)

On the other hand, the following graphs do not represent functions:

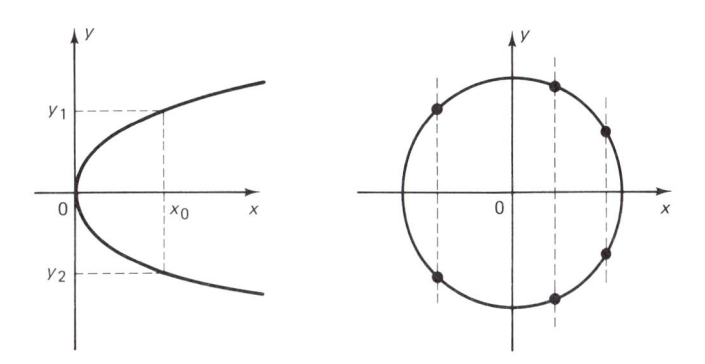

The reason is that there are vertical lines that meet the graphs in more than one point. Thus, corresponding to the value $x = x_0$ on the first graph there are two values y_1 and y_2 for y. In such a case the value of x does not determine a *single* value of y. The same happens for the dashed lines on the second graph. This leads to the following **vertical line test**:

> Any given curve (*or set of points*) in the (x, y) plane is the graph of some function (with y as dependent variable) provided that any vertical line meets the graph in at most one point.

On the graph of a function, the values along the x axis at which the graph is defined constitute the domain of the function. Correspondingly, the

Sec. 6.1   Functions

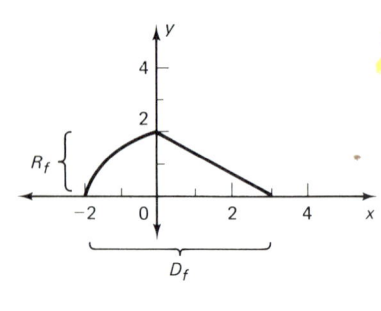

values along the y axis at which the graph has points constitute the range of the function. This is illustrated in the adjoining figure. In this figure $D_f = \{x \mid -2 \leq x \leq 3\}$ and $R_f = \{y \mid 0 \leq y \leq 2\}$.

Often the domain of a function is not stated explicitly. In such cases it is understood to be the set of all values of the argument for which the given rule makes sense. For a function f defined by an algebraic expression, the domain of f is the set of all real numbers x for which $f(x)$ is a well-defined real number. For example, the domain of the function $f(x) = \sqrt{x}$ is the set of nonnegative real numbers, since the square root makes sense only for $x \geq 0$. Similarly, in the case of the function $g(x) = x^2/(x - 3)$ the domain is the set of all real numbers except $x = 3$, since when $x = 3$ the denominator becomes zero and $g(3)$ is not defined.

In general, when finding the domain of a function we must keep these two conditions in mind: *any expression underneath a square-root sign cannot be negative and the denominator of any fraction cannot be zero.* (More generally, any expression underneath a radical of even index, such as $\sqrt[4]{}$ or $\sqrt[6]{}$, cannot be negative.)

Example 5 Find the domain of g, where

$$g(x) = \frac{x+3}{x-2}.$$

Solution Clearly $g(x)$ is not a well-defined real number for $x = 2$. For any other value of x, $g(x)$ *is* a well-defined real number. Thus the domain of g is the set of all real numbers except 2:

$$D_g = \{x \mid x \neq 2\}.$$

Example 6 Find the domain of f where

$$f(x) = \sqrt{x-4}.$$

Solution The domain of f is the set of all values of x for which the expression under the radical sign is nonnegative—that is,

$$x - 4 \geq 0 \quad \text{or} \quad x \geq 4.$$

For $x < 4$, the value of $f(x)$ is not a real number, since the quantity $(x - 4)$ beneath the square-root sign is negative. Thus,

$$D_f = \{x \mid x \geq 4\}.$$

The graph of f is shown in the adjoining figure, in which a few explicit points are plotted. [Note that the range of f is the set $\{y \mid y \geq 0\}$.]

In applied problems it is often necessary to construct an algebraic function from certain verbal information.

Example 7 Express the area of an equilateral triangle as a function of its perimeter.

Solution Let x denote the perimeter of the triangle and A its area. Each side then has length $x/3$, since the triangle is equilateral.

The formula for the area of a triangle is $A = \frac{1}{2}(\text{base})(\text{height})$. We have

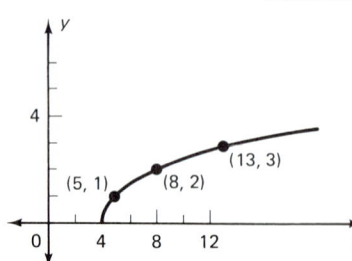

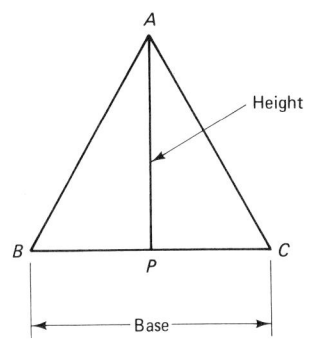

the base equal to $x/3$. For the height we must use Pythagoras's theorem applied to the triangle ABP in the adjacent figure:
$$AB^2 = AP^2 + BP^2.$$
Now $AB = x/3$ and $BP = x/6$ (being half the base). Therefore
$$\left(\frac{x}{3}\right)^2 = AP^2 + \left(\frac{x}{6}\right)^2$$
or
$$AP^2 = \left(\frac{x}{3}\right)^2 - \left(\frac{x}{6}\right)^2 = \frac{x^2}{9} - \frac{x^2}{36} = \frac{x^2}{12}.$$
Thus, the height $AP = x/\sqrt{12} = x/2\sqrt{3}$. Finally, therefore,
$$A = \frac{1}{2}\left(\frac{x}{3}\right)\left(\frac{x}{2\sqrt{3}}\right) = \frac{x^2}{12\sqrt{3}}.$$

Example 8 It is desired to construct a telephone link between two towns A and B situated on opposite banks of a river. The width of the river is 1 kilometer, and B lies 2 kilometers downstream from A. It costs \$$c$ per kilometer to construct a line overland and \$$2c$ per kilometer underwater. The telephone line will follow the river bank from A for a distance x (kilometers) and then will cross the river diagonally in a straight line directly to B. Determine the total cost of the line as a function of x.

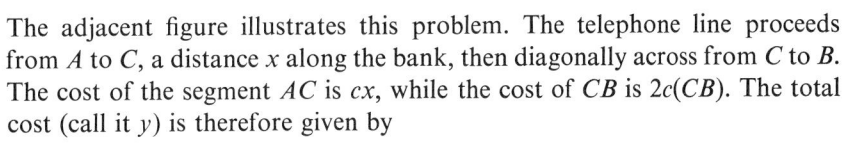

Solution The adjacent figure illustrates this problem. The telephone line proceeds from A to C, a distance x along the bank, then diagonally across from C to B. The cost of the segment AC is cx, while the cost of CB is $2c(CB)$. The total cost (call it y) is therefore given by 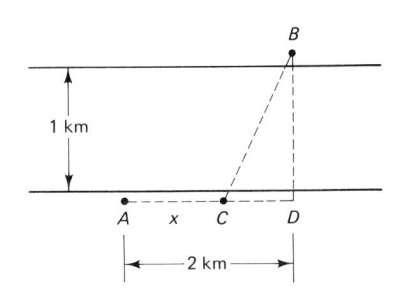
$$y = cx + 2c(CB).$$
In order to complete the problem we must express CB in terms of x. Now from Pythagoras's theorem applied to the triangle BCD,
$$BC^2 = BD^2 + CD^2.$$
But $BD = 1$ (the width of the river) and
$$CD = AD - AC = 2 - x.$$
Therefore,
$$BC^2 = 1^2 + (2 - x)^2 = 1 + (4 - 4x + x^2)$$
$$= x^2 - 4x + 5.$$
Thus the cost is given by
$$y = cx + 2c\sqrt{x^2 - 4x + 5}.$$
This is the required expression, giving y as a function of x.

In the preceding examples we have been concerned with functions that are defined by a single algebraic expression for all values of the independent variable throughout the domain of the function. It sometimes happens that we need to use functions that are defined by more than one expression.

Example 9 Electricity is charged to consumers at the rate of 10¢ per unit for the first 50 units and 3¢ per unit thereafter. Find the function $c(x)$ that gives the cost of using x units of electricity.

Solution For $x \leq 50$, each unit costs 10¢, so the total cost of x units is $10x$ cents. So $c(x) = 10x$ for $x \leq 50$. When $x = 50$, we get $c(50) = 500$: the cost of the first 50 units is equal to 500¢. When $x > 50$, the total cost is equal to the cost of the first 50 units (i.e., 500¢) plus the cost of the rest of the units used. The number of these excess units is $(x - 50)$, and they cost 3¢ each, so their total cost is $3(x - 50)$ cents. Thus the total bill when $x > 50$ comes to

$$c(x) = 500 + 3(x - 50) = 500 + 3x - 150$$
$$= 350 + 3x.$$

We can write $c(x)$ in the form

$$c(x) = \begin{cases} 10x & (x \leq 50) \\ 350 + 3x & (x > 50). \end{cases}$$

The graph of $y = c(x)$ is shown in the following figure. Note the way in which the nature of the graph changes at $x = 50$, where one formula takes over from the other.

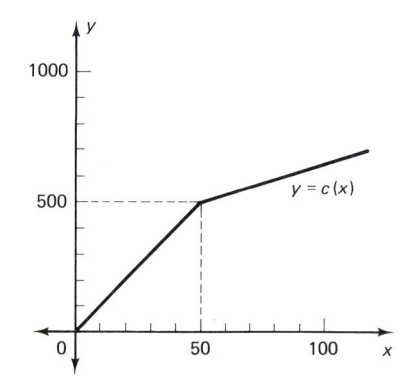

Example 10 Draw the graph of the following function:

$$f(x) = \begin{cases} 4 - x & (0 \leq x \leq 4) \\ \sqrt{x - 4} & (x > 4). \end{cases}$$

Solution The domain of this function is the set of all nonnegative real numbers. For $0 \leq x \leq 4$ the function is defined by the algebraic expression $f(x) = 4 - x$, while for $x > 4$ it is defined by the expression

$$f(x) = \sqrt{x - 4}.$$

Some values of $f(x)$ are set out in the accompanying table, and the graph of this function is shown in the following figure. It consists of two segments. For x

x	0	2	4	5	8	13
$y = f(x)$	4	2	0	1	2	3

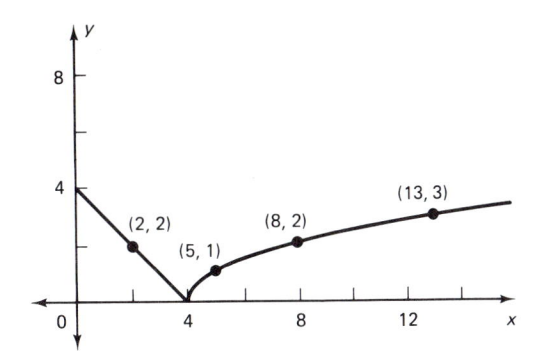

between 0 and 4 the graph consists of a straight line segment whose equation is $y = 4 - x$. For $x \geq 4$ the function is identical with that in Example 6.

In these examples, the function under consideration has been defined by two algebraic expressions. It is sometimes necessary to consider functions defined by three or even more different expressions.

VARIATION

We say that **y varies directly as x** or **y is directly proportional to x** if $y = kx$ for some constant k. The constant k is called the **constant of proportionality**.

If y is directly proportional to x, then when x is increased by any factor, y increases by the same factor. For example, doubling x also doubles y.

For example, hardware stores sell nails by weight. The number of nails you get is directly proportional to the number of pounds you buy (for each size of nail, of course). Suppose that one pound of a certain size contains 150 nails; then the number of nails, N, in x pounds is given by $N = 150x$. N is directly proportional to x, and the constant of proportionality is 150.

More generally, we say that **y varies directly as x^n** or **y is directly proportional to x^n** if $y = kx^n$ for some constant k. (Again k is the constant of proportionality.)

For example, the area of a circle is directly proportional to the square of its radius, and the constant of proportionality is π (since $A = \pi r^2$).

Example 11 The period T of a simple pendulum (that is, the time it takes to make one complete swing backward and forward) is directly proportional to the square root of its length, l. If $T = 2.84$ seconds when $l = 2$ meters, calculate the value of T when $l = 5$ meters.

Solution We are given that $T = k\sqrt{l}$. Substituting the given values, therefore,

$$2.84 = k\sqrt{2}$$

$$k = \frac{2.84}{\sqrt{2}}.$$

Sec. 6.1 Functions

When $l = 5$, the period is then

$$T = k\sqrt{5} = \frac{2.84}{\sqrt{2}}\sqrt{5} = 2.84\sqrt{\frac{5}{2}} = 2.84(1.58)$$
$$= 4.49$$

A pendulum of length 5 meters therefore has a period of 4.49 seconds.

We say that *y varies inversely as x* or *y is inversely proportional to x* if $y = k/x$ for some constant k. We say *y is inversely proportional to x^n* if $y = k/x^n$ for some constant k.

For example, if a pie is divided equally among several individuals, the amount that each receives is inversely proportional to the number of recipients. As a second example, the time t taken for a moving object to move a distance d is inversely proportional to its velocity V, since we have $t = d/V$.

Example 12 The maximum load P that can be supported by a column of length L without causing the column to buckle is inversely proportional to the square of L. If $P = 100$ kilograms when $L = 4$ meters, express P as a function of L.

Solution We are given that $P = k/L^2$. In order to determine the constant of proportionality, substitute the given values $P = 100$ and $L = 4$:

$$100 = \frac{k}{4^2} = \frac{k}{16}$$
$$k = 1600.$$

Therefore

$$P = \frac{1600}{L^2}$$

which expresses P as a function of L as required.

Diagnostic Test 6.1

Fill in the blanks.

1. The value of the function $x^2 - 2x$ when $x = -1$ is _____.
2. If $f(x) = 2x^3 - 3x^2$ then $f(-2) = $ _____.
3. If $f(x) = -7$, then $f(3) = $ _____ and $f(x^2) = $ _____.
4. If $f(x) = 1/x$, then $f(x + a) = $ _____.
5. The domain of $f(x) = \sqrt{x + 2}$ is $\{x \mid $ _____ $\}$.
6. The domain of $f(x) = 1/(x\sqrt{x - 1})$ is $\{x \mid $ _____ $\}$.
7. The domain of $f(x) = x^{-1}\sqrt[4]{x - 1}$ is $\{x \mid $ _____ $\}$.
8. If $f(x) = (2x + 7)/(x - 1)$, then the domain of f is _____.
9. If $f(x) = \sqrt{x - 1}/(x - 2)$, then the domain of f is _____.

10. The range of $f(x) = (x-1)^2$ is _____.

11. A given curve is the graph of some function if any _____ line meets the curve in _____ one point.

Exercises 6.1

1. Given $f(x) = 3x - 4$, evaluate $f(2), f(0)$, and $f(-2)$.

2. Given $g(t) = 3 - \frac{1}{2}t$, evaluate $g(10), g(6)$, and $g(-6)$.

3. Given $h(p) = 2 - \frac{1}{2}p^2$, evaluate $h(2), h(4)$, and $h(-4)$.

4. Given $h(y) = 2y^2 - 3$, evaluate $h(2), h(-1)$, and $h(0)$.

5. Given $f(x) = -1$, evaluate $f(-1), f(0)$, and $f(1)$.

6. Given $h(t) = 0$, evaluate $h(1), h(t^2)$, and $h(t-1)$.

7. Given $f(x) = \dfrac{2x-1}{x+1}$, evaluate $f(2), f(-2)$, and $f(-1)$.

8. Given $g(x) = \dfrac{x^2+2}{x^2-1}$, evaluate $g(3), g(\sqrt{2})$, and $g(1)$.

9. Given $f(t) = \sqrt{2-t}$, evaluate $f(0), f(3)$, and $f(-3)$.

10. Given $f(y) = y^{-1}\sqrt{y+1}$, evaluate $f(0), f(3)$, and $f(-3)$.

11. Given $f(x) = 3x^2 - 5x + 7$, find:
   (a) $f(0)$. (b) $f(2)$.
   (c) $f(-3)$. (d) $f(\frac{1}{2})$.
   (e) $f(-\frac{4}{3})$. (f) $f(c)$.
   (g) $f(c+h)$. (h) $\dfrac{f(c+h) - f(c)}{h}$.

12. Given $g(x) = 2x^2 + 3x - 5$, evaluate:
   (a) $g(1)$. (b) $g(-2)$.
   (c) $g(\frac{2}{3})$. (d) $g(a)$.
   (e) $g(a+h)$. (f) $\dfrac{g(a+h) - g(a)}{h}$.

13. Given
$$f(x) = \begin{cases} 2x - 3 & \text{if } x \geq 5 \\ 6 - 3x & \text{if } x < 5 \end{cases}$$
find:
(a) $f(0)$. (b) $f(7)$. (c) $f(-2)$.
(d) $f(5+h)$ and $f(5-h)$ where $h > 0$.

14. Given
$$h(x) = \begin{cases} 2 - 3x & \text{if } x > -1 \\ x + 5 & \text{if } x \leq -1 \end{cases}$$
find:
(a) $h(1)$. (b) $h(-1)$. (c) $h(-2)$.
(d) $h(2)$.
(e) $h(-1+p)$ and $h(-1-p)$ if $p > 0$.

15. Given
$$g(x) = \begin{cases} 4x + 3 & \text{if } -2 \leq x < 0 \\ 1 + x^2 & \text{if } 0 \leq x \leq 2 \\ 7 & \text{if } x > 2 \end{cases}$$
evaluate:
(a) $g(1)$. (b) $g(3)$. (c) $g(-1)$.
(d) $g(0)$. (e) $g(-3)$.
(f) $g(2+h)$ and $g(2-h)$ if $h > 0$.

16. Given
$$f(x) = \begin{cases} x^2 - 2 & \text{if } 1 < x \leq 2 \\ 1 & \text{if } 0 < x < 1 \\ 1 - x & \text{if } x \leq 0 \end{cases}$$
evaluate:
(a) $f(0)$. (b) $f(2)$. (c) $f(3)$.
(d) $f(-3)$. (e) $f(1)$. (f) $f(\frac{1}{4})$.

(17–32) Find the domains of the following functions.

17. $f(x) = 3 - x$.

18. $f(x) = 4$.

19. $f(x) = 2x + 3$.

20. $g(x) = 2x^2 - 3x + 7$.

21. $h(x) = \dfrac{x-1}{x-2}$.

22. $D(p) = \dfrac{2p+3}{p-1}$.

23. $f(x) = \sqrt{x-2}$.

24. $F(y) = -\sqrt{3y-2}$.

25. $g(t) = \dfrac{1}{\sqrt{2t-3}}.$

26. $G(u) = \dfrac{2}{\sqrt{3-2u}}.$

27. $g(x) = \dfrac{\sqrt{x}}{x+1}.$

28. $g(y) = \dfrac{\sqrt{1-y}}{y}.$

29. $f(t) = \dfrac{1}{(t-1)\sqrt{t}}.$

30. $h(p) = \dfrac{p+1}{(p-1)\sqrt{2-p}}.$

*31. $h(x) = \sqrt{1-x^2}.$

*32. $g(x) = \dfrac{1}{\sqrt{4-x^2}}.$

33. Express the area A of a circle as a function of the length L of its circumference.

34. Express the area A of a square as a function of the length p of its perimeter.

35. An object is traveling at a speed of 20 kilometers per hour. Express the distance D traveled as a function of time t.

36. An object starts from the point $(x, y) = (0, 2)$ and begins to move parallel to the x axis with a speed of 5 units per second. Express its distance D from the origin as a function of time t.

37. A and B start walking from the same point. A walks due north at a speed of 4 miles per hour and B walks due east at a speed of 3 miles per hour. Express the distance D between them as a function of time t.

38. One ship travels due west at a speed of 20 miles per hour and a second travels due north at 15 miles per hour. The second ship is initially 5 miles due south of the first ship. Calculate their distance apart D as a function of time.

39. A radio manufacturing firm has fixed costs of $3000, and the cost of labor and material is $15 per radio. Determine the cost function—that is, the total cost C as a function of the number x of radios manufactured. If each radio is sold for $25, determine the revenue function $R(x)$ and the profit function $P(x)$.

40. A manufacturer can sell 300 units of his product in a month if he charges $20 per unit and 500 units if he charges $15 per unit. Express the market demand x (the number of units that can be sold each month) as a function of the price p per unit, assuming it to be a linear function—that is, a function of the form $x = ap + b$, where a and b are constants to be determined.

41. A farmer has a fence of 200 yards length, with which he wants to enclose a rectangular field. Express the area A of the field as a function of the length x of one side of it.

42. A rectangle is inscribed in a circle of radius 3 centimeters. Express the area A of rectangle as a function of the length x of one of its sides.

43. A cistern is constructed to hold 300 cubic feet of water. The cistern has a square base and four vertical sides, all made of concrete, and a square top made of steel. If the concrete costs $1.50 per square foot and steel $4 per square foot, determine the total cost C as a function of the length x of the side of the square base.

44. Repeat Exercise 43 for a cistern with circular base and top. Express the cost C as a function of radius r of the base of the cylinder.

45. Sugar costs 25¢ per pound for amounts up to 50 pounds and 20¢ per pound for amounts over 50 pounds. If $C(x)$ denotes the cost of x pounds of sugar, express $C(x)$ by means of suitable algebraic expressions and sketch its graph.

46. A retailer can buy oranges from the wholesaler at the following prices: 20¢ per pound if 20 pounds or less are purchased; 15¢ per pound for amounts over 20 and up to 50 pounds, and 12¢ per pound for amounts over 50 pounds. Determine the cost $C(x)$ of purchasing x pounds of oranges.

(47–52) State whether or not the following graphs represent functions in which y is the dependent variable.

47.

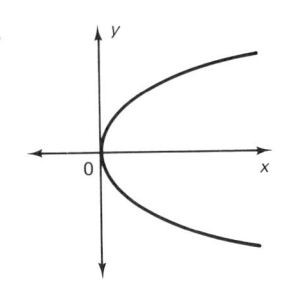

48.

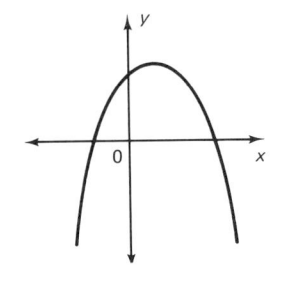

49.

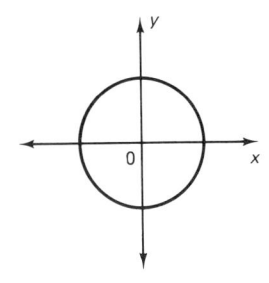

50.

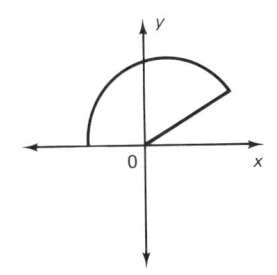

51.

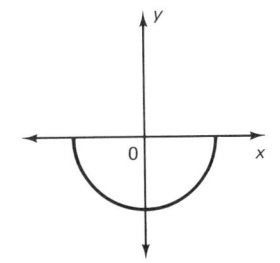

52.

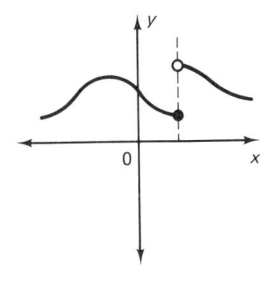

53. P is directly proportional to Q. If $P = 5$ when $Q = 2$, find P when $Q = 6$.

54. A varies directly as b. If $A = \frac{1}{4}$ when $b = 2$, find A when $b = \frac{1}{2}$.

55. W varies directly as x^3. If $W = 3$ when $x = 3$, find the constant of proportionality.

56. A is proportional to l^2. If $A = 5$ when $l = 4$, find the constant of proportionality.

57. S varies inversely as p^2. If $S = 1$ when $p = 2$, express S as a function of p and find S when $p = \frac{1}{2}$.

58. p is inversely proportional to x. If $p = 3$ when $x = 2$, express p as a function of x and find p when $x = 12$.

59. The length of a cassette tape is directly proportional to the playing time. If a 45-minute tape has length $421\frac{7}{8}$ feet, how long is a 60-minute tape?

60. The recording time of a tape reel is inversely proportional to the speed with which it is played. A certain reel will record for 80 minutes at a speed of $1\frac{7}{8}$ inches per second. How long will it record at 15 inches per second?

61. The terminal velocity V of a spherical droplet falling through the atmosphere is proportional to the square of its radius a. If $V = 10$ meters per second when $a = 10^{-3}$ meter, express V as a function of a.

62. The rate R at which a viscous liquid flows through a pipe of given length is proportional to the fourth power of r, the radius of the pipe. If $R = 4$ cubic centimeters per second when $r = 0.2$ centimeters, express R as a function of r.

63. The wavelength λ of a radio wave is inversely proportional to its frequency f. If $\lambda = 30$ meters when $f = 10$ megahertz (10 million cycles per second) express λ as a function of f and find λ when $f = 50$ megahertz.

64. The electrical resistance R of a wire of given length varies inversely as the square of its radius r. If $R = 2$ ohms when $r = 0.01$ centimeters, express R as a function of r. Evaluate R when $r = 0.005$ centimeters.

Sec. 6.1 Functions

6.2

Some Special Functions

CONSTANT FUNCTIONS AND LINEAR FUNCTIONS

A function of the form

$$f(x) = b$$

where b is some constant, is called a *constant function*. The value of this function is b no matter what the value of its argument. For example, $f(0) = b$, $f(1) = b, f(-2) = b$, and so on.

The graph of a constant function, $y = f(x)$ or $y = b$, is a straight line parallel to the x axis (see Section 5.3). The line is at a distance $|b|$ from the x axis and is above or below the x axis according as $b > 0$ or $b < 0$ (see the adjacent figure, drawn for the case $b > 0$).

In this case the domain of the function is the set of all real numbers and the range is the set $\{b\}$ containing only the one number, b.

A function of the form

$$f(x) = mx + b \qquad (m \neq 0)$$

where m and b are constants, is called a *linear function*. The graph of such a function, $y = f(x)$ or $y = mx + b$, is a straight line with slope m and with y intercept b (see Section 5.3). For example, the graph of $f(x) = \tfrac{1}{2}x - 1$ has slope $\tfrac{1}{2}$ and intersects the y axis at the point $(0, -1)$ as illustrated in the following figure.

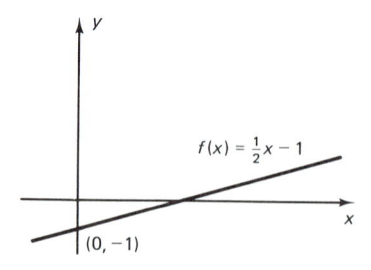

The domain and range of a linear function are both equal to the set of all real numbers.

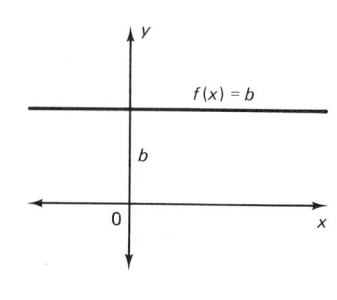

POWER FUNCTIONS

A function of the form

$$f(x) = ax^n$$

where a and n are constants, is called a ***power function***. For example, $f(x) = 2x^3$, $f(x) = \sqrt{x}$, and $f(x) = -3/x^2$ are all power functions with $n = 3$, $n = \frac{1}{2}$, and $n = -2$, respectively.

Example 1 Draw the graphs of the functions $f(x) = x^2$ and $g(x) = x^3$.

Solution The values of the two functions for several values of x are set out in the table below. In the two following figures the corresponding points are plotted and

x	-2.5	-2	-1.5	-1	-0.5	0	0.5	1	1.5	2	2.5
$f(x) = x^2$	6.25	4	2.25	1	0.25	0	0.25	1	2.25	4	6.25
$g(x) = x^3$	-15.6	-8	-3.38	-1	-0.13	0	0.13	1	3.38	8	15.6

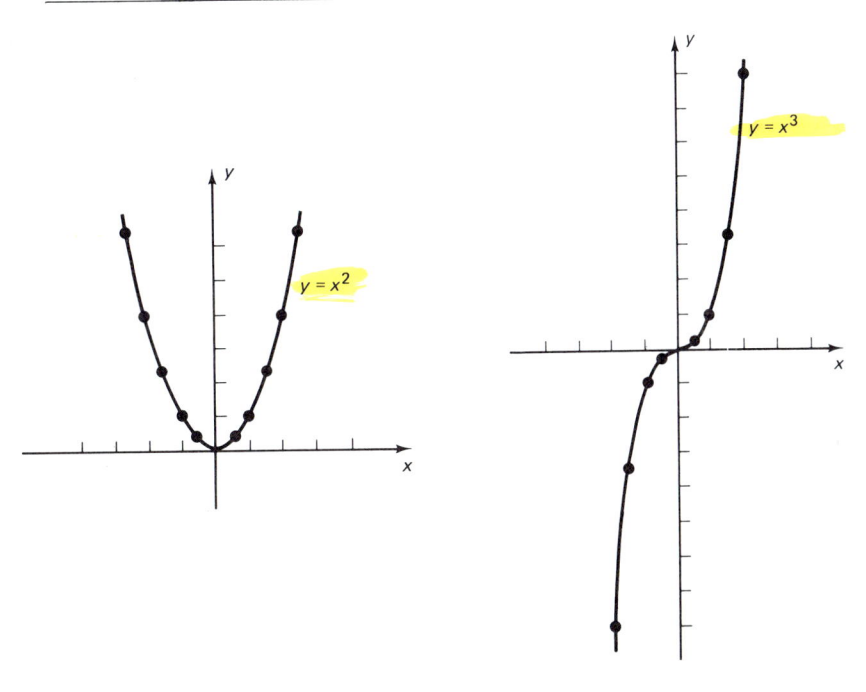

the graphs are drawn as smooth curves through these points. Observe that these graphs are symmetric about the y axis and the origin respectively.

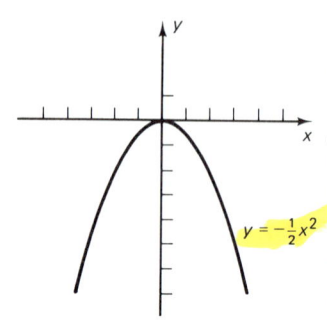

The function $f(x) = ax^2$ for $a > 0$ has a graph that is qualitatively similar to the graph of $y = x^2$. That is, the graph is a bowl-shaped figure. It is an example of a parabola (see Section 7-2). When $a < 0$, the function $f(x) = ax^2$ is again similar in form except that the bowl is upside down—that is, it opens downward. The graph of $y = -\frac{1}{2}x^2$ is shown in the adjacent figure and is typical of the form of the graph when $a < 0$.

Similarly, the graph of $y = ax^3$ for $a > 0$ has the same qualitative shape as the graph of $y = x^3$. The following figures illustrate typical graphs of $y = ax^3$ in the cases when $a > 0$ and when $a < 0$.

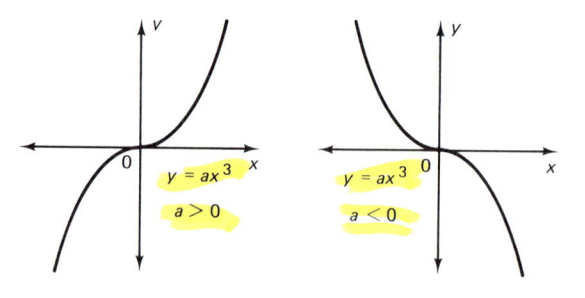

Example 2 Draw the graph of the function $f(x) = x^{1/2}$.

Solution The table of values of this function is given below. Note that $f(x) = x^{1/2}$ is defined only for $x \geq 0$. These points are plotted in the adjacent figure and

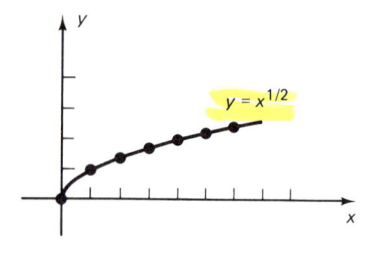

x	0	1	2	3	4	5	6
$f(x) = x^{1/2}$	0	1	1.41	1.73	2	2.24	2.45

connected by a smooth curve. The graph (as will be shown in Section 7.2) is half a parabola, opening to the right.

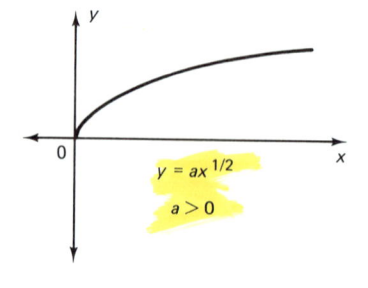

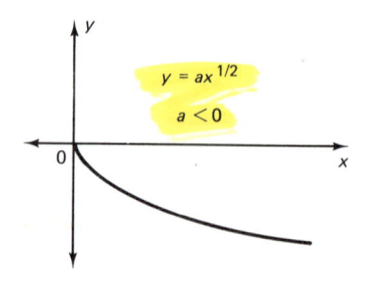

Again the graph in Example 2 is qualitatively typical of the graph of $y = ax^{1/2}$ when $a > 0$. In each case the graph is the upper half of a parabola, opening toward the right, and the graph just touches the y axis at the origin. When $a < 0$, the graph is the lower half of the parabola. The figure below illustrates the typical forms of the graphs of $y = ax^{1/2}$ in the two cases $a > 0$ and $a < 0$.

Example 3 Sketch the graph of the function $f(x) = x^{-1}$.

Solution The domain of this function is the set of all real numbers except zero. A table giving some values of the function is shown below. These values are

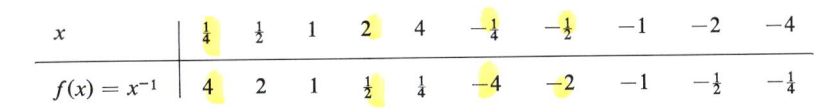

x	$\frac{1}{4}$	$\frac{1}{2}$	1	2	4	$-\frac{1}{4}$	$-\frac{1}{2}$	-1	-2	-4
$f(x) = x^{-1}$	4	2	1	$\frac{1}{2}$	$\frac{1}{4}$	-4	-2	-1	$-\frac{1}{2}$	$-\frac{1}{4}$

plotted on the following figure and a smooth curve drawn through them. Note that as x moves closer and closer to zero, the value of x^{-1} becomes numerically very large, being positive when x is positive and negative when x is negative. Thus the graph approaches closer and closer to the y axis with

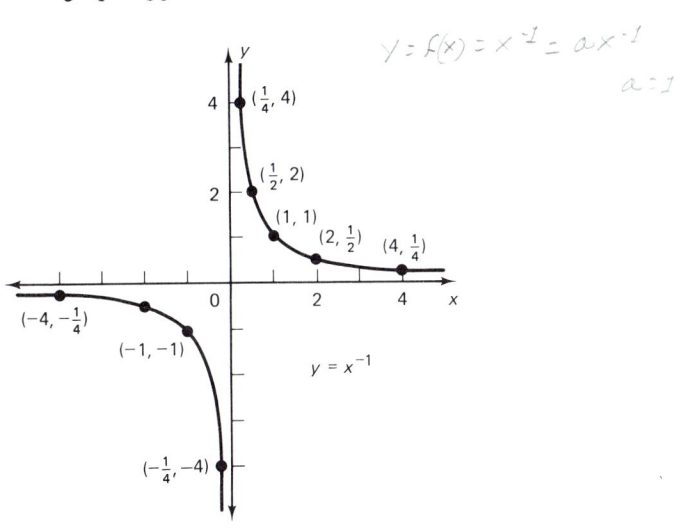

large positive and negative values of y, but it never actually crosses that axis. We say that the y axis is a *vertical asymptote* to the graph. Similarly as x becomes very large in magnitude, either positive or negative, x^{-1} gets closer and closer to zero and the graph approaches closer and closer to the x axis. We say that the x axis is a *horizontal asymptote* to the graph.

The graph of $y = ax^{-1}$ for $a > 0$ is qualitatively similar to the graph of $y = x^{-1}$. The graph lies in the first and third quadrants and has the x and y axes as horizontal and vertical asymptotes, respectively. When $a < 0$, the graph is again similar in form except that it lies in the second and fourth quadrants. The following figure shows the graph of $y = -x^{-1}$, which is typical of these graphs for $a < 0$.

Sec. 6.2 *Some Special Functions*

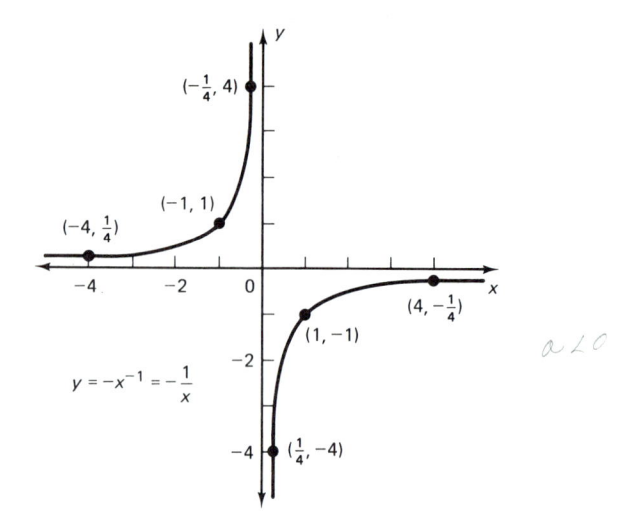

The graph of $y = ax^{-1}$ is a curve called a **rectangular hyperbola** (see Section 7.3).

ABSOLUTE-VALUE FUNCTIONS

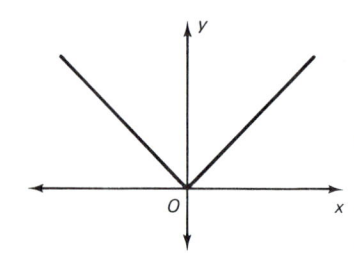

If x is a real number, the absolute value of x, denoted by $|x|$, is given by

$$|x| = \begin{cases} x & \text{if } x \geq 0 \\ -x & \text{if } x < 0. \end{cases}$$

Clearly, *the absolute value of a real number is always nonnegative.*

The function $f(x) = |x|$ is called the **absolute-value function**. The domain of f is the set of all real numbers, and the range is the set of all nonnegative real numbers. The graph of $y = |x|$ is shown in the adjoining figure.

Example 4 Find the domain and range and draw the graph of the function
$$f(x) = |x - 2|.$$

Solution The domain of f is the set of all real numbers and the range is the set of all nonnegative real numbers. Setting $y = f(x)$, we have
$$y = |x - 2|$$
or, using the above definition of absolute value,

$$y = x - 2 \quad \text{if } x - 2 \geq 0 \quad (\text{i.e., if } x \geq 2)$$
$$y = -(x - 2) \quad \text{if } x - 2 < 0 \quad (\text{i.e., if } x < 2).$$

Therefore, the graph of $f(x)$ consists of portions of the two straight lines
$$y = x - 2 \quad \text{and} \quad y = -(x - 2) = 2 - x$$

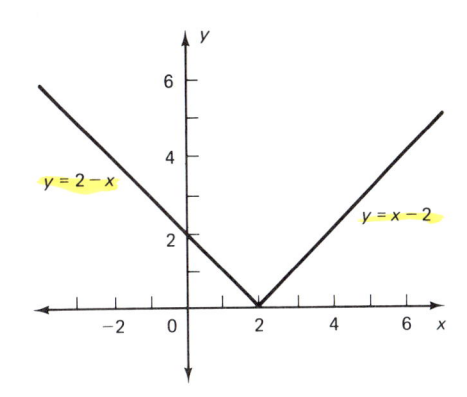

for $x \geq 2$ and $x < 2$, respectively. It is shown in the figure above. Note that $y \geq 0$ for all x.

Example 5 Find the domain and range and draw the graph of the function
$$f(x) = \frac{|x|}{x}.$$

Solution Clearly, the function is not defined for $x = 0$, since for this value of x the denominator becomes zero. Thus, the domain of f is the set of all real numbers except zero. If $x > 0$,
$$f(x) = \frac{|x|}{x} = \frac{x}{x} = 1.$$
If $x < 0$,
$$f(x) = \frac{|x|}{x} = \frac{-x}{x} = -1.$$
[For example, $f(-3) = |-3|/(-3) = 3/(-3) = -1$.] Thus the range consists of only two numbers, 1 and -1.

The graph of f consists of two horizontal straight lines, one above and one below the x axis as shown in the adjoining figure. Note that the endpoints where the two lines meet the y axis are *not* included in the graph. *We indicate this by drawing small open circles at the ends of the two lines.*

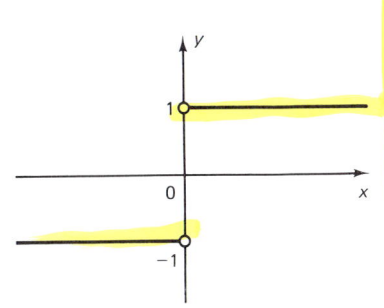

SOME CLASSES OF FUNCTIONS

A function f defined by the relation
$$f(x) = a_n x^n + a_{n-1} x^{n-1} + \cdots + a_1 x + a_0 \quad (a_n \neq 0)$$
where $a_0, a_1, \ldots, a_n$ are constants and n is a nonnegative integer, is said to be a ***polynomial function of degree n***. For example, the functions f and g defined by
$$f(x) = 3x^7 - 5x^4 + 2x - 1, \quad g(x) = x^3 + 7x^2 - 5x + 3$$
are polynomial functions of degree 7 and 3, respectively.

If $n = 0$, we obtain simply $f(x) = a_0$, a constant. Thus a polynomial function of degree zero is a constant function. If $n = 1$, we have $f(x) = a_1 x + a_0$ ($a_1 \neq 0$). Thus a polynomial function of degree 1 is a linear function.

A polynomial function of degree 2 is called a **quadratic function**. It has the general form

$$f(x) = a_2 x^2 + a_1 x + a_0 \qquad (a_2 \neq 0).$$

This type of function will be discussed in detail in Section 7.2. Its graph is always a parabola, which opens upward if $a_2 > 0$ and downward if $a_2 < 0$.

Similarly, a polynomial function of degree 3 is called a **cubic**. For example, the function f defined by

$$f(x) = 2x^3 - 5x^2 + 7x + 1$$

is a cubic function.

If a function can be expressed as the quotient of two polynomial functions, then it is called a **rational function**. Examples of rational functions are

$$f(x) = \frac{x^2 - 9}{x - 4}, \qquad g(x) = \frac{2x^3 - 7x + 1}{5x^2 - 2}.$$

In general, any rational function has the form $f(x) = p(x)/q(x)$, where $p(x)$ and $q(x)$ are polynomials in x.

A function f whose value $f(x)$ is found by a finite number of algebraic operations is called an **algebraic function**. The algebraic operations are addition, subtraction, multiplication, division, raising to powers, and extracting roots. For example, the functions f and g defined by

$$f(x) = \frac{(2x + 1)^2 - \sqrt{x^3 + 1}}{(x^2 + 1)^4}, \qquad g(x) = (2x^2 - 1)^{-1/7} + 5x^{3/4}$$

are algebraic functions.

Apart from algebraic functions, there are other functions called **transcendental functions**. Examples of transcendental functions are logarithmic functions and exponential functions, which we shall discuss in Chapter VIII. There are many other transcendental functions—for example, the trigonometric and inverse trigonometric functions.

Diagnostic Test 6.2

Fill in the blanks.

1. The function $f(x) = mx + b$ ($m \neq 0$) is called a _____ function.
2. The graph of a linear function is always a _____.
3. The graph of a constant function is a _____ straight line.
4. The graph of $y = x^2$ is a curve called a _parabola_.
5. The graph of $y = x^{-1}$ is a curve called a _hyperbola_.

6. The function $f(x) = |x|$ is called the ___absolute___.
7. The range of $f(x) = |x - 2|$ is the set ___≥ 0___.
8. The range of $f(x) = |x| - 2$ is the set ___-2 → +∞___.
9. If $f(x) = x/|x|$, then $f(x) = $ ___-1___ for $x < 0$.
10. $f(x) = x^6 - x^7$ is a ___7___ function of degree _____.

Exercises 6.2

(1–10) State whether the following functions are polynomials (give their degree) or rational functions or algebraic functions or none of these.

1. $x^{3/2} + 2x^{1/2} + 1$.
2. $\sqrt{x}$.
3. $(x^2 - 1)/(x - 2)$.
4. $\sqrt{x^2 + 2x + 1}$.
5. $|x^2 + 3x + 2|$.
6. $x + |x|$.
7. $10x^9 - 2x$.
8. $x^7 + x^{-7}$.
9. $\dfrac{x^2 - 4}{x - 2}$.
10. $\sqrt{x^4 + 6x^2 + 9}$.

(11–16) Draw the graphs of the following functions by plotting a few points and joining by a smooth curve.

11. $f(x) = 1 - \frac{1}{2}x^2$.
12. $f(x) = \frac{1}{2}x^2 - 2$.
13. $f(x) = \sqrt{x + 1}$.
14. $f(x) = \sqrt{4 - x}$.
15. $f(x) = (x - 2)^{-1}$.
16. $f(x) = (x + 1)^{-1}$.

(17–34) Find the domains of the following functions and sketch their graphs.

17. $f(x) = \frac{1}{3}x^3$.
18. $f(x) = 1 - x^3$.
19. $f(x) = 2\sqrt{x}$.
20. $g(x) = \sqrt{x - 2}$.
21. $g(x) = -\sqrt{3 - x}$.
22. $f(y) = 2\sqrt{-1 - y}$.
23. $h(t) = \dfrac{1}{t + 2}$.
24. $f(x) = \dfrac{-3}{x - 2}$.
25. $F(y) = \dfrac{1}{2 - y}$.
26. $H(t) = \dfrac{-1}{t + 1}$.
27. $f(x) = 2 - |x|$.
28. $g(x) = |x| + 3$.
29. $F(x) = |x + 3|$.
30. $F(x) = -|x - 2|$.
31. $g(t) = |t| + t$.
32. $g(t) = |t| - t$.
33. $f(x) = \dfrac{|x - 3|}{x - 3}$.
34. $g(x) = \dfrac{2 - x}{|x - 2|}$.

35. When an object is thrown vertically into the air with a speed of 80 feet per second, its height above the ground after t seconds is given by the formula $h = 80t - 16t^2$. Find h when $t = \frac{1}{2}, 1, \frac{3}{2}$, and 2. Draw the graph of h as a function of t. What can you say about $t = \frac{5}{2}$ and $t = 5$?

36. If in Exercise 35 the initial velocity is 120 feet per second, then $h = 120t - 16t^2$. Find h when $t = \frac{1}{2}, 1, \frac{3}{2}, 2, \frac{5}{2}, 3$ and draw the graph of h as a function of t. What can you say about $t = \frac{15}{4}$ and $t = \frac{15}{2}$?

37. Express the perimeter P of a square as a function of its area A. Sketch the graph of this function.

38. Express the surface area S of a sphere as a function of its volume V. Sketch the graph of this function.

*39. Using the absolute-value function, write a single formula that gives $f(x)$ for all $x \geq 0$ when
$$f(x) = \begin{cases} 2 + x & \text{for } 0 \leq x \leq 2 \\ 3x - 2 & \text{for } x \geq 2. \end{cases}$$

*40. Repeat Exercise 39 for the function
$$f(x) = \begin{cases} x & \text{for } 0 \leq x \leq 1 \\ 2 - x & \text{for } x \geq 1. \end{cases}$$

6.3

Combinations of Functions

A variety of situations arise in which we have to combine two or more functions in one of several ways to get new functions. For example, let $f(t)$ and $g(t)$ denote a person's income from two different sources at time t; then his combined income is $f(t) + g(t)$. From the two functions f and g we have in this way obtained a third function, the **sum** of f and g. If $C(x)$ denotes the cost of producing x units of a certain commodity and $R(x)$ the revenue obtained from the sale of x units, the profit $P(x)$ obtained by producing and selling x units is given by $P(x) = R(x) - C(x)$. The new function P so obtained is the **difference** of the two functions R and C.

As well as the sum and difference of two functions, we also encounter products and quotients of functions. For example, let $P(t)$ be the population size of a certain fish stock as a function of time and let $W(t)$ be the average weight per fish; then the total weight of the whole stock is given by the **product** of these two functions, $P(t)W(t)$. Alternatively, if $P(t)$ is the population of a certain country and $N(t)$ the national income at time t, then the per capita income is $N(t)/P(t)$. Here the new function is formed as the **quotient** of the two original functions.

These kinds of examples lead us to the following abstract definitions. Given two functions f and g, the sum, difference, product, and quotient functions are defined as follows:

Sum: $\qquad (f+g)(x) = f(x) + g(x)$

Difference: $\qquad (f-g)(x) = f(x) - g(x)$

Product: $\qquad (f \cdot g)(x) = f(x) \cdot g(x)$

Quotient: $\qquad \left(\dfrac{f}{g}\right)(x) = \dfrac{f(x)}{g(x)}, \quad$ provided $g(x) \neq 0$.

The domains of the sum, difference, and product functions are all equal to the common part of the domains of f and g—that is, the set of x at which both f and g are defined. In the case of the quotient function, the domain is the common part of the domains of f and g except those values of x for which $g(x) = 0$.

Example 1 If $f(x) = x^2 - 4$ and $g(x) = x - 2$, evaluate
(a) $(f+g)(1)$.
(b) $(f-g)(1)$.
(c) $(fg)(1)$.
(d) $(f/g)(1)$.
(e) $(f/g)(2)$.

Solution We have that
$$f(1) = 1^2 - 4 = -3 \quad \text{and} \quad g(1) = 1 - 2 = -1.$$
(a) $(f+g)(1) = f(1) + g(1) = -3 + (-1) = -4.$
(b) $(f-g)(1) = f(1) - g(1) = -3 - (-1) = -2.$
(c) $(fg)(1) = f(1) \cdot g(1) = (-3)(-1) = 3.$
(d) $(f/g)(1) = f(1)/g(1) = (-3)/(-1) = 3.$
(e) $f(2) = 2^2 - 4 = 0$ and $g(2) = 2 - 2 = 0.$
Since $g(2)$ is zero, $(f/g)(2)$ is not defined. The fact that $f(2)$ is also zero does not alter this situation.

Example 2 Let $f(x) = 1/(x-1)$ and $g(x) = \sqrt{x}$. Find $f+g$, $f-g$, $f \cdot g$, f/g, and g/f. Determine also their domains.

Solution We have:
$$(f+g)(x) = f(x) + g(x) = \frac{1}{x-1} + \sqrt{x}$$

$$(f-g)(x) = f(x) - g(x) = \frac{1}{x-1} - \sqrt{x}$$

$$(f \cdot g)(x) = f(x) \cdot g(x) = \frac{1}{x-1} \cdot \sqrt{x} = \frac{\sqrt{x}}{x-1}$$

$$\left(\frac{f}{g}\right)(x) = \frac{f(x)}{g(x)} = \frac{\frac{1}{x-1}}{\sqrt{x}} = \frac{1}{\sqrt{x}(x-1)}.$$

$$\left(\frac{g}{f}\right)(x) = \frac{g(x)}{f(x)} = \frac{\sqrt{x}}{\frac{1}{x-1}} = \sqrt{x}(x-1).$$

$f(x)$ is not defined for $x = 1$, since for this value of x the denominator is zero; thus, the domain of f is the set of all real numbers except 1. Similarly $g(x)$ is defined for values of x for which the expression under the radical sign is nonnegative—that is, $x \geq 0$. Thus,
$$D_f = \{x \mid x \neq 1\}, \qquad D_g = \{x \mid x \geq 0\}.$$
The common part of D_f and D_g is
$$\{x \mid x \geq 0 \quad \text{and} \quad x \neq 1\} \tag{i}$$
and this set provides the domain of $f+g$, $f-g$, and $f \cdot g$. $g(x) = \sqrt{x}$ is zero when $x = 0$, so this point must be excluded from the domain of f/g. Thus, the domain of f/g is
$$\{x \mid x > 0 \quad \text{and} \quad x \neq 1\}.$$
Since $f(x)$ is never zero, the domain of g/f is again the common part of D_f and D_g, namely the set (i) above. It appears from the formula $(g/f)(x) = \sqrt{x}(x-1)$ that this function is well defined when $x = 1$. In spite of this, it is still necessary to exclude $x = 1$ from the domain of this function, since g/f is defined only at points where both g and f are defined.

Sec. 6.3   Combinations of Functions

COMPOSITION OF TWO FUNCTIONS

A rather different way in which two functions can be combined to yield a third function is the *composition* of functions. In order to illustrate this, let us consider a specific example.

Let $f(x) = x^2$ and $g(x) = x + 2$. Obviously, then, $g(1) = 3$: the value of the function g when its argument is 1 is equal to 3. Suppose we use this value as the argument of f. We get $f(3) = 3^2 = 9$. In this way, we have used the original value of x as the argument of g and then the value of $g(x)$ as the argument of f. The final value, 9, represents the value at $x = 1$ of a new function, called the composition of f and g.

We can repeat this calculation with other values of x. For example, with $x = 2$ we have $g(2) = 4$, and, using this as the argument of f, we get $f(4) = 4^2 = 16$. With $x = 3$, we have $g(3) = 5$ and then $f(5) = 25$. With $x = -3$, we have $g(x) = -1$ and then $f(-1) = (-1)^2 = 1$.

You may like to check some of the other values in the following table.

x	-6	-5	-4	-3	-2	-1	0	1	2	3	4
$g(x)$	-4	-3	-2	-1	0	1	2	3	4	5	6
Value of the composition $= [g(x)]^2$	16	9	4	1	0	1	4	9	16	25	36

It is clear that the procedure we have described does define a perfectly good function. For each value of x, there is a single, uniquely defined value of the composition.

DEFINITION Let f and g be two functions. Let x belong to the domain of g and be such that $g(x)$ belongs to the domain of f. Then the *composite function* $f \circ g$ (read as "f circle g") is defined by

$$(f \circ g)(x) = f(g(x)).$$

Note that the value $g(x)$ is used as the argument of f. (Or, in other words, the output of the function g is used as input for the function f.)

Example 3 If $f(x) = x - 3\sqrt{x}$ and $g(x) = x^2$, evaluate
    (a) $(f \circ g)(2)$.   (b) $(f \circ g)(-2)$.   (c) $(f \circ g)(4)$.

Solution (a) We have $g(2) = 2^2 = 4$. Therefore,

$$(f \circ g)(2) = f(g(2)) = f(4) = 4 - 3\sqrt{4} = 4 - 6 = -2.$$

    (b) We have $g(-2) = (-2)^2 = 4$. Therefore,

$$(f \circ g)(-2) = f(g(-2)) = f(4) = -2$$

as in part (a).

(c) We have $g(4) = 16$. Therefore,
$$(f \circ g)(4) = f(g(4)) = f(16) = 16 - 3\sqrt{16} = 16 - 12 = 4.$$

We can similarly define $(g \circ f)(x) = g(f(x))$. The value $f(x)$ is here used as the argument of g. In general, $f \circ g$ and $g \circ f$ are quite different functions.

Example 4 Let $f(x) = 1/(x - 2)$ and $g(x) = \sqrt{x}$. Evaluate
(a) $(f \circ g)(9)$.
(b) $(f \circ g)(4)$.
(c) $(f \circ g)(x)$.
(d) $(g \circ f)(6)$.
(e) $(g \circ f)(1)$.
(f) $(g \circ f)(x)$.

Solution (a) $g(9) = \sqrt{9} = 3$. Therefore,
$$(f \circ g)(9) = f(g(9)) = f(3) = \frac{1}{3 - 2} = 1.$$

(b) $g(4) = \sqrt{4} = 2$.
$$(f \circ g)(4) = f(g(4)) = f(2) = \frac{1}{2 - 2}$$
which is not defined. The value $x = 4$ does not belong to the domain of $f \circ g$, so that $(f \circ g)(4)$ cannot be found.

(c) $g(x) = \sqrt{x}$.
$$(f \circ g)(x) = f(g(x)) = \frac{1}{g(x) - 2} = \frac{1}{\sqrt{x} - 2}.$$

(d) $f(6) = \frac{1}{6 - 2} = \frac{1}{4}$.
$$(g \circ f)(6) = g(f(6)) = g(\tfrac{1}{4}) = \sqrt{\tfrac{1}{4}} = \tfrac{1}{2}.$$

(e) $f(1) = \frac{1}{1 - 2} = -1$.
$$(g \circ f)(1) = g(f(1)) = g(-1) = \sqrt{-1}$$
which is not a real number. We cannot evaluate $(g \circ f)(1)$, because 1 does not belong to the domain of $g \circ f$.

(f) $f(x) = \frac{1}{x - 2}$.
$$(g \circ f)(x) = g(f(x)) = \sqrt{f(x)} = \sqrt{\frac{1}{x - 2}} = \frac{1}{\sqrt{x - 2}}.$$

The composition $(f \circ g)(x) = f(g(x))$ is defined, provided that $g(x)$ is defined (that is, x is in D_g) and $g(x)$ is in the domain of f. Thus
$$D_{f \circ g} = \{x \mid x \in D_g \text{ and } g(x) \in D_f\}.$$

It can be shown that, for the functions in Example 4,
$$D_{f \circ g} = \{x \mid x \geq 0 \text{ and } x \neq 4\}$$
and
$$D_{g \circ f} = \{x \mid x > 2\}.$$

Example 5 Assume that the size of police force needed to provide effective law enforcement in a city whose population is P (thousands) is given by the formula

$$F = 0.2P^{5/4}.$$

The population of the city is increasing according to the formula $P = 500 + 40t$, where t is time measured in years from 1980. How does the police force vary as a function of time? What size police force will be required in 1990?

Solution We are given F as a function of P and P as a function of t. To express F as a function of t we need the composition of these two given functions. We obtain this very simply by substituting for P in the formula for F:

$$F = 0.2P^{5/4} = 0.2(500 + 40t)^{5/4}.$$

In 1990, $t = 10$ and therefore

$$F = 0.2[500 + (40)(10)]^{5/4} = 0.2(900)^{5/4} = 985.9.$$

Thus a police force of 986 will be needed in 1990.

Diagnostic Test 6.3

Fill in the blanks.

1. By definition, $(f + g)(x) =$ _____, $(f \cdot g)(x) =$ _____, $(g/f)(x) =$ _____ provided _____.
2. If f and g are any two functions, then the domain of $f + g$ is _____ the domain of $f \cdot g$.
3. If $f(x) = 3$ and $g(x) = x^2$, then domain of $f + g$ is the set _____.
4. If $f(x) = x^2 + 7x - 1$ and $g(x) = 1/(x - 2)$, then the domain of $f \cdot g$ is the set _____.
5. If $f(x) = (x - 3)^2$ and $g(x) = 1/(x - 3)$, then the domain of $f \cdot g$ is the set _____.
6. By definition, $(f \circ g)(x) =$ _____.
7. If $f(x) = 3$ and $g(x) = 7$, then $(f \circ g)(x) =$ _____ and $(g \circ f)(x) =$ _____.
8. If $f(x) = 2$ and $g(x) = x^2$, then $(f \circ g)(x) =$ _____ and $(g \circ f)(x) =$ _____.

Exercises 6.3

(1–10) If $f(x) = x^2 - 3x$ and $g(x) = \sqrt{x - 1}$, evaluate the following.

1. $(f + g)(1)$.
2. $(g + f)(5)$.
3. $(g - f)(5)$.
4. $(f - g)(10)$.
5. $(fg)(0)$.
6. $(fg)(2)$.
7. $(f/g)(2)$.
8. $(f/g)(-1)$.
9. $(g/f)(3)$.
10. $(g/f)(5)$.

(11–18) If $f(x) = \sqrt{x}/(x-1)$ and $g(x) = (x-1)/\sqrt{x}$, evaluate the following.

11. $(f+g)(0)$. **12.** $(g-f)(1)$. **13.** $(fg)(2)$. **14.** $(fg)(-1)$.

15. $(f/g)(3)$. **16.** $(g/f)(5)$. **17.** $(f/g)(-1)$. **18.** $(g/f)(1)$.

(19–24) Find the sum, difference, product, and both quotients of the two functions f and g in each of the following exercises. Determine the domains of the resulting functions.

19. $f(x) = x^2, g(x) = \dfrac{1}{x-1}$.

20. $f(x) = x^2 + 1, g(x) = \sqrt{x}$.

21. $f(x) = \sqrt{x-1}, g(x) = \dfrac{1}{x+2}$.

22. $f(x) = 1 + \sqrt{x}, g(x) = \dfrac{2x+1}{x+2}$.

23. $f(x) = (x+1)^2, g(x) = \dfrac{1}{x^2-1}$.

24. $f(x) = \sqrt{x+1}, g(x) = \dfrac{1}{\sqrt{x-1}}$.

(25–32) Given $f(x) = x^2$ and $g(x) = \sqrt{x-1}$, evaluate.

25. $(f \circ g)(5)$. **26.** $(g \circ f)(3)$. **27.** $(f \circ g)(\tfrac{5}{4})$. **28.** $(g \circ f)(-2)$.

29. $(f \circ g)(\tfrac{1}{2})$. **30.** $(g \circ f)(\tfrac{1}{3})$. **31.** $(f \circ g)(2)$. **32.** $(g \circ f)(1)$.

(33–40) Given $f(x) = 1 + \sqrt{x}$ and $g(x) = x^2 - 3x$, evaluate.

33. $(f \circ g)(0)$. **34.** $(g \circ f)(0)$. **35.** $(f \circ g)(3)$. **36.** $(g \circ f)(1)$.

37. $(f \circ g)(2)$. **38.** $(g \circ f)(-1)$. **39.** $(f \circ g)(4)$. **40.** $(g \circ f)(4)$.

(41–48) Determine $(f \circ g)(x)$ and $(g \circ f)(x)$ in the following exercises.

41. $f(x) = x^2, g(x) = 1 + x$.

42. $f(x) = \sqrt{x} + 1, g(x) = x^2$.

43. $f(x) = \dfrac{1}{x+1}, g(x) = \sqrt{x} + 1$.

44. $f(x) = 2 + \sqrt{x}, g(x) = (x-2)^2$.

45. $f(x) = x^2 + 2, g(x) = x - 3$.

46. $f(x) = \dfrac{1}{x+1}, g(x) = \dfrac{1}{x^2+1}$.

47. $f(x) = 2x - 3, g(x) = 4$.

48. $f(x) = -2, g(x) = x^3 - 1$.

49. If $F(t) = \dfrac{t}{1+t}$ and $G(t) = \dfrac{t}{1-t}$, show that
$$F(t) - G(t) = -2G(t^2).$$

50. If $f(x) = x^2$ and $g(x) = \sqrt{x}$, find $(f \circ g)(x)$ and $(g \circ f)(x)$.

51. The rate at which a chemical is produced in a certain reaction depends on temperature T according to the formula $R = T^5 + 3\sqrt{T}$. If T varies with time t according to $T = 3(t+1)$, express R as a function of t and evaluate R when $t = 2$.

52. The velocity of a falling body varies with distance traveled according to the formula $v = 8\sqrt{y}$ (v = velocity in feet per second, y = distance in feet). The distance fallen varies with time t (in seconds) according to the formula $y = 16t^2$. Express v as a function of t.

53. The demand x for a certain commodity is given by $x = 2000 - 15p$, where p is the price per unit of the commodity. The monthly revenue R obtained from the sales of this commodity is given by $R = 2000p - 15p^2$. How does R depend on x?

54. A manufacturer can sell q units of his product at a price p per unit where $20p + 3q = 600$. As a function of quantity q demanded in the market, the total weekly revenue R is given by $R = 30q - 0.15q^2$. How does R depend on the price p?

55. If $f(x) = ax - 4$ and $g(x) = bx + 3$, find the condition on a and b such that $(f \circ g)(x) = (g \circ f)(x)$ for all x.

56. If $f(x) = x + a$ and $g(t) = t + b$, show that $(f \circ g)(x) = (g \circ f)(x)$.

Sec. 6.3 Combinations of Functions

6.4

Functions as Mappings; Inverse Functions

Given a function f and $y = f(x)$, we say that f *maps* each number x in the domain into the corresponding value y. For example, if $f(x) = x^2$, then f maps the value 2 into 4, it maps 3 into 9, it maps -1 into 1, and so on. Each number x in the domain is mapped by f into a certain number y in the range.

We can represent the action of f by the accompanying picture. The domain and range of f are represented by shaded regions. Each point in the shaded region representing X is imagined to correspond to a certain number

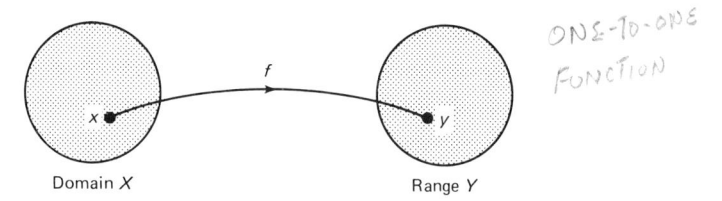

ONE-TO-ONE FUNCTION

Domain X — f → Range Y

x in the domain of f, and similarly for the points in the shaded region representing the range Y. If $y = f(x)$, then the action of f is represented by an arrow going from the point x in X to the point y in Y. Although only one such arrow is drawn in the above picture, there is a similar arrow starting from every number x in the domain X.

Example 1 Consider a small class of five students, Sue, Liz, Pam, Joe, and Fred. Let f be the function that assigns to each student his or her grade. Suppose the table of values of this function is as follows:

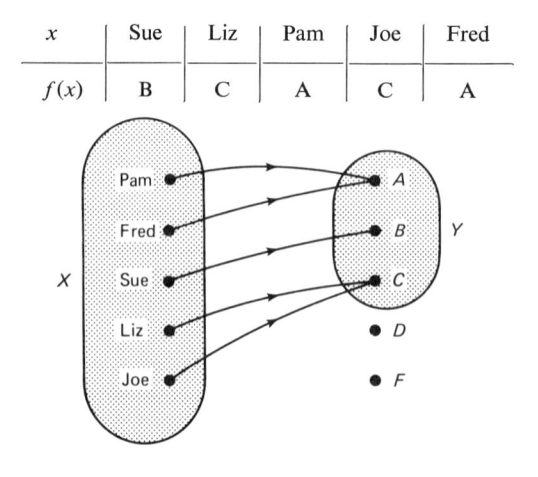

x	Sue	Liz	Pam	Joe	Fred
$f(x)$	B	C	A	C	A

X: Pam, Fred, Sue, Liz, Joe
Y: A, B, C, D, F

232   *Ch. 6   Functions and Graphs*

(that is, Sue gets grade B, Liz gets C, and so on). The domain is the set consisting of the five students, and the range is the set of their grades. The function f can be represented by the arrows in the diagram above. The grades D and F are not in the range of this particular function.

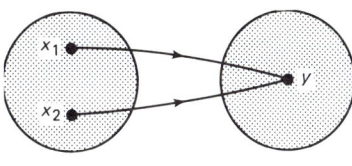

It is quite possible, as in this example, for two such arrows to end at the same value y in the range. If x_1 and x_2 are such that $f(x_1) = f(x_2)$, then f maps both x_1 and x_2 into the same y (see the adjacent figure). For example, if $f(x) = x^2$, then $f(-2) = 4$ and $f(2) = 4$. Thus f maps both 2 and -2 into the number 4.

If a function f is such that only one arrow ends at each and every value y in the range, then we say that f is a *one-to-one mapping* or a *one-to-one function*.

DEFINITION A function f is *one-to-one* if each value y in the range corresponds to exactly one value of x in the domain. In other words, if $x_1 \neq x_2$, then $f(x_1) \neq f(x_2)$.

Examples 2 (a) The function $f(x) = x^2$ is not one-to-one. For example, both $x = 2$ and $x = -2$ correspond to the value 4 in the range.
(b) The constant function $f(x) = b$ is not one-to-one. Every value of x corresponds to the same value b in the range.
(c) The linear function $f(x) = x - 2$ is one-to-one. For, if $x_1 \neq x_2$ then $x_1 - 2 \neq x_2 - 2$. In other words, $f(x_1) \neq f(x_2)$.
(d) The function $f(x) = x^7$ is one-to-one. For, if $x_1 \neq x_2$, then $x_1^7 \neq x_2^7$. In other words, $f(x_1) \neq f(x_2)$.

Note If $y = f(x)$ is a one-to-one function whose domain and range are subsets of the real numbers, then any horizontal line in the xy plane meets the graph of f in exactly one point, or else not at all. The reason is that, if a horizontal line on which $y = y_1$ were to meet the graph in two or more points, then there would be two or more values of x at which $f(x) = y_1$.

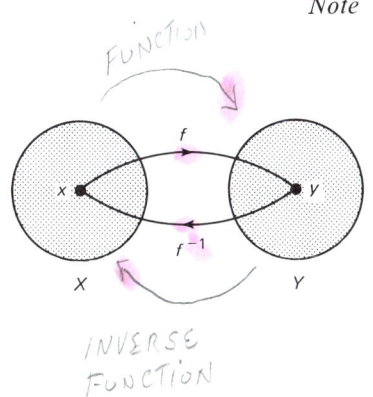

Let $f(x)$ be a one-to-one function. Then, given any number y in the range of f, there exists exactly one value of x in the domain such that $y = f(x)$. This means that there exists a second function that maps each y in the range of f into the corresponding x in the domain of f. This function is called the *inverse* of the function f and is denoted by the symbol f^{-1}.

The situation is illustrated in the adjoining figure. The original function f maps x to y; the inverse function f^{-1} maps y back to x.

Warning f^{-1} is not to be confused with the inverse power,

$$[f(x)]^{-1} = \frac{1}{f(x)}.$$

Note The domain of the function f^{-1} is equal to the range Y of the original function f. The range of f^{-1} is the domain X of f.

Sec. 6.4 Functions as Mappings; Inverse Functions

Examples 3 (a) Let $f(x) = 2x - 3$. Then $f(1) = -1$; that is, f maps the number 1 to -1. Then f^{-1} must map -1 back to 1. That is, $f^{-1}(-1) = 1$. Also $f(4) = 5$, so that f maps the number 4 to 5. Then f^{-1} must map 5 back to 4—that is, $f^{-1}(5) = 4$.

(b) Let $f(x) = x^3 - 3$. Then $f(2) = 5$. Since f maps 2 to 5, f^{-1} must map 5 back to 2—that is, $f^{-1}(5) = 2$. Similarly

$$f(3) = 24 \quad \text{so} \quad f^{-1}(24) = 3$$
$$f(-2) = -11 \quad \text{so} \quad f^{-1}(-11) = -2$$
$$f(0) = -3 \quad \text{so} \quad f^{-1}(-3) = 0, \quad \text{and so on.}$$

If $y = f(x)$, then f maps x to y. Then f^{-1} must map y back to x; that is, $f^{-1}(y) = x$. We see that the inverse function expresses x as a function of y. Consequently, *we can obtain the algebraic expression for f^{-1} by solving the equation $y = f(x)$ for x as a function of y.*

To summarize: For a one-to-one function f,

$$\boxed{y = f(x) \quad \text{if and only if} \quad x = f^{-1}(y).}$$

Example 4 Find the inverse of the function $f(x) = 2x + 1$.

Solution Setting $y = f(x) = 2x + 1$, we must solve for x as a function of y.

$$2x = y - 1$$
$$x = \frac{y-1}{2}.$$

Therefore the inverse function is given by

$$f^{-1}(y) = \frac{y-1}{2}.$$

The graphs of $y = f(x)$ and $x = f^{-1}(y)$ are shown in the following figures. Both graphs in this case are straight lines. Observe that when plotting the graph of $x = f^{-1}(y)$, we take the y axis horizontal and the x axis vertical, because y is the independent variable.

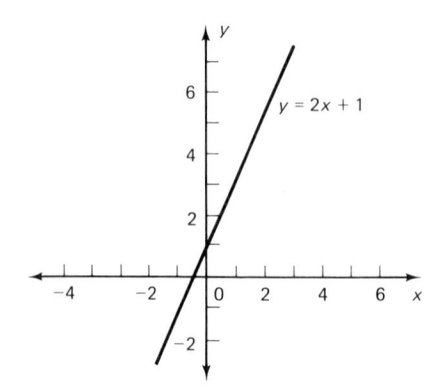

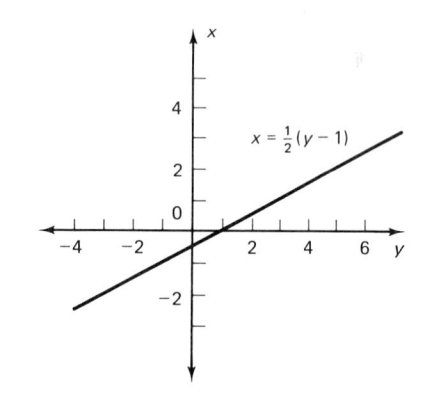

Example 5 Find the inverse of the function $f(x) = x^3$ and sketch its graph.

Solution Setting $y = f(x) = x^3$, we solve for x, obtaining
$$x = f^{-1}(y) = y^{1/3}.$$
The graphs of $y = f(x)$ and $x = f^{-1}(y)$ are shown in the following figures.

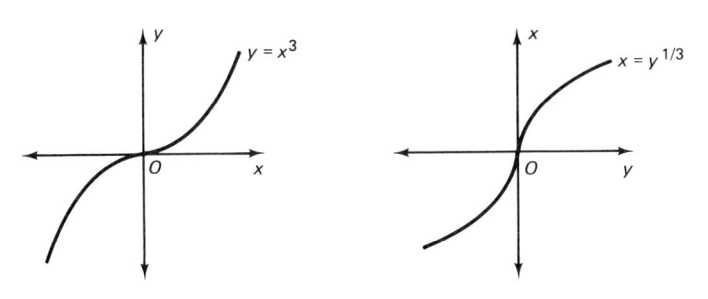

It is perhaps evident from these two examples that the graphs of a function $y = f(x)$ and its inverse function $x = f^{-1}(y)$ are closely related. In fact, the graph of the inverse function is obtained by flipping over the graph of the original function so that the coordinate axes become interchanged. For example, you might try holding the graph of $y = x^3$ in front of a mirror in such a way that the y axis is horizontal and points to the right and the x axis points vertically upward. The reflection you will see will be just the graph $x = y^{1/3}$, which is shown in the right-hand figure above.

The graphs of $y = f(x)$ and $x = f^{-1}(y)$ consist of precisely the same pairs of values of x and y. The difference rests only in that the axes are drawn in different directions in the two cases. Always the independent variable is given the horizontal axis.

Example 6 If $f(x) = (2x - 1)/(4x - 3)$, find $f^{-1}(y)$ and $f^{-1}(x)$.

Solution Setting $y = f(x)$, we obtain
$$y = \frac{2x - 1}{4x - 3}.$$
We must solve for x in terms of y. So, multiplying both sides by $(4x - 3)$ to eliminate the fraction, we get
$$y(4x - 3) = 2x - 1$$
or
$$4xy - 3y = 2x - 1.$$
Next we collect all the x terms on the left:
$$4xy - 2x = 3y - 1$$
$$x(4y - 2) = 3y - 1$$
and therefore
$$x = \frac{3y - 1}{4y - 2}.$$

Sec. 6.4 *Functions as Mappings; Inverse Functions*

Since this solution must be $x = f^{-1}(y)$, we have
$$f^{-1}(y) = \frac{3y-1}{4y-2}.$$
We are also asked to compute $f^{-1}(x)$. But this just means renaming the variable. For example, $f^{-1}(u) = (3u-1)/(4u-2)$, $f^{-1}(z) = (3z-1)/(4z-2)$, and so on. In particular,
$$f^{-1}(x) = \frac{3x-1}{4x-2}.$$

Another way of seeing the relationship between the two graphs is to plot the graphs of f and f^{-1} on the same axes. As in the preceding example, we construct $f^{-1}(x)$ and plot the graph of $y = f^{-1}(x)$ on the same xy axes as $y = f(x)$. It turns out that the two graphs are always reflections of one another in the line $y = x$.

The following figure shows the graphs of the function $f(x) = x^3$ and its inverse function, $f^{-1}(x) = x^{1/3}$, drawn on the same axes. Clearly the two graphs are reflections of one another in the line $y = x$. (See Exercise 45.)

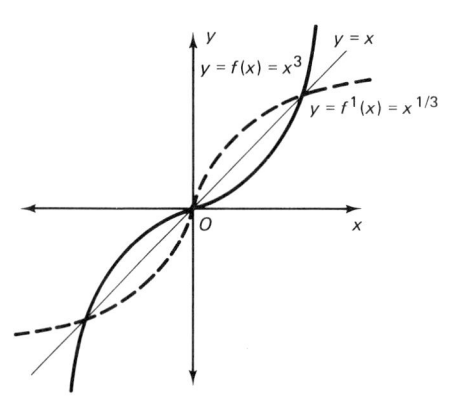

Not every function has an inverse. Consider, for example, the function $y = x^2$. Solving for x in terms of y, we obtain
$$x^2 = y \quad \text{—that is,} \quad x = \pm\sqrt{y}.$$
For any value of y in the region $y > 0$, then, there are two possible values of x, and so we cannot say that x is a function of y. The function $y = x^2$ is not one-to-one.

This example is shown graphically in the figures at the top of page 237. The left figure shows the graph of $y = x^2$ which is a parabola opening upward. The right figure shows the same graph but with the axes flipped over, the y axis being horizontal and the x axis vertical. For each $y > 0$ we have the two values of x, $x = +\sqrt{y}$ and $x = -\sqrt{y}$. For example, when $y = 1$, x has the values $+1$ and -1, both satisfying the relation $y = x^2$.

The graph in the right-hand figure corresponds to two functions rather than to one. The upper branch of the parabola is the graph of $x = +\sqrt{y}$, while the lower branch is the graph of $x = -\sqrt{y}$. Thus we can say that the function $y = x^2$ has two possible inverse functions, one given by $x = +\sqrt{y}$ and the other by $x = -\sqrt{y}$.

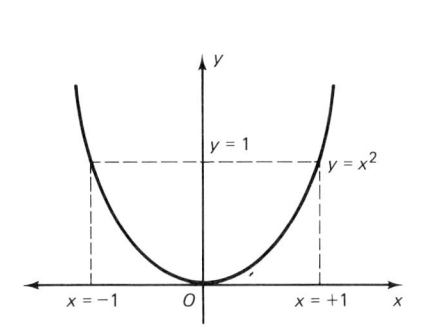

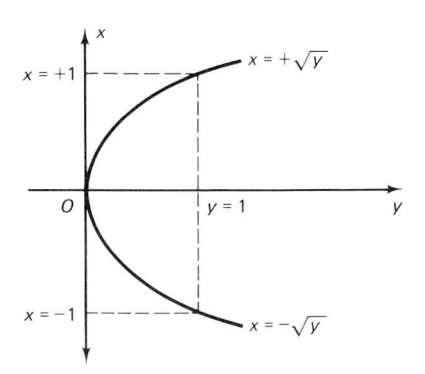

In a case such as this, we can make the definition of f^{-1} unambiguous by restricting the values of x. For example, if x is restricted to the region $x \geq 0$, then $y = x^2$ has the unique inverse $x = +\sqrt{y}$. On the other hand, if x is restricted to the region $x \leq 0$, then the inverse is given by $x = -\sqrt{y}$. Placing a restriction on x in this way means restricting the domain of the original function f. We conclude, therefore, that in cases where a function $y = f(x)$ is not one-to-one, an inverse function can be defined if we place a suitable restriction on the domain of f.

It is worth observing that *a function $f(x)$ has a unique inverse whenever any horizontal line intersects its graph in at most one point.*

The composition of two functions, $f \circ g$, has an elegant interpretation in terms of mappings. Let $y = g(x)$, so that g maps x to y, and let $z = f(y)$, so that f maps y to z, as shown in the following figure. Then

$$z = f[g(x)] = (f \circ g)(x).$$

That is, $f \circ g$ is the function that maps x to z. So the mapping $f \circ g$ corresponds to the mapping g followed by the mapping f.

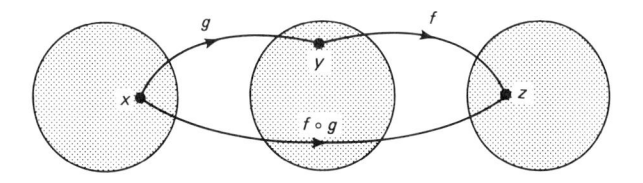

If a mapping f is followed by its inverse mapping f^{-1}, we get back to the original value of the argument. That is,

$(f^{-1} \circ f)(x) = x$	for all x in the domain of f
$(f \circ f^{-1})(y) = y$	for all y in the range of f.

Sec. 6.4   *Functions as Mappings; Inverse Functions*

These are readily proved. For, we have $y = f(x)$ and $x = f^{-1}(y)$, and therefore
$$(f^{-1} \circ f)(x) = f^{-1}[f(x)] = f^{-1}(y) = x$$
$$(f \circ f^{-1})(y) = f[f^{-1}(y)] = f(x) = y.$$
Thus the composition of f and f^{-1} in either order gives the identity function— the function that leaves the variable unchanged.

Diagnostic Test 6.4

Fill in the blanks.

1. A function $f(x)$ is one-to-one if _____ whenever $x_1 \neq x_2$.
2. If $v = g(u)$ and g is one-to-one, then $u =$ _____.
3. If $x = f^{-1}(t)$, then $t =$ _____.
4. If $f(x) = 1/x$, then $f^{-1}(y) =$ _____.
5. If $f(x) = x - 2$, then $f^{-1}(x) =$ _____.
6. If $f(x) = 3$, then $f^{-1}(x) =$ _____.
7. If $f(x) = x^2 + 1$ ($x > 0$), then $f^{-1}(y) =$ _____.
8. $f(x)$ has a unique inverse if any _____ line meets the graph of f in _____ one point.
9. The graph of $y = f^{-1}(x)$ is the reflection of the graph of $y = f(x)$ in the line _____.
10. $(f \circ f^{-1})(x) =$ _____ for all x in the _____ of f.
11. $(f^{-1} \circ f)(x) =$ _____ for all x in the _____ of f.

Exercises 6.4

(1–6) Using the following table of values, calculate the stated quantity.

x	-3	-2	-1	0	1	2	3
$f(x)$	0	1	2	$\frac{5}{2}$	3	$-\frac{1}{2}$	-1

1. $f^{-1}(0)$. 2. $f^{-1}(-1)$. 3. $f^{-1}(3)$.
4. $f^{-1}(-\frac{1}{2})$. 5. $f^{-1}(-2)$. 6. $f^{-1}(-3)$.

(7–10) Given that $f(x) = x - 2$, calculate the following.

7. $f^{-1}(-1)$. 8. $f^{-1}(-2)$. 9. $f^{-1}(0)$. 10. $f^{-1}(2)$.

(11–24) Find the inverse of the following functions. Draw the graphs of the function and its inverse in each case.

11. $y = -3x - 4$. 12. $y = x - 1$. 13. $p = 4 - \frac{2}{5}x$. 14. $q = 3p + 6$.
15. $y = \sqrt{3x - 4}$. 16. $y = \sqrt{\frac{1}{4}x + 2}$. 17. $y = x^5$. 18. $y = \sqrt{x}$.
19. $y = \sqrt{4 - x}$. 20. $y = -\sqrt{2 - x}$. 21. $y = \dfrac{1}{x - 2}$. 22. $y = \dfrac{1}{x + 2}$.

23. $y = \dfrac{3 - x}{2x + 3}$.

24. $y = \dfrac{2x + 1}{4x - 3}$.

25. Given $f(x) = 3x + 2$ find $f^{-1}(x)$.

26. Given $f(x) = \sqrt{2x - 1}$ find $f^{-1}(x)$.

27. Given $f(x) = \dfrac{3x + 5}{5x + 3}$ find $f^{-1}(x)$.

28. Given $g(x) = \dfrac{2x - 1}{1 - 7x}$ find $g^{-1}(x)$.

(29–34) By placing a suitable restriction on the domain of each of the following functions, find an inverse function.

29. $y = (x + 1)^2$.

30. $y = (3 - 2x)^2$.

31. $y = x^{2/3}$.

32. $y = \sqrt{x^2 + 1}$.

*33. $y = x^2 + 3x + 2$.

34. $y = x^4 - 1$.

35. Can a constant function have an inverse function?

36. Show that any linear function has an inverse function (that is, it is one-to-one).

*37. Show that any quadratic function is not one-to-one.

*38. Show that the inverse of $f \circ g$ is $g^{-1} \circ f^{-1}$.

39. If $s = 16t^2$ expresses distance traveled as a function of time, find an expression for the time needed to travel a distance s.

40. Repeat Exercise 39 if $s = \sqrt{t^2 + 2}$ $(t \geq 0)$.

41. A moving object is accelerating so that its velocity at time t is given by $v = (2t + 3)/(t + 4)$ $(t > 0)$. Find an expression for the time needed to reach a velocity v.

42. Using Example 7 of Section 6.1, express the perimeter of an equilateral triangle as a function of its area.

*43. Using Example 9 of Section 6.1, find an expression for the number of units x as a function of their cost c.

*44. f is a one-to-one function with domain X and range Y. What are the domains and ranges of $f \circ f^{-1}$ and $f^{-1} \circ f$?

*45. Show that if the point (a, b) lies on the graph of $y = f(x)$ then (b, a) lies on the graph of $y = f^{-1}(x)$. Show that (a, b) and (b, a) are reflections of one another in the line $y = x$.

6.5

Horizontal and Vertical Asymptotes

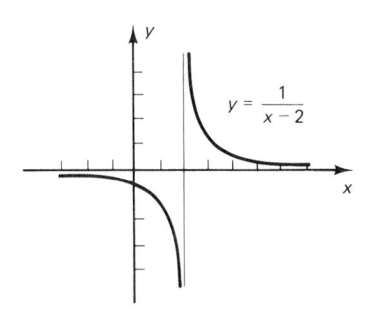

We have already discussed the graph of the function $y = 1/x$ in Section 6.2. The function $y = 1/(x - 2)$ has a similar graph, shown in the adjacent figure. Like the graph of $1/x$, this graph has the x axis as a horizontal asymptote. However, the vertical asymptote is now the line $x = 2$, rather than the y axis.

In general, a horizontal asymptote is determined by the way in which the values of the function behave as x becomes numerically large. The values of the function $1/(x - 2)$ for a sequence of increasing large values of x are given in the following table.

x	4	10	40	100
$1/(x - 2)$	0.5	0.125	0.0263	0.0102

Clearly, as x becomes very large, the value of $1/(x-2)$ becomes very small, which means that the graph must approach closer and closer to the x axis. Observe that $1/(x-2)$ remains positive, even though small, so that the graph always lies a little above the x axis.

In the following table the values of $1/(x-2)$ are given for a sequence of negative values of x that grow increasingly larger in magnitude.

x	-4	-10	-40	-100
$1/(x-2)$	-0.167	-0.0833	-0.0238	-0.0098

Once again the values are seen to get closer and closer to zero as x grows more and more negative. This means that the x axis is again a horizontal asymptote at the left-hand side of the graph. In this case, however, $1/(x-2)$ is negative when x is numerically large and negative, so that the graph always lies a little below the x axis for negative values of x.

Thus the graph of $y = 1/(x-2)$ has the line $y = 0$ (x axis) as its horizontal asymptote. It approaches this line from above as x becomes large and positive, and it approaches the asymptote from below as $|x|$ becomes large with $x < 0$.

It is convenient to have a shorthand notation for these ideas. We write

$$f(x) \to 0^+ \quad \text{as} \quad x \to \infty$$

to mean "$f(x)$ becomes closer and closer to zero, keeping always positive values, as x takes larger and larger positive values." Similarly

$$f(x) \to 0^- \quad \text{as} \quad x \to -\infty$$

means that "$f(x)$ becomes closer and closer to zero with negative values as x becomes numerically larger and negative." We read $x \to \infty$ as "x approaches infinity," and so on.

Examples 1 The following examples illustrate this notation further.
(a) $1/x \to 0^+$ as $x \to \infty$
$1/x \to 0^-$ as $x \to -\infty$
(b) $5/(x+4) \to 0^+$ as $x \to \infty$
$5/(x+4) \to 0^-$ as $x \to -\infty$
(c) $-1/(x-1) \to 0^-$ as $x \to \infty$
$-1/(x-1) \to 0^+$ as $x \to -\infty$
(d) $1/x^2 \to 0^+$ as $x \to \infty$
$1/x^2 \to 0^+$ as $x \to -\infty$

The function $1/(x-2)$ has a *vertical* asymptote at $x = 2$, because as x approaches close to 2, the value of the function increases in magnitude without bound. The following tables illustrate this. In the first table a sequence of values of x are chosen that approach closer and closer to 2 but always remain above 2. The values of $1/(x-2)$ are seen to become larger and larger. In the second table the values of x are all less than 2, and we see that as x approaches 2, the values of $1/(x-2)$ become numerically very large but are negative. These properties are apparent from the graph of $y = 1/(x-2)$.

x	2.1	2.01	2.001	2.0001
$1/(x-2)$	10	100	1000	10,000
x	1.9	1.99	1.999	1.9999
$1/(x-2)$	-10	-100	-1000	$-10,000$

We write $x \to 2^+$ to indicate that x approaches 2 from above—that is, through values that are always greater than 2. Correspondingly $x \to 2^-$ indicates that x approaches 2 from below—that is, through values that are always less than 2. In the case of the function $1/(x-2)$, we write

$$\frac{1}{x-2} \to \infty \quad \text{as} \quad x \to 2^+$$

$$\frac{1}{x-2} \to -\infty \quad \text{as} \quad x \to 2^-.$$

The first statement indicates that $1/(x-2)$ increases without bound as x approaches 2 from above. The second states that $1/(x-2)$ becomes negatively large without bound as x approaches 2 from below.

Example 2 Find the horizontal and vertical asymptotes of the function

$$y = \frac{2}{3-x}.$$

Solution For the horizontal asymptotes we must examine the behavior of the function for large numerical values of x. Now when x is large and positive, $(3-x)$ will be large and negative (for example, if $x = 100$ then $3 - x = -97$). Then $2/(3-x)$ will be small and negative. This leads to the conclusion that

$$\frac{2}{3-x} \to 0^- \quad \text{as} \quad x \to \infty.$$

Similarly when $x < 0$ and $|x|$ is large, $(3-x)$ will be large and positive (for example, if $x = -100$, then $3 - x = 103$). Then $2/(3-x)$ will be small but positive. Thus we conclude that

$$\frac{2}{3-x} \to 0^+ \quad \text{as} \quad x \to -\infty.$$

Therefore, the x axis is the horizontal asymptote for the graph of $y = 2/(3-x)$. As $x \to \infty$, the graph lies slightly below the x axis, while as $x \to -\infty$, the graph lies slightly above this axis. Examine the graph below and you will see these properties demonstrated.

The graph has a vertical asymptote at $x = 3$, because as x approaches 3, the denominator $(3-x)$ approaches 0 and so $2/(3-x)$ becomes numerically large. When x approaches 3 from above, $(3-x)$ is small and negative, so

$2/(3-x)$ is large and negative. For example, when $x = 3.1$, we have $(3-x) = -0.1$ and so $2/(3-x) = 2/(-0.1) = -20$. In general, we can see that

$$\frac{2}{3-x} \to -\infty \quad \text{as} \quad x \to 3^+.$$

Similarly we have

$$\frac{2}{3-x} \to +\infty \quad \text{as} \quad x \to 3^-.$$

For example, when $x = 2.9$, $(3-x) = 0.1$, and so

$$\frac{2}{3-x} = \frac{2}{0.1} = 20.$$

Thus as x approaches 3 from below, the graph approaches the vertical asymptote with large positive values. As x approaches 3 from above, the graph approaches the asymptote with numerically large negative values. Again these properties are apparent from the adjacent figure.

Example 3 Find the horizontal and vertical asymptotes of the function

$$y = \frac{2x-4}{x-1}.$$

Use them to sketch its graph.

Solution The vertical asymptote is at $x = 1$, since as x approaches 1, the denominator approaches 0. When x is a little greater than 1, $(x-1)$ is small and positive and $(2x-4)$ in the numerator is close to -2. Thus $y \approx (-2)/(x-1) =$ (negative) $\div$ (small and positive), and so y is large and negative. [For example, if $x = 1.1$, then $(x-1) = 0.1$, small and positive; and $2x - 4 = -1.8$, close to -2. Then $y = (-1.8)/(0.1) = -18$, large and negative.]

Similarly, when x is a little less than 1, $(x-1)$ is small and negative and $(2x-4)$ is still close to -2. Therefore, $y \approx (-2)/(x-1) =$ (negative) $\div$ (small and negative), and so y is large and positive. (For example, when $x = 0.9$, you can verify that $y = +18$.)

To summarize, then,

$$y \to -\infty \quad \text{as} \quad x \to 1^+$$
$$y \to +\infty \quad \text{as} \quad x \to 1^-.$$

This information allows us to sketch the position of the graph in relation to the vertical asymptote. We obtain the segments of curve shown in the adjoining figure.

For the horizontal asymptote, we divide numerator and denominator by x:

$$y = \frac{2x-4}{x-1} = \frac{(2x-4)/x}{(x-1)/x} = \frac{\left(2 - \dfrac{4}{x}\right)}{\left(1 - \dfrac{1}{x}\right)}.$$

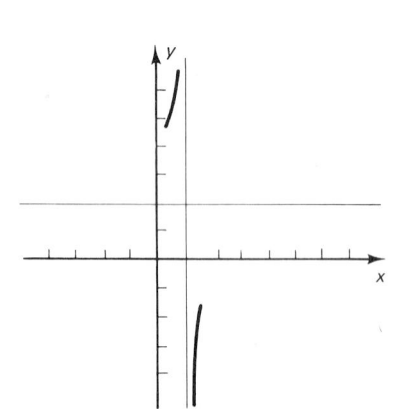

As $x \to \pm\infty$, the terms $4/x$ and $1/x$ become very small. The numerator becomes close to 2, the denominator becomes close to 1, and y becomes closer and closer to $2/1 = 2$. Thus we can write

$$y \to 2 \quad \text{as} \quad x \to \pm\infty$$

and $y = 2$ is the horizontal asymptote.

In order to decide on which side of the horizontal asymptote the curve lies, we consider the difference $(y - 2)$:

$$y - 2 = \frac{2x - 4}{x - 1} - 2$$

$$= \frac{2x - 4 - 2(x - 1)}{x - 1}$$

$$= \frac{-2}{x - 1}.$$

Now when x approaches $+\infty$, $-2/(x - 1)$ is small and negative, so $y - 2 < 0$. Thus $y \to 2^-$ as $x \to +\infty$, and the graph lies below the asymptote, as shown in the adjacent figure. On the other hand, when x approaches $-\infty$, $-2/(x - 1)$ is small and positive, so $y - 2 > 0$. Thus $y \to 2^+$ as $x \to -\infty$, and the graph lies above the asymptote.

Note that in the figure the two points $(2, 0)$ and $(0, 4)$ are marked. In sketching a graph you should always plot one or two explicit points in order to position the graph correctly. The intersections with the coordinates axes are often the easiest points to plot. In this example, when $x = 0$ we have $y = (2 \cdot 0 - 4)/(0 - 1) = 4$, so the intersection with the y axis is at $(0, 4)$. Setting $y = 0$, we must have $2x - 4 = 0$ or $x = 2$, so the intersection with the x axis is at $(2, 0)$.

In the preceding examples the graph always lay on opposite sides of the asymptote at its two extremities. This is not always the case.

Example 4 Sketch the graph of the function

$$y = \frac{1}{(x + 1)^2}.$$

Solution As x becomes numerically large, whether positive or negative, $(x + 1)^2$ becomes very large and positive, so y becomes small and is positive. Therefore $y \to 0^+$ as $x \to \infty$ and as $x \to -\infty$. The x axis is the horizontal asymptote, and the graph lies above the x axis at both ends. This allows us to position the two segments of the graph shown in the left figure on page 244.

The vertical asymptote is at $x = -1$, since as x approaches -1 the denominator $(x + 1)^2$ becomes very small. Since the denominator is squared, it is always positive, so that $y \to +\infty$ as $x \to -1^+$ and as $x \to -1^-$. The complete graph is shown in the right-hand figure that follows.

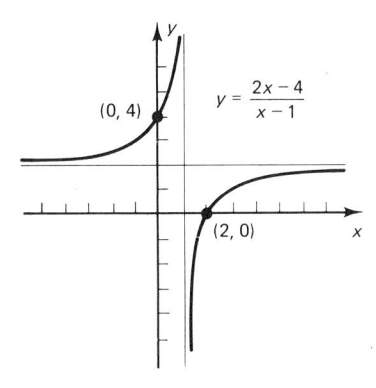

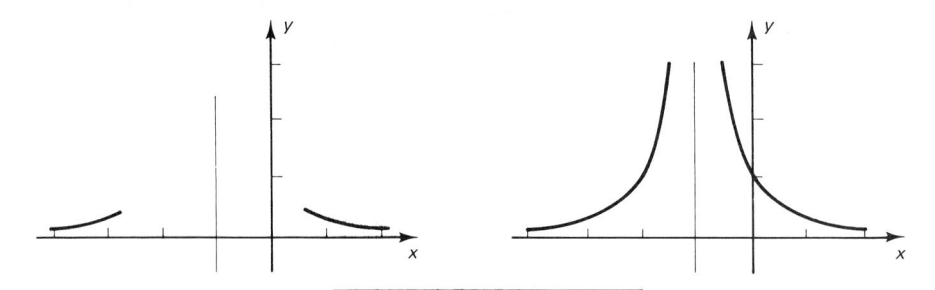

The techniques described above can often be used to sketch the graphs of rational functions. The general procedure is to locate the horizontal and vertical asymptotes and find which sides of the asymptotes the graph lies on. This is frequently sufficient information to allow a rough sketch to be drawn. Plotting a few specific points, such as the intersections with the coordinate axes, is then usually enough to position the graph reasonably accurately.

The following statements summarize the approach to finding asymptotes for rational functions.

> *Vertical asymptotes* are found at values of x for which the denominator is equal to zero. The *horizontal asymptote* is found by dividing numerator and denominator by the *highest power of x in the denominator*. We then use the fact that all the inverse powers, $1/x$, $1/x^2$, and so on become very small as $x \to \pm\infty$.

The approach is illustrated in the following (more difficult) example.

Example 5 Sketch the graph of the function

$$y = \frac{x^2}{x^2 - 4x + 3}.$$

Solution We can write

$$y = \frac{x^2}{(x-1)(x-3)}.$$

Therefore the vertical asymptotes are at $x = 1$ and $x = 3$.

Let us examine which sides of the asymptote $x = 1$ the graph lies on. We rewrite y in the form

$$y = \left(\frac{x^2}{x-3}\right)\left(\frac{1}{x-1}\right).$$

The reason is to separate out the factor that becomes unbounded as x approaches 1. The first factor, $x^2/(x-2)$, will always be fairly close to the value $1^2/(1-3) = -\frac{1}{2}$ when x is close to 1. When x is just below 1, $(x - 1)$ is small and negative while $x^2/(x - 2)$ is negative, since it is close to $-\frac{1}{2}$. Therefore $y = $ (negative)/(small and negative), and so y is large and positive. That is, $y \to +\infty$ as $x \to 1^-$.

When x is just above 1, $(x - 1)$ is small and positive, while $x^2/(x - 2)$

is still negative ($\approx -\frac{1}{2}$). Therefore $y = $ (negative)/(small and positive), and so y is numerically large and negative. That is, $y \to -\infty$ as $x \to 1^+$.

Similarly, for the asymptote at $x = 3$ we write

$$y = \left(\frac{x^2}{x-1}\right)\left(\frac{1}{x-3}\right).$$

For x close to 3, the first factor is approximately $3^2/(3-1) = \frac{9}{2}$. Therefore for x just below 3, $(x-3)$ is small and negative, so $y = $ (positive)/(small and negative). Thus $y \to -\infty$ as $x \to 3^-$. Similarly we find that $y \to +\infty$ as $x \to 3^+$. This information allows us to position segments of the graph in relation to its vertical asymptotes, as shown in the first figure below.

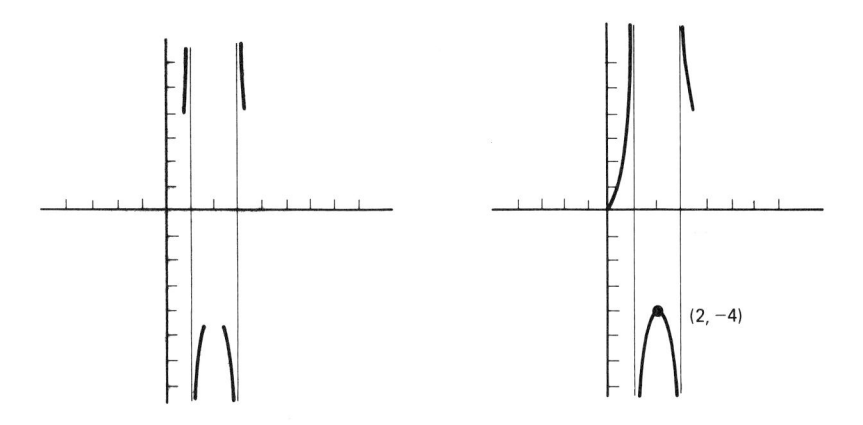

Note also that when $x = 0$, $y = 0$ also. Furthermore, setting $y = 0$, we have $x^2 = 0$, or $x = 0$. Thus the only intersection with either of the coordinate axes is at the origin. We can conclude, therefore, that the left-hand branch of the graph must connect down to the origin, as shown in the above right-hand figure.

In addition, the two lower branches cannot cross the x axis, hence they must be joined together. To position them correctly we can plot the point corresponding to $x = 2$ halfway between the two asymptotes. When $x = 2$,

$$y = \frac{2^2}{2^2 - 4 \cdot 2 + 3} = -4.$$

So the point is $(2, -4)$, as plotted in the figure.

To complete the sketch we need the horizontal asymptote. Dividing numerator and denominator by x^2 (the highest power of x in the denominator), we obtain

$$y = \frac{x^2/x^2}{(x^2 - 4x + 3)/x^2} = \frac{1}{1 - \dfrac{4}{x} + \dfrac{3}{x^2}}.$$

The terms $4/x$ and $3/x^2$ become very small as $x \to \pm\infty$, and therefore

$$y \to 1.$$

In order to determine on which side of the horizontal asymptote $y = 1$ the graph lies, we consider the difference $y - 1$:

$$y - 1 = \frac{x^2}{x^2 - 4x + 3} - 1$$
$$= \frac{x^2 - (x^2 - 4x + 3)}{x^2 - 4x + 3}$$
$$= \frac{4x - 3}{x^2 - 4x + 3}.$$

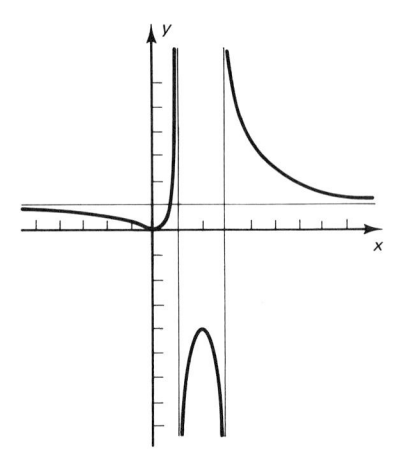

Now when $|x|$ is large, the numerator has the same sign as $4x$, since $4x$ is numerically much larger than 3. The denominator has the same sign as x^2, since the x^2 term is much larger in magnitude than either $4x$ or 3. Thus the denominator is always positive when $|x|$ is very large. So $y - 1$ has the same sign as the numerator—that is, the same sign as x. Therefore,

$$y \to 1^+ \quad \text{as} \quad x \to +\infty$$
$$y \to 1^- \quad \text{as} \quad x \to -\infty.$$

We conclude, therefore, that the graph lies above the asymptote as $x \to +\infty$ and below the asymptote as $x \to -\infty$.

The graph is shown in the above figure.

Diagnostic Test 6.5

Fill in the blanks.

(1–4) If $f(x) = 2x/(x - 1)$, then:

1. $f(x) \to$ _____ as $x \to +\infty$.
2. $f(x) \to$ _____ as $x \to -\infty$.
3. $f(x) \to$ _____ as $x \to 1^+$.
4. $f(x) \to$ _____ as $x \to 1^-$.

5. The horizontal and vertical asymptotes of $y = x/(2x + 1)$ are $y = $ _____ and $x = $ _____, respectively.

6. The vertical asymptotes of $y = x/(x^2 - 1)$ are _____ and _____.

7. The graph of $y = 1/[x^2(x + 1)]$ has vertical asymptotes _____.

Exercises 6.5

(1–12) Specimen example: For the function $y = (x + 1)/(2x - 3)$,
$$y \to \tfrac{1}{2} \quad \text{as} \quad x \to \pm\infty$$
$$y \to -\infty \quad \text{as} \quad x \to (\tfrac{3}{2})^-$$
$$y \to +\infty \quad \text{as} \quad x \to (\tfrac{3}{2})^+.$$

Make corresponding statements about the following functions.

1. $y = \dfrac{1}{x - 4}$.
2. $y = \dfrac{3}{x + 1}$.
3. $y = \dfrac{2}{3 - x}$.
4. $y = \dfrac{1}{5 - 2x}$.
5. $y = \dfrac{x}{x + 2}$.
6. $y = \dfrac{x}{2 - x}$.
7. $y = \left(\dfrac{x}{x + 1}\right)^2$.
8. $y = \left(\dfrac{x + 1}{x - 1}\right)^2$.
9. $y = \dfrac{x^2}{x^2 + 3x + 2}$.
10. $y = \dfrac{x^2 + 1}{x^2 - x - 2}$.
11. $y = \dfrac{x^2 - x - 6}{x^2 - 1}$.
12. $y = \dfrac{4x^2 - 1}{x^2 - 4}$.

(13–30) Sketch the graphs of the following functions.

13. $y = \dfrac{1}{x - 1}$.
14. $y = \dfrac{2}{x + 2}$.
15. $y = \dfrac{-2}{x + 2}$.
16. $y = \dfrac{1}{5 - 2x}$.
17. $y = \dfrac{2x + 1}{x + 1}$.
18. $y = \dfrac{3x - 6}{x - 1}$.
19. $y = \dfrac{x - 2}{2x + 3}$.
20. $y = \dfrac{2 - x}{x + 2}$.
21. $y = \dfrac{2}{(x + 1)^2}$.
22. $y = \dfrac{1}{(x - 1)^3}$.
23. $y = \dfrac{x^2}{x^2 - 1}$.
24. $y = \dfrac{x^2}{x^2 - x - 6}$.
25. $y = \dfrac{x^2 + 1}{x^2 - 1}$.
26. $y = \dfrac{x^2 - 2}{x^2 - 3x + 2}$.
27. $y = \dfrac{x}{x^2 - 1}$.
28. $y = \dfrac{x}{x^2 - 3x - 4}$.
29. $y = \dfrac{x}{x^2 - 3x + 2}$.
30. $y = \dfrac{x - 1}{(x + 1)(x + 3)}$.

Chapter Review 6

1. Define and/or explain the following terms.
   (a) Function.
   (b) Domain and range of a function.
   (c) Argument and value of a function.
   (d) Graph of a function.
   (e) Constant function.
   (f) Linear function.
   (g) Power function.

2. Describe the following.
   (a) Two criteria you would use in finding the domain of an algebraic function.
   (b) The graphs of $y = x^n$ for different values of n.
   (c) How you would calculate the domains of the sum and the quotient of two functions.
   (d) How you would calculate the inverse of a given function.

(h) Absolute-value function.
(i) Polynomial, rational, and algebraic functions.
(j) Quadratic and cubic functions.
(k) The sum, difference, product, and quotient of two functions.
(l) The composition of two functions.
(m) One-to-one function.
(n) Inverse function.
(o) Horizontal asymptote.
(p) Vertical asymptote.

(e) The relationship between the graphs of a function and its inverse.

Review Exercises on Chapter 6

1. Given $g(x) = x^3 - 3x^{-1}$, evaluate $g(2)$ and $g(-1)$.

2. Given $h(y) = \sqrt{2y + 5}$, evaluate $h(2)$ and $h(-2)$.

3. If
$$f(x) = \begin{cases} 2x - 1 & \text{if } x \geq 1 \\ 4 - x & \text{if } x < 1 \end{cases}$$
evaluate:
(a) $f(-1)$.
(b) $f(1)$.
(c) $f(10)$.
(d) $f(0)$.

4. If
$$p(t) = \begin{cases} t & \text{for } t > 2 \\ 1 & \text{for } t = 2 \\ t^2 - 1 & \text{for } t < 2 \end{cases}$$
evaluate:
(a) $p(3)$.
(b) $p(2)$.
(c) $p(0)$.
(d) $p(-2)$.

(5–8) Find the domains of the following functions.

5. $f(x) = \dfrac{x + 1}{2x + 3}$.

6. $g(x) = \dfrac{\sqrt{2x - 3}}{3x - 2}$.

7. $h(x) = \dfrac{1}{\sqrt{|x|}}$.

8. $F(x) = \dfrac{x}{\sqrt{4 - x}}$.

9. Express the volume V of a sphere as a function of its surface area S.

10. A and B start walking from the same point. A walks due west at a speed of 3 miles per hour. B walks due north at a speed of 4 miles per hour for one hour, then turns and walks due east at the same speed. Express the distance d between A and B as a function of time t.

11. In Exercise 10, after an additional hour, B turns and retraces his steps at the same speed. Express the distance d between A and B as a function of time t.

12. Pacific Airways charges $6 to transport each pound of merchandise 900 miles and $10 to transport each pound 1700 miles. Determine the cost function $C(x)$, assuming it to be a linear function of distance x.

13. Give an example of a function f that satisfies each property for all values of x and y.
(a) $f(x) = f(-x)$ (such a function is called an even function).
(b) $f(-x) = -f(x)$ (such a function is called an odd function).
(c) $f(x + y) = f(x) + f(y)$.

14. Two functions f and g are said to be *equal* if $f(x) = g(x)$ for all x in the domain and $D_f = D_g$. Use this criterion to determine which of the following functions are equal to $f(x) = (2x^2 + x)/x$.
(a) $g(x) = 2x + 1$.
(b) $h(x) = \sqrt{1 + 4x + 4x^2}$.
(c) $F(x) = \dfrac{2x^3 + x^2}{x^2}$.
(d) $G(x) = \dfrac{(x^3 + 2x)(1 + 2x)}{x(x^2 + 2)}$.

(15–20) Find the domains of the following functions and sketch their graphs.

15. $f(x) = 1 - \frac{1}{2}x^3$.
16. $f(x) = \sqrt{x+2}$.
17. $g(x) = -\sqrt{4-x}$.
18. $h(x) = |x+2| - 2$.
19. $F(x) = |x-1| + x$.
20. $G(x) = \dfrac{x}{x+|x|}$.

(21–28) If $f(x) = \sqrt{x+1}$ and $g(x) = x^2 - 1$, evaluate the following.

21. $(f+g)(-2)$.
22. $(f \circ g)(-2)$.
23. $(gf)(3)$.
24. $(g/f)(-1)$.
25. $(f/g)(-1)$.
26. $(g \circ f)(1)$.
27. $(g \circ f)(x)$.
28. $(f \circ g)(x)$.

29. If $f(x) = x - \sqrt{x}$ and $g(x) = x^2 + 2x + 1$, find $(f \circ g)(x)$ and $(g \circ f)(x)$.

*30. If $(g \circ f)(x) = g(x)$ for all x and g is one-to-one, show that $f(x) = x$.

(31–36) Find the inverses of the following functions.

31. $y = 2x + 2$.
32. $y = \sqrt{2x - 3}$.
33. $y = \dfrac{3}{2x-1}$.
34. $y = \dfrac{2x-1}{2x+3}$.
35. $y = (2x-3)^4$ $(x > \frac{3}{2})$.
36. $y = \sqrt{2x^2 + 3}$ $(x > 0)$.

*(37–38) Find the inverses of the following functions. Draw the graph of each inverse function.

37. $y = \begin{cases} x+2 & (x \geq 1) \\ 2x+1 & (x < 1). \end{cases}$
38. $y = \begin{cases} x^2 - 1 & (x \geq 3) \\ 2x + 2 & (x < 3). \end{cases}$

(39–42) Make the same type of statements about the following functions as in Exercises 1–12 of Section 6.5.

39. $y = \dfrac{3x-1}{x-3}$.
40. $y = \dfrac{2x+1}{x+2}$.
41. $y = \dfrac{x^2-2}{x^2+2x+1}$.
42. $y = \dfrac{x^2+x+1}{(x+2)(x-3)}$.

(43–48) Sketch the graphs of the following functions.

43. $y = \dfrac{x-2}{2x-3}$.
44. $y = \dfrac{x}{3-2x}$.
45. $y = \dfrac{x}{(x-1)^2}$.
46. $y = \dfrac{x^2}{(x-1)^2}$.
47. $y = \dfrac{x^2-1}{x^2-4}$.
48. $y = \dfrac{x^2-4}{x^2-1}$.

49. If $f(x) = px - q$ and $g(u) = ru + s$, where p, q, r, s are constants, find the relation between p, q, r, and s such that $(f \circ g)(x) = (g \circ f)(x)$.

7 CONICS

In this chapter we shall look at a group of curves called **conic sections** or simply **conics**—so named because they can all be obtained by the intersection of a plane with a double-ended right circular cone. The various cases are shown in the following figure.

If the intersecting plane is perpendicular to the central axis of the cone, as in part (a) of the figure, the curve of intersection is a circle. If, as in part (b), the plane is not perpendicular to the axis of the cone but cuts completely across one half of the cone, the curve of intersection is an elongated closed curve called an ellipse. In part (c) the plane intersects only one half of the cone but does not cut across it completely; here the curve of intersection is an

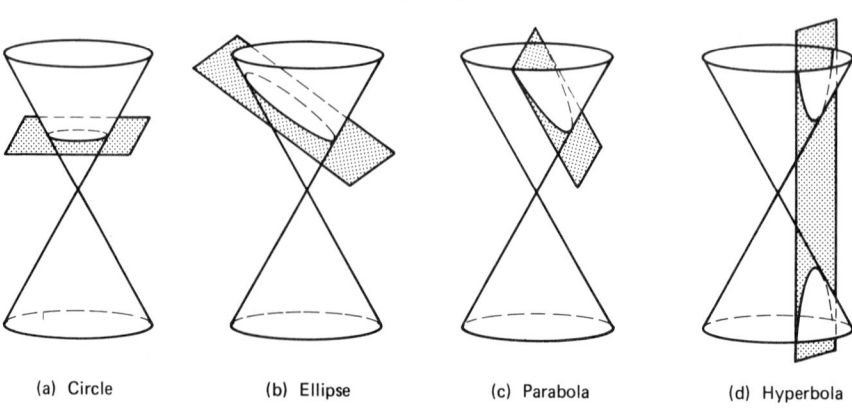

(a) Circle   (b) Ellipse   (c) Parabola   (d) Hyperbola open curve called a parabola. Finally, in part (d), the plane intersects both halves of the cone; in this case the curve of intersection is an open curve with two branches and is called a hyperbola.

These curves possess many fascinating geometrical properties, some of which we shall encounter later. What makes the study of the conics so interesting is the interplay between their geometry and algebra.

Conics occur in many important applications. For example, the path of a planet or comet around the sun is an ellipse, as is the path of a satellite about the earth. The path of a projectile close to the earth's surface is a parabola,* and the properties of parabolas form a basic part of the study of ballistics. Everyone is familiar with the parabolic mirror; the parabolic dish antenna is a common feature of our landscapes; and every spy, we are led to believe, has a parabolic microphone in his back pocket. The curve traced by the cables of a suspension bridge is also a parabola.† When a circle is viewed at an angle, it appears as an ellipse, so ellipses play a central role in engineering drawing and other artwork. These are just a few of the many applications of these curves.

7.1

The Circle

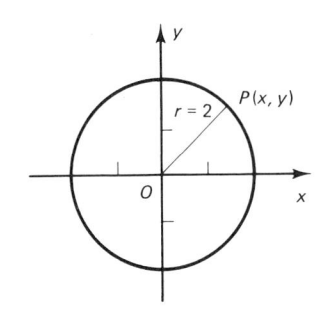

Let C be a given point in the xy plane and let r be a given nonnegative number. Then the set of points P in the plane such that the distance $CP = r$ is called a *circle*. C is called the **center** and r is called the **radius** of the circle.

Consider as an example the circle whose center is at the origin and whose radius is 2 units (see the adjacent figure). Let P be an arbitrary point on the circle with coordinates (x, y). Then the distance $OP = 2$ units.

From the distance formula, $OP = \sqrt{x^2 + y^2}$. Therefore,

$$OP = \sqrt{x^2 + y^2} = 2.$$

Squaring both sides of this equation to remove the radical, we get

$$x^2 + y^2 = 4.$$

This equation is satisfied by every point (x, y) lying on the circle. We refer to it as *the equation of the circle* whose radius is 2 and center is at the origin.

Now let us find the equation of the circle with center at the point (h, k) and radius r. Let (x, y) be any point on the circle. Then the distance between this point (x, y) and the center (h, k) is given by the distance formula to be

$$\sqrt{(x - h)^2 + (y - k)^2}.$$

---

*When we ignore the effects of air resistance.
†When the load is uniform across the span.

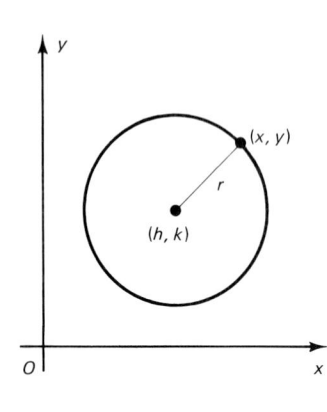

Setting this equal to the given **radius** r, we obtain the equation
$$\sqrt{(x-h)^2 + (y-k)^2} = r.$$
Squaring both sides, we obtain
$$(x-h)^2 + (y-k)^2 = r^2. \tag{1}$$

This is called the ***center-radius form of the equation of the circle***.

In particular, if the center is at the origin, $h = k = 0$ and the equation (1) of the circle reduces to
$$x^2 + y^2 = r^2. \tag{2}$$

This is the ***equation of the circle of radius r with center at the origin***. Our opening example corresponded to the case $r = 2$.

Example 1 Find the equation of the circle with center at $(2, -3)$ and radius 5.

Solution Here $h = 2$, $k = -3$, and $r = 5$. Using the standard equation (1) of the circle, we have
$$(x-2)^2 + [y-(-3)]^2 = 5^2$$
$$(x-2)^2 + (y+3)^2 = 25.$$
Expanding the squares, we reduce this to
$$x^2 - 4x + 4 + y^2 + 6y + 9 = 25$$
$$x^2 + y^2 - 4x + 6y - 12 = 0.$$

The left side of the final equation in this example can be seen to have the following general form: the terms $x^2 + y^2$ plus a term proportional to x plus a term proportional to y plus a constant term. In fact, we can see that the general equation (1) can always be written in this form. For, expanding the two squares, we have
$$x^2 - 2hx + h^2 + y^2 - 2ky + k^2 = r^2$$
and so
$$x^2 + y^2 - 2hx - 2ky + (h^2 + k^2 - r^2) = 0.$$
This is of the general form
$$x^2 + y^2 + Bx + Cy + D = 0 \tag{3}$$
where B, C, and D are constants given by
$$B = -2h, \quad C = -2k, \quad D = h^2 + k^2 - r^2. \tag{4}$$

In Example 1 these constants have the values $B = -4$, $C = 6$, and $D = -12$.

Equation (3) is called the ***general form*** of the equation of the circle. Often we are given the equation of a circle in this general form and we would like to work out its radius and the position of its center. In order to do this we can use the method of completing the square. This is best illustrated by an example.

Example 2 Find the coordinates of the center and the radius of the circle whose equation is
$$x^2 + y^2 - 3x + 6y - 1 = 0.$$

Solution We first rewrite the given equation in the form
$$(x^2 - 3x \quad) + (y^2 + 6y \quad) = 1.$$

We have left blank spaces in the parentheses in order to suggest completing each expression to make it a perfect square. The technique was described in Section 3.4. In order to make $(x^2 - 3x \quad)$ a perfect square, we add $(-\frac{3}{2})^2 = \frac{9}{4}$ to it. To make $(y^2 + 6x \quad)$ a perfect square we add $(\frac{6}{2})^2 = 9$ to it. We must also add corresponding quantities to the right side. Thus we write the given equation as
$$(x^2 - 3x + \tfrac{9}{4}) + (y^2 + 6y + 9) = 1 + \tfrac{9}{4} + 9$$

—that is,
$$(x - \tfrac{3}{2})^2 + (y + 3)^2 = \tfrac{49}{4}.$$

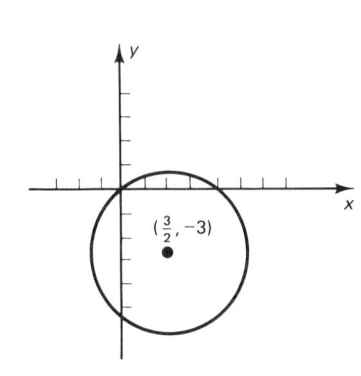

Comparing this with Equation (1),
$$(x - h)^2 + (y - k)^2 = r^2$$

we see that $h = \frac{3}{2}$, $k = -3$, and $r^2 = \frac{49}{4}$. Hence the center of the circle is the point $(\frac{3}{2}, -3)$ and the radius is $r = \frac{7}{2}$. The circle is shown in the adjoining figure.

Example 3 Determine whether the graph of the equation
$$x^2 + y^2 + 2x + 4y + 10 = 0 \qquad \text{(i)}$$
is a circle. If it is, find the center and radius.

Solution We proceed to rearrange the terms as in the previous example:
$$(x^2 + 2x \quad) + (y^2 + 4y \quad) = -10.$$

Completing the squares in the two sets of parentheses, we get
$$(x^2 + 2x + 1) + (y^2 + 4y + 4) = -10 + 1 + 4$$
$$(x + 1)^2 + (y + 2)^2 = -5. \qquad \text{(ii)}$$

Comparing with Equation (1), we see that $r^2 = -5$. But r^2 cannot be negative, so we conclude that Equation (i) is *not* the equation of a circle. In fact there are no values of x and y that satisfy Equation (i). We can see this from Equation (ii), in which the left side is always nonnegative for any values of x and y (being the sum of two squares) and cannot possibly be equal to -5.

We see from this example that an equation of the general form of Equation (3) may not represent the equation of a circle because the value of r^2 may turn out to be negative.

Example 4 Determine whether the equation
$$x^2 + y^2 - 4x + 8y + 20 = 0$$
represents a circle. If so, find its center and radius.

Solution Completing the squares as usual, we find
$$(x^2 - 4x \quad) + (y^2 + 8y \quad) = -20$$
$$(x^2 - 4x + 4) + (y^2 + 8y + 16) = -20 + 4 + 16$$
$$(x - 2)^2 + (y + 4)^2 = 0.$$

Comparing with the center-radius form (1), we conclude that $(h, k) = (2, -4)$ and $r = 0$. The graph is a circle with zero radius, which means it consists of the one point $(2, -4)$ alone.

There are other ways in which we can specify a circle other than by giving its center and radius. One way is to give the center and one point on the circle. Another way is to specify three points that lie on the circle, since one and only one circle passes through three given points.

Example 5 Determine the equation of the circle whose center is $(2, 3)$ and that passes through the point $(5, -1)$. Determine where this circle meets the y axis.

Solution The radius of the circle must equal the distance between the center $(2, 3)$ and the point $(5, -1)$ (see the adjoining figure). That is, 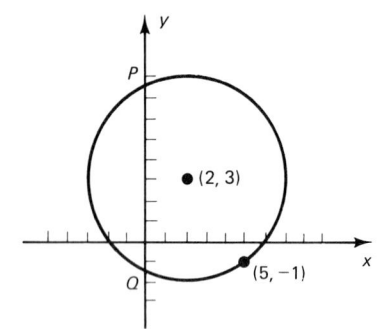
$$r^2 = (2 - 5)^2 + [3 - (-1)]^2 = (-3)^2 + 4^2 = 25.$$
Thus $r = 5$.

Using the center-radius form of the equation, we therefore get
$$(x - h)^2 + (y - k)^2 = r^2$$
$$(x - 2)^2 + (y - 3)^2 = 25$$
$$(x^2 - 4x + 4) + (y^2 - 6y + 9) = 25$$
$$x^2 + y^2 - 4x - 6y - 12 = 0 \qquad \text{(i)}$$
which is the required equation.

We wish to find the coordinates of P and Q, the points where the circle cuts the y axis. At these points, $x = 0$. When $x = 0$, Equation (i) becomes
$$y^2 - 6y - 12 = 0.$$
We must use the quadratic formula to solve this equation:
$$y = \frac{6 \pm \sqrt{6^2 - 4(1)(-12)}}{2}$$
$$= \frac{6 \pm \sqrt{84}}{2}$$
$$= 3 \pm \sqrt{21}.$$
Thus P is $(0, 3 + \sqrt{21}) \approx (0, 7.58)$ and Q is $(0, 3 - \sqrt{21}) \approx (0, -1.58)$.

Example 6 Determine the equation of the circle that passes through the three points $(1, 2)$, $(3, -1)$, and $(2, -2)$.

Solution The circle must have an equation of the form
$$x^2 + y^2 + Bx + Cy + D = 0. \qquad (i)$$
Substituting $x = 1$ and $y = 2$ [since the point $(1, 2)$ lies on the circle], we obtain
$$1^2 + 2^2 + B + 2C + D = 0$$
$$B + 2C + D = -5.$$
Similarly, since $(3, -1)$ lies on the circle,
$$3^2 + (-1)^2 + 3B - C + D = 0$$
$$3B - C + D = -10.$$
Finally, since $(2, -2)$ lies on the circle,
$$2^2 + (-2)^2 + 2B - 2C + D = 0$$
$$2B - 2C + D = -8.$$
Thus we have three equations:
$$B + 2C + D = -5$$
$$3B - C + D = -10 \qquad (ii)$$
$$2B - 2C + D = -8.$$
These form a system of three simultaneous algebraic equations to be solved for the unknown coefficients B, C, and D.

The solution is actually not hard, because if we subtract the first equation from each of the other two, we immediately eliminate D and have two equations to solve for B and C:
$$2B - 3C = -5$$
$$B - 4C = -3.$$
The solution of this pair of equations is $B = -\frac{11}{5}$ and $C = \frac{1}{5}$. Then from the first equation in system (ii) we have
$$D = -5 - B - 2C = -5 - (-\frac{11}{5}) - 2(\frac{1}{5}) = -\frac{16}{5}.$$
Thus the required equation (i) for the circle is
$$x^2 + y^2 - \frac{11}{5}x + \frac{1}{5}y - \frac{16}{5} = 0.$$

Diagnostic Test 7.1

1. The graph of the equation $x^2 + y^2 = 1$ is a _____.
2. The graph of the equation $x^2 + y^2 + 9 = 0$ is _____.
3. The graph of the equation $(x - 2)^2 + (y - 3)^2 + 16 = 0$ is _____.
4. The graph of the equation $x^2 + y^2 = 0$ is _____.
5. The graph of the equation $(x - 2)^2 + (y + 5)^2 = 0$ is _____.

6. If A, B, C are nonzero constants, then the graph of the equation $Ax^2 + By^2 + C = 0$ is a circle only if _____.

7. The equation of a circle with center at (h, k) and radius r is _____.

8. The graph of the equation $(x - 2)^2 + (y + 3)^2 = 16$ is a _____ of radius _____ and center at _____.

Exercises 7.1

(1–10) Find the equation of the circle whose center is the given point and that has the given radius.

1. $(0, 2)$, $r = 5$. **2.** $(2, 5)$, $r = 3$. **3.** $(-3, 0)$, $r = 4$. **4.** $(0, 0)$, $r = 7$.

5. $(-2, -5)$, $r = 1$. **6.** $(4, -1)$, $r = 3$. **7.** $(6, -6)$, $r = 6$. **8.** $(-3, 4)$, $r = 5$.

9. $(-2, 8)$, $r = 3$. **10.** $(-\frac{3}{2}, -2)$, $r = \frac{5}{2}$.

(11–14) Find the equation of the circle that has the first point as center and that passes through the second point.

11. $(0, 3)$; $(-1, 2)$. **12.** $(-3, 0)$; $(1, 4)$. **13.** $(-2, 4)$; $(2, -4)$. **14.** $(5, 1)$; $(4, 2)$.

(15–26) Determine whether each of the following equations represents a circle. If it does, find its center and radius.

15. $x^2 + y^2 + 2x + 2y + 1 = 0$.
16. $x^2 + y^2 - 2x + 6y + 6 = 0$.
17. $x^2 + y^2 - 4x - 8y + 4 = 0$.
18. $x^2 + y^2 + 6x - 3y + 12 = 0$.
19. $x^2 + y^2 + 3x - 5y + 1 = 0$.
20. $x^2 + y^2 - 7x - y - \frac{7}{2} = 0$.
21. $2x^2 + 2y^2 - 5x + 4y - 1 = 0$.
22. $3x^2 + 3y^2 + 6x + y - 2 = 0$.
23. $3x^2 + 3y^2 - 2x + 4y - \frac{11}{9} = 0$.
24. $4x^2 + 4y^2 - 5x - 7y - 3 = 0$.
25. $x^2 + y^2 + 2x + 6y + 10 = 0$.
26. $x^2 + y^2 - 10x + 12y + 61 = 0$.

(27–36) For each equation in Exercises 15–24 that represents a circle, determine the coordinates of the points where the circle meets the two coordinate axes.

37. Determine the equation of the circle that has center $(2, -3)$ and that passes through the point $(-3, 2)$.

38. Determine the equation of the circle that has center $(-3, 4)$ and that passes through the origin.

39. Determine the equation of the circle that has center $(2, 3)$ and just touches the x axis. Find the points where this circle meets the y axis.

40. Determine the equation of the circle that has center $(-4, 1)$ and just touches the y axis. Find the points where this circle meets the x axis.

(41–42) Determine the equation of the circle that passes through the origin and the two given points.

41. $(3, 0)$ and $(1, 2)$. **42.** $(-2, 2)$ and $(4, 1)$.

(43–46) Determine the equation of the circle that passes through the three given points.

43. $(1, 1)$, $(-1, 1)$, $(0, 2)$. **44.** $(2, 0)$, $(0, 3)$, $(4, 1)$.

45. $(-1, 3)$, $(-2, 4)$, $(-3, 4)$. **46.** $(2, 4)$, $(-1, 2)$, $(1, -3)$.

47. At a price of p dollars per unit a manufacturer can sell x units of his product, where x are p are related by

$$x^2 + p^2 + 400x + 300p = 60{,}000.$$

Plot the demand curve. What is the price above which no sales are possible?

48. A car dealer can sell x cars of a particular model when he charges $\$p$ per car, where

$$x^2 + p^2 + 400x + 250p = 19{,}437{,}500.$$

Plot the demand curve for this model of car. What is the highest price at which sales are still possible?

49. The owner of an apple orchard can produce either apples or apple cider. The possible amounts x of apples (in kilograms) and y of apple cider (in liters) are related by the equation

$$x^2 + y^2 + 8x + 250y = 6859.$$

Plot the graph of this relation (called the product transformation curve) and determine the maximum amounts of apples and apple cider that can be produced.

50. The Atlas Cycle Company manufactures two types of bicycles, *Coronado* and *Eastern Star*. The possible quantities x and y (in thousands) that can be produced per year are related by

$$x^2 + y^2 + 6x + 10y = 47.$$

Sketch the product transformation curve for this company. What maximum numbers of cycles of each type can be produced?

7.2

The Parabola

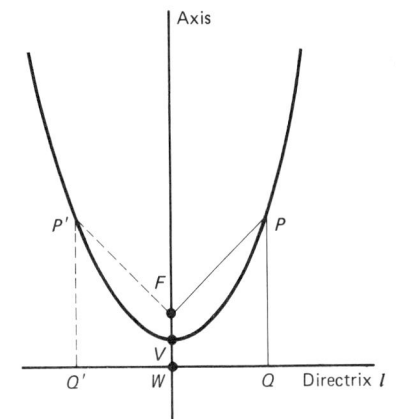

One of the geometrical definitions of a parabola is as follows. A *parabola* is the set of points that are equidistant from a given point and a given line. The given point is called the *focus* and the given line the *directrix*.

Let F be the focus and the line l be the directrix, as in the adjacent figure. For any point P on the parabola, the distance FP is equal to the perpendicular distance PQ from P onto the directrix. The line through the focus perpendicular to the directrix is called the *axis* of the parabola. The parabola is symmetrical about the axis, because to each point P on the right of the axis there will be a symmetrical point P' on the left. The point V where the parabola crosses its axis is called the *vertex*. Clearly $FV = VW$ (see the adjacent figure), so the vertex is halfway between the focus and the directrix.

In order to find the equation of the parabola, let us choose the axis of the parabola as the y axis and its vertex as the origin as shown on the following figure. If F is the point $(0, c)$, then the directrix must be the line $y = -c$.

Let $P(x, y)$ be any point on the parabola. Then

$$FP^2 = (x - 0)^2 + (y - c)^2 = x^2 + y^2 - 2yc + c^2.$$

The distance of P from the directrix is $PQ = y + c$. Therefore, since

$$FP^2 = PQ^2$$

$$x^2 + y^2 - 2yc + c^2 = (y + c)^2 = y^2 + 2yc + c^2$$

$$x^2 = 4yc$$

Setting $4c = 1/a$ and multiplying both sides by a, we obtain the equation in the form

$$\boxed{y = ax^2, \quad a = \frac{1}{4c}.}$$

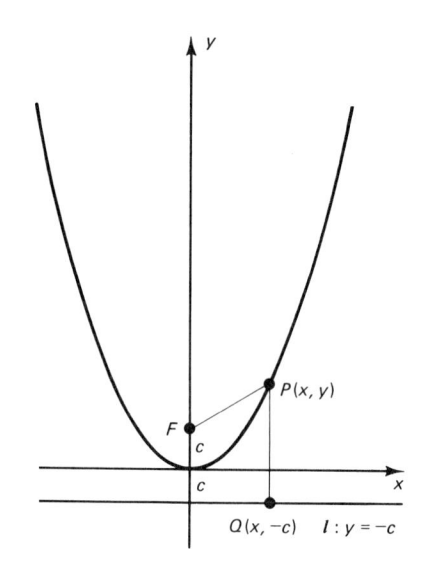

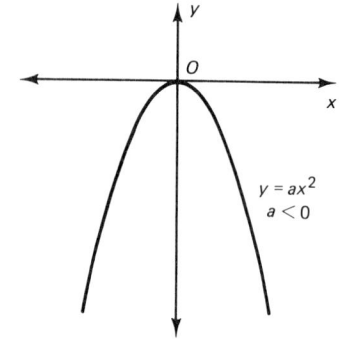

Note that this equation remains unchanged if x is replaced by $-x$, consistent with the symmetry about the y axis that we noted earlier.

We see then that the graph of the function $y = ax^2$ is a parabola with vertex at the origin. The parabola opens upward if $a > 0$ and downward if $a < 0$ (as in the adjacent figure). The focus is at the point $(0, 1/4a)$ and the directrix is $y = -1/4a$.

The graph of the function $y = a(x - 2)^2$ is also a parabola, having exactly the same shape as the graph of $y = ax^2$ but with its vertex at the point $(2, 0)$. Its axis is the vertical line $x = 2$. The adjacent figure illustrates that the graph of $y = a(x - 2)^2$ can be obtained by simply translating (i.e., moving) the graph of $y = ax^2$ by 2 units in the x direction.

We can generalize as follows: The graph of $y = a(x - h)^2$ is a parabola having the same shape as $y = ax^2$ but with vertex at the point $(h, 0)$. It is obtained by translating the graph of $y = ax^2$ by h units in the x direction.

If we simply add a constant k, obtaining $y = a(x - h)^2 + k$, the effect is to move the whole graph in the y direction by k units. Thus, the graph of

$$y = a(x - h)^2 + k$$

is a parabola having the same shape and size as $y = ax^2$ but with its vertex shifted from the origin to the point (h, k). This is known as the **standard equation of the parabola** with axis parallel to the y axis.

Example 1 Find the equation of the parabola that has the same shape as $y = \frac{1}{2}x^2$ but has vertex at (a) $(3, 2)$ and (b) $(-2, -1)$.

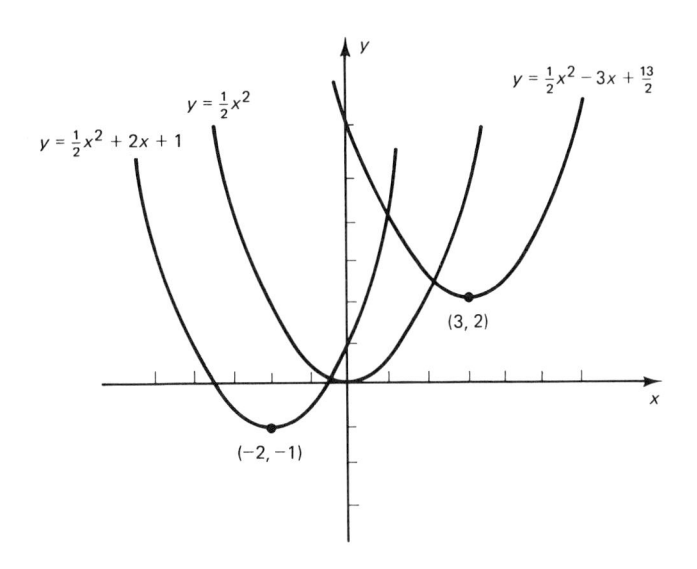

Solution (a) When the vertex is (h, k), we have the equation
$$y = a(x - h)^2 + k.$$
For $a = \frac{1}{2}$ and $(h, k) = (3, 2)$ this becomes
$$y = \tfrac{1}{2}(x - 3)^2 + 2$$
$$= \tfrac{1}{2}(x^2 - 6x + 9) + 2$$
$$= \tfrac{1}{2}x^2 - 3x + \tfrac{13}{2}.$$

(b) When $(h, k) = (-2, -1)$, we have
$$y = \tfrac{1}{2}[x - (-2)]^2 + (-1)$$
$$= \tfrac{1}{2}(x + 2)^2 - 1$$
$$= \tfrac{1}{2}(x^2 + 4x + 4) - 1$$
$$= \tfrac{1}{2}x^2 + 2x + 1$$

which is the required equation. The three parabolas are illustrated in the above figure.

Notice that in the above example the parabolas are graphs of the *quadratic functions** $y = \tfrac{1}{2}x^2 - 3x + \tfrac{13}{2}$ and $y = \tfrac{1}{2}x^2 + 2x + 1$. In both cases the coefficient of x^2 is $\tfrac{1}{2}$, the same as in the original parabola $y = \tfrac{1}{2}x^2$. We can generalize this result; the general quadratic function $y = ax^2 + bx + c$ has a graph that is a parabola identical in shape and size to the graph of $y = ax^2$ but with its vertex shifted away from the origin.

---

*See Section 6.2(d).

Sec. 7.2   The Parabola

For a given quadratic function it is often important to locate the position of the vertex of the corresponding parabola. We can do this by the method of completing the square, which enables us to write the given quadratic function in the form $y = a(x - h)^2 + k$. We can then readily extract the values of h and k.

Example 2 Find the vertex of the parabola whose equation is

$$y = x^2 + 6x + 10.$$

Solution We write

$$y = (x^2 + 6x) + 10$$

grouping the x^2 and x terms together. We wish to add a suitable number into the parentheses so that the expression inside becomes a perfect square of the form $(x - h)^2$. According to the method of completing the square, we must add $(\frac{6}{2})^2 = 3^2 = 9$:

$$y = (x^2 + 6x + 9) + 10 - 9.$$

Note that the 9 must also be subtracted outside the parentheses. Thus we can write

$$y = (x^2 + 6x + 9) + 1$$
$$= (x + 3)^2 + 1.$$

Comparing this with the general form

$$y = a(x - h)^2 + k$$

we see that $a = 1$, $h = -3$, and $k = 1$. In particular the vertex is at the point $(h, k) = (-3, 1)$. Since $a = 1$, the graph has the same shape and size as that of $y = x^2$. The two graphs are shown in the following figure.

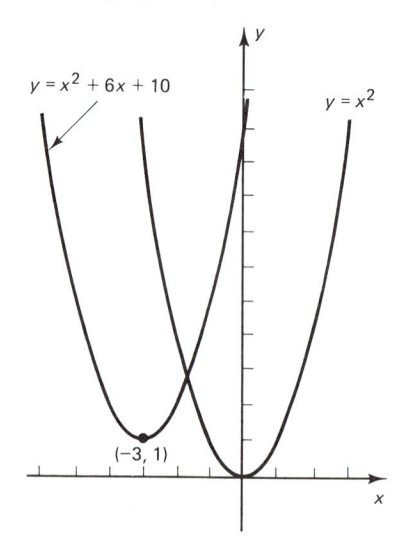

Example 3 Find the vertex of the parabola whose equation is
$$y = 1 + 10x - 2x^2.$$

Solution We first extract the coefficient of x^2 as a factor from the x^2 and x terms.
$$y = -2(x^2 - 5x) + 1.$$

Completing the square inside the parentheses, we get
$$y = -2\{[x^2 - 5x + (\tfrac{5}{2})^2] - (\tfrac{5}{2})^2\} + 1$$
$$= -2[(x - \tfrac{5}{2})^2 - \tfrac{25}{4}] + 1$$
$$= -2(x - \tfrac{5}{2})^2 + \tfrac{25}{2} + 1$$
$$= -2(x - \tfrac{5}{2})^2 + \tfrac{27}{2}.$$

This has the standard form
$$y = a(x - h)^2 + k$$
with $a = -2$, $h = \tfrac{5}{2}$, and $k = \tfrac{27}{2}$. The vertex is therefore at the point $(\tfrac{5}{2}, \tfrac{27}{2})$, as illustrated in the following figure. Since a is negative, the parabola opens downward.

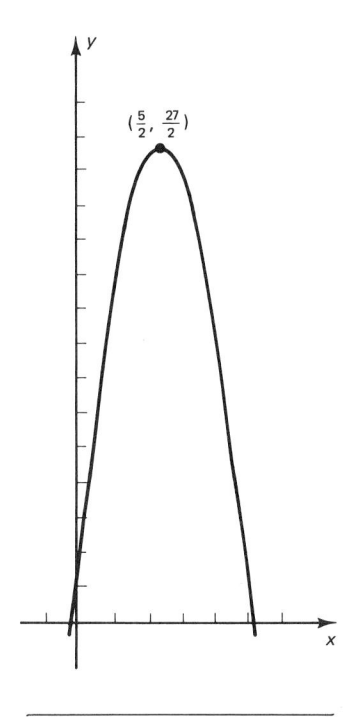

When sketching the graph of a quadratic function we would normally compute the coordinates of its vertex as the first step. In addition we would plot a few other points on the graph. The easiest points to find are usually the points of intersection with the coordinate axes.

Sec. 7.2   The Parabola

Example 4 Sketch the graph of
$$y = \tfrac{1}{2}x^2 + x - 4.$$

Solution We observe immediately that the graph is a parabola opening upward, having the same shape and size as the graph of $y = \tfrac{1}{2}x^2$. We find the vertex by completing the square as usual:
$$y = \tfrac{1}{2}(x^2 + 2x) - 4$$
$$= \tfrac{1}{2}(x + 1)^2 - \tfrac{9}{2}.$$

The vertex is therefore $(-1, -\tfrac{9}{2})$.

The graph meets the y axis where $x = 0$, and so
$$y = \tfrac{1}{2}(0)^2 + 0 - 4 = -4.$$

Thus the point $(0, -4)$ lies on the graph.

The graph meets the x axis where $y = 0$. Thus
$$\tfrac{1}{2}x^2 + x - 4 = 0$$

and we have a quadratic equation for x. Multiplying all the terms by 2, we get
$$x^2 + 2x - 8 = 0$$

which can be factored to give
$$(x + 4)(x - 2) = 0.$$

Thus $x = -4$ or $x = 2$. The intersections with the x axis are at $(-4, 0)$ and $(2, 0)$.

We can plot these four points as shown in the first of the following figures. The dotted vertical line through the vertex is the axis of the parabola

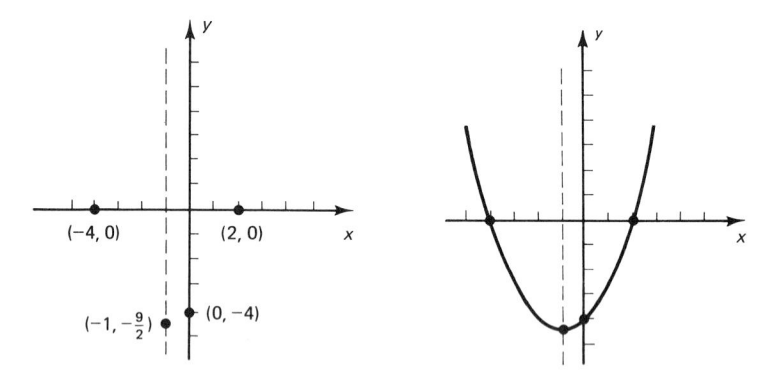

about which the graph must be symmetrical. The second figure shows how the graph can be sketched with this information.

If you feel that the vertex and the intersections with the coordinate axes do not allow you to draw a sufficiently accurate sketch, you can always plot two or three more points. This may be necessary when the vertex is close to the origin. It will certainly be necessary when the parabola lies entirely above or below the x axis.

Example 5 Sketch the graph of $y = 2x^2 - 4x + 7$.

Solution Proceeding as usual, we find that
$$y = 2(x-1)^2 + 5 \tag{i}$$
so the vertex is (1, 5). The curve meets the y axis at (0, 7) as shown in the figure below. From (i) it is immediately clear that y is never less than 5; therefore the graph does not intersect the x axis. (If we put $y = 0$, we would find that the quadratic equation for x had no real roots.)

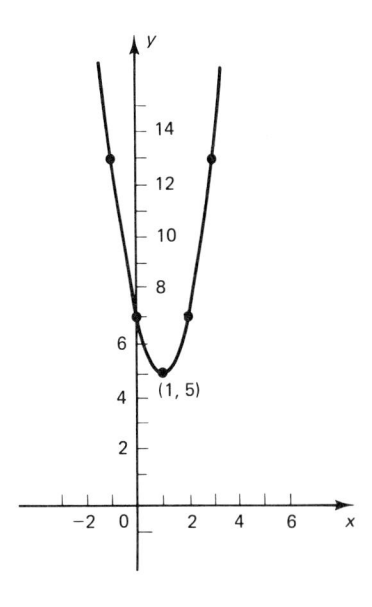

As an aid in sketching the graph, we therefore compute values of y corresponding to a few values of x:

x	-1	0	1	2	3
y	13	7	5	7	13

Plotting these points, we are able to draw the graph shown in the figure above.

MAXIMUM AND MINIMUM VALUES

The vertex of the parabola $y = ax^2 + bx + c$ represents the lowest point when $a > 0$ or the highest point when $a < 0$. It follows therefore that for $a > 0$, the function $f(x) = ax^2 + bx + c$ takes its minimum value at the vertex of the corresponding parabola. Correspondingly, when $a < 0$, the function $f(x) = ax^2 + bx + c$ takes its largest value at the vertex of the parabola.

Problems in which we are required to calculate the maximum and minimum values of certain functions arise very frequently in applications. They are studied at length in most courses of calculus. However, we can solve some of these problems by making use of the properties of parabolas.

Example 6 A farmer has 200 yards of fencing with which he wishes to enclose a rectangular field. One side of the field can make use of a fence that already exists. What is the maximum area he can enclose?

Solution Let the sides of the field be denoted by x and y, as shown in the adjacent figure, the y side being parallel to the fence that already exists. Then the length of new fence is $2x + y$, which must equal the available 200 yards:
$$2x + y = 200.$$
The area enclosed is $A = xy$. But $y = 200 - 2x$, and so
$$A = x(200 - 2x) = 200x - 2x^2 \qquad \text{(i)}$$
Comparing this with $f(x) = ax^2 + bx + c$, we see that A is a quadratic function of x, with $a = -2$, $b = 200$, $c = 0$. Therefore, since $a < 0$, the quadratic function has a maximum value at the vertex of the corresponding parabola.

Completing the square as usual, we find that
$$\begin{aligned} A &= -2(x^2 - 100x) \\ &= -2(x^2 - 100x + 2500) + 5000 \\ &= -2(x - 50)^2 + 5000. \end{aligned}$$
Thus the coordinates of the vertex are $x = 50$ and $A = 5000$. The largest area that can be enclosed is 5000 square yards. This is achieved by making $x = 50$ yards. The other side has length $y = 200 - 2x = 100$ yards.

Example 7 An object is projected vertically into the air with an initial speed of 80 feet per second (about 54 miles per hour). After t seconds its height above the ground in feet is given by the formula
$$H = 80t - 16t^2.$$
Find the maximum height it reaches.

Solution The height H is a quadratic function of t. Since the coefficient of t^2 is negative, the maximum value of H will occur at the vertex of the corresponding parabola. We have
$$\begin{aligned} H &= -16(t^2 - 5t) \\ &= -16[(t^2 - 5t + (\tfrac{5}{2})^2) - (\tfrac{5}{2})^2] \\ &= -16[(t - \tfrac{5}{2})^2 - \tfrac{25}{4}] \\ &= -16(t - \tfrac{5}{2})^2 + 100. \end{aligned}$$
The vertex is at $t = \tfrac{5}{2}$, $H = 100$. Thus the maximum height reached is 100 feet, and it takes $\tfrac{5}{2}$ seconds to reach this height. The graph of H as a function of t is shown in the following figure. (Note the different scales on the two axes.)

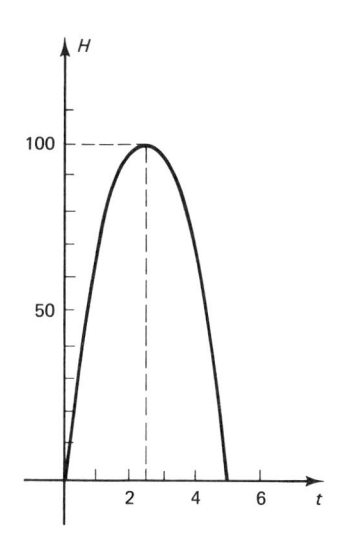

Example 8 Mr. Woolhouse owns an apartment building that has 60 suites. He can rent all the suites if he charges a monthly rent of $200 per suite. At a higher rent some of the suites will remain vacant. On average, for each increase in rent of $5, one suite remains vacant with no possibility of being rented. Determine the functional relationship between the total monthly revenue and the number of vacant units. What monthly rent will make the total revenue maximum? What is this maximum revenue?

Solution Let x denote the number of vacant units. The number of rented apartments is then $(60 - x)$ and the monthly rent per suite is $(200 + 5x)$ dollars. If R denotes the total monthly revenue (in dollars), then

$$R = (\text{rent per unit})(\text{number of units rented})$$
$$= (200 + 5x)(60 - x)$$
$$= -5x^2 + 100x + 12{,}000.$$

The total monthly revenue R is a quadratic function of x with $a = -5 < 0$. The graph is therefore a parabola that opens downward, and the maximum value of R is obtained at the vertex. We have

$$R = -5(x^2 - 20x) + 12{,}000$$
$$= -5(x - 10)^2 + 12{,}500.$$

Thus the maximum value of R is $12,500 per month. This is achieved by setting the rent at $(200 + 5x) = \$250$ per month, in which case 10 suites will be vacant.

Example 9 Find the point on the line $2x + y = 4$ that is closest to the point $(3, 2)$.

Solution The square of the distance between the point (x, y) and $(3, 2)$ is

$$d^2 = (x - 3)^2 + (y - 2)^2.$$

If (x, y) lies on the given line, then
$$y = 4 - 2x.$$
Therefore
$$d^2 = (x - 3)^2 + (4 - 2x - 2)^2$$
$$= 5x^2 - 14x + 13.$$

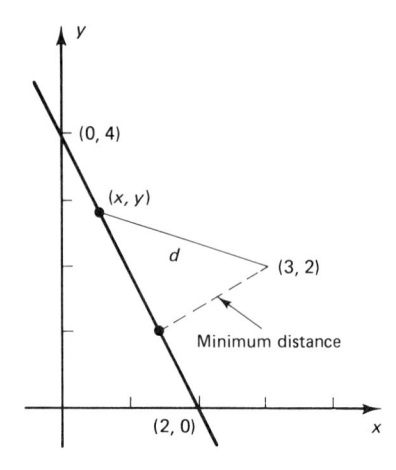

Since d^2 is a quadratic function of x with positive coefficient of x^2, the minimum value of d^2 occurs at the vertex of the corresponding parabola. We find, as usual,
$$d^2 = 5(x - \tfrac{7}{5})^2 + \tfrac{16}{5}$$
and so $x = \tfrac{7}{5}$ at the vertex. Then $y = 4 - 2x = \tfrac{6}{5}$. So the point on the line closest to $(3, 2)$ is $(\tfrac{7}{5}, \tfrac{6}{5})$.

HORIZONTAL PARABOLAS

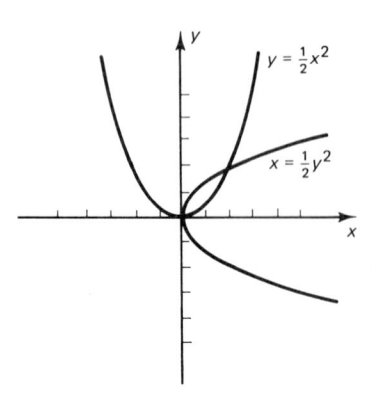

If we interchange x and y in the equation $y = ax^2$, we obtain $x = ay^2$. This latter equation must also have a parabola as its graph, but this time with the x axis as its axis of symmetry. The adjoining figure shows the graphs of the two parabolas $y = \tfrac{1}{2}x^2$ and $x = \tfrac{1}{2}y^2$. One curve can be obtained by rotating the other through 90 degrees.

The equation $x = \tfrac{1}{2}y^2$ can be solved for y. There are two solutions: $y = \sqrt{2x}$ and $y = -\sqrt{2x}$. The graph of $y = \sqrt{2x}$ is the upper half of the horizontal parabola and the graph of $y = -\sqrt{2x}$ is the lower half. (See Section 6.2.)

In general, the graph of $x = ay^2$ is a horizontal parabola with vertex at the origin, opening to the right if $a > 0$ and to the left if $a < 0$. The focus is on the x axis, at $(1/4a, 0)$, while the directrix is the vertical line $x = -1/4a$. This parabola is *not* the graph of a function, but each half of it (upper or lower) is the graph of a function ($y = +\sqrt{x/a}$ or $y = -\sqrt{x/a}$).

The parabola that has the same shape and size as $x = ay^2$ but has vertex at (h, k) is given by the equation
$$x = a(y - k)^2 + h.$$

Diagnostic Test 7.2

1. The graph of $y = ax^2$ is a _____ with vertex at _____ and that opens _____ if $a > 0$ and opens _____ if $a < 0$.
2. The equation $y - h = a(x - k)^2$ represents a _____ with vertex at _____ and that opens _____ if $a > 0$ and opens _____ if $a < 0$.
3. The equation $(x - 2)^2 + 4y - 8 = 0$ represents a _____ with vertex at _____ and that opens _____.
4. The equation $x + (y + 3)^2 = 0$ represents a _____ with vertex at _____ and that opens _____.
5. The equation $x = ay^2$ represents a _____ with vertex at _____ and that opens _____ if $a > 0$ and opens _____ if $a < 0$.
6. The graph of $y = ax^2 + bx + c$ $(a \neq 0)$ is a _____ that opens _____ if $a > 0$ and opens _____ if $a < 0$.
7. The graph of $y = a\sqrt{x}$ is a _____ with vertex at _____, that opens _____, lies above the x axis if _____, and lies below the x axis if _____.
8. The graph of $y = a\sqrt{-x}$ is a _____ with vertex at _____, that opens _____, lies above the x axis if _____, and lies below the x axis if _____.
9. The graph of $y - k = a\sqrt{x - h}$ is a _____ with vertex at _____.
10. The axis of any parabola is _____ to its directrix.
11. The focus of any parabola lies on the _____ of the parabola.
12. The _____ of a parabola is the middle point of the perpendicular from the focus onto the directrix.

Exercises 7.2

(1–12) Find the vertices of the following parabolas.

1. $y = -3x^2$.
2. $y = \frac{1}{4}x^2$.
3. $y = x^2 + 2$.
4. $y = x^2 - 5$.
5. $y = -\frac{1}{2}x^2 - 2$.
6. $y = 6 - 3x^2$.
7. $y = x^2 - 4x + 1$.
8. $y = x^2 + 10x - 4$.
9. $y = 3x^2 + 2x - 2$.
10. $y = 5x^2 + 2x$.
11. $y = 4x - 6x^2$.
12. $y = 3x - 2x^2 + 2$.

(13–24) Find the points where the following parabolas intersect the x and y coordinate axes.

13. $y = x^2 - 4$.
14. $y = 2x^2 - 2$.
15. $y = 3x^2 + 1$.

16. $y = -x^2 - 5$.
17. $y = x^2 + 3x + 2$.
18. $y = 2x^2 - 2x - 12$.
19. $y = 3x - x^2$.
20. $y = 2x^2 + 5x$.
21. $y = 2x^2 + 2x + 4$.
22. $y = \frac{1}{2}x^2 - x - 84$.
23. $y = 2 + 2x - x^2$.
24. $y = 3 - 5x - x^2$.

(25–34) Sketch the graphs of the following parabolas.

25. $y = 1 - x^2$.
26. $y = x^2 - 16$.
27. $y = 2x^2 + 3x - 1$.
28. $y = 4x - x^2$.
29. $y = x^2 + 2x$.
30. $y = \frac{1}{2}x^2 - x - 1$.
31. $y = 3 - x - 3x^2$.
32. $y = x^2 + 4x + 1$.
33. $y = 2x - \frac{1}{2}x^2 - 2$.
34. $y = \frac{1}{3}x^2 + 2x + 3$.

35. Find two numbers whose sum is 16 and whose product is as large as possible.

36. Find two numbers whose difference is 8 and whose product is as small as possible.

37. Find two numbers whose difference is 5 and the sum of whose squares is as small as possible.

38. Find two numbers whose sum is 16, such that the sum of the square of one and three times the square of the other is as small as possible.

39. Find the point on the line $2x + 3y = 6$ that is closest to the origin.

40. Find the point on the line $x - 4y = 8$ that is closest to the point (6, 0).

41. A farmer has 500 yards of fencing with which to enclose a rectangular paddock. What is the largest area he can enclose?

42. The yield of apples from each tree in an orchard is $(500 - 5x)$ pounds, where x is the density with which the trees are planted (the number of trees per acre). Find the value of x that makes the total yield per acre a maximum.

43. If rice plants are sown at a density of x plants per square foot, the yield of rice from a certain location is $x(10 - 0.5x)$ bushels per acre. What value of x maximizes the yield per acre?

44. If apple trees are planted at 30 per acre, the value of the crop produced by each tree is $180. For each additional tree planted per acre, the value of the crop falls by $3. What number of trees must be planted per acre to obtain the maximum value of the crop? What is this maximum value per acre of the crop?

45. If a publisher prices one of his books at $20 per copy, he will sell 10,000 copies. For every dollar increase in the price his sales fall by 400 copies. What should he charge per copy to obtain the maximum revenue? What is the value of this maximum revenue?

46. In the preceding question, the cost of producing and selling each copy is $13. What price should the publisher charge for each book so as to gain a maximum profit?

47. Chou-ching Realty has built a new rental unit of 40 apartments. It is known from market research that if a rent of $150 per month is charged, all the suites will be occupied. For each $5 increase in the rent one unit will remain vacant. What monthly rent should be charged for each unit to obtain the maximum monthly rental revenue? Find this maximum revenue.

48. At a price of $10 each, a company expects to sell 5000 items of its product per week. For each increase of $1 in the price the weekly sales will fall by 250. What price should be charged in order to generate the maximum weekly revenue?

49. The monthly revenue from selling x units of a certain commodity is given by $R(x) = 12x - 0.01x^2$ dollars. Determine the number of units that must be sold each month to maximize the revenue. What is the corresponding maximum revenue?

50. The profit $P(x)$ obtained by manufacturing and selling x units of a certain product is given by
$$P(x) = 60x - x^2.$$
Determine the number of units that must be produced and sold so as to maximize the profit. What is the maximum profit?

51. (a) The revenue R obtained by selling x units is given by $R(x) = 60x - 0.01x^2$. Determine the number of units that must be sold each month so as to maximize the revenue. What is this maximum revenue?
(b) The cost of producing x units is equal to $2000 + 25x$ dollars. How many units must be produced and sold each month to obtain a maximum profit? What is this maximum profit?

*52. The market demand for a certain product is x units when the price charged to consumers is $\$p$, where
$$15p + 2x = 720.$$
The cost (in dollars) of producing x units is given by $C(x) = 200 + 6x$. What price p per unit should be charged to consumers to obtain a maximum profit?

*53. Find a formula for the coordinates of the vertex of the parabola $y = ax^2 + bx + c$.

*54. If $a > 0$, show that $ax^2 + bx + c > 0$ for all x if and only if $b^2 - 4ac < 0$.
If $a < 0$, show that $ax^2 + bx + c < 0$ for all x if and only if $b^2 - 4ac < 0$.
Interpret these results in terms of the graph of $y = ax^2 + bx + c$.

*55. Find the condition on a, b, and c such that the graph of $y = ax^2 + bx + c$ just touches the x axis at its vertex.

7.3

The Ellipse and Hyperbola

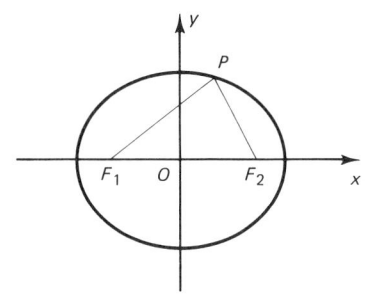

Let F_1 and F_2 be two given points and a a given number. Then the set of all points the sum of whose distances from F_1 and F_2 is equal to $2a$ is a curve called an *ellipse*. F_1 and F_2 are its two *foci*. If P is any point on the ellipse, we have
$$PF_1 + PF_2 = 2a. \tag{1}$$

Let us choose coordinate axes such that the foci lie on the x axis at equal distances from the origin, as in the adjacent figure. Let F_1 and F_2 have coordinates $(-c, 0)$ and $(c, 0)$, respectively. Then if $P(x, y)$ is a general point on the ellipse satisfying (1), it can be shown (see Exercise 47) that

$$\boxed{\frac{x^2}{a^2} + \frac{y^2}{b^2} = 1} \tag{2}$$

where $b^2 = a^2 - c^2$. This equation is called the ***standard form of the equation of the ellipse.***

If $c = 0$, then $b^2 = a^2$ and Equation (2) reduces to

$$\frac{x^2}{a^2} + \frac{y^2}{a^2} = 1, \quad \text{that is,} \quad x^2 + y^2 = a^2.$$

This is the equation of a circle centered at the origin and radius a. But if $c = 0$, then F_1 and F_2 both coincide at the origin, and the condition (1) reduces to the condition $2PO = 2a$, which does define such a circle. Thus a circle is a special case of an ellipse.

Example 1 Draw the graph of the equation

$$\frac{x^2}{9} + \frac{y^2}{4} = 1 \tag{i}$$

that corresponds to the values $a = 3$, $b = 2$ in Equation (1).

Solution First let us find the points of intersection with the coordinate axes. On the x axis we have $y = 0$, and (i) becomes $x^2/9 = 1$; that is, $x = \pm 3$. Thus there are two points of intersection, $(\pm 3, 0)$.

Similarly, when $x = 0$, we have $y^2/4 = 1$ and $y = \pm 2$. So again there are two points on the y axis, $(0, \pm 2)$.

Next let us solve for y. We have

$$\frac{y^2}{4} = 1 - \frac{x^2}{9} = \frac{1}{9}(9 - x^2)$$

and so, taking the square root of both sides, we get

$$\frac{y}{2} = \pm \frac{1}{3}\sqrt{9 - x^2}$$

$$y = \pm \frac{2}{3}\sqrt{9 - x^2}.$$

We observe two things about this solution. First, for each value of x there are two values of y, which are of equal magnitude but of opposite sign. This means that the graph must be *symmetrical* about the x axis.

Second, we observe that if the value of y is to be a real number, we must have $9 - x^2 \geq 0$. This means $x^2 \leq 9$; that is, $-3 \leq x \leq 3$. The graph has no points if $x < -3$ or if $x > 3$.

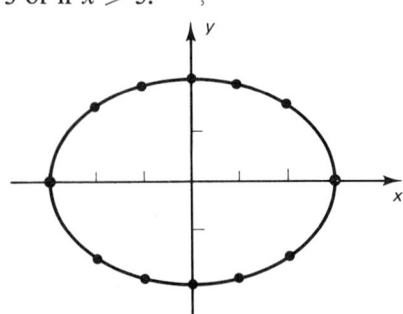

We have the following table of values.

x	-3	-2	-1	0	1	2	3
y	0	$\pm\frac{2}{3}\sqrt{5}$	$\pm\frac{2}{3}\sqrt{8}$	± 2	$\pm\frac{2}{3}\sqrt{8}$	$\pm\frac{2}{3}\sqrt{5}$	0

These points are plotted on the figure above (note that $\frac{2}{3}\sqrt{5} \approx 1.49$ and $\frac{2}{3}\sqrt{8} \approx 1.89$).

In the example above the graph is symmetrical about the x axis because the original equation $(x^2/9) + (y^2/4) = 1$ involves only even powers of y (see Section 5.2). But this equation also involves only even powers of x, implying that the graph must be symmetrical about the y axis as well. These symmetries about both coordinate axes are apparent from the above figure. Furthermore, the graph is also symmetric about the origin.

The graph of Equation (2) for general a and b is qualitatively similar to that in Example 1. First we observe that, since the equation involves only even powers of x and y, the graph is symmetrical about both the x axis and the y axis for any values of the constants a and b.

Next let us calculate the points of intersection with the coordinate axes. On the x axis, $y = 0$, and setting $y = 0$ in Equation (2) we get

$$\frac{x^2}{a^2} = 1; \quad \text{that is,} \quad x = \pm a.$$

So the graph intersects the x axis at the two points $(\pm a, 0)$.

On the y axis, $x = 0$, and Equation (1) becomes

$$\frac{y^2}{b^2} = 1; \quad \text{that is,} \quad y = \pm b.$$

So the graph intersects the y axis at the two points $(0, \pm b)$.

Next let us solve Equation (2) for y in terms of x. As in Example 1, we get

$$\frac{y^2}{b^2} = 1 - \frac{x^2}{a^2} = \frac{1}{a^2}(a^2 - x^2)$$

and, taking the square root,

$$\frac{y}{b} = \pm\frac{1}{a}\sqrt{a^2 - x^2}.$$

Therefore,

$$y = \pm\frac{b}{a}\sqrt{a^2 - x^2}.$$

In order for y to be a real number, the expression under the radical must be nonnegative, $a^2 - x^2 \geq 0$. Thus $x^2 \leq a^2$; that is, $-a \leq x \leq a$. The graph therefore has no points for $x < -a$ or for $x > a$.

In a similar way, we can solve for x in terms of y. We obtain that

$-b \leq y \leq b$, and so there are no points on the graph for which $y < -b$ or $y > b$.

The figures below illustrate the graph of Equation (2) in the two cases $a > b$ and $a < b$. In both cases the curve is an ellipse. The x and y axes, about which the curve is symmetrical, are called the **axes of the ellipse**. The ellipse cuts off a length $2a$ on the x axis and a length $2b$ on the y axis. The greater of these two lengths is called the **major axis** of the ellipse and the lesser its **minor axis**. For example, the ellipse in Example 1 has major axis equal to 6 and minor axis equal to 4. The origin is called the **center** of the ellipse. (Note that when $b > a$, the foci of the ellipse are on the y axis.)

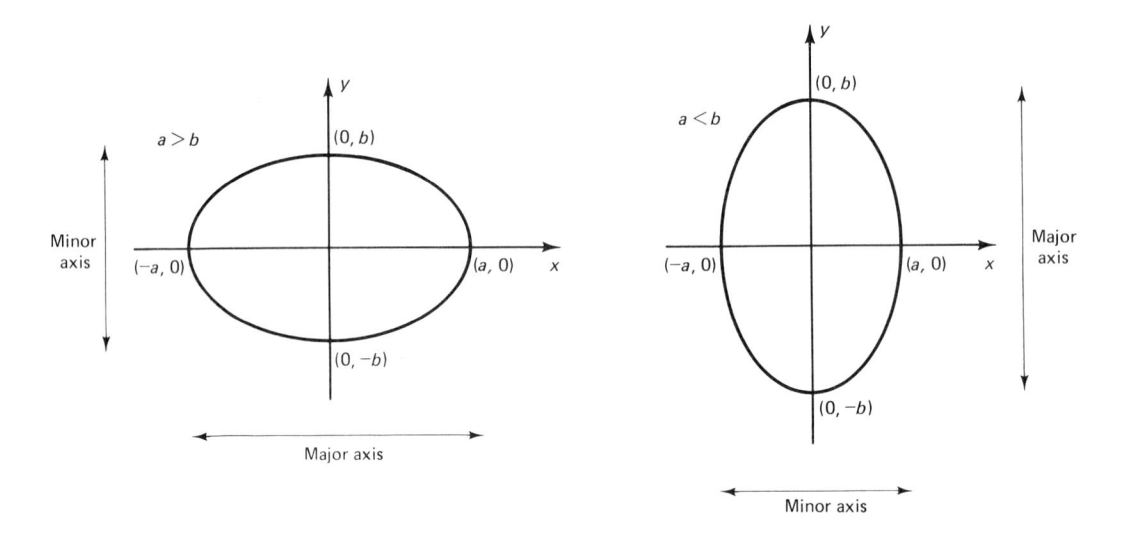

Example 2 Draw the graph of the equation $4x^2 + y^2 = 25$.

Solution We write the given equation in the standard form (1) by dividing through by 25:

$$\frac{4x^2}{25} + \frac{y^2}{25} = 1$$

$$\frac{x^2}{\frac{25}{4}} + \frac{y^2}{25} = 1.$$

Thus, $a^2 = \frac{25}{4}$ and $b^2 = 25$, so $a = \frac{5}{2}$ and $b = 5$. Since $a < b$, the graph is an elongated ellipse with major axis equal to $2b = 10$ (on the y axis) and minor axis equal to $2a = 5$.

The intercepts with the coordinate axes are at the points $(\pm a, 0)$ and $(0, \pm b)$—that is, at $(\pm \frac{5}{2}, 0)$ and $(0, \pm 5)$. We note that when $x = \pm 1$, $y = \pm \sqrt{21} \approx 4.6$. And when $x = \pm 2$, $y = \pm 3$. These few extra points enable us to draw the graph shown in the following figure.

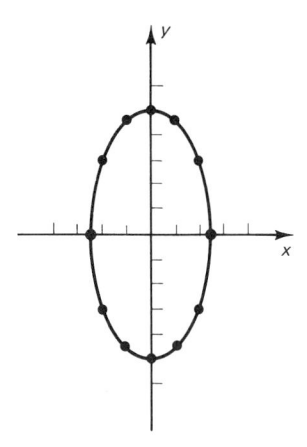

THE HYPERBOLA

Let F_1 and F_2 be two given points and a a given number. Then the set of all points whose distances from F_1 and F_2 *differ* by $2a$ is called a **hyperbola**. F_1 and F_2 are its two *foci*. If P is any point on the hyperbola, we have

$$PF_1 - PF_2 = \pm 2a \tag{3}$$

(the plus or minus depending on whether PF_1 is greater or less than PF_2).

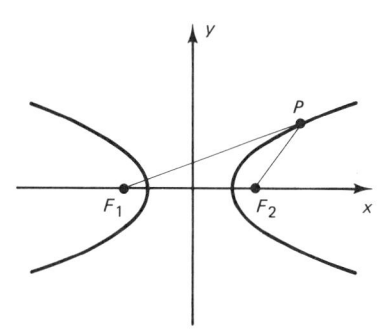

Again take coordinate axes such that F_1 and F_2 lie on the x axis at equal distances from the origin. Let F_1 be $(-c, 0)$ and F_2 be $(c, 0)$. Then if $P(x, y)$ is a general point on the hyperbola satisfying (3), we can show that

$$\boxed{\frac{x^2}{a^2} - \frac{y^2}{b^2} = 1} \tag{4}$$

where $b^2 = c^2 - a^2$ (see Exercise 48). This is called the **standard equation of the hyperbola**.

Example 3 Draw the graph of the equation

$$\frac{x^2}{9} - \frac{y^2}{4} = 1 \tag{i}$$

that corresponds to $a = 3$ and $b = 2$ in Equation (4).

Sec. 7.3 *The Ellipse and Hyperbola*

Solution We notice first that Equation (4), like the equation of the ellipse, involves only the squares x^2 and y^2. It follows that the graph is symmetrical about the x and y axes for any values of a and b (see Section 5.2.)

To find the points of intersection with the x axis, we put $y = 0$ in Equation (i); we get $x^2/9 = 1$, $x = \pm 3$, giving the points $(\pm 3, 0)$. Similarly, to find the intersections with the y axis we put $x = 0$; we get $-y^2/4 = 1$, which has no real solution. Therefore the graph does not cross the y axis.

Let us solve Equation (i) for y. We get

$$\frac{y^2}{4} = \frac{x^2}{9} - 1 = \frac{1}{9}(x^2 - 9)$$

and, taking the square root,

$$\frac{y}{2} = \pm \frac{1}{3}\sqrt{x^2 - 9}.$$

Thus

$$y = \pm \frac{2}{3}\sqrt{x^2 - 9}.$$

The two solutions for y correspond to the symmetry of the graph about the x axis that we noted earlier.

We observe that the solution for y is a real number if and only if $x^2 - 9 \geq 0$. This is satisfied if $x \leq -3$ or if $x \geq 3$. Thus the graph of Equation (i) has points only in the regions $x \leq -3$ and $x \geq 3$ and has no points for $-3 < x < 3$.

We have the following table of values.

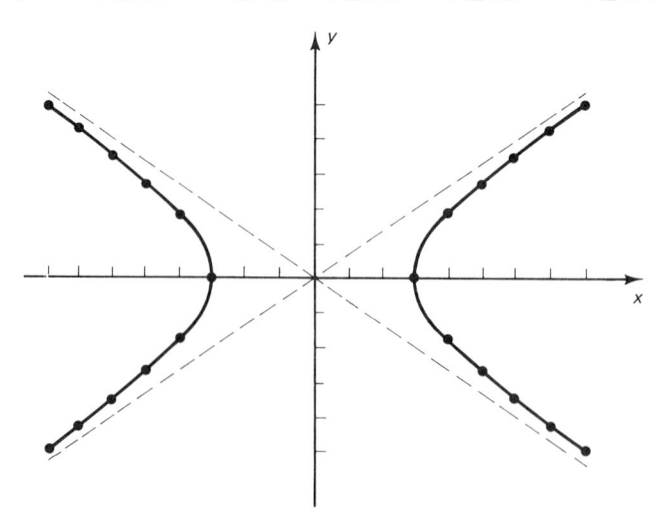

x	± 3	± 4	± 5	± 6	± 7	± 8
y	0	$\pm \frac{2\sqrt{7}}{3}$	$\pm \frac{8}{3}$	$\pm 2\sqrt{3}$	$\pm \frac{4}{3}\sqrt{10}$	$\pm \frac{2\sqrt{55}}{3}$
		$\approx \pm 1.76$	$\approx \pm 2.67$	$\approx \pm 3.46$	$\approx \pm 4.22$	$\approx \pm 4.94$

These points are plotted on the figure on page 274 and the graph is sketched through them. The resulting curve is a hyperbola. (The dotted lines in the figure will be explained later.)

In the general case of Equation (4) the graph intersects the x axis (i.e., $y = 0$) where $x^2/a^2 = 1$. Thus the points of intersection are $(\pm a, 0)$. On the other hand there are no intersections with the y axis because when we put $x = 0$ we get $-y^2/b^2 = 1$—that is, $y^2 = -b^2$, which has no solutions.

Solving the general equation for y, we get

$$\frac{y^2}{b^2} = \frac{x^2}{a^2} - 1 = \frac{1}{a^2}(x^2 - a^2)$$

and so

$$\frac{y}{b} = \pm\frac{1}{a}\sqrt{x^2 - a^2}.$$

Therefore

$$y = \pm\frac{b}{a}\sqrt{x^2 - a^2}. \tag{5}$$

In this case the value of y is a real number only when $x^2 - a^2 \geq 0$—that is, $x^2 \geq a^2$. This is satisfied if $x \leq -a$ or if $x \geq a$. Thus there are no points on the graph for $-a < x < a$.

It is important to know the form of the graph when x is numerically large. When x is very large in magnitude, then x^2 will be much greater than a^2 and the quantity $(x^2 - a^2)$ in (5) will be almost the same as x^2 itself. Then the solution for y is approximately given by

$$y = \pm\frac{b}{a}\sqrt{x^2 - a^2} \approx \pm\frac{b}{a}\sqrt{x^2}$$

$$y \approx \pm\frac{b}{a}x.$$

Thus when x is numerically very large, the graph of Equation (4) becomes very close to the two straight lines $y = \pm(b/a)x$.

Returning to Example 3, in which $a = 3$, $b = 2$, we find that these lines are $y = \pm\frac{2}{3}x$. These two straight lines have been drawn dotted in the figure in Example 3, and we can see quite clearly that as x increases in magnitude, the graph of the hyperbola comes closer and closer to one or the other of the lines. For example, when $x = 100$, the values of y on the hyperbola are

$$y = \pm(\tfrac{2}{3})\sqrt{9991} \approx \pm 66.63$$

while on the two straight lines the values of y are $\pm\frac{200}{3} = \pm 66.67$, almost the same.

In the general case the graph of Equation (4) has the form shown in the following figure. The two lines $y = (b/a)x$ and $y = (-b/a)x$ are called the *asymptotes* of the hyperbola. The word asymptote has the same sense as in Section 6.5, in that *the graph approaches closer and closer to the asymptote as it recedes further and further from the origin.* The asymptotes of the hyperbola

Sec. 7.3   The Ellipse and Hyperbola

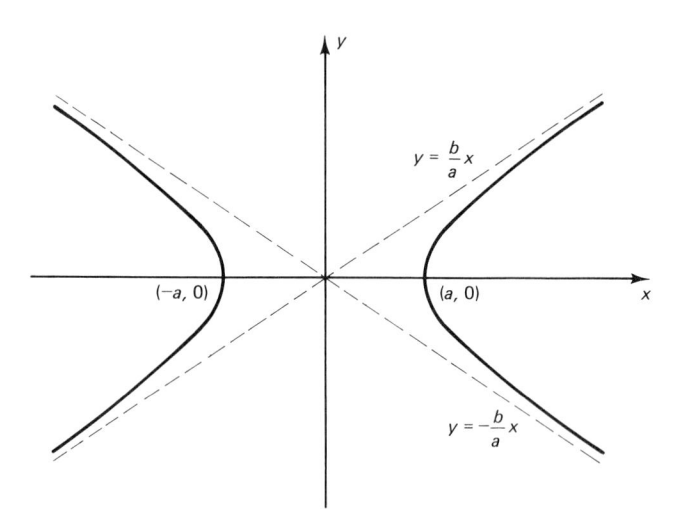

[Figure: Hyperbola with vertices $(-a, 0)$ and $(a, 0)$, and asymptotes $y = \frac{b}{a}x$ and $y = -\frac{b}{a}x$.]

are, however, inclined to the coordinate axes rather than being horizontal or vertical as in the earlier graphs.

Example 4 Find the asymptotes of the graph of $9x^2 - 16y^2 = 25$.

Solution We first note that the graph of this equation *is* a hyperbola, since we can write the equation in the same form as Equation (4) by dividing through by 25:

$$\frac{9x^2}{25} - \frac{16y^2}{25} = 1$$

or

$$\frac{x^2}{(\frac{5}{3})^2} - \frac{y^2}{(\frac{5}{4})^2} = 1.$$

Therefore $a = \frac{5}{3}$ and $b = \frac{5}{4}$.

The equations of the asymptotes can be obtained from the general formula $y = \pm(b/a)x$:

$$y = \pm\left(\frac{\frac{5}{4}}{\frac{5}{3}}\right)x = \pm\frac{3}{4}x.$$

There is, however, a quicker way of getting the asymptotes from the original equation itself: we simply replace the constant on the right side of the equation with zero and keep only x^2 and y^2 terms. We get

$$9x^2 - 16y^2 = 0$$

$$y^2 = \frac{9x^2}{16}$$

$$y = \pm\frac{3}{4}x.$$

The procedure used in this example works in general. Returning to Equation (4) and replacing the constant on the right with zero, we get

$$\frac{x^2}{a^2} - \frac{y^2}{b^2} = 0.$$

This gives $y = \pm(b/a)x$, which, as we know, are the equations of the asymptotes.

Example 5 Sketch the graph of the equation $3y^2 - x^2 = 12$.

Solution This equation cannot be written in quite the same form as Equation (4) because the signs of the coefficients of x^2 any y^2 are wrong way round. The graph is still a hyperbola, however. It does not intersect the x axis, but it intersects the y axis at the points $(0, \pm 2)$. Its asymptotes are given by

$$3y^2 - x^2 = 0$$

$$y = \pm\left(\frac{1}{\sqrt{3}}\right)x.$$

The graph is sketched in the following figure.*

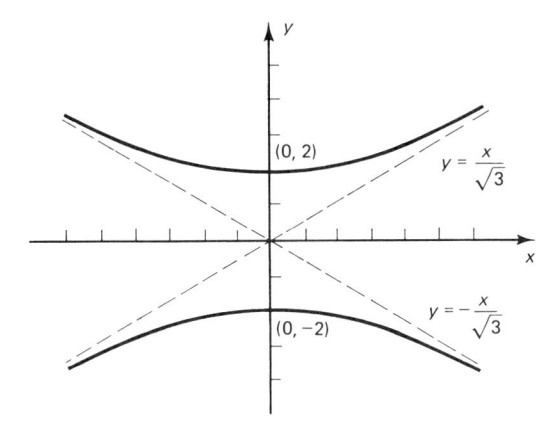

Example 6 Find the equation of the hyperbola that has the lines $y = \pm 2x$ as asymptotes and that meets the y axis at the points $(0, \pm 3/2)$.

Solution The equation must take one of the following two forms:

$$\text{either} \quad \frac{x^2}{a^2} - \frac{y^2}{b^2} = 1 \quad \text{or} \quad \frac{y^2}{b^2} - \frac{x^2}{a^2} = 1.$$

However, a hyperbola of the first type intersects the x axis and not the y axis. Therefore the equation must be of the second type,

$$\frac{y^2}{b^2} - \frac{x^2}{a^2} = 1. \tag{i}$$

---

*The foci of this hyperbola are on the y axis rather than the x axis.

Putting $x = 0$, we get the intersections with the y axis at $y = \pm b$. These points are given to be $(0, \pm \frac{3}{2})$, and therefore $b = \frac{3}{2}$. The asymptotes are $y = \pm(b/a)x$, which are given to be $y = \pm 2x$. Therefore $b/a = 2$. Hence $a = b/2 = \frac{3}{4}$: The equation (i) therefore becomes

$$\frac{y^2}{(\frac{3}{2})^2} - \frac{x^2}{(\frac{3}{4})^2} = 1$$

—that is,

$$4y^2 - 16x^2 = 9.$$

When $b = a$, the asymptotes become the lines $y = \pm x$, having slopes ± 1. The two asymptotes are then perpendicular to one another. In this case the graph is called a **rectangular hyperbola**.

Setting $b = a$ in Equation (4) and multiplying through by a^2, we obtain

$$x^2 - y^2 = a^2.$$

This is the equation of the rectangular hyperbola that meets the x axis at the points $(\pm a, 0)$. Similarly the equation $y^2 - x^2 = a^2$ describes the rectangular hyperbola that meets the y axis at the points $(0, \pm a)$.

Diagnostic Test 7.3

Fill in the blanks.

1. The graph of $x^2/2 + y^2/3 = 1$ is _____ that crosses the x axis at _____ and the y axis at _____.

2. The equation of the ellipse that crosses the x axis at $(\pm 2, 0)$ and the y axis at $(0, \pm 5)$ is _____.

3. The equations of the ellipses that are symmetrical about the x and y axes and have major axis 6 and minor axis 2 are _____ and _____.

4. The ellipse $x^2/3 + y^2/4 = 1$ has major axis _____ and minor axis _____.

5. The graph of $x^2/2 - y^2/3 = 1$ is _____ that meets the coordinate axes at _____.

6. The _____ of $x^2/2 - y^2/8 = 1$ are $y = \pm 2x$.

7. The graphs of $x^2 - 2y^2 = 1$ and $x^2 - 2y^2 = -1$ have the same _____.

8. The graph of $2x^2 - 2y^2 = 1$ is a _____.

9. The asymptotes of $x^2/2 - y^2/3 = 1$ are _____.

10. The asymptotes of $3y^2 - 2x^2 = 1$ are _____.

11. If P is any point on an ellipse, then the _____ of the distances of P from the two foci is constant.

12. The set of points the difference of whose distances from two fixed points is constant is a _____.

Exercises 7.3

(1–6) Find the points of intersection of the following ellipses with the coordinate axes, and calculate the major and minor axes.

1. $\dfrac{x^2}{16} + \dfrac{y^2}{9} = 1.$
2. $\dfrac{x^2}{25} + \dfrac{y^2}{49} = 1.$
3. $5x^2 + 20y^2 = 1.$
4. $\tfrac{2}{3}x^2 + y^2 = 1.$
5. $3x^2 + y^2 = 3.$
6. $4x^2 + 6y^2 = 5.$

(7–14) Sketch the graphs of the following equations.

7. $x^2 + \dfrac{y^2}{9} = 1.$
8. $\dfrac{x^2}{9} + \dfrac{y^2}{16} = 1.$
9. $4x^2 + 16y^2 = 64.$
10. $x^2 + 25y^2 = 25.$
11. $x^2 + 9y^2 = 25.$
12. $4x^2 + 9y^2 = 1.$
13. $4x^2 + 3y^2 = 12.$
14. $2x^2 + 5y^2 = 20.$

(15–18) Write down the equations of the ellipses that have the coordinate axes as axes of symmetry and that intersect these axes at the following points.

15. $(\pm 2, 0), (0, \pm 4).$
16. $(\pm \tfrac{2}{3}, 0), (0, \pm \tfrac{1}{3}).$
17. $(\pm \tfrac{1}{2}, 0), (0, \pm \tfrac{1}{3}).$
18. $(\pm 4, 0), (0, \pm \tfrac{4}{5}).$

(19–28) Find the intersections with the coordinate axes and the equations of the asymptotes of the graphs of the following equations.

19. $\dfrac{x^2}{5} - y^2 = 1.$
20. $\dfrac{x^2}{3} - \dfrac{y^2}{12} = 1.$
21. $2x^2 - 4y^2 = 1.$
22. $x^2 - 8y^2 = 1.$
23. $3x^2 - 2y^2 = 8.$
24. $5x^2 - 2y^2 = 6.$
25. $\dfrac{y^2}{4} - \dfrac{x^2}{3} = 1.$
26. $\dfrac{y^2}{2} - \dfrac{x^2}{8} = 1.$
27. $5y^2 - 6x^2 = 15.$
28. $3y^2 - x^2 = 4.$

(29–36) Sketch the graphs of the following equations.

29. $\dfrac{x^2}{4} - \dfrac{y^2}{16} = 1.$
30. $\dfrac{x^2}{4} - \dfrac{y^2}{9} = 1.$
31. $9x^2 - 4y^2 = 4.$
32. $x^2 - 4y^2 = 16.$
33. $3x^2 - 3y^2 = 4.$
34. $2x^2 - 3y^2 = 4.$
35. $y^2 - 3x^2 = 1.$
36. $2y^2 - 2x^2 = 3.$

(37–42) Find the equations of the hyperbolas that have the given lines as asymptotes and that intersect the coordinate axes at the given points.

37. $y = \pm x; (\pm 2, 0).$
38. $y = \pm 2x; (\pm 1, 0).$
39. $y = \pm \tfrac{2}{3}x; (\pm 1, 0).$
40. $y = \pm \tfrac{5}{4}x; (\pm 4, 0).$
41. $y = \pm 3x; (0, \pm 2).$
42. $y = \pm \tfrac{1}{2}x; (0, \pm \tfrac{1}{3}).$

(43–46) Find the equations of the rectangular hyperbolas that are symmetrical about the coordinate axes and that meet these axes at the given points.

43. $(\pm 2, 0).$
44. $(\pm \tfrac{1}{2}, 0).$
45. $(0, \pm \sqrt{2}/3).$
46. $(0, \pm \sqrt{2}).$

*47. The point $P(x, y)$ is such that the sum of its distances from the two points $(\pm c, 0)$ is equal to $2a$. Show that x and y satisfy the equation $x^2/a^2 + y^2/b^2 = 1$, where $b^2 = a^2 - c^2$. [*Hint:* $\sqrt{(x+c)^2 + y^2} + \sqrt{(x-c)^2 + y^2} = 2a$.]

*48. The point $P(x, y)$ is such that the difference between its distances from the two points $(\pm c, 0)$ is equal to $2a$. Show that x and y satisfy the equation $x^2/a^2 - y^2/b^2 = 1$, where $b^2 = c^2 - a^2$.

49. The arch of a bridge over a highway is in the form of half an ellipse. The width of the arch at the bottom is 50 feet and its height at the center is 18 feet. Write down the equation of the arch, taking the origin at the center of the ellipse and the x axis horizontal.

50. A vertical pipe of circular cross section passes through a sloping roof. If the diameter of the pipe is 20 centimeters and the roof slopes at 45 degrees, find the major and minor axes of the elliptical hole that must be cut in the roof to allow passage of the pipe.

*51. Repeat Exercise 50 when the roof has slope m.

7.4

General Equation of Conics

Let $F(x, y) = 0$ be some given equation relating x and y whose graph is a certain curve. Suppose that this curve is translated by h units in the x direction and then by k units in the y direction. Then the new curve so obtained is the graph of the equation

$$F(x - h, y - k) = 0.$$

We obtain this equation from the original equation $F(x, y) = 0$ by replacing x with $x - h$ and replacing y with $y - k$.

We have already seen examples of this general result. The equation $x^2 + y^2 - r^2 = 0$ represents a circle of radius r centered at the origin. Making the above replacements, we obtain

$$(x - h)^2 + (y - k)^2 - r^2 = 0 \tag{1}$$

which is the equation of the circle of radius r centered at (h, k).

Similarly, $y = ax^2$ is the equation of a parabola with vertex at the origin and the y axis as its axis. Replacing x with $x - h$ and y with $y - k$, we obtain

$$y - k = a(x - h)^2. \tag{2}$$

The graph of this equation is a similar parabola with vertex at (h, k).

As another example, the equation $x = ay^2$ represents a horizontal parabola with vertex at the origin. A similar parabola with vertex (h, k) has the equation

$$x - h = a(y - k)^2. \tag{3}$$

We can make similar replacements in the equation

$$\frac{x^2}{a^2} + \frac{y^2}{b^2} = 1$$

of the ellipse whose center is the origin. It follows that the equation

$$\frac{(x - h)^2}{a^2} + \frac{(y - k)^2}{b^2} = 1 \tag{4}$$

represents an ellipse whose center is at the point (h, k). The axes of the ellipse are still parallel to the coordinate axes, and the ellipse has the same size and shape as does the graph of $x^2/a^2 + y^2/b^2 = 1$. (See the adjoining figure.)

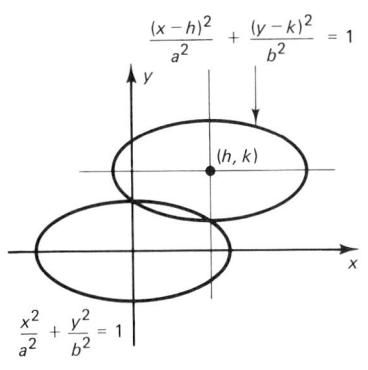

In the same way, from the equations

$$\frac{x^2}{a^2} - \frac{y^2}{b^2} = \pm 1$$

which represent hyperbolas centered at the origin, we obtain the equations

$$\frac{(x-h)^2}{a^2} - \frac{(y-k)^2}{b^2} = \pm 1 \tag{5}$$

which represent similar hyperbolas with center (h, k).

Example 1 Find the equation of the ellipse with center $(-2, 1)$ and with major axis 6 (in the x direction) and minor axis 2.

Solution In Equation (4) we are given $2a = 6$, $2b = 2$, and the center $(h, k) = (-2, 1)$. Therefore the equation of the ellipse is

$$\frac{[x-(-2)]^2}{3^2} + \frac{(y-1)^2}{1^2} = 1.$$

After simplification, this becomes

$$x^2 + 9y^2 + 4x - 18y + 4 = 0.$$

Example 2 The hyperbola whose equation is $3x^2 - y^2 = 4$ is translated so that its center is at the point $(2, -3)$. Find the equation of the new curve.

Solution In accordance with the general principle, in order to translate the hyperbola by 2 units in the x direction and by (-3) units in the y direction, we replace x with $x - 2$ and y with $y - (-3) = y + 3$. We obtain the equation

$$3(x-2)^2 - (y+3)^2 = 4$$

or, after simplification,

$$3x^2 - y^2 - 12x - 6y - 1 = 0.$$

The equations obtained in Examples 1 and 2 were both of the form

$$Ax^2 + Cy^2 + Dx + Ey + F = 0 \tag{6}$$

where A, C, D, E, and F are five constants. In fact, whatever the values of these constants, as long as A and C are not both zero, the graph of Equation (6) is always a conic section. By completing the square so as to combine the x and y terms, respectively, with x^2 and y^2 terms, we can always rewrite such an equation in one of the standard forms (1) through (5).

Example 3 Write the equation

$$2y^2 - x - 4y + 1 = 0$$

in one of the standard forms. Hence describe its graph.

Solution There is no x^2 term, so we combine the y^2 and y terms only:

$$2(y^2 - 2y) - x + 1 = 0.$$

Completing the square as usual, we get
$$2(y^2 - 2y + 1) - 2 - x + 1 = 0$$
$$2(y - 1)^2 - x - 1 = 0.$$
This equation can be rewritten in the form (3);
$$x + 1 = 2(y - 1)^2.$$
Therefore its graph is a horizontal parabola (similar to $x = 2y^2$) with vertex at $(h, k) = (-1, 1)$.

Example 4 Write the equation
$$3x^2 - 2y^2 - 12x + 6y = 0$$
in one of the standard forms and hence describe its graph.

Solution The given equation can be written
$$3(x^2 - 4x) - 2(y^2 - 3y) = 0$$
$$3(x - 2)^2 - 2(y - \tfrac{3}{2})^2 = \tfrac{15}{2}$$
$$\frac{(x - 2)^2}{\tfrac{5}{2}} - \frac{(y - \tfrac{3}{2})^2}{\tfrac{15}{4}} = 1.$$

Comparing with (5), we see that the graph is a hyperbola with center $(h, k) = (2, \tfrac{3}{2})$.

For the general equation (6) we can determine the type of conic it represents by inspection of the coefficients of x^2 and y^2. The following table summarizes the various cases.

Coefficients	Conic	Standard form	Properties
$A = 0, C \neq 0$	Horizontal parabola	$x - h = a(y - k)^2$	Vertex (h, k); opens right if $a > 0$, left if $a < 0$
$A \neq 0, C = 0$	Vertical parabola	$y - k = a(x - h)^2$	Vertex (h, k); opens up if $a > 0$, down if $a < 0$
$A = C$	Circle*	$(x - h)^2 + (y - k)^2 = r^2$	Center (h, k); radius r
$A \neq C$, A, C same sign	Ellipse*	$\dfrac{(x - h)^2}{a^2} + \dfrac{(y - k)^2}{b^2} = 1$	Center (h, k); semi-axes a, b
$A \neq C$, A, C of opposite signs	Hyperbola	$\dfrac{(x - h)^2}{a^2} - \dfrac{(y - k)^2}{b^2} = \pm 1$	Center (h, k); asymptotes parallel to $y = \pm(b/a)x$

*In these cases it is possible that the graph does not exist. See Example 3 in Section 7.1.

Diagnostic Test 7.4

Fill in the blanks.

1. The graph of the equation $x^2 + 2x - y + 7 = 0$ is a _____.
2. The graph of the equation $y^2 = 3x - 4y + 5$ is a _____.
3. The graph of the equation $y = ax^2 + bx + c$ $(a \neq 0)$ is a _____.
4. The graph of the equation $x = py^2 + qy + r$ $(p \neq 0)$ is a _____.
5. The graph of $3x^2 + 3y^2 + 2x - 4y = 0$ is a _____.
6. The graph of $2x^2 + y^2 - 3x + 4y = 5$ is _____.
7. The graph of $3x^2 - 4y^2 + 6x + 8y = 4$ is _____.
8. The equation $Ax^2 + Cy^2 + Dx + Ey + F = 0$ represents a parabola if _____.
9. The equation $Ax^2 + By^2 + Cx + Dy + E = 0$ represents a hyperbola if _____.
10. If the equation $Ax^2 + By^2 + Cx + Dy + E = 0$ represents a circle, then _____.
11. If the equation $Ax^2 + By^2 + Cx + Dy + E = 0$ represents an ellipse, then _____.

Exercises 7.4

(1–2) Find the equation of the ellipse with the given semi-axes and center. (The first semi-axis is parallel to the x axis in each case.)

1. $a = 4, b = 2;\ (h, k) = (3, -2)$.
2. $a = 3, b = 5;\ (h, k) = (-2, -1)$.

(3–4) Find the equation of the hyperbola obtained by translating the given hyperbola so that its center is at the given point (h, k).

3. $2x^2 - 3y^2 = 1;\ (h, k) = (-2, 2)$.
4. $4x^2 - y^2 = 2;\ (h, k) = (-1, -4)$.

(5–10) Find the equation of the parabola obtained by translating the given parabola so that its vertex is at the given point (h, k).

5. $y = 3x^2;\ (h, k) = (-2, 1)$.
6. $y = -2x^2;\ (h, k) = (-3, -4)$.
7. $y^2 + 4x = 0;\ (h, k) = (-4, -2)$.
8. $2y^2 - 3x = 0;\ (h, k) = (3, -1)$.
9. $2x^2 + 2x + 3y = 0;\ (h, k) = (-2, 2)$.
10. $y^2 + 3x + 4y - 2 = 0;\ (h, k) = (3, 2)$.

(11–20) Write each equation in one of the standard forms (1) through (5) and hence describe its graph.

11. $2x^2 - y^2 + 8x - 6y + 1 = 0$.
12. $3x^2 + 2y^2 + 12x - 8y = 0$.
13. $2x^2 + 3x + y + 1 = 0$.
14. $x^2 - 3y^2 + 4x - 2 = 0$.
15. $2x^2 + 4y^2 + 16(x + y) - 2 = 0$.
16. $3y^2 - 2x - 6y + 2 = 0$.
17. $4x^2 + 4y^2 - x + y - 1 = 0$.
18. $2x^2 - y^2 + 4x + 10y + 8 = 0$.
19. $2x^2 + y^2 - 6x - 6y + 12 = 0$.
20. $x^2 - 6x + 2y - 1 = 0$.

Chapter Review 7

Review the following topics.

(a) The equation of a circle centered at the origin.

(b) The equation of a circle with a given point as center.

(c) The calculation of the center and radius of a circle from the general form of its equation.

(d) The graph of the function $y = ax^2$.

(e) The graph of a general quadratic function.

(f) The calculation of the vertex of the parabola for a given quadratic function.

(g) The calculation of the maximum or minimum value of a quadratic function.

(h) The equations of the ellipse and hyperbola.

(i) The procedure for finding the asymptotes of a hyperbola.

(j) The procedure for determining whether the graph of a given equation is symmetrical about either of the coordinate axes.

(k) The equation of a horizontal or vertical parabola with vertex at (h, k).

(l) The equation of an ellipse or hyperbola with center at (h, k).

(m) Reduction of $Ax^2 + Cy^2 + Dx + Ey + F = 0$ to one of the standard forms and the classification of its graph.

Review Exercises on Chapter 7

(1–6) Determine the equation of the circle that has the given description.

1. Center at $(-2, 4)$ and radius 4.
2. Center at $(-3, -1)$ and radius 2.
3. Center at $(1, -2)$ and passing through $(-1, 2)$.
4. Center at $(-2, -3)$ and passing through the origin.
5. Center at $(\frac{1}{2}, \frac{3}{2})$ and touching the x axis.
6. Center at $(-3, 2)$ and touching the y axis.

(7–12) Determine whether each of the following equations represents a circle. If it does, find its center and radius and the points (if any) where it meets the coordinate axes.

7. $x^2 + y^2 + 4x + 4y - 1 = 0$.
8. $x^2 + y^2 - 2x + 6y + 6 = 0$.
9. $x^2 + y^2 - 3x + 4y + 8 = 0$.
10. $x^2 + y^2 - 6x - 4y - 3 = 0$.
11. $x^2 + y^2 + 10x - 4y + 4 = 0$.
12. $x^2 + y^2 + 2x - 8y + 17 = 0$.
13. A circle of radius 5 units has its center at $(p, -1)$ and passes through the point $(1, 2)$. Determine p.
14. Find the equation of the circle of radius r that lies in the first quadrant and touches both coordinate axes.

15. A manufacturer can sell x units of its product at p per unit, where x and p are related by
$$x^2 + p^2 + 200x + 150p = 49{,}400.$$
Plot the demand curve. What is the price above which no sales are possible?

16. A shoe manufacturing firm can produce men's and women's shoes by varying the production process. The possible amounts x and y (in hundreds of pairs) are related by the equation
$$x^2 + y^2 + 40x + 30y = 975.$$
Plot the product transformation curve for this firm (that is, the graph relating the quantities x and y of the two products).

(17–22) Find the vertices of the following parabolas and the points where they intersect the coordinate axes.

17. $y = \frac{1}{2}x^2 - 2$.

18. $y = 2x^2 + 1$.

19. $y = x^2 - 8x + 12$.

20. $y = 4x^2 - 4x - 3$.

21. $y = 15 + x - 2x^2$.

22. $y = -3x^2 + 6x - 4$.

(23–26) Sketch the graphs of the following parabolas.

23. $y = \frac{1}{2}(x^2 - 1)$.

24. $y = 3x - 2x^2$.

25. $y = \frac{5}{4} - 2x - x^2$.

26. $y = 2x^2 + 4x + 3$.

27. The owner of an apartment building can rent all his 60 suites if he charges $120 per month for each suite. If he increases the rent by $5, two of the suites remain unoccupied with no possibility of being rented. Assume the relation between the number of vacant suites and the rent to be linear.
  (a) Find the revenue R as a function of monthly rent p per unit.
  (b) Find the revenue as a function of the number y of occupied suites.
  (c) Find the rent that maximizes the monthly revenue.

28. Find two numbers whose difference is 5 and whose product is as small as possible.

29. Find the point on the line $3x + 4y = 12$ that lies nearest to the point $(1, -1)$.

30. A projectile is fired vertically upward at 240 feet per second. Its height after t seconds is given by $H = 240t - 16t^2$. Determine the maximum height reached.

31. A power cable suspended between two pylons has the shape of a parabola. The pylons are 80 feet high and 300 feet apart. In the center the cable hangs 70 feet above the ground. Find the height of the cable at a point 75 feet from one pylon.

32. The cable of a suspension bridge is parabolic in shape. In the center it is 10 feet above the roadway, and three-quarters of the way across it is 40 feet above the roadway. Assuming the roadway is level, calculate the height of the cable at the ends of the bridge.

(33–38) State whether the graph of each of the following equations is an ellipse or hyperbola and give the intersections with the coordinate axes.

33. $3x^2 + 5y^2 = 2$.

34. $-5x^2 + 2y^2 = -1$.

35. $2x^2 - 3y^2 = -1$.

36. $2x^2 + 3y^2 = -1$.

37. $y^2 = 2 + \frac{1}{3}x^2$.

38. $x^2 = 5(2 - y^2)$.

(39–40) Give the asymptotes of the following hyperbolas.

39. $2x^2 - 9y^2 = -1$.

40. $5y^2 - 2x^2 = 10$.

(41–42) Find the equations of the ellipses that are symmetrical about the coordinate axes and meet these axes at the following points.

41. $(\pm \frac{3}{2}, 0), (0, \pm 4)$.

42. $(\pm \frac{2}{3}, 0), (0, \pm \frac{1}{6})$.

(43–44) Find the equations of the hyperbolas that have the given lines as asymptotes and that intersect the coordinate axes at the given points.

43. $y = \pm(1/\sqrt{2})x; (\pm 3, 0)$.

44. $y = \pm 2x; (0, \pm \frac{1}{2})$.

(45–48) Write each equation in one of the standard forms. Hence describe its graph.

45. $3x^2 + 2y^2 + x + y - 1 = 0$.

46. $x^2 + 6x + 2y - 3 = 0$.

47. $4y^2 + 8(x + y) = 0$.

48. $x^2 - 3y^2 - 6x - 12y + 2 = 0$.

8
EXPONENTIALS AND LOGARITHMS

8.1

Exponential Functions

In Section 1.8 we showed how a^x can be defined when x is any rational number and a is any positive real number. Just to remind you how this goes, taking $a = 2$, we have for example that $2^{7/5}$ is the fifth root of 2^7, that is, the fifth root of 128; or $2^{4/9}$ is the ninth root of 2^4, that is, of 16; or $2^{-3/8}$ is the eighth root of 2^{-3}, that is, the eighth root of $\frac{1}{8}$. In general, if x is any rational number, we can write $x = p/q$, where p and q are integers with $q > 0$; then $a^x = a^{p/q}$ is equal to the qth root of a^p.

The question arises as to whether we can define a^x when x is irrational; for example, can we define $a^{\sqrt{2}}$ or a^π? Such quantities can in fact be defined by taking approximations to the irrational exponent x by means of rational numbers. For example, $\sqrt{2}$ can be approximated by the rational number $1.4 = \frac{14}{10}$; therefore we can take $a^{1.4} = \sqrt[10]{a^{14}}$ as an approximation to $a^{\sqrt{2}}$. Or a better approximation is $\sqrt{2} \approx 1.41 = \frac{141}{100}$, so we take $a^{1.41} = \sqrt[100]{a^{141}}$ as a better approximation to $a^{\sqrt{2}}$. Continuing in this way, we get a succession of numbers, $a^{1.4}, a^{1.41}, a^{1.414}, a^{1.4142}, \ldots$, which provide better and better approximations to $a^{\sqrt{2}}$. Each number in this sequence is well defined (in terms of powers and roots), and we can calculate $a^{\sqrt{2}}$ as accurately as we like by going far enough down the sequence.

This definition is correct but incomplete. We cannot be sure (unless we prove it) that the sequence of numbers used in the definition do in fact get closer and closer to a single real number that could be taken as $a^{\sqrt{2}}$. The proof is beyond the scope of this book, and we shall simply say that a completely rigorous definition of a^x can be given for all real exponents x and $a > 0$.

Example 1 Construct the graph of $y = 2^x$. Use it to estimate $2^{\sqrt{2}}$.

Solution We have the following table of values.

x	-3	-2	-1	0	1	2	3
y	$\frac{1}{8}$	$\frac{1}{4}$	$\frac{1}{2}$	1	2	4	8

The corresponding points are plotted on the figure below and connected by a smooth curve. 2^x can be defined for all real values of x, and the smooth curve is the graph of $y = 2^x$ for *all* x. When $x = \sqrt{2} \approx 1.41$ we read off from the graph that $y \approx 2.67$. Thus, $2^{\sqrt{2}} \approx 2.67$.

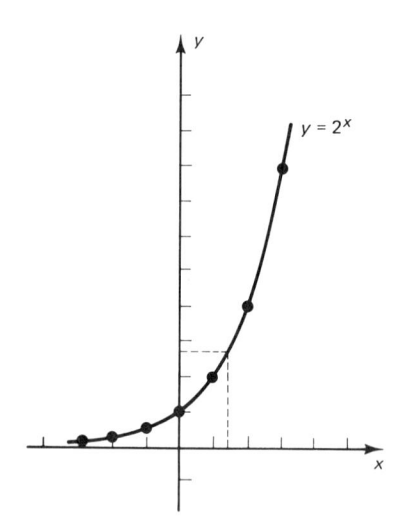

Example 2 Construct the graphs of $f(x) = (\frac{3}{2})^x$ and $g(x) = (\frac{2}{3})^x$.

Solution The following table gives a number of values of both $(\frac{3}{2})^x$ and $(\frac{2}{3})^x$.

x	-3	-2	-1	0	1	2	3
$y = (\frac{3}{2})^x$	$\frac{8}{27}$	$\frac{4}{9}$	$\frac{2}{3}$	1	$\frac{3}{2}$	$\frac{9}{4}$	$\frac{27}{8}$
$y = (\frac{2}{3})^x$	$\frac{27}{8}$	$\frac{9}{4}$	$\frac{3}{2}$	1	$\frac{2}{3}$	$\frac{4}{9}$	$\frac{8}{27}$

The two graphs are drawn in the figure below. Note that one graph is the mirror image of the other in the y axis.

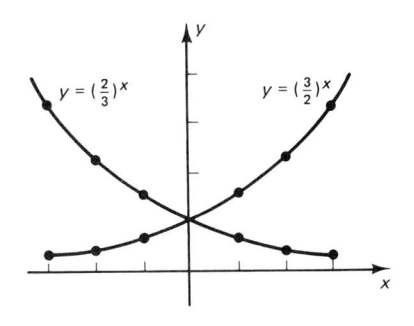

DEFINITION If $a > 0$, then the function $f(x) = a^x$ is called the **exponential function with base a**. Since a^x is defined for all real x, its domain is the set of all real numbers.

The graphs obtained in Examples 1 and 2 are characteristic of exponential functions. If $a > 1$ [such as $y = 2^x$ or $y = (\frac{3}{2})^x$], the graph of $y = a^x$ approaches the x axis as a horizontal asymptote as x becomes numerically large and negative (i.e., as $x \to -\infty$). As x takes larger and larger positive values, however, the graph rises more and more steeply. On the other hand, if $0 < a < 1$ [such as $y = (\frac{2}{3})^x$], the graph of $y = a^x$ approaches the x axis as a horizontal asymptote as x becomes large and positive, and y becomes large and positive as x becomes increasingly negative.

We note that a^x is positive for any real values of a and x ($a > 0$). Therefore the graph of $y = a^x$ is always situated above the x axis. Furthermore $a^0 = 1$ for any $a > 0$, so that the graph always crosses the y axis at $(0, 1)$.

The figure below illustrates the graphs of two typical exponential functions $y = a^x$ and $y = b^x$ in the case $a > b > 1$. The two graphs cross as they cross the y axis at $(0, 1)$.

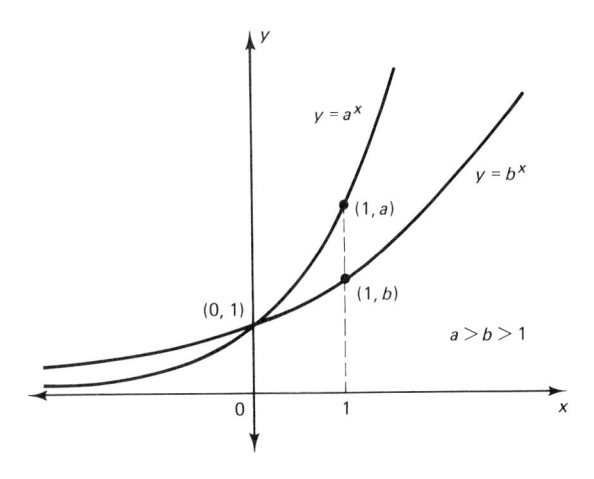

Sec. 8.1   Exponential Functions

When $a > 1$, the function $y = a^x$ is said to be a ***growing exponential function***, whereas when $0 < a < 1$ it is said to be a ***decaying exponential function***. The graphs of $y = a^x$ and $y = b^x$ when $0 < b < a < 1$ are in the figure below. These functions decrease as x increases.

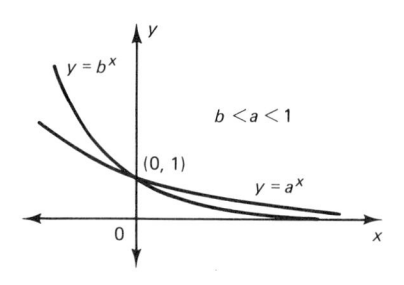

SOME APPLICATIONS

Exponential functions find many applications, some of which we shall see later. One application is to the growth of populations: a population growing in an environment that does not limit its expansion will tend to grow as an exponential function of time. The following example illustrates such a case.

Example 3 A population of bacteria growing in a laboratory is doubling in size every 15 minutes. Initially there are 1000 organisms in the population. How many will there be after 1 hour, after 3 hours, and after t hours?

Solution After 15 minutes the population will have doubled to 2000. After a further 15 minutes it will have doubled again to 4000. After 45 minutes it will be 8000, and after an hour the population will be 16,000.

Each 15 minutes the population is multiplied by 2. An hour contains four such periods, so the population is multiplied by 2 four times; that is, it is multiplied by $2^4 = 16$. This accounts for the increase from 1000 to 16,000.

During 3 hours there are twelve periods of 15 minutes, so the population will be multiplied by 2 twelve times. After 3 hours, therefore, the population will be $1000 \cdot 2^{12} = 1000 \cdot 4096 = 4,096,000$.

During t hours the number of 15-minute periods is $4t$. Therefore the population is doubled $4t$ times—that is, it is multiplied by 2^{4t}. Therefore:

$$\text{population after } t \text{ hr} = (1000)(2^{4t}).$$

Putting $t = 1$ and $t = 3$ in this formula gives the answers to the first two parts of the example. The graph is shown in the adjacent figure (note the different scales).

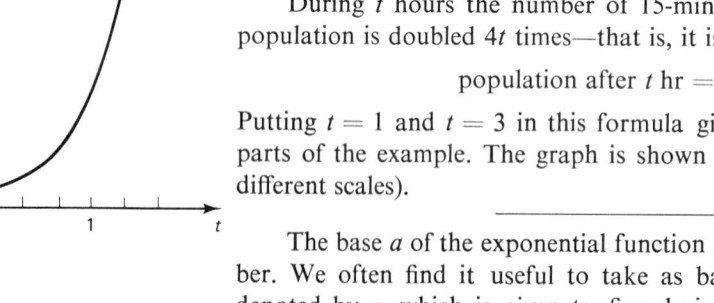

The base a of the exponential function a^x can be any positive real number. We often find it useful to take as base a certain irrational number denoted by e, which is given to five decimal places by: $e = 2.71828\ldots$. The corresponding exponential function, written e^x, is called the ***natural exponential function***.

The reason for the importance of this particular exponential function cannot fully be explained without the use of calculus. However, later in this chapter (Section 4) we give a partial explanation.

The function e^x is in fact so important that a table giving some of its values is included in the Appendix. Values of this function can also be obtained from many pocket calculators.

Examples 4 Using the table in the Appendix or a pocket calculator, find the approximate values of the following:
(a) e^2. (b) $e^{3.55}$. (c) $e^{-0.24}$.

Solutions Directly from the table, the approximate (rounded off) values are
(a) $e^2 = 7.3891$. (b) $e^{3.55} = 34.813$. (c) $e^{-0.24} = 0.7866$.
[In case (c) the value is read from the column headed by e^{-x} adjacent to the value 0.24 in the x column.]

Example 5 The population of a certain developing nation is found to be given in millions by the formula

$$P = 15e^{0.02t}$$

where t is the number of years measured from 1960 (1960 means $t = 0$, 1961 means $t = 1$, and so on). Determine the population in 1980 and the projected population in 1990, assuming this formula continues to hold until then.

Solution In 1980, $t = 20$, and so

$$P = 15e^{(0.02)(20)} = 15e^{0.4}$$
$$= 15(1.4918) = 22.4 \quad \text{(after rounding off)}.$$

So in 1980 the population is 22.4 million. After a further 10 years, $t = 30$, and so

$$P = 15e^{(0.02)(30)} = 15e^{0.6}$$
$$= 15(1.8221) = 27.3 \quad \text{(after rounding off)}.$$

Thus in 1990 the projected population will be 27.3 million.

Another case in which exponential functions arise concerns investments made at compound interest. Consider a sum of money, say $100, invested at a fixed rate of interest, say 6% per annum. After one year the investment will have increased in value by 6%, to $106. If the interest is compounded, then during the second year this whole sum of $106 earns interest at 6%. Thus the value of the investment at the end of the second year will consist of the $106 plus 6% of $106 in interest, giving a total value of

$$\$106 + (0.06)(\$106) = \$106(1 + 0.06)$$
$$= \$100(1.06)(1.06)$$
$$= \$100(1.06)^2$$
$$= 112.36.$$

Sec. 8.1 Exponential Functions

During the third year the value will increase by an amount of interest equal to 6% of $112.36, giving a total value at the end of the year equal to

$$\$112.36 + \$112.36(0.06) = \$112.36(1 + 0.06)$$
$$= \$100(1.06)^2(1 + 0.06)$$
$$= \$100(1.06)^3.$$

In general, the investment increases by a factor of 1.06 with each year that passes. After n years its value is $\$100(1.06)^n$.

Let us consider the general case. Let a sum P be invested at a rate of interest of $R\%$ per annum. Then the interest in the first year is $(R/100)P$, so the value of the investment after one year is

$$P + \left(\frac{R}{100}\right)P = P\left(1 + \frac{R}{100}\right) = P(1 + i)$$

where we have set $i = R/100$.

The interest in the second year will be $R\%$ of the value $P(1 + i)$ at the beginning of that year: Interest $= (R/100)P(1 + i)$. Thus the value after two years is

$$P(1 + i) + \left(\frac{R}{100}\right)P(1 + i) = P(1 + i)\left(1 + \frac{R}{100}\right) = P(1 + i)^2.$$

We see that each year the value of the investment is multiplied by a factor $(1 + i)$ from its value the previous year. After n years the value is given by the formula:

$$\boxed{\text{value after } n \text{ years} = P(1 + i)^n, \quad i = \frac{R}{100}.} \quad (1)$$

This is equivalent to the exponential function Pa^n, where $a = (1 + i)$ is the base and n is the variable.

Example 6 A sum of $200 is invested at 5% interest compounded annually. Find the value of the investment after 10 years.

Solution In this case $R = 5$ and $i = R/100 = 0.05$. Then after n years the value of the investment is

$$P(1 + i)^n = 200(1.05)^n.$$

When $n = 10$, this is $200(1.05)^{10}$. In order to complete this calculation, we need to evaluate $(1.05)^{10}$. If your calculator has a y^x key, you can make this evaluation immediately. The answer is 1.6289. An alternative method of evaluating such expressions is to use logarithms. We shall describe this method in Section 8.4. Meanwhile, taking the value from the calculator, we see that the value of the investment after 10 years is

$$200(1.6289) = 325.78 \text{ dollars.}$$

In some cases, interest is compounded more than once per year—for example, semiannually (twice per year), quarterly (four times per year), or monthly (twelve times per year). In such a case the annual rate of interest $R\%$ that is usually quoted is the so-called **nominal rate**. If compounding occurs k times per year and if the nominal rate of interest is $R\%$, this means that the interest rate at each compounding is equal to $(R/k)\%$. In n years the number of compoundings is kn. For example, at 8% nominal interest compounded quarterly, an investment is increased by 2% every three months. In five years there would be 20 such compoundings.

From the compound-interest formula, the value after n years is given by

$$P\left(1 + \frac{R/k}{100}\right)^{nk} = P\left(1 + \frac{i}{k}\right)^{nk}.$$

Thus for compounding k times per year,

$$\boxed{\text{value after } n \text{ years} = P\left(1 + \frac{i}{k}\right)^{nk}, \quad i = \frac{R}{100}.} \quad (2)$$

Example 7 A sum of $2000 is invested at a nominal rate of interest of 9% compounded monthly. Find the value of the investment after three years.

Solution Here $R = 9$, $k = 12$, and $n = 3$. The investment is compounded monthly, and the interest rate at each compounding is $R/k = \frac{9}{12} = 0.75\%$. Thus at each compounding the value is increased by a factor

$$1 + \frac{R/k}{100} = \left(1 + \frac{0.75}{100}\right) = 1.0075.$$

During three years there will be 36 such compoundings. Hence the value will be

$$2000(1.0075)^{36} \approx 2000(1.3086) = 2617.20 \text{ dollars.}$$

(Again the y^x key proved useful.)

Further applications of exponential functions are given in the examples that follow.

Example 8 The population of the earth at the beginning of 1976 was 4 billion and was growing at 2% per year. What will be the population in the year 2000, assuming that the rate of growth remains unchanged?

Solution When a population grows by 2% a year, its size at any time is 1.02 times what it was one year earlier. Thus at the beginning of 1977 the population was 1.02 (4 billion). Furthermore, at the beginning of 1978 it was 1.02 times the population at the beginning of 1977—that is, $(1.02)^2 \cdot$ (4 billion). And so we continue, with the population multiplying by a factor of 1.02 for every year that passes. The population at the beginning of the year 2000—that is, after 24 years—will therefore be $(1.02)^{24}$(4 billion), or

$$1.608 \cdot (4 \text{ billion}) = 6.43 \text{ billion.}$$

[We used the y^x key to evaluate $(1.02)^{24}$.]

Example 9 What rate of annual interest doubles the value of an investment in five years when compounded annually?

Solution An investment P increases to $P(1+i)^5$ after five years, where $i = R/100$. For doubling, we have
$$P(1+i)^5 = 2P$$
$$(1+i)^5 = 2$$
$$1 + i = 2^{1/5} \approx 1.1487.$$

Therefore $i = 0.1487 = R/100$, and so $R = 14.87$. The required interest rate is 14.87%.

Example 10 The population of a certain developing nation is increasing by 3% per year. By how much per year must the gross national product increase if the income per capita is to double in 20 years?

Solution Let the present population be denoted by P_0. Then, since the population increases by a factor (1.03) each year, the size of the population after 20 years is given by
$$P = P_0(1.03)^{20}.$$

Let the present gross national product (GNP) be I_0. Then the present income per capita is obtained by dividing this quantity I_0 by the population size, hence is equal to (I_0/P_0). Let the GNP increase each year by a factor k. Then after one year it will equal $I_0 k$, after two years $I_0 k^2$, and so on, until after 20 years the GNP will be
$$I = I_0 k^{20}.$$

The income per capita after 20 years is therefore
$$\frac{I}{P} = \frac{I_0 k^{20}}{P_0(1.03)^{20}} = \frac{I_0}{P_0}\left(\frac{k}{1.03}\right)^{20}.$$

We wish to find the value of k for which this income per capita is twice its present value. That is,
$$\frac{I}{P} = \frac{I_0}{P_0}\left(\frac{k}{1.03}\right)^{20} = 2\frac{I_0}{P_0}.$$

It follows that
$$\left(\frac{k}{1.03}\right)^{20} = 2.$$

Therefore,
$$\frac{k}{1.03} = 2^{1/20} \approx 1.0353$$
$$k = (1.03)(1.0353) \approx 1.066$$

The GNP must increase by a factor 1.066 each year. This corresponds to a yearly increase of 6.6%.

The exponential-growth equation must be applied to population problems with caution. It is a law of ecology that a population will grow exponentially, provided that no environmental factors limit or otherwise influence the growth. For instance, if the supply of food available to the population of a certain species is limited, then exponential growth must eventually cease as the food supply becomes insufficient to support the ever-growing population. Among other factors that inhibit indefinite growth are the supply of shelter for the species, which typically is limited, the interaction with predator species, sociological factors that might slow the expansion of the population in overcrowded circumstances, and, in the last resort, simply the limited availability of physical space for the population. Such factors eventually operate to end the exponential growth of the population, causing it to level off at some value that is the maximum population the given habitat can support.

It is becoming apparent that such constraints are being imposed on the human population, and it seems likely that human populations will not grow exponentially during the coming decades. Thus projections of the human population far into the future based on the exponential-growth equation are unlikely to prove accurate. They indicate what will happen if present trends continue, which may be different from what will happen in fact. Nevertheless, the exponential population model *is* applicable in a number of cases—for example, for bacteria or insect proliferation.

Diagnostic Test 8.1

Fill in the blanks.

1. $y = b^x$ is an _____ function and b is its _____.
2. $y = a^x$ is a decaying exponential function if _____.
3. $y = e^x$ is the _____ function.
4. $e =$ _____ to four significant figures.
5. The domain of the exponential function $f(x) = a^x$ is the set _____ and the range is the set _____.
6. If $b > a > 0$, the graph of $f(x) = b^x$ lies _____ the graph of $f(x) = a^x$ for $x > 0$ and lies _____ the graph of $f(x) = a^x$ for $x < 0$.
7. If $0 < a < 1$, then a^x _____ 1 for $x > 0$.
8. If $k > 0$, then $e^{kt} > e^{kt'}$ whenever t _____ t'.
9. If $k < 0$, then $e^{kt} < e^{kt'}$ whenever t _____ t'.
10. The graph of $f(x) = a^x$ has an asymptote given by the equation _____.
11. The graph of $f(x) = 2^{-x}$ _____ (rises or falls) to the right.
12. The graph of $f(x) = (0.3)^{-x}$ _____ (rises or falls) to the right.
13. The graph of $f(x) = -2^x$ _____ (rises or falls) to the right.

Exercises 8.1

(1–8) Construct the graphs of the following exponential functions by calculating and plotting a few points on them.

1. $y = 3^x$.
2. $y = (\frac{5}{2})^x$.
3. $y = (\frac{1}{3})^x$.
4. $y = (\frac{2}{5})^x$.
5. $y = (\frac{3}{4})^x$.
6. $y = (\frac{1}{2})^x$.
7. $y = (\frac{1}{2})^{-x}$.
8. $y = (\frac{2}{3})^{-x}$.

(9–18) Evaluate the following, using the table in the Appendix or the e^x key on your calculator.

9. e^3.
10. e^4.
11. e^{-2}.
12. e^{-5}.
13. $e^{0.41}$.
14. $e^{2.75}$.
15. $e^{-1.05}$.
16. $e^{-0.68}$.
17. $e^{-5.2}$.
18. $e^{-2.35}$.

19. Using values from the table, draw the graph of $y = e^x$.

20. Using values from the table, draw the graph of $y = e^{-0.5x}$.

21. In Example 5 of this section the population was given by $P = 15e^{0.02t}$. Using the table, draw the graph of P for $0 \leq t \leq 40$ (i.e., between 1960 and 2000). Use your graph to estimate the date at which the population hits 30 million.

22. A certain population is changing according to the formula
$$P = 8e^{0.05t}.$$
Draw the graph of P as a function of t for the interval $0 \leq t \leq 20$. What is the value of P when $t = 15$? Use your graph to estimate the value of t at which (a) $P = 12$ and (b) $P = 16$.

23. A certain depressed economic region has a population that is in decline. In 1970 its population was 500,000, and thereafter its population was given by the formula
$$P = 500,000e^{-0.02t}$$
where t is time in years. Find the region's population in 1980. Assuming this trend continues, find the projected population in the year 2000.

24. The population of a certain city at time t (measured in years) is given by the formula
$$P = 50,000e^{0.05t}.$$
Calculate the population (a) when $t = 10$, (b) when $t = 15$.

*25. In Exercise 23, calculate the percentage decline in population per year.

*26. In Exercise 24, calculate the percentage growth in population per year.

27. A machine is purchased for $10,000 and depreciates continuously from the date of purchase, its value after t years being given by the formula
$$V = 10,000e^{-0.2t}.$$
(a) Find the value of the machine after eight years.
*(b) Find the percentage decline in value each year.

28. A population of microorganisms doubles every 20 minutes. If 200 organisms are present initially, find a formula for the population size after t hours.

29. A population of microorganisms is doubling every 45 minutes. If 5000 organisms are present initially, how many will there be after (a) 3 hours, (b) 6 hours, (c) t hours?

30. During the autumn, half of a population of flies die off on average every three days. If initially the population size is one million, find the number of survivors after (a) 3 weeks, (b) t weeks.

(31–51) In order to complete the remaining exercises, you will need to use the y^x key of a calculator. Most of the answers are given in incomplete form as a formula. Alternatively you might prefer to postpone these exercises until after you have covered Section 8.4.

(31–34) If $2000 is invested at 6% compound interest per annum, find the following:

31. The value of the investment after 2 years.

32. The value of the investment after 4 years.

33. The value of the investment after 8 years.

34. The value of the investment after 12 years.

(35–38) If $100 is invested at 8% compound interest per annum, find the following:

35. The value of the investment after 2 years.

36. The value of the investment after 5 years.

37. The value of the investment after 10 years.

38. The value of the investment after 20 years.

39. What rate of compound interest doubles the value of an investment in 10 years when compounded annually?

40. What rate of compound interest triples the value of an investment in 10 years when compounded annually?

(41–44) A sum of $2000 is invested at a nominal interest rate of 12%. Calculate its value:

41. After 1 year if compounding is quarterly.

42. After 1 year if compounding is monthly.

43. After 4 years if compounding occurs every six months.

44. After 6 years with quarterly compounding.

45. A sum of $1000 is invested for 5 years. Calculate its value for (a) monthly compounding at the nominal interest rate of 8%, and (b) annual compounding at 8.25%. Which is better?

46. Which is better for the investor—monthly compounding at the nominal rate of 10% or quarterly compounding at the nominal rate of 10.2%?

47. What nominal rate of interest doubles an investment in 8 years when compounded quarterly?

48. What nominal rate of interest triples an investment in 10 years when compounded monthly?

49. The population of the earth at the beginning of 1976 was 4 billion. If the growth rate continues at 2% per year, what will be the population in the year 2026?

50. In Exercise 49, what will be the population in the year 2076 if the present trend continues?

51. The profits of a certain company have been increasing by an average of 12% per year between 1975 and 1980. In 1980 they were $5.2 million. Assuming that this growth rate continues, find the profits in 1985.

8.2

Logarithms

Let $a > 0$ be a given base ($a \neq 1$) and y any positive real number. If x is such that $a^x = y$, then we call x the **logarithm of y with base a.** We write this $x = \log_a y$. Thus x is the logarithm of y with base a if $a^x = y$; that is,

> $\log_a y$ is the power to which a must be raised in order to get y.

For example, take $a = 3$ and $y = 9$. Now 3 must be raised to the power 2 in order to get 9, and therefore $\log_3 9 = 2$.

Summarizing the definition:

$$x = \log_a y \quad \text{if and only if} \quad y = a^x.$$

The two statements $x = \log_a y$ and $y = a^x$ mean exactly the same thing. For example, the statement $x = \log_4 2$ means the same as the statement $2 = 4^x$. (Here $a = 4$, $y = 2$. These statements are true for $x = \frac{1}{2}$, since $2 = 4^{1/2}$.) Of these two equivalent statements, $x = \log_a y$ is said to be the **logarithmic form** and $y = a^x$ the **exponential form**.

Examples 1 Write the following statements in logarithmic form:
(a) $2^4 = 16$.   (b) $(\frac{1}{2})^{-3} = 8$.
Write the following statements in exponential form and hence verify them:
(c) $\log_4 8 = \frac{3}{2}$.   (d) $\log_{27}(\frac{1}{9}) = -\frac{2}{3}$.

Solutions (a) We have $2^4 = 16$. Comparing this with the equation $a^x = y$, we see that $a = 2$, $x = 4$, and $y = 16$. The logarithmic form, $\log_a y = x$, is

$$\log_2 16 = 4.$$

(b) Comparing $(\frac{1}{2})^{-3} = 8$ with $a^x = y$, we have $a = \frac{1}{2}$, $x = -3$, and $y = 8$. The logarithmic form is

$$\log_{1/2} 8 = -3.$$

(c) Comparing $\log_4 8 = \frac{3}{2}$ with the equation $\log_a y = x$, we have $a = 4$, $y = 8$, and $x = \frac{3}{2}$. The exponential form, $y = a^x$, is

$$8 = 4^{3/2}.$$

This is clearly a true statement, since $4^{3/2} = (4^{1/2})^3 = 2^3 = 8$. It follows that the given logarithmic form must also be true.

(d) Here $a = 27$, $y = \frac{1}{9}$, and $x = -\frac{2}{3}$ and the exponential form is

$$27^{-2/3} = \frac{1}{9}.$$

Again this is easily verified: $27^{-2/3} = (27^{1/3})^{-2} = 3^{-2} = \frac{1}{9}$.

Examples 2 Find the values:
(a) $\log_2 16$.   (b) $\log_{1/3} 243$.   (c) $\log_8 32$.

Solutions (a) Let $x = \log_2 16$. Then from the definition of logarithms it follows that $16 = 2^x$. But $16 = 2^4$, and so $x = 4$. Therefore $\log_2 16 = 4$.
(b) Let $x = \log_{1/3} 243$. Then from the definition

$$243 = (\tfrac{1}{3})^x.$$

But $243 = 3^5 = (\frac{1}{3})^{-5}$. Therefore $x = -5$.
(c) Let $x = \log_8 32$. Then from the definition

$$32 = 8^x$$

—that is,

$$2^5 = (2^3)^x$$
$$= 2^{3x}.$$

Therefore $3x = 5$, and $x = \frac{5}{3}$. Thus
$$\log_8 32 = \tfrac{5}{3}.$$

Example 3 Evaluate $2^{\log_4 9}$.

Solution Let $x = \log_4 9$, so that $9 = 4^x$. The quantity we wish to calculate is therefore
$$2^{\log_4 9} = 2^x = (4^{1/2})^x$$
$$= (4^x)^{1/2} = 9^{1/2} = 3.$$

LOGARITHMIC FUNCTIONS

The logarithm of any positive real number is uniquely defined by the above procedure. Therefore it defines a function, called the **logarithmic function with base a**, whose domain is the set of positive real numbers.

Example 4 Construct the graph of the logarithmic function with base 2.

Solution Let us use x as the independent variable and write
$$y = \log_2 x.$$
According to the definition this means the same thing as
$$x = 2^y.$$
(Note that x and y have been interchanged from the definition of logarithm, and $a = 2$.) So we can construct the following table of values. For each point (x, y) we first give y a value and then calculate* the corresponding value of x.

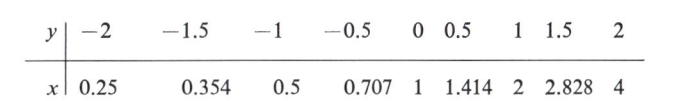

y	-2	-1.5	-1	-0.5	0	0.5	1	1.5	2
x	0.25	0.354	0.5	0.707	1	1.414	2	2.828	4

The tabulated values of (x, y) are plotted in the adjacent figure and joined by a smooth curve that provides the required graph of $y = \log_2 x$.

Note that when we write $y = \log_2 x$, we are regarding x as the independent variable, so that x axis is drawn horizontally. In spite of this, we can obtain the various plotted points by giving y particular values and calculating the corresponding values of x.

You may have realized already that the logarithmic function is the inverse of the exponential function. The construction of the inverse of a function f was described in Section 6.4: If f is a one-to-one function, then the equation $y = f(x)$ can be solved for x whenever y is in the range of f, and we write the solution in the form $x = f^{-1}(y)$. But this is precisely how

---

*For the irrational values we have $2^{-0.5} = \sqrt{2}/2 \approx 1.414/2$, and so on. Alternatively the y^x key of a calculator can be used.

we construct the logarithmic function. We start with the exponential function $y = a^x$ and solve for x in terms of y: the solution is $x = \log_a y$. So if y is the exponential function of x with base a, then x is the logarithmic function of y with base a.

As with any other inverse function, the graph of the function $x = \log_a y$ for general base a can be obtained from the graph of the exponential function $y = a^x$ simply by flipping the axes. The two graphs are illustrated in the figure shown below for $a > 1$. If you hold the first graph ($y = a^x$) in front of a mirror and rotate it so that the y axis is horizontal and the x axis points vertically upward, you will see a curve identical to that shown in the second figure.

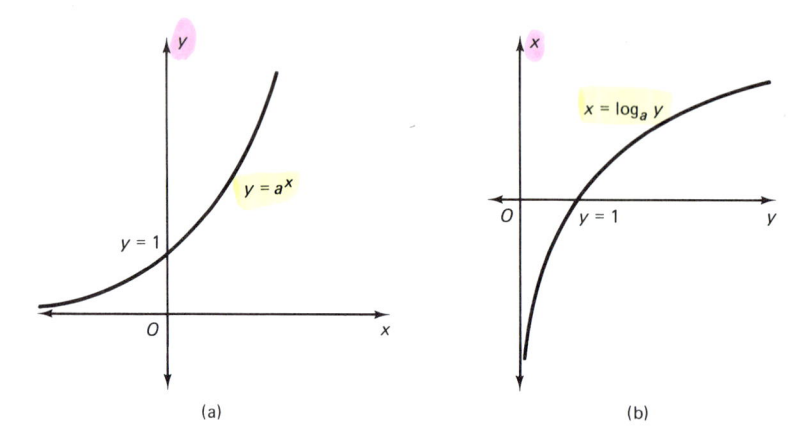

Alternatively, we can draw the graphs of $y = a^x$ and $y = \log_a x$ on the same xy plane, as in the figure below. The two graphs are reflections of one another in the line $y = x$. (Recall that this is always true of the graphs of a function and its inverse.)

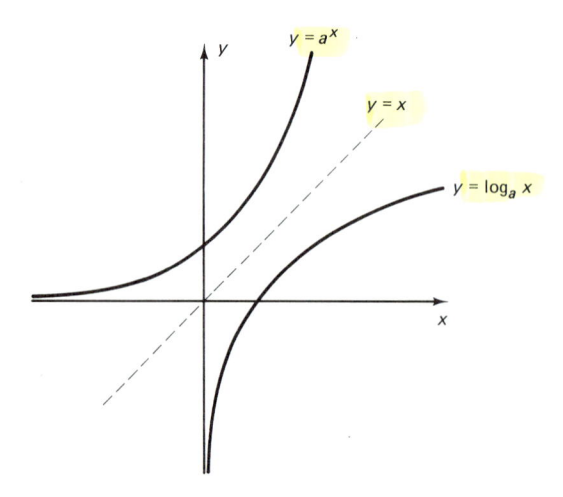

Note The value $a = 1$ for the base must be excluded. The graph of $y = a^x$ becomes the horizontal line $y = 1$ when $a = 1$, and no inverse function exists in this case. (A constant function has no inverse, since it is not one-to-one.)

ELEMENTARY PROPERTIES

We note that $\log_a y$ is only defined for $y > 0$ and for $a > 0$, $a \neq 1$. For $y = 1$, we have

$$\boxed{\log_a 1 = 0}$$

for any base a. This is because in the exponential form, $a^0 = 1$ for any $a > 0$. Thus: *the logarithm of 1 with any positive base (not equal to 1) is zero.*

If $y = a^x$, then $x = \log_a y$. Substituting for x in the first equation, we then find

$$\boxed{y = a^{\log_a y}.}$$

This equation is often summarized as follows: *a number y is equal to the base raised to the power of the logarithm of y.*

We can also substitute $y = a^x$ in the equation $x = \log_a y$. We get

$$\boxed{x = \log_a(a^x).}$$

This identity holds for any real number x.

The following equations illustrate these two identities:

$$3^{\log_3 4} = 4, \qquad \left(\tfrac{1}{2}\right)^{\log_{1/2} 5} = 5$$

$$\log_{10}(10^{-2}) = -2, \qquad \log_{3/2}\left[\left(\tfrac{3}{2}\right)^4\right] = 4.$$

If we take $x = 1$ in the equation $y = a^x$, we get

$$y = a^x = a^1 = a.$$

Therefore the statement $\log_a y = x$ becomes

$$\boxed{\log_a a = 1.}$$

So *the logarithm of any positive number with itself as base is always equal to 1.* For example, $\log_{10} 10 = 1$; $\log_{1/2}\left(\tfrac{1}{2}\right) = 1$.

LOGARITHMIC EQUATIONS

A logarithmic equation is one in which the unknown quantity appears in the argument or in the base of a logarithmic function. Equations of this type can often be solved by writing them in the corresponding exponential form.

Example 5 Solve for x:
$$\log_2 (3 - 4x) = 4.$$

Solution In exponential form, the given logarithmic equation is
$$(3 - 4x) = 2^4 = 16$$
Thus
$$-4x = 13$$
$$x = -\tfrac{13}{4}.$$

Example 6 Solve for x:
$$\log_x (x + 2) = 2. \qquad (i)$$

Solution The given equation when rewritten in exponential form becomes
$$(x + 2) = x^2.$$
Thus we obtain a quadratic equation for x:
$$x^2 - x - 2 = 0$$
$$(x - 2)(x + 1) = 0.$$

The solutions are $x = 2$ and $x = -1$. However, $x = -1$ cannot be a solution of the original equation (i) because the base x cannot be negative. The only solution is $x = 2$. [When $x = 2$, equation (i) becomes $\log_2 (2 + 2) = 2$, which is a true statement.]

Note When solving logarithmic equations, always check that each solution you find satisfies the *original* equation. Bear in mind that the base of any logarithm must be positive and different from 1 and that the argument of any logarithm must be positive.

Diagnostic Test 8.2

Fill in the blanks.

1. If $y = a^x$, then $x = $ _____ .
2. If $2^3 = 8$, then $3 = $ _____ .
3. If $\log_a u = v$, then $u = $ _____ .
4. If $\log_2 p = q$, then $p = $ _____ .
5. $\log_7 7 = $ _____ .
6. $\log_3 1 = $ _____ .
7. $\log_2 (-1) = $ _____ .
8. $\log_{0.5} (\tfrac{1}{2}) = $ _____ .
9. $10^{\log_{10} x} = $ _____ .
10. $3^{\log_3 4} = $ _____ .
11. $\log_a a = $ _____ for all a _____ .
12. $\log_a y$ is the _____ to which _____ must be raised to get _____ .

13. The domain of $f(x) = \log_a x$ is the set _____.
14. The range of $f(x) = \log_a x$ is the set _____.
15. The graph of $y = \log_a x$ always passes through the point _____ for all values of a for which $\log_a x$ is defined.
16. The domain of $f(x) = \log_a |x|$ is the set _____, and the range is the set _____.
17. The domain of $f(x) = \log_a (t + 3)$ is the set _____, and the range is the set _____.
18. If $f(x) = 10^x$, then $f^{-1}(x) = $ _____.
19. If $f(t) = \log_a t$, then $f^{-1}(x) = $ _____.

Exercises 8.2

(1–12) Verify the following statements and rewrite them in logarithmic form with an appropriate base.

1. $2^3 = 8$.
2. $3^4 = 81$.
3. $16^{1/2} = 4$.
4. $27^{1/3} = 3$.
5. $8^{-1/3} = \frac{1}{2}$.
6. $81^{-1/2} = \frac{1}{9}$.
7. $27^{-4/3} = \frac{1}{81}$.
8. $16^{3/4} = 8$.
9. $125^{2/3} = 25$.
10. $8^{-5/3} = \frac{1}{32}$.
11. $(\frac{8}{27})^{-1/3} = \frac{3}{2}$.
12. $(\frac{625}{16})^{-3/4} = \frac{8}{125}$.

(13–20) Write the following equations in exponential form and hence verify them.

13. $\log_3 27 = 3$.
14. $\log_5 25 = 2$.
15. $\log_{1/2} 8 = -3$.
16. $\log_7 (\frac{1}{7}) = -1$.
17. $\log_{1/9} (\frac{1}{243}) = \frac{5}{2}$.
18. $\log_8 (32) = \frac{5}{3}$.
19. $\log_4 (\frac{1}{2}) = -\frac{1}{2}$.
20. $\log_2 (\frac{1}{4}) = -2$.

(21–40) Evaluate the following by using the definition of the logarithm.

21. $\log_2 512$.
22. $\log_3 81$.
23. $\log_{27} 243$.
24. $\log_{16} 64$.
25. $\log_{\sqrt{2}} 16$.
26. $\log_{16} \sqrt{2}$.
27. $\log_8 128$.
28. $\log_2 (0.25)$.
29. $10^{\log_{10} 100}$.
30. $10^{\log_{10} 2}$.
31. $\log_4 (2^p)$.
32. $\log_2 (4^p)$.
33. $2^{\log_2 x}$.
34. $3^{2 \log_3 x}$.
35. $2^{\log_{1/2} (3)}$.
36. $2^{\log_4 3}$.
37. $3^{\log_9 (2)}$.
38. $3^{\log_{1/3} (5)}$.
*39. $\log_a 32 \div \log_a 4$.
*40. $(\log_3 2)(\log_2 3)$.

(41–46) Draw the graphs of the following functions.

41. $y = \log_3 x$.
42. $y = \log_2 (x - 1)$.
43. $y = \log_{1/2} x$.
44. $y = \log_{2/3} x$.
45. $y = \log_2 |x|$.
46. $y = \log_{1/2} |x|$.

(47–64) Find the solutions of the following logarithmic equations.

47. $\log_3 (2x + 1) = 3$.
48. $\log_4 (1 - 5x) = -1$.
49. $\log_{1/2} (1 - x) = -2$.
50. $\log_{2/3} (4 - 5x) = 2$.
51. $\log_9 |x + 2| = \frac{1}{2}$.
52. $\log_2 |1 - 2x| = -1$.
53. $\log_x (2x + 3) = 2$.
54. $\log_x (5x - 6) = 2$.
55. $\log_{x+1} (2x - 3) = 2$.
56. $\log_{x+1} (3x + 7) = 2$.
57. $\log_{x+3} (3x + 7) = 2$.
58. $\log_x (6 - 5x) = 2$.
59. $\log_x 4x = 3$.
60. $\log_{|x|} 2x = 3$.
*61. $\log_5 x = \log_x 5$.
*62. $\log_2 x = \log_x 16$.
*63. $\log_4 x^2 = \log_x 2$.
*64. $\log_a (x/2) = \log_a (2/x)$.

8.3

Properties of Logarithms

Logarithms were introduced in the seventeenth century for the purpose of simplifying arithmetical calculations. By using logarithms, we can replace the processes of multiplication and division of pairs of numbers by the much easier processes of addition and subtraction. In the next section we describe how this is done. The use of logarithms for this purpose (and others) rests on four basic properties:

$$
\begin{aligned}
&\text{I} \quad \log_a (uv) = \log_a u + \log_a v \\
&\text{II} \quad \log_a \left(\frac{u}{v}\right) = \log_a u - \log_a v \\
&\text{III} \quad \log_a \left(\frac{1}{u}\right) = -\log_a u \\
&\text{IV} \quad \log_a (u^n) = n \log_a u.
\end{aligned}
$$

These properties will be proved at the end of this section. Let us illustrate them with some examples.

Examples 1 (a) Since $8 = 2^3$, it follows that $\log_2 8 = 3$. Since $16 = 2^4$, it follows that $\log_2 16 = 4$. From Property I, therefore,

$$\log_2 (8 \cdot 16) = \log_2 8 + \log_2 16$$
$$\log_2 (128) = 3 + 4 = 7.$$

This statement has the exponential form $128 = 2^7$, which is obviously true.

(b) We can also calculate $\log_2 (128)$ by making use of the last property of logarithms:

$$\log_2 (128) = \log_2 (2^7) = 7 \log_2 2.$$

But $\log_a a = 1$ for any $a > 0$, so in particular $\log_2 2 = 1$. Therefore, $\log_2 (128) = 7$.

(c) Since $81 = (\frac{1}{3})^{-4}$, it follows that

$$\log_{1/3} (81) = -4.$$

Therefore, using Property III,

$$\log_{1/3} (\tfrac{1}{81}) = -\log_{1/3} (81) = -(-4) = 4.$$

This result in exponential form corresponds to the true statement $\frac{1}{81} = (\frac{1}{3})^4$.

Examples 2 Express the following in terms of $\log_a x$ and $\log_a y$:
(a) $\log_a (x^3/y^2)$. (b) $\log_a \sqrt{xy^3}$.

Solutions (a) $\log_a (x^3/y^2) = \log_a x^3 - \log_a y^2$ (by Property II)
$= 3 \log_a x - 2 \log_a y$ (by Property IV).
(b) $\log_a \sqrt{xy^3} = \log_a (xy^3)^{1/2}$
$= \tfrac{1}{2} \log_a (xy^3)$ (by Property IV)
$= \tfrac{1}{2}(\log_a x + \log_a y^3)$ (by Property I)
$= \tfrac{1}{2}(\log_a x + 3 \log_a y)$ (by Property IV)
$= \tfrac{1}{2} \log_a x + \tfrac{3}{2} \log_a y.$

Examples 3 If $x = \log_2 3$, express the following quantities in terms of x.
(a) $\log_2 (\tfrac{1}{3})$. (b) $\log_2 (\tfrac{2}{3})$. (c) $\log_2 \sqrt{\tfrac{27}{2}}$.

Solutions (a) $\log_2 (\tfrac{1}{3}) = -\log_2 3$ (Property III)
$= -x.$
(b) $\log_2 (\tfrac{2}{3}) = \log_2 2 - \log_2 3$ (Property II)
$= 1 - x.$
(Here we used the fact that $\log_2 2 = 1$, since $\log_a a = 1$ for any $a > 0$.)
(c) $\log_2 \sqrt{\tfrac{27}{2}} = \log_2 (\tfrac{27}{2})^{1/2}$
$= \tfrac{1}{2} \log_2 (\tfrac{27}{2})$ (Property IV)
$= \tfrac{1}{2}(\log_2 27 - \log_2 2)$ (Property II)
$= \tfrac{1}{2}(\log_2 3^3 - 1)$
$= \tfrac{1}{2}[3 \log_2 3 - 1]$ (Property IV)
$= \tfrac{1}{2}(3x - 1).$

Remark Properties I and II extend to products and quotients in which several factors are involved. For example,

$$\log_a (uvwz) = \log_a u + \log_a v + \log_a w + \log_a z$$

$$\log_a \left(\frac{u}{vw}\right) = \log_a u - \log_a v - \log_a w$$

and so on. In general, the logarithm of a quotient in which several factors appear in the numerator and/or denominator is the sum of the logarithms of all the factors in the numerator minus the sum of the logarithms of all the factors in the denominator.

COMMON LOGARITHMS

Logarithms used to be used extensively in carrying out arithmetical calculations involving multiplication, division, and the calculation of powers and roots. Nowadays, with electronic calculators widely available, such usage has considerably diminished, although in some areas logarithms are still used. (For example, the ship's navigator must still learn how to use logarithm tables and other tables to guard against the possibility that his electronics may fail.) In the next section we shall describe the use of logarithm tables in arithmetical calculations.

The logarithms ordinarily used for these calculations are called *common logarithms* and are obtained by using the number 10 as base (that is, $a = 10$). Thus the common logarithm of a number y is $\log_{10} y$, which for convenience is usually denoted by $\log y$, the base being omitted.* So when the base is not written, it should be understood to be 10.

The following statements are then equivalent:

$$x = \log y, \qquad y = 10^x.$$

The following examples demonstrate the application of the Properties I through IV to common logarithms.

Examples 4 In the general relations

$$y = 10^x, \qquad x = \log y$$

we shall take some particular values of x and y.
(a) $x = 1$. Then $y = 10^1 = 10$, and so $\log 10 = 1$.
(b) $x = 2$. Then $y = 10^2 = 100$, and so $\log 100 = 2$.
(c) $x = -1$. Then $y = 10^{-1} = 0.1$, and so $\log 0.1 = -1$.
(d) To four figures† $\log 3 = 0.4771$. This means that $3 = 10^{0.4771\cdots}$. Using the first property of logarithms proved above, it follows that

$$\log 30 = \log 3 + \log 10 = 1.4771$$
$$\log 300 = \log 3 + \log 100 = 2.4771$$
$$\log (0.3) = \log 3 + \log 0.1 = 0.4771 - 1 = -0.5229.$$

(e) Also to four figures $\log 2 = 0.3010$. Thus

$$\log 6 = \log (3 \cdot 2) = \log 3 + \log 2 = 0.4771 + 0.3010 = 0.7781$$
$$\log 4 = \log (2^2) = 2 \log 2 = 0.6020.$$

Examples 5 Given that $\log 2 = 0.3010$ and $\log 3 = 0.4771$ to four figures, evaluate (a) $\log 5$, (b) $\log 18$, (c) $\log \sqrt{54}$.

Solutions (a) $\log 5 = \log (10 \div 2) = \log 10 - \log 2$
$\qquad\qquad = 1 - 0.3010 = 0.6990$.
(*Note:* $\log 2 + \log 3$ equals not $\log 5$ but $\log 6$.)
(b) $\log 18 = \log (2 \cdot 3^2) = \log 2 + \log 3^2$
$\qquad\quad = \log 2 + 2 \log 3$
$\qquad\quad = 0.3010 + 2(0.4771) = 1.2552$.
(c) $\log \sqrt{54} = \frac{1}{2} \log 54$
$\qquad\qquad = \frac{1}{2} \log (2 \cdot 3^3) = \frac{1}{2} [\log 2 + 3 \log 3]$
$\qquad\qquad = \frac{1}{2} [0.3010 + 3(0.4771)] = 0.8662$.

---

*In some books $\log y$ means the natural logarithm of y (see later in this section).
†The use of tables to find common logarithms is described in Section 8.4. You may also be able to use your calculator for this purpose.

Example 6 Simplify (without using log tables or calculators)
$$E = \log 2 + 16 \log \left(\frac{16}{15}\right) + 12 \log \left(\frac{25}{24}\right) + 7 \log \left(\frac{81}{80}\right).$$

Solution
$$E = \log 2 + 16 \log \left(\frac{2^4}{3 \cdot 5}\right) + 12 \log \left(\frac{5^2}{2^3 \cdot 3}\right) + 7 \log \left(\frac{3^4}{2^4 \cdot 5}\right)$$
$$= \log 2 + 16(\log 2^4 - \log 3 - \log 5) + 12(\log 5^2 - \log 2^3 - \log 3)$$
$$+ 7(\log 3^4 - \log 2^4 - \log 5)$$
$$= \log 2 + 16(4 \log 2 - \log 3 - \log 5) + 12(2 \log 5 - 3 \log 2 - \log 3)$$
$$+ 7(4 \log 3 - 4 \log 2 - \log 5)$$
$$= (1 + 64 - 36 - 28) \log 2 + (-16 - 12 + 28) \log 3$$
$$+ (-16 + 24 - 7) \log 5$$
$$= \log 2 + \log 5 = \log (2 \cdot 5) = \log 10 = 1.$$

Example 7 Solve the following logarithmic equation for x:
$$2 \log (2x + 2) = \log \left(1 + \frac{12x}{25}\right) + 2. \qquad (i)$$

Solution Since $\log (100) = 2$, we can write the given equation in the form
$$2 \log (2x + 2) = \log \left(1 + \frac{12x}{25}\right) + \log 100$$

—that is,
$$\log (2x + 2)^2 = \log \left[100 \left(1 + \frac{12x}{25}\right)\right] \quad \text{(using Properties I and IV)}$$
$$= \log (100 + 48x).$$

Therefore,
$$(2x + 2)^2 = 100 + 48x.$$

This can be simplified to the quadratic equation
$$x^2 - 10x - 24 = 0$$
whose roots are $x = -2$ and $x = 12$.

Now remember that with logarithmic equations we must always check to see that the solutions we find really do satisfy the original equation. With $x = -2$, in fact, the left side of (i) becomes $2 \log (2x + 2) = 2 \log (-2)$, which is undefined. Therefore $x = -2$ is *not* a solution. When $x = 12$, equation (i) becomes
$$2 \log 26 = \log \tfrac{169}{25} + 2$$
which can be seen to be true. Therefore $x = 12$ *is* a solution.

NATURAL LOGARITHMS

We can also form logarithms with base e. (Recall that $e = 2.71828\ldots$.) They are called ***natural logarithms*** (or sometimes Napierian logarithms). They are denoted by the symbol ln:

$$y = e^x, \quad x = \log_e y = \ln y.$$

Note that the following identities are obtained by eliminating either x or y from these two equations:

$$e^{\ln y} = y, \quad \ln(e^x) = x.$$

The natural logarithm is important because of certain mathematical properties it has in connection with calculus. This type of logarithm *can* be used to perform arithmetical calculations instead of the common logarithm, but much less conveniently. Its importance arises in other ways, some of which we shall see later.

The natural logarithm has all the properties discussed earlier for logarithms with a general base a. In particular we note that $\ln e = 1$.

Those with a suitable pocket calculator can find the natural logarithm of any number by pressing the appropriate key. Alternatively, we can use the table of natural logarithms in the Appendix.

Examples 8 Using the table in the Appendix, find the values to four decimal places of:
(a) ln 3.4.
(b) ln 100.
(c) ln 340.
(d) ln 0.34.

Solutions (a) From the table we find directly that

$$\ln 3.4 = 1.2238.$$

(b) At the bottom of the table we find the value

$$\ln 10 = 2.3026.$$

Therefore,

$$\ln 100 = \ln 10^2 = 2 \ln 10 = 2(2.3026)$$
$$= 4.6052.$$

(c) $\ln(340) = \ln[(3.4)(100)]$
$= \ln 3.4 + \ln 100$
$= 1.2238 + 4.6052 = 5.8290.$

Observe that the table provides the natural logarithms of numbers lying between 1 and 10. For numbers lying outside this range, an appropriate power of 10 has to be extracted (100 in this example).

(d) $\ln(0.34) = \ln[(3.4)(10^{-1})]$
$= \ln 3.4 + \ln(10^{-1})$
$= \ln 3.4 - \ln 10$
$= 1.2238 - 2.3026 = -1.0788.$

PROOFS OF THE BASIC PROPERTIES OF LOGARITHMS

I Let $x = \log_a u$ and $t = \log_a v$. Then, from the definition of the logarithm,
$$u = a^x, \quad v = a^t.$$
It follows that
$$uv = (a^x)(a^t) = a^{x+t}$$
after we use one of the fundamental properties of exponents. Consequently, from the definition of the logarithm, it follows that $(x + t)$ must be the logarithm of (uv) with base a:
$$x + t = \log_a(uv).$$
Substituting for x and t, we get the required formula
$$\log_a(uv) = \log_a u + \log_a v.$$

II We may obtain the second result by considering u/v.
$$\frac{u}{v} = \frac{a^x}{a^t} = a^{x-t}.$$
Thus $(x - t) = \log_a(u/v)$, or equivalently
$$\log_a\left(\frac{u}{v}\right) = \log_a u - \log_a v.$$

III The particular example of this result obtained by setting $u = 1$ gives the third formula. When $u = 1$, $\log_a u = 0$, and we get
$$\log_a\left(\frac{1}{v}\right) = -\log_a v.$$
If we now replace v by u, we get the required result.

IV Fourth, let $x = \log_a u$, so that $u = a^x$. Then $u^n = (a^x)^n = a^{xn}$. Thus $xn = \log_a(u^n)$, or
$$\log_a u^n = n \log_a u.$$

Diagnostic Test 8.3

Fill in the blanks.

1. If $x = \log y$, then $y =$ _____.
2. If $t = \ln u$, then $u =$ _____.

3. If $f(x) = \ln x$, then $f^{-1}(x) = $ _____.

4. If $g^{-1}(t) = \log t$, then $g(x) = $ _____.

5. If $f(x) = 2^x$, then $f^{-1}(x) = $ _____.

6. If $f(u) = e^u$, then $f^{-1}(x) = $ _____.

7. $10^{\log x} = $ _____.

8. $e^{\ln u} = $ _____.

9. $\log 10 = $ _____.

10. $\log 1 = $ _____.

11. $\ln 1 = $ _____.

12. $\ln e = $ _____.

13. $\ln 6 = \ln 2 + $ _____.

14. $\log x + \log y = \log ($ _____ $)$.

15. $\log x - \log y = \log ($ _____ $)$.

16. $\log (13.4) = \log 1.34 + $ _____.

17. $\ln 8 = $ _____ $\ln 2$.

18. $\ln (0.25) = $ _____ $\ln 2$.

19. $\log (0.1) = $ _____ 1.

20. $\ln e^3 = $ _____.

Exercises 8.3

(1–10) Express the following in terms of $\log x$, $\log y$, and $\log z$.

1. $\log (xy/z)$.
2. $\log (x/yz)$.
3. $\log (x^2 z)$.
4. $\log (z/y^2)$.
5. $\log (\sqrt{xz}/y)$.
6. $\log (z\sqrt{xy})$.
7. $\log \sqrt{x^3/yz}$.
8. $\log (y\sqrt{x/z})$.
9. $\log (x^{1/2} y^{1/3} z^{1/4})$.
10. $\log (x^{-3} y^{-4} z^{-5})$.

(11–20) Express each of the following as a single logarithm.

11. $\log x - 2 \log y$.
12. $-\log z - \log x$.
13. $\frac{1}{2} \log y + 2 \log z$.
14. $2 \log x + \frac{1}{2}(\log y + \log z)$.
15. $\frac{1}{2} \log y + \log z + 1$.
16. $2 - \frac{1}{2} \log z$.
17. $2 \log_2 y - \frac{1}{3} \log_2 z - 4$.
18. $2 - \frac{1}{2} \log_3 z$.
19. $\log_3 x + \frac{1}{2} \log_3 z - 3$.
20. $2 \log_5 x + 3 \log_5 y - 2$.

(21–26) Given that $x = \log_6 2$, express the following in terms of x.

21. $\log_6 3$.
22. $\log_6 4$.
23. $\log_6 12$.
24. $\log_6 18$.
25. $\log_6 \sqrt{54}$.
26. $\log_6 \sqrt[3]{\frac{2}{9}}$.

(27–34) Given that $\log 5 = a$ and $\log 9 = b$, express the following in terms of a and b.

27. $\log 2$.
28. $\log 3$.
29. $\log 12$.
30. $\log 75$.
31. $\log \sqrt{30}$.
32. $\log \sqrt{60}$.
33. $\log (\frac{2}{27})$.
34. $\log (20/\sqrt{3})$.

(35–46) Evaluate the following, using the table in the Appendix.

35. $\ln 3.41$.
36. $\ln 2.68$.
37. $\ln 1.25$.
38. $\ln 8.69$.
39. $\ln 84.2$.
40. $\ln 71.4$.
41. $\ln 593$.
42. $\ln 2550$.
43. $\ln 0.341$.
44. $\ln 0.569$.
45. $\ln 0.00917$.
46. $\ln 0.0213$.

STUDY RULES OF EXPONENTS, AYARD + LARDNER ... PP $\underline{27}$-39

$$5^n = \underbrace{5_1 \cdot 5_2 \cdot 5_3 \cdots 5_n}_{n}$$

∴
$$5^2 = 5 \cdot 5$$
$$5^1 = 5 \cdot 1 = 5$$
$$5^0 = 5 \div 5 = 1 \longrightarrow \boxed{a^0 = 1} \quad \text{where } a \neq 0$$
$$5^{-1} = 1 \div 5 = 1/5$$
$$5^{-2} = 1 \div 5^2 = 1/5^2$$
⋮
$$5^{-n} = 1 \div 5^n = 1/5^n \longrightarrow \boxed{a^{-m} = 1/a^m} \quad \text{see more rules}$$

$$\left.\begin{array}{l} q = 3 \cdot 3 \\ q = 3^2 \end{array}\right\} \longrightarrow \underset{\text{BASE}}{\log_3} \underset{\text{RESULT}}{q} = \underset{\text{LOGARITHM/EXPONENT}}{2}$$

$$\left.\begin{array}{l} q = \underbrace{a \cdot a \cdot a \cdots a}_{x \text{ TIMES}} \\ y = a^x \end{array}\right\} \quad \text{EXPONENTIAL FOR } y$$

"y IS a MULTIPLIED x TIMES"

$$\underset{\text{BASE}}{\log_a} \underset{\text{VARIABLE where } y > 0}{y} = \underset{\text{LOGARITHM/EXPONENT}}{x} \quad \text{LOGARITHM FOR } y$$
$$a > 0, \neq 1$$

"YOU NEED TO MULTIPLY a x TIMES TO GET y"

IF $a^0 = 1$
$$1 = a^0$$
$$\boxed{\log_a 1 = 0}$$

The LOGARITHM OF 1 with ANY POSITIVE BASE (NOT EQUAL TO 1) IS ZERO

(1) $y = a^x$
(2) $x = \log_a y$
SUBSTITUTE (2) INTO 1
(1) $y = a^x$
$$y = a^{\log_a y}$$
$$\boxed{y = a^{\log_a y}}$$

SOLVING FOR y IN LOG FOR y

$a^1 \cdot a^{-1} = a^{1-1} = a^0$
$a^1 \cdot a^{-1} = a \cdot \frac{1}{a} = \frac{a}{a} = 1$

$$X = \log_a i$$
$$X = \log_a \left(\frac{a}{x}\right)$$

SOLVING FOR X IN LOG FORM

(47–48) Verify the following (without using tables or calculators).

47. $7 \log \left(\frac{16}{15}\right) + 5 \log \left(\frac{25}{24}\right) + 3 \log \left(\frac{81}{80}\right) = \log 2.$ **48.** $3 \log \left(\frac{36}{25}\right) + \log \left(\frac{6}{27}\right)^3 - 2 \log \left(\frac{16}{125}\right) = \log 2.$

(49–56) Solve the following equations for x. (Do not use logarithm tables or calculators.)

49. $\log_3 (x + 4) = \log_3 (1 - x) + 1.$ **50.** $\log_2 (1 - x) = \log_2 (x + 1) - 1.$

51. $\log (10x + 5) - \log (x - 4) = \log 2.$ **52.** $\log_3 3 + \log_3 (x + 1) - \log_3 (2x - 7) = 4.$

53. $\log x = \log 3 + 2 \log 2 - \frac{3}{4} \log 16.$ **54.** $\log (4x - 3) = \log (x + 1) + \log 3.$

55. $2 \ln (x + 1) = \ln 2 + \ln (3x - 1).$ **56.** $2 \log_3 (2x - 1) = \log_3 (x - 1) + 2.$

57. Show that $\log (\sqrt{x^2 + 1} - x) = -\log (\sqrt{x^2 + 1} + x).$

8.4

The Use of Common Logarithms

In this section we describe the way logarithms are used in performing certain types of arithmetical calculation. The logarithms used for this purpose are *common logarithms* or *logarithms with base 10*. A short table of common logarithms is given in Appendix I, and we begin by explaining how this table is used.

COMMON LOGARITHM TABLES

The values given in the tables are the logarithms of numbers lying between 1.00 and 9.99. As a rule the logarithms are really irrational numbers, and the values in the table are rounded off to four decimal places.

Examples 1   Look up in the tables the following common logarithms:
(a)   log 1.6.   (b)   log 2.34.

Solutions   An extract from the table of common logarithms is shown on page 312.

(a) To find log 1.6 we simply look down the left-hand column until we reach the number 1.6. In the first column, headed by zero, we find the value

$$\log 1.6 = 0.2041.$$

You will see this value shaded in the figure.

(b) To find log 2.34 we look down the left-hand column until we reach the number 2.3. Then we move across to the vertical column headed by 4, the third significant figure in 2.34. We find

$$\log 2.34 = 0.3692.$$

Again this value has been shaded in the figure.

FOUR-PLACE COMMON LOGARITHMS

N	0	1	2	3	4	5	6	7	8	9
1.0	.0000	.0043	.0086	.0128	.0170	.0212	.0253	.0294	.0334	.0374
1.1	.0414	.0453	.0492	.0531	.0569	.0607	.0645	.0682	.0719	.0755
1.2	.0792	.0828	.0864	.0899	.0934	.0969	.1004	.1038	.1072	.1106
1.3	.1139	.1173	.1206	.1239	.1271	.1303	.1335	.1367	.1399	.1430
1.4	.1461	.1492	.1523	.1553	.1584	.1614	.1644	.1673	.1703	.1732
1.5	.1761	.1790	.1818	.1847	.1875	.1903	.1931	.1959	.1987	.2014
1.6	.2041	.2068	.2095	.2122	.2148	.2175	.2201	.2227	.2253	.2279
1.7	.2304	.2330	.2355	.2380	.2405	.2430	.2455	.2480	.2504	.2529
1.8	.2553	.2577	.2601	.2625	.2648	.2672	.2695	.2718	.2742	.2765
1.9	.2788	.2810	.2833	.2856	.2878	.2900	.2923	.2945	.2967	.2989
2.0	.3010	.3032	.3054	.3075	.3096	.3118	.3139	.3160	.3181	.3201
2.1	.3222	.3243	.3263	.3284	.3304	.3324	.3345	.3365	.3385	.3404
2.2	.3424	.3444	.3464	.3483	.3502	.3522	.3541	.3560	.3579	.3598
2.3	.3617	.3636	.3655	.3674	.3692	.3711	.3729	.3747	.3766	.3784
2.4	.3802	.3820	.3838	.3856	.3874	.3892	.3909	.3927	.3945	.3962
2.5	.3979	.3997	.4014	.4031	.4048	.4065	.4082	.4099	.4116	.4133
2.6	.4150	.4166	.4183	.4200	.4216	.4232	.4249	.4265	.4281	.4298
2.7	.4314	.4330	.4346	.4362	.4378	.4393	.4409	.4425	.4440	.4456
2.8	.4472	.4487	.4502	.4518	.4533	.4548	.4564	.4579	.4594	.4609
2.9	.4624	.4639	.4654	.4669	.4683	.4698	.4713	.4728	.4742	.4757

Although the table gives the logarithms only of numbers that lie between 1 and 10, we can easily use it to find the logarithms of numbers lying outside that range. In order to do this, we find it useful first to express the given number in scientific notation, as illustrated in the following examples.

Examples 2 Express the following numbers in scientific notation:
(a) 3790.
(b) 0.075.

Solutions In scientific notation a number is expressed as the product of two parts— (1) a number lying between 1 and 10 and (2) a power of 10. Thus:
(a) $3790 = 3.79 \times 10^3$.
(b) $0.075 = 7.5 \times 10^{-2}$.

Examples 3 Use the tables to find:
(a) log 23.4.
(b) log 3790.
(c) log 0.734.

Solutions (a) log 23.4 = log (2.34 × 10¹)
         = log 2.34 + log 10
         = 0.3692 + 1   (from Example 1(b))
         = 1.3692.
(b) log 3790 = log (3.79 × 10³)
         = log 3.79 + log 10³
         = 0.5786 + 3 log 10

where log 3.79 was found from the table. Therefore,
$$\log 3790 = 0.5786 + 3$$
$$= 3.5786.$$

(c) $\log 0.734 = \log(7.34 \times 10^{-1})$
$= \log 7.34 + \log 10^{-1}$
$= \log 7.34 + (-1)\log 10$
$= 0.8657 + (-1)$
$= -0.1343.$

Observe that in this case the answer obtained is negative. It is usually inconvenient to deal with negative numbers, and we write instead

$$\log 0.734 = \bar{1}.8657.$$

This is read "bar one point eight six five seven." The notation $\bar{1}.8657$ is used as an abbreviation for $-1 + 0.8657$. When doing arithmetical calculations with a number such as $\bar{1}.8657$, we must remember that the integer part is negative (i.e., -1) but the decimal part is positive.

Examples 4 Find (a) log 0.016 and (b) log 0.000379.

Solutions (a) $\log 0.016 = \log(1.6 \times 10^{-2})$
$= \log 1.6 + \log(10^{-2})$
$= 0.2041 + (-2)$ [from Example 1(a)]
$= \bar{2}.2041.$

(b) $\log 0.000379 = \log(3.79 \times 10^{-4})$
$= 0.5786 + (-4)$
$= \bar{4}.5786.$

DEFINITION The decimal part of a common logarithm—that is, the part found from the table in the Appendix—is called the ***mantissa***. The integer part is the ***characteristic***. For example, log 23.4 has a characteristic of $+1$ and a mantissa of 0.3692. Log 0.016 has a mantissa of 0.2041 and a characteristic of -2.

When a given number is written in scientific notation, the characteristic of its logarithm is equal to the power of 10 that appears. For example, $0.016 = 1.6 \times 10^{-2}$ in scientific notation, so log 0.016 has characteristic -2.

It is useful to remember that if $x > 10$, then $\log x > 1$; while if $x < 1$, then log x is negative.

MULTIPLICATION AND DIVISION USING LOGARITHMS

Logarithms can be used as an aid when multiplying or dividing numbers. Their use in multiplication is based on the following property of logarithms: for any two positive real numbers a and b,

$$\log(ab) = \log a + \log b.$$

Sec. 8.4 *The Use of Common Logarithms*

Example 5 Use logarithms to evaluate (3.79)(1.6).

Solution Let $x = (3.79)(1.6)$. Then

$$\log x = \log [(3.79)(1.6)]$$
$$= \log 3.79 + \log 1.6$$
$$= 0.5786 + 0.2041 \quad \text{(from log tables)}$$
$$\log x = 0.7827.$$

Now, using the table of logarithms backward, we find that

$$\log 6.06 = 0.7825 \quad \text{and} \quad \log 6.07 = 0.7832.$$

Taking the nearest of these, we conclude that, to two decimal places, $x = 6.06$—that is,

$$(3.79)(1.6) = 6.06.$$

Example 6 Evaluate (0.734)(0.379).

Solution Let $x = (0.734)(0.379)$. Then

$$\log x = \log [(0.734)(0.379)]$$
$$= \log 0.734 + \log 0.379$$
$$= \log (7.34 \times 10^{-1}) + \log (3.79 \times 10^{-1})$$
$$= \bar{1}.8657 + \bar{1}.5786 \quad \text{(from tables)}$$
$$= (-1 + 0.8657) + (-1 + 0.5786)$$
$$= -2 + 1.4443 \qquad (*)$$
$$= -1 + 0.4443$$
$$\log x = \bar{1}.4443.$$

Observe that, in simplifying $-2 + 1.4443$ in the step (*) above, we have kept the decimal part (mantissa) positive. From the tables we find that $\log 2.78 = 0.4440$ is the nearest value to the mantissa 0.4443. Therefore, taking account of the characteristic, we conclude that $\bar{1}.4443$ is the logarithm of 2.78×10^{-1} or 0.278. Thus to three figures, $x = 0.278$ or

$$(0.734)(0.379) = 0.278.$$

The process of using the logarithm tables backward in order to find the number that has a given logarithm is known as ***finding the antilogarithm***. Thus, in the preceding examples, the antilogarithm of 0.7827 is 6.06 and antilog $(\bar{1}.4443) = 0.278$. Tables of antilogarithms exist by means of which this step can be performed directly, but it is just as easy to use the logarithm table backward.

The use of logarithms for division is based on the property that, for any pair of positive numbers a and b,

$$\log \frac{a}{b} = \log a - \log b.$$

Thus to calculate $a \div b$ we subtract the logarithm of b from the logarithm of a and then find the antilogarithm of the result.

Example 7 Evaluate $7.34 \div 3.79$.

Solution Let $x = 7.34/3.79$. Then
$$\log x = \log \frac{7.34}{3.79}$$
$$= \log 7.34 - \log 3.79$$
$$= 0.8657 - 0.5786 \quad \text{(from tables)}$$
$$\log x = 0.2871.$$

From the logarithm tables, we find that the closest logarithm to 0.2871 is $\log 1.94$, which is 0.2878. Thus to three figures
$$x = \text{antilog}\,(0.2871) = 1.94$$
or
$$7.34 \div 3.79 = 1.94.$$

Example 8 Evaluate $0.379 \div 0.00734$.

Solution Letting $x = 0.379 \div 0.00734$ and taking logarithms on both sides, we have
$$\log x = \log(0.379 \div 0.00734)$$
$$= \log 0.379 - \log 0.00734$$
$$= \bar{1}.5786 - \bar{3}.8657 \quad \text{(from tables)}$$
$$= (-1 + 0.5786) - (-3 + 0.8657)$$
$$= -1 + 0.5786 + 3 - 0.8657$$
$$= 2 + 0.5786 - 0.8657$$

—that is,
$$\log x = 2.5786 - 0.8657 = 1.7129.$$

From the tables, we find that $\text{antilog}(1.7129) = 51.6$ to three figures. Therefore,
$$0.379 \div 0.00734 = 51.6.$$

Example 9 Evaluate $(0.00734)(1.17)/0.951$.

Solution Let x denote the required quantity. Then
$$\log x = \log \frac{(0.00734)(1.17)}{0.951}$$
$$= \log 0.00734 + \log 1.17 - \log 0.951$$
$$= \bar{3}.8657 + 0.0682 - \bar{1}.9782$$
$$= \underbrace{(-3 + 0.8657) + 0.0682}_{= 0.9339} - (-1 + 0.9782)$$
$$= -2 + 0.9339 - 0.9782 \qquad\qquad (*)$$
$$= -3 + 1.9339 - 0.9782$$
$$= -3 + 0.9557$$
$$\log x = \bar{3}.9557.$$

Observe that in the step following line (∗) we increased 0.9339 by 1 before subtracting 0.9782. The purpose was to *keep the mantissa positive*.

Taking the antilogarithm, we find $x = 9.03 \times 10^{-3} = 0.00903$. That is, to three significant figures

$$\frac{(0.00734)(1.17)}{0.951} = 0.00903.$$

POWERS, ROOTS, AND EXPONENTS USING LOGARITHMS

The third use to which logarithms can be put is in evaluating powers and roots of numbers. It rests on the following property, which holds for any real numbers n and $a > 0$:

$$\log(a^n) = n \log a.$$

Example 10 Evaluate $(3.79)^{5/2}$.

Solution Let $x = (3.79)^{5/2}$. Then

$$\begin{aligned}
\log x &= \log(3.79)^{5/2} \\
&= \tfrac{5}{2} \log 3.79 \\
&= \tfrac{5}{2}(0.5786) \\
\log x &= 1.4465.
\end{aligned}$$

The antilogarithm of 1.4465 is 28.0, so

$$x = (3.79)^{5/2} = 28.0.$$

Example 11 Evaluate $(0.734)^9$.

Solution Let $x = (0.734)^9$. Then

$$\begin{aligned}
\log x &= \log(0.734)^9 \\
&= 9 \log 0.734 \\
&= 9(\bar{1}.8657) \\
&= 9(-1 + 0.8657) \\
&= -9 + 7.7913 \\
&= -2 + 0.7913 \\
\log x &= \bar{2}.7913.
\end{aligned}$$

Therefore,

$$\begin{aligned}
x &= \text{antilog}\,(\bar{2}.7913) \\
&= 6.18 \times 10^{-2} = 0.0618.
\end{aligned}$$

Thus, to three figures,

$$x = (0.734)^9 = 0.0618.$$

Example 12 Evaluate the fifth root of 0.0734.

Solution Let $x = (0.0734)^{1/5}$. Then

$$\log x = \log (0.0734)^{1/5}$$
$$= \tfrac{1}{5} \log 0.0734$$
$$= \tfrac{1}{5}(\bar{2}.8657).$$

In carrying out the necessary division by 5, we must be sure to keep the characteristic a whole number. The rule in this situation is to *decrease* the characteristic so that it becomes exactly divisible by the divisor. In this case, for example, we write

$$\bar{2}.8657 = -2 + 0.8657 = -5 + 3.8657.$$

Then

$$\log x = \tfrac{1}{5}(-5 + 3.8657)$$
$$= -1 + 0.7731$$
$$= \bar{1}.7731.$$

Taking the antilogarithm, we find to three figures $x = 5.93 \times 10^{-1} = 0.593$. That is,

$$(0.0734)^{1/5} = 0.593.$$

Example 13 Find the value of n for which

$$(3.79)^n = 7.34.$$

Solution Taking logarithms of both sides of the given equation, we get

$$\log (3.79)^n = \log 7.34.$$

and so, using Property IV of logarithms,

$$n \log 3.79 = \log 7.34.$$
$$n = \frac{\log 7.34}{\log 3.79}$$
$$n = \frac{0.8657}{0.5786}$$
$$= \frac{0.866}{0.579}$$

where we have rounded off to three figures. To find the value of this ratio, we again take logarithms on both sides to get

$$\log n = \log \frac{0.866}{0.579}$$
$$= \log 0.866 - \log 0.579$$
$$= \bar{1}.9375 - \bar{1}.7627 \quad \text{(from the table)}$$
$$= 0.1748.$$

Therefore,
$$n = \text{antilog}(0.1748)$$
$$= 1.50.$$
Thus to three figures, $n = 1.50$, or
$$(3.79)^n = 7.34 \quad \text{when } n = 1.50.$$

Diagnostic Test 8.4

Fill in the blanks.

1. If $\log x = 2.3478$, then the mantissa of $\log x$ is _____ and the characteristic of $\log x$ is _____.
2. If $\log t = \bar{3}.2573$, then the mantissa of $\log t$ is _____ and its characteristic is _____.
3. If $\log u = -3.7800$, then the mantissa of $\log u$ is _____ and the characteristic is _____.
4. $\bar{2}.37 + \bar{1}.79 = $ _____.
5. $\bar{4}.45 - \bar{5}.28 = $ _____.
6. $3.20 - \bar{5}.28 = $ _____.
7. $3.50 \div \bar{1}.75 = $ _____.

Exercises 8.4

(1–12) Find the common logarithms of the following numbers.

1. 7.5.
2. 6.3.
3. 1.71.
4. 8.64.
5. 21.4.
6. 343.
7. 2570.
8. 3710.
9. 0.0271.
10. 0.350.
11. 0.0015.
12. 0.000236.

(13–20) Find the antilogarithms of the following numbers to the nearest three significant digits.

13. 0.7371.
14. 0.6666.
15. 1.2345.
16. 2.3478.
17. $\bar{1}.2540$.
18. $\bar{1}.4609$.
19. $\bar{3}.2313$.
20. $\bar{4}.0955$.

(21–58) Use logarithms to evaluate the following quantities, giving the answer to three significant figures.

21. (3.4)(5.8).
22. (6.2)(9.7).
23. (1.76)(4.12).
24. (8.13)(1.42).
25. (12.7)(28.0).
26. (405)(83.1).
27. (5170)(0.0257).
28. (0.431)(0.892).
29. (0.325)(0.0281).
30. (0.602)(0.011).
31. $4.27 \div 2.83$.
32. $1.87 \div 4.61$.
33. $28.4 \div 0.208$.
34. $0.646 \div 0.00293$.
35. $0.427 \div 20.6$.
36. $0.037 \div 12.8$.
37. $0.0148 \div 255$.
38. $0.015 \div 0.732$.
39. $0.0407 \div 6080$.
40. $0.00281 \div 0.0902$.
41. (1.35)(47.2)(0.551).
42. (0.42)(0.146)(0.0118).
43. $\dfrac{(0.462)(1.39)}{55.6}$.
44. $\dfrac{1.57}{(42.3)(8.9)}$.
45. $\dfrac{0.0523}{(2.41)(0.804)}$.
46. $\dfrac{(2.08)(0.00681)}{0.0546}$.
47. $(4.27)^4$.
48. $(6.16)^7$.
49. $(0.157)^3$.
50. $(0.782)^6$.

51. $(0.851)^{3/2}$.   **52.** $(0.0253)^5$.   **53.** $\sqrt{2740}$.
54. $\sqrt[3]{4160}$.   **55.** $(0.461)^{1/3}$.   **56.** $(0.0321)^{1/6}$.
57. $\sqrt{\frac{346}{902}}$.   **58.** $\sqrt[3]{29.5}/\sqrt{0.295}$.

(59–64) Find the value of n that satisfies each of the following equations.

59. $(1.5)^n = 2.43$.   **60.** $(7.6)^n = 2.15$.   **61.** $(0.23)^n = 0.46$.
62. $(0.32)^n = 0.087$.   **63.** $(0.5)^n = 3(0.2)^n$.   **64.** $(0.471)^n = 1.36^n/2$.

8.5

Applications of Logarithms

EXPONENTIAL EQUATIONS

An important application of logarithms is to solve certain types of equations in which the unknown variable appears as an exponent. Consider the following examples.

Example 1 In 1970 the population of a certain city was 2 million and was increasing at the rate of 5% each year. Assuming this growth rate continues, when will the population pass the 5 million mark?

Solution At a rate of increase of 5%, the population is multiplied by a factor 1.05 each year. After n years, starting from 1970, the population level is therefore

$$2(1.05)^n \text{ million.}$$

We require the value of n for which this level is 5 million, so we have

$$2(1.05)^n = 5$$
$$(1.05)^n = 2.5.$$

Observe that in this equation the unknown quantity n appears as an exponent. We can solve it by taking logarithms of both sides. It does not matter which base we use for the logarithms, but usually common logarithms are the most convenient. We obtain

$$\log (1.05)^n = \log 2.5$$

or, using Property IV of logarithms,

$$n \log 1.05 = \log 2.5.$$

Therefore,

$$n = \frac{\log 2.5}{\log 1.05}$$

$$= \frac{0.3979}{0.0212} \quad \text{(from the table of common logarithms in the Appendix)}$$

$$\approx 18.8.$$

It takes 18.8 years for the population to climb to 5 million. This level will be reached during 1988.

Example 2 The sum of $100 is invested at 6% compound interest per annum. How long does it take to increase in value to $150?

Solution At 6% interest per annum, the investment grows by a factor 1.06 each year. Therefore, after n years, the value is $100(1.06)^n$ dollars. Setting this equal to 150, we obtain the following equation for n:

$$100(1.06)^n = 150$$
$$(1.06)^n = 1.5.$$

Taking logarithms of both sides, we obtain

$$\log (1.06)^n = n \log 1.06 = \log 1.5$$

and so

$$n = \frac{\log 1.5}{\log 1.06}$$
$$= \frac{0.1761}{0.0253}$$
$$= 6.96.$$

Thus it takes almost seven years for the investment to increase in value to $150.

These two examples led to an equation of the type

$$a^x = b$$

where a and b are two given positive constants and x is the unknown variable. Such an equation can always be solved by taking logarithms of both sides:

$$\log (a^x) = \log b$$
$$x \log a = \log b$$
$$x = \frac{\log b}{\log a}.$$

There is no difference in principle here between problems involving growing exponential functions ($a > 1$) and those involving decaying exponentials ($a < 1$). The following example involves a decaying exponential function.

Example 3 Shortly after a person consumes a substantial dose of whiskey, the alcohol level in her blood rises to a level of 0.3 milligrams per milliliter. Thereafter this level decreases according to the formula $(0.3)(0.5)^t$, where t is the time measured in hours from the moment at which the peak level is reached. How long is it before the person can legally drive her automobile? (In her locality the legal limit is 0.08 milligrams per milliliter of blood alcohol.)

Solution We wish to find the value of t at which

$$(0.3)(0.5)^t = 0.08.$$

That is,
$$(0.5)^t = \frac{0.08}{0.3} = 0.267.$$

Taking logarithms, we obtain that
$$\log (0.5)^t = \log 0.267$$
$$t \log (0.5) = \log 0.267.$$

Therefore,
$$t = \frac{\log 0.267}{\log 0.5}$$
$$= \frac{-0.5735}{-0.3010} \quad \text{(from the Appendix table*)}$$
$$= 1.91.$$

It therefore takes 1.91 hours before the person is legally fit to drive.

RADIOACTIVE DECAY

There are certain elements, such as radium, uranium, and plutonium, whose atoms are unstable. The nucleus can spontaneously emit radiation and thereby change the atom into a different chemical element. Elements with this property are said to be radioactive, and the process by which they change into new elements is called radioactive decay.

Some radioactive elements are so unstable that they decay in fractions of a microsecond. Other elements are almost stable, decaying over periods of thousands or millions of years.

In a given piece of radioactive material, the atoms do not decay all at the same time but rather in a random and unpredictable manner. Over a period of time, as more and more of its atoms decay, the remaining amount of undecayed radioactive material gradually reduces to zero. It is well established that the amount of undecayed material is an exponential function of time. If y is the amount of radioactive material that remains at time t and if y_0 is the amount that was present at the initial time $t = 0$, then we have

$$y = y_0 e^{-kt}$$

where k is a constant called the *decay constant*. The value of k depends on the particular element in question.

Example 4 The decay constant for radium is 4.36×10^{-4} when time is measured in years. If we initially have 10 grams of radium, how many grams will remain after 1500 years? After how many years will only 4 grams remain?

---

*From the table we would normally express the two logarithms in characteristic-mantissa form as: $\log 0.267 = \bar{1}.4265$ and $\log 0.5 = \bar{1}.6990$. In order to calculate their ratio, however, we must reexpress them as negative numbers. For example, $\log 0.267 = \bar{1}.4265 = -1 + 0.4265 = -0.5735$.

Sec. 8.5 Applications of Logarithms

Solution Since $y_0 = 10$, the number of grams of radium that remain after t years is given by
$$y = 10e^{-kt} = 10e^{-(0.000436)t}.$$
When $t = 1500$, this is
$$y = 10e^{-(0.000436)(1500)}$$
$$= 10e^{-0.65}$$
$$= 10(0.52)$$
$$= 5.2.$$
(Note that we have rounded off to two significant figures.) Thus 5.2 grams remain after 1500 years.

We also want the value of t at which $y = 4$. We have
$$y = 4 = 10e^{-kt}$$
$$e^{-kt} = 0.4.$$
In this equation the unknown t appears in an exponent. So we take logarithms of both sides. Since the base e appears already, it is best to use natural logarithms:
$$\ln(e^{-kt}) = \ln 0.4$$
$$-kt \ln e = \ln 0.4$$
$$-kt = \ln 0.4 \quad (\ln e = 1)$$
$$t = \frac{-\ln 0.4}{k}$$
$$= \frac{-(-0.9163)}{0.000436}$$
$$= 2102.$$
Thus it takes 2102 years for the amount of radium to decay to 4 grams.

We define the **half-life** of a radioactive isotope to be the amount of time for half of the amount that was originally present to decay. Denoting this time by T, we have that when $t = T$, $y = \frac{1}{2}y_0$, since only half the original amount y_0 remains. Therefore,
$$y = y_0 e^{-kT} = \frac{1}{2}y_0$$
$$e^{-kT} = \frac{1}{2}.$$
Taking natural logarithms of both sides again, we get
$$\ln e^{-kT} = \ln \frac{1}{2}$$
$$-kT = \ln \frac{1}{2}$$
$$= -\ln 2$$
by Property III of logarithms. Therefore
$$T = \frac{\ln 2}{k}$$

This equation establishes the relationship between the half-life T and the decay constant k for any radioactive material.

Example 5 Calculate the half-life for radium.

Solution From Example 4, we have $k = 0.000436$. Therefore
$$T = \frac{\ln 2}{k} = \frac{0.6931}{k} = \frac{0.6931}{0.000436}$$
$$= 1590.$$
Therefore the half-life for radium is 1590 years.

CONTINUOUS COMPOUNDING OF INTEREST

Consider a sum P invested at a nominal interest rate of $R\%$ compounded k times per annum. We saw in Section 8.1 that the value after n years is given by the formula
$$A = P\left(1 + \frac{i}{k}\right)^{nk}, \quad i = \frac{R}{100}.$$

Suppose as an example that $100 is invested for four years at the nominal interest rate of 8% p.a. (per annum) ($P = 100$, $n = 4$, $R = 8$). Then
$$A = 100\left(1 + \frac{0.08}{k}\right)^{4k}.$$

k	Value after four years
1	$136.05
2	$136.86
4	$137.28
12	$137.57
52	$137.68
365	$137.71
1000	$137.71

The adjacent table sets out the value of the investment after four years for different values of k. For annual compounding $k = 1$, while the other values refer to semiannual ($k = 2$), quarterly ($k = 4$), monthly ($k = 12$), weekly ($k = 52$) and daily ($k = 365$) compounding. We can see that as the frequency of compounding is increased, the value of the investment also increases; however it does not increase indefinitely, but rather approaches closer and closer to a certain value. To the nearest cent, there is no difference between compounding 365 times a year and 1000 times a year: the value of the investment after four years would still be $137.71.

Because of this we can envisage the possibility of what is called continuous compounding. By this is meant that the number k is allowed to approach infinity; we write $k \to \infty$. This corresponds to compounding the interest infinitely often during the year. With our $100 invested at the nominal rate of 8% per annum, continuous compounding gives a value of $137.71 after four years, the same value as daily compounding.

In the above expression for the value of the investment after four years, let us write $k = (0.08)q$. Then $4k = (0.32)q$, and the value after four years takes the form
$$\$100\left(1 + \frac{0.08}{k}\right)^{4k} = \$100\left(1 + \frac{1}{q}\right)^{0.32q} = \$100\left[\left(1 + \frac{1}{q}\right)^{q}\right]^{0.32}.$$

The reason for writing it in this form is as follows. As $k \to \infty$ then $q = k/0.08$ also becomes arbitrarily large. The quantity inside the square bracket,

q	$\left(1+\dfrac{1}{q}\right)^q$
1	2
2	2.25
10	2.594
100	2.705
1000	2.717
10,000	2.718

$(1 + 1/q)^q$, gets closer and closer to a certain value as $q \to \infty$. We can see this from the adjacent table, which gives the values of $(1 + 1/q)^q$ for a series of increasingly large values of q. The eventual value to which $(1 + 1/q)^q$ gets closer and closer as q increases indefinitely is the number that we previously denoted by the letter e ($e = 2.71828$ to five places of decimals). This is one of the reasons why e is such an important number.

Returning to the above example of continuous compounding we see that as $q \to \infty$, the value of the investment after four years gets closer and closer to $\$100e^{0.32}$.

This example can be generalized (see Exercise 59). It can be shown that *if a sum P is compounded continuously at a nominal annual interest rate of R%, its value after n years is given by*

$$A(n) = Pe^{ni}, \qquad i = \frac{R}{100}.$$

We see therefore that the natural exponential function arises in a very basic way in problems concerning continuous compounding.

Example 6 An investment of $250 is compounded continuously at a nominal rate of interest of $7\frac{1}{2}\%$. What will be the value of the investment after six years?

Solution We must use the formula $A(n) = Pe^{ni}$ for the value after n years. In this example, $P = 250$, $n = 6$, and $i = 7.5/100 = 0.075$. Therefore $ni = (0.075)(6) = 0.45$, and the value is

$$A(6) = Pe^{6i} = 250e^{0.45} \text{ dollars.}$$

The value of $e^{0.45}$ can be found in the table in the Appendix, and we get

$$A(6) = 250e^{0.45} = 250(1.5683) = 392.08.$$

Thus the value of the investment after six years is $392.08.

Example 7 What nominal rate of interest, when compounded continuously, gives the same growth over a whole year as a 10% annual rate of interest?

Solution A sum P invested at a nominal rate of interest $R\%$ compounded continuously has a value $A(1) = Pe^i$ after one year ($i = R/100$). (Take $n = 1$ in the formula for continuous compounding.) If invested at 10% per annum it would increase by a factor (1.1) during each year. Therefore we must set

$$Pe^i = (1.1)P \quad \text{or} \quad e^i = 1.1.$$

If we take natural logarithms of this equation, we get

$$\ln(e^i) = \ln(1.1).$$

But $\ln(e^i) = i$ for any real number i, so

$$i = \ln(1.1) = 0.0953.$$

Therefore $R = 100i = 9.53$. So 10% interest compounded annually is equivalent to the annual growth provided by a 9.53% nominal rate of interest compounded continuously.

LOGARITHMIC SCALES

Logarithmic scales are used to measure a number of physical quantities. The purpose is to reduce the scale of variation where a natural physical variable covers a very wide range. Examples are the decibel scale for loudness, stellar magnitudes for star brightness, the Richter scale for earthquakes, and the pH scale for acidity.

Loudness. We can measure the intensity I of sound in watts per square meter. The intensity I_0 which is the threshold of hearing for the average human ear is at about 10^{-12} watts per square meter when the sound frequency is 100 hertz (1 hertz = 1 cycle per second). The range of intensities I that are typically experienced by the ear vary from I_0 up to the threshold of pain, which occurs at an intensity of about 100 watts per square meter. Thus I can cover an immense range.

For this reason the *loudness* L of a sound is defined in *decibels* by the equation

$$L = 10 \log \frac{I}{I_0}.$$

When $I = I_0$, the loudness is zero. At the pain threshold, $I/I_0 \approx 10^{14}$, and the loudness is 140 decibels. Thus in terms of loudness the scale of variation of audible sounds is reduced to the manageable range of 0–140 decibels.

Example 8 A quiet room has a loudness of background noise of 32 decibels. A loud conversation has a noise level of 65 decibels. Calculate the ratio of the two sound intensities.

Solution Let I be the sound intensity for the quiet room. Then the corresponding loudness is

$$10 \log \frac{I}{I_0} = 32$$

$$\log \frac{I}{I_0} = 3.2.$$

Thus

$$\frac{I}{I_0} = \text{antilog}(3.2)$$

$$= 1580.$$

This means that the sound intensity in a quiet room is 1580 times the threshold of hearing: $I = 1580 I_0$. Let I' be the sound intensity for the loud conversation. Then

$$10 \log \frac{I'}{I_0} = 65$$

$$\log \frac{I'}{I_0} = 6.5$$

$$\frac{I'}{I_0} = \text{antilog } (6.5)$$

$$= 3.16 \times 10^6.$$

I' is 3.16 million times the threshold: $I' = 3.16 \times 10^6 I_0$. Therefore the ratio of the two intensities is

$$\frac{I'}{I} = \frac{3.16 \times 10^6 I_0}{1580 I_0}$$

$$= 2000.$$

The decibel scale of loudness can be interpreted as follows. If a sound increases in loudness by 10 decibels, then its intensity increases by a *factor* of 10. For example, if a certain acoustic insulation reduces the sound transmitted through a wall by 20 decibels, then the intensity of sound transmitted (the amount of sound energy transmitted) is reduced by a factor of 10^2 or 100.

Richter Scale. The **intensity** of an earthquake, A, is defined as the amplitude of movement measured on a standard seismograph located 100 kilometers from the earthquake's epicenter. The **magnitude** R of the earthquake on the Richter scale is defined as

$$\boxed{R = \log \frac{A}{A_0}}$$

where A_0 is a certain minimum intensity used for comparison (in fact A_0 is taken as 1 micron amplitude). Thus if one earthquake has 10 times the intensity of another, its magnitude on the Richter scale is greater by 1. For example, if one earthquake has magnitude 5 and a second has magnitude 6, then the second is 10 times as intense as the first.

Example 9 The 1964 earthquake in Alaska measured 8.5 on the Richter scale. Calculate A/A_0 for this earthquake. The largest earthquakes ever recorded measured 8.9 in magnitude. How much more intense were these earthquakes than the Alaskan one?

Solution For the Alaskan earthquake,

$$\log \frac{A}{A_0} = 8.5$$

$$\frac{A}{A_0} = \text{antilog } 8.5$$

$$= 3.16 \times 10^8.$$

For the largest earthquakes, having intensity A',

$$\log \frac{A'}{A_0} = 8.9.$$

$$\frac{A'}{A_0} = \text{antilog } 8.9$$

$$= 7.94 \times 10^8.$$

The ratio of the intensities of these strongest earthquakes to that of the Alaskan earthquake is therefore

$$\frac{A'}{A} = \frac{A'}{A_0} \div \frac{A}{A_0} = \frac{7.94 \times 10^8}{3.16 \times 10^8} = \frac{7.94}{3.16} = 2.51.$$

The strongest earthquakes were therefore 2.5 times as intense as the 1964 Alaskan earthquake.

Stellar Magnitudes. If B is the apparent brightness of a star (that is, its brightness as seen from the earth), then its ***magnitude*** M is defined by

$$\boxed{M = -\frac{5}{2} \log \frac{B}{B_0} = \frac{5}{2} \log \frac{B_0}{B}}$$

where B_0 is a certain standard brightness used for comparison. The value of B_0 is fixed such that the faintest stars just visible to the human eye have magnitude $M = 6$.

Because of the minus sign in the definition of M, the brighter the star the lower its magnitude. The two brightest stars (Sirius and Canopus) have negative magnitudes, while the other visible stars range in brightness from 0 to 6. The brightest heavenly objects, apart from the sun and moon, are the planets Venus and Jupiter, which have average magnitudes of -3.9 and -2.0, respectively. (The brightness of the planets varies considerably as they change position relative to the sun.)

From the definition we can see that if one star is 10 times as bright as another, the second will have magnitude 2.5 greater than the first. Consequently a star of first magnitude ($M = 1$) is 100 times as bright as a star of sixth magnitude.

Example 10 How much brighter is Venus on average than the star Polaris (the pole star), whose magnitude is 2.1?

Solution For Venus, we have the average magnitude -3.9:

$$M = -\frac{5}{2} \log \frac{B}{B_0} = -3.9$$

—that is,

$$\frac{B}{B_0} = \text{antilog}\left[\frac{2}{5}(3.9)\right] = \text{antilog}(1.56)$$

$$= 36.3.$$

For Polaris, taking the brightness as B', we have
$$M = \frac{5}{2} \log \frac{B_0}{B'} = 2.1$$
—that is,
$$\frac{B_0}{B'} = \text{antilog}\left[\frac{2}{5}(2.1)\right] = \text{antilog}(0.84)$$
$$= 6.92.$$
Thus
$$\frac{B}{B'} = \left(\frac{B}{B_0}\right)\left(\frac{B_0}{B'}\right) = (36.3)(6.92) \approx 250.$$
Therefore Venus is 250 times as bright as Polaris.

Diagnostic Test 8.5

Fill in the blanks.

1. If $(1.08)^n = 2$, then $n = $ _____.
2. If $a^x = b$, then $x = $ _____.
3. If $e^{kt} = y$, then $t = $ _____.
4. $(1 + 1/p)^p \to $ _____ as $p \to \infty$.
5. A sum P compounded k times per year at nominal interest rate R grows to _____ after n years, where $i = R/100$.
6. A sum P compounded continuously at nominal interest rate R grows to _____ after n years, where $i = R/100$.
7. $e^{kT} = \frac{1}{2}$, where $T = $ _____.

Exercises 8.5

(1–16) Solve the following equations for x.

1. $10^x = 25$.
2. $10^{x/2} = 50$.
3. $2^x = 25$.
4. $3^x = 2$.
5. $4^{x/4} = 3$.
6. $2^{3x} = 10$.
7. $3^x 2^{3x} = 4$.
8. $3^x 2^{1-x} = 10$.
9. $3^x = 2^{2-x}$.
10. $(3^x)^2 = 2\sqrt{2^x}$.
11. $4^{2x} = 2(\sqrt{3})^x$.
12. $5^x 2^{1-2x} = \sqrt{3^x}$.
13. $(2^x)^x = 25$.
14. $(2^x)^x = 3^x$.
15. $a^x = cb^x$.
16. $(a^x)^2 = b^{x+1}$.

17. The population of the earth in 1976 was 4 billion and was growing at 2% per year. If this rate of growth continues, calculate when the population will reach 10 billion.

18. The population of China in 1970 was 750 million and was growing at 4% per year. When would this population reach 2 billion, assuming that the same growth rate continued? (This growth rate is now very much out of date.)

19. Using the data of the preceding questions, calculate when the population of China would become equal to half the population of the earth.

20. Country A has a population 50% greater than that of country B. A's population is growing at 1.5% per year and B's at 3.5%. How many years will it take for the two populations to be equal?

21. In Exercise 20, how many years will it take for B's population to be 50% greater than A's?

22. The profits of a company have been increasing at an average of 12% per year between 1975 and 1980, and in 1980 they reached the level of $5.2 million. Assuming this rate of growth continues, how long will it be before they reach $8 million per year?

23. Two competing newspapers have circulations of 1 million and 2 million, respectively. If the circulation of the first is increasing by 2% each month while that of the second is declining by 1% each month, calculate how long it will be before the two have equal circulations.

24. In Exercise 23, calculate how long it will be before the expanding newspaper has twice the circulation of the declining one.

(25–26) $500 is invested at 12% compounded annually.

25. How long does it take to increase to $1000?

26. How long does it take to increase to $2000?

(27–28) $1000 is invested at 8% interest compounded annually.

27. How long does it take to increase to $1500?

28. How long does it take to increase to $3000?

(29–31) $1000 is invested at 12% nominal interest.

29. How long does it take to increase to $2000 if compounded quarterly?

30. How long does it take to increase to $2000 if compounded monthly?

31. How long does it take to increase to $1500 if compounded monthly?

32. The following rule of thumb is often employed by people in finance: if the rate of interest is R% per annum, then the number of years, n, for an investment to double is given by dividing R into 70 (i.e., $n = 70/R$). Calculate n *exactly* for the following values of R: 4, 8, 12, 16, and 20. Compare your answer with those obtained from the above formula, and hence assess the accuracy of the rule $n = 70/R$.

33. Calculate the nominal rate of interest that compounded semiannually is equivalent to an 8% annual interest.

34. Calculate the nominal rate of interest that compounded monthly is equivalent to an 8% annual interest.

35. The decay constant for radium is 4.36×10^{-4}. If 5 grams of radium are left for 1000 years, how much radium will remain?

36. The decay constant for radioactive beryllium is 1.5×10^{-7} (time is measured in years). How much of 1 gram of beryllium will remain after 100 million years?

37. In Exercise 35, how long will it be before the amount of radium has decayed to 2 grams?

38. In Exercise 36, how long will it be before only 0.1 gram of beryllium remains?

39. The isotope C^{14} is used extensively for carbon dating in archeology. Its half-life is 5570 years. Calculate its decay constant.

40. In Exercise 36, calculate the half-life of radioactive beryllium.

41. An organic archeological specimen has a C^{14} content that is 40% of the C^{14} content found in living matter. How long is it since the specimen was part of a living organism? (See Exercise 39.)

42. A bone from an ancient burial site has a C^{14} content that is 25% of the C^{14} content of living organisms. How many years ago did the owner of the bone pass away?

43. An investment is compounded continuously at a nominal rate of 8% per annum ($i = 0.08$). How long does it take the investment to double in value?

44. An investment of $100 is compounded continuously for four years, increasing in value to $150. Calculate the nominal rate of interest.

45. An investment is compounded continuously for two years at a nominal rate of R% and for a further four years at the nominal rate of $2R$%. Find R if the value exactly doubles.

46. An investment of $100 is compounded continuously for two years at a nominal rate of interest of 9% and then for a further five years at nominal rate of interest of 11%. Calculate the value of the investment after the seven-year period.

47. Which is better for the investor, continuous compounding with a nominal annual rate of 8% or quarterly compounding with a nominal annual rate of 8.2%?

48. Calculate the nominal rate of interest that doubles an investment in 10 years when compounded continuously.

49. A normal conversation has a sound intensity of $10^6 I_0$. Calculate the loudness in decibels.

50. A jet engine at 100 feet has a sound intensity of $10^{13} I_0$. Calculate the loudness in decibels.

51. A conversation has a loudness of 60 decibels, and an electric typewriter has a loudness of 80 decibels. What is the ratio of their intensities?

52. A heavy truck causes sound of 90 decibels to fall on the wall of a building. If the acoustic insulation reduces sound by 25 decibels, what will be the intensity of the sound transmitted into the building?

53. Two earthquakes measure 5.0 and 6.5 on the Richter scale. Calculate the ratio of their intensities.

54. The San Francisco earthquake of 1906 measured 8.3 on the Richter scale. The Los Angeles earthquake of 1971 measured 6.7. How many times more intense was the San Francisco earthquake?

(55–58) Venus has magnitude -3.9, Jupiter has magnitude -2.0, the star Sirius has magnitude -1.6, and the stars Vega and Alpheratz have magnitudes 0.1 and 2.2.

55. How much brighter is Venus than Sirius?

56. How much brighter is Sirius than Alpheratz?

57. How much brighter is Jupiter than Alpheratz?

58. How much brighter is Venus than Vega?

*59. Prove the formula $A(n) = Pe^{ni}$. [*Hint:* Start from the formula $A = P(1 + i/k)^{nk}$, set $k = iq$ and let $q \to \infty$.]

8.6

Base-change Formulas

Consider the function $y = e^{kx}$, where k is a given constant. The base of this exponential function is e. We can, however, rewrite it in the following way:

$$y = e^{kx} = (e^k)^x.$$

Therefore, if we define $a = e^k$, we have $y = a^x$. In this way we have changed the base of the exponential function from e to a.

We see from this that the function $y = e^{kx}$ can be expressed in the alternative form $y = a^x$, where $a = e^k$. Conversely, any exponential function $y = a^x$ can be reexpressed in terms of the natural exponential function. For, we can write $a = e^{\ln a}$, and so
$$y = (e^{\ln a})^x = e^{(\ln a)x}.$$
Thus the function $y = a^x$ can be written in the form $y = e^{kx}$, where $k = \ln a$.

Example 1 Express the following in the form $y = e^{kx}$:
(a) $y = 3^x$. (b) $y = 4^{x/4}$.

Solution (a) Here the base is 3. We have
$$y = 3^x = (e^{\ln 3})^x$$
$$= e^{(1.0986)x}.$$

(b) Here $a = 4$:
$$y = 4^{x/4} = (e^{\ln 4})^{x/4}$$
$$= e^{[(1/4)\ln 4]x}$$
$$= e^{(1.3863/4)x}$$
$$= e^{(0.3466)x}.$$

Example 2 A population of microorganisms has an initial size of 100 and is doubling every 20 minutes. Write an expression for the population size in the form $y = be^{kt}$, where t is time measured in hours.

Solution The population doubles three times in each hour. After t hours there will be $3t$ such doublings and therefore the population size will be
$$y = 100(2^{3t})$$
$$= 100(e^{\ln 2})^{3t}$$
$$= 100e^{(3\ln 2)t}$$
$$= 100e^{(\ln 8)t}$$
$$= 100e^{(2.0794)t}.$$
This formula has the required form $y = be^{kt}$, with $b = 100$ and $k = 2.0794$.

It is the usual practice to write any growing exponential function a^x in the form e^{kx}, where $k = \ln a$. A decaying exponential function, defined by a^x with $a < 1$, would normally be written as e^{-kx}, where k is a positive constant given by $k = -\ln a$. The constant k is known as the **specific growth rate** for the function e^{kx} and as the **specific decay rate** for the decaying exponential e^{-kx}.

Example 3 In Example 3 of Section 8.4 the alcohol level in a person's blood at a time t was given by the formula $(0.3)(0.5)^t$ milligrams per milliliter. We can write this in terms of a natural exponential:
$$(0.5)^t = e^{kt}$$

where
$$k = \ln(0.5) = -0.69$$
to two figures. Therefore the alcohol level after t hours is
$$(0.3)e^{-(0.69)t}.$$
This is an example of a decaying exponential function, and its specific decay rate is 0.69 per hour.

It is possible to express logarithms with respect to one base in terms of logarithms with respect to any other base. This is done by means of the so-called *base-change formula*, which states that

$$\log_b y = (\log_a y)(\log_b a).$$

A proof is outlined in Exercise 37. Let us examine certain special cases of this formula. First, take $b = 10$, so that $\log_b y = \log y$ and $\log_b a = \log a$. Then the base-change formula can be written

$$(\log_a y)(\log a) = \log y$$

or

$$\log_a y = \frac{\log y}{\log a}.$$

So *the logarithm of y with base a is equal to the common logarithm of y divided by the common logarithm of a.*

Example 4 Taking $a = 2$, we obtain that
$$\log_2 y = \frac{\log y}{\log 2} = \frac{\log y}{0.3010}.$$
For example,
$$\log_2 3 = \frac{\log 3}{\log 2} = \frac{0.4771}{0.3010} = 1.5850.$$

Second, in the base-change formula let us take $y = b$. Then the left-hand side becomes $\log_b b$, which equals 1. Therefore we obtain the result that

$$(\log_a b)(\log_b a) = 1$$
$$\text{or} \quad \log_b a = \frac{1}{\log_a b}.$$

Example 5 Take $b = 10$, and we get the relation
$$\log_a 10 = \frac{1}{\log a}.$$
For example,
(a) $\log_2 10 = 1/\log 2 = 1/0.3010 = 3.3219$,

(b) $\log_3 10 = 1/\log 3 = 1/0.4771 = 2.0959$,
(c) $\ln 10 = \log_e 10 = 1/\log e$.

The values of these two logarithms are to four decimal places:

$$\log e = \log(2.7183) = 0.4343$$
$$\ln 10 = 2.3026$$

and these two are reciprocals of one another.

The base-change formula allows us to relate the logarithm with a general base a to the common logarithm. In particular, taking $a = e$, we can express the natural logarithm in terms of the common logarithm:

$$\log_e y = \frac{\log y}{\log e} = \frac{\log y}{0.4343}$$

—that is,

$$\boxed{\ln y = 2.3026 \log y.}$$

Thus, in order to find the natural logarithm of y we can look up the common logarithm of y and multiply it by 2.3026. This method of finding the natural logarithm of a number is actually not very convenient when compared to the use of a separate table. The relationship between the two logarithms, however, is of some theoretical importance.

Example 6 From the tables of common logarithms we find that $\log 2 = 0.3010$ and hence $\log 0.2 = 0.3010 - 1 = -0.6990$. The natural logarithms of 2 and 0.2 are therefore given as follows:

$$\ln 2 = 2.3026 \log 2 = (2.3026)(0.3010) = 0.6931$$
$$\ln 0.2 = 2.3026 \log 0.2 = (2.3026)(-0.6990) = -1.6095.$$

You can check these with the values obtained from the natural logarithm tables. There may be small differences in the final digit, owing to round-off errors.

Diagnostic Test 8.6

Fill in the blanks.

1. If $a^x = e^{kx}$, then $k = $ _____.
2. If $200(0.6)^t = be^{-kt}$, then $b = $ _____ and $k = $ _____.
3. $\log x / \log y = $ _____.
4. $(\log_b a)(\log_a b) = $ _____.
5. $(\log e)(\ln 10) = $ _____.
6. $\ln x = (\log x)($_____$)$.
7. $(\log_2 10)^{-1} = \log$ _____.
8. $(\log_3 e)^{-1} = \ln$ _____.
9. $\log_a x = (\ln x)($_____$)$.
10. $\ln x / \ln a = \log$ _____.

Exercises 8.6

(1–4) Express the following functions in the form $y = a^t$ for a suitable base a.

1. $y = e^{2t}$.   **2.** $y = e^{t/2}$.   **3.** $y = e^{-t/5}$.   **4.** $y = e^{-3t}$.

(5–8) Express the following functions in the form $y = e^{kx}$ for a suitable constant k.

5. $y = (2.5)^x$.   **6.** $y = (\frac{5}{4})^x$.   **7.** $y = (\frac{4}{5})^x$.   **8.** $y = (\frac{1}{2})^x$.

(9–12) Express the following functions in the form $y = be^{kt}$.

9. $y = 3(2^t)$.   **10.** $y = (1000)2^{t/3}$.   **11.** $y = 5(1.04)^t$.   **12.** $6 \times 10^8(1.05)^t$.

13. A population of microorganisms is doubling every 45 minutes. If the initial size is 5000, express the population P after t hours in the form $P = be^{kt}$.

14. The earth's population is at present 4 billion and is increasing by 2% each year. Express the population y at a time t years from now in the form $y = ae^{kt}$.

15. A company purchases a machine for $10,000. Each year the value of the machine decreases by 20%. Express the value V in the form be^{kt}, where b and k are constants and time $t = 0$ corresponds to the date of purchase.

16. Between January 1975 and January 1980 the consumer price index I rose from 121 to 196.
(a) Calculate the average percentage increase per annum during this period.
(b) Express I in the form be^{kt}, with $t = 0$ corresponding to January 1975.
(c) Assuming this growth rate continues, determine when I will reach 250.

17. Given that $\log 5 = 0.6990$ and $\log 11 = 1.0414$, find $\log_5 11$.

18. Given that $\log 3 = 0.4771$ and $\log 6 = 0.7782$, find $\log_3 6$.

(19–20) Given $\log 3 = 0.4771$ and $\log 6 = 0.7782$, find the following.

19. $\log_2 6$.   **20.** $\log_2 5$.

21. Given $\log 4 = 0.6021$, find $\log_4 10$.   **22.** Given $\log 7 = 0.8451$, find $\log_7 10$.

(23–26) If $x = \log_2 10$, express the following in terms of x.

23. $\log_5 10$.   **24.** $\log_4 10$.   **25.** $\log_{20} 10$.   **26.** $\log_{\sqrt{2}} 10$.

(27–30) If $z = \log_{12} 6$, express the following in terms of z.

27. $\log_2 6$.   **28.** $\log_{18} 6$.   **29.** $\log_{\sqrt{3}} 6$.   **30.** $\log_{1/3} \frac{1}{6}$.

(31–32) If $x = \log 2$ and $y = \log 3$, express the following in terms of x and y.

31. $\log_{12} 18$.   **32.** $\log_{\sqrt{6}} 9$.

33. Prove that
(a) $(\log_a b)(\log_b a) = 1$.
(b) $(\log_a b)(\log_b c)(\log_c a) = 1$.
(c) $(\log_a b)(\log_b c)(\log_c d)(\log_d a) = 1$.

34. Prove that
$$\log_{ab} x = \left(\frac{1}{\log_a x} + \frac{1}{\log_b x}\right)^{-1}.$$

35. Prove that $a^{px} = b^{qx}$, where
$$q = p \log_b a.$$
(This is the general base-change formula for exponential functions.)

(36–38) Prove the following identities.

36. $\log_3 x = \log_9 x^2$.   **37.** $\log_a t = \log_{a^n} t^n$.

38. $\log_a (1/x) = \log_{1/a} x$.

*39. Prove the base-change formula for logarithms given on p. 332. [*Hint:* Let $y = a^x$, $a = b^c$ so that $x = \log_a y$, $c = \log_b a$. Then prove that $y = b^{cx}$ and write it in logarithmic form.]

Chapter Review 8

1. Define and/or explain the following.
   (a) a^x where x is irrational.
   (b) The exponential function with base a.
   (c) Growing and decaying exponential functions.
   (d) The natural exponential function.
   (e) The compound-interest formula with compounding once per year or k times per year.
   (f) The logarithm of x with base a.
   (g) The relationship between logarithmic and exponential functions.
   (h) The properties of logarithms.
   (i) Common logarithms and natural logarithms.
   (j) Characteristic and mantissa of a common logarithm.
   (k) Decay constant and half-life of a radioactive isotope.
   (l) Continuous compounding of an investment.
   (m) The base-change formula for logarithms.

2. Describe how you would perform the following operations.
   (a) Find the common logarithm of a number from the tables.
   (b) Find the natural logarithm of a number from the tables.
   (c) Use common logarithms to:
      (1) multiply two numbers together.
      (2) divide one number by another.
      (3) take powers and roots of a number.
   (d) Solve an equation in which the unknown was in the exponent.
   (e) Change the function a^x to a natural exponential function.

Review Exercises on Chapter 8

(1–2) Construct the graphs of the following functions.

1. $y = 1 + 2(\tfrac{1}{4})^x$.

2. $y = 1 - (\tfrac{4}{3})^x$.

(3–6) Evaluate the following.

3. $e^{-1.75}$.

4. $e^{2.9}$.

5. $e^{-4.4} \div e^{-2.5}$.

6. $(e^{-1.65})(e^{2.4})$.

7. If $500 is invested at 7% interest per annum compounded annually, what is its value after seven years?

8. If, in Exercise 7, $100 is added to the investment after each year, calculate the new value after three years.

9. The sum of $1000 is borrowed at the rate of interest of 10%, compounded annually. The loan is to be repaid in two equal installments, at the end of one year and at the end of the second year. How much must the installments be?

10. Repeat Exercise 9 in the case where the loan is repaid in three equal annual installments.

11. The population of Britain in 1600 is believed to have been about 5 million. Three hundred fifty years later it had increased to 50 million. What was the average percentage growth per year during that period? (Assume a uniform exponential growth.)

12. If a population increases from 5 million to 200 million over a period of 200 years, what is the average percentage growth per year?

(13–14) Rewrite in logarithmic form.

13. $(\frac{1}{9})^{-3/2} = 27$.

14. $(\sqrt{2})^{-6} = \frac{1}{8}$.

(15–16) Verify the following by writing them in exponential form.

15. $\log_{4/3} (3\sqrt{3}/8) = -\frac{3}{2}$.

16. $\log_{1/2\sqrt{2}} 32 = -\frac{10}{3}$.

(17–22) Evaluate.

17. $\log_{\sqrt{27}} 81$.

18. $\log_{36} (1/\sqrt{6})$.

19. $9^{\log_3 (1/2)}$.

20. $16^{\log_8 27}$.

21. $\ln 0.0126$.

22. $\ln 204$.

(23–26) If $\log_{12} 2 = x$, express the following in terms of x.

23. $\log_{12} 3$.

24. $\log_{12} \sqrt{108}$.

25. $\log_{27} 12$.

26. $\log_2 6$.

(27–32) Use common logarithms to evaluate.

27. $(60.5)(0.0798)$.

28. $81.4 \div 92.0$.

29. $27(0.023)^{1/3}$.

30. $(0.57)^3 \div (0.32)^{1/2}$.

31. The value of n for which $(0.081)^n = 0.24$.

32. The value of p for which $(1.08)^p = (1.04)^{2p+3}$.

(33–36) Solve for x.

33. $\log_3 (x + 2) + \log_3 (2x + 7) = 3$.

34. $\log_2 x + \log_8 x = 2$.

35. $2^{x+1} = 3^{2-2x}$.

36. $(2^x)^x = 4^{1-x}$.

37. A population of bacteria is doubling every 25 minutes. How long does it take the population to increase by a factor of 10?

38. The half-life of radioactive carbon is 5570 years. How long does it take for 30% of a specimen to decay away?

39. A sum of $100 is invested at the nominal rate of interest of 12% per annum. How much is the investment worth after five years if compounded (a) annually, (b) quarterly, (c) continuously?

40. At what nominal rate of interest does money triple in value in 10 years if compounded continuously?

41. Population A is initially 20% of population B. Population A is growing at 4% per year and B is growing at 1.5% per year. After how many years will A be 40% of B?

42. The GNP of nation A increased from 0.5 to 1.1 billion dollars between 1970 and 1980.
   (a) Calculate the average percentage growth per annum.
   (b) Express the GNP at time t in the form be^{kt}.
   (c) Assuming this growth rate continues, calculate when the GNP will reach $1.5 billion.

43. The GNP of nation B during the same period as in Exercise 42 increased from 1.0 to 1.5 billion dollars.
   (a) Calculate the average percentage growth per annum for B.
   (b) Express the GNP in the form $b'e^{k't}$.
   (c) Calculate when nation A's GNP overtakes B's.

44. A new product was introduced onto the market at $t = 0$, and thereafter its monthly sales grew according to the formula
$$S = 4000(1 - e^{-kt})^3.$$
If $S = 2000$ when $t = 10$ (i.e., after 10 months), find the value of k.

45. A population is growing according to the formula

$$P = 5 \times 10^6 e^{0.06t}$$

where t is in years. Calculate the percentage growth per annum. How long does it take the population to increase by 50%?

46. A population has a size given by the formula

$$P = P_0 e^{kt}.$$

Find an expression for the percentage growth per unit of time and for the length of time necessary for the population (a) to double in size, (b) to triple in size.

47. The weight of a culture of bacteria is given by

$$y = \frac{2}{1 + 3(2^{-t})}$$

when t is measured in hours. What are the weights when $t = 0, 1, 2,$ and 4?

*48. A new improved strain of rice is developed. After t years the proportion of rice farmers who have switched to the new strain is found to be given by

$$p = [1 + Ce^{-kt}]^{-1}$$

At $t = 0$, 2% of farmers are using the new strain. Four years later, 50% are doing so. Evaluate C and k and calculate how many years it is before 90% have switched.

(49–50) Express the following functions in the form $y = be^{kx}$.

49. $y = 4000(1.09)^x$.

50. $y = P(1 + i)^x$.

(51–52) If $x = \log_3 12$, express the following in terms of x.

51. $\log_2 3$.

52. $\log_{\sqrt{6}} 3$.

POLYNOMIALS

9.1

Division of Polynomials

A *polynomial function* of x is an expression each of whose terms consists of a nonnegative integral power of x multiplied by a numerical coefficient. The highest power of x that appears is called the ***degree*** of the polynomial. A *rational function* is a quotient whose numerator and denominator are polynomials. Thus, for example,

$$\frac{2x^3 + x - 7}{x^2 + 4x}$$

is a rational function whose numerator has degree 3 and whose denominator has degree 2.

When the degree of the numerator is greater than or equal to the degree of the denominator, a rational function can be divided out using long division. The process is exactly analogous to long division in arithmetic. To remind yourself how this goes, work through the following example, in which $625 \div 23$ is simplified by long division.

$$\begin{array}{r} 27 \\ 23\,)\,\overline{625} \\ 46 \\ \overline{165} \\ 161 \\ \overline{4} \end{array}$$ ← $2 \times 23 = 46$
     ← subtract 46 from 62, then bring down 5
     ← $7 \times 23 = 161$
     ← subtract 161 from 165, giving remainder 4

We write the result of this calculation in the form

$$\frac{625}{23} = 27 + \frac{4}{23}.$$

We call 23 the denominator or *divisor* in the original fraction, 27 the **quotient**, and 4 the **remainder**. In general we have

$$\boxed{\frac{\text{NUMERATOR}}{\text{DIVISOR}} = \text{QUOTIENT} + \frac{\text{REMAINDER}}{\text{DIVISOR}}.} \qquad (1)$$

If we multiply both sides of this equation by the DIVISOR, we get the equivalent statement

$$\boxed{\text{NUMERATOR} = (\text{DIVISOR})(\text{QUOTIENT}) + \text{REMAINDER}.} \qquad (2)$$

In the arithmetic example above, this equation is

$$625 = (23)(27) + 4.$$

The truth of this equation is easily verified, and it provides a check on our long division. Equation (2) is often used as a check, as we shall see.

Now let us use long division to simplify rational functions.

Example 1 Simplify $\dfrac{2x^2 - x - 1}{x - 2}$ by long division.

Solution Here the numerator is $(2x^2 - x - 1)$ and the divisor is $(x - 2)$. The first step is to write the numerator and divisor side by side as in arithmetic long division. Then the x term in the divisor is divided into the highest-degree term $(2x^2)$ in the numerator. The result, $2x^2 \div x = 2x$, becomes the first term in the quotient:

$$\begin{array}{r} 2x\phantom{{}-x-1} \\ x - 2\,)\,\overline{2x^2 - x - 1} \\ 2x^2 \div x = 2x \end{array}$$

Next the divisor, $x - 2$, is multiplied by this $2x$ and the result, $2x^2 - 4x$, set out below the numerator. This $(2x^2 - 4x)$ must be subtracted from the numerator.

$$\begin{array}{r} \text{Multiply} 2x\phantom{{}-x-1} \\ x - 2\,)\,\overline{2x^2 - x - 1} \\ 2x^2 - 4x \end{array}$$

Sec. 9.1   *Division of Polynomials*

It is easier, however, if we change the signs and add it to the numerator. The $2x^2$ terms cancel and we are left with $3x$. The next term (-1) is brought down.

$$\begin{array}{r} 2x \\ x-2 \overline{\smash{\big)}\, 2x^2 - x - 1} \\ -\,2x^2 + 4x \\ \hline 3x - 1 \end{array}$$

Change signs
Add

Now the process is repeated. The x term in the divisor is divided into the highest-degree term remaining. The result, $3x \div x = 3$, becomes the next term in the quotient.

$$\begin{array}{r} 2x + 3 \\ x-2 \overline{\smash{\big)}\, 2x^2 - x - 1} \\ -2x^2 + 4x \\ \hline 3x - 1 \end{array}$$

$3x \div x = 3$

The divisor is now multiplied by this 3 and the result, $3x - 6$, set below. Its signs are changed (to $-3x + 6$) and it is added to the $3x - 1$. The result is 5.

$$\begin{array}{r} 2x + 3 \\ (x-2) \overline{\smash{\big)}\, 2x^2 - x - 1} \\ -2x^2 + 4x \\ \hline 3x - 1 \\ -3x + 6 \\ \hline 5 \end{array}$$

Multiply
Change signs
Add

Since this remainder has degree less than the degree of the divisor, the division terminates. The final quotient is $2x + 3$ and the remainder is 5. The result can be written in the form of Equation (1):

$$\frac{2x^2 - x - 1}{x - 2} = 2x + 3 + \frac{5}{x - 2}.$$

In order to check the result in Example 1, we write it in the form (2):

numerator = (divisor)(quotient) + remainder
$$2x^2 - x - 1 = (x - 2)(2x + 3) + 5.$$

We easily verify this by multiplying out the parentheses on the right.

Before starting a long-division calculation, we must arrange both numerator and divisor in descending powers of the variable—that is with highest powers on the left. If any powers are missing, they should be filled in with zero coefficients. *The division always terminates when we reach a remainder whose degree is smaller than that of the divisor.*

Example 2 Simplify $\dfrac{11 - 6x^3}{2x^2 + 3 - 2x}$ by long division.

Solution Rearranging the numerator and divisor in descending powers, we obtain

$$\dfrac{-6x^3 + 0x^2 + 0x + 11}{2x^2 - 2x + 3}.$$

Note the inclusion of the $0x^2$ and $0x$ terms to fill in the missing powers in the numerator.

The details of the division are set out below, with brief explanatory notes beneath.

$$
\begin{array}{r}
-3x - 3 \quad \leftarrow \text{quotient} \\
2x^2 - 2x + 3 \,\overline{)\,-6x^3 + 0x^2 + 0x + 11} \quad \leftarrow \text{numerator} \\
\underline{-6x^3 + 6x^2 - 9x} \quad \leftarrow \text{(A)} \\
-6x^2 + 9x + 11 \quad \leftarrow \text{(B)} \\
\underline{-6x^2 + 6x - 9} \quad \leftarrow \text{(C)} \\
3x + 20 \quad \leftarrow \text{remainder (D)}
\end{array}
$$

(A) $(-6x^3) \div (2x^2) = -3x$, which becomes the first term in the quotient. $(-3x)(2x^2 - 2x + 3) = -6x^3 + 6x^2 - 9x$.

(B) Change the signs in $(-6x^3 + 6x^2 - 9x)$, then add it to the original numerator, leaving $(-6x^2 + 9x + 11)$, the first remainder.

(C) $(-6x^2) \div (2x^2) = -3$, which becomes the second term in the quotient. $(-3)(2x^2 - 2x + 3) = -6x^2 + 6x - 9$.

(D) Change the signs in $(-6x^2 + 6x - 9)$, then add it to the first remainder, leaving $(3x + 20)$. This is the final remainder, since it has degree 1 whereas the divisor has degree 2.

We write the result as follows [see Equation (1)]:

$$\dfrac{11 - 6x^3}{2x^2 - 2x + 3} = -3x - 3 + \dfrac{3x + 20}{2x^2 - 2x + 3}.$$

As a check, we use the form (2):

numerator = (divisor)(quotient) + remainder
$$11 - 6x^3 = (2x^2 - 2x + 3)(-3x - 3) + (3x + 20).$$

We easily verify this by multiplying out the parentheses on the right.

SYNTHETIC DIVISION

When the divisor has the form $(x - c)$, where c is a constant, the process of long division can be shortened considerably. The method is called **synthetic division**. We first show how it works in terms of an example.

Sec. 9.1 *Division of Polynomials*

Example 3 Simplify $\dfrac{3x^3 - 5x^2 + 5}{x - 2}$ using synthetic division.

Solution We first set down the coefficients in the numerator in order of descending powers of x and place the constant c ($= 2$ here) to the left:

$$\underline{2\ |}\ \ 3\ \ -5\ \ 0\ \ 5$$

Note that the x term has zero coefficient. This must not be omitted. Next, missing a line, we copy the leading coefficient 3, then multiply 3 by c ($= 2$), placing the result beneath the -5:

$$\begin{array}{r|rrrr} 2 & 3 & -5 & 0 & 5 \\ \text{copy} & & 6 & & \\ \hline & 3 & 1 & & \end{array} \qquad \begin{array}{l} (3 \cdot 2 = 6) \\ \\ (-5 + 6 = 1) \end{array}$$

The second column is added to give 1. This 1 is multiplied by c ($= 2$) and the result placed under 0 in the third column. The third column is then added:

$$\begin{array}{r|rrrr} 2 & 3 & -5 & 0 & 5 \\ & & 6 & 2 & \\ \hline & 3 & 1 & 2 & \end{array} \qquad \begin{array}{l} (1 \cdot 2 = 2) \\ \\ (0 + 2 = 2) \end{array}$$

This third column sum (2) is multiplied by c ($= 2$) and the result placed below 5. The fourth column is then added:

$$\begin{array}{r|rrrr} 2 & 3 & -5 & 0 & 5 \\ & & 6 & 2 & 4 \\ \hline & 3 & 1 & 2 & 9 \end{array} \qquad \begin{array}{l} (2 \cdot 2 = 4) \\ \\ (5 + 4 + 9) \end{array}$$

In the last line, the final entry (9 in this case) gives the remainder. The other entries (3, 1, and 2) give the coefficients in the quotient:

$$\text{quotient} = 3x^2 + x + 2, \qquad \text{remainder} = 9.$$

(Note that when dividing by $x - c$, the quotient always has degree one less than the degree of the numerator. Here the numerator has degree 3, so the quotient has degree 2.) Thus we have

$$\frac{3x^3 - 5x^2 + 5}{x - 2} = 3x^2 + x + 2 + \frac{9}{x - 2}.$$

You should verify this result by writing it in the form (2) and multiplying out the right side.

Example 4 Simplify $\dfrac{2x^4 - 11x^2 + 4x}{x + 3}$ by synthetic division.

Solution The details are as follows, with explanatory notes below.

$$\begin{array}{r|rrrrr}
-3 & 2 & 0 & -11 & 4 & 0 \\
  &   & -6 & 18 & -21 & 51 \\
\hline
  & 2 & -6 & 7 & -17 & 51 \\
  & (A) & (B) & (C) & (D) &
\end{array}$$

First, $(x + 3)$ is of the form $x - c$ with $c = -3$.
(A) $(2)(-3) = -6$; $\quad 0 + (-6) = -6$
(B) $(-6)(-3) = 18$; $\quad -11 + 18 = 7$
(C) $(7)(-3) = -21$; $\quad 4 + (-21) = -17$
(D) $(-17)(-3) = 51$; $\quad 0 + 51 = 51$.

Thus, since the quotient has degree 3, one less than the numerator,

$$\text{quotient} = 2x^3 - 6x^2 + 7x - 17, \quad \text{remainder} = 51$$

and we have

$$\frac{2x^4 - 11x^2 + 4x}{x + 3} = 2x^3 - 6x^2 + 7x - 17 + \frac{51}{x + 3}.$$

We can illustrate the justification of synthetic division by setting out Example 3 as an ordinary long division.

$$\begin{array}{r}
3x^2 + x + 2 \\
x - 2 \overline{\smash{)}3x^3 - 5x^2 + 0x + 5} \\
\underline{3x^3 - 6x^2 } \\
x^2 + 0x \\
\underline{x^2 - 2x } \\
2x + 5 \\
\underline{2x - 4} \\
9
\end{array}$$

If we omit all the x's, retaining only the coefficients, this calculation can be abbreviated as follows (only the encircled signs are retained from now on):

$$\begin{array}{r}
3 1 2 \\
-2 \overline{\smash{)}3 -5 0 5} \\
\underline{-\,\textcircled{3} +6 } \\
1 0 \\
\underline{-\,\textcircled{1} +2 } \\
2 5 \\
\underline{-\,\textcircled{2} +4} \\
9
\end{array}$$

Sec. 9.1 Division of Polynomials

In this calculation the encircled numbers 3, 1, and 2 are simply repetitions of the numbers directly above them. And the numbers 0 and 5 indicated by arrows are repetitions of numbers in the original numerator. If we omit all these repeated numbers, we get the following array:

$$
\begin{array}{r|rrrr}
 & 3 & 1 & 2 & \\
-2\,)\,& 3 & -5 & 0 & 5 \\
 & & +6 & & \\ \hline
 & & 1 & & \\
 & & & +2 & \\ \hline
 & & & 2 & \\
 & & & & +4 \\ \hline
 & & & & 9
\end{array}
$$

This array can be compressed vertically, and we get

$$
\begin{array}{r|rrrr}
 & 3 & 1 & 2 & \\
-2\,)\,& 3 & -5 & 0 & 5 \\
 & & 6 & 2 & 4 \\ \hline
 & ③ & 1 & 2 & 9
\end{array}
$$

The extra encircled 3 has been copied from the top row. With this extra number, the bottom row actually duplicates the top row, plus one additional number, the remainder. Therefore we can omit the top row entirely, and we get

$$
\begin{array}{r|rrrr}
-2\,)\,& 3 & -5 & 0 & 5 \\
 & & 6 & 2 & 4 \\ \hline
 & 3 & 1 & 2 & 9
\end{array}
$$

We observe that in this array, each element in the middle row is formed by multiplying the preceding element in the bottom row by 2, as indicated by the arrows below. The bottom row is formed by adding the second row to the first.

$$
\begin{array}{r|rrrr}
+2\,)\,& 3 & -5 & 0 & 5 \\
 & & 6 & 2 & 4 \\ \hline
 & 3 \nearrow & 1 \nearrow & 2 \nearrow & 9
\end{array}
$$

If we drop the negative sign in front of the 2 in the divisor, we have exactly the method of synthetic division (see Example 3). In the last row, the final element (9) is the remainder and the other elements are the coefficients in the quotient (which in this case is $3x^2 + x + 2$).

Important Note The remainder must have degree less than the divisor. In the present case the divisor $(x - c)$ has degree 1. Therefore the remainder must be a polynomial of degree zero—that is, *the remainder must always be just a constant.*

Diagnostic Test 9.1

Fill in the blanks.

1. $\dfrac{\text{Numerator}}{\text{divisor}} = \underline{\hspace{2cm}} + \dfrac{\underline{\hspace{1cm}}}{\text{divisor}}$.

2. In the equation $\dfrac{2x}{x+1} = 2 - \dfrac{2}{x+1}$ the remainder is _____.

3. The degree of the remainder is always less than the degree of the _____.

4. When the divisor has the form $(x - c)$ the remainder is a _____.

5. For the following synthetic division array,

$$\begin{array}{r|rrrr} 3 & 3 & -10 & 0 & 6 \\ & & 9 & -3 & -9 \\ \hline & 3 & -1 & -3 & -3 \end{array}$$

the numerator = _____, divisor = _____, quotient = _____, and remainder = _____.

6. The synthetic division array for $(x^2 + 1) \div (x + 1)$ is _____. The quotient is _____ and the remainder is _____.

7. In the division $(x^2 + 3x + 2) \div (x + 2)$ the remainder is _____.

Exercises 9.1

(1–16) Simplify the following by long division.

1. $(x^2 - 5x + 7) \div (x - 2)$.
2. $(6x^2 + x - 4) \div (3x - 1)$.
3. $(t^2 + 1) \div (t - 1)$.
4. $t^2 \div (t + 1)$.
5. $(2x^2 + x - 1) \div (2x - 1)$.
6. $(12x^2 + 10x - 8) \div (3x + 4)$.
7. $(2x^2 + 1) \div (x^2 - x + 2)$.
8. $(x^3 + x) \div (x^3 - x^2 - 1)$.
9. $(x^3 + 2x^2 + 3x + 4) \div (x + 1)$.
10. $(2x^3 - x^2 + 2) \div (2x - 1)$.
11. $x^4 \div (x - 2)$.
12. $x^3 \div (x + 1)$.
13. $(6x^4 + 2x^3 - 2x) \div (2x^2 - 3)$.
14. $(2x^4 + x^3 - 7x^2) \div (x^2 + 1)$.
15. $(2x^4 - 3x^2 + x + 5) \div (2x^2 + x + 1)$.
16. $(4x^4 - 6x^3 + 2x - 1) \div (2x^2 + x - 2)$.

(17–32) Use synthetic division to simplify the following.

17. $(2x - 5) \div (x - 2)$.
18. $(6x + 5) \div (x - 5)$.
19. $(x^2 + 6x - 1) \div (x - 3)$.
20. $(2x^2 + x - 3) \div (x - 2)$.
21. $(2x^2 + 3x + 4) \div (x + 3)$.
22. $(x^2 + 7x - 11) \div (x + 2)$.
23. $(2x^3 + x - 7) \div (x - 2)$.
24. $(3x^3 + 2x^2 - 5) \div (x - 4)$.
25. $(x^3 + 7x^2 + x + 1) \div (x + 4)$.
26. $(x^3 + 1) \div (x - 2)$.
27. $x^4 \div (x + 2)$.
28. $(x^4 + x) \div (x + 3)$.
29. $(2x^4 + 4x^3 + x + 3) \div (x + 1)$.
30. $(x^4 - 3x^2 - x - 2) \div (x - 2)$.
31. $(x^5 + x^3 + x) \div (x - 2)$.
32. $(2x^5 + x^4 - 3) \div (x + 2)$.

9.2

Complex Numbers

ROOTS OF QUADRATIC EQUATIONS

The values of x at which a polynomial has zero value are called the **zeros** of the polynomial. The zeros of the quadratic function $ax^2 + bx + c$ (polynomial of degree 2) are therefore the solutions of the quadratic equation

$$ax^2 + bx + c = 0.$$

They are given by the quadratic formula (see page 000):

$$x = \frac{-b \pm \sqrt{b^2 - 4ac}}{2a}.$$

If $b^2 - 4ac > 0$, the quadratic function $ax^2 + bx + c$ has two zeros, which are distinct real numbers. If $b^2 - 4ac = 0$, there is only one zero (at $x = -b/2a$) and it is a real number. If $b^2 - 4ac < 0$, the quadratic function has no real zeros. The quantity $b^2 - 4ac$ is called the **discriminant** of the quadratic function $ax^2 + bx + c$.

Example 1 Investigate the zeros of the polynomial $x^2 - 2x + c$ for different values of c. Interpret the results in terms of the graph of the polynomial.

Solution Comparing $x^2 - 2x + c$ with the general form $ax^2 + bx + c$, we have $a = 1$ and $b = -2$. From the quadratic formula, therefore, the solutions of $x^2 - 2x + c = 0$ are given by

$$x = \frac{-(-2) \pm \sqrt{(-2)^2 - 4(1)c}}{2(1)}.$$

$$= \tfrac{1}{2}(2 \pm \sqrt{4 - 4c})$$

$$= \tfrac{1}{2}[2 \pm \sqrt{4(1 - c)}]$$

$$= 1 \pm \sqrt{1 - c}.$$

There are two real and distinct zeros if $1 - c > 0$—that is, if $c < 1$. If $c = 1$, there is only one distinct zero (at $x = 1$). If $1 - c < 0$ (i.e., if $c > 1$), there are no real zeros.

The graph of $y = x^2 - 2x + c$ is a parabola opening upward. We can find the position of the vertex by completing the square on the right (see Section 7.2). We have

$$y = x^2 - 2x + c$$
$$= (x - 1)^2 + (c - 1).$$

Therefore the vertex is at the point $x = 1$, $y = c - 1$.

The figure below shows the graph of $y = x^2 - 2x + c$ in the three cases, $c = 1$, $c < 1$, and $c > 1$. The zeros of the quadratic function $x^2 - 2x + c$ are the values of x where the graph crosses the x axis, since on the x axis $y = 0$—that is, $x^2 - 2x + c = 0$.

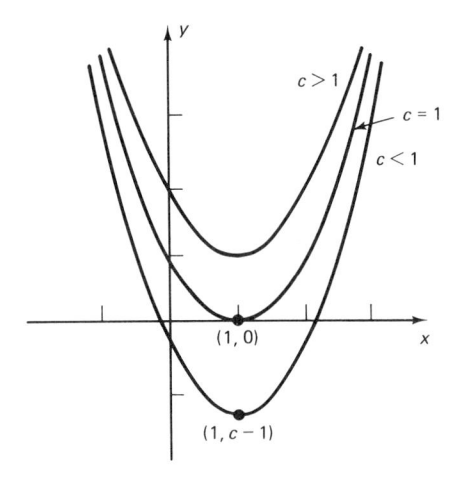

When $c < 1$, the parabola cuts the x axis at two distinct real points, giving two distinct real zeros. Let us imagine that c is increased toward 1. The parabola is shifted upward, and the zeros move closer together. When $c = 1$, the two zeros coincide: the parabola touches the x axis at the point $(1, 0)$ and there is only one distinct real zero, at $x = 1$. (We speak of this zero as a **double zero**, or a zero of **multiplicity** 2, since it corresponds to the coincidence of the two distinct roots.) When $c > 1$, the parabola does not intersect the x axis. In this case the quadratic has no real zeros.

The graphs in Example 1 are typical of the general quadratic function $y = ax^2 + bx + c$. The graph is a parabola that opens upward if $a > 0$ and downward if $a < 0$. When $b^2 - 4ac > 0$, the parabola cuts the x axis at two distinct points, so the quadratic function $ax^2 + bx + c$ has two distinct real zeros. When $b^2 - 4ac = 0$, the parabola touches the x axis (at its vertex) and the two zeros become coincident at this point. When $b^2 - 4ac < 0$, the parabola lies entirely on one side of the x-axis so there are no real zeros in this case.

The basic problem with the quadratic formula when $b^2 - 4ac < 0$ is that it involves the square root of a negative number. Such a square root cannot exist, of course, as a real number. Perhaps, however, it does exist as some other sort of number. In fact, let us suppose that square roots of negative numbers do exist, and let us call them *imaginary numbers*. We shall also suppose that many of the operations we are accustomed to perform with real numbers can also be performed with imaginary numbers. This will all be made more precise later on, but for the moment let us see where it takes us.

The imaginary number $\sqrt{-1}$ is called the **imaginary unit** and is denoted by i. If $i = \sqrt{-1}$, we have

$$i^2 = -1.$$

Any imaginary number can be expressed in terms of i. For example,

$$\sqrt{-4} = \sqrt{(4)(-1)} = \sqrt{4}\sqrt{-1} = 2i.$$

In general,*

$$\text{if } a > 0, \quad \sqrt{-a} = (\sqrt{a})i.$$

Example 2 Find the solutions of the following quadratic equations:

(a) $x^2 + 4 = 0.$
(b) $x^2 + 4x + 13 = 0.$

Solution (a) The given equation can be written

$$x^2 = -4.$$

Taking the square root of both sides, we get

$$x = \pm\sqrt{-4}$$
$$= \pm 2i.$$

The solutions are therefore the imaginary numbers $2i$ and $-2i$. Notice that there are two solutions.

(b) We could use the quadratic formula for the solutions. However, let us use completing the square instead:

$$x^2 + 4x = -13$$
$$x^2 + 4x + 4 = -13 + 4$$
$$(x + 2)^2 = -9$$
$$x + 2 = \pm\sqrt{-9} = \pm\sqrt{(9)(-1)}$$
$$= \pm 3i$$

Therefore,

$$x = -2 \pm 3i.$$

The roots are $-2 + 3i$ and $-2 - 3i$.

In Example 2(b) the roots have turned out to be neither real nor imaginary numbers, but rather each root is the sum of a real number and an imaginary number. Clearly, then, if imaginary numbers are to be of much use to us, we will have to include such combinations of real and imaginary numbers in our considerations. They are called **complex numbers**. Specifically, a complex number is a number of the form $p + qi$, where p and q are real

---

*The general rule $\sqrt{ab} = \sqrt{a}\sqrt{b}$ should be used only when $a \geq 0$ and $b \geq 0$. For example, $\sqrt{-4}\sqrt{-9} = (2i)(3i) = 6i^2 = -6$, whereas $\sqrt{(-4)(-9)} = \sqrt{36} = 6$.

numbers and $i = \sqrt{-1}$. For example, the following are all complex numbers:

$$2 - 3i, \quad -1 - i, \quad \tfrac{3}{4} + \tfrac{1}{2}i, \quad -1 + i, \quad 3 + 0i, \quad 0 - 2i.$$

The introduction of complex numbers allows us to make the following general statement:

The quadratic function $ax^2 + bx + c$, where a, b, c are real numbers ($a \neq 0$), always has exactly two zeros. If $b^2 - 4ac > 0$, the two zeros are distinct real numbers. If $b^2 - 4ac = 0$, the two zeros are coincident but still real: there is a zero of multiplicity 2 at $x = -b/2a$. (We count this double zero twice.) And if $b^2 - 4ac < 0$, there are two zeros, which are complex numbers.

COMPLEX NUMBERS

A complex number has the form $a + bi$, where a and b are two real numbers. The number a is called the **real part** and bi is called the *imaginary part.*

The plus sign in $a + bi$ is perhaps the most confusing thing about it. You should think of it as simply a connection, holding the real and imaginary parts together, not as an addition sign in the ordinary arithmetic sense. For example, think of the complex number $2 + 3i$ as being "2 real units and 3 imaginary units." The real and imaginary parts are different types of numbers and cannot be added in the ordinary sense; they are simply adjoined as two parts of the complex number.

This distinction between the two parts of a complex number is behind the rule for equality of complex numbers. Two complex numbers are *equal* if and only if their real parts are equal *and* their imaginary parts are equal. That is:

$$\boxed{a + bi = c + di \quad \text{if and only if} \quad a = c \quad \text{and} \quad b = d.}$$

For example, $2 + yi = x - 5i$ if and only if $x = 2$ and $y = -5$. The two complex numbers $a + i$ and $3 - 2i$ can never be equal, even for $a = 3$, since their imaginary parts are different.

We want to use complex numbers to solve quadratic equations (and later on other equations). To do so, we have to be able to add and multiply complex numbers together. This means that we need *rules* for adding and multiplying pairs of complex numbers. In fact, the rules permit us to add, multiply, or otherwise simplify complex numbers by treating i as if it were an ordinary variable and replacing i^2 by -1 wherever it appears.

First we consider the sum of two complex numbers, $(a + bi)$ and $(c + di)$. The rule is

$$\boxed{(a + bi) + (c + di) = (a + c) + (b + d)i.} \tag{1}$$

In other words, to add two complex numbers we add their two real parts and their two imaginary parts separately.

Examples 3 (a) $(2 + 5i) + (3 + i) = (2 + 3) + (5 + 1)i$
$$= 5 + 6i.$$
(b) $(-4 + 2i) + (2 - i) = (-4 + 2) + (2 - 1)i$
$$= -2 + 1i$$
$$= -2 + i.$$

The rule for forming the product of two complex numbers $a + bi$ and $c + di$ is as follows:

$$\boxed{(a + bi)(c + di) = (ac - bd) + (ad + bc)i.} \qquad (2)$$

This is the same result we would obtain by blindly applying the familiar algebraic methods to multiply out the parentheses on the right. For example, by using the method of arcs, we would obtain that

$$(a + bi)(c + di) = ac + a(di) + (bi)c + (bi)(di)$$
$$= ac + adi + bci + bdi^2$$
$$= ac + (ad + bc)i + bd(-1)$$
$$= (ac - bd) + (ad + bc)i.$$

Example 4 Evaluate $(2 - 5i)(6 + 2i)$.

Solution When evaluating products of complex numbers, it is generally better to multiply out the parentheses using the same methods you would for real expressions (e.g., the methods of arcs) rather than to rely on remembering Equation (2). We have

$$(2 - 5i)(6 + 2i) = (2)(6) + (2)(2i) - (5i)(6) - (5i)(2i)$$
$$= 12 + 4i - 30i - 10i^2$$
$$= 12 - 26i - 10(-1)$$
$$= 22 - 26i.$$

Example 5 For what (real) values of a and b is the following equation satisfied?
$$(a + bi)(2 + i) = (a + bi) + (2 + i).$$

Solution We simplify both sides of the equation, using the rules for addition and multiplication of complex numbers:

$$(a)(2) + (a)(i) + (bi)(2) + (bi)(i) = (a + 2) + bi + i$$
$$2a + ai + 2bi + bi^2 = (a + 2) + (b + 1)i$$
$$(2a - b) + (a + 2b)i = (a + 2) + (b + 1)i.$$

Now two complex numbers are equal if and only if their real and imaginary parts are equal. Thus we must have

$$2a - b = a + 2 \quad \text{and} \quad a + 2b = b + 1.$$

These simplify to
$$a - b = 2, \quad a + b = 1$$
and the solution is readily seen to be $a = \frac{3}{2}$, $b = -\frac{1}{2}$.

Note The rules (1) and (2) are the **axioms** (i.e., the basic laws) for complex numbers. *They cannot be proved*; they are simply *assumed*.

We would like to be able to say that a complex number whose imaginary part is zero is a real number. This statement is almost true. Let $R = \{a + 0i\}$ denote the set of all complex numbers with zero imaginary parts. Then we can define a one-to-one correspondence by letting the element $a + 0i$ in R correspond to the real number a. If we add the two members $(a + 0i)$ and $(c + 0i)$ in R, we get
$$(a + 0i) + (c + 0i) = (a + c) + 0i.$$
The sum belongs to the set R, since its imaginary part is zero. It corresponds to the real number $(a + c)$. Furthermore, the product of these two numbers in R, according to Formula (2), is given by
$$(a + 0i)(c + 0i) = ac + 0i.$$
Thus the product of $(a + 0i)$ and $(c + 0i)$ is still in R, and it corresponds to the real number ac.

We conclude, therefore, the the sum (or product) of any two numbers in R corresponds to the sum (or product) of the two corresponding real numbers. It is in this sense of one-to-one correspondence that $a + 0i$ can be identified with the real number a.

Example 6 Show that $x = -2 + 3i$ is a solution of the equation $x^2 + 4x + 13 = 0$ [see Example 2(b)].

Solution The coefficients 4 and 13 in the given equation must be reinterpreted as the complex numbers $4 + 0i$ and $13 + 0i$, respectively. For example, $4x$ makes no sense if x is a complex number, since the product of a real number and a complex number is not defined. However, $(4 + 0i)x$ is the product of two complex numbers and is well defined.

We have, with $x = -2 + 3i$,
$$\begin{aligned}
x^2 + 4x + 13 &= (-2 + 3i)^2 + (4 + 0i)(-2 + 3i) + (13 + 0i) \\
&= (-2 + 3i)(-2 + 3i) + (4)(-2) + (4)(3i) \\
&\quad + (0i)(-2) + (0i)(3i) + 13 + 0i \\
&= (-2)^2 + 2(-2)(3i) + (3i)^2 + (-8) + 12i \\
&\quad + 0i + 0 + 13 + 0i \\
&= 4 - 12i + 9i^2 - 8 + 12i + 0i + 0 + 13 + 0i \\
&= (4 - 9 - 8 + 0 + 13) + (-12i + 12i + 0i + 0i) \\
&= 0 + 0i.
\end{aligned}$$

Thus the equation is satisfied by the given value of x.

Note that in Example 6, the fact that we changed 4 to $4 + 0i$ and 13 to $13 + 0i$ made no difference to the end result. It meant only that we had to carry along a lot of extra 0 and $0i$ terms. In practice we would not bother with these unnecessary terms. For example, we would write
$$4x = 4(-2 + 3i) = 4(-2) + 4(3i) = -8 + 12i.$$
This gives the right answer, but keep in mind that the coefficient 4 is really the complex number $4 + 0i$.

We see, then, that a complex number $a + 0i$ with zero imaginary part can be denoted by the real number a without causing any error. Similarly a complex number $0 + bi$ with zero real part can be denoted simply by the imaginary number bi.

Let us use the product formula (2) to evaluate i^2. We have
$$i^2 = (0 + 1i)(0 + 1i).$$
Thus in (2) we have $a = 0, b = 1, c = 0, d = 1$, and so
$$i^2 = (0.0 - 1.1) + (0.1 + 1.0)i$$
$$= -1 + 0i.$$
Thus the "definition" $i^2 = -1$ with which we introduced imaginary numbers is really an abbreviation for the property $(0 + 1i)^2 = -1 + 0i$ that holds for complex numbers. It is all right to write $i^2 = -1$, provided we interpret this equation as a statement about complex numbers.

We have also the following properties of i:
$$i^3 = i^2 \cdot i = (-1)i = -i$$
$$i^4 = (i^2)^2 = (-1)^2 = 1$$
and continuing, we find $i^5 = i, i^6 = -1, i^7 = -i, i^8 = 1$, and so on. These equivalences are useful sometimes in simplifying complex numbers.

Example 7 Evaluate i^{57} and i^{83}.

Solution We note that $57 = (4)(14) + 1$ and so
$$i^{57} = i^{(4)(14)+1} = (i^4)^{14} \cdot i^1$$
$$= 1^{14} \cdot i = i.$$
Here we used the fact that $i^4 = 1$. This is why we expressed 57 as a multiple of 4 plus a remainder.

Similarly $83 = (4)(20) + 3$. Therefore
$$i^{83} = (i^4)^{20} \cdot i^3 = 1^{20} \cdot i^3 = i^3 = i^2 \cdot i = (-1)i = -i.$$

Example 8 Simplify $3i(7 + 13i - 2i^2) + 3(2 + 3i)(2 - 3i)$.

Solution
$$3i(7 + 13i - 2i^2) + 3(2 + 3i)(2 - 3i)$$
$$= 21i + 39i^2 - 6i^3 + 3(4 - 9i^2)$$
$$= 21i + 39(-1) - 6(-i) + 3[4 - 9(-1)]$$
$$= 21i - 39 + 6i + 12 + 27$$
$$= 0 + 27i$$
$$= 27i.$$

Subtraction of complex numbers is very easy to define.
$$(a + bi) - (c + di) = (a - c) + (b - d)i.$$
For example,
$$(3 + i) - (2 - 2i) = (3 - 2) + [1 - (-2)]i$$
$$= 1 + 3i.$$

On the other hand, division is not so easy. We can best do it using conjugate complex numbers.

CONJUGATE COMPLEX NUMBERS

DEFINITION If $(a + bi)$ is a complex number, its **conjugate** is the complex number $(a - bi)$. It is often denoted by $\overline{a + bi}$. Note that the conjugate is formed by reversing the sign of the imaginary part.

Example 9 Write down the conjugates of $2 + 3i$, $2 - 3i$, 5, $4i$.

Solution
$$\overline{2 + 3i} = 2 - 3i.$$
$$\overline{2 - 3i} = 2 - (-3i) = 2 + 3i.$$
$$\overline{5} = \overline{5 + 0i} = 5 - 0i = 5.$$
$$\overline{4i} = \overline{0 + 4i} = 0 - 4i = -4i.$$

We can extend the results of Examples 9 to certain general properties of conjugates:

The conjugate of a real number (i.e., a complex number with zero imaginary part) is the same as the number itself.

The conjugate of a purely imaginary number is the negative of that number.

The conjugate of the conjugate of a complex number is equal to the original number.

It can be seen that the sum of a complex number and its conjugate is equal to twice the real part of the number, hence is always a real number. Another useful property of conjugates is that *the product of a complex number and its conjugate is also always real*. We have

$$(a + bi)(a - bi) = a^2 - b^2 i^2 \quad \text{(difference-of-squares formula)}$$
$$= a^2 - b^2(-1).$$
$$(a + bi)(a - bi) = a^2 + b^2.$$

For example,
$$(3 + 4i)(3 - 4i) = 3^2 - 4^2 i^2$$
$$= 3^2 - 4^2(-1)$$
$$= 3^2 + 4^2$$
$$= 25.$$

We can use this property to divide one complex number by another. In order to simplify any fraction in which the denominator is a complex number, the method is to *multiply both numerator and denominator by the conjugate of the denominator.*

Examples 10 Express the following in the form $a + bi$:

(a) $\dfrac{1}{3 + 4i}$.

(b) $\dfrac{5 + 2i}{1 - 4i}$.

Solutions (a) Multiply numerator and denominator by $(3 - 4i)$, the conjugate of the denominator $(3 + 4i)$:

$$\frac{1}{3 + 4i} = \frac{3 - 4i}{(3 + 4i)(3 - 4i)}$$

$$= \frac{3 - 4i}{3^2 + 4^2}$$

$$= \frac{3 - 4i}{25}$$

$$= \frac{3}{25} - \frac{4}{25}i.$$

(b) Multiply numerator and denominator by $(1 + 4i)$:

$$\frac{5 + 2i}{1 - 4i} = \frac{(5 + 2i)(1 + 4i)}{(1 - 4i)(1 + 4i)}$$

$$= \frac{(5)(1) + (5)(4i) + (2i)(1) + (2i)(4i)}{1^2 + 4^2}$$

$$= \frac{5 + 20i + 2i + 8i^2}{17}$$

$$= \frac{-3 + 22i}{17}$$

$$= -\frac{3}{17} + \frac{22}{17}i.$$

THEOREM 1 The conjugate of the sum (or product) of two complex numbers is equal to the sum (or product) of the conjugates of the two numbers. That is,

(a) $\overline{(a + bi) + (c + di)} = \overline{(a + bi)} + \overline{(c + di)}$.
(b) $\overline{(a + bi)(c + di)} = \overline{(a + bi)}\,\overline{(c + di)}$.

Proof The proof of part (a) is almost trivial, so we prove part (b) only. Now

$$\overline{(a + bi)(c + di)} = \overline{(ac - bd) + (bc + ad)i}$$

$$= (ac - bd) - (bc + ad)i. \qquad \text{(i)}$$

Also

$$\overline{(a + bi)}\,\overline{(c + di)} = (a - bi)(c - di)$$

$$= ac - adi - (bi)c + (bi)(di)$$

$$= (ac - bd) - (ad + bc)i. \qquad \text{(ii)}$$

Clearly (i) and (ii) agree, and so we have proved part (b) of the theorem.

Diagnostic Test 9.2

Fill in the blanks.

1. A quadratic function has real zeros if and only if its _____ is _____.
2. The function $ax^2 + x + 1$ has complex zeros if $a >$ _____.
3. $i = \sqrt{-1}$ is called the _____.
4. The complex number $2 - 5i$ has real part _____ and imaginary part _____.
5. The complex number $a + bi$ is purely imaginary if _____.
6. $i^2 =$ _____, $i^3 =$ _____, $i^4 =$ _____, $i^5 =$ _____.
7. If n is any integer, then $i^{4n} =$ _____ and $i^{4n+1} =$ _____.
8. $(2 - 3i) + (-3 + 2i) =$ _____.
9. $x - 2i = 3 + yi$ if and only if $x =$ _____ and $y =$ _____.
10. $(a + bi)(c + di) =$ _____ + _____ i.
11. $i(2 - 3i) =$ _____.
12. $4 + 2i - 3i^2 + 5i^3 =$ _____.
13. $(2 + i)(2 - i) =$ _____.
14. The imaginary part of $(a + bi)(a - bi)$ is _____.
15. The complex conjugate of $-3 + 7i$ is _____.
16. The complex conjugate of $-3i$ is _____.
17. $(a + bi) - \overline{(a + bi)} =$ _____.
18. $(a + bi)\overline{(a + bi)} =$ _____.
19. $1/(1 + i) =$ _____ + _____ i.
20. $(a + bi)/(c + di) = (a + bi)($ _____ $)/(c^2 + d^2)$.

Exercises 9.2

(1–8) Find the solutions of the following quadratic equations.

1. $x^2 + 3 = 0$.
2. $x^2 + 7 = 0$.
3. $x^2 - 2x + 2 = 0$.
4. $x^2 + 2x + 3 = 0$.
5. $2x^2 + 6x + 5 = 0$.
6. $3x^2 - 5x + 3 = 0$.
*7. $x^2 + (4 - i)x + (4 - 2i) = 0$.
*8. $x^2 + (1 + 6i)x + (-9 + 3i) = 0$.

[*Hint:* For 7 and 8 use the quadratic formula.]

(9–12) Investigate the zeros of the following polynomials for different values of the constant c. Interpret the results in terms of the graph of the polynomial.

9. $x^2 + c$.
10. $x^2 - cx$.
11. $c - 4x - x^2$.
12. $x^2 + 4x + c$.

(13–42) Express the following complex numbers in the form $a + bi$.

13. $5 + (3 - 2i)$.
14. $4i - (2 - i)$.
15. $3 + i - (-1 + 4i)$.
16. $2 + 5i + (-4 - 3i)$.
17. $4 - 2i + (-2 - i)$.
18. $3(1 - 2i) + 2(2 + 2i)$.
19. $3 + 2i - 2i^2$.
20. $4 - i + 5i^2$.
21. $i(3 - i)$.
22. $2i(2 + 3i)$.
23. $(1 + i)^2$.
24. $\left(\dfrac{1}{\sqrt{2}} - \dfrac{1}{\sqrt{2}}i\right)^2$.
25. $(3 + 2i)(5 - i)$.
26. $(4 + i)(4 + 3i)$.
27. $(2 - i)(2 + 3i)$.
28. $(3 - 2i)(1 - 4i)$.
29. $\dfrac{3 + 2i}{i}$.
30. $\dfrac{4 - i}{2i}$.
31. $\dfrac{1}{1 + i}$.
32. $\dfrac{1}{2 - 3i}$.
33. $\dfrac{2 - 3i}{2 + 3i}$.
34. $\dfrac{4 + i}{4 - i}$.
35. $\dfrac{2 - 5i}{2 - 3i}$.
36. $\dfrac{4 - 3i}{1 - 2i}$.
37. $1 + 2i + 3i^2 + 4i^3$.
38. $3i - 5i^3 + 7i^5$.
39. $\dfrac{2 - 4i}{i^3}$.
40. $\dfrac{i^4 + 2i^6}{i^8}$.
41. $\dfrac{2i + 3i^2}{1 - i^3}$.
42. $\dfrac{2 + 3i^5}{i + 2i^4}$.

(43–46) Write down the conjugates of the following complex numbers.

43. $-2 - 4i$.
44. $-3 + 2i$.
45. $16i$.
46. 0.

47. Show that the cubes of the following complex numbers are all equal to 1:

$$1, \quad -\dfrac{1}{2} + \dfrac{\sqrt{3}}{2}i, \quad -\dfrac{1}{2} - \dfrac{\sqrt{3}}{2}i.$$

(These numbers are known as the three cube roots of unity.)

48. Show that the fourth powers of the following complex numbers are all equal to -1:

$$\dfrac{1}{\sqrt{2}} + \dfrac{1}{\sqrt{2}}i, \quad \dfrac{1}{\sqrt{2}} - \dfrac{1}{\sqrt{2}}i,$$
$$-\dfrac{1}{\sqrt{2}} + \dfrac{1}{\sqrt{2}}i, \quad -\dfrac{1}{\sqrt{2}} - \dfrac{1}{\sqrt{2}}i.$$

(49–52) For what values of a and b does the complex number $a + bi$ satisfy the following equations?

49. $3(a + bi) = (2 + 4i) - (a + bi)$.
50. $2i(a + bi) = i + (a + bi)$.
51. $(1 + i)(a + bi) = (1 + i) + (a + bi)$.
52. $(2 + 3i)(a + bi) = (2 + 3i) + (a + bi)$.

(53–58) Simplify the following.

53. $\dfrac{(1 + 2i)(3 - i)}{2 + i}$.
54. $\dfrac{(3 - 2i)(4 + i)}{3 + 2i}$.
55. $\dfrac{1}{(1 + 2i)^2}$.
56. $\dfrac{1 + \sqrt{3}\,i}{(1 - \sqrt{3}\,i)^2}$.
57. $\dfrac{3 + 4i}{1 + i} + \dfrac{2 - i}{1 - i}$.
58. $\dfrac{2 - 5i}{2 - i} - \dfrac{1 - 3i}{2 + i}$.

9.3

Factorization Theory

In this section we shall be concerned with certain properties of the zeros of polynomials. Recall that a zero of a polynomial is a value of the variable at which the polynomial is zero.

Let $f(x)$ be a given polynomial of degree n and consider the fraction

$f(x)/(x - c)$, where c is a constant. This fraction can be divided out, either by synthetic division or full long division, and let us suppose that the quotient obtained is $q(x)$ and the remainder is r. [The remainder is just a constant when dividing by $(x - c)$.] Then we have the usual equation

$$\frac{f(x)}{x - c} = \text{quotient} + \frac{\text{remainder}}{x - c} = q(x) + \frac{r}{x - c}$$

or, after multiplying both sides by $(x - c)$,

$$f(x) = (x - c)q(x) + r. \tag{1}$$

This equation is an identity; that is, it is true for all values of x. Let us put $x = c$. Then we obtain that

$$f(c) = (c - c)q(c) + r = 0 + r = r.$$

This result is called the **remainder theorem**. It can be stated as follows:

> If a polynomial $f(x)$ is divided by $(x - c)$, then the remainder is $f(c)$.

Substituting $r = f(c)$ into Equation (1), we obtain the algebraic form of this theorem:

$$f(x) = (x - c)q(x) + f(c).$$

Example 1 Verify the remainder theorem for $f(x) = x^3 - 3x^2 + 1$ and the divisor $(x + 2)$.

Solution We begin by carrying out the synthetic division to find the quotient and remainder:

$$\begin{array}{r|rrrr} -2 & 1 & -3 & 0 & 1 \\ & & -2 & 10 & -20 \\ \hline & 1 & -5 & 10 & -19 \end{array}$$

Thus $q(x) = x^2 - 5x + 10$ and $r = -19$. Next let us evaluate $f(c)$—that is, $f(-2)$:

$$f(-2) = (-2)^3 - 3(-2)^2 + 1 = -8 - 12 + 1 = -19.$$

Clearly $r = f(-2)$, so the theorem is verified.

Example 2 Use the remainder theorem to evaluate the remainder when $f(x) = 2x^3 + 4x - 5$ is divided by $(x + 1)$.

Solution The divisor $(x - c)$ becomes $(x + 1)$ when $c = -1$. Therefore, the remainder will be $f(c)$—that is, $f(-1)$:

$$\text{remainder} = 2(-1)^3 + 4(-1) - 5$$
$$= -11.$$

If $(x - c)$ is a *factor* of $f(x)$, then we will obtain no remainder on dividing out the fraction $f(x)/(x - c)$. This leads to the so-called **factor theorem**:

> $(x - c)$ is a factor of the polynomial $f(x)$ if and only if $f(c) = 0$.

This theorem is commonly used in factoring polynomials.

Example 3 Use the factor theorem to show that $(x - 3)$ is a factor of $2x^2 - 5x - 3$. Verify the result by factoring this quadratic.

Solution Let $f(x) = 2x^2 - 5x - 3$. By the factor theorem, $(x - 3)$ is a factor if and only if $f(3) = 0$. We have

$$f(3) = 2(3)^2 - 5 \cdot 3 - 3 = 18 - 15 - 3 = 0$$

and therefore $(x - 3)$ must be a factor of $f(x)$.

From the usual method of factoring quadratic polynomials,

$$(2x^2 - 5x - 3) = (x - 3)(2x + 1)$$

showing that $(x - 3)$ is indeed a factor.

Observe in this example that we can write

$$f(x) = 2x^2 - 5x - 3 = 2(x - 3)(x + \tfrac{1}{2}).$$

Therefore $(x + \tfrac{1}{2})$ is also a factor of $f(x)$. By the factor theorem, therefore, $f(-\tfrac{1}{2}) = 0$. This is easily verified:

$$f(-\tfrac{1}{2}) = 2(-\tfrac{1}{2})^2 - 5(-\tfrac{1}{2}) - 3 = \tfrac{1}{2} + \tfrac{5}{2} - 3 = 0.$$

Example 4 Use the factor theorem to determine whether $(x + 1)$, $(x - 1)$, and $(x - 2)$ are factors of $x^3 - 4x^2 + x + 6$.

Solution Let $f(x) = x^3 - 4x^2 + x + 6$. Then

$$f(-1) = (-1)^3 - 4(-1)^2 + (-1) + 6 = -1 - 4 - 1 + 6 = 0.$$

Therefore $(x + 1)$ is a factor.

$$f(1) = 1^3 - 4(1)^2 + 1 + 6 = 1 - 4 + 1 + 6 = 4 \neq 0.$$

Therefore $(x - 1)$ is not a factor.

$$f(2) = 2^3 - 4 \cdot 2^2 + 2 + 6 = 8 - 16 + 2 + 6 = 0.$$

Therefore $(x - 2)$ is a factor.

Check As a check, let us divide $f(x)$ by $(x - 2)$. By synthetic division or long division we obtain the quotient $(x^2 - 2x - 3)$. We can then write

$$x^3 - 4x^2 + x + 6 = (x - 2)(x^2 - 2x - 3)$$
$$= (x - 2)(x + 1)(x - 3)$$

where in the last step we factored the quadratic $x^2 - 2x - 3$ in the ordinary way. In the final expression both $(x - 2)$ and $(x + 1)$ are revealed as factors. There is also a third factor, $(x - 3)$, and it is easy to verify that $f(3) = 0$, as required by the factor theorem.

COMPLETE FACTORIZATION

In Examples 3 and 4 we saw two polynomials broken up into factors of the type $(x - c)$:
$$2x^2 - 5x - 3 = 2(x - 3)(x + \tfrac{1}{2})$$
$$x^3 - 4x^2 + x + 6 = (x - 2)(x + 1)(x - 3).$$

These are examples of what is called the **complete factorization** of a polynomial.

Example 5 Verify that $(x + 2)$ is a factor of
$$f(x) = 2x^4 - 38x^2 - 60x$$
and then completely factor this polynomial.

Solution We have
$$f(-2) = 2(-2)^4 - 38(-2)^2 - 60(-2) = 32 - 152 + 120 = 0.$$

Therefore, by the factor theorem, $(x + 2)$ is a factor of $f(x)$. The synthetic division array is as follows:

$$\begin{array}{r|rrrrr} -2 & 2 & 0 & -38 & -60 & 0 \\ & & -4 & 8 & 60 & 0 \\ \hline & 2 & -4 & -30 & 0 & 0 \end{array}$$

So the quotient is $2x^3 - 4x^2 - 30x$ (and the remainder is zero, as expected). Therefore
$$f(x) = (x + 2)(2x^3 - 4x^2 - 30x).$$
$$= 2x(x + 2)(x^2 - 2x - 15).$$

The remaining quadratic can be factored, and we obtain the complete factorization
$$f(x) = 2x(x + 2)(x + 3)(x - 5).$$

[The x factor is of the form $(x - c)$ with $c = 0$.]

Example 6 Completely factor the polynomial $f(x) = 2x^2 + x + 1$.

Solution In this case the quadratic cannot be factored by the usual trial-and-error methods. By the factor theorem, however, we know that $(x - c)$ is a factor if and only if $f(c) = 0$. So we should investigate the zeros of $f(x)$ to find the possible values of c. Using the quadratic formula, we have
$$2x^2 + x + 1 = 0$$
$$x = \frac{-1 \pm \sqrt{(-1)^2 - 4(2)(1)}}{2(2)}$$
$$= \frac{-1 \pm \sqrt{-7}}{4}$$
$$= -\frac{1}{4} \pm \frac{\sqrt{7}}{4}i.$$

Sec. 9.3 *Factorization Theory*

Thus the possible values of c are the complex numbers $-\frac{1}{4} \pm (\sqrt{7}/4)i$. The two factors $(x - c)$ are therefore

$$x - \left(-\frac{1}{4} + \frac{\sqrt{7}}{4}i\right) = x + \frac{1}{4} - \frac{\sqrt{7}}{4}i$$

and

$$x - \left(-\frac{1}{4} - \frac{\sqrt{7}}{4}i\right) = x + \frac{1}{4} + \frac{\sqrt{7}}{4}i.$$

We can easily verify, using the difference-of-squares formula, that

$$\left(x + \frac{1}{4} - \frac{\sqrt{7}}{4}i\right)\left(x + \frac{1}{4} + \frac{\sqrt{7}}{4}i\right) = \left(x + \frac{1}{4}\right)^2 - \left(\frac{\sqrt{7}}{4}i\right)^2$$

$$= x^2 + \frac{1}{2}x + \frac{1}{2}.$$

Therefore, we have the complete factorization

$$2x^2 + x + 1 = 2\left(x + \frac{1}{4} - \frac{\sqrt{7}}{4}i\right)\left(x + \frac{1}{4} + \frac{\sqrt{7}}{4}i\right).$$

It turns out that *every* polynomial can be completely factored into the product of a number of factors of the type $(x - c)$ and a constant. The proof of this result rests on a theorem called the fundamental theorem of algebra, which we shall simply quote here, because the proof requires some quite advanced techniques.

THEOREM 1 (FUNDAMENTAL THEOREM OF ALGEBRA)

If $f(x)$ is a polynomial of degree greater than zero, then $f(x)$ has at least one zero. That is, there is at least one value of x, which may be a real or complex number, for which $f(x) = 0$.

For example, if $f(x)$ has degree 1, then $f(x)$ has the form $ax + b$ for some constants a and b and there is one and only one zero, at $x = -b/a$. If $f(x)$ has degree 2, then there are two zeros given by the quadratic formula, and they may be real or complex numbers. They may also be coincident, in which case there is only one value of x at which $f(x) = 0$.

Let $f(x)$ be a polynomial of degree n. Let c_1 be such that $f(c_1) = 0$. (Such a c_1 exists by the above theorem.) Then by the factor theorem, $(x - c_1)$ is a factor of $f(x)$, and so we can write

$$f(x) = (x - c_1)q_1(x). \tag{2}$$

The quotient, $q_1(x)$, is also a polynomial and its degree must be $(n - 1)$, one less than the degree of $f(x)$.

But now we can apply the fundamental theorem to $q_1(x)$, provided that its degree $n - 1 > 0$. There exists a value c_2 such that $q_1(c_2) = 0$. But then $(x - c_2)$ must be a factor of $q_1(x)$, and we can write

$$q_1(x) = (x - c_2)q_2(x).$$

The quotient $q_2(x)$ is a polynomial whose degree is $(n-2)$, one less than the degree of $q_1(x)$.

Substituting for $q_1(x)$ into (2), we then have that for $n > 1$,

$$f(x) = (x - c_1)(x - c_2)q_2(x)$$

where $q_2(x)$ has degree $(n-2)$.

We can continue this process. At each stage we extract a factor of the type $(x - c)$ and reduce the degree of the quotient by 1. After n steps we will have extracted n such factors and will have reduced the degree of the quotient to zero. That is, the nth quotient q_n will be a constant. So we have

$$f(x) = (x - c_1)(x - c_2)(\ldots)(x - c_n)q_n. \qquad (3)$$

If we multiply out all the parentheses on the right in Equation (3), we will obviously get a polynomial of degree n. The x^n term will be obtained from the product of the x terms from all the factors $(x - c_1) \ldots (x - c_n)$, and therefore its coefficient will be q_n. Consequently we have the result that q_n *must equal the coefficient of x^n in $f(x)$*. This is called the **leading coefficient** in $f(x)$: it is the coefficient of the term of highest degree.

We have proved the following.

THEOREM 2 If $f(x)$ is a polynomial of degree $n > 0$ with leading coefficient a, then there exist n numbers $c_1, c_2, \ldots, c_n$ (which may be real or complex) such that

$$\boxed{f(x) = a(x - c_1)(x - c_2)(\ldots)(x - c_n).} \qquad (4)$$

The numbers $c_1, c_2, \ldots, c_n$ are zeros of $f(x)$. This is obvious when we substitute into the right side of Equation (4). For example, substituting $x = c_2$, we have

$$f(c_2) = a(c_2 - c_1)(c_2 - c_2)(\ldots)(c_2 - c_n) = 0$$

since the third factor here is zero. Moreover, these numbers $c_1, c_2, \ldots, c_n$ are the *only* zeros of $f(x)$. Because if x does not equal any of these numbers, every one of the factors $(x - c_1), (x - c_2), \ldots, (x - c_n)$ on the right of Equation (4) is nonzero and hence $f(x)$ must be nonzero, since it is equal to the product of nonzero factors.

It follows immediately that *any polynomial of degree n can have at most n distinct zeros*. If $c_1, c_2, \ldots, c_n$ are all different from one another, there will be exactly n zeros, but if some of them are equal, there will be less than n.

If a given factor $(x - c)$ occurs k times in the complete factorization (4), then we say that c is a **zero of multiplicity k** of $f(x)$. For example, suppose that

$$f(x) = 3x^2(x - 2)(x - 4)^4. \qquad (5)$$

The degree of f is 7, but there are only three distinct zeros, namely $x = 0, 2,$ and 4. The zero at $x = 0$ has multiplicity 2, that at $x = 2$ has multiplicity 1, and that at $x = 4$ has multiplicity 4.

Although there may be fewer distinct zeros than the degree of the polynomial, *if each zero is counted k times, where k is its multiplicity*, then the number of zeros is equal to the degree of the polynomial. For example, for the polynomial (5) above, if we count two zeros at $x = 0$, one at $x = 2$, and four at $x = 4$, the total number of zeros is seven, the same as the degree of $f(x)$.

Example 7 Give the complete factorizations of

(a) $f(x) = x^3 + 4x^2 + 4x$.
(b) $f(x) = 3x^7 - 3x^3$.

Identify all the zeros according to their multiplicities in each case.

Solution (a) We have

$$f(x) = x(x^2 + 4x + 4)$$
$$= x(x + 2)^2.$$

This is the complete factorization. There is a zero at $x = 0$ of multiplicity 1 and a zero of multiplicity 2 at $x = -2$.

(b) We have

$$f(x) = 3x^3(x^4 - 1)$$
$$= 3x^3(x^2 - 1)(x^2 + 1) \qquad \text{(difference of squares)}$$
$$= 3x^3(x - 1)(x + 1)(x^2 - i^2) \qquad \text{(since } i^2 = -1\text{)}$$
$$= 3x^3(x - 1)(x + 1)(x - i)(x + i).$$

From the complete factorization we see that there is a zero of multiplicity 3 at $x = 0$ and four zeros of multiplicity 1 at $x = 1, -1, i,$ and $-i$.

Diagnostic Test 9.3

Fill in the blanks.

1. The remainder theorem states that if a polynomial $f(x)$ is divided by $(x - c)$ the remainder is _____.

2. In the division $(x^2 + 3x + 3) \div (x + 2)$ the remainder is _____.

3. In the division $(x^3 + 4x^2 - 5x - 1) \div (x - 1)$ the remainder is _____.

4. For $f(x) = x^3 + 4x^2 + 4x$ we have $f(-2) = 0$ and therefore _____ is a factor of $f(x)$.

5. The factor theorem states that $(x - c)$ is a factor of a polynomial $f(x)$ if and only if _____.

6. $(x + 1)$ _____ a factor of $x^4 - x^3 + x^2 - x$.

7. If α, β are the zeros of $ax^2 + bx + c$, then the complete factorization of $ax^2 + bx + c$ is _____.

8. If α, β, γ are the zeros of $ax^3 + bx^2 + cx + d$, then $ax^3 + bx^2 + cx + d =$ _____.

9. A polynomial of degree n has _____ different zeros.

10. The complete factorization of $2x^2 - 5x + 2$ is _____.

11. If $f(x) = 3x^4(x - 2)^2(x - 3)$, then the zero 0 has multiplicity _____, 2 has multiplicity _____, and 3 has multiplicity _____.

12. If $f(x) = 2x(x^2 + 1)^2$, the zeros of $f(x)$ are _____ and they have multiplicities _____, respectively.

13. For $f(x) = 3x^2(2x^2 + 1)^2$ the complete factorization is $f(x) = $ _____.

Exercises 9.3

(1–6) Verify the remainder theorem for the following polynomials $f(x)$ and factors $(x - c)$.

1. $f(x) = x^2 - 3x - 10$; $(x + 2)$.
2. $f(x) = 3x^2 + 5x - 2$; $(x - \frac{1}{3})$.
3. $f(x) = x^3 - x^2 + x - 1$; $(x - 1)$.
4. $f(x) = 2x^3 - 5x^2 + 2x$; $(x - 2)$.
5. $f(x) = x^4 + x^3 + 2x^2 + x + 1$; $(x - i)$.
6. $f(x) = x^4 + 1$; $\left(x - \frac{1}{\sqrt{2}} + \frac{1}{\sqrt{2}}i\right)$.

(7–10) Use the remainder theorem to evaluate the remainder when the given $f(x)$ is divided by the given $(x - c)$.

7. $f(x) = 2x^4 + 3x + 2$; $(x + 2)$.
8. $f(x) = x^5 - 3x^3 + 3x$; $(x + 1)$.
9. $f(x) = x^{53} - 1$; $(x + 1)$.
10. $f(x) = 2x^{99} - 1$; $(x - 1)$.

(11–14) Use synthetic division to evaluate the stated $f(c)$.

11. $f(x) = 3x^4 - 2x^3 - 2x^2 + 3$; $f(2)$.
12. $f(x) = x^4 + 2x^3 - 4x + 1$; $f(-3)$.
13. $f(x) = 2x^5 - x^4 + x^2 - 4x + 3$; $f(-\frac{3}{2})$.
14. $f(x) = 3x^5 + x^4 + 4x^3 - 10x^2 + x + 4$; $f(\frac{2}{3})$.

(15–20) Use the factor theorem to determine whether $(x - 1)$ and/or $(x + 1)$ is a factor of the following polynomials.

15. $f(x) = x^8 - x^6 + x^2 - 1$.
16. $f(x) = x^9 + 1$.
17. $f(x) = x^{11} - 2x^6 + x$.
18. $f(x) = x^{13} + x^7 - x^6 - 1$.
19. $f(x) = x^{999} + 1$.
20. $f(x) = x^{50} - 1$.

(21–24) Determine whether the given $(x - c)$ is a factor of the given polynomial $f(x)$.

21. $(x - \frac{1}{2})$; $f(x) = 2x^3 + x^2 + 3x - 2$.
22. $(x + 2)$; $f(x) = x^3 - 3x^2 - 4x + 12$.
23. $(x + 3)$; $f(x) = x^3 + 5x^2 + 8x + 6$.
24. $(x - \frac{1}{3})$; $f(x) = 3x^3 - x^2 + 2x - 1$.

(25–30) Determine the values of the constant k for which the given $(x - c)$ is a factor of the given $f(x)$.

25. $(x - 1)$; $f(x) = x^3 + x^2 + 3x + k$.
26. $(x + 1)$; $f(x) = x^4 + x^2 + k$.
27. $(x + \frac{1}{2})$; $f(x) = 2x^3 + x^2 + kx + 1$.
28. $(x - 3)$; $f(x) = x^3 - kx^2 - 15$.
29. $(x - 1 - i)$; $f(x) = x^3 - 2x + k$.
30. $(x + 2i)$; $f(x) = x^3 + kx^2 + kx - 1$.

(31–38) Find all the zeros of the following polynomials. Completely factor each polynomial.

31. $x^2 - 3x - 2$.
32. $x^2 - 8x$.
33. $x^3 - 4x$.
34. $x^3 - 2x^2 + x$.
35. $x^4 - 4$.
36. $x^4 + x^2$.
37. $x^3 - 1$.
38. $x^3 + 1$.

(39–44) List the zeros of the following polynomials together with their multiplicities. Give also the leading coefficient of each polynomial.

39. $f(x) = (x - 1)^2(x + 1)$.
40. $f(x) = (x - 2)^2(x - 3)^3$.

41. $f(x) = (2x + 1)^2(x + 2)(x^2 + 3)^3$.

42. $f(x) = x^4(2x^2 - 3)^2$.

43. $f(x) = 2x^3(4x^2 - 3)^2$.

44. $f(x) = (x^2 + 1)^3(x^2 - 4)^2$.

(45–52) Show that the given $(x - c)$ is a factor of the given cubic polynomial $f(x)$. Hence completely factor $f(x)$.

45. $(x - 2); f(x) = x^3 - 3x - 2$.

46. $(x + 3); f(x) = 2x^3 + 5x^2 - 4x - 3$.

47. $(x + \tfrac{1}{2}); f(x) = 4x^3 - 4x^2 - 7x - 2$.

48. $(x - \tfrac{1}{4}); f(x) = 4x^3 - 29x^2 + 31x - 6$.

49. $(x - 1); f(x) = x^3 - 2x^2 + 2x - 1$.

50. $(x + 2); f(x) = x^3 + 4x^2 + 6x + 4$.

51. $(x - \tfrac{1}{2}); f(x) = 2x^3 + 7x^2 + 6x - 5$.

52. $(x + \tfrac{1}{3}); f(x) = 6x^3 + 11x^2 + 9x + 2$.

9.4

Polynomials with Real Coefficients

The results of Section 9.3 are true for polynomials whose coefficients are complex numbers as well as those with real coefficients. When the coefficients are all real numbers, certain further properties can be found.

Consider the zeros of the polynomial $f(x) = x^2 - 6x + 13$, which is a polynomial of degree 2 whose coefficients (1, −6, and 13) are all real numbers. From the quadratic formula we find that the zeros are

$$x = \frac{6 \pm \sqrt{6^2 - 4 \cdot 1 \cdot 13}}{2} = \frac{6 \pm \sqrt{-16}}{2} = 3 \pm 2i.$$

The important thing to note about these two zeros is that they are complex conjugates of one another.

This result is generally true: Any quadratic function with real coefficients either has two real zeros or one real zero of multiplicity 2 or two complex zeros that are complex conjugates of one another. We can easily prove the last part of this statement by using the quadratic formula (for the case when $b^2 - 4ac < 0$). It also follows, however, from the following general theorem.

THEOREM 1 Let $f(x)$ be a polynomial with real coefficients. If $p + qi$ ($q \neq 0$) is a complex zero of $f(x)$, then $p - qi$ is also a zero of $f(x)$. That is, *the complex zeros occur in conjugate pairs.*

Proof We shall prove this theorem only for the cubic polynomial

$$f(x) = ax^3 + bx^2 + cx + d$$

where a, b, c, and d are real. Since $f(p + qi) = 0$, we have

$$a(p + qi)^3 + b(p + qi)^2 + c(p + qi) + d = 0.$$

Taking the conjugate of both sides of this equation and using part (a) of Theorem 1 of Section 9.2, we have

$$\overline{a(p+qi)^3} + \overline{b(p+qi)^2} + \overline{c(p+qi)} + \bar{d} = 0.$$

From part (b) of the same theorem, then,

$$\bar{a}(\overline{p+qi})^3 + \bar{b}(\overline{p+qi})^2 + \bar{c}(\overline{p+qi}) + \bar{d} = 0.$$

But $\bar{a} = a$, $\bar{b} = b$, and so on, and $\overline{p+qi} = p - qi$. Therefore,

$$a(p-qi)^3 + b(p-qi)^2 + c(p-qi) + d = 0.$$

This proves that $(p - qi)$ is a zero of $f(x)$. The extension of this proof to higher-degree polynomials is straightforward.

Example 1 Find the polynomial of smallest degree with real coefficients that has zeros at $x = 2 + i$ and $x = 1$.

Solution Since the polynomial has real coefficients, it must also have a zero at $x = 2 - i$. By Theorem 1 of Section 9.3, therefore,

$$f(x) = a(x - 1)[x - (2 + i)][x - (2 - i)].$$

[If $f(x)$ had any more zeros than these three, it would be of higher degree.] The constant a cannot be determined on the basis of the information given. Multiplying out the parentheses, we get

$$f(x) = a(x - 1)[(x - 2)^2 - i^2]$$
$$= a(x - 1)(x^2 - 4x + 5)$$
$$= a(x^3 - 5x^2 + 9x - 5).$$

Since the complex roots occur in conjugate pairs, it follows that the total number of complex roots is always even. It is also true (although we shall not prove it) that the multiplicities of the conjugate zeros $p + qi$ and $p - qi$ are always equal. Thus when the complex zeros are counted according to their multiplicities, the total number of them is still even. Recall that the total number of zeros altogether is equal to the degree of the polynomial. This leads to the following result.

THEOREM 2 Let $f(x)$ be a polynomial of degree n with real coefficients. Then if n is even, the number of real zeros of $f(x)$ is also even (where each zero is counted according to its multiplicity). If n is odd, the number of real zeros of $f(x)$ is also odd.

COROLLARY A polynomial of odd degree having real coefficients always has at least one real zero.

The following examples illustrate cubic polynomials with various combinations of zeros. In each case the coefficients are real. The graphs are shown in each case.

Example 2 (a) The cubic
$$f(x) = x^3 - x$$
$$= x(x-1)(x+1)$$

has three distinct real zeros, at $x = -1, 0,$ and $+1$.

(b) The cubic
$$f(x) = x^3 - x^2 - x + 1$$
$$= (x-1)^2(x+1)$$

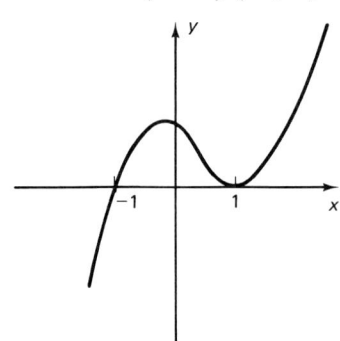

has two real zeros, one at $x = -1$ and one with multiplicity 2 at $x = 1$.

(c) The cubic
$$f(x) = x^3 - 3x^2 + 3x - 1$$
$$= (x-1)^3$$

has one real zero of multiplicity 3 at $x = 1$.

(d) The cubic
$$f(x) = x^3 - x^2 + x - 1$$
$$= (x-1)(x-i)(x+i)$$

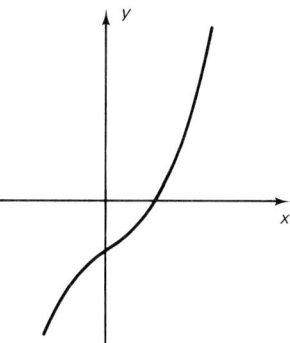

has one real zero at $x = 1$ and a pair of complex conjugate zeros at $x = \pm i$.

POLYNOMIALS WITH INTEGER COEFFICIENTS

We have said quite a lot about the qualitative properties of the zeros of polynomials but very little about how you can calculate them. As a matter of fact this is in general not very easy. Of course for degree 2 polynomials we have the quadratic formula, which enables us to calculate the zeros, but for higher-degree polynomials we do not know a formula for the zeros. Such formulas do exist for polynomials of degree 3 or 4, but they are not very convenient to use. No such formula exists for general polynomials of degree 5 or more.

Nonetheless, certain techniques can be used to find zeros of polynomials. One of these techniques relates to polynomials whose coefficients are integers. Such polynomials very often have one or more of their zeros as rational numbers, and such rational zeros can be found by trial and error. The method is based on the following theorem.

THEOREM 3

Let
$$f(x) = a_n x^n + a_{n-1} x^{n-1} + \cdots + a_1 x + a_0$$
be a polynomial of degree n whose coefficients $a_0, a_1, \ldots, a_n$ are integers. Suppose that $f(x)$ has a rational zero at $x = p/q$ where the fraction p/q is reduced to its lowest terms (i.e., p and q have no common factors and $q > 0$). Then p must be a factor of a_0 and q must be a factor of a_n.

A proof will be given later. First we shall give some examples illustrating the use of this theorem to locate zeros.

Example 3 Find the zeros of the polynomial
$$f(x) = 2x^3 - 3x^2 + 4x - 6.$$

Solution In the notation of Theorem 3 we have $n = 3$, and the coefficients are $a_3 = 2$, $a_2 = -3$, $a_1 = 4$, $a_0 = -6$.

If a rational zero p/q exists, then, according to the theorem, p must be a factor of a_0 and q must be a factor of a_3. That is, p must be a factor of -6 and q must be a factor of 2. Thus the possibilities are: $p = \pm 1, \pm 2, \pm 3,$ or ± 6 and $q = 1$ or 2. (We need consider $q > 0$ only.) Bearing in mind also that p and q can have no common factors, we are left with the following possible choices for $p, q,$ and x ($x = p/q$):

$$p = \pm 1, q = 1 \rightarrow x = \pm 1$$
$$p = \pm 1, q = 2 \rightarrow x = \pm \tfrac{1}{2}$$
$$p = \pm 2, q = 1 \rightarrow x = \pm 2$$
$$p = \pm 3, q = 1 \rightarrow x = \pm 3$$
$$p = \pm 3, q = 2 \rightarrow x = \pm \tfrac{3}{2}$$
$$p = \pm 6, q = 1 \rightarrow x = \pm 6.$$

What we must now do is evaluate $f(x)$ at each of these values of x. If $f(x)$ is zero for any one, then we have found a zero. If $f(x) \neq 0$ for every one of these values, then $f(x)$ has no rational zeros. We find the following values:

$$f(1) = -3, \quad f(-1) = -15, \quad f(\tfrac{1}{2}) = -\tfrac{9}{2}, \quad f(-\tfrac{1}{2}) = -9$$
$$f(2) = 6, \quad f(-2) = -42, \quad f(3) = 33, \quad f(-3) = -99$$
$$f(\tfrac{3}{2}) = 0, \quad f(-\tfrac{3}{2}) = -\tfrac{51}{2}, \quad f(6) = 342, \quad f(-6) = -570.$$

It follows that the only rational zero is at $x = \tfrac{3}{2}$.

To find the other zeros, we must divide $f(x)$ by $(x - \tfrac{3}{2})$. The quotient turns out to be $2x^2 + 4$. Therefore,

$$f(x) = (x - \tfrac{3}{2})(2x^2 + 4)$$
$$= 2(x - \tfrac{3}{2})(x^2 + 2)$$
$$= 2(x - \tfrac{3}{2})(x^2 - 2i^2)$$
$$= 2(x - \tfrac{3}{2})(x + \sqrt{2}\,i)(x - \sqrt{2}\,i).$$

The three zeros are therefore at $x = \tfrac{3}{2}, \sqrt{2}\,i,$ and $-\sqrt{2}\,i$.

In the last example we listed the values of $f(x)$ at all the possible rational zeros. In practice, however, we can usually avoid much of the work that this entails. In this example, for instance, it soon becomes clear that when x is negative, all the terms in $f(x)$ have negative values and therefore $f(x)$ cannot be zero. Thus we can avoid calculating $f(x)$ for any of the negative values of x. Also we might realize that when $x > 2$, the leading term $2x^3$ is always much bigger than all the rest of the terms, so again $f(x)$ cannot have any zeros greater than 2. Thus we are able to reduce the number of calculations very considerably.

Example 4 Find the zeros of the polynomial
$$f(x) = 3x^4 - 2x^3 + 3x^2 - 8x + 4.$$

Solution Suppose $x = p/q$ is a rational zero ($q > 0$). Then p must be a factor of 4 and q must be a factor of 3. The possibilities are $p = \pm 1, \pm 2$, or ± 4 and $q = 1$ or 3. Thus the possible values of x are $\pm 1, \pm \frac{1}{3}, \pm 2, \pm \frac{2}{3}, \pm 4$, or $\pm \frac{4}{3}$.

We note that when $x < 0$, all the terms in $f(x)$ are positive. Therefore $f(x)$ never vanishes when $x < 0$, and we need consider only the positive values of x. We find

$$f(1) = 0, \qquad f(2) = 32, \qquad f(4) = 660$$
$$f(\tfrac{1}{3}) = \tfrac{44}{27}, \qquad f(\tfrac{2}{3}) = 0, \qquad f(\tfrac{4}{3}) = \tfrac{92}{27}.$$

Thus there are two rational zeros, at $x = 1$ and $x = \frac{2}{3}$.

Using synthetic division twice to remove the factors $(x - 1)$ and $(x - \frac{2}{3})$, we find that

$$f(x) = (x - 1)(3x^3 + x^2 + 4x - 4)$$
$$= (x - 1)(x - \tfrac{2}{3})(3x^2 + 3x + 6)$$
$$= 3(x - 1)(x - \tfrac{2}{3})(x^2 + x + 2).$$

From the quadratic formula, we find that $x^2 + x + 2 = 0$ when $x = \frac{1}{2}(-1 \pm \sqrt{-7}) = \frac{1}{2}(-1 \pm \sqrt{7}\,i)$. Thus the zeros of $f(x)$ are $x = 1$, $x = \frac{2}{3}$, $x = -\frac{1}{2} + (\sqrt{7}/2)i$ and $x = -\frac{1}{2} - (\sqrt{7}/2)i$.

Proof of Theorem 3 We shall prove Theorem 3 for the cubic polynomial
$$f(x) = a_3 x^3 + a_2 x^2 + a_1 x + a_0.$$

The proof for general degree is along the same lines.

Remember that $a_0, a_1, a_2,$ and a_3 are integers. Let $x = p/q$ be a rational zero of $f(x)$, where p and q have no common factors. Then

$$f\!\left(\frac{p}{q}\right) = a_3\!\left(\frac{p}{q}\right)^3 + a_2\!\left(\frac{p}{q}\right)^2 + a_1\!\left(\frac{p}{q}\right) + a_0 = 0.$$

Multiply both sides of this equation by q^3:

$$a_3 p^3 + a_2 p^2 q + a_1 p q^2 + a_0 q^3 = 0. \tag{i}$$

We can write this as

$$a_3 p^3 = -a_2 p^2 q - a_1 p q^2 - a_0 q^3$$
$$= q(-a_2 p^2 - a_1 p q - a_0 q^2).$$

Clearly q is a factor of the right side and therefore must also be a factor of $a_3 p^3$. But q and p have no common factors. Therefore q must be a factor of a_3.

Similarly we can write (i) in the form

$$a_0 q^3 = p(-a_3 p^2 - a_2 pq - a_1 q^2)$$

and from this we deduce that p must be a factor of a_0. Thus the proof is complete.

DESCARTES' RULE OF SIGNS

Let a polynomial be written in order of descending powers. Then a variation in sign is said to occur whenever two adjacent terms have opposite signs, ignoring terms with zero coefficients. For example, the polynomial

$$3x^6 - x^4 - 2x^3 + 2x^2 + 4 \qquad (1)$$

has two variations in sign as indicated by the arrows.

Descartes' rule of signs allows us to obtain information on the zeros of a polynomial from the number of variations in sign. It is as follows:

> The number of positive real zeros of a polynomial with real coefficients is at most equal to the number of variations in sign. The difference between the number of positive zeros and the number of variations in sign is always even.

(Each positive zero is counted according to its multiplicity.)

For example, the above polynomial (1) has two variations in sign. Therefore it has either two positive zeros or none at all.

We can use the rule for negative zeros also by replacing x by $-x$ in the given polynomial and counting the variations in sign in the polynomial so obtained. For if the polynomial $f(x)$ has a certain number of negative zeros, then the polynomial $f(-x)$ has the same number of positive zeros. Again taking (1) as an example,

$$f(x) = 3x^6 - x^4 - 2x^3 + 2x^2 + 4$$
$$f(-x) = 3(-x)^6 - (-x)^4 - 2(-x)^3 + 2(-x)^2 + 4$$
$$= 3x^6 - x^4 + 2x^3 + 2x^2 + 4.$$

Since $f(-x)$ has two variations in sign, as shown, the original $f(x)$ has either two negative zeros or none.

Example 5 Use Descartes' rule of signs for the polynomial

$$f(x) = 2x^4 - x^3 - 3x^2 - 2x - 1.$$

Solution
$$f(x) = 2x^4 - x^3 - 3x^2 - 2x - 1$$

$$f(-x) = 2x^4 + x^3 - 3x^2 + 2x - 1.$$

$f(x)$ has one variation in sign, therefore there is exactly one positive zero. $f(-x)$ has three variations in sign, so the number of negative zeros of $f(x)$ is either three or one.

APPROXIMATE CALCULATION OF ZEROS

There exist several methods of computing approximate values of the zeros of polynomials. The most widely used is called Newton's method, which you will probably learn about when you take calculus. Other methods are simpler, and we shall describe one of these here. The method is called *successive linear interpolation*. It is less efficient than Newton's method but quite easy to understand. We shall illustrate it with an example.

Example 6 Compute an approximate value for the zero of the polynomial

$$f(x) = x^3 - 2x - 2$$

that lies between 1 and 2.

Solution First, how do we know there is a zero between 1 and 2? We have

$$f(1) = 1^3 - 2(1) - 2 = -3$$
$$f(2) = 2^3 - 2(2) - 2 = 2.$$

As x increases from 1 to 2, therefore, $f(x)$ changes from negative to positive, so somewhere between it must be zero.

The two points $(1, -3)$ and $(2, 2)$ lie on the graph of $y = f(x)$. In order to estimate where the graph crosses the x axis, we approximate it as a straight line through these two points (as in the adjacent figure). The slope of this line is

$$m = \frac{2 - (-3)}{2 - 1} = 5$$

and its equation (point-slope formula) is

$$y - (-3) = 5(x - 1)$$
$$y = 5x - 8.$$

At the point A where this crosses the x axis (see above figure) we have $y = 0$ —that is, $5x - 8 = 0$, $x = 1.6$. This gives us an approximation to the zero of $f(x)$. This method is called *linear interpolation*.

The approximation $x \approx 1.6$ cannot be expected to be very accurate, so we should carry the computation at least one stage further. We have in fact

$$f(1.6) = (1.6)^3 - 2(1.6) - 2 = -1.104.$$

Since $f(1.6)$ is still negative, the zero must in fact lie closer to 2. Let us try some neighboring values of x. We find

$$f(1.7) = -0.487, \quad f(1.8) = 0.232.$$

Thus $f(x)$ changes sign between 1.7 and 1.8, so the zero must lie in this interval. We can again use linear interpolation to estimate its position.

The calculation is illustrated in the figure on page 372. The two points $(1.7, -0.487)$ and $(1.8, 0.232)$ lie on the graph of $y = f(x)$, and between

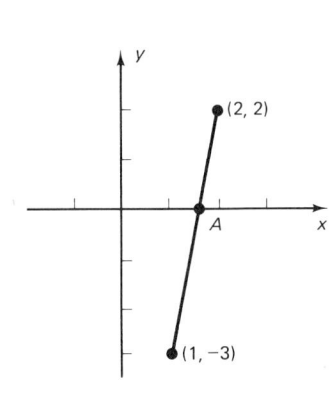

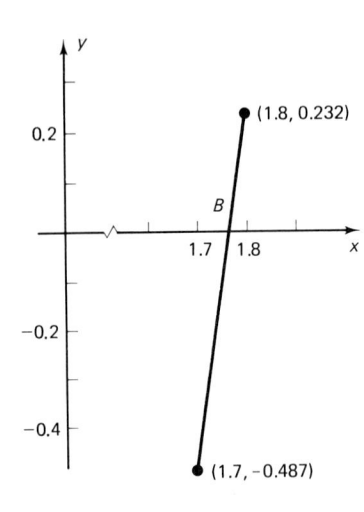

them the graph is approximated by a straight line. The slope is
$$\frac{0.232 - (-0.487)}{1.8 - 1.7} = 7.19$$
and its equation is
$$y - (-0.487) = 7.19(x - 1.7)$$
$$y = 7.19x - 12.71.$$

This crosses the x axis at B, where $y = 0$ and $x = 12.71/7.19 \approx 1.77$.

For a more accurate value of the zero we can repeat this procedure as many times as we desire. Let us run through one more cycle. We find that
$$f(1.77) = 0.00523, \quad f(1.76) = -0.06822.$$
Therefore the slope of the linear interpolation is
$$\frac{0.00523 - (-0.06822)}{1.77 - 1.76} = 7.345.$$
Its equation is
$$y - 0.00523 = 7.345(x - 1.77)$$
$$y = 7.345x - 12.9954.$$

When $y = 0$, we then have $x = 12.9954/7.345 = 1.7693$. This is quite accurate—you can check that $f(1.7693) = 0.000057$, which is quite close to zero.

In order to start the calculation of the zeros of a polynomial by this method, we need to find an interval in which the polynomial changes sign. This is usually done by trial and error. We would usually compute the values of the polynomial at a succession of integer values of x in order to locate the required interval.

Example 7 Find intervals that contain the zeros of the polynomial
$$f(x) = x^4 - 3x^3 - 2.$$

Solution We find
$$f(0) = -2, \quad f(1) = -4, \quad f(2) = -10, \quad f(3) = -2, \quad f(4) = 62.$$

Thus there is one zero between 3 and 4. As a matter of fact, since $f(4)$ is so large, the zero must be very close to 3. Let us compute $f(3.1)$. We find that $f(3.1) = 0.98$, and therefore the zero lies between 3 and 3.1. We can now proceed to calculate a more accurate value by using successive interpolations.

If $x \geq 4$, the x^4 term in $f(x)$ is much larger than the other two terms put together, so there can be no zero in this region. This same argument also applies for $x < -4$. In the negative region we have
$$f(0) = -2, \quad f(-1) = 2, \quad f(-3) = 160, \quad f(-4) = 446.$$

There is therefore a zero between -1 and 0. The first linear interpolation for this zero would be -0.5.

In using this method, it is obviously advisable to have a calculator. A programmable calculator is even better, because you need to repeat the calculation of $f(x)$ for a number of values of x.

The method of Example 7 is far from foolproof, so beware. Consider, for example, the function $f(x) = 8x^2 - 8x + 1$. Clearly $f(0) = 1$ and $f(1) = 1$, so you might be led to conclude that $f(x)$ has no zeros between 0 and 1. In fact, however, it has two zeros, at $x = (2 \pm \sqrt{2})/4 \approx 0.146$ and 0.854. Its graph is shown in the figure on the left below.

On the other hand, if $f(a)$ and $f(b)$ have opposite signs, certainly $f(x)$ has one zero between a and b, but it may have more than one. The two figures on the right indicate cases in which more zeros than one occur. In both figures, $f(a) < 0$ and $f(b) > 0$.

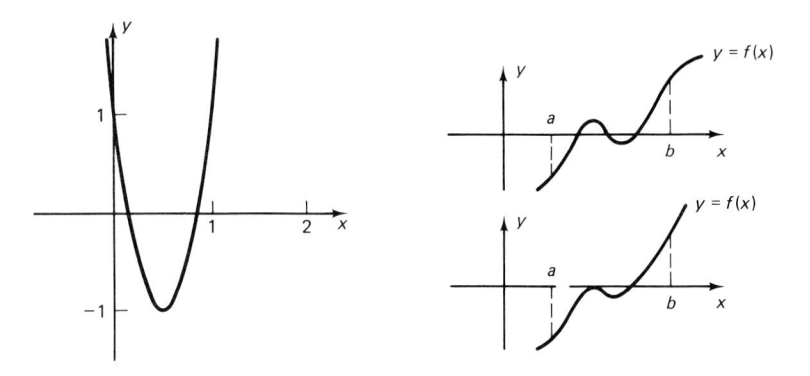

These difficulties arise particularly in cases where the zeros of $f(x)$ are close together.

The approximation techniques we have described can be used to find the zeros of functions other than simply polynomials—for example, of functions involving exponentials or logarithms.

We conclude with an application problem to illustrate the use of these techniques.

Example 8 A firm can sell all the units it can produce at a price of 14 per unit. From experience it is known that the cost of producing x units per day is given by

$$C(x) = x^3 - x^2 + 4x + 5.$$

Find the values of x at which the firm breaks even. [Each unit of production actually consists of 1000 items and prices and costs are in units of $1000.]

Solution The revenue from producing and selling x units at a price of 14 each is $R = 14x$. At the break-even point this revenue exactly equals $C(x)$, the cost of production. Thus we have the equation

$$C(x) = R$$
$$x^3 - x^2 + 4x + 5 = 14x$$
$$x^3 - x^2 - 10x + 5 = 0.$$

Sec. 9.4 Polynomials with Real Coefficients

The value of x at which the firm breaks even is thus a zero of the polynomial $f(x) = x^3 - x^2 - 10x + 5$.

First we observe that $f(x)$ has two variations in sign, so by Descartes' rule of signs there are either two or no positive zeros.

Next, computing the value of $f(x)$ at a few positive integers, we find $f(0) = 5, f(1) = -5, f(2) = -11, f(3) = -7, f(4) = 13$. Thus the positive zeros lie between 0 and 1 and between 3 and 4.

Consider first the zero between 3 and 4. The straight line joining the points $(3, -7)$ and $(4, 13)$ has equation $y = 20x - 67$. It crosses the x axis at $x = \frac{67}{20} = 3.35$, which is the first interpolation.

We find $f(3.35) = -2.127$. Then $f(3.5) = 0.625$. Interpolating once again, we obtain the line $y = 18.35x - 63.60$ and the zero $x \approx 3.466$.

Proceeding in a similar way with the other zero, we find a first interpolation of 0.5 and a second interpolation (between 0.5 and 0.45) of 0.488.

There are thus two break-even points, $x = 0.488$ and $x = 3.466$.

Diagnostic Test 9.4

Fill in the blanks.

1. If $2 + 3i$ is a zero of $f(x) = x^5 - 4x^4 + 13x^3 + x^2 - 4x + 13$, then _____ is also a zero of $f(x)$.

2. Given that $-3 + 4i$ is a zero of the polynomial $g(x)$ with real coefficients, then _____ is also a zero of $g(x)$.

3. If $x = p/q$ is a zero of $15x^5 + bx^3 + cx^2 + dx - 6$, then p is a factor of _____ and q is a factor of _____.

4. The polynomial $3x^5 - x + 1$ has _____ positive zeros.

5. The polynomial $5x^6 - 3x^2 + 7x - 2$ has _____ positive zeros.

6. If $f(2) = 6$ and $f(3) = -4$, then the first linear interpolation for the zero of $f(x)$ is $x =$ _____.

Exercises 9.4

(1–8) Find the polynomials with real coefficients that have the given zeros and the given degree.

1. $2 - i$, degree 2.
2. $3 + 4i$, degree 2.
3. $-2, 2 + 2i$, degree 3.
4. $1, 3 + 2i$, degree 3.
5. $1 + i, -1 + 2i$, degree 4.
6. $2i, 3 - 5i$, degree 4.
7. $0, -1, 2 + 3i$, degree 4.
8. $0, 3i, 2 - i$, degree 5.

(9–16) What is the smallest degree of a polynomial with real coefficients that has the following zeros?

9. $-1, 2 + 2i$.
10. $3, 4i, 2$.
11. $6i, -1 + i$.
12. $0, 3, 4, 3i$.
13. $0, 2, 2i, 4i$.
14. $2i, 2 + i, -2 + i$.
15. $1 + i, -1 + i, 1 - i$.
16. $2 - 3i, -2 - 3i, 2 + 3i, -2 + 3i$.

(17–20) Find the polynomials with real coefficients and smallest degree that have the given zeros.

17. $i, 1 + i$.   18. $0, 2i, 2 - i$.   19. $0, 2, -i, 2i$.   20. $-i, -2i, -3i$.

(21–34) Find all the zeros of the following polynomials.

21. $x^3 - 2x^2 + x - 2$.
22. $x^3 + 4x^2 + x + 4$.
23. $2x^3 - 7x^2 + 7x - 2$.
24. $2x^3 - 9x^2 + 12x - 4$.
25. $2x^3 + 3x^2 + 5x + 2$.
26. $4x^3 + 6x^2 + 4x + 1$.
27. $3x^3 + 2x^2 + 4x - 16$.
28. $4x^3 - 8x^2 + 9x - 9$.
29. $2x^4 - 11x^3 + 19x^2 - 13x + 3$.
30. $x^4 - 12x^3 + 52x^2 - 96x + 64$.
31. $2x^4 - 9x^3 + 5x^2 + 14x - 12$.
32. $8x^4 - 32x^3 + 22x^2 + 26x - 15$.
33. $8x^4 + 2x^3 + 15x^2 + 4x - 2$.
34. $x^4 - 4x^3 + 5x^2 - 14x + 24$.

(35–42) Use successive linear interpolations to find the zeros of the following polynomials in the given intervals correct to two decimal places. For Exercises 35–38 check your result by using the quadratic formula.

35. $f(x) = x^2 - 2x - 2$; between 2 and 3.
36. $f(x) = x^2 + 4x - 3$; between 0 and 1.
37. $f(x) = 2x^2 + 4x - 5$; between -2 and -3.
38. $f(x) = 3x^2 + x - 6$; between -1 and -2.
39. $f(x) = x^3 + 3x^2 - 4x - 6$; between 1 and 2.
40. $f(x) = x^3 - 8x + 2$; between 2 and 3.
41. $f(x) = x^4 + 3x^3 + 2x + 7$; between -2 and -3.
42. $f(x) = x^5 + 8x^2 - 2x + 20$; between -2 and -3.

(43–50) Find intervals between adjacent integers in which the following polynomials have zeros.

43. $f(x) = 2x^2 - 3x - 4$.
44. $f(x) = x^2 + 4x - 7$.
45. $f(x) = x^3 - 6x + 1$.
46. $f(x) = x^3 - 8x - 2$.
47. $f(x) = x^3 - 2x^2 - 5x + 2$.
48. $f(x) = x^3 - 4x^2 + 3x - 3$.
49. $f(x) = x^3 + 3x^2 + 5x + 1$.
50. $f(x) = x^3 - 4x^2 - 3x + 3$.

Chapter Review 9

Review the following topics.

(a) Long division of polynomials.
(b) Synthetic division.
(c) The terms divisor, quotient, and remainder.
(d) The terms real part and imaginary part of a complex number.
(e) The rule for equality of two complex numbers.
(f) The rules for addition and multiplication of complex numbers.
(g) The sense in which the real numbers are a subset of the complex numbers.
(h) The correct meaning of the equation $i^2 = -1$.
(i) The conjugate of a complex number.
(j) The method of division of one complex number by another.
(k) The remainder theorem.
(l) The factor theorem.
(m) The complete factorization of a polynomial.
(n) The multiplicity of a zero.
(o) The relationship between the number of zeros and the degree of a polynomial.
(p) Properties of the zeros when the polynomial has real coefficients.
(q) The method of finding the rational zeros when the polynomial has integer coefficients.
(r) The method of successive linear interpolation to approximate the real zeros.

Review Exercises on Chapter 9

(1–4) Simplify the following by long division.

1. $(x^3 - 3x^2 + 2) \div (x^2 - 3)$.
2. $(x^3 + 4x^2 - x - 1) \div (x^2 - 2x + 1)$.
3. $(6x^4 + 3x^3 - x - 4) \div (2x^2 + x + 2)$.
4. $(12x^4 - 3x^2 + x - 2) \div (3x^2 + 3x - 1)$.

(5–8) Use synthetic division to simplify the following.

5. $(x^3 + x - 5) \div (x - 4)$.
6. $(2x^3 - 3x^2 + x + 2) \div (x + 2)$.
7. $(2x^6 + 5x^4 + x^3 - 2) \div (x + 3)$.
8. $(3x^5 - x^4 + 2x^3 - x^2 - x + 4) \div (x - 2)$.

9. Sketch the graph of $f(x) = (x - 2)^2 + c$ for $c = 0$, $c > 0$, and $c < 0$. Show that for $c > 0$, $f(x)$ has two complex zeros, while for $c < 0$ it has two distinct real zeros.

*10. Sketch the graph of $f(x) = (x + 1)^2(x - 2) + c$ for $c = 0$. Deduce from this the form of its graph in the cases $c < 0$, $0 < c < 4$, $c = 4$, and $c > 4$. What can you say about the zeros of $f(x)$ in these various cases?

(11–18) Express the following in the form $a + bi$.

11. $(3 - 4i)(2 + 5i)$.
12. $(5 - i)(4 - 3i)$.
13. $(1 + i)(2 + i)(3 + i)$.
14. $(2 + i)(3 + 2i) + 5(2 - i)^2$.
15. $\dfrac{3 - i}{3 + 2i} + \dfrac{4}{3 - 2i}$.
16. $\dfrac{4 - 2i}{5 + 3i}$.
17. $\dfrac{2}{3 - 2i} + \dfrac{4}{3 + 2i}$.
18. $\dfrac{3 - 2i}{i(1 + 4i)} + 3i^3(1 - i)$.

(19–21) For what values of $a + bi$ are the following equations satisfied?

19. $i(a + bi) = (4 - 3i) + (a + bi)$.
20. $a + bi = 2 - i + \overline{(a + bi)}$.
21. $(a + bi)\overline{(a + bi)} = 2(a + bi) + 3$.

*22. Show that $(a + bi)^2 = p + qi$ if and only if $a^2 - b^2 = p$ and $2ab = q$. Hence by eliminating b show that $a^2 = \tfrac{1}{2}(p + \sqrt{p^2 + q^2})$. Therefore $b^2 = \tfrac{1}{2}(-p + \sqrt{p^2 + q^2})$. Hence find $(a + bi)$ such that $(a + bi)^2 = 3 + 4i$. We can regard $a + bi$ as the square root of $p + qi$.

(23–24) Evaluate the remainder when the given $f(x)$ is divided by the given $(x - c)$.

23. $f(x) = x^{50} - 2x^4 + 3$; $(x + 1)$.
24. $f(x) = x^{32} - 4x^{30} + 3x - 1$; $(x - 2)$.

(25–26) Determine whether the given $(x - c)$ is a factor of the given $f(x)$.

25. $(x + 1)$; $f(x) = x^{500} - 2x^{250} + 1$.
26. $(x - 2)$; $f(x) = x^{10} - 4x^9 + 4x^8 + x^3 - 3x + 2$.

(27–28) Determine the values of k for which the given $(x - c)$ is a factor of the given $f(x)$.

27. $(x + 3)$; $f(x) = x^3 - kx^2 - 2x - 4$.
28. $(x - \tfrac{1}{2})$; $f(x) = kx^5 - 3x + 4$.

(29–36) Completely factor the following polynomials.

29. $x^4 + 2x^3 + 5x^2$.
30. $x^7 - x^3$.
31. $4x^3 - 4x^2 - x - 3$.
32. $4x^4 - 5x^2 - 9$.
33. $6x^3 + 11x^2 + 9x + 2$.
34. $2x^3 - 3x^2 - 4$.
35. $6x^4 - 13x^3 + 5x^2 - 8x + 4$.
36. $6x^4 - 5x^3 - 38x^2 - 5x + 6$.

37. Find a polynomial with real coefficients of smallest degree that has a zero of multiplicity 2 at $(2 + 3i)$.

(38–40) Find the polynomials of smallest degree with real coefficients that have the following zeros.

38. $2, 1 + i$. **39.** $0, 3, 2i$. **40.** $2 + 4i, 3i$.

(41–42) Find all the zeros of the following polynomials.

41. $6x^3 + 19x^2 + x - 6$. **42.** $9x^4 + 15x^3 - 20x^2 - 20x + 16$.

(43–44) Use successive linear interpolations to find the zeros of the given $f(x)$ in the given intervals correct to two decimal places.

43. $f(x) = 2x^3 - 8x^2 + x + 6$; between 3 and 4. **44.** $f(x) = 2 + 4x - x^3$; between -1 and -2.

45. Locate all the zeros of $f(x) = x^3 - 3x - 1$ correct to two decimal places.

10 TOPICS IN ALGEBRA

10.1

Arithmetic Progressions

A *sequence* is an ordered list of numbers. For example the two lists

(a) 2, 5, 8, 11, 14, 17

(b) 3, 6, 12, 24, 48, ...

are examples of sequences. In sequence (a) the *first term* is 2, the *second term* is 5, and so on. Each term is obtained by adding 3 to the preceding term. In sequence (b) the first term is 3, the fourth term is 24, and so on, and any term can be obtained by doubling the preceding term. Sequences of these types arise in many problems, particularly in the mathematics of finance.

A sequence is said to be *finite* if it contains a limited number of terms—that is, if it has a last term. If it has no last term, it is called an *infinite* sequence. Sequence (a) is a finite sequence, its last term being 17. The ellipsis in sequence (b) is used to indicate that this sequence continues indefinitely.

The first, second, third, and subsequent terms in a sequence will be denoted by T_1, T_2, T_3, and so on. Thus T_7 denotes the seventh term, T_{10} the tenth term, and T_n the nth term. The following sequence illustrates this notation:

$$2, \quad 4, \quad 6, \quad 8, \quad \ldots, \quad 2n, \quad \ldots$$
$$\uparrow \quad \uparrow \quad \uparrow \quad \uparrow \qquad\qquad \uparrow$$
$$T_1 \quad T_2 \quad T_3 \quad T_4 \qquad\qquad T_n$$

The nth term of a sequence is commonly called its ***general term***.

Suppose a man borrows a sum of $5000 from the bank at 1% interest per month. He agrees to pay each month $200 toward the principal plus the interest on the balance. At the end of the first month he pays $200 plus the interest on $5000 at 1% per month, which is $50. Thus his first payment is $250, and he owes only $4800 to the bank. At the end of the second month he pays $200 toward the principal plus the interest on $4800, which at 1% for one month is $48. Thus his second payment is $248. Continuing in this way, his successive payments (in dollars) are

$$250, \quad 248, \quad 246, \quad 244, \quad \ldots.$$

This sequence is an example of an arithmetic progression.

DEFINITION A sequence is said to be an ***arithmetic progression*** (***A.P.***) if the difference between any term and the preceding one is the same throughout the sequence. The algebraic difference between each term and the preceding one is called the ***common difference*** and is denoted by d.

The above sequence (of payments) is in A.P. because the difference between any term and its preceding term is -2. This A.P. has 250 as its first term and $-2 (= 248 - 250)$ as its common difference. Similarly the sequence

$$2, \quad 5, \quad 8, \quad 11, \quad 14, \quad \ldots$$

is an A.P. with first term 2 and common difference 3.

If T_n denotes the nth term in an arithmetic progression, then $T_2 - T_1 = d$, $T_3 - T_2 = d$, and so on. In general, for any n,

$$\boxed{T_{n+1} - T_n = d.}$$

THE nTH TERM OF AN A.P.

If a is the first term and d the common difference of an A.P., then the successive terms of the A.P. are

$$a, \quad a + d, \quad a + 2d, \quad a + 3d, \quad \ldots.$$

The nth term is given by the formula

$$\boxed{T_n = a + (n - 1)d.} \tag{1}$$

Sec. 10.1   *Arithmetic Progressions*

For example, setting $n = 1, 2, 3, \ldots$ in this formula, we find
$$T_1 = a + (1-1)d = a$$
$$T_2 = a + (2-1)d = a + d$$
$$T_3 = a + (3-1)d = a + 2d$$
$$T_4 = a + (4-1)d = a + 3d$$

and in general
$$T_n = a + (n-1)d.$$

Equation (1) contains four numbers, a, d, n, and T_n. If any three of them are given, we can find the fourth.

Example 1 Find (a) the fifteenth term (b) the nth term of the sequence
$$1, \ 5, \ 9, \ 13, \ \ldots.$$

Solution The given sequence is an A.P. because
$$5 - 1 = 9 - 5 = 13 - 9 = 4.$$

Thus the common difference is 4; that is, $d = 4$. Also $a = 1$.
(a) Using Equation (1) above with $n = 15$,
$$T_{15} = a + (15-1)d = a + 14d$$
$$= 1 + (14)(4) = 57.$$
(b) $T_n = a + (n-1)d$
$$= 1 + (n-1)4 = 4n - 3.$$

Thus the fifteenth term is 57 and the nth term is $4n - 3$.

Example 2 George's monthly payments to the bank toward his loan are in A.P. If his sixth and tenth payments are $345 and $333, what will be his fifteenth payment?

Solution Let a be the first term and d the common difference of the monthly payments. Then successive payments (in dollars) are
$$a, \ a + d, \ a + 2d, \ \ldots.$$

Since the sixth and tenth payments (in dollars) are 345 and 333, $T_6 = 345$ and $T_{10} = 333$. Using Equation (1) for $n = 6$ and $n = 10$,
$$T_6 = a + 5d = 345 \quad \text{(given)} \qquad \text{(i)}$$
$$T_{10} = a + 9d = 333 \quad \text{(given)} \qquad \text{(ii)}$$

Subtracting (i) from (ii), we get
$$4d = 333 - 345 = -12$$
$$d = -3.$$

Substituting this value of d in (i), we obtain
$$a - 15 = 345 \quad \text{or} \quad a = 360.$$
Now
$$T_{15} = a + 14d$$
$$= 360 + 14(-3) = 318.$$

Thus his fifteenth payment to the bank will be $318.

SUM OF n-TERMS OF AN A.P.

We wish to find an expression for the sum of the first n terms of an A.P. Consider the following example with five terms:

$$S = 2 + 5 + 8 + 11 + 14$$
Also: $\quad S = 14 + 11 + 8 + 5 + 2 \quad$ (same sequence written backwards)
Add: $\quad 2S = 16 + 16 + 16 + 16 + 16 = 5(16).$

When the same sequence is written backward and added term-by-term to the original sequence, each pair of terms produces a sum of 16. We then obtain $S = 5(16)/2 = 40$.

This method works in general. Consider the A.P.
$$a, \quad a+d, \quad a+2d, \quad \ldots.$$

If the sequence contains n terms and if l denotes the last term (i.e., the nth term), then
$$l = a + (n-1)d. \tag{2}$$

The next to last term will be $(l - d)$, the second from last term will be $(l - 2d)$, and so on. Let S_n denote the sum of all these n terms. Then
$$S_n = a + (a+d) + (a+2d) + \cdots + (l-2d) + (l-d) + l. \tag{3}$$
Also:
$$S_n = l + (l-d) + (l-2d) + \cdots + (a+2d) + (a+d) + a \tag{4}$$
(same sequence written backward). Adding (3) and (4) term by term vertically, we get
$$2S_n = (a+l) + (a+d+l-d) + (a+2d+l-2d)$$
$$\quad + \cdots + (l-d+a+d) + (l+a)$$
$$= (a+l) + (a+l) + \cdots + (a+l) + (a+l) \quad (n \text{ terms})$$
$$= n(a+l).$$
Thus
$$S_n = \frac{n}{2}(a+l). \tag{5}$$

Substituting the value of l from (2) into (5), we get

$$S_n = \frac{n}{2}[a + a + (n-1)d]$$

or

$$S_n = \frac{n}{2}[2a + (n-1)d].$$

These results can be summarized as below:

THEOREM 1 The sum of n terms of an A.P. whose first term is a and the common difference d, is given by

$$\boxed{S_n = \frac{n}{2}[2a + (n-1)d]}$$

or

$$\boxed{S_n = \frac{n}{2}(a + l) \\ \text{where } l = a + (n-1)d.}$$

Example 3 Find the sum of the progression $2 + 5 + 8 + 11 + 14 + \cdots$ to 20 terms.

Solution The given sequence is an A.P. because

$$5 - 2 = 8 - 5 = 11 - 8 = 14 - 11 = 3.$$

Thus, the common difference is 3; that is, $d = 3$. Also $a = 2$ and $n = 20$. Therefore,

$$S_n = \frac{n}{2}[2a + (n-1)d]$$

$$S_{20} = \frac{20}{2}[2(2) + (20-1)3]$$

$$= 10(4 + 57) = 610.$$

Example 4 A man borrows $5000 from the bank at 1% per month. Each month he pays back $200 toward the principal plus the monthly interest on the outstanding balance. How much will his total payments amount to by the time he has repaid the loan?

Solution As discussed at the beginning of this section, the sequence of repayments is

$$250, \quad 248, \quad 246, \quad \ldots.$$

These form an A.P. with $a = 250$ and $d = -2$. Since $200 of principal is repaid each month, the total number of payments is $n = 5000/200 = 25$. The last term is therefore

$$l = T_{25} = a + 24d = 250 + 24(-2) = 202.$$

The total payment made is given by the sum of all 25 terms:

$$S_n = \frac{n}{2}(a + l) = \frac{25}{2}(250 + 202) = 5650.$$

The total amount paid to the bank will be $5650, which means that the interest paid will amount to $650.

Example 5 A man agrees to pay an interest-free debt of $5800 in a number of installments, each succeeding installment exceeding the previous one by $20. If the first installment is $100, find how many installments will be necessary to wipe out the loan completely.

Solution Since the first installment is $100 and each succeeding installment increases by $20, the various installments (in dollars) are

$$100, \quad 120, \quad 140, \quad 160, \quad \ldots.$$

These numbers form an arithmetic progression with $a = 100, d = 20$. Let n installments be necessary to wipe out the loan of $5800 completely. Then the sum of the first n terms of the above sequence must equal 5800; that is, $S_n = 5800$. Using the formula for the sum of an A.P., we get

$$S_n = \frac{n}{2}[2a + (n - 1)d]$$

or

$$5800 = \frac{n}{2}[200 + (n - 1)20]$$

$$= \frac{n}{2}(20n + 180)$$

$$= 10n^2 + 90n.$$

Dividing throughout by 10 and rearranging, we get

$$n^2 + 9n - 580 = 0$$
$$(n - 20)(n + 29) = 0.$$

This gives $n = 20$ or $n = -29$. Since a negative value of n is not permissible, we have $n = 20$. Thus, 20 installments will be necessary to wipe out the loan completely.

Example 6 The nth term of a sequence is given by the formula $T_n = 5n + 15$. Find the sum of the first 12 terms.

Solution First let us examine the first few terms. Putting $n = 1, 2, 3$, and so on, we find $T_1 = 5 \cdot 1 + 15 = 20, T_2 = 25, T_3 = 30$, and so on. The sequence looks like an arithmetic progression with $a = 20$ (first term) and $d = 5$ (common difference). To make sure, we evaluate the nth term of the A.P. with $a = 20$ and $d = 5$:

$$a + (n - 1)d = 20 + (n - 1)5 = 20 + 5n - 5 = 5n + 15$$

agreeing with the given formula for T_n.

Sec. 10.1 *Arithmetic Progressions*

The sum of 12 terms is then given by

$$S_n = \tfrac{1}{2}n[2a + (n-1)d]$$
$$S_{12} = \tfrac{1}{2}(12)[2(20) + (11)(5)]$$
$$= 570.$$

Diagnostic Test 10.1

Fill in the blanks.

1. A sequence $T_1, T_2, T_3, \ldots$ is in A.P. if _____.
2. If the first term and the common difference of an A.P. are x and y, respectively, then the first three terms are _____.
3. The nth term of the sequence $a, a + d, a + 2d, \ldots$ is _____.
4. The pth term of the sequence $a, a - d, a - 2d, \ldots$ is _____.
5. The sum of n terms of the sequence $d, d + a, d + 2a, \ldots$ is given by _____.
6. If the nth term of a sequence is $3n + 7$, then its pth term is _____.
7. If the pth term of a sequence is $5 - 2p$, then the third term of the sequence is _____.
*8. If T_n denotes the nth term of an A.P., then $T_{n-2} + T_2 = T_5 + $ _____.
9. If S_n denotes the sum to n terms of a sequence $T_1, T_2, T_3, \ldots$, then
$$S_n - S_{n-1} = $$ _____.

Exercises 10.1

(1–8) Find the indicated terms of the given sequences.

1. Tenth and fifteenth terms of $3, 7, 11, 15, \ldots$.
2. Sixth and tenth terms of $7, 6, 5, 4, \ldots$.
3. Seventh and nth terms of $5, 3, 1, -1, \ldots$.
4. Fourteenth and pth terms of $60, 58, 56, \ldots$.
5. The nth term of $4, 4\tfrac{1}{3}, 4\tfrac{2}{3}, 5, \ldots$.
6. The nth term of $5, 4.8, 4.6, 4.4, \ldots$.
7. The rth term of $7, 6\tfrac{3}{4}, 6\tfrac{1}{2}, 6\tfrac{1}{4}, \ldots$.
8. The pth term of $9, 8.7, 8.4, 8.1, \ldots$.
9. If the third and seventh terms of an A.P. are 18 and 30, respectively, find the fifteenth term.
10. If the fifth and tenth terms of an A.P. are 38 and 23, respectively, find the nth term.
11. The eighth and thirteenth terms of an A.P. are 2.9 and 1.4, respectively. Find the nth term.
12. If the ninth term of an A.P. is zero, show that the twenty-ninth term is twice the nineteenth term.
13. Which term of the sequence $5, 14, 23, 32, \ldots$ is 239?
14. Which term of the sequence $72, 70, 68, 66, \ldots$ is 40?
15. The last term of the sequence $20, 18, 16, \ldots$ is -4. Find the number of terms in the sequence.
16. The last term of the sequence $3.2, 3.9, 4.6, \ldots$ is 13. Find the number of terms in the sequence.

(17–26) Find the sum of the following progressions.

17. $1 + 4 + 7 + 10 + \cdots$ to 30 terms.
18. $70 + 68 + 66 + 64 + \cdots$ to 15 terms.

19. The sum of the first n natural numbers.
20. The sum of the first n odd natural numbers.
21. $20 + 17 + 14 + \cdots$ to n terms.
22. $58 + 53 + 48 + 43 + \cdots$ to n terms.
23. $51 + 48 + 45 + 42 + \cdots + 18$.
24. $15 + 17 + 19 + 21 + \cdots + 55$.
25. $7 + 6.7 + 6.4 + 6.1 + \cdots + 4$.
26. $3.8 + 4.2 + 4.6 + \cdots + 10.2$.
27. How many terms of the sequence 9, 12, 15, ... will make the sum 306?
28. How many terms of the sequence $-12, -7, -2, 3, 8, \ldots$ must be added to obtain a sum of 105?
29. The seventh term of an A.P. is 15 and the sum to 15 terms is 255. Find the tenth term of the sequence.
30. The sum of 20 terms of an A.P. is 70. If the first term of the sequence is 13, find the common difference of the A.P.
31. If S denotes the sum of a given sequence that is an A.P., a, l the first and the last terms, and n the number of terms in the sequence, show that

$$l = \frac{2S}{n} - a.$$

32. If $5, x, y, z, 17$ are in A.P., find x, y, z.

33. Find three numbers in A.P. whose sum is 15 and whose product is 105. [*Hint:* Take the three numbers as $a - d, a, a + d$.]

34. Find the three numbers in A.P. whose sum is 15 and such that the sum of the squares of the two extremes is 58.

35. Find four numbers in A.P. whose sum is 8 and whose product is -15. [*Hint:* Take the four numbers as $a - 3d, a - d, a + d, a + 3d$.]

36. Find four numbers in A.P. whose sum is 10 and such that the sum of the squares of the two extremes is 53.

37. A manufacturing company installs a machine at a cost of $1500. At the end of nine years the machine has a value of $420. Assuming that yearly depreciation is a constant amount, find the annual depreciation.

38. It cost $2000 to install a machine that depreciated annually by $160. What was the life of the machine if its scrap value was $400?

39. Steve's monthly payments to the bank toward his loan form a sequence in A.P. If his eighth and fifteenth payments are $213 and $185, respectively, what will be his twentieth payment?

40. The monthly salary of a person is increased annually in A.P. It is known that he was drawing $440 a month during the seventh year of his service and $1160 a month during the twenty-fifth year.
    (a) Find his starting salary and the annual increment.
    (b) What should be his salary at the time of retirement, on completion of 38 years of service?

41. In Exercise 39 suppose that Steve paid a total of $5490 to the bank.
    (a) Find the number of payments he made.
    (b) What was his last payment?

42. A debt of $1800 is to be repaid in one year by making a payment of $150 at the end of each month plus interest at the rate of 1% per month on the outstanding balance. Find the total interest paid.

43. A person deposits $50 at the beginning of every month into a savings account in which interest is allowed at 0.5% per month on the minimum monthly balance. Find the balance of the account at the end of the second year, calculating at simple interest.

44. A tube-well is bored 600 feet deep. The first foot costs $15, and the cost per foot increases by $2 for every subsequent foot. Find the cost of boring the five-hundredth foot and the total cost.

45. A man agrees to pay a debt of $1800 in a number of installments, each installment (beginning with the second) being less than the previous one by $10. If his fifth installment is $200, find how many installments will be necessary to wipe out the debt completely.

46. On November 1 every year a person buys Canada savings bonds to a value exceeding his previous year's purchase by $50. After 10 years he finds that the total cost of the bonds purchased by him has been $4250. Find the value of the bonds purchased by him (a) in the first year (b) in the seventh year.

10.2

Geometric Progressions

Suppose $1000 is deposited with a bank that calculates interest at the rate of 10% compounded annually. The value of this investment (in dollars) at the end of one year is equal to $1000(1.1) = $1100. At the end of two years it is $1100(1.1) = $1000(1.1)^2$, and so on. Thus the values of the investment (in dollars) at the end of 0, 1, 2, 3, ... years form the sequence

$$1000, \quad 1000(1.1), \quad 1000(1.1)^2, \quad 1000(1.1)^3, \quad \ldots \quad (1)$$

This sequence is an example of a geometric progression.

DEFINITION A sequence of terms is said to be a *geometric progression (G.P.)* if the ratio of each term to the preceding one is the same throughout. This constant ratio is called the *common ratio* of the G.P.

Thus the sequence 2, 6, 18, 54, 162, ... is a G.P. because

$$\frac{6}{2} = \frac{18}{6} = \frac{54}{18} = \frac{162}{54} = 3.$$

The common ratio is 3. Similarly the sequence $\frac{1}{3}, -\frac{1}{6}, \frac{1}{12}, -\frac{1}{24}, \ldots$ is a G.P. with common ratio $-\frac{1}{2}$.

If $T_1, T_2, T_3, \ldots, T_n, \ldots$ are the terms in a geometric progression with common ratio r, then $T_2/T_1 = r$, $T_3/T_2 = r$, and so on. In general,

$$\boxed{\frac{T_{n+1}}{T_n} = r.}$$

THE nTH TERM OF A G.P.

Each term in a G.P. is obtained by multiplying the preceding term by the common ratio. If a is the first term and r the common ratio, then successive terms of the G.P. are

$$a, \quad ar, \quad ar^2, \quad ar^3, \quad \ldots.$$

Observe that the sequence (1) above fits this pattern with $a = 1000$ and $r = 1.1$. In this G.P. we observe that the power of r in any term is one less than the number of the term. Thus the nth term is given by

$$T_n = ar^{n-1}. \tag{1}$$

Example 1 Find the fifth term and the nth term of the sequence 2, 6, 18, 54, ...

Solution The given sequence is a G.P., because

$$\frac{6}{2} = \frac{18}{6} = \frac{54}{18} = 3.$$

Thus, successive terms have a constant ratio of 3; that is, $r = 3$. Also $a = 2$. Therefore,

$$T_5 = ar^4 = 2(3^4) = 162.$$

and

$$T_n = ar^{n-1} = 2(3^{n-1}).$$

Example 2 The fourth and ninth terms of a G.P. are $\frac{1}{2}$ and $\frac{16}{243}$. Find the sixth term.

Solution Let a be the first term and r the common ratio of the given G.P. Then

$$T_4 = ar^3 = \tfrac{1}{2} \quad \text{(given)} \tag{i}$$

and

$$T_9 = ar^8 = \tfrac{16}{243} \quad \text{(given)} \tag{ii}$$

Dividing (ii) by (i), we get

$$\frac{ar^8}{ar^3} = \frac{\frac{16}{243}}{\frac{1}{2}}$$

$$r^5 = \tfrac{16}{243} \cdot \tfrac{2}{1} = \tfrac{32}{243} = (\tfrac{2}{3})^5.$$

Therefore $r = \tfrac{2}{3}$. Substituting this value of r in (i), we have

$$a(\tfrac{2}{3})^3 = \tfrac{1}{2}.$$

Thus,

$$a = \tfrac{1}{2} \cdot \tfrac{27}{8} = \tfrac{27}{16}.$$

Then,

$$T_6 = ar^5$$

$$= \tfrac{27}{16}(\tfrac{2}{3})^5 = \tfrac{27}{16} \cdot \tfrac{32}{243} = \tfrac{2}{9}.$$

Hence, the sixth term is $\tfrac{2}{9}$.

Example 3 Which term of the sequence 512, 256, 128, 64, ... is $\tfrac{1}{8}$?

Solution The given sequence is a G.P., because

$$\tfrac{256}{512} = \tfrac{128}{256} = \tfrac{64}{128} = \tfrac{1}{2}.$$

Thus the common ratio is $r = \tfrac{1}{2}$. Also the first term is $a = 512$. Let $\tfrac{1}{8}$ be the nth term in the sequence—that is, $T_n = \tfrac{1}{8}$. Using the formula for the nth term

of a G.P., we have
$$T_n = ar^{n-1}$$
$$\tfrac{1}{8} = 512(\tfrac{1}{2})^{n-1}$$
$$2^{-3} = 2^9 \cdot \frac{1}{2^{n-1}}$$
$$2^{-3} = 2^{9-(n-1)}$$
$$= 2^{10-n}.$$

Since the bases on the two sides are equal, the powers must also be equal. Thus,
$$-3 = 10 - n$$
$$n = 10 + 3 = 13.$$

Hence $\tfrac{1}{8}$ is the thirteenth term of the sequence.

SUM OF n TERMS OF A G.P.

THEOREM 1   If a is the first term and r the common ratio of a G.P., then the sum S_n of the first n terms is given by

$$\boxed{S_n = \frac{a(1-r^n)}{1-r}.} \qquad (2)$$

Proof   The n terms of the given sequence in G.P. are
$$a, \quad ar, \quad ar^2, \quad \ldots, \quad ar^{n-2}, \quad ar^{n-1}.$$
Therefore, their sum is
$$S_n = a + ar + ar^2 + \cdots + ar^{n-2} + ar^{n-1}. \qquad \text{(i)}$$
Multiplying both sides by $-r$ we have
$$-rS_n = -ar - ar^2 - \cdots - ar^{n-1} - ar^n. \qquad \text{(ii)}$$
Adding (i) and (ii) vertically, we obtain

$$\begin{aligned}
S_n &= a + ar + ar^2 + \cdots + ar^{n-1} \\
-rS_n &= - ar - ar^2 - \cdots - ar^{n-1} - ar^n \\
\hline
S_n - rS_n &= a \phantom{+ ar + ar^2 + \cdots + ar^{n-1}} - ar^n
\end{aligned}$$

Therefore,
$$S_n(1 - r) = a(1 - r^n)$$
$$S_n = \frac{a(1 - r^n)}{1 - r}$$

which proves the result.

Notes   (a)   The above formula for S_n is valid only when $r \neq 1$. When $r = 1$, the given G.P. becomes
$$a + a + a + \cdots + a \quad (n \text{ terms})$$
which has the sum na.

(b) Multiplying numerator and denominator of (2) by (-1), we obtain the alternative formula

$$S_n = \frac{a(r^n - 1)}{r - 1}.$$

This formula is generally used when $r > 1$, whereas formula (2) is more useful when $r < 1$.

Example 4 Find the sum of eight terms of the sequence $2 - 4 + 8 - 16 + \cdots$.

Solution The given sequence is a G.P. with $a = 2$ and $r = -2$. Here $n = 8$. Therefore, using the formula

$$S_n = \frac{a(1 - r^n)}{1 - r}$$

we have

$$S_8 = \frac{2[1 - (-2)^8]}{1 - (-2)}$$
$$= \tfrac{2}{3}(1 - 2^8) = \tfrac{2}{3}(1 - 256) = \tfrac{2}{3}(-255) = -170.$$

(Check: $S_8 = 2 - 4 + 8 - 16 + 32 - 64 + 128 - 256 = -170$.)

Example 5 How many terms of the sequence $486, 162, 54, 18, \ldots$ should be added to obtain a sum of 728?

Solution The given sequence is in G.P. with $r = \tfrac{1}{3}$ and $a = 486$. Let n terms be added to obtain a sum of 728; that is, $S_n = 728$. Now,

$$S_n = \frac{a(1 - r^n)}{1 - r}.$$

Substituting the given values, we have

$$728 = \frac{486\left(1 - \dfrac{1}{3^n}\right)}{1 - \tfrac{1}{3}}$$
$$= 729(1 - 3^{-n})$$

—that is,

$$\frac{728}{729} = 1 - 3^{-n}.$$

Therefore,

$$3^{-n} = 1 - \frac{728}{729}$$
$$= \frac{1}{729} = \frac{1}{3^6} = 3^{-6}$$

Consequently $n = 6$. Six terms added together will give a sum of 728. (Check: $S_6 = 486 + 162 + 54 + 18 + 6 + 2 = 728$.)

SUM OF AN INFINITE G.P.

The sum of the first n terms of the geometric sequence

$$a + ar + ar^2 + \cdots$$

is given by

$$S_n = \frac{a(1 - r^n)}{1 - r}. \qquad (2)$$

Let us consider the behavior of r^n for large n when $-1 < r < 1$. For the sake of definiteness, let $r = \frac{1}{2}$. Then the following table gives the values of r^n for several different values of n:

n	1	2	3	4	5	6	7
r^n	0.5	0.25	0.125	0.0625	0.03125	0.015625	0.0078125

From this table we observe that as n gets larger and larger, r^n gets smaller and smaller; ultimately, when n approaches infinity, r^n approaches zero. In fact, whenever $-1 < r < 1$, r^n gets closer and closer to zero as n becomes larger and larger. Thus from (2) above we can say that the sum of an infinite number of terms in the G.P. is given by

$$S_\infty = \frac{a(1 - 0)}{1 - r} = \frac{a}{1 - r}.$$

This leads us to the following theorem.

THEOREM 2 (SUM OF AN INFINITE G.P.)

The sum S of an infinite geometric sequence

$$a + ar + ar^2 + \cdots$$

is given by

$$\boxed{S = \frac{a}{1 - r}, \qquad \text{provided } -1 < r < 1.} \qquad (3)$$

Example 6 Find the sum to infinity of the sequence $1 - \frac{1}{3} + \frac{1}{9} - \frac{1}{27} + \cdots$.

Solution The given sequence is a G.P., with $a = 1$ and $r = -\frac{1}{3}$. The sum to infinity is given by

$$S = \frac{a}{1 - r}$$

$$= \frac{1}{1 - (-\frac{1}{3})} = \frac{1}{\frac{4}{3}} = \frac{3}{4}.$$

Example 7 Write the following repeating decimals as fractions.
(a) $0.55555\ldots$
(b) $2.345454545\ldots$

Solution (a) We can write 0.55555... as

$$0.5555... = 0.5 + 0.05 + 0.005 + 0.0005 + \cdots.$$

The sequence on the right side is an infinite geometric progression with first term $a = 0.5$ and the common ratio $r = 0.1$. Using Equation (3) for the sum of an infinite G.P., we have

$$0.555... = 0.5 + 0.05 + 0.005 + 0.0005 + \cdots$$
$$= \frac{0.5}{1 - 0.1} = \frac{0.5}{0.9} = \frac{5}{9}.$$

(b) We can write the repeating decimal as

$$2.3454545... = 2.3 + 0.045 + 0.00045 + 0.0000045 + \cdots.$$

Ignoring the first term, the sequence on the right is a G.P. This G.P. has first term 0.045 and common ratio 0.01. Thus, using the formula for the sum of an infinite G.P. again, we have

$$2.3454545... = 2.3 + \underbrace{0.045 + 0.00045 + 0.0000045 + \cdots}_{\text{G. P.}}$$

$$= 2.3 + \frac{0.045}{1 - 0.01}$$

$$= 2.3 + \frac{0.045}{0.99}$$

$$= \frac{23}{10} + \frac{45}{990}$$

$$= \frac{23}{10} + \frac{1}{22}$$

$$= \frac{516}{220}.$$

Diagnostic Test 10.2

Fill in the blanks:

1. A sequence $a_1, a_2, a_3, a_4, \ldots$ is in G.P. if _____.
2. The sequence 27, 18, 12, 8, ... is in _____ (A.P. or G.P.).
3. The sequence 5, 4.7, 4.4, 4.1, ... is in _____ (A.P. or G.P.).
4. The sequence $e^x, e^{2x}, e^{3x}, \ldots$ is in _____ (A.P. or G.P.).
5. The sequence log 3, log 9, log 27, ... is in _____ (A.P. or G.P.).
6. The sequence $7 + 77 + 777 + \cdots$ is _____.
7. If the nth term of a sequence is $T_n = 2n + 3$, then the sequence is in _____ (A.P. or G.P.).
8. If the pth term of a sequence is 3^{p+2}, then the sequence is in _____ (A.P. or G.P.).

9. The nth term of the sequence $a, ar, ar^2, \ldots$ is _____.

10. The sum to n terms of the sequence $a, ar, ar^2, \ldots$ is _____.

11. The sum of the infinite G.P. $a + ar + ar^2 + \cdots$ is given by _____ provided that _____.

12. The sum of the sequence $1 + \frac{1}{2} + \frac{1}{4} + \frac{1}{8} + \cdots$ is _____.

13. The sum of the infinite G.P. $1 + 2 + 4 + 8 + \cdots$ is _____.

14. $0.777\ldots$ can be expressed as a sum of a geometric progression as _____.

15. The nth term of the sequence

$$(2 + 1), (3 + \tfrac{1}{2}), (4 + \tfrac{1}{4}), (5 + \tfrac{1}{8}) \cdots$$

is _____.

Exercises 10.2

(1–8) Find the indicated terms of the following sequences.

1. Ninth term of $3, 6, 12, 24, \ldots$.
2. Tenth term of $2, -6, 18, -54, \ldots$.
3. Eighth term of $\sqrt{3}, 3, 3\sqrt{3}, 9, \ldots$.
4. Seventh term of $8, -4\sqrt{2}, 4, -2\sqrt{2}, \ldots$.
5. The nth term of $\frac{2}{9}, \frac{1}{3}, \frac{1}{2}, \ldots$.
6. The pth term of $\frac{2}{5}, -\frac{1}{2}, \frac{5}{8}, \ldots$.
7. The nth term of $\frac{3}{4}, -\frac{1}{2}, \frac{1}{3}, \ldots$.
8. The nth term of $\frac{5}{4}, \frac{1}{2}, \frac{1}{5}, \ldots$.

(9–12) Which term of the sequence

9. $96, 48, 24, 12, \ldots$ is $\frac{3}{16}$?
10. $18, 12, 8, \ldots$ is $\frac{512}{729}$?
11. $3, -6, 12, -24, \ldots$ is -384?
12. $\frac{3}{4}, \frac{1}{2}, \frac{1}{3}, \ldots$ is $\frac{8}{81}$?

13. The second term of a G.P. is 24 and the fifth term is 81. Find the first three terms of the sequence and its tenth term.

14. The third and seventh terms of a G.P. are $\frac{3}{4}$ and $\frac{243}{64}$, respectively. Find the first term of the G.P.

15. The fifth, eighth, and eleventh terms of a G.P. are x, y, and z, respectively. Show that $y^2 = xz$.

16. If $3, x, y, -24$ are in G.P., find the values of x, y.

(17–32) Find the sum of each sequence.

17. $2 + 6 + 18 + 54 + \cdots$ to 12 terms.
18. $3 - 6 + 12 - 24 + \cdots$ to 9 terms.
19. $\sqrt{3} - 3 + 3\sqrt{3} - 9 + \cdots$ to 10 terms.
20. $\sqrt{2} + \dfrac{1}{\sqrt{2}} + \dfrac{1}{2\sqrt{2}} + \cdots$ to 6 terms.
21. $1 + 2 + 4 + 8 + \cdots$ to n terms.
22. $3 + 1.5 + 0.75 + 0.375 + \cdots$ to p terms.
23. $\frac{2}{9} + \frac{1}{3} + \frac{1}{2} + \cdots$ to n terms.
24. $\frac{5}{4} - \frac{1}{2} + \frac{1}{5} - \frac{2}{25} + \cdots$ to n terms.
25. $\frac{3}{4} + \frac{1}{4} + \frac{1}{3} + \cdots + \frac{16}{243}$.
26. $8 - 4\sqrt{2} + 4 - 2\sqrt{2} + \cdots + \frac{1}{2}$.
27. $1 + \frac{1}{2} + \frac{1}{4} + \frac{1}{8} + \cdots$.
28. $1 - \frac{1}{3} + \frac{1}{9} - \frac{1}{27} + \cdots$.
29. $\frac{3}{4} - \frac{1}{2} + \frac{1}{3} \cdots$.
30. $3 + \sqrt{3} + 1 + 1/\sqrt{3} + \cdots$.
31. $a + br + ar^2 + br^3 + ar^4 + br^5 + \cdots$ $(-1 < r < 1)$.
32. $x(x + y) + x^2(x^2 + y^2) + x^3(x^3 + y^3) + \cdots$ $(|x| < 1, |y| < 1)$.
33. Prove that $9^{1/3} \cdot 9^{1/9} \cdot 9^{1/27} \cdots = 3$.
34. Evaluate $4^{1/3} \cdot 4^{-1/9} \cdot 4^{1/27} \cdot 4^{-1/81} \cdots$.

35. If $y = 1 + x + x^2 + x^3 + \cdots$, $(-1 < x < 1)$, then show that
$$x = \frac{y-1}{y}.$$

36. If $v = \frac{1}{1+i} (i > 0)$, show that
$$v + v^2 + v^3 + \cdots = \frac{1}{i}.$$

37. If a, r, l, and S denote the first term, common ratio, last term, and sum of a sequence in G.P., then show that
$$S = \frac{a - rl}{1 - r}.$$

38. Use the result in Exercise 37 to determine the common ratio of a G.P. whose first term is 96, last term is 3, and sum is 189.

(39–44) Express the following infinite decimals as fractions.

39. 0.4444....
40. 0.77777....
41. 0.23232323....
42. 1.343434....
43. 0.8272727....
44. 2.5717171....

45. The nth term of a sequence is 3^n. Find the sum of the first p terms.

46. The nth term of an infinite sequence is 2^{-n}. Find the sum of the sequence.

47. An article costs $512 when new, but by usage it loses one-fourth of its value yearly. If the article is worth y dollars after x years of usage, show that
$$\log y = (9 - 2x) \log 2 + x \log 3.$$

48. A machine is depreciated annually at the rate of 10% on its declining value. The original cost was $10,000 and the ultimate scrap value was $5314.41. Find the effective life of the machine.

49. The sum of $2000 is invested in a savings account at 8% interest compounded annually. Find its value after five years.

50. The sum of $5000 is invested in a savings account that is compounded quarterly at a nominal interest rate of 8% per annum. Find its value after three years.

51. A yeast colony initially has five cells and is doubling in size every 15 minutes. If T_n denotes the number of cells after n hours, show that $T_1, T_2, \ldots$ are in G.P. and find the common ratio and first term.

52. Let T_n denote the number of cancer cells in a growth after a period of n months from its first detection. Approximate measurements show that $T_1, T_2, \ldots$ form a geometric progression with $T_1 = 10^4$ and $T_6 = 5 \times 10^4$. At what value of n will the number of cells reach 1 million?

53. The frequencies of notes on a well-tempered musical scale form a G.P. If middle C and high C have frequencies of 256 and 512 cycles per second and there are 11 notes between them, calculate the frequencies of E (the fifth note) and G sharp (the ninth note).

10.3

Mathematical Induction

Mathematical induction is an important method of proof used in mathematics. It is commonly used when we need to prove the general validity of a formula or statement that involves a positive integer n. As an example, consider the sum of the first n consecutive positive even integers. Suppose that we observe the following pattern:

$$
\begin{aligned}
n = 1: &\qquad 2 = 2 = 1 \cdot 2 \\
n = 2: &\qquad 2 + 4 = 6 = 2 \cdot 3 \\
n = 3: &\qquad 2 + 4 + 6 = 12 = 3 \cdot 4 \\
n = 4: &\qquad 2 + 4 + 6 + 8 = 20 = 4 \cdot 5
\end{aligned}
\qquad (1)
$$

After observing this pattern, we may suspect that the sum of the first n even positive integers is $n(n + 1)$; that is,

$$2 + 4 + 6 + 8 + \cdots + 2n = n(n + 1). \qquad (2)$$

The question is: how can we prove that the statement (2) is true for *all* positive integral values of n? For any specific value of n, say $n = 10$ or 20, we can actually compute the two sides of (2) and check their equality as we did for $n = 1, 2, 3$, and 4 in (1). To verify in this manner the validity of (2) for *every* positive integral value of n would involve an infinite number of computations—an impossible task. This is where the method of mathematical induction can be used.

Informally, according to this method the statement or formula such as (2) is proved if we can verify the following two conditions:

(i) The formula is true for $n = 1$.
(ii) If the formula is true for *any* given positive integer $n = k$, then it is also true for $n = k + 1$.

These two conditions are sufficient to guarantee that the given statement is true for every integer n. For by (i) it is true for $n = 1$. Then, taking $k = 1$ in condition (ii), the statement must also be true for $n = k + 1 = 2$. Next take $k = 2$ in condition (ii), and we see that the statement must be true for $n = k + 1 = 3$. And so we continue, using (ii) repeatedly to show that the given statement must be true in turn for $n = 4, 5, 6$, and so on.

The formal principle underlying this method of proof is as follows.

PRINCIPLE OF MATHEMATICAL INDUCTION.

Let P_n be a statement or a formula involving a natural number n. Then P_n is true for every positive integer n if the following two conditions are met.

(a) P_n is true for $n = 1$.
(b) If P_n is true for any positive integer $n = k$, then it is also true for the next integer $n = k + 1$. That is, the truth of P_k for any k implies the truth of P_{k+1}.

Condition (b) is often referred to as the *inductive step*.

In the use of the principle of mathematical induction, the following two steps should be followed.

Step 1. Verify that the statement P_n is true for $n = 1$.

Step 2. *Assume* that P_n is true for some positive integer $n = k$ and prove that this implies that P_n is also true for $n = k + 1$.

Example 1 Use mathematical induction to prove that the sum of the first n even positive integers is $n(n + 1)$; that is,

$$2 + 4 + 6 + \cdots + 2n = n(n + 1). \tag{i}$$

Solution Let P_n denote the statement given in (i).

Step 1. Verify that the statement is true for $n = 1$. When $n = 1$, P_1 becomes

$$2 = 1(1 + 1)$$

which is obviously true.

Step 2. We assume that P_n is true for $n = k$; that is,

$$2 + 4 + 6 + \cdots + 2k = k(k + 1) \tag{ii}$$

and then we shall try to prove that P_n must also be true for $n = k + 1$. That is, we shall try to show that

$$2 + 4 + 6 + \cdots + 2k + 2(k + 1) = (k + 1)(k + 1 + 1)$$

or

$$2 + 4 + 6 + \cdots + 2k + (2k + 2) = (k + 1)(k + 2). \tag{iii}$$

We can achieve this by adding the next even integer $(2k + 2)$ to both sides of (ii). Doing so gives

$$\underbrace{[2 + 4 + 6 + \cdots + 2k]}_{= k(k+1) \text{ from (ii)}} + (2k + 2) = k(k + 1) + (2k + 2)$$

$$= k^2 + 3k + 2$$

$$= (k + 1)(k + 2)$$

which is precisely (iii).

Since both conditions of mathematical induction are satisfied, the given statement is true for all positive integers n.

Example 2 Use mathematical induction to prove that for any positive integer n,

$$3 + 7 + 11 + \cdots + (4n - 1) = n(2n + 1). \tag{i}$$

Solution Let P_n denote the statement given in (i). It may be noted that there are exactly n terms on the left side.

Step 1. Show that the statement P_n is true for $n = 1$. When $n = 1$, P_1 becomes

$$3 = 1(2 \cdot 1 + 1)$$

which is true.

Step 2. Assume that P_n is true for $n = k$; that is,

$$3 + 7 + 11 + \cdots + (4k - 1) = k(2k + 1). \tag{ii}$$

Sec. 10.3   *Mathematical Induction*

Then we want to prove that P_n is true for $n = k + 1$. We can achieve this by adding $4(k + 1) - 1 = (4k + 3)$ to both sides of (ii). Doing so gives

$$3 + 7 + 11 + \cdots + (4k - 1) + (4k + 3)$$
$$= k(2k + 1) + (4k + 3)$$
$$= 2k^2 + 5k + 3$$
$$= (2k + 3)(k + 1)$$

which is precisely the statement P_{k+1}. Thus, we have proved that if P_n is true for $n = k$, then P_n is also true for $n = k + 1$.

Since both conditions of mathematical induction are satisfied, the given statement is true for all positive integers n.

Example 3 Use mathematical induction to prove that 3 is a factor of $4^n - 1$ for every positive integer n.

Solution Let P_n denote the given statement—that is,

P_n: 3 is a factor of $4^n - 1$.

Step 1. For $n = 1$, P_1 becomes

3 is a factor of $4^1 - 1 = 3$

which is true.

Step 2. Assume P_n is true for $n = k$; that is,

3 is a factor of $4^k - 1$.

Then we can write

$$4^k - 1 = 3Q \qquad \text{(i)}$$

where Q is the other factor of $4^k - 1$. We want to prove that P_n is true for $n = k + 1$; that is, that 3 is a factor of $4^{k+1} - 1$. Now

$$4^{k+1} - 1 = 4(4^k - 1) + 3$$
$$= 4(3Q) + 3 \qquad \text{[using (i)]}$$
$$= 3(4Q + 1)$$

which shows that 3 is also a factor of $4^{k+1} - 1$. Thus we have proved that if P_k is true then so is P_{k+1}.

Therefore, both conditions of mathematical induction are met, and 3 is a factor of $4^n - 1$ for every positive integer n.

Often we encounter statements P_n involving a natural number n that we wish to prove not for *all* positive integers n but for $n \geq m$, where $m \neq 1$ is some given positive integer. In such cases we use the principle of extended mathematical induction.

> PRINCIPLE OF EXTENDED MATHEMATICAL INDUCTION.
>
> Let P_n be a statement involving a natural number n. Let m be a positive integer. Then P_n is true for every integer $n \geq m$, provided that
>
> (a) P_n is true for $n = m$, and
> (b) if P_n is true for any positive integer $k \geq m$, then it is also true for $n = k + 1$.

Example 4 Use mathematical induction to prove that

$$n^3 \geq (n+1)^2 \quad \text{for } n \geq 3. \tag{i}$$

Solution Let P_n denote the statement given in (i). For $n = 3$, P_3 becomes

$$3^3 \geq (3+1)^2 \quad \text{or} \quad 27 \geq 16$$

which is true. Let us assume that P_n is true for $n = k$, where $k \geq 3$; that is,

$$k^3 \geq (k+1)^2. \tag{ii}$$

Now we wish to prove that P_n is true for $n = k + 1$; that is,

$$(k+1)^3 \geq (k+1+1)^2$$

or

$$(k+1)^3 \geq (k+2)^2 \tag{iii}$$

Now

$$\begin{aligned}(k+1)^3 &= k^3 + 3k^2 + 3k + 1 \\ &\geq (k+1)^2 + 3k^2 + 3k + 1 \quad \text{[using (ii)]} \\ &= 4k^2 + 5k + 2 \\ &= (k+2)^2 + 3k^2 + (k-2) \\ &\geq (k+2)^2\end{aligned}$$

because $3k^2$ and $(k-2)$ are both positive for $k \geq 3$. Thus we have proved the truth of (iii), and by mathematical induction the given statement is true for all $n \geq 3$.

Exercises 10.3

(1–26) Use mathematical induction to prove that the following statements are true for each positive integer n.

1. $1 + 3 + 5 + 7 + \cdots + (2n-1) = n^2$.
2. $1 + 2 + 3 + 4 + \cdots + n = \frac{1}{2}n(n+1)$.
3. $1 + 7 + 13 + 19 + \cdots + (6n-5) = n(3n-2)$.
4. $2 + 5 + 8 + 11 + \cdots + (3n-1) = \frac{1}{2}n(3n+1)$.
5. $1 + 2 + 4 + 8 + \cdots + 2^{n-1} = 2^n - 1$.
6. $1 + 5 + 5^2 + 5^3 + \cdots + 5^{n-1} = \frac{1}{4}(5^n - 1)$.
7. $\frac{1}{2} + \frac{1}{2^2} + \frac{1}{2^3} + \cdots + \frac{1}{2^n} = 1 - \frac{1}{2^n}$.
8. $\frac{1}{5} + \frac{1}{5^2} + \frac{1}{5^3} + \cdots + \frac{1}{5^n} = \frac{1}{4}\left(1 - \frac{1}{5^n}\right)$.

9. $\dfrac{1}{1\cdot 2}+\dfrac{1}{2\cdot 3}+\dfrac{1}{3\cdot 4}+\cdots+\dfrac{1}{n(n+1)}=\dfrac{n}{n+1}$.

10. $\dfrac{1}{1\cdot 3}+\dfrac{1}{3\cdot 5}+\dfrac{1}{5\cdot 7}+\cdots+\dfrac{1}{(2n-1)(2n+1)}=\dfrac{n}{2n+1}$.

11. $1\cdot 2+2\cdot 3+3\cdot 4+\cdots+n(n+1)=\tfrac{1}{3}n(n+1)(n+2)$.

12. $1\cdot 2+3\cdot 4+5\cdot 6+\cdots+(2n-1)(2n)=\tfrac{1}{3}n(n+1)(4n-1)$.

13. $1^2+2^2+3^2+4^2+\cdots+n^2=\tfrac{1}{6}n(n+1)(2n+1)$.

14. $1^3+2^3+3^3+4^3+\cdots+n^3=\tfrac{1}{4}n^2(n+1)^2$.

15. $1\cdot 2^0+2\cdot 2^1+3\cdot 2^2+\cdots+n\cdot 2^{n-1}=1+(n-1)2^n$.

16. $1+\dfrac{2}{3}+\dfrac{3}{3^2}+\dfrac{4}{3^3}+\cdots+\dfrac{n}{3^{n-1}}=\dfrac{9}{4}\left(1-\dfrac{2n+3}{3^{n+1}}\right)$.

17. $1+3n\leq 4^n$. 18. $1+2n\leq 3^n$. 19. 2 is a factor of n^2+n.

20. 3 is a factor of n^3-n+3. 21. 2 is a factor of 3^n-1. 22. 4 is a factor of 5^n-1.

23. For $r\neq 1$, $a+ar+ar^2+\cdots+ar^{n-1}=a(1-r^n)/(1-r)$.

24. $a+(a+d)+(a+2d)+\cdots+[a+(n-1)d]=\dfrac{n}{2}[2a+(n-1)d]$.

25. $(1-x)(1+x+x^2+x^3+\cdots+x^{n-1})=1-x^n$.

26. $(x-y)(x^{n-1}+x^{n-2}y+x^{n-3}y^2+\cdots+xy^{n-2}+y^{n-1})=x^n-y^n$.

(27–30) Use mathematical induction to prove the following statements.

27. $2^n+3^n<4^n$ for $n\geq 2$.

28. For every real number $x>-1$,
$$(1+x)^n>1+nx \quad \text{for } n\geq 2.$$

29. $2^n>2n$ for $n\geq 3$.

30. $3^n>2n+1$ for $n\geq 2$.

31. Use mathematical induction to prove the compound interest formula $A(n)=P(1+i)^n$.

10.4

Permutations and Combinations

Let two different roads A and B connect the cities X and Y and three different roads P, Q, R connect the cities Y and Z. In how many ways can we travel by road from city X to city Z via city Y?

We can travel from X to Y in two different ways. With each of these two choices we have three choices of roads from Y to Z. Thus we have $2\cdot 3=6$ routes from city X to city Z via city Y, as illustrated in the tree diagram.

```
                          P    ⓩ  = AP
                       ⓨ  Q
                    A     R  ⓩ  = AQ
                             ⓩ  = AR
          ⓧ
                          P  ⓩ  = BP
                    B     Q
                       ⓨ     ⓩ  = BQ
                          R  ⓩ  = BR
```

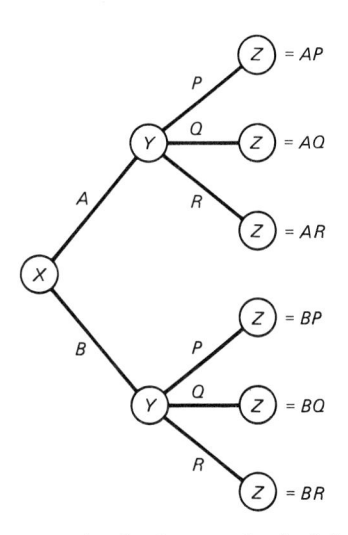

This example illustrates the *fundamental principle of counting*.

FUNDAMENTAL PRINCIPLE OF COUNTING.

Suppose that we wish to make n selections, or choices, one after another. If the first selection can be made in p_1 ways and then the second selection can be made in p_2 ways, and so on, until the nth selection can be made in p_n ways, then the n selections together in the order indicated can be made in

$$p_1 \cdot p_2 \cdot \cdots \cdot p_n \text{ ways.}$$

Example 1 A student has four shirts, six pairs of pants, and two ties. How many different combinations of shirts, pants and ties can she wear?

Solution The student can select one of the four shirts in 4 different ways. She can select one pair of pants out of her six pairs in 6 ways. Finally she can select one of the two ties in 2 ways. Thus by the principle of counting she can select a shirt, pants, and a tie in $(4)(6)(2) = 48$ ways. In other words, she can wear 48 different combinations of pants, shirts, and ties.

Example 2 A North American corporation is in the process of expanding its business into Europe and intends to send three representatives, one to Brussels, one to Frankfurt, and one to Zurich. If there are ten representatives to choose from, in how many ways can the three be selected?

Solution There are 10 ways of selecting the representative to be sent to Brussels. After that choice is made, however, there are only nine representatives left from which to make the next choice. Thus, if the Frankfurt representative is chosen next, this second choice can be made in 9 ways. Similarly, once the other two have been selected, the Zurich representative can be chosen in only 8 ways.

Consequently, the number of ways in which the three representatives can be chosen is given by the product $(10)(9)(8) = 720$.

Example 2 provides an example of what are called *ordered selections* or *permutations*. We have a group of ten persons and we wish to select three of them for a certain purpose. Since each of the three is needed for a different purpose, the *order* in which the selection is made is important. On the other hand, if the corporation intended to send all three representatives to the same place, then the order would not matter, and we would obtain a different answer.

Consider the general case in which we have n different objects or individuals and we wish to make an ordered selection of r objects from among them.

Since there are n objects to choose from, the first object may be chosen in n ways. However, since the first object is not replaced, there are only $(n - 1)$ objects from which the second can be chosen. Hence the second object can be selected in $(n - 1)$ ways. Similarly, once the first two have been chosen, the third object can be selected in $(n - 2)$ ways, and so on. The last (rth) object can be chosen in $(n - r + 1)$ ways. Thus, the number of ways in which the whole permutation can be selected is given by the product

$$n(n - 1)(n - 2)(...)(n - r + 1).$$

This product is denoted by the symbol $_nP_r$ and is called the **number of permutations of r objects from among n**. Thus,

$$_nP_r = n(n - 1)(n - 2) \ldots (n - r + 1). \tag{1}$$

Note that $_nP_r$ is given by *the product of r consecutive integers, starting with n and decreasing to $(n - r + 1)$*. For example,

$$_8P_3 = 8(8 - 1)(8 - 2) = 8 \cdot 7 \cdot 6 = 336$$

$$_{15}P_2 = 15 \cdot 14 = 210.$$

Example 3 From a group of four people, it is required to select individuals to participate in three different psychological tests. In how many ways can the selection be made?

Solution Since the tests are *different*, the order in which the three are chosen is significant. The number of ways in which the choice can be made is therefore the number of permutations of four people taken three at a time—that is,

$$_4P_3 = 4 \cdot 3 \cdot 2 = 24.$$

When $r = n$, we have from (1),

$$_nP_n = n(n - 1)(n - 2) \cdot \cdots \cdot 2 \cdot 1. \tag{2}$$

The right side is the product of first n natural numbers. This product is called *factorial n* and denoted by $n!$. Thus

$$n! = 1 \cdot 2 \cdot 3 \cdot \cdots \cdot (n-1)n. \tag{3}$$

For example,

$$1! = 1, \quad 2! = 1 \cdot 2 = 2, \quad 3! = 1 \cdot 2 \cdot 3 = 6$$
$$4! = 1 \cdot 2 \cdot 3 \cdot 4 = 24, \quad 5! = 1 \cdot 2 \cdot 3 \cdot 4 \cdot 5 = 120.$$

As a matter of convenience we also define $0! = 1$. From (2) it follows that

$$_nP_n = n!. \tag{4}$$

From (3) we have

$$n! = \underbrace{1 \cdot 2 \cdot 3 \cdot \cdots \cdot (n-1)}_{= (n-1)!} \cdot n$$
$$= n \cdot (n-1)!.$$

Thus we have the following *reduction formula* for $n!$:

$$n! = n \cdot (n-1)!. \tag{5}$$

For example, with $n = 5$, we have $5! = (5)(4!)$.

We can express $_nP_r$ in terms of factorials. For example,

$$_8P_3 = 8 \cdot 7 \cdot 6$$
$$= \frac{8 \cdot 7 \cdot 6 \cdot 5 \cdot 4 \cdot 3 \cdot 2 \cdot 1}{5 \cdot 4 \cdot 3 \cdot 2 \cdot 1} = \frac{8!}{5!}$$
$$_{15}P_2 = 15 \cdot 14$$
$$= \frac{15 \cdot 14 \cdot 13 \cdot 12 \cdot \ldots \cdot 2 \cdot 1}{13 \cdot 12 \cdot \ldots \cdot 2 \cdot 1} = \frac{15!}{13!}.$$

In general,

$$_nP_r = n(n-1)(\ldots)(n-r+1)$$
$$= \frac{n(n-1)(\ldots)(n-r+1)(n-r)(n-r-1)(\ldots) \cdot 2 \cdot 1}{(n-r)(n-r-1)(\ldots) \cdot 2 \cdot 1}.$$

In this step we have introduced the factors $(n-r), (n-r-1), \ldots, 2, 1$ into both numerator and denominator, which leaves the value unchanged. But now the numerator is equal to $n!$ and the denominator is equal to $(n-r)!$, so that

$$_nP_r = \frac{n!}{(n-r)!}. \tag{6}$$

Example 4 How many six-digit numbers can be formed by using the digits 4, 5, 6, 7, 8, 9 with no digit being repeated? How many of them are not divisible by 5?

Solution (a) To form a six-digit number, we have to arrange the given six digits in a row to fill up the six places (units, tens, hundreds, and so on). Since the order of the digits in the number is important, we need the number of permutations of all the six digits. Thus the number can be formed in $_6P_6 = 6!$ ways. The number of six-digit numbers that can be formed is

$$6! = 720.$$

(b) Let us first find how many of these numbers are divisible by 5. For such numbers, the last digit must be 5. Once this last place has been filled with the 5, the remaining five places can be filled with the remaining five digits in $_5P_5 = 5! = 120$ ways. Thus there are 120 six-digit numbers that are divisible by 5. Therefore, the number of six-digit numbers that are not divisible by 5 is

$$720 - 120 = 600.$$

Example 5 How many different words can be formed out of the letters of the word DAUGHTER so that the vowels are all together?

Solution There are eight different letters in the word DAUGHTER, of which three are vowels: A, E, U. Since the vowels are always to be together, we may enclose them in a bracket as (AEU) and regard this whole bracket as a single letter. Now we have six "letters" namely D, G, H, T, R, (AEU), which are to be arranged in a row to form words. Since the order in which these "letters" occur in the words is important, they can be arranged in $_6P_6 = 6! = 720$ different ways. But the three vowels A, E, U in the bracket can be arranged themselves within the bracket in $_3P_3 = 3! = 6$ ways. Thus, by the fundamental principle of counting, the total number of words that can be formed so that the vowels are together is

$$(720)(6) = 4320.$$

Example 6 In how many ways can six boys and four girls be arranged in a row so that no two girls are together?

Solution Since no two girls are to be together, it is necessary that every two girls be separated by at least one boy. First let us position the six boys. They can be arranged in a row in $_6P_6 = 6!$ ways. The four girls then must be placed in four places out of seven marked X in the diagram below. This will ensure that no two girls are together.

$$\text{X B X B X B X B X B X B X}$$

But four places out of the seven marked X can be filled in $_7P_4$ ways. Thus, the required number of ways in which the six boys and four girls can be arranged is

$$_7P_4(6!) = (7 \cdot 6 \cdot 5 \cdot 4)(720) = 604{,}800.$$

Now let us consider the case when r objects are selected from n but when the order in which the objects are chosen is of no significance. For instance, in Example 3 we could suppose that the three individuals were chosen from the available four in order to participate in the *same* test. Then the order in which the three were chosen would be immaterial, and all we would need to know would be which three formed the chosen group.

Let us label the available four by the letters a, b, c, d. If the order in which the choice is made is important, then the number of choices is equal to $_4P_3 = 4 \cdot 3 \cdot 2 = 24$. We can list these 24 choices as follows:

$$\begin{array}{cccccc}
bcd, & bdc, & cbd, & cdb, & dcb, & dbc \\
acd, & adc, & cad, & cda, & dac, & dca \\
abd, & adb, & bad, & bda, & dab, & dba \\
abc, & acb, & bac, & bca, & cab, & cba
\end{array}$$

We observe from this list that each group of three individuals appears six times, corresponding to the six different ways in which the three individuals can be ordered. If the order in which the three individuals are selected is of no importance, all these six permutations of each group are equivalent to one another; for example, bcd is equivalent to bdc, to cbd, and so on. When order is immaterial, there are only four different choices, corresponding to the four rows in the above table.

Now consider the general problem in which r objects are chosen from among n, where the order of selection is of no significance. Each such choice is called a **combination** of r objects from among n, and the **number of combinations** is denoted by the symbol $_nC_r$.

Any *permutation* of r objects from among n can be formed by first choosing the combination of r objects and then arranging these r objects in an appropriate order. The number of permutations, $_nP_r$ is therefore equal to the number of ways of choosing particular combinations of r objects from among n multiplied by the number of ways in which each combination can be ordered. That is,

$$_nP_r = \,_nC_r \cdot N(r)$$

where $N(r)$ is the number of ordered arrangements of the chosen r objects. But $N(r)$ must be equal to $_rP_r = r!$. Therefore

$$_nP_r = \,_nC_r(r!)$$

and so

$$\boxed{_nC_r = \frac{_nP_r}{r!} = \frac{n!}{(n-r)!\,r!}.} \qquad (6)$$

We can also write the number of combinations in the form

$$\boxed{_nC_r = \frac{n(n-1)\ldots(n-r+1)}{r(r-1)\cdot\ldots\cdot 2 \cdot 1}.} \qquad (7)$$

Note that in this fraction both numerator and denominator contain r consecutive integers. The first factor is n in the numerator and r in the denominator. For example,

$$_8C_3 = \frac{8 \cdot 7 \cdot 6}{3 \cdot 2 \cdot 1} \quad \begin{array}{l} \leftarrow \text{ 3 factors starting with 8} \\ \leftarrow \text{ 3 factors starting with 3} \end{array}$$

$$= 56$$

$$_7C_5 = \frac{7 \cdot 6 \cdot 5 \cdot 4 \cdot 3}{5 \cdot 4 \cdot 3 \cdot 2 \cdot 1} \quad \begin{array}{l} \leftarrow \text{ 5 factors starting with 7} \\ \leftarrow \text{ 5 factors starting with 5} \end{array}$$

$$= 21.$$

Setting $r = n$ and $r = 0$ in (6), we have, respectively,

$$_nC_n = \frac{n!}{(n-n)!n!} = \frac{n!}{0!n!} = 1$$

and

$$_nC_0 = \frac{n!}{(n-0)!0!} = \frac{n!}{n!} = 1.$$

(Recall that $0! = 1$.) Thus

$$\boxed{_nC_0 = {_nC_n} = 1 \quad \text{for all } n.} \tag{8}$$

Again replacing r by $n - r$ in (6), we have

$$_nC_{n-r} = \frac{n!}{[n-(n-r)]!(n-r)!}$$

$$= \frac{n!}{r!(n-r)!}$$

$$= {_nC_r}.$$

That is,

$$\boxed{_nC_{n-r} = {_nC_r} \quad \text{for all } r \leq n.} \tag{9}$$

Example 7 Evaluate $_{50}C_{48}$.

Solution Using Equation (9), we have

$$_{50}C_{48} = {_{50}C_{50-48}} = {_{50}C_2}$$

$$= \frac{50 \cdot 49}{2 \cdot 1} = 1225$$

where we have used (7). (In $_{50}C_2$, two factors occur in the numerator and denominator. If $_{50}C_{48}$ were written out directly, it would have 48 factors in numerator and denominator!)

Example 8 There are 10 points in a plane, no three of which are collinear except for 4 that are all in a straight line. How many straight lines can be formed by joining pairs of the points?

Solution When 2 points are joined, we get a straight line. There are 10 points in all, and if no 3 of them were in a straight line, the number of straight lines formed would be

$$_{10}C_2 = \frac{10 \cdot 9}{2 \cdot 1} = 45.$$

But 4 of the points lie on one straight line. Thus these 4 points give only one straight line instead of the $_4C_2 = 6$ straight lines that they would have given if no 3 of them were collinear. Thus the number of lines formed by joining the given 10 points is equal to

$$45 - 6 + 1 = 40.$$

Example 9 A committee of five persons is to be formed out of eight men and four women. Find the number of ways in which this can be done if the committee contains:

(a) Exactly two women. (b) At least two women.

Solution (a) The committee of five in this case consists of two women and therefore three men. Now three men can be selected out of eight in $_8C_3$ ways, and two women can be selected out of four in $_4C_2$ ways. Therefore, by the principle of counting, the number of different committees consisting of two women and three men is

$$_8C_3 \cdot {}_4C_2 = \frac{8 \cdot 7 \cdot 6}{3 \cdot 2 \cdot 1} \cdot \frac{4 \cdot 3}{2 \cdot 1} = 336.$$

(b) Since at least two women are included in the committee of five persons, the committee can consist of one of the following combinations:

(1) two women and three men
(2) three women and two men
(3) four women and one man.

As in part (a) above, the numbers of committees formed in the above cases are

(1) $_4C_2 \cdot {}_8C_3 = 6 \cdot 56 = 336$
(2) $_4C_3 \cdot {}_8C_2 = 4 \cdot 28 = 112$
(3) $_4C_4 \cdot {}_8C_1 = 1 \cdot 8 = 8.$

Therefore the total number of committees that will include at least two women is

$$336 + 112 + 8 = 456.$$

Note Can you see what is wrong with the following "solution" to part (b)? (It is not so easy). Since 2 of the committee must definitely be women, we choose them first. This can be done in $_4C_2$ ways. The remaining 3 committee members can be either men or women, so they can be chosen freely from among the remaining 10 individuals. This can be done in $_{10}C_3$ ways. By the principle of counting, the number of ways of making both selections is therefore

$$_4C_2 \cdot {}_{10}C_3 = 6 \cdot 120 = 720.$$

This shows that counting problems can often be quite tricky!

Example 10 How many different poker hands of 5 cards can be chosen from a deck of 52 cards? How many of those hands contain 3 aces and 2 kings?

Solution The total number of poker hands is $_{52}C_5$ ($\approx$ 2.6 million). The number of ways of choosing 3 aces from among the 4 available is $_4C_3$. The number of ways of choosing 2 kings from among the 4 available is $_4C_2$. The number of hands that contain 3 aces and 2 kings is therefore

$$_4C_3 \cdot {}_4C_2 = \frac{4 \cdot 3 \cdot 2}{3 \cdot 2 \cdot 1} \cdot \frac{4 \cdot 3}{2 \cdot 1} = 24.$$

Before closing this section, we summarize the important formulas.

---

(1) $n! = 1 \cdot 2 \cdot 3 \cdot \ldots \cdot (n-1)n$ and $0! = 1$

$n! = n \cdot (n-1)!$

$= n(n-1) \cdot (n-2)!$ etc.

(2) $_nP_r = n(n-1)(n-2) \ldots$ to r factors

$= n(n-1)(n-2) \ldots (n-r+1)$

$= \dfrac{n!}{(n-r)!}$

$_nP_n = n!$

(3) $_nC_r = \dfrac{_nP_r}{r!} = \dfrac{n!}{r!(n-r)!}$

$= \dfrac{n(n-1)(n-2) \ldots \text{to } r \text{ factors}}{r(r-1)(r-2) \ldots \text{to } r \text{ factors}}$

$_nC_r = {}_nC_{n-r}$

$_nC_0 = {}_nC_n = 1$

Diagnostic Test 10.4

Fill in the blanks.

1. $_nP_2 = $ _____ and $_8P_3 = $ _____.
2. $_nC_2 = $ _____ and $_8C_3 = $ _____.
3. $_nC_r = {}_nC$_____ and $_8C_5 = {}_8C$_____.
4. $_nC_n = $ _____ and $_nC_0 = $ _____ for all positive integers n.

5. $_nP_r = $ _____ .

6. $_nC_r = $ _____ .

7. $_nP_r = $ _____ $(_nC_r)$.

8. $n! = $ _____ $\cdot (n-2)!$

9. $0! = $ _____ .

10. $\dfrac{n!}{r!(n-r)!} = $ _____ and $\dfrac{n!}{(n-r)!} = $ _____ .

11. Three vacant positions can be filled from among five available persons in _____ ways.

12. Three persons can occupy five vacant seats in _____ ways.

13. Five persons can be arranged in a row for a photograph in _____ ways.

14. A group of five persons can be selected out of eight in _____ ways.

Exercises 10.4

(1–6) Evaluate the following.

1. (a) $\dfrac{20!}{18!}$. (b) $\dfrac{10!}{9!+8!}$. (c) $\dfrac{8!}{7!-6!}$.   2. (a) $\dfrac{50!}{48!}$. (b) $\dfrac{8!}{8!-7!}$. (c) $\dfrac{9!}{8!+7!}$.

3. (a) $_{10}P_2$. (b) $_6P_4$. (c) $_5P_5$.   4. (a) $_nP_3$. (b) $_4P_6$. (c) $_nP_0$.

5. (a) $_{10}C_2$. (b) $_8C_3$. (c) $_{20}C_{18}$.   6. (a) $_{35}C_0$. (b) $_{30}C_{28}$. (c) $_{70}C_{70}$.

(7–20) Find the value of n so that the following equations are satisfied.

7. $(n+1)! = 20\, n!$.   8. $(n+2)! = 42\, n!$.   9. $_nP_2 = 56$.   10. $_nP_2 = 90$.

11. $_{11}P_n = 990$.   12. $_6P_n = 360$.   13. $_nC_2 = 21$.   14. $_nC_2 = 45$.

15. $_nC_{12} = {_nC_8}$.   16. $_nC_7 = {_nP_1}$.   17. $_{n+5}P_4 \div {_{n+4}P_2} = 54$.   18. $_{n+3}P_6 \div {_{n+2}P_4} = 14$.

19. $_{n+5}C_3 \div {_{n+4}C_3} = \tfrac{4}{3}$.   20. $_{n+1}C_4 \div {_nC_2} = \tfrac{9}{2}$.

21. Show that $(_{10}C_6)(_6C_2) = (_{10}C_2)(_8C_4)$.

22. Show that $(_{12}C_5)(_7C_4) = (_{12}C_3)(_9C_5)$.

23. Show that $_9C_3 + {_9C_4} = {_{10}C_4}$.

*24. Show that $_nC_r + {_nC_{r+1}} = {_{n+1}C_{r+1}}$.

25. Four persons enter a bus in which there are three vacant seats. In how many ways can the seats be filled?

26. In how many ways can five prizes be awarded to seven students when
    (a) No student can receive more than one prize?
    *(b) There is no restriction to the number of prizes each student may receive?

27. How many numbers less than 3000 can be formed by using the digits 7, 3, 5, 2 when no digit is repeated?

28. How many numbers each consisting of four different digits can be formed from the digits 0, 1, 2, 3?

29. How many different words can be formed out of the letters of the word PECULIAR when no letter is repeated? How many of these words will begin with P?

30. In how many ways can the letters of the word VOWEL be arranged so that the vowels are always together?

31. In how many ways can five books on chemistry and four books on physics be arranged on a shelf so that the books on the same subject are always together?

32. In how many ways can five boys and eight girls be seated in a row so that the boys are always together and the girls are always together?

33. Find the number of ways in which seven boys and five girls can be arranged in a row so that no two girls are together.

34. In how many ways can the letters of the word MALENKOV be arranged so that no two vowels are together?

35. A father has eight children and takes three of them at a time to the zoo as often as he can without the same three children being together more than once. How often will the father go and how often will each child go?

36. In a mathematics examination, 10 questions are presented. In how many ways can a student select 6 questions to answer? If question 1 is made compulsory, in how many ways can a student now select a total of 6 questions?

37. There are 16 points in a plane, no 3 of them in a straight line. Find the number of (a) straight lines that can be formed by joining the given points, (b) triangles that have the given points as vertices.

38. There are 12 points in a plane, no 3 of them collinear except that 5 are on the same straight line. Find the number of triangles that can be formed with the given points as vertices.

39. A person has 15 acquaintances, of whom 10 are relatives. In how many ways can he invite 9 guests so that 7 of them will be relatives?

40. In how many ways can a committee of seven persons be formed among eight men and six women so that the committee contains exactly three women?

41. In how many ways may 12 hockey players, of whom at least 2 must be goal tenders, be chosen from 15 players, it being known that only 4 of the players can play in goal?

42. In an examination paper there are two parts each consisting of seven questions. A candidate is required to answer nine questions but not more than five questions from each part. In how many ways can the nine questions be selected?

43. In how many ways can we select 6 students out of a class of 20 so that the tallest student is always included?

44. In how many ways can we select 6 persons from a group of 30 so that the selection always includes the tallest person and excludes the smallest person?

45. In how many ways can 4 cards be drawn from a pack of 52 so that one of the drawn cards is an ace?

46. How many different four-letter code words can be formed by using the letters of the word EQUATION, when no letter is repeated?

47. A box contains seven black and five white balls, all of different sizes. In how many ways can we draw two balls so that
    (a) Both the balls drawn are black?
    (b) The two drawn balls are of different colors?
    (c) The two drawn balls are of the same color?

48. A box contains five red, four black, and three white balls. Three balls are drawn from the box. In how many ways can the following happen?
    (a) All the balls drawn are red.
    (b) The balls drawn are two red and one black.
    (c) The three balls drawn are of different colors.

49. How many different poker hands are there that contain two aces and two kings (the fifth card being neither ace nor king)?

50. How many different poker hands are there that contain four of a kind?

51. Two dice are rolled. Show that there are 36 different outcomes. How many of these outcomes have a total score of eight on the two dice?

52. In Exercise 51, how many outcomes have a total score of more than eight?

10.5

The Binomial Theorem

An expression consisting of only two terms is called a *binomial expression*. For instance, $2x + 3y$, $x^2 - 4a$, and $3a + 5/b$ are all binomial expressions.

By actual multiplication, we can show that

$$(x + y)^1 = x + y$$
$$(x + y)^2 = x^2 + 2xy + y^2$$
$$(x + y)^3 = x^3 + 3x^2y + 3xy^2 + y^3 \qquad (1)$$
$$(x + y)^4 = x^4 + 4x^3y + 6x^2y^2 + 4xy^3 + y^4$$
$$(x + y)^5 = x^5 + 5x^4y + 10x^3y^2 + 10x^2y^3 + 5xy^4 + y^5.$$

The expression on the right side is called the *expansion* of the binomial on the left side. Now, suppose we are interested in the expansion of $(x + y)^{20}$. To obtain this expansion by actual multiplication of $(x + y)$ by itself 20 times would involve lengthy and tedious calculations. Thus, it would be useful if we had a general formula for the expansion of $(x + y)^n$, where n is any positive integer. The development of such a formula is the purpose of this section.

Let us first make the following observations from the expansions of $(x + y)^n$ for $n = 1, 2, 3, 4,$ and 5 listed above in (1).

(a) The number of terms in the expansion on the right is one more than the exponent n. For example, the expansion of $(x + y)^2$ contains 3 terms, $(x + y)^3$ contains 4 terms, and so on. In general, the expansion of $(x + y)^n$ will consist of $(n + 1)$ terms.

(b) The first term in the expansion of $(x + y)^n$ is x^n. For example, the first term in the expansion of $(x + y)^3$ is x^3, and so on. Similarly the last term in the expansion of $(x + y)^n$ is y^n.

(c) As we move from one term to the next in the expansion, the exponent of x decreases by 1 while the exponent of y increases by 1. The sum of the powers of x and y in any term is equal to the exponent n.

In view of the above observations, we can say that apart from the numerical coefficients, the successive terms of $(x + y)^n$ will consist of

$$x^n, \quad x^{n-1}y, \quad x^{n-2}y^2, \quad x^{n-3}y^3, \quad \ldots, \quad xy^{n-1}, \quad y^n. \qquad (2)$$

Let us now study the pattern of the coefficients of the various terms in the expansion of $(x + y)^n$ in (1). We read off the following coefficients:

BINOMIAL	COEFFICIENTS
$(x+y)^0$	1
$(x+y)^1$	1 1
$(x+y)^2$	1 2 1
$(x+y)^3$	1 3 3 1
$(x+y)^4$	1 4 6 4 1
$(x+y)^5$	1 5 10 10 5 1

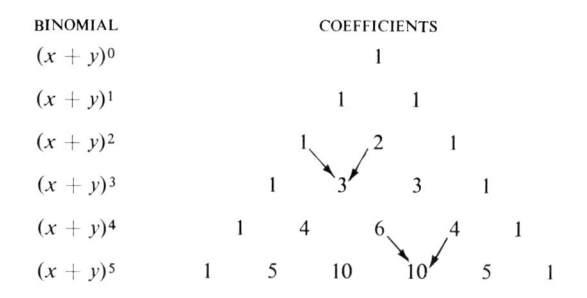

The triangle of numbers formed in this manner is called *Pascal's triangle* after the French mathematician Blaise Pascal (1623–1662). It may be observed that the end numbers in any row are 1 and any number not on the end is the sum of the two closest numbers (one to the left and one to the right) in the preceding row. For example, the 3 in the fourth row is the sum of the 2 and 1 in the third row; and the 10 in the sixth row is the sum of the 6 and 4 in the fifth row, as shown.

This property of the coefficients enables us to write down the expansion of $(x+y)^{n+1}$ whenever the expansion of $(x+y)^n$ is known.

Example 1 Given the expansion of $(x+y)^5$ in (1), use the Pascal triangle method to write down the expansion of $(x+y)^6$.

Solution The coefficients of $(x+y)^5$ form the bottom row in the triangle printed above. The next row has a 1 at each end, and the numbers in the middle are obtained by adding pairs in the $(x+y)^5$ row:

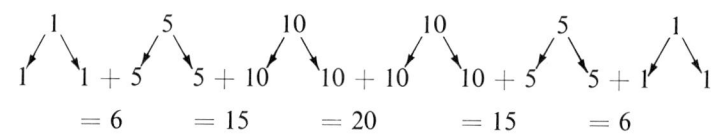

$$1 \quad 1+5=6 \quad 5+10=15 \quad 10+10=20 \quad 10+5=15 \quad 5+1=6 \quad 1$$

These numbers are precisely the coefficients of successive terms in the expansion of $(x+y)^6$. Thus,

$$(x+y)^6 = x^6 + 6x^5y + 15x^4y^2 + 20x^3y^3 + 15x^2y^4 + 6xy^5 + y^6.$$

From Pascal's triangle, the coefficients of successive terms in the expansion of $(x+y)^4$ are

$$1 \quad 4 \quad 6 \quad 4 \quad 1.$$

In combinatorial notation, these coefficients are

$$_4C_0, \quad _4C_1, \quad _4C_2, \quad _4C_3, \quad _4C_4$$

because $_4C_0 = 1$, $_4C_1 = 4/1 = 4$, $_4C_2 = (4 \cdot 3)/(1 \cdot 2) = 6$, $_4C_3 = (4 \cdot 3 \cdot 2)/(3 \cdot 2 \cdot 1) = 4$, and $_4C_4 = (4 \cdot 3 \cdot 2 \cdot 1)/(4 \cdot 3 \cdot 2 \cdot 1) = 1$.

In the same manner, the coefficients 1, 5, 10, 10, 5, 1 in the expansion of $(x+y)^5$ are $_5C_0, _5C_1, _5C_2, _5C_3, _5C_4$, and $_5C_5$, respectively. These observations lead us to speculate that the coefficients of successive terms in the expansion of $(x + y)^n$, where n is any positive integer, are

$$_nC_0, \ _nC_1, \ _nC_2, \ _nC_3, \ \ldots, \ _nC_{n-1}, \ _nC_n.$$

This is in fact the case, and combining this sequence with the list (2), we can write down the expansion of $(x + y)^n$ as stated in the following theorem.

THEOREM (THE BINOMIAL THEOREM)

If n is a positive integer, then
$$(x + y)^n = {_nC_0}x^n + {_nC_1}x^{n-1}y + {_nC_2}x^{n-2}y^2 + {_nC_3}x^{n-3}y^3$$
$$+ \cdots + {_nC_{n-1}}xy^{n-1} + {_nC_n}y^n. \tag{3}$$

We shall prove this theorem at the end of this section.

The right side of (3) is called the **binomial expansion** of $(x + y)^n$. The $(r + 1)$th term in this expansion is called the **general term** and is given by

$$T_{r+1} = {_nC_r}x^{n-r}y^r. \tag{4}$$

The coefficients $_nC_0, \ _nC_1, \ _nC_2, \ldots, \ _nC_n$ of the various terms in the expansion of $(x + y)^n$ are called the **binomial coefficients**.

The binomial expansion may be written in full as follows.

$$(x + y)^n = x^n + \frac{n}{1}x^{n-1}y + \frac{n(n-1)}{1 \cdot 2}x^{n-2}y^2 + \frac{n(n-1)(n-2)}{1 \cdot 2 \cdot 3}x^{n-3}y^3$$
$$+ \cdots + \frac{n(n-1)(n-2)\ldots(n-r+1)}{1 \cdot 2 \cdot 3 \cdot \ldots \cdot r}x^{n-r}y^r + \cdots + y^n.$$

Example 2 Find the binomial expansion of $\left(2x + \dfrac{y}{2}\right)^6$.

Solution Using the binomial theorem (3) with $n = 6$ and x replaced by $2x$ and y by $y/2$, we have

$$\left(2x + \frac{y}{2}\right)^6 = {_6C_0}(2x)^6 + {_6C_1}(2x)^5\left(\frac{y}{2}\right) + {_6C_2}(2x)^4\left(\frac{y}{2}\right)^2 + {_6C_3}(2x)^3\left(\frac{y}{2}\right)^3$$
$$+ {_6C_4}(2x)^2\left(\frac{y}{2}\right)^4 + {_6C_5}(2x)\left(\frac{y}{2}\right)^5 + {_6C_6}\left(\frac{y}{2}\right)^6$$

$$= 1(64x^6) + \frac{6}{1}(32x^5)\left(\frac{y}{2}\right) + \frac{6 \cdot 5}{1 \cdot 2}(16x^4)\left(\frac{y^2}{4}\right)$$
$$+ \frac{6 \cdot 5 \cdot 4}{1 \cdot 2 \cdot 3}(8x^3)\left(\frac{y^3}{8}\right) + \frac{6 \cdot 5 \cdot 4 \cdot 3}{1 \cdot 2 \cdot 3 \cdot 4}(4x^2)\left(\frac{y^4}{16}\right)$$
$$+ \frac{6 \cdot 5 \cdot 4 \cdot 3 \cdot 2}{1 \cdot 2 \cdot 3 \cdot 4 \cdot 5}(2x)\left(\frac{y^5}{32}\right) + 1\left(\frac{y^6}{64}\right)$$

$$= 64x^6 + 96x^5y + 60x^4y^2 + 20x^3y^3 + \tfrac{15}{4}x^2y^4 + \tfrac{3}{8}xy^5 + \tfrac{1}{64}y^6.$$

Example 3 Expand $(2x - 3)^5$.

Solution We can write $(2x - 3)^5 = [2x + (-3)]^5$. Thus, using the binomial theorem (3) with $n = 5$, x replaced by $2x$, and y by -3, we obtain

$$(2x - 3)^5 = {}_5C_0(2x)^5 + {}_5C_1(2x)^4(-3) + {}_5C_2(2x)^3(-3)^2$$
$$+ {}_5C_3(2x)^2(-3)^3 + {}_5C_4(2x)(-3)^4 + {}_5C_5(-3)^5$$
$$= 1(32x^5) + 5(16x^4)(-3) + 10(8x^3)(9)$$
$$+ 10(4x^2)(-27) + 5(2x)(81) + 1(-243)$$
$$= 32x^5 - 240x^4 + 720x^3 - 1080x^2 + 810x - 243.$$

Example 4 Find the seventh term in the expansion of $[x + (2/x^2)]^9$.

Solution We know that the $(r + 1)$th term in $(x + y)^n$ is given by

$$T_{r+1} = {}_nC_r x^{n-r} y^r.$$

In this case we want $T_7 = T_{6+1}$, so that $r = 6$. Also y is replaced by $2/x^2$ and $n = 9$. Thus,

$$T_7 = {}_9C_6 x^{9-6} \left(\frac{2}{x^2}\right)^6$$
$$= 84x^3 \cdot \frac{64}{x^{12}} = 5376x^{-9}.$$

Example 5 Find the middle term in the expansion of $(x/a + a/x)^{10}$.

Solution The total number of terms in the expansion is $n + 1 = 10 + 1 = 11$. Leaving five terms each at the beginning and the end, the sixth term is the middle term. Thus,

$$T_6 = T_{5+1} = {}_{10}C_5 \left(\frac{x}{a}\right)^{10-5} \left(\frac{a}{x}\right)^5$$
$$= {}_{10}C_5 = 252.$$

Example 6 Find the coefficient of x^{16} in the expansion of $(x^2 - 2x)^{10}$.

Solution Let x^{16} occur in the $(r + 1)$th term in the expansion. Now, using (4), we have

$$T_{r+1} = {}_{10}C_r(x^2)^{10-r}(-2x)^r$$
$$= {}_{10}C_r x^{20-2r}(-2)^r x^r$$
$$= (-2)^r {}_{10}C_r x^{20-r}. \quad \text{(i)}$$

Since T_{r+1} is required to be the term that contains x^{16}, we must have

$$20 - r = 16 \quad \text{or} \quad r = 4.$$

Substituting $r = 4$ in (i), we have

$$T_5 = (-2)^4 \, {}_{10}C_4 x^{16}.$$

Therefore, the coefficient of x^{16} is

$$(-2)^4 \, {}_{10}C_4 = 16 \cdot \frac{10 \cdot 9 \cdot 8 \cdot 7}{4 \cdot 3 \cdot 2 \cdot 1} = 3360.$$

PROOF OF BINOMIAL THEOREM

Let us now prove the binomial theorem, given in (3):

$$(x+y)^n = {}_nC_0 x^n + {}_nC_1 x^{n-1}y + {}_nC_2 x^{n-2}y^2 + \cdots + {}_nC_{n-1}xy^{n-1} + {}_nC_n y^n. \tag{5}$$

We use mathematical induction to prove this result. Let P_n denote the statement given in (5). For $n = 1$, P_1 becomes

$$(x+y)^1 = {}_1C_0 x^1 + {}_1C_1 y^1$$

—that is,

$$(x+y)^1 = x + y$$

because ${}_1C_0 = {}_1C_1 = 1$. Thus P_n is true for $n = 1$.

Next, let us assume that P_n is true for $n = k$; that is,

$$(x+y)^k = {}_kC_0 x^k + {}_kC_1 x^{k-1}y + {}_kC_2 x^{k-2}y^2 + \cdots + {}_kC_{k-1}xy^{k-1} + {}_kC_k y^k \tag{6}$$

and prove that P_n is also true for $n = k+1$:

$$(x+y)^{k+1} = {}_{k+1}C_0 x^{k+1} + {}_{k+1}C_1 x^k y + {}_{k+1}C_2 x^{k-1}y^2$$
$$+ \cdots + {}_{k+1}C_{k+1} y^k. \tag{7}$$

To obtain (7) from (6), we multiply both sides of (6) by $(x+y)$. This gives

$$(x+y)^{k+1} = (x+y)[{}_kC_0 x^k + {}_kC_1 x^{k-1}y + {}_kC_2 x^{k-2}y^2 + \cdots + {}_kC_r x^{k-r}y^r$$
$$+ \cdots + {}_kC_{k-1}xy^{k-1} + {}_kC_k y^k]$$
$$= {}_kC_0 x^{k+1} + {}_kC_1 x^k y + {}_kC_2 x^{k-1}y^2 + \cdots + {}_kC_r x^{k+1-r}y^r + \cdots + {}_kC_k xy^k$$
$$+ {}_kC_0 x^k y + {}_kC_1 x^{k-1}y^2 + \cdots + {}_kC_{r-1} x^{k+1-r}y^r + \cdots + {}_kC_{k-1}xy^k$$
$$+ {}_kC_k y^{k+1}.$$

Here the terms in the first line are obtained by multiplying the square bracket by x and the terms in the second line by multiplying by y. Combining the like terms by adding the two lines vertically, we have

$$(x+y)^{k+1} = {}_kC_0 x^{k+1} + ({}_kC_1 + {}_kC_0)x^k y + ({}_kC_2 + {}_kC_1)x^{k-1}y^2$$
$$+ \cdots + ({}_kC_r + {}_kC_{r-1})x^{k+1-r}y^r$$
$$+ \cdots + ({}_kC_k + {}_kC_{k-1})xy^k + {}_kC_k y^{k+1}. \tag{8}$$

Since ${}_nC_0 = {}_nC_n = 1$ for all positive integers n, we have

$${}_kC_0 = {}_{k+1}C_0 \quad \text{and} \quad {}_kC_k = {}_{k+1}C_{k+1}.$$

Therefore the first and last terms in (8) agree with those in (7). Now

$${}_kC_1 + {}_kC_0 = \frac{k}{1} + 1 = k + 1 = {}_{k+1}C_1$$

$${}_kC_2 + {}_kC_1 = \frac{k(k-1)}{1 \cdot 2} + \frac{k}{1} = \frac{k}{2}(k-1+2) = \frac{k(k+1)}{1 \cdot 2} = {}_{k+1}C_2$$

and so on. In general, it can be shown in a similar way that

$$_kC_r + {_kC_{r-1}} = {_{k+1}C_r}.$$

When we use these results, (8) becomes

$$(x+y)^{k+1} = {_{k+1}C_0}x^{k+1} + {_{k+1}C_1}x^k y + {_{k+1}C_2}x^{k-1}y^2$$
$$+ \cdots + {_{k+1}C_r}x^{k+1-r}y^r + \cdots + {_{k+1}C_k}xy^k + {_{k+1}C_{k+1}}y^{k+1}$$

and this is nothing but (7). Thus, we have satisfied both conditions of mathematical induction, and the result (5) is therefore true for all positive integer values of n.

Diagnostic Test 10.5

Fill in the blanks.

1. According to the binomial theorem, for any positive integer n,
   $(x + a)^n =$ _____.
2. The pth term in the expansion of $(x + a)^n$ is _____.
3. The seventh term in the expansion of $(x - a)^9$ is _____.
4. The fifth term from the end in the expansion of $(x + y)^{10}$ is _____.
5. The rth term from the end in the expansion of $(x + y)^n$ is the _____ th term from the beginning.
6. The binomial coefficients of any two terms in the expansion of $(x + a)^n$ that are at the same distances, respectively, from the beginning and end are _____.
7. The binomial expansion of $(x + a)^n$ has _____ terms.
8. The sum of powers of x and y in each term in the expansion of $(x + y)^n$ is _____.
9. The binomial expansion of $(x + y)^{13}$ has _____ middle term(s).
10. The binomial expansion of $(a + b)^{20}$ has _____ middle term(s).

Exercises 10.5

(1–12) Use the binomial theorem to expand the following, and simplify.

1. $(a + b)^7$.
2. $(2x + y)^4$.
3. $\left(2x + \dfrac{y}{2}\right)^5$.
4. $\left(x^2 + \dfrac{1}{x}\right)^6$.
5. $\left(\dfrac{2x}{3} + \dfrac{3}{2x}\right)^5$.
6. $(x^2 + 3x)^4$.
7. $(a - 2b)^6$.
8. $\left(2a - \dfrac{b}{2a}\right)^7$.
9. $(2p - 3q)^5$.
10. $\left(x - \dfrac{1}{x}\right)^7$.
11. $\left(p^2 - \dfrac{q}{p}\right)^5$.
12. $(1 - x)^8$.

(13–22) Find the indicated terms in simplified form in the given binomial expansions.

13. The seventh term in $\left(3x + \dfrac{y}{2}\right)^9$.
14. The eighth term in $\left(\dfrac{x}{2} - \dfrac{4}{x}\right)^9$.

15. The nth term in $\left(x - \dfrac{1}{x^2}\right)^{3n}$.

16. The pth term in $\left(x + \dfrac{1}{2x}\right)^{2p}$.

17. The fifth term from the end in $\left(\dfrac{x^3}{2} + \dfrac{2}{x^2}\right)^9$.

18. The fourth term from the end in $\left(\dfrac{4x}{3} - \dfrac{3}{2x}\right)^7$.

19. The middle term in $\left(x - \dfrac{1}{x}\right)^{10}$.

20. The middle term in $\left(\dfrac{3}{x} - 2x\right)^{12}$.

21. The two middle terms in $\left(3x - \dfrac{x^3}{6}\right)^9$.

22. The two middle terms in $\left(x + \dfrac{2}{x}\right)^7$.

(23–26) Find in simplified form the coefficient of the indicated term.

23. x^{15} in the expansion of $(x + x^2)^{10}$.

24. x^5 in the expansion of $(2 + 3x)^{12}$.

25. x^{-2} in the expansion of $\left(2x^5 - \dfrac{1}{x^2}\right)^8$.

26. x^{-3} in the expansion of $\left(\dfrac{3x^2}{2} - \dfrac{1}{3x}\right)^9$.

(27–32) Find the term independent of x in the binomial expansions of the following.

27. $\left(x^2 + \dfrac{1}{2x}\right)^{12}$.

28. $\left(\dfrac{x^3}{2} - \dfrac{3}{x^2}\right)^{10}$.

29. $\left(\dfrac{4}{3}x^2 - \dfrac{3}{2x}\right)^9$.

30. $\left(\dfrac{3}{2}x^2 - \dfrac{1}{3x}\right)^9$.

31. $\left(x^3 + \dfrac{2}{x}\right)^{10}$.

32. $\left(x^2 - \dfrac{1}{x^3}\right)^{11}$.

*33. Show that the middle term in the expansion of $(1 + x)^{2n}$ is
$$\dfrac{1 \cdot 3 \cdot 5 \cdot 7 \cdots (2n - 1)}{n!} 2^n x^n$$
where n is a positive integer.

*34. If P is the sum of the odd terms and Q the sum of the even terms in the expansion of $(x + y)^n$, then prove that
  (a) $P^2 - Q^2 = (x^2 - y^2)^n$.
  (b) $4PQ = (x + y)^{2n} - (x - y)^{2n}$.
  (c) $2(P^2 + Q^2) = (x + y)^{2n} + (x - y)^{2n}$.

*35. Given that the coefficients of x^{12} and x^{13} in the expansion of $(3x^2 + ax)^9$ are equal, find the value of a.

*36. The first three terms in the expansion of $(1 + ax)^n$ are $1 + 4x + 7x^2$. Find n and a.

37. Evaluate $(2 + \sqrt{3})^5 + (2 - \sqrt{3})^5$.

38. Evaluate $(\sqrt{2} + 1)^6 - (\sqrt{2} - 1)^6$.

*(39–40) If $_nC_0, {}_nC_1, {}_nC_2, \ldots, {}_nC_n$ are the binomial coefficients in the expansion of $(x + y)^n$, prove that

39. $_nC_0 + {}_nC_1 + {}_nC_2 + \cdots + {}_nC_n = 2^n$.

40. $_nC_1 + {}_nC_2 + {}_nC_3 + \cdots + {}_nC_n = 1 + 2 + 2^2 + \cdots + 2^{n-1}$.

Chapter Review 10

Explain (with examples where possible) the following terms:

(a) Sequence.

(b) Arithmetic progression (formulas for the nth term and the sum of n terms).

(c) Geometric progression (formulas for the nth term and the sum of n terms).

(d) Infinite geometric progression (formula for the sum and the condition when this sum can be found).

(e) Principle of mathematical induction.

(f) Principle of extended mathematical induction.

(g) Fundamental principle of counting.
(h) Permutations (formulas for $_nP_r$).
(i) Combinations (formulas for $_nC_r$).
(j) Factorial n.
(k) Reduction formula for factorial n.
(l) Binomial theorem for positive integral exponent.
(m) Binomial coefficients.
(n) Pascal's triangle.

Review Exercises on Chapter 10

1. Find the nth term of the sequence
$$-15, \ -12, \ -9, \ -6, \ \ldots$$

2. Find an arithmetic progression of six terms whose first term is $\frac{2}{3}$ and last term is $\frac{22}{3}$.

3. Find an A.P. of four terms whose first term is p and last term is q.

(4–6) Find the sum of the following progressions.

4. $0.9 + 0.91 + 0.92 + 0.93 + \cdots$ to 100 terms.

5. $20 + 18 + 16 + 14 + \cdots$ to n terms.

6. $\log 2 + \log 4 + \log 8 + \cdots$ to n terms.

7. If the third and tenth terms of an A.P. are 7 and 21, respectively, find the sum to n terms of this A.P.

8. The sum of 10 terms of an A.P. is 65 and the sum of 20 terms is 230. Find the sum of n terms of this A.P.

9. In a given A.P. the pth term is q and the qth term is p. Show that the $(p+q)$th term is zero.

10. How many terms of the sequence $-12 - 9 - 6 - 3 - \cdots$ will make the sum 54?

*11. If the sum of p terms of an A.P. is q and the sum of q terms is p, show that the sum of $(p+q)$ terms is $-(p+q)$.

12. Find the thirteenth term of the sequence $3, -\sqrt{3}, 1, -\frac{1}{\sqrt{3}} + \cdots$.

13. Which term of the sequence 96, 48, 24, 12 $\cdots$ is $\frac{3}{16}$?

14. Find a geometric progression of seven terms whose first term is $\frac{2}{9}$ and last term is $\frac{81}{32}$.

15. Find the nth term of the sequence $(a+b)$, $(2a+b^2)$, $(3a+b^3)$, $\cdots$.

16. The fourth and seventh terms of a G.P. are 12 and $\frac{3}{2}$, respectively. Find the sum of six terms of this G.P.

17. The second and fifth terms of a G.P. are 18 and $\frac{16}{3}$, respectively. Find the sum to infinity of this G.P.

18. Express $0.3272727\ldots$ as a fraction.

19. A bicycle is worth \$243 when new, but by usage it loses one-third of its value yearly. If the bicycle is worth y dollars after x years of usage, show that
$$\log y = (5-x)\log 3 + x \log 2.$$

20. A man borrows \$5115 to be repaid in 10 monthly installments. If each installment is double the value of the preceding one, find the value of the first and last installments.

(21–22) Find the sums of the following sequences.

21. $18 + 6\sqrt{3} + 6 + 2\sqrt{3} + \cdots$ to p terms.

22. $e^x + e^{2x} + e^{3x} + \cdots$ to n terms.

(23–30) Use mathematical induction to prove that the following statements are true for all positive integers n.

23. $2 + 6 + 10 + 14 + \cdots + (4n - 2) = 2n^2$.

24. $1 + 3 + 3^2 + 3^3 + \cdots + 3^{n-1} = \frac{1}{2}(3^n - 1)$.

25. $5 + 5 \cdot 6 + 5 \cdot 6^2 + 5 \cdot 6^3 + \cdots + 5 \cdot 6^{n-1} = 6^n - 1$.

26. $\dfrac{1}{2 \cdot 5} + \dfrac{1}{5 \cdot 8} + \dfrac{1}{8 \cdot 11} + \cdots + \dfrac{1}{(3n - 1)(3n + 2)} = \dfrac{n}{3n + 2}$.

27. 8 is a factor of $3^{2n} - 1$.

28. If $0 < a < 1$, then $0 < a^n < 1$.

29. $n^2 < 2^n$ for $n \geq 5$.

30. $3^n > 2n + 5$ for $n \geq 3$.

31. If $_nP_6 = (30)_nP_4$, find the value of n.

32. Find n if $_{2n+1}P_{n-1} \div {}_{2n-1}P_n = \frac{3}{5}$.

33. Find r if $_{2n}C_r = {}_{2n}C_{r+2}$.

*34. Find n if $_{n+2}C_8 \div {}_{n-2}P_4 = \frac{57}{16}$.

35. Show that the sum of the first n natural numbers is given by $_{n+1}C_{n-1}$.

36. Prove that $(n + 1)_nC_r = (r + 1)_{n+1}C_{r+1}$.

37. How many different permutations of four letters can be formed by using the letters of the word MALENKOV?

38. How many four-digit numbers can be formed by using the digits 0, 1, 2, 3, 4, 5, no digit being repeated?

39. How many three-digit numbers can be formed by using the digits 0, 1, 3, 5, 7, 9 when no digit is repeated? How many of these numbers are divisible by 5?

40. In how many ways can five persons be seated in a row so that two particular individuals x and y are side by side?

41. In how many ways can the letters of the word EQUATION be arranged so that no two consonants are together?

42. Automobile license plates consist of three letters from the English alphabet followed by three digits selected from 0, 1, 2, ..., 9. The digits or letters may be repeated, except that the three digits cannot all be zeros. Find the number of different license plate numbers that are possible.

43. A class consists of 10 boys and 7 girls. A committee of 3 persons is to be formed out of this class. In how many ways can we do this, if the committee contains
   (a) Exactly one girl?
   (b) At least one girl?

44. A box contains 4 black and 5 red balls, all of different sizes. A pair of balls are drawn from the box. In how many ways can the following happen?
   (a) The two balls are of different colors.
   (b) Both the balls drawn are black.
   (c) Both the balls drawn are red.
   (d) The balls drawn are of the same color.

45. There are 15 points in a plane, no three of them collinear except for 6 that are in a straight line. Find the number of
   (a) Straight lines that can be formed by joining these points.
   (b) Triangles with the given points as vertices.

46. Prove that $_nP_{n-1} = {}_nP_n$ for all positive integers.

47. Given $(x + y)^5 = x^5 + 5x^4y + 10x^3y^2 + 10x^2y^3 + 5xy^4 + y^5$, use Pascal's triangle to write down the binomial expansion of $(x - y)^6$.

48. Find the middle term in the expansion of $(2x + y/2)^{12}$.

49. Find the two middle terms in the binomial expansion of $(a - b)^{15}$.

50. Find the coefficient of x^{-4} in the expansion of $(3x^2 - 2/x)^7$.

51. Find the coefficient of x^{20} in the expansion of $(x^3 + 2x)^{10}$.

52. Find the term independent of x in the expansion of $(2x^2 + 1/x^3)^{10}$.

53. Find the fifth term from the end in the expansion of $(3x - y)^9$.

*54. If $_nC_0, {_nC_1}, {_nC_2}, \ldots, {_nC_n}$ are the binomial coefficients in the expansion of $(x + y)^n$, show that

$$_nC_0 + {_nC_2} + {_nC_4} + \cdots$$
$$= {_nC_1} + {_nC_3} + {_nC_5} + \cdots.$$

11
MATRICES AND LINEAR EQUATIONS

11.1

Solution of Linear Systems by Row Reduction

In Section 5.5 we described how systems of linear equations in two variables can be solved by the method of substitution or the method of addition. In this section we shall develop a method of solving systems of linear equations that can be used regardless of the number of equations or unknowns involved. Let us first illustrate the principles of the method by solving the following simple system of two equations:

$$\begin{aligned} 2x + 3y &= 3 \\ x - 2y &= 5. \end{aligned} \quad (1)$$

If we interchange the two equations (for reasons that will become clear later), we get the equivalent system

$$\begin{aligned} x - 2y &= 5 \\ 2x + 3y &= 3. \end{aligned} \quad (2)$$

If we multiply the first of these equations by -2, we obtain $-2x + 4y =$

-10. Adding this equation to the second equation in (2), we get
$$2x + 3y + (-2x + 4y) = 3 + (-10)$$
$$0x + 7y = -7.$$

The system (2) then becomes
$$x - 2y = 5$$
$$0x + 7y = -7. \qquad (3)$$

Multiplying both sides of the second equation by $\frac{1}{7}$, we get the equivalent system
$$x - 2y = 5$$
$$0x + y = -1. \qquad (4)$$

From the second equation in (4), we have $y = -1$. Thus $2y = -2$. Adding this to the first equation in (4), we get the system
$$x + 0y = 3$$
$$0x + y = -1. \qquad (5)$$

Therefore $x = 3$ and $y = -1$, and we have solved the given system of equations (1).

In the method illustrated above we perform certain operations on the original system (1), transforming it into the system (5) from which the values of the unknowns x and y can be read off directly. At each operation the system is transformed into a new system that is equivalent to the old one. The operations consist of the following basic types:

> 1. Interchanging two equations.
> 2. Multiplying or dividing an equation by a nonzero constant.
> 3. Adding (or subtracting) a constant multiple of one equation to (or from) another equation.

(6)

If we keep track of the positions of the variables and the equals signs, then we can write a system of linear equations as an array of numbers with the variables omitted. For example, the system (1) above,
$$2x + 3y = 3$$
$$x - 2y = 5$$
can be abbreviated as
$$\begin{bmatrix} 2 & 3 & | & 3 \\ 1 & -2 & | & 5 \end{bmatrix}.$$

This array of numbers is called the *augmented matrix* for the given system. Note that, we have written the coefficients of x and y to the left of the vertical line and the constants appearing on the right of the equations to the right of the vertical line. The augmented matrix is simply a way of writing down the system of equations without writing down the variables every time.

Example 1 If the unknown variables are x, y, z, t in that order, the augmented matrix

$$\begin{bmatrix} 2 & -1 & 3 & 4 & | & 5 \\ 1 & 3 & -2 & 0 & | & 7 \\ -4 & 0 & 5 & 1 & | & -3 \end{bmatrix}$$

corresponds to the linear system

$$\begin{aligned} 2x - y + 3z + 4t &= 5 \\ x + 3y - 2z &= 7 \\ -4x + 5z + t &= -3. \end{aligned}$$

Since each row of the augmented matrix corresponds to an equation in the linear system, the three operations listed in (6) correspond to the following three *row operations* of the augmented matrix.

> 1. Interchanging two rows.
> 2. Multiplying or dividing a row by a nonzero constant.
> 3. Adding (or subtracting) a constant multiple of one row to (or from) another row.

(7)

We shall illustrate the use of row operations on an augmented matrix to solve the following system:

$$\begin{aligned} 3x - 2y &= 4 \\ x + 3y &= 5. \end{aligned}$$

The augmented matrix in this case is

$$\begin{bmatrix} 3 & -2 & | & 4 \\ 1 & 3 & | & 5 \end{bmatrix}.$$

For illustrative purposes, we shall solve the system by showing operations on the equations side-by-side with the corresponding operations on the augmented matrix.

SYSTEM

$$\begin{aligned} 3x - 2y &= 4 \\ x + 3y &= 5. \end{aligned}$$

AUGMENTED MATRIX

$$\begin{bmatrix} 3 & -2 & | & 4 \\ 1 & 3 & | & 5 \end{bmatrix}.$$

Interchange the first and second equations to obtain

$$\begin{aligned} x + 3y &= 5 \\ 3x - 2y &= 4 \end{aligned}$$

Interchange the first and second rows to obtain

$$\begin{bmatrix} 1 & 3 & | & 5 \\ 3 & -2 & | & 4 \end{bmatrix}.$$

Add -3 times the first equation to the second equation:

$$\begin{aligned} x + 3y &= 5 \\ 0x - 11y &= -11. \end{aligned}$$

Add -3 times the first row to the second row:

$$\begin{bmatrix} 1 & 3 & | & 5 \\ 0 & -11 & | & -11 \end{bmatrix}.$$

Divide both sides of the second equation by -11 to obtain

$$x + 3y = 5$$
$$0x + y = 1$$

Subtract 3 times the second equation from the first equation to obtain

$$x + 0y = 2$$
$$0x + y = 1.$$

Divide the second row by -11 to obtain

$$\begin{bmatrix} 1 & 3 & | & 5 \\ 0 & 1 & | & 1 \end{bmatrix}.$$

Subtract 3 times the second row from the first to obtain

$$\begin{bmatrix} 1 & 0 & | & 2 \\ 0 & 1 & | & 1 \end{bmatrix}.$$

The solution is therefore $x = 2$ and $y = 1$.

Observe that in the final augmented matrix the values of x and y are given by the entries in the last column. The entries in the remaining part of the augmented matrix also have a very special form called a **unit matrix**, in which the *first* row has 1 in the *first* position, the *second* row has 1 in the *second* position, and the other entries are zero. Thus *to solve a given system we use row operations to change its augmented matrix so that the entries to the left of the vertical line form a unit matrix*. This may not always be possible, but if it is, the solution is then given by the final entries in the last column.

The final form of the augmented matrix is called the **reduced matrix**, and this method of solving linear systems is called the **method of row reduction**.

Before we explain further, we introduce some notations to avoid repeating lengthy expressions. We shall use the symbol R_p for the pth row of the augmented matrix. Thus R_1 denotes the first row, R_2 the second row, and so on. When we say, "Apply $R_2 - 2R_1$," this means "Subtract twice the first row from the second row"; the operation $R_3 + 4R_2$ consists of adding four times the second row to the third row; the operation $R_2 + R_3$ means adding the third row to the second row (*not* the second row to the third row). The operation $2R_3$ means multiplying the third row of the augmented matrix by 2; $-\frac{1}{2}R_1$ means multiplying the first row by $(-\frac{1}{2})$. Finally notation such as $R_1 \leftrightarrow R_3$ means the operation of interchanging the first and third rows. We shall also use notation such as

$$\text{augmented matrix } \mathbf{A} \xrightarrow{R_1 - 2R_2} \text{augmented matrix } \mathbf{B}$$

which means that the augmented matrix $\mathbf{B}$ is obtained by applying the operation $R_1 - 2R_2$ (that is, subtracting twice the second row from the first row) on the matrix $\mathbf{A}$.

Now we are in a position to explain the method of row reduction in detail. We shall do this through an example.

Example 2 Use the method of row reduction to solve the following system of linear equations:

$$2x - 3y + 4z = 13$$
$$x + y + 2z = 4$$
$$3x + 5y - z = -4.$$

Solution The augmented matrix for this system is

$$\begin{bmatrix} 2 & -3 & 4 & | & 13 \\ 1 & 1 & 2 & | & 4 \\ 3 & 5 & -1 & | & -4 \end{bmatrix}.$$

Our purpose is to apply row operations on this matrix until we obtain its reduced form. The best method, generally, is to attack the columns one by one. We first change the first column so that its first element is 1 and other two elements are zero. Then we change the second column so that its second element is 1 and other two elements are zero. Finally we change the third column to have 1 in the third position and zeros elsewhere.

In the first column here, the first entry is 2. To change this entry to 1, we could divide R_1 by 2, or we could interchange R_1 and R_2. If we apply $\frac{1}{2}R_1$, we shall immediately introduce fractions, whereas if we interchange R_1 and R_2 (i.e., apply $R_1 \leftrightarrow R_2$), we shall avoid fractions (at least for the time being). Thus it is preferable to apply $R_1 \leftrightarrow R_2$ and obtain

$$\begin{bmatrix} 1 & 1 & 2 & | & 4 \\ 2 & -3 & 4 & | & 13 \\ 3 & 5 & -1 & | & -4 \end{bmatrix}.$$

Now that we have obtained the first entry 1 in the first column, we *use the first row to change the other elements in the first column to zero*. We can do this by applying the operations $R_2 - 2R_1$ and $R_3 - 3R_1$. This gives us the augmented matrix:

$$\begin{bmatrix} 1 & 1 & 2 & | & 4 \\ 2-2(1) & -3-2(1) & 4-2(2) & | & 13-2(4) \\ 3-3(1) & 5-3(1) & -1-3(2) & | & -4-3(4) \end{bmatrix} = \begin{bmatrix} 1 & 1 & 2 & | & 4 \\ 0 & -5 & 0 & | & 5 \\ 0 & 2 & -7 & | & -16 \end{bmatrix}$$

We now attack the second column. In this column we must have 1 in the second row and zero in the first and third rows. While achieving this goal, *we must be careful not to change the first column*. (This means, for instance, that we cannot add six times the first row to the second row because this will change the first column entries.) There are many ways of making the second entry in column 2 equal to 1. For example, we can apply $-\frac{1}{5}R_2$ or $R_2 + 3R_3$. However, application of $-\frac{1}{5}R_2$ is simpler in this case and leads to the augmented matrix:

$$\begin{bmatrix} 1 & 1 & 2 & | & 4 \\ 0 & 1 & 0 & | & -1 \\ 0 & 2 & -7 & | & -16 \end{bmatrix}.$$

We now use the *second row* to make the other two entries in the second column zero. Thus we apply the operations $R_1 - R_2$ and $R_3 - 2R_2$, which gives

$$\begin{bmatrix} 1-0 & 1-1 & 2-0 & | & 4-(-1) \\ 0 & 1 & 0 & | & -1 \\ 0-2(0) & 2-2(1) & -7-2(0) & | & -16-2(-1) \end{bmatrix} = \begin{bmatrix} 1 & 0 & 2 & | & 5 \\ 0 & 1 & 0 & | & -1 \\ 0 & 0 & -7 & | & -14 \end{bmatrix}.$$

Notice that these operations have not changed the first column. Finally we attack the third column. We must make the third entry in this column equal to 1, and we can do so by applying $-\frac{1}{7}R_3$. This leads to

$$\begin{bmatrix} 1 & 0 & 2 & | & 5 \\ 0 & 1 & 0 & | & -1 \\ 0 & 0 & 1 & | & 2 \end{bmatrix}.$$

We also require that in the third column, the entries in the first and second rows be zero so we apply the operation $R_1 - 2R_3$. This gives

$$\begin{bmatrix} 1 & 0 & 2-2(1) & | & 5-2(2) \\ 0 & 1 & 0 & | & -1 \\ 0 & 0 & 1 & | & 2 \end{bmatrix} = \begin{bmatrix} 1 & 0 & 0 & | & 1 \\ 0 & 1 & 0 & | & -1 \\ 0 & 0 & 1 & | & 2 \end{bmatrix}.$$

Thus we have attained our goal—that is, changed the first three columns of the augmented matrix of the system to a unit matrix. The final augmented matrix represents the system

$$1x + 0y + 0z = 1 \qquad\qquad x = 1$$
$$0x + 1y + 0z = -1 \quad\text{or}\quad y = -1$$
$$0x + 0y + 1z = 2 \qquad\qquad z = 2$$

from which the required solution is read off directly.

In light of the above example, we may summarize the steps involved in changing the augmented matrix to its reduced form as follows.* Each step is carried out by means of one or more of the row operations listed in (7) above.

1. Use row operations to obtain the top entry in the first column equal to 1.
2. Add or subtract appropriate multiples of the first row to the other rows so that the remaining entries in the first column become zero.
3. Without disturbing the first column, use row operations to make the second entry in the second column equal to 1. Then add or subtract suitable multiples of the second row to the other rows to obtain zeros in the rest of the second column.
4. Without disturbing the first two columns, make the third entry in the third column equal to 1. Then, by adding suitable multiples of the third row, obtain zeros in the rest of the third column.
5. Continue the process column by column until the reduced form of the matrix is obtained—that is, until the augmented matrix has a unit matrix to the left of the vertical line. The solutions for the variables are then given by the entries in the last column.

---

*The procedure does not always work, and in certain cases it must be modified, as we shall see below.

As discussed in Chapter 5, a system of linear equations may have a unique solution, an infinite number of solutions, or no solution at all. Let us see what happens with the method of row reduction in the latter two cases.

Example 3 Find the solution of the system
$$-2x + 3y = 4$$
$$6x - 9y = -12.$$

Solution We obtain the following sequence of augmented matrices:

$$\begin{bmatrix} -2 & 3 & | & 4 \\ 6 & -9 & | & -12 \end{bmatrix} \xrightarrow{-\frac{1}{2}R_1} \begin{bmatrix} 1 & -\frac{3}{2} & | & -2 \\ 6 & -9 & | & -12 \end{bmatrix}$$

$$\xrightarrow{R_2 - 6R_1} \begin{bmatrix} 1 & -\frac{3}{2} & | & -2 \\ 0 & 0 & | & 0 \end{bmatrix}.$$

At this stage the method of row reduction grinds to a halt because the second row has both elements zero to the left of the vertical line. This second row corresponds to the equation $0x + 0y = 0$. This equation is a trivial identity and can be dropped. Thus there is only one independent equation in the system, corresponding to the first row in the augmented matrix,

$$x - \frac{3}{2}y = -2.$$

Any pair of values of x and y that satisfies this equation is a solution of the given system.

Geometrically the pair of equations in Example 3 have as their graphs one and the same straight line. (See Section 5.5.) Thus there is not a unique point of intersection, and the number of solutions is infinite.

This situation can also arise when there are more than two equations.

Example 4 Solve the system
$$x + y - z = 4$$
$$3x - 2y + 4z = 9$$
$$9x - y + 5z = 30.$$

Solution We reduce the augmented matrix for this system as follows:

$$\begin{bmatrix} 1 & 1 & -1 & | & 4 \\ 3 & -2 & 4 & | & 9 \\ 9 & -1 & 5 & | & 30 \end{bmatrix} \xrightarrow[R_3 - 9R_1]{R_2 - 3R_1} \begin{bmatrix} 1 & 1 & -1 & | & 4 \\ 0 & -5 & 7 & | & -3 \\ 0 & -10 & 14 & | & -6 \end{bmatrix}$$

$$\xrightarrow{-\frac{1}{5}R_2} \begin{bmatrix} 1 & 1 & -1 & | & 4 \\ 0 & 1 & -\frac{7}{5} & | & \frac{3}{5} \\ 0 & -10 & 14 & | & -6 \end{bmatrix}$$

$$\xrightarrow[R_3 + 10R_2]{R_1 - R_2} \begin{bmatrix} 1 & 0 & \frac{2}{5} & | & \frac{17}{5} \\ 0 & 1 & -\frac{7}{5} & | & \frac{3}{5} \\ 0 & 0 & 0 & | & 0 \end{bmatrix}.$$

So far we have obtained the first two columns in the desired form. However, the third row now consists entirely of zeros, and therefore we are unable to obtain 1 in the third row third column without disturbing the first and second columns. Thus we cannot continue the process of row reduction any further. The augmented matrix we have obtained corresponds to the following equations:

$$x + \tfrac{2}{5}z = \tfrac{17}{5}$$
$$y - \tfrac{7}{5}z = \tfrac{3}{5}. \qquad (i)$$

The third equation is $0x + 0y + 0z = 0$ or $0 = 0$, which is satisfied identically for all values of x, y, and z and can be ignored. We see therefore that the given system of three equations can be reduced to only two independent equations.

The two equations in (i) can be solved for x and y in terms of z:

$$x = \tfrac{17}{5} - \tfrac{2}{5}z = \tfrac{1}{5}(17 - 2z)$$
$$y = \tfrac{3}{5} + \tfrac{7}{5}z = \tfrac{1}{5}(3 + 7z). \qquad (ii)$$

The variable z is arbitrary and can take any value. For example, if $z = 1$, then $x = \tfrac{1}{5}(17 - 2) = 3$, $y = \tfrac{1}{5}(3 + 7) = 2$. Thus $x = 3$, $y = 2$, and $z = 1$ is one solution. By changing the value of z, we get different values of x and y from (ii) and therefore a different solution of the given system. Thus the system has *an infinite number of solutions*. The general form of solution is: $x = \tfrac{1}{5}(17 - 2z)$, $y = \tfrac{1}{5}(3 + 7z)$, z arbitrary.

Example 5 Solve the following system:

$$x + y + 2z = 9$$
$$3x - 2y + 7z = 20$$
$$2x + 7y + 3z = 27.$$

Solution We reduce the augmented matrix for the system as follows:

$$\begin{bmatrix} 1 & 1 & 2 & | & 9 \\ 3 & -2 & 7 & | & 20 \\ 2 & 7 & 3 & | & 27 \end{bmatrix} \xrightarrow{\substack{R_2 - 3R_1 \\ R_3 - 2R_1}} \begin{bmatrix} 1 & 1 & 2 & | & 9 \\ 0 & -5 & 1 & | & -7 \\ 0 & 5 & -1 & | & 9 \end{bmatrix}$$

$$\xrightarrow{-\tfrac{1}{5}R_2} \begin{bmatrix} 1 & 1 & 2 & | & 9 \\ 0 & 1 & -\tfrac{1}{5} & | & \tfrac{7}{5} \\ 0 & 5 & -1 & | & 9 \end{bmatrix}$$

$$\xrightarrow{\substack{R_1 - R_2 \\ R_3 - 5R_2}} \begin{bmatrix} 1 & 0 & \tfrac{11}{5} & | & \tfrac{38}{5} \\ 0 & 1 & -\tfrac{1}{5} & | & \tfrac{7}{5} \\ 0 & 0 & 0 & | & 2 \end{bmatrix}.$$

We have obtained the first two columns in the desired form of a unit matrix. However, we cannot put 1 in the third column and third row without affecting these two columns, so the reduction cannot proceed any further.

The third row represents the equation

$$0x + 0y + 0z = 2 \quad \text{or} \quad 0 = 2.$$

Clearly this equation is absurd. This means that the *system does not have a solution*; that is, there are no values of x, y, and z that satisfy all the three equations of the system.

> In general, *a system will have no solution if a row is obtained in which all the entries except the last one are zero.*

It is clear from these examples that the procedure of row reduction outlined earlier is not sufficiently general to cope with all cases. We cannot always reduce an augmented matrix to a form in which there is a unit matrix to the left of the vertical line. In these exceptional cases, we can reduce it to a form that has the following properties.

1. The first nonzero entry in each row is 1.
2. In the column in which such a 1 appears all other entries are zero.
3. The first nonzero entry in any row is to the right of the first nonzero entry in every preceding row.
4. Any rows consisting entirely of zeros are below the rows with nonzero entries.

In this form the method of row reduction can be used for any system, regardless of the numbers of equations and variables. If the final reduced form contains a row in which only the last entry is nonzero, then the system is inconsistent (i.e., it has no solution). Otherwise it is consistent.

All the systems we have considered had the same number of equations as variables. The method of row reduction is also useful in cases where the number of equations is different from the number of unknowns involved.

If a system has fewer equations than the number of variables, it will always have more than one solution provided it is not inconsistent. We use the method of Example 4 above and try to obtain a unit matrix in the columns corresponding to some of the variables. This then gives the solution for the corresponding variables in terms of the others.

We conclude with an application.

Example 6 A fertilizer plant produces three types of fertilizer. Type A contains 25% potash, 45% nitrate, and 30% phosphate. Type B contains 15% potash, 50% nitrate, and 35% phosphate. Type C contains no potash, 75% nitrate, and 25% phosphate. The plant has supplies of 1.5 tons per day of potash, 5 tons per day of nitrates, and 3 tons per day of phosphate. How much of each type of fertilizer should be produced so as to use up exactly the supplies of ingredients?

Solution Let x tons per day of type A, y tons per day of type B, and z tons per day of type C be produced. The amount of potash consumed will be

$$\tfrac{25}{100}x + \tfrac{15}{100}y + 0z = 1.5.$$

Similarly the amounts of nitrate and phosphate are

$$\frac{45}{100}x + \frac{50}{100}y + \frac{75}{100}z = 5$$

$$\frac{30}{100}x + \frac{35}{100}y + \frac{25}{100}z = 3.$$

Clearing the fractions, we obtain the system

$$5x + 3y + 0z = 30$$
$$9x + 10y + 15z = 100$$
$$6x + 7y + 5z = 60.$$

The row reduction then proceeds as follows

$$\begin{bmatrix} 5 & 3 & 0 & | & 30 \\ 9 & 10 & 15 & | & 100 \\ 6 & 7 & 5 & | & 60 \end{bmatrix} \xrightarrow{\frac{1}{5}R_1} \begin{bmatrix} 1 & \frac{3}{5} & 0 & | & 6 \\ 9 & 10 & 15 & | & 100 \\ 6 & 7 & 5 & | & 60 \end{bmatrix}$$

$$\xrightarrow[R_3 - 6R_1]{R_2 - 9R_1} \begin{bmatrix} 1 & \frac{3}{5} & 0 & | & 6 \\ 0 & \frac{23}{5} & 15 & | & 46 \\ 0 & \frac{17}{5} & 5 & | & 24 \end{bmatrix}$$

$$\xrightarrow{\frac{5}{23}R_2} \begin{bmatrix} 1 & \frac{3}{5} & 0 & | & 6 \\ 0 & 1 & \frac{75}{23} & | & 10 \\ 0 & \frac{17}{5} & 5 & | & 24 \end{bmatrix}$$

$$\xrightarrow[R_3 - \frac{17}{5}R_2]{R_1 - \frac{3}{5}R_2} \begin{bmatrix} 1 & 0 & -\frac{45}{23} & | & 0 \\ 0 & 1 & \frac{75}{23} & | & 10 \\ 0 & 0 & -\frac{140}{23} & | & -10 \end{bmatrix}$$

$$\xrightarrow{-\frac{23}{140}R_3} \begin{bmatrix} 1 & 0 & -\frac{45}{23} & | & 0 \\ 0 & 1 & \frac{75}{23} & | & 10 \\ 0 & 0 & 1 & | & \frac{23}{14} \end{bmatrix}$$

$$\xrightarrow[R_2 - \frac{75}{23}R_3]{R_1 + \frac{45}{23}R_3} \begin{bmatrix} 1 & 0 & 0 & | & \frac{45}{14} \\ 0 & 1 & 0 & | & \frac{65}{14} \\ 0 & 0 & 1 & | & \frac{23}{14} \end{bmatrix}.$$

The solution is therefore $x = \frac{45}{14}$, $y = \frac{65}{14}$, $z = \frac{23}{14}$. The plant should produce $\frac{45}{14} \approx 3.21$ tons of type A, $\frac{65}{14} \approx 4.64$ tons of type B, and $\frac{23}{14} \approx 1.64$ tons of type C per day.

Exercises 11.1

(1–40) Use the method of row reduction to find the solutions (if they exist) of the following systems of linear equations.

1. $2x + 3y = 7$
$3x - y = 5.$

2. $x + 2y = 1$
$3y + 2x = 3.$

3. $u + 3v = 1$
$2u - v = 9.$

4. $3p + 2q = 5$
$p - 3q + 2 = 0.$

5. $2x - 3y = 4$
$6y = 4(x - 5).$

6. $x = 10 - 2y$
$y = 5 - \dfrac{x}{2}.$

7. $x + y = 3$
$y + z = 5$
$x + z = 4.$

8. $x + 2y = 1$
$3y + 5z = 7$
$2x - y = 7.$

9. $x + y + z = 6$
$2x - y + 3z = 9$
$-x + 2y + z = 6.$

10. $x + 2y - z = -3$
$3y + 4z = 5$
$2x - y + 3z = 9.$

11. $3x_1 + 2x_2 + x_3 = 6$
$2x_1 - x_2 + 4x_3 = -4$
$x_1 + x_2 - 2x_3 = 5.$

12. $2u - 3v + 4w = 13$
$u + v + w = 6$
$-3u + 2v + w + 1 = 0.$

13. $p - q + r = -1$
$3p - 2r = -7$
$r + 4q = 10.$

14. $b = 3 - a$
$c = 4 - a - b$
$3a + 2b + c = 8.$

15. $x + y + z = 5$
$-x + y + 3z = 1$
$x + 2y + 3z = 8.$

16. $x + y = 3$
$2x + y + z = 4$
$2x + 2y - 2z = 5.$

17. $x + y + z = 3$
$-x - y + z = -1$
$3x + 3y + 4z = 8.$

18. $6x - 5y + 6z = 7$
$2x + y + 6z = 5$
$2x - y + 3z = 3.$

19. $u - v + 2w = 5$
$4u + v + 3w = 15$
$5u - 2v + 7w = 31.$

20. $-x + y + z = 4$
$3x - y + 2z = -3$
$4x - 2y + z = 3.$

21. $2x + y - z = 2$
$3x + 2y + 4z = 8$
$5x + 4y + 14z = 20.$

22. $a + b - 2c = 3$
$2a + 3b + c = 13$
$7a + 9b - 4c = 35.$

23. $x + 2y + z - t = 0$
$y - 2z + 2t = 13$
$2x + 4y - z + 2t = 17$
$y - z - 3t = 0.$

24. $p - q - r = 4$
$q - r - s = -5$
$r - s - p = -8$
$p + 2q + 2r + s = -5.$

25. $x + y + z = 1$
$2x + 3y - w = 3$
$-x + 2z + 3w = 3$
$2y - z + w = 5.$

26. $x_1 + x_2 + x_3 + x_4 = 2$
$x_1 - x_2 + x_3 + 2x_4 = -4$
$2x_1 + x_2 - x_3 + x_4 = 1$
$-x_1 + x_2 + x_3 - x_4 = 4.$

27. $x + 2y - 3z - t = 2$
$2x + 4y + z - t = 1$
$3x + 6y + 2z + t = -7$
$x + 2y + z + t = 6.$

28. $p + 2q - r + 2s = 6$
$-2p + q + 2r + 3s = 6$
$3p + 5q - 3r + s = 0$
$p + 2q - r + s = 2.$

29. $u + v - w = 4$
$3u - v + 2w = -1$
$2u + 3v + w = 7$
$u + 2v + 3w = 2.$

30. $3x + 2y + z = 10$
$2x - y + 3z = 9$
$x + y - 2z = -3$
$2x + 3y + 4z = 20.$

31. $x + y - 2z = -3$
$2x + 3y + z = 10$
$-x + 2y + 3z = 9$
$3x + y - z = 4$
$x - 2y - z = 2.$

32. $x_1 + 2x_2 - x_3 = 2$
$3x_1 + x_2 + 4x_3 = 17$
$-2x_1 + 3x_2 + 5x_3 = 19$
$x_1 + x_2 + 2x_3 = 9$
$4x_1 - x_2 + x_3 = 4.$

33. $2x - y + 3z = 9$
$3y - 6x - 9z = 12.$

34. $u - 2v + w = 7$
$5u - 10v + 5w = 36.$

35. $x + y - z = 2$
$2x - 3y + 4z = -3.$

36. $2x + y - 3z = 10$
$3x + 2y + z = 11.$

37. Find $a, b,$ and c such that the parabola $y = ax^2 + bx + c$ passes through the points $(2, 0)$, $(4, 2)$, and $(-2, 3)$.

38. Find the equation of the circle that passes through the three points $(-1, 2), (2, -2),$ and $(3, -1)$. (See Section 7.1.)

39. A firm produces three products, A, B, and C, that require processing by three machines. The time (in hours) required for processing one unit of each product by the three machines is given below.

$$\begin{array}{c} & \begin{array}{ccc} A & B & C \end{array} \\ \begin{array}{c} \text{Machine I} \\ \text{Machine II} \\ \text{Machine III} \end{array} & \begin{bmatrix} 3 & 1 & 2 \\ 1 & 2 & 4 \\ 2 & 1 & 1 \end{bmatrix} \end{array}$$

Machine I is available for 850 hours, machine II for 1200 hours, and machine III for 550 hours. How many units of each product should be produced to make use of all the available time on the machines?

40. A shipping company loaded three types of cargo on its light transport plane. The space required by each unit of the three types of cargo was 5, 2, and 4 cubic feet, respectively. Each unit of the three types of cargo weighed 2, 3, and 1 kilograms, respectively, whereas the unit values of the three types of cargo were $10, $40, and $60, respectively. Determine the number of units of each type of cargo loaded if the cargo had a total value of $13,500, occupied 1050 cubic feet of space, and weighed 550 kilograms.

41. A person invested a total of $20,000 in three different investments at 6%, 8%, and 10%. The total annual return was $1624, and the return from the 10% investment was twice the return from the 6% investment. How much was invested in each?

42. A contractor has 5000 work-hours of labor available for three projects. The costs per work-hour of the three projects are $8, $10, and $12, respectively, and the total cost is $53,000. If the number of work-hours for the third project is equal to the sum of the work-hours for the first two projects, find the number of work-hours that can be used for each project.

(43–44) In the electrical circuit shown below, R_1, R_2, and R_3 are resistances (in ohms), E_1 and E_2 are the voltages of the cells shown, and i_1, i_2, and i_3 are the currents shown (in amperes). Kirchoff's laws lead to the following equations:

$$R_1 i_1 + R_3 i_3 = E_1$$
$$R_2 i_2 + R_3 i_3 = E_2$$
$$i_1 + i_2 - i_3 = 0.$$

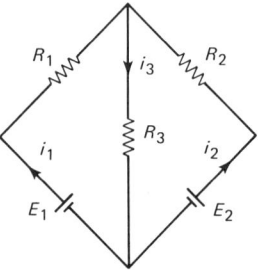

43. Find i_1, i_2, and i_3 if $R_1 = 3$, $R_2 = 2$, $R_3 = 4$, $E_1 = 0$, and $E_2 = 8$.

44. Find i_1, i_2, and i_3 if $R_1 = R_2 = 5$, $R_3 = 8$, $E_1 = 4$, $E_2 = 6$.

11.2

Matrices

Often we encounter data that can be arranged in a natural way as a rectangular array of numbers. One example is the augmented matrix of a system of linear equations. Other examples arise in various fields such as business and economics, engineering, and the biological and physical sciences.

Consider the following example from agriculture. Two strains of wheat (called S1 and S2) are tested for their average yield (in bushels per acre). Each strain can be treated with one of three types of fertilizer (called F1, F2, and F3). The results are shown in the following table.

	F1	F2	F3
S1	48	54	59
S2	49	57	52

For example, strain S1 treated with fertilizer F2 yields 54 bushels per acre. Strain S2 treated with fertilizer F1 yields 49 bushels per acre, and so on. We see that the six numbers in the table naturally form a rectangular array.

DEFINITIONS A *matrix* (plural *matrices*) is a rectangular array of numbers. The array is enclosed in large brackets, as shown in the examples below. Matrices are generally denoted by boldface capital letters **A**, **B**, **C**, and so on. Some examples of matrices follow:

$$\mathbf{A} = \begin{bmatrix} 2 & -3 & 7 \\ 1 & 0 & 4 \end{bmatrix}, \quad \mathbf{B} = \begin{bmatrix} 3 & 4 & 5 & 6 \\ 7 & 8 & 9 & 1 \\ 5 & 4 & 3 & 2 \end{bmatrix}, \quad \mathbf{C} = \begin{bmatrix} 4 \\ 2 \\ 3 \\ 1 \end{bmatrix}$$

$$\mathbf{D} = [1 \quad 2 \quad 3 \quad 5 \quad 6], \quad \mathbf{E} = [3].$$

The real numbers that form the array are called the *entries* or *elements* of the matrix. The elements in any horizontal line form a *row* and those in any vertical line form a *column* of the matrix. For example, the matrix **B** above has three rows and four columns. The elements of first row are 3, 4, 5, 6 and those of third column are 5, 9, 3.

If a matrix has m rows and n columns, then it is said to be of *size $m \times n$* (which is read "m by n"). Of the matrices given above, **A** is of size 2×3, **B** of size 3×4, and **C** of size 4×1.

A matrix of size $1 \times n$ has only one row and is often called a *row matrix* or a *row vector*. Similarly a matrix of size $m \times 1$ has only one column and is called a *column matrix* or a *column vector*. In the above examples, **D** is a row vector and **C** is a column vector.

It is often convenient to use a double subscript notation for the elements of a matrix, a_{ij} being used to denote the element of the matrix **A** that lies in the ith row and the jth column. For example a_{24} denotes the entry that lies in the second row and fourth column of **A**. If A is the 2×3 matrix

$$\mathbf{A} = \begin{bmatrix} 2 & -3 & 7 \\ 1 & 0 & 4 \end{bmatrix}$$

then $a_{11} = 2$, $a_{12} = -3$, $a_{13} = 7$, $a_{21} = 1$, $a_{22} = 0$, and $a_{23} = 4$.

In general, if **A** is an $m \times n$ matrix, we can write

$$\mathbf{A} = \begin{bmatrix} a_{11} & a_{12} & a_{13} & \cdots & a_{1j} & \cdots & a_{1n} \\ a_{21} & a_{22} & a_{23} & \cdots & a_{2j} & \cdots & a_{2n} \\ \vdots & \vdots & \vdots & & \vdots & & \vdots \\ a_{i1} & a_{i2} & a_{i3} & \cdots & a_{ij} & \cdots & a_{in} \\ \vdots & \vdots & \vdots & & \vdots & & \vdots \\ a_{m1} & a_{m2} & a_{m3} & \cdots & a_{mj} & \cdots & a_{mn} \end{bmatrix} \begin{matrix} \leftarrow \text{first row} \\ \leftarrow \text{second row} \\ \\ \leftarrow i\text{th row} \\ \\ \leftarrow m\text{th row} \end{matrix}$$

↑ ↑ ↑ ↑ ↑
first second third *j*th   *n*th
column column column column column

The matrix **A** can be denoted by $[a_{ij}]$, where its size is understood. If the size is also to be specified, then we can write $\mathbf{A} = [a_{ij}]_{m \times n}$.

Example 1 Write down the 2×3 matrix $[a_{ij}]$ for which

$$a_{ij} = 2i + j - 3.$$

Solution A 2×3 matrix $[a_{ij}]$ has the following form:

$$\mathbf{A} = \begin{bmatrix} a_{11} & a_{12} & a_{13} \\ a_{21} & a_{22} & a_{23} \end{bmatrix}. \tag{i}$$

Now we have

$$a_{ij} = 2i + j - 3. \tag{ii}$$

Substituting $i = j = 1$ in (ii) gives

$$a_{11} = 2(1) + 1 - 3 = 0.$$

Similarly,

$$a_{12} = 2(1) + 2 - 3 = 1$$
$$a_{13} = 2(1) + 3 - 3 = 2$$
$$a_{21} = 2(2) + 1 - 3 = 2$$
$$a_{22} = 2(2) + 2 - 3 = 3$$
$$a_{33} = 2(2) + 3 - 3 = 4.$$

Therefore from (i), we have

$$\mathbf{A} = \begin{bmatrix} 0 & 1 & 2 \\ 2 & 3 & 4 \end{bmatrix}.$$

SOME SPECIAL MATRICES

1. A matrix all of whose elements are zero is called a ***zero matrix*** and is denoted by **0**. Thus, the zero matrix of size 2×3 is

$$\mathbf{0} = \begin{bmatrix} 0 & 0 & 0 \\ 0 & 0 & 0 \end{bmatrix}.$$

2. A matrix having the same number of rows and columns is called a *square matrix*. The following are examples of square matrices:

$$\mathbf{P} = \begin{bmatrix} 1 & 2 \\ 3 & 4 \end{bmatrix}, \quad \mathbf{Q} = \begin{bmatrix} 2 & 1 & 3 \\ 4 & -2 & 1 \\ 3 & 0 & 2 \end{bmatrix}, \quad \mathbf{R} = [2].$$

The size of **P** is 2×2, of **Q** is 3×3, and of **R** is 1×1.

If $\mathbf{A} = [a_{ij}]$ is a square matrix, then the elements a_{ij} for which $i = j$ (that is, the elements a_{11}, a_{22}, a_{33}, and so on) are called the *diagonal elements* of the matrix.

3. A square matrix is called an *identity matrix* or *unit matrix* if its diagonal elements are all equal to 1 and its nondiagonal elements are all equal to zero. The identity matrices of sizes 2×2 and 3×3 are therefore as follows:

$$\begin{bmatrix} 1 & 0 \\ 0 & 1 \end{bmatrix}, \quad \begin{bmatrix} 1 & 0 & 0 \\ 0 & 1 & 0 \\ 0 & 0 & 1 \end{bmatrix}.$$

The identity matrix is usually denoted by **I** when its size is understood without ambiguity. Sometimes the identity matrix of size $n \times n$ is denoted by $\mathbf{I}_n$ when we need to make the size clear.

EQUAL MATRICES

DEFINITION Two matrices **A** and **B** are said to be *equal* if (i) they are of the same size and (ii) their corresponding elements are equal.

Example 2 Given

$$\mathbf{A} = \begin{bmatrix} 2 & x & 3 \\ y & -1 & 4 \end{bmatrix} \quad \text{and} \quad \mathbf{B} = \begin{bmatrix} a & 5 & 3 \\ 0 & b & 4 \end{bmatrix}.$$

Find x, y, a, and b if $\mathbf{A} = \mathbf{B}$.

Solution Clearly **A** and **B** are of the same size, 2×3. Thus condition (i) is satisfied. From condition (ii), the corresponding entries in the two matrices must be equal—that is,

$$2 = a, \quad x = 5, \quad y = 0, \quad \text{and} \quad -1 = b.$$

Thus, if $\mathbf{A} = \mathbf{B}$, then $x = 5, y = 0, a = 2, b = -1$.

SCALAR MULTIPLICATION OF A MATRIX

Scalar multiplication of a matrix refers to the operation of multiplying the matrix by a real number. If $\mathbf{A} = [a_{ij}]$ is an $m \times n$ matrix and c is any real number, then the product $c\mathbf{A}$ is an $m \times n$ matrix obtained by multiplying each element of **A** by the constant c. In other words, $c\mathbf{A} = [ca_{ij}]$.

Example 3 Given $\mathbf{A} = \begin{bmatrix} 1 & 0 & -1 \\ 0 & -2 & 4 \end{bmatrix}$, find $2\mathbf{A}$.

Solution

$$2\mathbf{A} = 2\begin{bmatrix} 1 & 0 & -1 \\ 0 & -2 & 4 \end{bmatrix}$$

$$= \begin{bmatrix} 2(1) & 2(0) & 2(-1) \\ 2(0) & 2(-2) & 2(4) \end{bmatrix}$$

$$= \begin{bmatrix} 2 & 0 & -2 \\ 0 & -4 & 8 \end{bmatrix}.$$

ADDITION AND SUBTRACTION OF MATRICES

Two matrices $\mathbf{A}$ and $\mathbf{B}$ of the *same size* can be added (or subtracted) by adding (or subtracting) their corresponding elements. In other words, if $\mathbf{A} = [a_{ij}]$ and $\mathbf{B} = [b_{ij}]$ are two matrices of the same size, then $\mathbf{A} + \mathbf{B} = [a_{ij} + b_{ij}]$ and $\mathbf{A} - \mathbf{B} = [a_{ij} - b_{ij}]$.

Examples 4 (a) $\begin{bmatrix} 2 & 0 & -1 \\ 3 & 4 & 5 \\ 1 & -2 & 3 \end{bmatrix} + \begin{bmatrix} 3 & 1 & 2 \\ 2 & 0 & -3 \\ 3 & 2 & -4 \end{bmatrix}$

$= \begin{bmatrix} 2+3 & 0+1 & -1+2 \\ 3+2 & 4+0 & 5+(-3) \\ 1+3 & -2+2 & 3+(-4) \end{bmatrix} = \begin{bmatrix} 5 & 1 & 1 \\ 5 & 4 & 2 \\ 4 & 0 & -1 \end{bmatrix}.$

(b) $\begin{bmatrix} 2 & 0 & -1 \\ 3 & 4 & 5 \end{bmatrix} - \begin{bmatrix} 4 & 1 & -2 \\ 2 & 6 & 1 \end{bmatrix} = \begin{bmatrix} -2 & -1 & 1 \\ 1 & -2 & 4 \end{bmatrix}.$

Example 5 Given

$$\mathbf{A} = \begin{bmatrix} 3 & 1 & 4 \\ 2 & -3 & 5 \end{bmatrix} \text{ and } \mathbf{B} = \begin{bmatrix} 1 & 2 & 3 \\ 4 & 5 & 6 \end{bmatrix}$$

determine the matrix $\mathbf{X}$ such that $\mathbf{X} + \mathbf{A} = 2\mathbf{B}$.

Solution We have $\mathbf{X} + \mathbf{A} = 2\mathbf{B}$ or $\mathbf{X} = 2\mathbf{B} - \mathbf{A}$. Thus,

$$\mathbf{X} = 2\begin{bmatrix} 1 & 2 & 3 \\ 4 & 5 & 6 \end{bmatrix} - \begin{bmatrix} 3 & 1 & 4 \\ 2 & -3 & 5 \end{bmatrix}$$

$$= \begin{bmatrix} 2 & 4 & 6 \\ 8 & 10 & 12 \end{bmatrix} - \begin{bmatrix} 3 & 1 & 4 \\ 2 & -3 & 5 \end{bmatrix}$$

$$= \begin{bmatrix} -1 & 3 & 2 \\ 6 & 13 & 7 \end{bmatrix}.$$

Example 6 A manufacturing firm makes three models of TV sets, each model in three different sizes. The production capacity (in thousands) at its New York plant is given by the matrix:

$$\begin{array}{c} \text{Size 1 (20'')} \\ \text{Size 2 (23'')} \\ \text{Size 3 (26'')} \end{array} \begin{array}{ccc} \text{Model I} & \text{Model II} & \text{Model III} \\ \begin{bmatrix} 5 & 3 & 2 \\ 7 & 4 & 5 \\ 10 & 8 & 4 \end{bmatrix} \end{array} = \mathbf{A}.$$

(In other words, the plant capacity is 5000 sets of Model I 20-inch TV's; 8000 sets of Model II 26-inch TV's and so on.) The production capacity at its California plant is given by the matrix:

$$\begin{array}{c} \text{Size 1} \\ \text{Size 2} \\ \text{Size 3} \end{array} \begin{array}{ccc} \text{Model I} & \text{Model II} & \text{Model III} \\ \begin{bmatrix} 4 & 5 & 3 \\ 9 & 6 & 4 \\ 8 & 12 & 2 \end{bmatrix} \end{array} = \mathbf{B}.$$

(a) What is the total production matrix at the two plants?
(b) If the firm decides to increase production at its New York plant by 20%, what will be the new production matrix at that plant?

Solution (a) The combined production (in thousands) at the two plants is given by the sum of the two matrices **A** and **B**:

$$\mathbf{A} + \mathbf{B} = \begin{bmatrix} 5 & 3 & 2 \\ 7 & 4 & 5 \\ 10 & 8 & 4 \end{bmatrix} + \begin{bmatrix} 4 & 5 & 3 \\ 9 & 6 & 4 \\ 8 & 12 & 2 \end{bmatrix} = \begin{bmatrix} 9 & 8 & 5 \\ 16 & 10 & 9 \\ 18 & 20 & 6 \end{bmatrix}$$

(i.e., 9000 sets of Model I 20-inch TV's, and so on).

(b) If the production at New York increases by 20%, the new production (in thousands) will be given by the matrix 1.2**A**:

$$1.2\mathbf{A} = 1.2 \begin{bmatrix} 5 & 3 & 2 \\ 7 & 4 & 5 \\ 10 & 8 & 4 \end{bmatrix} = \begin{bmatrix} 6 & 3.6 & 2.4 \\ 8.4 & 4.8 & 6 \\ 12 & 9.6 & 4.8 \end{bmatrix}$$

(i.e., 4800 sets of Model II 23-inch TV's, and so on).

Diagnostic Test 11.2

Fill in the blanks.

1. If $\begin{bmatrix} x & 2 \\ 3 & y \end{bmatrix} = \begin{bmatrix} 1 & u \\ v & 5 \end{bmatrix}$, then $x =$ _____, $y =$ _____, $u =$ _____, and $v =$ _____.

2. If $\mathbf{A} = \begin{bmatrix} 3 & 4 \\ 5 & -7 \end{bmatrix}$, then $3\mathbf{A} =$ _____.

3. Two matrices of size $m \times n$ and $p \times q$ can be added or subtracted if and only if _____.

4. If $\mathbf{A} = \begin{bmatrix} 2 & 3 & 1 \\ 3 & -4 & 5 \end{bmatrix}$ and $\mathbf{B} = \begin{bmatrix} 1 & 2 \\ 3 & 4 \\ 5 & 6 \end{bmatrix}$, then $\mathbf{A} + \mathbf{B} =$ _____.

5. If $\begin{bmatrix} x & 2 & y \\ 3 & t & 4 \end{bmatrix} = \begin{bmatrix} 1 & 2 & 3 \\ 3 & 5 & 4 \end{bmatrix}$, then $x =$ _____, $y =$ _____, and $t =$ _____.

6. If $\mathbf{A} + \mathbf{B} = \mathbf{A}$ then $\mathbf{B} =$ _____.

7. If $\mathbf{A} + \mathbf{B} = \mathbf{A} + \mathbf{C}$ then $\mathbf{B} =$ _____.

Exercises 11.2

1. Write down the sizes of the following matrices.

$$\mathbf{A} = \begin{bmatrix} 1 & 0 \\ 2 & 3 \end{bmatrix}, \quad \mathbf{B} = \begin{bmatrix} 2 & 3 & 1 \\ -1 & 2 & 3 \end{bmatrix},$$

$$\mathbf{C} = \begin{bmatrix} 3 \\ 1 \\ 2 \end{bmatrix}, \quad \mathbf{D} = \begin{bmatrix} 1 & 2 & 3 \\ 4 & 5 & 6 \\ 9 & 8 & 7 \end{bmatrix}$$

$$\mathbf{E} = \begin{bmatrix} 3 & 4 & 5 \\ 1 & 0 & 2 \end{bmatrix}, \quad \mathbf{F} = \begin{bmatrix} 2 & -1 \\ -1 & 1 \end{bmatrix}$$

$$\mathbf{G} = [4 \; 1 \; 3], \quad \mathbf{H} = [1].$$

5. Give an example of a 3×3 matrix $[c_{ij}]$ for which $c_{ij} = -c_{ji}$.

2. In Exercise 1, if $\mathbf{B} = [b_{ij}]$, find $b_{12}, b_{22}, b_{21}, b_{23}, b_{32}$.

3. Write down the 2×2 matrix $\mathbf{A} = [a_{ij}]$ for which $a_{ij} = i + j - 2$.

4. Write down the 3×2 matrix $\mathbf{B} = [b_{ij}]$ for which $b_{ij} = 2i + 3j - 4$.

6. Write down the 3×4 matrix $\mathbf{A} = [a_{ij}]$ for which
$$a_{ij} = \begin{cases} i + j & \text{if } i \neq j \\ 0 & \text{if } i = j. \end{cases}$$

(7–14) Perform the indicated operations and simplify.

7. $3 \begin{bmatrix} 2 & 4 \\ 1 & 3 \end{bmatrix}.$

8. $4 \begin{bmatrix} 1 & -2 \\ -1 & 5 \end{bmatrix}.$

9. $-2 \begin{bmatrix} 1 & -2 & 3 \\ -2 & 1 & -4 \\ 3 & 0 & 2 \end{bmatrix}.$

10. $-3 \begin{bmatrix} 0 & 1 & -2 \\ 2 & -3 & 4 \\ 1 & 5 & 3 \end{bmatrix}.$

11. $\begin{bmatrix} 2 & 1 & 3 \\ -1 & 4 & 7 \end{bmatrix} + \begin{bmatrix} 0 & -1 & 2 \\ 1 & 2 & -8 \end{bmatrix}.$

12. $\begin{bmatrix} 1 & -2 \\ 3 & 4 \\ 5 & 1 \end{bmatrix} + \begin{bmatrix} 3 & 1 \\ -1 & 2 \\ 1 & 3 \end{bmatrix}.$

13. $\begin{bmatrix} 3 & 1 \\ -2 & 5 \\ 0 & -1 \end{bmatrix} - \begin{bmatrix} 1 & -2 \\ 2 & -1 \\ -3 & 2 \end{bmatrix}.$

14. $\begin{bmatrix} 2 & 5 & 7 \\ 1 & -2 & 3 \\ 3 & 4 & -1 \end{bmatrix} - \begin{bmatrix} 1 & -1 & 2 \\ 2 & 1 & -1 \\ 3 & 2 & 1 \end{bmatrix}.$

15. $2 \begin{bmatrix} 1 & 2 \\ -1 & 3 \end{bmatrix} + 3 \begin{bmatrix} -2 & 3 \\ 1 & 0 \end{bmatrix}.$

16. $3 \begin{bmatrix} 2 & 1 \\ -1 & 3 \\ 4 & 7 \end{bmatrix} - 2 \begin{bmatrix} 1 & -2 \\ 2 & 3 \\ -3 & 0 \end{bmatrix}.$

17. $2 \begin{bmatrix} 1 & 2 & 3 \\ 2 & -1 & 0 \\ 4 & 5 & 6 \end{bmatrix} + 3 \begin{bmatrix} 0 & -1 & 2 \\ 3 & 2 & -4 \\ -1 & 0 & 3 \end{bmatrix}.$

18. $4 \begin{bmatrix} 1 & 0 & -3 & 4 \\ 2 & -1 & 5 & 1 \\ 3 & 2 & 0 & -2 \end{bmatrix} - 5 \begin{bmatrix} 2 & -1 & 2 & 3 \\ 1 & 0 & -3 & 4 \\ 3 & 1 & 0 & -5 \end{bmatrix}.$

(19–28) Determine the values of the unknowns so that the following matrix equations are true.

19. $\begin{bmatrix} x & 2 \\ 3 & y \end{bmatrix} = \begin{bmatrix} 1 & 2 \\ 3 & 4 \end{bmatrix}.$

20. $\begin{bmatrix} 3 & -1 \\ x & 0 \end{bmatrix} = \begin{bmatrix} y+2 & z \\ 4 & t-1 \end{bmatrix}.$

21. $\begin{bmatrix} 4 & x & 3 \\ y & -1 & 2 \end{bmatrix} = \begin{bmatrix} y-1 & 2-x & 3 \\ 5 & z+1 & 2 \end{bmatrix}.$

22. $\begin{bmatrix} x+2 & 5 & y-3 \\ 4 & z-6 & 7 \end{bmatrix} = \begin{bmatrix} 3 & t+1 & 2y-5 \\ 4 & 2 & z-1 \end{bmatrix}.$

23. $\begin{bmatrix} 1 & -2 & x \\ y & 3 & 4 \\ 2 & z & 3 \end{bmatrix} = \begin{bmatrix} 1 & t & 6 \\ 5 & 3 & 4 \\ u & 2 & v \end{bmatrix}.$

24. $\begin{bmatrix} x+1 & 2 & 3 \\ 4 & y-1 & 5 \\ u & -1 & z+2 \end{bmatrix} = \begin{bmatrix} 2x-1 & t+1 & 3 \\ v+1 & -3 & 5 \\ -4 & w-1 & 2z-1 \end{bmatrix}.$

25. $\begin{bmatrix} x & 3 & 4 \\ 2 & -1 & y \\ 1 & z & -3 \end{bmatrix} + \begin{bmatrix} 1 & t & -1 \\ 3 & 4 & x \\ u & y & 2 \end{bmatrix} = \begin{bmatrix} 2 & 7 & v+1 \\ 5 & w-2 & 3 \\ 0 & 5 & -1 \end{bmatrix}.$

26. $\begin{bmatrix} x+1 & -2 & 3 \\ 4 & 1 & z+2 \\ -1 & y & 2 \end{bmatrix} + 2\begin{bmatrix} 3 & -1 & 2 \\ 1 & 2 & -3 \\ 4 & -1 & 0 \end{bmatrix} = \begin{bmatrix} 6 & u+2 & 7 \\ v+1 & 5 & -7 \\ 7 & 0 & w \end{bmatrix}.$

27. $3\begin{bmatrix} x & 1 & -1 \\ 0 & -2 & 3 \\ 1 & y & 2 \end{bmatrix} + 2\begin{bmatrix} -2 & t & 0 \\ z & 1 & -1 \\ u & 2 & v \end{bmatrix} = \begin{bmatrix} w-4 & 1 & -v \\ 4 & 2u & 2v+y \\ -1 & x+7 & 12 \end{bmatrix}.$

28. $2\begin{bmatrix} 1 & x+1 & 0 \\ 0 & -2 & y-1 \\ z & 1 & 2 \end{bmatrix} - 3\begin{bmatrix} u & -1 & 2 \\ 1 & v+2 & 3 \\ 0 & -3 & 1 \end{bmatrix} = \begin{bmatrix} 8 & 7 & 2v-2z \\ u+y & -7 & 1-7z \\ 4 & w+11 & t \end{bmatrix}.$

29. A company has plants at three locations X, Y, Z and four warehouses at the locations A, B, C, and D. The cost (in dollars) of transporting each unit of its product from plants to warehouses is given by the following matrix:

FROM

	X	Y	Z
A	10	12	15
B	13	10	12
C	8	15	6
D	16	9	10

TO

(a) If the transportation costs are increased by $1 per unit uniformly, what is the new matrix?
(b) If the transportation costs go up by 20%, write the new costs in matrix form.

30. The trade between three countries I, II, and III during 1976 (in millions of U.S. dollars) is given by the matrix $\mathbf{A} = [a_{ij}]$, where a_{ij} represents the exports from the ith country to the jth country,

$$\mathbf{A} = \begin{bmatrix} 0 & 16 & 20 \\ 17 & 0 & 18 \\ 21 & 14 & 0 \end{bmatrix}.$$

The trade between these three countries during the year 1977 (in millions of U.S. dollars) was given by the matrix

$$\mathbf{B} = \begin{bmatrix} 0 & 17 & 19 \\ 18 & 0 & 20 \\ 24 & 16 & 0 \end{bmatrix}.$$

(a) Write a matrix representing the total trade between the three countries for the two-year period 1976 and 1977.
(b) If in 1976 and 1977 one U.S. dollar was equal to 5 Hong Kong dollars, write the matrix representing the total trade for the two years in Hong Kong dollars.

11.3

Products of Matrices

In this section we explain when and how two matrices can be multiplied. We begin by considering the multiplication of a row vector and a column vector.

DEFINITION Let **A** be a $1 \times n$ row vector and **B** be an $n \times 1$ column vector. Then the matrix product **AB** is defined to be the sum of the products of corresponding elements in the two vectors. That is, if

$$\mathbf{A} = [a_1 \quad a_2 \quad \ldots \quad a_n] \quad \text{and} \quad \mathbf{B} = \begin{bmatrix} b_1 \\ b_2 \\ \vdots \\ b_n \end{bmatrix}$$

then

$$\mathbf{AB} = [a_1 \quad a_2 \quad \ldots \quad a_n] \begin{bmatrix} b_1 \\ b_2 \\ \vdots \\ b_n \end{bmatrix} = [a_1 b_1 + a_2 b_2 + \cdots + a_n b_n].$$

Note that the product matrix **AB** is of size 1×1.

Notes
1. If **A** is a $1 \times n$ matrix and **B** is an $m \times 1$ matrix, then the product **AB** *is not defined* when $n \neq m$. Thus the product **AB** is defined only when the row vector **A** and the column vector **B** have the same number of elements.
2. In forming this type of product of a row vector and a column vector, the row vector must always be on the left.

Examples 1 (a) $[2 \quad 3] \begin{bmatrix} -1 \\ 4 \end{bmatrix} = [2(-1) + 3(4)] = [-2 + 12] = [10]$.

(b) $[4 \quad 5 \quad 6] \begin{bmatrix} 2 \\ 0 \\ -1 \end{bmatrix} = [4(2) + 5(0) + 6(-1)] = [2]$.

Let us now extend our definition of the product when the two given matrices are not row and column vectors. Let **A** and **B** be two matrices of given size. The product **AB** is defined only *if the number of columns in* **A** *is equal to the number of rows in* **B**. If this condition is met, then the product **AB** can be defined as follows:

DEFINITION Let $\mathbf{A} = [a_{ij}]$ be an $m \times n$ matrix and $\mathbf{B} = [b_{ij}]$ be an $n \times p$ matrix. Then the product **AB** is an $m \times p$ matrix $\mathbf{C} = [c_{ij}]$, where ij element c_{ij} is obtained by multiplying the elements in the ith row of **A** by the corresponding elements in the jth column of **B** and summing the products.

The ith row in **A** and jth column in **B** are shown below:

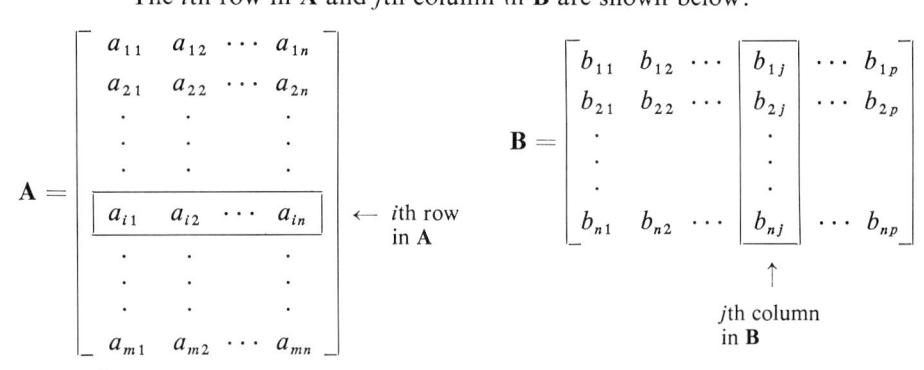

$$\mathbf{A} = \begin{bmatrix} a_{11} & a_{12} & \cdots & a_{1n} \\ a_{21} & a_{22} & \cdots & a_{2n} \\ \vdots & \vdots & & \vdots \\ \boxed{a_{i1} \quad a_{i2} \quad \cdots \quad a_{in}} \\ \vdots & \vdots & & \vdots \\ a_{m1} & a_{m2} & \cdots & a_{mn} \end{bmatrix} \leftarrow i\text{th row in } \mathbf{A}$$

$$\mathbf{B} = \begin{bmatrix} b_{11} & b_{12} & \cdots & b_{1j} & \cdots & b_{1p} \\ b_{21} & b_{22} & \cdots & b_{2j} & \cdots & b_{2p} \\ \vdots & \vdots & & \vdots & & \vdots \\ b_{n1} & b_{n2} & \cdots & b_{nj} & \cdots & b_{np} \end{bmatrix}$$

$\uparrow$ jth column in **B**

Then

$$c_{ij} = a_{i1}b_{1j} + a_{i2}b_{2j} + \cdots + a_{in}b_{nj}.$$

The following diagram may help you remember the product **AB** of two matrices **A** and **B**.

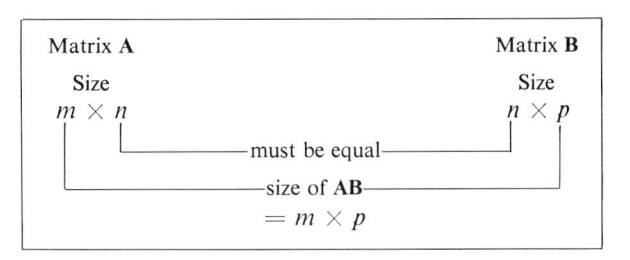

Matrix **A**   Matrix **B**
Size   Size
$m \times n$ $n \times p$
— must be equal —
— size of **AB** —
$= m \times p$

Example 2 Let

$$\mathbf{A} = \begin{bmatrix} 2 & 3 \\ 4 & 1 \end{bmatrix}, \quad \mathbf{B} = \begin{bmatrix} 3 & 1 & 0 \\ 2 & -3 & 4 \end{bmatrix}.$$

Find **AB** and **BA** if they exist.

Solution Here **A** is 2×2 and **B** is 2×3. Since the number of columns in **A** is equal to the number of rows in **B**, the product **AB** is defined. It is of size 2×3. If $\mathbf{C} = \mathbf{AB}$, then we can write

$$\mathbf{C} = \begin{bmatrix} c_{11} & c_{12} & c_{13} \\ c_{21} & c_{22} & c_{23} \end{bmatrix}.$$

The element c_{ij} is found by multiplying the ith row of **A** and the jth column of **B**. For example, to obtain the element c_{12} in the first row and second column, we add the products of the elements in the first row of **A** and the second column of **B**:

Row 1 of **A**	Column 2 of **B**		Product
2	1	$\longrightarrow$	2
3	-3	$\longrightarrow$	-9
		Sum	$-7 = c_{12}.$

Thus, in full,

$$\mathbf{AB} = \begin{bmatrix} 2 & 3 \\ 4 & 1 \end{bmatrix} \begin{bmatrix} 3 & 1 & 0 \\ 2 & -3 & 4 \end{bmatrix}$$

$$= \begin{bmatrix} 2(3) + 3(2) & 2(1) + 3(-3) & 2(0) + 3(4) \\ 4(3) + 1(2) & 4(1) + 1(-3) & 4(0) + 1(4) \end{bmatrix}$$

$$= \begin{bmatrix} 12 & -7 & 12 \\ 14 & 1 & 4 \end{bmatrix}.$$

The product **BA** is not defined because the number of columns of **B** is not equal to the number of rows of **A**.

Example 3 Given

$$\mathbf{A} = \begin{bmatrix} 1 & 2 \\ 4 & 5 \end{bmatrix} \quad \text{and} \quad \mathbf{B} = \begin{bmatrix} -2 & 1 \\ 3 & 2 \end{bmatrix}.$$

Find **AB** and **BA**.

Solution Here **A** and **B** are both of size 2×2. Thus **AB** and **BA** are both defined and both have size 2×2. We have

$$\mathbf{AB} = \begin{bmatrix} 1 & 2 \\ 4 & 5 \end{bmatrix} \begin{bmatrix} -2 & 1 \\ 3 & 2 \end{bmatrix}$$

$$= \begin{bmatrix} 1(-2) + 2(3) & 1(1) + 2(2) \\ 4(-2) + 5(3) & 4(1) + 5(2) \end{bmatrix}$$

$$= \begin{bmatrix} 4 & 5 \\ 7 & 14 \end{bmatrix}.$$

$$\mathbf{BA} = \begin{bmatrix} -2 & 1 \\ 3 & 2 \end{bmatrix} \begin{bmatrix} 1 & 2 \\ 4 & 5 \end{bmatrix}$$

$$= \begin{bmatrix} -2(1) + 1(4) & -2(2) + 1(5) \\ 3(1) + 2(4) & 3(2) + 2(5) \end{bmatrix}$$

$$= \begin{bmatrix} 2 & 1 \\ 11 & 16 \end{bmatrix}.$$

Clearly $\mathbf{AB} \neq \mathbf{BA}$, even though both products are defined.

It may be remarked that *if* **A** *is an* $m \times n$ *matrix, then the products* **AB** *and* **BA** *are both defined only if* **B** *is of size* $n \times m$. In particular, *if* **A** *is a square matrix, then* **AB** *and* **BA** *are both defined provided* **B** *is also a square matrix of the same size as* **A**. In general **AB** $\neq$ **BA** (in contrast to the addition of matrices, where the commutative property **A** + **B** = **B** + **A** does hold true).

If **A**, **B**, and **C** are three matrices of sizes $m \times n, n \times p$, and $p \times q$, respectively, then the products **AB**, **BC**, (**AB**)**C** and **A**(**BC**) are all defined. It can be shown that

$$(\mathbf{AB})\mathbf{C} = \mathbf{A}(\mathbf{BC}) \qquad \text{ASSOCIATIVE LAW.}$$

In such triple products we can therefore omit the parentheses and write simply **ABC**. The product matrix **ABC** is of size $m \times q$.

Example 4 Determine the matrix **X** such that the matrix equation **AX** = **C** holds true, where

$$\mathbf{A} = \begin{bmatrix} 1 & -1 \\ 0 & 3 \\ 2 & 0 \end{bmatrix}, \quad \mathbf{C} = \begin{bmatrix} -2 & 3 \\ 9 & -3 \\ 2 & 4 \end{bmatrix}.$$

Solution The product **AX** is defined if the number of columns in **A** is equal to the number of rows in **X**. Thus **X** must have two rows. Moreover, the number of columns in the product matrix **C** = **AX** is the same as the number of columns in **X**. Since **C** has two columns, **X** also must have two columns. In other words, **X** must be of size 2×2. Let

$$\mathbf{X} = \begin{bmatrix} a & b \\ c & d \end{bmatrix}.$$

Then the given equation **C** = **AX** becomes

$$\begin{bmatrix} -2 & 3 \\ 9 & -3 \\ 2 & 4 \end{bmatrix} = \begin{bmatrix} 1 & -1 \\ 0 & 3 \\ 2 & 0 \end{bmatrix} \begin{bmatrix} a & b \\ c & d \end{bmatrix}$$

$$= \begin{bmatrix} a-c & b-d \\ 3c & 3d \\ 2a & 2b \end{bmatrix}.$$

This implies

$$a - c = -2 \quad \text{and} \quad b - d = 3$$
$$3c = 9 \qquad\qquad\qquad 3d = -3$$
$$2a = 2 \qquad\qquad\qquad 2b = 4.$$

Solving these equations, we have $a = 1, c = 3, b = 2$, and $d = -1$. Thus,

$$\mathbf{X} = \begin{bmatrix} a & b \\ c & d \end{bmatrix} = \begin{bmatrix} 1 & 2 \\ 3 & -1 \end{bmatrix}.$$

Example 5 Let
$$A = \begin{bmatrix} a & b \\ c & d \end{bmatrix}.$$

Find **AI** and **IA** where **I** denotes the identity matrix.

Solution The products **AI** and **IA** are both defined if **A** and **I** are square matrices of the same size. Since **A** is of size 2×2, the identity matrix **I** must also be taken of size 2×2; that is,
$$I = \begin{bmatrix} 1 & 0 \\ 0 & 1 \end{bmatrix}.$$

Thus,
$$AI = \begin{bmatrix} a & b \\ c & d \end{bmatrix} \begin{bmatrix} 1 & 0 \\ 0 & 1 \end{bmatrix}$$
$$= \begin{bmatrix} a(1) + b(0) & a(0) + b(1) \\ c(1) + d(0) & c(0) + d(1) \end{bmatrix}$$
$$= \begin{bmatrix} a & b \\ c & d \end{bmatrix} = A.$$

Similarly,
$$IA = \begin{bmatrix} 1 & 0 \\ 0 & 1 \end{bmatrix} \begin{bmatrix} a & b \\ c & d \end{bmatrix}$$
$$= \begin{bmatrix} a & b \\ c & d \end{bmatrix} = A.$$

Therefore, $AI = IA = A$.

We see from this example that when any 2×2 matrix is multiplied by the identity matrix, it remains unchanged. This result is easily seen to hold for square matrices of any size. In other words, **I** behaves the same way in matrix multiplication as the number 1 behaves in multiplication of real numbers. This justifies the name *identity matrix* for **I**. If **A** is a *square matrix* of any size, then it is always true that

$$\boxed{AI = IA = A.}$$

If **A** is a square matrix of size $n \times n$, we can form the product **AA** of **A** with itself. This product is of the same size, $n \times n$, as **A**, so we can multiply it again by **A**, forming **AAA**. We use the notation A^2 to denote the product **AA**, A^3 to denote the product **AAA**, and so on. Note that $A^2, A^3, \ldots$ are defined only if A is a square matrix.

Example 6 Given
$$A = \begin{bmatrix} 0 & 1 \\ 0 & 0 \end{bmatrix}, \quad B = \begin{bmatrix} 1 & 2 \\ 3 & 4 \end{bmatrix}, \quad C = \begin{bmatrix} 5 & 7 \\ 3 & 4 \end{bmatrix}.$$

Prove that $A^2 = 0$ and $AB = AC$.

Solution

$$A^2 = AA = \begin{bmatrix} 0 & 1 \\ 0 & 0 \end{bmatrix} \begin{bmatrix} 0 & 1 \\ 0 & 0 \end{bmatrix}$$

$$= \begin{bmatrix} 0(0) + 1(0) & 0(1) + 0(0) \\ 0(0) + 0(0) & 0(1) + 0(0) \end{bmatrix}$$

$$= \begin{bmatrix} 0 & 0 \\ 0 & 0 \end{bmatrix} = \mathbf{0}.$$

$$AB = \begin{bmatrix} 0 & 1 \\ 0 & 0 \end{bmatrix} \begin{bmatrix} 1 & 2 \\ 3 & 4 \end{bmatrix}$$

$$= \begin{bmatrix} 0(1) + 1(3) & 0(2) + 1(4) \\ 0(1) + 0(3) & 0(2) + 0(4) \end{bmatrix}$$

$$= \begin{bmatrix} 3 & 4 \\ 0 & 0 \end{bmatrix}.$$

$$AC = \begin{bmatrix} 0 & 1 \\ 0 & 0 \end{bmatrix} \begin{bmatrix} 5 & 7 \\ 3 & 4 \end{bmatrix}$$

$$= \begin{bmatrix} 0(5) + 1(3) & 0(7) + 1(4) \\ 0(5) + 0(3) & 0(7) + 0(4) \end{bmatrix}$$

$$= \begin{bmatrix} 3 & 4 \\ 0 & 0 \end{bmatrix}.$$

Clearly $AB = AC$.

If a is a real number such that $a^2 = 0$, then a must be zero. This is not true in the case of matrices. That is, if $A^2 = 0$, then A may not be a zero matrix, as shown by Example 6 above. Also for real numbers, if $ab = ac$ and $a \neq 0$, then $b = c$. This again is not true in the case of matrices: in Example 6 we have $AB = AC$ and $A \neq 0$ but $B \neq C$.

SYSTEMS OF LINEAR EQUATIONS

By using the idea of matrix multiplication, we can write systems of linear equations in the form of matrix equations. Consider, for example, the system

$$2x - 3y = 7$$
$$4x + y = 21$$

consisting of two simultaneous linear equations for the variables x and y. Now, we have the following matrix product:

$$\begin{bmatrix} 2 & -3 \\ 4 & 1 \end{bmatrix} \begin{bmatrix} x \\ y \end{bmatrix} = \begin{bmatrix} 2x - 3y \\ 4x + y \end{bmatrix} = \begin{bmatrix} 7 \\ 21 \end{bmatrix}$$

from the the given simultaneous equations. Therefore

$$\begin{bmatrix} 2 & -3 \\ 4 & 1 \end{bmatrix} \begin{bmatrix} x \\ y \end{bmatrix} = \begin{bmatrix} 7 \\ 21 \end{bmatrix}.$$

If we define matrices as follows,

$$A = \begin{bmatrix} 2 & -3 \\ 4 & 1 \end{bmatrix}, \quad X = \begin{bmatrix} x \\ y \end{bmatrix}, \quad B = \begin{bmatrix} 7 \\ 21 \end{bmatrix}$$

then this matrix equation can be written simply as

$$AX = B.$$

Observe that the matrices **A** and **B** have elements whose values are given numbers. The matrix **X** contains the unknown quantities x and y. The column matrix **X** is commonly called the *variable vector*, **A** the *coefficient matrix*, and **B** the *value vector*.

By introducing appropriate matrices **A**, **B**, and **X**, we can express any system of linear equations as a matrix equation.

Example 7 Express the following system of equations in matrix form:

$$2x + 3y + 4z = 7$$
$$4y = 2 + 5z$$
$$3z - 2x + 6 = 0.$$

Solution We first rearrange the equations so that the constant terms appear on the right, and on the left the x, y, and z terms are aligned in the same column:

$$2x + 3y + 4z = 7$$
$$0x + 4y - 5z = 2$$
$$-2x + 0y + 3z = -6.$$

Observe that the missing terms are written in as $0x$ and $0y$ in the second and third equations. Now, if we define

$$A = \begin{bmatrix} 2 & 3 & 4 \\ 0 & 4 & -5 \\ -2 & 0 & 3 \end{bmatrix}, \quad X = \begin{bmatrix} x \\ y \\ z \end{bmatrix}, \quad B = \begin{bmatrix} 7 \\ 2 \\ -6 \end{bmatrix}$$

we can write the given system in the form $AX = B$. Again, **A** and **B** are known matrices of numbers and **X** is the matrix whose elements consist of unknown variables.

Suppose now that we are given a general system of m linear equations involving n variables. We denote the variables by $x_1, x_2, \ldots, x_n$ and suppose that the system takes the form

$$a_{11}x_1 + a_{12}x_2 + \cdots + a_{1n}x_n = b_1$$
$$a_{21}x_1 + a_{22}x_2 + \cdots + a_{2n}x_n = b_2$$
$$a_{31}x_1 + a_{32}x_2 + \cdots + a_{3n}x_n = b_3 \quad (1)$$
$$\vdots$$
$$a_{m1}x_1 + a_{m2}x_2 + \cdots + a_{mn}x_n = b_m.$$

Here the coefficients a_{ij} are certain given numbers, a_{ij} being the coefficient of x_j in the ith equation, and $b_1, b_2, \ldots, b_m$ are given right sides of the equations.

Let us introduce the $m \times n$ matrix $\mathbf{A}$ whose elements consist of the coefficients of $x_1, x_2, \ldots, x_n$: $\mathbf{A} = [a_{ij}]$. The first column of $\mathbf{A}$ contains all the coefficients of x_1, the second column contains all the coefficients of x_2, and so on. Let $\mathbf{X}$ be the column vector formed by the n variables $x_1, x_2, \ldots, x_n$ and $\mathbf{B}$ be the column vector formed by the m constants on the right side of the equations. Thus

$$\mathbf{X} = \begin{bmatrix} x_1 \\ x_2 \\ x_3 \\ \cdot \\ \cdot \\ \cdot \\ x_n \end{bmatrix} \quad \text{and} \quad \mathbf{B} = \begin{bmatrix} b_1 \\ b_2 \\ b_3 \\ \cdot \\ \cdot \\ \cdot \\ b_m \end{bmatrix}.$$

Now consider the product $\mathbf{AX}$. This product is defined because the number of columns in $\mathbf{A}$ is equal to the number of rows in $\mathbf{X}$. We have

$$\mathbf{AX} = \begin{bmatrix} a_{11} & a_{12} & \cdots & a_{1n} \\ a_{21} & a_{22} & \cdots & a_{2n} \\ a_{31} & a_{32} & \cdots & a_{3n} \\ \cdot & \cdot & & \cdot \\ \cdot & \cdot & & \cdot \\ \cdot & \cdot & & \cdot \\ a_{m1} & a_{m2} & \cdots & a_{mn} \end{bmatrix} \begin{bmatrix} x_1 \\ x_2 \\ x_3 \\ \cdot \\ \cdot \\ \cdot \\ x_n \end{bmatrix}$$

$$= \begin{bmatrix} a_{11}x_1 + a_{12}x_2 + \cdots + a_{1n}x_n \\ a_{21}x_1 + a_{22}x_2 + \cdots + a_{2n}x_n \\ a_{31}x_1 + a_{32}x_2 + \cdots + a_{3n}x_n \\ \cdot \\ \cdot \\ \cdot \\ a_{m1}x_1 + a_{m2}x_2 + \cdots + a_{mn}x_n \end{bmatrix} = \begin{bmatrix} b_1 \\ b_2 \\ b_3 \\ \cdot \\ \cdot \\ \cdot \\ b_m \end{bmatrix} = \mathbf{B}$$

where we have used Equations (1). Thus the system of Equations (1) is again equivalent to the single matrix equation $\mathbf{AX} = \mathbf{B}$.

Diagnostic Test 11.3

Fill in the blanks.

1. If $\mathbf{A}$ is an $m \times n$ matrix and $\mathbf{B}$ is a $p \times q$ matrix, then the product $\mathbf{AB}$ is defined if _____ and the product $\mathbf{BA}$ is defined if _____.

2. If $\mathbf{A}$ is an $m \times n$ matrix, then $\mathbf{A}^2$ is defined only if _____.

3. If $\mathbf{A}$ is of size $m \times n$ and $\mathbf{B}$ is of size $p \times q$ such that $\mathbf{AB}$ is defined, then the size of $\mathbf{AB}$ is _____.

4. If **A** is of size $p \times q$ and **B** is another matrix such that **AB** and **BA** are both defined, then **B** must be of size _____.

5. If **A** is of size $n \times n$ and $\mathbf{AB} = \mathbf{BA}$, then **B** must be of size _____.

6. If $\mathbf{A} = \mathbf{0}$, then $\mathbf{A}^2 =$ _____.

7. If $\mathbf{A}^2 = \mathbf{0}$, then **A** _____ a zero matrix.

8. If **A** is an identity matrix of order n, then $\mathbf{A}^3$ is an identity matrix of order _____.

*9. If **A** is an $m \times n$ matrix and **I** an identity matrix, then $\mathbf{AI} =$ _____ only if **I** is of size _____.

10. The system of equations
$$2x + 3y = 7$$
$$5y - 4x = 3$$
is equivalent to the matrix equation $\mathbf{AX} = \mathbf{B}$ if

$\mathbf{A} =$ _____, $\mathbf{X} =$ _____, $\mathbf{B} =$ _____.

*11. If **A** and **B** are two square matrices of the same size, then $(\mathbf{A} + \mathbf{B})(\mathbf{A} - \mathbf{B}) = \mathbf{A}^2 - \mathbf{B}^2$, provided _____.

Exercises 11.3

(1–6) If **A** is a 3×4, **B** is a 4×3, **C** is a 2×3, and **D** is a 4×5 matrix, find the sizes of the following product matrices.

1. **AB**.
2. **BA**.
3. **CA**.
4. **AD**.
5. **CAD**.
6. **CBA**.

(7–22) Perform the indicated operations and simplify.

7. $[2 \ 3]\begin{bmatrix} 4 \\ 5 \end{bmatrix}$.

8. $\begin{bmatrix} 3 \\ 1 \end{bmatrix}[2 \ 5 \ 4]$.

9. $[1 \ 2]\begin{bmatrix} 3 & 1 & 4 \\ 1 & -2 & 0 \end{bmatrix}$.

10. $[2 \ 0 \ 1]\begin{bmatrix} 0 & 2 \\ 1 & -1 \\ 3 & 0 \end{bmatrix}$.

11. $\begin{bmatrix} 3 & 0 & 1 \\ 2 & 4 & 0 \end{bmatrix}\begin{bmatrix} 4 \\ 5 \\ 6 \end{bmatrix}$.

12. $\begin{bmatrix} 1 & -2 \\ -3 & 4 \\ 5 & 6 \end{bmatrix}\begin{bmatrix} 2 \\ 0 \end{bmatrix}$.

13. $\begin{bmatrix} 2 & 1 \\ 3 & -2 \end{bmatrix}\begin{bmatrix} 1 & -2 \\ 4 & 3 \end{bmatrix}$.

14. $\begin{bmatrix} 3 & 1 \\ 1 & 5 \end{bmatrix}\begin{bmatrix} 2 & -1 & 3 \\ 1 & 4 & -2 \end{bmatrix}$.

15. $\begin{bmatrix} 2 & 3 & 1 \\ -1 & 2 & -3 \\ 4 & 5 & 6 \end{bmatrix}\begin{bmatrix} 1 \\ 2 \\ 3 \end{bmatrix}$.

16. $\begin{bmatrix} 1 & 0 & 2 \\ 0 & 2 & -1 \\ -2 & 1 & 0 \end{bmatrix}\begin{bmatrix} 3 & -2 \\ 2 & 1 \\ -1 & 3 \end{bmatrix}$.

17. $\begin{bmatrix} 2 & 1 & 4 \\ 5 & 3 & 6 \end{bmatrix}\begin{bmatrix} 1 & 0 & 2 & 4 \\ 3 & -1 & 0 & 1 \\ 0 & 2 & 1 & 3 \end{bmatrix}$.

18. $\begin{bmatrix} 2 & -1 & 0 \\ 1 & 3 & 2 \\ 4 & 0 & -3 \end{bmatrix}\begin{bmatrix} 1 & 0 & 2 \\ 0 & 2 & 1 \\ 2 & 1 & 0 \end{bmatrix}$.

19. $\begin{bmatrix} 1 & 2 & 3 \\ 4 & 5 & 6 \end{bmatrix} \begin{bmatrix} -1 & 0 \\ 2 & 4 \\ 0 & 3 \end{bmatrix} \begin{bmatrix} 3 & -1 \\ -2 & 1 \end{bmatrix}.$

20. $\begin{bmatrix} 1 & 0 & 2 \\ 0 & 2 & -1 \\ 3 & 1 & 0 \end{bmatrix} \begin{bmatrix} 2 & -1 \\ 1 & 0 \\ 0 & 3 \end{bmatrix} \begin{bmatrix} 0 & 1 & -2 \\ 3 & 0 & 1 \end{bmatrix}.$

21. $\begin{bmatrix} 4 & 1 & -2 \\ -3 & 2 & 1 \end{bmatrix} \left(\begin{bmatrix} 5 & 6 \\ 1 & 0 \\ 2 & -3 \end{bmatrix} + \begin{bmatrix} -4 & 2 \\ 3 & 1 \\ -2 & 3 \end{bmatrix} \right).$

22. $\begin{bmatrix} 2 & 1 \\ 0 & 2 \\ 3 & -1 \end{bmatrix} \left(\begin{bmatrix} 1 & -2 \\ 2 & -1 \end{bmatrix} + 3 \begin{bmatrix} 2 & 0 \\ 1 & 2 \end{bmatrix} \right).$

23. Given $\mathbf{A} = \begin{bmatrix} 1 & 2 \\ 2 & 3 \end{bmatrix}$, evaluate $\mathbf{A}^2 + 2\mathbf{A} - 3\mathbf{I}$.

24. Given $\mathbf{A} = \begin{bmatrix} 1 & 0 & 0 \\ 0 & 2 & 1 \\ 0 & 0 & 3 \end{bmatrix}$, evaluate $\mathbf{A}^2 - 5\mathbf{A} + 2\mathbf{I}$.

25. Given
$$\mathbf{A} = \begin{bmatrix} 1 & 2 \\ 3 & 4 \end{bmatrix}, \quad \mathbf{B} = \begin{bmatrix} 2 & -1 \\ -3 & -2 \end{bmatrix}.$$
(a) Evaluate $(\mathbf{A} + \mathbf{B})^2$.
(b) Evaluate $\mathbf{A}^2 + 2\mathbf{AB} + \mathbf{B}^2$.
(c) Is $(\mathbf{A} + \mathbf{B})^2 = \mathbf{A}^2 + 2\mathbf{AB} + \mathbf{B}^2$?

26. Given
$$\mathbf{A} = \begin{bmatrix} 2 & 3 \\ 1 & 2 \end{bmatrix}, \quad \mathbf{B} = \begin{bmatrix} 1 & 0 \\ 2 & -1 \end{bmatrix}.$$
Compute $\mathbf{A}^2 - \mathbf{B}^2$ and $(\mathbf{A} - \mathbf{B})(\mathbf{A} + \mathbf{B})$ and show that $\mathbf{A}^2 - \mathbf{B}^2 \neq (\mathbf{A} - \mathbf{B})(\mathbf{A} + \mathbf{B})$.

*(27–32) Determine the matrix **A** such that the following matrix equations hold true.

27. $\mathbf{A} \begin{bmatrix} 2 & 1 \\ 1 & 0 \end{bmatrix} = [5 \ \ 3].$

28. $\mathbf{A}[1 \ \ 3] = \begin{bmatrix} 2 & 6 \\ 3 & 9 \end{bmatrix}.$

29. $\begin{bmatrix} 1 & 2 \\ 0 & -1 \\ 2 & 3 \end{bmatrix} \mathbf{A} = \begin{bmatrix} -1 \\ 1 \\ -1 \end{bmatrix}.$

30. $\begin{bmatrix} 1 & 0 & 2 \\ 2 & -1 & 0 \\ 0 & 1 & 3 \end{bmatrix} \mathbf{A} = \begin{bmatrix} 7 \\ 0 \\ 11 \end{bmatrix}.$

31. $\begin{bmatrix} 2 & 0 \\ 1 & -1 \\ 0 & 1 \end{bmatrix} \mathbf{A} = \begin{bmatrix} 6 & 0 \\ 3 & -1 \\ 0 & 1 \end{bmatrix}.$

32. $\mathbf{A} \begin{bmatrix} 1 & 2 \\ 3 & 4 \end{bmatrix} = \begin{bmatrix} 7 & 10 \\ 15 & 22 \end{bmatrix}.$

(33–38) Express the following systems of linear equations in matrix form.

33. $2x + 3y = 7$
$x + 4y = 5.$

34. $3x - 2y = 4$
$4x + 5y = 7.$

35. $x + 2y + 3z = 8$
$2x - y + 4z = 13$
$3y - 2z = 5.$

36. $2x - y = 3$
$3y + 4z = 7$
$5z + x = 9.$

37. $2x + y - u = 0$
$3y + 2z + 4u = 5$
$x - 2y + 4z + u = 12.$

38. $2x_1 - 3x_2 + 4x_3 = 5$
$3x_3 + 5x_4 - x_1 = 7$
$x_1 + x_2 = x_3 + 2x_4.$

39. Given $\mathbf{A} = \begin{bmatrix} 1 & 2 \\ 3 & 6 \end{bmatrix}$, find a 2×2 nonzero matrix **B** such that **AB** is a zero matrix. (There is more than one answer.)

40. Give an example of two nonzero matrices **A** and **B** of different sizes such that the product **AB** is defined and is a zero matrix. (There are many possible answers.)

(41–44) Determine the matrix $\mathbf{A}^n$ for a general positive integer n, where **A** is as given.

41. $\mathbf{A} = \begin{bmatrix} 1 & 0 \\ 0 & 1 \end{bmatrix}.$

42. $\mathbf{A} = \begin{bmatrix} 0 & 1 \\ 1 & 0 \end{bmatrix}.$

Sec. 11.3 Products of Matrices

*43. $A = \begin{bmatrix} 1 & 0 \\ \frac{1}{2} & \frac{1}{2} \end{bmatrix}$.

*44. $A = \begin{bmatrix} 0 & 1 & 0 \\ 1 & 0 & 0 \\ 0 & 0 & 1 \end{bmatrix}$.

11.4

Inverse of a Matrix (Optional Section)

DEFINITION Let **A** be a square matrix of size $n \times n$. Then a matrix **B** is said to be an *inverse* of *A* if it satisfies the two matrix equations

$$\mathbf{AB} = \mathbf{I} \quad \text{and} \quad \mathbf{BA} = \mathbf{I} \tag{1}$$

where **I** is the identity matrix of size $n \times n$. In other words, the product of the matrices **A** and **B** in either order is the identity matrix.

It is clear from this definition that **B** must be a square matrix of the same size as **A**, otherwise one or both of the products **AB** and **BA** will not be defined.

Example 1 Show that $\mathbf{B} = \begin{bmatrix} -2 & 1 \\ \frac{3}{2} & -\frac{1}{2} \end{bmatrix}$ is an inverse of $\mathbf{A} = \begin{bmatrix} 1 & 2 \\ 3 & 4 \end{bmatrix}$.

Solution To show that **B** is an inverse of **A**, all we need prove is that $\mathbf{AB} = \mathbf{I}$ and $\mathbf{BA} = \mathbf{I}$. Now,

$$\mathbf{AB} = \begin{bmatrix} 1 & 2 \\ 3 & 4 \end{bmatrix} \begin{bmatrix} -2 & 1 \\ \frac{3}{2} & -\frac{1}{2} \end{bmatrix}$$

$$= \begin{bmatrix} 1(-2) + 2(\frac{3}{2}) & 1(1) + 2(-\frac{1}{2}) \\ 3(-2) + 4(\frac{3}{2}) & 3(1) + 4(-\frac{1}{2}) \end{bmatrix} = \begin{bmatrix} 1 & 0 \\ 0 & 1 \end{bmatrix} = \mathbf{I}$$

and

$$\mathbf{BA} = \begin{bmatrix} -2 & 1 \\ \frac{3}{2} & -\frac{1}{2} \end{bmatrix} \begin{bmatrix} 1 & 2 \\ 3 & 4 \end{bmatrix}$$

$$= \begin{bmatrix} -2(1) + 1(3) & -2(2) + 1(4) \\ \frac{3}{2}(1) - \frac{1}{2}(3) & \frac{3}{2}(2) - \frac{1}{2}(4) \end{bmatrix} = \begin{bmatrix} 1 & 0 \\ 0 & 1 \end{bmatrix} = \mathbf{I}.$$

Thus **B** is an inverse of **A**, as required

Not every square matrix has an inverse. This is illustrated by the following example.

Example 2 Find an inverse of the matrix $\mathbf{A} = \begin{bmatrix} 1 & 2 \\ 2 & 4 \end{bmatrix}$, if such an inverse exists.

Solution Let **B** be an inverse of **A**. If **B** exists, it is a square matrix of the same size as **A** and so must be of the form

$$\mathbf{B} = \begin{bmatrix} a & b \\ c & d \end{bmatrix}$$

where *a*, *b*, *c*, and *d* are certain entries.

Now the equation $\mathbf{AB} = \mathbf{I}$ implies that

$$\begin{bmatrix} 1 & 2 \\ 2 & 4 \end{bmatrix} \begin{bmatrix} a & b \\ c & d \end{bmatrix} = \begin{bmatrix} 1 & 0 \\ 0 & 1 \end{bmatrix}$$

—that is,

$$\begin{bmatrix} a + 2c & b + 2d \\ 2a + 4c & 2b + 4d \end{bmatrix} = \begin{bmatrix} 1 & 0 \\ 0 & 1 \end{bmatrix}.$$

Therefore, comparing entries in these matrices, we obtain that

$$\begin{matrix} a + 2c = 1 \\ 2a + 4c = 0 \end{matrix} \quad \text{and} \quad \begin{matrix} b + 2d = 0 \\ 2b + 4d = 1. \end{matrix} \tag{i}$$

These systems of equations are inconsistent, as we can easily see by dividing the lower two equations by 2. Thus the systems (i) *have no solution*, and therefore there is no matrix $\mathbf{B}$ that satisfies the condition that $\mathbf{AB} = \mathbf{I}$. The matrix $\mathbf{A}$ does not have an inverse.

DEFINITION A matrix $\mathbf{A}$ is said to be ***invertible*** (or ***nonsingular***) if it has an inverse. If the matrix $\mathbf{A}$ does not have an inverse, then it is said to be a ***singular matrix***.

It can be shown that the inverse of any nonsingular matrix is unique. That is, if $\mathbf{A}$ has an inverse at all, then it has only one inverse. For this reason we denote the inverse of $\mathbf{A}$ by $\mathbf{A}^{-1}$ (read "A inverse"). Thus we have the two equations

$$\boxed{\mathbf{AA}^{-1} = \mathbf{I} \quad \text{and} \quad \mathbf{A}^{-1}\mathbf{A} = \mathbf{I}.}$$

Let us now turn to the problem of finding the inverse of a nonsingular matrix. As an example, suppose we want to find the inverse of

$$\mathbf{A} = \begin{bmatrix} 1 & 3 \\ 2 & 5 \end{bmatrix}.$$

Let $\mathbf{B} = \begin{bmatrix} a & b \\ c & d \end{bmatrix}$ denote the inverse of $\mathbf{A}$. Then $\mathbf{B}$ must satisfy the two equations

$$\mathbf{AB} = \mathbf{I} \quad \text{and} \quad \mathbf{BA} = \mathbf{I}.$$

The matrix equation $\mathbf{AB} = \mathbf{I}$, when written out in full, is

$$\begin{bmatrix} 1 & 3 \\ 2 & 5 \end{bmatrix} \begin{bmatrix} a & b \\ c & d \end{bmatrix} = \begin{bmatrix} 1 & 0 \\ 0 & 1 \end{bmatrix}$$

—that is,

$$\begin{bmatrix} a + 3c & b + 3d \\ 2a + 5c & 2b + 5d \end{bmatrix} = \begin{bmatrix} 1 & 0 \\ 0 & 1 \end{bmatrix}.$$

Therefore,

$$\begin{matrix} a + 3c = 1, & b + 3d = 0 \\ 2a + 5c = 0, & 2b + 5d = 1. \end{matrix} \tag{2}$$

Note that the two equations on the left form a system of equations for the unknown entries a and c, while the two equations on the right form a system of equations for b and d. To solve these two systems of equations, we must transform the corresponding augmented matrices

$$\begin{bmatrix} 1 & 3 & | & 1 \\ 2 & 5 & | & 0 \end{bmatrix} \text{ and } \begin{bmatrix} 1 & 3 & | & 0 \\ 2 & 5 & | & 1 \end{bmatrix} \tag{3}$$

to their reduced forms. You can verify that these reduced forms are, respectively,

$$\begin{bmatrix} 1 & 0 & | & -5 \\ 0 & 1 & | & 2 \end{bmatrix} \text{ and } \begin{bmatrix} 1 & 0 & | & 3 \\ 0 & 1 & | & -1 \end{bmatrix}.$$

Therefore, $a = -5$, $c = 2$ and $b = 3$, $d = -1$. Consequently the matrix $\mathbf{B}$ that satisfies the equation $\mathbf{AB} = \mathbf{I}$ is

$$\mathbf{B} = \begin{bmatrix} a & b \\ c & d \end{bmatrix} = \begin{bmatrix} -5 & 3 \\ 2 & -1 \end{bmatrix}.$$

It is now easy to verify that this matrix $\mathbf{B}$ also satisfies the equation $\mathbf{BA} = \mathbf{I}$.* Therefore $\mathbf{B}$ is the inverse of $\mathbf{A}$, and we can write

$$\mathbf{A}^{-1} = \begin{bmatrix} -5 & 3 \\ 2 & -1 \end{bmatrix}.$$

To find the inverse in this example we solved the two linear systems (2) by reducing their augmented matrices (3). Now we may observe that the two systems in (2) have the same coefficient matrix $\mathbf{A}$, and so we can reduce both augmented matrices in the same calculation, since both require the same sequence of row operations. The procedure we can use for this simultaneous reduction is to write down the coefficient matrix $\mathbf{A}$, draw a vertical line, and write the constants appearing in the right sides of systems (2) in two columns as shown below.

$$\begin{bmatrix} 1 & 3 & | & 1 & 0 \\ 2 & 5 & | & 0 & 1 \end{bmatrix}. \tag{4}$$

We now perform row operations in the usual way to reduce the left half of this augmented matrix to an identity matrix.

It may be observed that the elements to the right of the vertical line in the augmented matrix (4) form a 2×2 identity matrix. Thus, this augmented matrix can be abbreviated as simply $\mathbf{A}|\mathbf{I}$. If we transform $\mathbf{A}|\mathbf{I}$ to reduced form, we shall be simultaneously reducing the two augmented matrices (3) and thereby solving the two linear systems (2) at the same time.

In this example, $\mathbf{A}|\mathbf{I}$ is reduced by the following sequence of row operations:

---

*It can be shown (though the proof is quite difficult) that if either one of the two conditions $\mathbf{AB} = \mathbf{I}$, $\mathbf{BA} = \mathbf{I}$ is satisfied, then the other one must be satisfied automatically. This is why we need to use only one of these conditions in order to determine $\mathbf{B}$.

$$\mathbf{A}|\mathbf{I} = \begin{bmatrix} 1 & 3 & | & 1 & 0 \\ 2 & 5 & | & 0 & 1 \end{bmatrix} \xrightarrow{R_2 - 2R_1} \begin{bmatrix} 1 & 3 & | & 1 & 0 \\ 0 & -1 & | & -2 & 1 \end{bmatrix}$$

$$\xrightarrow{-R_2} \begin{bmatrix} 1 & 3 & | & 1 & 0 \\ 0 & 1 & | & 2 & -1 \end{bmatrix}$$

$$\xrightarrow{R_1 - 3R_2} \begin{bmatrix} 1 & 0 & | & -5 & 3 \\ 0 & 1 & | & 2 & -1 \end{bmatrix}.$$

This is now the required reduced matrix, having an identity matrix to the left of the vertical line. The elements to the right of this line are the solutions of the two systems (2); in other words, they form the matrix

$$\begin{bmatrix} a & b \\ c & d \end{bmatrix} = \begin{bmatrix} -5 & 3 \\ 2 & -1 \end{bmatrix}.$$

But this matrix is the inverse of **A**. We conclude therefore that the reduced form of the augmented matrix $\mathbf{A}|\mathbf{I}$ contains the inverse of **A** on the right of the vertical line. To summarize:

*Let **A** be an invertible square matrix of size $n \times n$, and **I** the identity matrix of the same size. Then the reduced form of $\mathbf{A}|\mathbf{I}$ is $\mathbf{I}|\mathbf{A}^{-1}$.*

Example 3 Given $\mathbf{A} = \begin{bmatrix} 1 & 2 & 3 \\ 2 & 5 & 7 \\ 3 & 7 & 8 \end{bmatrix}$. Find $\mathbf{A}^{-1}$.

Solution We have

$$\mathbf{A}|\mathbf{I} = \begin{bmatrix} 1 & 2 & 3 & | & 1 & 0 & 0 \\ 2 & 5 & 7 & | & 0 & 1 & 0 \\ 3 & 7 & 8 & | & 0 & 0 & 1 \end{bmatrix} \xrightarrow[R_3 - 3R_1]{R_2 - 2R_1} \begin{bmatrix} 1 & 2 & 3 & | & 1 & 0 & 0 \\ 0 & 1 & 1 & | & -2 & 1 & 0 \\ 0 & 1 & -1 & | & -3 & 0 & 1 \end{bmatrix}$$

$$\xrightarrow[R_3 - R_2]{R_1 - 2R_2} \begin{bmatrix} 1 & 0 & 1 & | & 5 & -2 & 0 \\ 0 & 1 & 1 & | & -2 & 1 & 0 \\ 0 & 0 & -2 & | & -1 & -1 & 1 \end{bmatrix}$$

$$\xrightarrow{-\frac{1}{2}R_3} \begin{bmatrix} 1 & 0 & 1 & | & 5 & -2 & 0 \\ 0 & 1 & 1 & | & -2 & 1 & 0 \\ 0 & 0 & 1 & | & \frac{1}{2} & \frac{1}{2} & -\frac{1}{2} \end{bmatrix}$$

$$\xrightarrow[R_2 - R_3]{R_1 - R_3} \begin{bmatrix} 1 & 0 & 0 & | & \frac{9}{2} & -\frac{5}{2} & \frac{1}{2} \\ 0 & 1 & 0 & | & -\frac{5}{2} & \frac{1}{2} & \frac{1}{2} \\ 0 & 0 & 1 & | & \frac{1}{2} & \frac{1}{2} & -\frac{1}{2} \end{bmatrix}.$$

This matrix is now in the reduced form $\mathbf{I}|\mathbf{A}^{-1}$. Thus,

Sec. 11.4 *Inverse of a Matrix (Optional Section)*

$$\mathbf{A}^{-1} = \begin{bmatrix} \frac{9}{2} & -\frac{5}{2} & \frac{1}{2} \\ -\frac{5}{2} & \frac{1}{2} & \frac{1}{2} \\ \frac{1}{2} & \frac{1}{2} & -\frac{1}{2} \end{bmatrix} = \frac{1}{2} \begin{bmatrix} 9 & -5 & 1 \\ -5 & 1 & 1 \\ 1 & 1 & -1 \end{bmatrix}.$$

You can easily verify that this is in fact the inverse matrix of **A** by checking the two equations

$$\mathbf{A}\mathbf{A}^{-1} = \mathbf{I} \quad \text{and} \quad \mathbf{A}^{-1}\mathbf{A} = \mathbf{I}.$$

How do we know whether a given matrix **A** is invertible or not? If we carry out the process of transforming $\mathbf{A} | \mathbf{I}$ to its reduced form and if at any step we find that one of the rows on the left of the vertical line consists entirely of zeros, then it can be shown that $\mathbf{A}^{-1}$ does not exist.

Example 4 Given $\mathbf{A} = \begin{bmatrix} 1 & 2 & 3 \\ 2 & 5 & 7 \\ 3 & 7 & 10 \end{bmatrix}$, find $\mathbf{A}^{-1}$ if it exists.

Solution We have

$$\mathbf{A} | \mathbf{I} = \begin{bmatrix} 1 & 2 & 3 & | & 1 & 0 & 0 \\ 2 & 5 & 7 & | & 0 & 1 & 0 \\ 3 & 7 & 10 & | & 0 & 0 & 1 \end{bmatrix} \xrightarrow[R_3 - 3R_1]{R_2 - 2R_1} \begin{bmatrix} 1 & 2 & 3 & | & 1 & 0 & 0 \\ 0 & 1 & 1 & | & -2 & 1 & 0 \\ 0 & 1 & 1 & | & -3 & 0 & 1 \end{bmatrix}$$

$$\xrightarrow{R_3 - R_2} \begin{bmatrix} 1 & 2 & 3 & | & 1 & 0 & 0 \\ 0 & 1 & 1 & | & -2 & 1 & 0 \\ 0 & 0 & 0 & | & -1 & -1 & 1 \end{bmatrix}.$$

Since the third row to the left of the vertical line consists entirely of zeros, the reduction cannot be completed. We must conclude that $\mathbf{A}^{-1}$ does not exist and that **A** is a *singular* matrix.

Inverses of matrices have many uses, one of which is in the solution of systems of equations. In the case when we have n equations in n variables, we can solve the system by finding the inverse of the coefficient matrix.

A system of equations can be written in matrix form as $\mathbf{AX} = \mathbf{B}$. If the coefficient matrix **A** is invertible, then $\mathbf{A}^{-1}$ exists. Multiplying both sides of the given matrix equation by $\mathbf{A}^{-1}$ on the left, we obtain

$$\mathbf{A}^{-1}(\mathbf{AX}) = \mathbf{A}^{-1}\mathbf{B}.$$

Using the associative property, we can write this as

$$(\mathbf{A}^{-1}\mathbf{A})\mathbf{X} = \mathbf{A}^{-1}\mathbf{B}$$

—that is,

$$\mathbf{IX} = \mathbf{A}^{-1}\mathbf{B}$$

$$\mathbf{X} = \mathbf{A}^{-1}\mathbf{B}.$$

So we have obtained an expression for the solution **X** of the given system of equations.

Example 5 Solve the following system of linear equations:
$$x + 2y + 3z = 3$$
$$2x + 5y + 7z = 6$$
$$3x + 7y + 8z = 5.$$

Solution The given system of equations in matrix form is
$$\mathbf{AX} = \mathbf{B} \tag{i}$$
where
$$\mathbf{A} = \begin{bmatrix} 1 & 2 & 3 \\ 2 & 5 & 7 \\ 3 & 7 & 8 \end{bmatrix}, \quad \mathbf{X} = \begin{bmatrix} x \\ y \\ z \end{bmatrix}, \quad \text{and} \quad \mathbf{B} = \begin{bmatrix} 3 \\ 6 \\ 5 \end{bmatrix}.$$

Then $\mathbf{A}^{-1}$ (as found in Example 3 above) is given by
$$\mathbf{A}^{-1} = \tfrac{1}{2} \begin{bmatrix} 9 & -5 & 1 \\ -5 & 1 & 1 \\ 1 & 1 & -1 \end{bmatrix}.$$

It follows that the solution of (i) is given by
$$\mathbf{X} = \mathbf{A}^{-1}\mathbf{B} = \tfrac{1}{2} \begin{bmatrix} 9 & -5 & 1 \\ -5 & 1 & 1 \\ 1 & 1 & -1 \end{bmatrix} \begin{bmatrix} 3 \\ 6 \\ 5 \end{bmatrix}$$
$$= \tfrac{1}{2} \begin{bmatrix} 27 - 30 + 5 \\ -15 + 6 + 5 \\ 3 + 6 - 5 \end{bmatrix} = \tfrac{1}{2} \begin{bmatrix} 2 \\ -4 \\ 4 \end{bmatrix} = \begin{bmatrix} 1 \\ -2 \\ 2 \end{bmatrix}$$

—that is,
$$\begin{bmatrix} x \\ y \\ z \end{bmatrix} = \begin{bmatrix} 1 \\ -2 \\ 2 \end{bmatrix}.$$

Therefore $x = 1$, $y = -2$ and $z = 2$.

At first glance it may appear that this method of solving a system of equations is much less convenient than the simple method of row reduction described in Section 11-1. In most cases this conclusion is correct. However the advantage of using the inverse matrix occurs in cases where several systems of equations must be solved all having the same coefficient matrix. For problems such as this, the solution of *all* the systems can be immediately written down once the inverse of the coefficient matrix has been found, and it is not necessary to use row reduction over and over again for each system. (See Exercises 42–44.)

Diagnostic Test 11.4

Fill in the blanks.

1. The matrix **Q** is said to be an inverse of a matrix **P** if and only if _____ .
2. If **A** is of size $m \times n$ and A^{-1} exists, then _____ .
3. If **A** is of size $n \times n$, and A^{-1} exists, then A^{-1} is of size _____ .
4. The inverse of an identity matrix is _____ .
5. The inverse of a zero matrix is _____ .
6. If $AX = B$, where **A** is a nonsingular matrix, then $X = $ _____ .
*7. If $XP = Q$, where **P** is a nonsingular matrix, then $X = $ _____ .
8. A matrix **P** is said to be nonsingular if _____ exists.
*9. $(A^{-1})^{-1} = $ _____ .
*10. If $AB = AC$, then $B = C$ provided _____ .
11. If $AB = AC$ and A^{-1} exists, then $B = $ _____ .

Exercises 11.4

(1–26) In the following exercises, find the inverse of the given matrix if it exists.

1. $\begin{bmatrix} 2 & 0 \\ 0 & 3 \end{bmatrix}$.
2. $\begin{bmatrix} 0 & 2 \\ 5 & 0 \end{bmatrix}$.
3. $\begin{bmatrix} 2 & 5 \\ 3 & 0 \end{bmatrix}$.
4. $\begin{bmatrix} 0 & 1 \\ 2 & -3 \end{bmatrix}$.
5. $\begin{bmatrix} 2 & 5 \\ 3 & 4 \end{bmatrix}$.
6. $\begin{bmatrix} 3 & 1 \\ 4 & 2 \end{bmatrix}$.
7. $\begin{bmatrix} 1 & -2 \\ -3 & 4 \end{bmatrix}$.
8. $\begin{bmatrix} 3 & 2 \\ -1 & 4 \end{bmatrix}$.
9. $\begin{bmatrix} 1 & -3 \\ -2 & 6 \end{bmatrix}$.
10. $\begin{bmatrix} 3 & -2 \\ -6 & 4 \end{bmatrix}$.
11. $\begin{bmatrix} 1 & 2 \\ 0 & 0 \end{bmatrix}$.
12. $\begin{bmatrix} 0 & 0 \\ 3 & 4 \end{bmatrix}$.
13. $\begin{bmatrix} 1 & 0 & 0 \\ 0 & 2 & 0 \\ 0 & 0 & 3 \end{bmatrix}$.
14. $\begin{bmatrix} 0 & 0 & 1 \\ 0 & 2 & 0 \\ 3 & 0 & 0 \end{bmatrix}$.
15. $\begin{bmatrix} 1 & 0 & 2 \\ 0 & 3 & 1 \\ 2 & -1 & 0 \end{bmatrix}$.
16. $\begin{bmatrix} 2 & 1 & 0 \\ 1 & 0 & 3 \\ 0 & 2 & 1 \end{bmatrix}$.
17. $\begin{bmatrix} 2 & 3 & 4 \\ 1 & 2 & 0 \\ 4 & 5 & 6 \end{bmatrix}$.
18. $\begin{bmatrix} 1 & 2 & 3 \\ -2 & 1 & 4 \\ -3 & -4 & 1 \end{bmatrix}$.
19. $\begin{bmatrix} 2 & 1 & -1 \\ 3 & 2 & 0 \\ 4 & 3 & 1 \end{bmatrix}$.
20. $\begin{bmatrix} 3 & -4 & 5 \\ 4 & -3 & 6 \\ 6 & -8 & 10 \end{bmatrix}$.
21. $\begin{bmatrix} -1 & 2 & -3 \\ 2 & -1 & 1 \\ 3 & 1 & 2 \end{bmatrix}$.

22. $\begin{bmatrix} -3 & 2 & 1 \\ 2 & -1 & 3 \\ 1 & -3 & 2 \end{bmatrix}.$

23. $\begin{bmatrix} 1 & 2 & 3 \\ 4 & 5 & 6 \\ 7 & 8 & 9 \end{bmatrix}.$

24. $\begin{bmatrix} 2 & 1 & 3 \\ 0 & 0 & 0 \\ 1 & 2 & -1 \end{bmatrix}.$

25. $\begin{bmatrix} 1 & -1 & 1 & 2 \\ 2 & -3 & 0 & 3 \\ 1 & 1 & 1 & 1 \\ 3 & 0 & -1 & 2 \end{bmatrix}.$

26. $\begin{bmatrix} 2 & 1 & 3 & 4 \\ 1 & 1 & 1 & -1 \\ -1 & 1 & -1 & 0 \\ 3 & 0 & 1 & 2 \end{bmatrix}.$

(27–34) Solve the following systems of equations by finding the inverse of the coefficient matrix.

27. $2x - 3y = 1$
    $3x + 4y = 10.$

28. $3x_1 + 2x_2 = 1$
    $2x_1 - x_2 = 3.$

29. $4u + 5v = 14$
    $2v - 3u = 1.$

30. $3y - 2z = -4$
    $5z + 4y = -13.$

31. $2x - y + 3z = -3$
    $x + y + z = 2$
    $3x + 2y - z = 8.$

32. $x + 2y - z = 1$
    $2z - 3x = 2$
    $3y + 2z = 5.$

33. $2u + 3v - 4w = -10$
    $w - 2u - 1 = 0$
    $u + 2v = 1.$

34. $p + 2q - 3r = 1$
    $q - 2p + r = 3$
    $2r + p - 2 = 0.$

35. If **A** is a nonsingular matrix and $\mathbf{AB} = \mathbf{AC}$, show that $\mathbf{B} = \mathbf{C}$.

36. If $\mathbf{AB} = \mathbf{A}$ and **A** is nonsingular, show that $\mathbf{B} = \mathbf{I}$.

37. Given

$$\mathbf{A} = \begin{bmatrix} 1 & 3 \\ 2 & 4 \end{bmatrix} \quad \text{and} \quad \mathbf{B} = \begin{bmatrix} 2 & -1 \\ -3 & 1 \end{bmatrix}$$

verify the result $(\mathbf{AB})^{-1} = \mathbf{B}^{-1}\mathbf{A}^{-1}$.

38. Use the matrices **A** and **B** in Exercise 37 above to verify that $(\mathbf{A}^{-1}\mathbf{B})^{-1} = \mathbf{B}^{-1}\mathbf{A}$.

*39. Show that $(\mathbf{A}^{-1})^{-1} = \mathbf{A}$ for any invertible matrix **A**.

*40. Show that for any two invertible $n \times n$ matrices **A** and **B**,

$$(\mathbf{AB})^{-1} = \mathbf{B}^{-1}\mathbf{A}^{-1}.$$

*41. Show that if two matrices **B** and **C** are both inverses of a matrix **A**, then $\mathbf{B} = \mathbf{C}$. [*Hint:* Consider **BAC**.]

42. Use the inverse matrix to find the solution of the following system for any values of a, b, and c:

$x - 3y + 2z = a$
$\quad\quad 2y - z = b$
$2x \quad\quad + 3z = c.$

43. A business firm is planning an expansion and it expects to have available K units of capital and L units of labor. The firm has two production lines. To produce one extra unit on the first production line requires 2 units of capital and 5 units of labor, while on the second production line it requires 3 units of capital and 4 units of labor. If all the available capital and labor is used, how many extra units should be produced on each line? (Use an inverse matrix.)

44. In Exercises 43 and 44 of Section 11.1, use an inverse matrix to express i_1, i_2, and i_3 as functions of E_1 and E_2: (a) when $R_1 = 3$, $R_2 = 2$, $R_3 = 4$; (b) for general R_1, R_2, and R_3.

11.5

Determinants and Cramer's Rule

Corresponding to any square matrix there is a real number called its determinant. We denote the determinant by enclosing the matrix in vertical bars. For example, if **A** is the 2×2 matrix given by

$$\mathbf{A} = \begin{bmatrix} 2 & 3 \\ 4 & 5 \end{bmatrix}$$

then its determinant is denoted by $|\mathbf{A}|$, or in full by

$$\begin{vmatrix} 2 & 3 \\ 4 & 5 \end{vmatrix}.$$

The determinant of an $n \times n$ matrix is said to be a **determinant of order n**. For example, $|\mathbf{A}|$ above is a determinant of order 2 and

$$\begin{vmatrix} 3 & 0 & -1 \\ 2 & 0 & 2 \\ -4 & 2 & 3 \end{vmatrix}$$

is a determinant of order 3.

The symbol Δ (capital delta) is also often used to denote a given determinant.

We shall begin by defining determinants of order 2.

DEFINITION A **determinant of order 2** is defined by the following expression:

$$\begin{vmatrix} a_1 & b_1 \\ a_2 & b_2 \end{vmatrix} = a_1 b_2 - a_2 b_1.$$

In other words, the determinant is given by the product of the elements a_1, b_2 on the main diagonal minus the product of the elements a_2, b_1 on the cross-diagonal. We can indicate these two diagonals by means of arrows in the following way:

$$\begin{matrix} (+) & (-) \end{matrix}$$
$$\begin{vmatrix} a_1 & b_1 \\ a_2 & b_2 \end{vmatrix} = a_1 b_2 - a_2 b_1.$$

The $(+)$ and $(-)$ signs indicate the signs associated with the two products.

Example 1 Evaluate the following determinants:

(a) $\begin{vmatrix} 2 & -3 \\ 4 & 5 \end{vmatrix}.$
(b) $\begin{vmatrix} 3 & 2 \\ 0 & 4 \end{vmatrix}.$

Solution (a)
$$\Delta = \begin{vmatrix} 2 & -3 \\ 4 & 5 \end{vmatrix} = 2(5) - 4(-3) = 10 + 12 = 22.$$

(b)
$$\Delta = \begin{vmatrix} 3 & 2 \\ 0 & 4 \end{vmatrix} = 3(4) - 0(2) = 12 - 0 = 12.$$

DEFINITION   A *determinant of order 3* is defined by the following expression:

$$\Delta = \begin{vmatrix} a_1 & b_1 & c_1 \\ a_2 & b_2 & c_2 \\ a_3 & b_3 & c_3 \end{vmatrix} = a_1 b_2 c_3 + a_2 b_3 c_1 + a_3 b_1 c_2 \\ - a_1 b_3 c_2 - a_3 b_2 c_1 - a_2 b_1 c_3. \quad (1)$$

The expression on the right is called the *complete expansion* of the third-order determinant Δ. Observe that it contains six terms, three positive and three negative, each term consisting of a product of three elements in the determinant.

Usually, when evaluating a third-order determinant we do not use the complete expansion, but instead we use what are called cofactors. Each element in the determinant has a cofactor, which is denoted by the corresponding capital letter; for example, A_2 denotes the cofactor of a_2, B_3 denotes the cofactor of b_3, and so on. They are defined as follows.

DEFINITION   Let Δ be a determinant of order 3. The *minor* of any element in Δ is the 2×2 determinant obtained by deleting the row and column that contain that element.

Example 2   Find the minor of the encircled element in the determinant

$$\begin{vmatrix} 2 & -1 & 0 \\ 3 & 0 & ④ \\ -3 & 2 & 6 \end{vmatrix}$$

Solution   We must delete the second row and the third column:

$$\begin{vmatrix} 2 & -1 & 0 \\ \cancel{3} & \cancel{0} & \cancel{4} \\ -3 & 2 & \cancel{6} \end{vmatrix} = \begin{vmatrix} 2 & -1 \\ -3 & 2 \end{vmatrix} = (2)(2) - (-1)(-3) = 1.$$

DEFINITION   If an element in Δ occurs in the ith row and jth column, then its *cofactor* is equal to its minor multiplied by $(-1)^{i+j}$.

Example 3 (a)   In the determinant Δ given in Equation (1), a_2 occurs in the second row and first column ($i = 2, j = 1$), so its cofactor is

$$A_2 = (-1)^{2+1} \begin{vmatrix} \cancel{a_1} & b_1 & c_1 \\ \cancel{a_2} & \cancel{b_2} & \cancel{c_2} \\ \cancel{a_3} & b_3 & c_3 \end{vmatrix} = (-1)^3 \begin{vmatrix} b_1 & c_1 \\ b_3 & c_3 \end{vmatrix} = -\begin{vmatrix} b_1 & c_1 \\ b_3 & c_3 \end{vmatrix} \\ = -(b_1 c_3 - b_3 c_1).$$

(b) c_3 occurs in the third row and third column ($i = j = 3$), so its cofactor is

$$C_3 = (-1)^{3+3} \begin{vmatrix} a_1 & b_1 & \cancel{c_1} \\ a_2 & b_2 & \cancel{c_2} \\ \cancel{a_3} & \cancel{b_3} & \cancel{c_3} \end{vmatrix} = (-1)^6 \begin{vmatrix} a_1 & b_1 \\ a_2 & b_2 \end{vmatrix} = \begin{vmatrix} a_1 & b_1 \\ a_2 & b_2 \end{vmatrix}$$
$$= a_1 b_2 - a_2 b_1.$$

The connection between a determinant and its various cofactors is given by the following theorem.

THEOREM 1 The value of a determinant can be found by multiplying the elements in any row (or column) by their cofactors and adding the products for all elements in the given row (or column).

Let us verify that this theorem is true for expansion by the first row. The theorem states that

$$\Delta = a_1 A_1 + b_1 B_1 + c_1 C_1. \tag{2}$$

The three cofactors here are as follows:

$$A_1 = (-1)^{1+1} \begin{vmatrix} b_2 & c_2 \\ b_3 & c_3 \end{vmatrix} = (b_2 c_3 - b_3 c_2)$$

$$B_1 = (-1)^{1+2} \begin{vmatrix} a_2 & c_2 \\ a_3 & c_3 \end{vmatrix} = -(a_2 c_3 - a_3 c_2)$$

$$C_1 = (-1)^{1+3} \begin{vmatrix} a_2 & b_2 \\ a_3 & b_3 \end{vmatrix} = (a_2 b_3 - a_3 b_2).$$

Substituting these into (2) above, we obtain

$$\Delta = a_1(b_2 c_3 - b_3 c_2) - b_1(a_2 c_3 - a_3 c_2) + c_1(a_2 b_3 - a_3 b_2).$$

We easily see that this expression agrees with the complete expansion given in the definition of the third-order determinant.

Example 4 Evaluate the determinant:

$$\Delta = \begin{vmatrix} 2 & 3 & -1 \\ 1 & 4 & 2 \\ -3 & 1 & 4 \end{vmatrix}.$$

Solution Expanding by the first row, we have

$$\Delta = a_1 A_1 + b_1 B_1 + c_1 C_1$$

$$= 2 \begin{vmatrix} 4 & 2 \\ 1 & 4 \end{vmatrix} - 3 \begin{vmatrix} 1 & 2 \\ -3 & 4 \end{vmatrix} + (-1) \begin{vmatrix} 1 & 4 \\ -3 & 1 \end{vmatrix} \tag{1}$$

$$= 2(4 \cdot 4 - 1 \cdot 2) - 3[1 \cdot 4 - (-3)2] + (-1)[1 \cdot 1 - (-3)4]$$

$$= 2(16 - 2) - 3(4 + 6) - 1(1 + 12) = -15.$$

Example 5 Evaluate the determinant

$$\Delta = \begin{vmatrix} 2 & -3 & 0 \\ 1 & 4 & 3 \\ -5 & 6 & 0 \end{vmatrix}$$

(a) By expanding by the second column.
(b) By expanding by the third column.

Solution (a) Expanding by the second column, we get

$$\Delta = b_1 B_1 + b_2 B_2 + b_3 B_3$$
$$= -(-3)\begin{vmatrix} 1 & 3 \\ -5 & 0 \end{vmatrix} + 4\begin{vmatrix} 2 & 0 \\ -5 & 0 \end{vmatrix} - 6\begin{vmatrix} 2 & 0 \\ 1 & 3 \end{vmatrix} \qquad (2)$$
$$= 3[(1(0) - 3(-5)] + 4[2(0) - (-5)(0)] - 6[2(3) - 1(0)]$$
$$= 3(15) + 4(0) - 6(6) = 9.$$

(b) Expanding by the third column, we get

$$\Delta = c_1 C_1 + c_2 C_2 + c_3 C_3$$
$$= (0)\begin{vmatrix} 1 & 4 \\ -5 & 6 \end{vmatrix} - 3\begin{vmatrix} 2 & -3 \\ -5 & 6 \end{vmatrix} + (0)\begin{vmatrix} 2 & -3 \\ 1 & 4 \end{vmatrix}. \qquad (3)$$

In this expansion, two of the terms are immediately zero, so:

$$\Delta = -3[2(6) - (-3)(-5)] = 9.$$

In this example, both methods gave the same answer, but the calculations involved in the second method were a little easier because the third column had two zeros, so two of the three terms in the expansion could immediately be set equal to zero. *When expanding a determinant, it is generally easier to choose a row or column that has the maximum number of zeros.*

We can expand a determinant of order 3 by any row or column. In such an expansion the terms alternate in sign, and each element in the given row or column multiplies the 2×2 determinant (the minor) that can be obtained by deleting from Δ the row and column that contains that element.

We observe that sometimes the sign of the first term in such an expansion is positive [as in (1) and (3)] and sometimes it is negative [as in (2)]. In fact, *the first term in an expansion is positive when expanded by the first or third row (or column) and is negative when expanded by the second row (or column).*

PARENTHETICAL REMARK

These rules extend in a natural way to determinants of orders higher than 3. Any such determinant can be evaluated by expansion along any row or column. The determinant is obtained by multiplying each element in the row (or column) by its cofactor and adding up all the products so obtained. The rule for evaluating the cofactors is exactly the same as for 3×3 determinants:

the cofactor of the element that lies in the ith row and jth column is equal to $(-1)^{i+j}$ multiplied by the smaller determinant that is obtained by deleting the ith row and the jth column.

CRAMER'S RULE

Determinants have many applications. One of the most important is to systems of linear equations in which the number of equations is equal to the number of unknowns. In fact the concept of determinants originated from the study of such systems.

Consider the following system of two equations in two unknowns:

$$a_1 x + b_1 y = k_1 \qquad (4)$$
$$a_2 x + b_2 y = k_2.$$

Let us solve this system by eliminating y. We multiply the first equation by b_2, multiply the second equation by $(-b_1)$, and add the results together:

$$a_1 b_2 x + b_1 b_2 y = k_1 b_2$$
$$-a_2 b_1 x - b_1 b_2 y = -k_2 b_1$$
$$\overline{(a_1 b_2 - a_2 b_1) x + 0y = k_1 b_2 - k_2 b_1}$$

Thus we obtain the solution

$$x = \frac{k_1 b_2 - k_2 b_1}{a_1 b_2 - a_2 b_1}. \qquad (5)$$

If we eliminate x and solve for y in a similar way, we obtain

$$y = \frac{k_2 a_1 - k_1 a_2}{a_1 b_2 - a_2 b_1}. \qquad (6)$$

The solutions (5) and (6) can be expressed in terms of determinants:

$$x = \frac{\Delta_1}{\Delta}, \qquad y = \frac{\Delta_2}{\Delta} \qquad (7)$$

where

$$\Delta = \begin{vmatrix} a_1 & b_1 \\ a_2 & b_2 \end{vmatrix}, \quad \Delta_1 = \begin{vmatrix} k_1 & b_1 \\ k_2 & b_2 \end{vmatrix}, \quad \Delta_2 = \begin{vmatrix} a_1 & k_1 \\ a_2 & k_2 \end{vmatrix}.$$

Notice that Δ is the *determinant of the matrix of coefficients* in the given system (4). Δ_1 is obtained by replacing the first column in Δ by the right sides of the system (4). Δ_2 is obtained by replacing the second column in Δ by these right sides. This form of the solution is called **Cramer's Rule**.

Example 6 Use Cramer's rule to solve the system

$$2x - 3y = 5$$
$$3x + 4y = -2.$$

Solution

$$\Delta = \begin{vmatrix} 2 & -3 \\ 3 & 4 \end{vmatrix} = (2)(4) - (-3)(3) = 17$$

$$\Delta_1 = \begin{vmatrix} 5 & -3 \\ -2 & 4 \end{vmatrix} = (5)(4) - (-3)(-2) = 14$$

$$\Delta_2 = \begin{vmatrix} 2 & 5 \\ 3 & -2 \end{vmatrix} = (2)(-2) - (5)(3) = -19.$$

The solution is therefore

$$x = \frac{\Delta_1}{\Delta} = \tfrac{14}{17}$$

$$y = \frac{\Delta_2}{\Delta} = -\tfrac{19}{17}.$$

An important aspect of Cramer's rule is that it allows us to tell immediately whether or not a system is consistent. In fact it follows immediately from (7) that the system (4) has a unique solution if and only if the determinant of coefficients Δ is nonzero.

Cramer's rule extends to general systems of n equations in n unknowns. The following theorem gives the rule for three unknowns.

THEOREM 2 (CRAMER'S RULE)

Consider the following system of three equations in three unknowns, x, y, and z:

$$\begin{aligned} a_1 x + b_1 y + c_1 z &= k_1 \\ a_2 x + b_2 y + c_2 z &= k_2 \\ a_3 x + b_3 y + c_3 z &= k_3. \end{aligned} \quad (8)$$

Let

$$\Delta = \begin{vmatrix} a_1 & b_1 & c_1 \\ a_2 & b_2 & c_2 \\ a_3 & b_3 & c_3 \end{vmatrix}$$

be the determinant of coefficients and let $\Delta_1, \Delta_2, \Delta_3$ be obtained by replacing respectively the first, second, and third columns in Δ by the constant terms. In other words,

$$\Delta_1 = \begin{vmatrix} k_1 & b_1 & c_1 \\ k_2 & b_2 & c_2 \\ k_3 & b_3 & c_3 \end{vmatrix}, \quad \Delta_2 = \begin{vmatrix} a_1 & k_1 & c_1 \\ a_2 & k_2 & c_2 \\ a_3 & k_3 & c_3 \end{vmatrix}, \quad \Delta_3 = \begin{vmatrix} a_1 & b_1 & k_1 \\ a_2 & b_2 & k_2 \\ a_3 & b_3 & k_3 \end{vmatrix}.$$

If $\Delta \neq 0$, then the system (8) has a unique solution given by

$$x = \frac{\Delta_1}{\Delta}, \quad y = \frac{\Delta_2}{\Delta}, \quad z = \frac{\Delta_3}{\Delta}.$$

If $\Delta = 0$ and $\Delta_1 = \Delta_2 = \Delta_3 = 0$, then the system has an infinite number of solutions. If $\Delta = 0$ and either $\Delta_1 \neq 0$ or $\Delta_2 \neq 0$ or $\Delta_3 \neq 0$, then the system has no solution.

Example 7 Use determinants to solve the following systems of equations:
$$2x - 3y + z = 5$$
$$x + 2y - z = 7$$
$$6x - 9y + 3z = 4.$$

Solution The determinant of coefficients is

$$\Delta = \begin{vmatrix} 2 & -3 & 1 \\ 1 & 2 & -1 \\ 6 & -9 & 3 \end{vmatrix} = 0$$

as we can see by expanding by the first row. Replacing the first-column elements in Δ by the constant terms, we have

$$\Delta_1 = \begin{vmatrix} 5 & -3 & 1 \\ 7 & 2 & -1 \\ 4 & -9 & 3 \end{vmatrix} = -11.$$

Since $\Delta = 0$ and $\Delta_1 \neq 0$, the given system has *no* solution.

Example 8 Use determinants to solve the following system of equations:
$$3x - y + 2z = -1$$
$$2x + y - z = 5$$
$$x + 2y + z = 4.$$

Solution The determinant of coefficients is

$$\Delta = \begin{vmatrix} 3 & -1 & 2 \\ 2 & 1 & -1 \\ 1 & 2 & 1 \end{vmatrix}.$$

Expanding by the first row, we obtain

$$\Delta = 3 \begin{vmatrix} 1 & -1 \\ 2 & 1 \end{vmatrix} - (-1) \begin{vmatrix} 2 & -1 \\ 1 & 1 \end{vmatrix} + 2 \begin{vmatrix} 2 & 1 \\ 1 & 2 \end{vmatrix}$$
$$= 3(1+2) + 1(2+1) + 2(4-1) = 18.$$

Since $\Delta \neq 0$, the system has a unique solution given by

$$x = \frac{\Delta_1}{\Delta}, \qquad y = \frac{\Delta_2}{\Delta}, \qquad z = \frac{\Delta_3}{\Delta}. \qquad \text{(i)}$$

Replacing the first, second, and third columns in Δ respectively by the constant terms, we get

$$\Delta_1 = \begin{vmatrix} -1 & -1 & 2 \\ 5 & 1 & -1 \\ 4 & 2 & 1 \end{vmatrix} = 18, \qquad \Delta_2 = \begin{vmatrix} 3 & -1 & 2 \\ 2 & 5 & -1 \\ 1 & 4 & 1 \end{vmatrix} = 36$$

$$\Delta_3 = \begin{vmatrix} 3 & -1 & -1 \\ 2 & 1 & 5 \\ 1 & 2 & 4 \end{vmatrix} = -18.$$

From (i), therefore, we have

$$x = \frac{\Delta_1}{\Delta} = \frac{18}{18} = 1$$

$$y = \frac{\Delta_2}{\Delta} = \frac{36}{18} = 2$$

$$z = \frac{\Delta_3}{\Delta} = -\frac{18}{18} = -1.$$

Hence the required solution is $x = 1, y = 2, z = -1$.

It should be pointed out that Cramer's rule is not usually the most efficient method of solving systems of equations. As a rule the method of row reduction described in Section 1 involves shorter calculations. Cramer's rule is important mainly from a theoretical standpoint. One of the most significant results arising from it is that *a system of n linear equations in n unknowns has a unique solution if and only if the determinant of coefficients is nonzero.*

Diagnostic Test 11.5

Fill in the blanks.

1. By definition, the determinant
$$\begin{vmatrix} a & b \\ c & d \end{vmatrix} = \underline{\qquad}.$$

2. If $\Delta = \begin{vmatrix} a & b \\ c & d \end{vmatrix}$ then the minor of
   (a) a in Δ is _____.
   (b) b in Δ is _____.

3. The cofactor of the element 3 in $\begin{vmatrix} 1 & 5 & 4 \\ 2 & 4 & 3 \\ 7 & 1 & 2 \end{vmatrix}$ is _____.

4. The cofactor of an element x in the ith row and jth column of a determinant is equal to _____ times the minor of x.

5. We can use Cramer's rule for solving a system of linear equations only if the number equations is _____ the number of unknowns.

6. If Δ denotes the determinant of coefficients of a given system, then by Cramer's rule, the system has a unique solution if and only if _____.

Exercises 11.5

(1–4) Write each of the following for the determinant $\Delta = \begin{vmatrix} a & b & c \\ p & q & r \\ l & m & n \end{vmatrix}$.

1. The minor of q.
2. The minor of n.
3. The cofactor of r.
4. The cofactor of m.

Sec. 11.5   *Determinants and Cramer's Rule*

(5–28) Evaluate the following determinants.

5. $\begin{vmatrix} 3 & -1 \\ 4 & 7 \end{vmatrix}.$

6. $\begin{vmatrix} 5 & 8 \\ 3 & -2 \end{vmatrix}.$

7. $\begin{vmatrix} -6 & -7 \\ -8 & -3 \end{vmatrix}.$

8. $\begin{vmatrix} 2 & x \\ 0 & 1 \end{vmatrix}.$

9. $\begin{vmatrix} a & -2 \\ b & 3 \end{vmatrix}.$

10. $\begin{vmatrix} 5 & a \\ -a & 4 \end{vmatrix}.$

11. $\begin{vmatrix} 32 & 2 \\ 64 & 5 \end{vmatrix}.$

12. $\begin{vmatrix} 274 & 3 \\ 558 & 7 \end{vmatrix}.$

13. $\begin{vmatrix} 59 & 3 \\ 64 & 0 \end{vmatrix}.$

14. $\begin{vmatrix} a+1 & 2-a \\ 2a+3 & 5-2a \end{vmatrix}.$

15. $\begin{vmatrix} a & b \\ a+b & b+c \end{vmatrix}.$

16. $\begin{vmatrix} x+2 & x-1 \\ 3 & x \end{vmatrix}.$

17. $\begin{vmatrix} 2 & 1 & 4 \\ 3 & 5 & -1 \\ 1 & 0 & 0 \end{vmatrix}.$

18. $\begin{vmatrix} 1 & -2 & 3 \\ 4 & 0 & 5 \\ 6 & 0 & 7 \end{vmatrix}.$

19. $\begin{vmatrix} 1 & 3 & -2 \\ 0 & 2 & 1 \\ 4 & -1 & 3 \end{vmatrix}.$

20. $\begin{vmatrix} 3 & 1 & 0 \\ 0 & -2 & 1 \\ 2 & 0 & 5 \end{vmatrix}.$

21. $\begin{vmatrix} 2 & 3 & 4 \\ 5 & 6 & 7 \\ 8 & 9 & 10 \end{vmatrix}.$

22. $\begin{vmatrix} 5 & 10 & 1 \\ 8 & 5 & 4 \\ 1 & 4 & 2 \end{vmatrix}.$

23. $\begin{vmatrix} 1 & 2 & 4 \\ 2 & 5 & 1 \\ 3 & 8 & 4 \end{vmatrix}.$

24. $\begin{vmatrix} 7 & 9 & 3 \\ 8 & 2 & 0 \\ 0 & 5 & 4 \end{vmatrix}.$

25. $\begin{vmatrix} a & b & c \\ 0 & d & e \\ 0 & 0 & f \end{vmatrix}.$

26. $\begin{vmatrix} x & 0 & 0 \\ a & y & 0 \\ b & c & z \end{vmatrix}.$

27. $\begin{vmatrix} 1 & 0 & -1 \\ 2 & 1 & 0 \\ 0 & -2 & 1 \end{vmatrix}.$

28. $\begin{vmatrix} 2 & 3 & 4 \\ 1 & 0 & -1 \\ 0 & -2 & 1 \end{vmatrix}.$

(29–32) Determine x such that:

29. $\begin{vmatrix} x & 3 \\ 2 & 5 \end{vmatrix} = 9.$

30. $\begin{vmatrix} x+3 & 2 \\ x & x+1 \end{vmatrix} = 3.$

31. $\begin{vmatrix} 1 & 0 & 0 \\ x^2 & x-2 & 3 \\ x & x+1 & x \end{vmatrix} = 3.$

32. $\begin{vmatrix} x+1 & 2 & x \\ x & x^2 & 2 \\ 0 & 1 & 0 \end{vmatrix} = 1.$

(33–50) Use Cramer's rule to solve the following systems of equations.

33. $3x + 2y = 1.$
    $2x - y = 3.$

34. $2x - 5y = 8$
    $3y + 7x = -13.$

35. $4x + 5y - 14 = 0$
    $3y = 7 - x.$

36. $2(x - y) = 5$
    $4(1 - y) = 3x.$

37. $x/3 + y/2 = 7$
    $x/2 - y/5 = 1.$

38. $\frac{2}{3}u + \frac{3}{4}v = 13$
    $\frac{5}{2}u + \frac{1}{3}v = 19.$

39. $2x + 3y = 13$
    $6x + 9y = 40.$

40. $3x = 2(2 + y)$
    $4y = 7 + 6x.$

41. $x + y + z = -1$
    $2x + 3y - z = 0$
    $3x - 2y + z = 4.$

42. $2x - y + z = 2$
    $3x + y - 2z = 9$
    $-x + 2y + 5z = -5.$

43. $2x + y + z = 0$
    $x + 2y - z = -6$
    $x + 5y + 2z = 0.$

44. $x + 3y - z = 0$
    $3x - y + 2z = 0$
    $2x - 5y + z = 5.$

45. $x + 2y = 5$
    $3y - z = 1$
    $2x - y + 3z = 11.$

46. $2v + 5w = 3$
    $4u - 3w = 5$
    $3u - 4v + 2w = 12.$

47. $2x - y + 3z = 4$
    $x + 3y - z = 5$
    $6x - 3y + 9z = 10.$

48. $4x + 2y - 6z = 7$
$3x - y + 2z = 12$
$6x + 3y - 9z = 10.$

49. $x + 2y - z = 2$
$2x - 3y + 4z = -4$
$3x + y + z = 0.$

50. $2p - r = 5$
$p + 3q = 9$
$3p - q + 5r = 12.$

51–53. Use Cramer's rule to repeat Exercises 42–44 of Section 11.4.

Chapter Review 11

1. Define the following:
   (a) The size of a matrix.
   (b) The equality of two matrices.
   (c) The sum and difference of two matrices.
   (d) Scalar multiplication of a matrix.
   (e) Identity matrix.
   (f) Zero matrix.
   (g) Row vector and column vector.
   (h) The product of two matrices.
   (i) The inverse of a matrix.
   (j) Singular and nonsingular matrices.
   (k) The minor and cofactor of an element in a determinant of order 3.

2. Describe the method of:
   (a) Finding the product of two matrices.
   (b) Solving a system of linear equations by row reduction.
   (c) Finding the inverse of a matrix if it exists.
   (d) Evaluating a determinant of order 2.
   (e) Expanding a determinant of order 3.
   (f) Solving a system of equations by Cramer's rule.
   (g) Using Cramer's rule to check whether a system has one solution, no solutions, or an infinity of solutions.

Review Exercises on Chapter 11

1. Write down the 2×3 matrix $\mathbf{A} = a_{ij}$, for which $a_{ij} = i - j + 2$.

2. Write down a 3×2 matrix $\mathbf{A} = a_{ij}$, for which
$$a_{ij} = \begin{cases} i - j & \text{if } i \neq j. \\ i + j & \text{if } i = j. \end{cases}$$

(3–6) Determine the values of the unknowns so that the following matrix equations are true.

3. $\begin{bmatrix} 2 & a \\ b & -1 \end{bmatrix} = \begin{bmatrix} c & 5 \\ -2 & d \end{bmatrix}.$

4. $\begin{bmatrix} x+3 & 5 \\ 2 & y-1 \end{bmatrix} = \begin{bmatrix} 4 & y+3 \\ x+1 & x \end{bmatrix}.$

5. $\begin{bmatrix} 2 & y & 1 \\ x & 3 & z \\ 0 & -4 & 5 \end{bmatrix} = \begin{bmatrix} u & 3 & 1 \\ 4 & v & 2 \\ 0 & w & v+2 \end{bmatrix}.$

6. $\begin{bmatrix} x & -2 \\ 1 & y \end{bmatrix} + 2 \begin{bmatrix} 1 & y \\ x-1 & 2 \end{bmatrix} = \begin{bmatrix} 1 & 0 \\ -3 & 5 \end{bmatrix}.$

(7–12) Perform the indicated matrix operations and simplify.

7. $\begin{bmatrix} 2 & -1 \\ 3 & 4 \end{bmatrix} + 2 \begin{bmatrix} 1 & -2 \\ 4 & 3 \end{bmatrix} - 3 \begin{bmatrix} 1 & 2 \\ -3 & 0 \end{bmatrix}.$

8. $\begin{bmatrix} 1 & 2 & 3 \\ 3 & -1 & 2 \\ -2 & 3 & 1 \end{bmatrix} - 3 \begin{bmatrix} 0 & 1 & -1 \\ -2 & 3 & 0 \\ 1 & 0 & 2 \end{bmatrix} + 5 \begin{bmatrix} 1 & 0 & 2 \\ -1 & 2 & 3 \\ 0 & -1 & 0 \end{bmatrix}.$

9. $\begin{bmatrix} 1 & 0 & -1 \\ 2 & 1 & 0 \end{bmatrix} \begin{bmatrix} 2 & -1 \\ 1 & 3 \\ -3 & 2 \end{bmatrix} + 2 \begin{bmatrix} 1 & 2 \\ 3 & 4 \end{bmatrix}.$

10. $\begin{bmatrix} 2 & 3 \\ 1 & 0 \\ -3 & 1 \end{bmatrix} - 2 \begin{bmatrix} -1 & 2 \\ 0 & -1 \\ 2 & 3 \end{bmatrix} \begin{bmatrix} 2 & 1 \\ 3 & -1 \end{bmatrix}.$

11. $\begin{bmatrix} 1 & 2 & 3 \\ 0 & -1 & 2 \end{bmatrix} \begin{bmatrix} 2 & -1 \\ 3 & 4 \\ 1 & 0 \end{bmatrix} - \begin{bmatrix} 3 & 1 \\ -1 & 2 \end{bmatrix} \begin{bmatrix} 0 & 1 \\ 2 & 3 \end{bmatrix}.$

12. $\begin{bmatrix} 1 & 0 & -1 \\ 0 & 1 & 2 \end{bmatrix} \left(\begin{bmatrix} 2 & 0 \\ 1 & -1 \\ 0 & 1 \end{bmatrix} + 3 \begin{bmatrix} 0 & -1 \\ 1 & 2 \\ 0 & 0 \end{bmatrix} \right) + \begin{bmatrix} 2 & 1 \\ 3 & 0 \end{bmatrix} \begin{bmatrix} -1 & 2 \\ 0 & 1 \end{bmatrix}.$

(13–22) Solve the following systems of equations by the method of row reduction.

13. $2x - 3y = 1$
$3x + 4y = 10.$

14. $3u - 4v = 11$
$5u + 2v = 1.$

15. $2x + y - 3z = -5$
$-x + 2y + z = -1$
$3x - y - 2z = 0.$

16. $x - y + z = 2$
$3x - 2y + 4z = 5$
$2x + y + 3z = 8.$

17. $x + y + 3z = -1$
$3x + 4y + 2z = 5$
$x + 2y = 4 + z.$

18. $2p - 3q + r = 7$
$-p + 2q + 3r = 3$
$3p + q - 2r = -2.$

19. $\begin{bmatrix} 2 & -1 \\ 1 & 3 \end{bmatrix} \begin{bmatrix} x \\ y \end{bmatrix} = \begin{bmatrix} -4 \\ 5 \end{bmatrix}.$

20. $\begin{bmatrix} 3 & 2 \\ 4 & -1 \end{bmatrix} \begin{bmatrix} x \\ y \end{bmatrix} + \begin{bmatrix} 5 \\ -3 \end{bmatrix} = \begin{bmatrix} 9 \\ 6 \end{bmatrix}.$

21. $x[3 \quad 1 \quad 2] + y[2 \quad -3 \quad 1] = [1 \quad 4 \quad 1].$

22. $\begin{bmatrix} 2 & -1 & 3 \\ 3 & 1 & -2 \\ 1 & 2 & 1 \end{bmatrix} \begin{bmatrix} x \\ y \\ z \end{bmatrix} = \begin{bmatrix} 4 \\ -3 \\ 7 \end{bmatrix}.$

(23–24) Determine the matrix **X** such that the following equations are satisfied.

*23. $\begin{bmatrix} 2 & 1 \\ 3 & 4 \end{bmatrix} \mathbf{X} = \begin{bmatrix} 2 & -1 \\ 3 & 1 \end{bmatrix}.$

*24. $\begin{bmatrix} 2 & 1 & 3 \\ 1 & 2 & -1 \\ -1 & 1 & 1 \end{bmatrix} \mathbf{X} = \begin{bmatrix} 7 & 14 \\ -3 & 1 \\ 0 & 2 \end{bmatrix}.$

(25–34) Find the inverses of the matrices given below, when they exist.

25. $\begin{bmatrix} 1 & -3 \\ 2 & 5 \end{bmatrix}.$

26. $\begin{bmatrix} 2 & 4 \\ 5 & -3 \end{bmatrix}.$

27. $\begin{bmatrix} a & b \\ -b & a \end{bmatrix}$ $(a, b \neq 0).$

28. $\begin{bmatrix} 1 & 1 \\ a & b \end{bmatrix}$ $(a \neq b).$

29. $\begin{bmatrix} 3 & 1 \\ 1 & -2 \\ 2 & 5 \end{bmatrix}.$

30. $\begin{bmatrix} 1 & 0 & 2 \\ 0 & 2 & 1 \\ 2 & 1 & 0 \end{bmatrix}.$

31. $\begin{bmatrix} 1 & 2 & 3 \\ 2 & 3 & 4 \\ 3 & 1 & 2 \end{bmatrix}.$

32. $\begin{bmatrix} 2 & -1 & 1 \\ 1 & 2 & -1 \\ -1 & 1 & 2 \end{bmatrix}.$

33. $\begin{bmatrix} 1 & -2 & 3 \\ 4 & 1 & 6 \\ 7 & 4 & 9 \end{bmatrix}.$

34. $\begin{bmatrix} 2 & -1 & -3 \\ 4 & 1 & -1 \\ 7 & -2 & -8 \end{bmatrix}.$

(35–38) Solve the following systems of equations by using the inverse of the coefficient matrix.

35. $4x - 3y = 1$
$3x + 2y = 5.$

36. $2u - 5v = 11$
$3u + 4v = 5.$

37. $x + y + z = 1$
$2x - y + 3z = -2$
$3x + 2y - z = 6.$

38. $3p - 2q + 4r = 13$
$p + q - 2r = 1$
$2p - 3q + 5r = 11.$

(39–44) Evaluate the determinants.

39. $\begin{vmatrix} a & b \\ c & d \end{vmatrix}.$

40. $\begin{vmatrix} x & y \\ -y & x \end{vmatrix}.$

41. $\begin{vmatrix} 2 & 0 & 3 \\ -1 & 1 & 4 \\ 0 & 2 & -2 \end{vmatrix}.$

42. $\begin{vmatrix} 1 & 2 & 3 \\ 5 & 4 & 3 \\ 5 & 6 & 7 \end{vmatrix}.$

43. $\begin{vmatrix} -1 & 3 & 2 \\ 4 & 0 & 4 \\ -2 & 3 & 1 \end{vmatrix}.$

44. $\begin{vmatrix} x & y & 0 \\ x & 0 & z \\ 0 & y & z \end{vmatrix}.$

(45–48) Use Cramer's rule to solve the following systems of equations.

45. $x - 3y = 2$
$2x + 4y = 4.$

46. $2x + 5y = 7$
$3x - 2y = -6.$

47. $3x - 4y + z = -2$
$x - 3y + 2z = 1$
$4x + y = 5.$

48. $2x - 3y = 1$
$-3x + 2z = 1$
$2y - 3z = 1.$

12 TRIGONOMETRY

12.1

Angles

"Trigonometry" is a Greek word that means "triangle measurement." Over two thousand years ago the Greeks used trigonometry to solve problems in land surveying, astronomy, and navigation. Today it is widely used in virtually all branches of science, engineering, and technology. In its most basic form, trigonometry is the study of relationships between the sides and angles of a right triangle.

We begin by defining angles. An *angle* consists of two line segments that have a common endpoint, called the *vertex* of the angle. The two lines forming the angle are called the *sides* of the angle, the *initial side* and the *terminal side*. We also identify the direction of the angle by a curved arrow that indicates the rotation from the initial side to the terminal side (see the left-hand figure at the top of the next page).

When an angle is in a plane where a Cartesian coordinate system is prescribed, it is convenient to take the vertex of the angle at the origin and the initial side of the angle along the positive x axis. In such a case, the angle is said to be *in standard position* (see the right-hand figure at the top of the next page).

(a) (b)

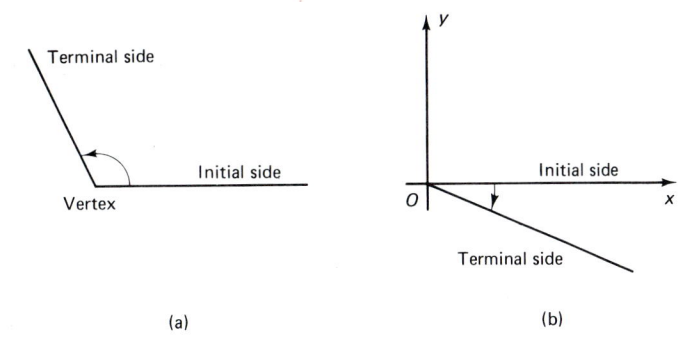

An angle in standard position is said to be in the first, second, third, or fourth quadrant if its terminal side lies in the first, second, third, or fourth quadrant, respectively, as shown below.

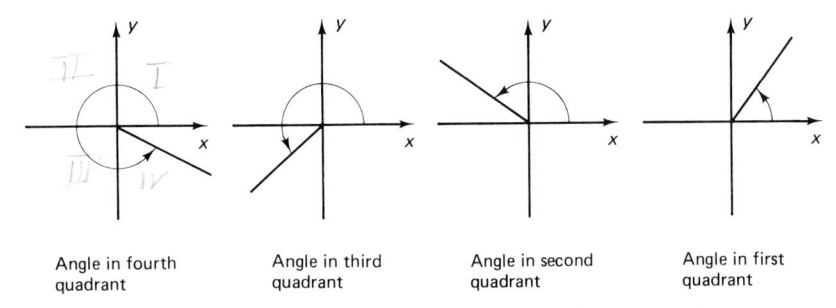

Angle in fourth quadrant   Angle in third quadrant   Angle in second quadrant   Angle in first quadrant

Two units called *degrees* and *radians* are generally used to measure angles. We shall study both of them.

As shown below, an angle formed by one complete counterclockwise revolution of the initial side has a measure of 360 degrees, which we write as 360°. Sometimes an angle of 360° is called a *round angle*. An angle of half

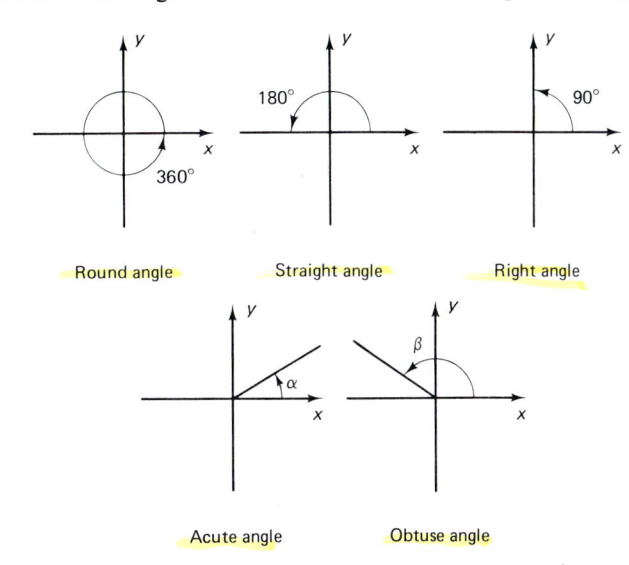

Round angle   Straight angle   Right angle

Acute angle   Obtuse angle

Sec. 12.1   Angles

a revolution, or 180°, is called a ***straight angle*** and an angle of 90° (a quarter of a revolution) a ***right angle***. An angle* α that is between 0° and 90° ($0° < \alpha < 90°$) is an ***acute angle*** while an angle β that is between 90° and 180° ($90° < \beta < 180°$) is an ***obtuse angle***.

An angle is given a negative measure if it is formed by clockwise rotation (see adjoining figure). Angles formed by counterclockwise rotation are given positive values.

For many purposes, such as the accurate location of stars, one must make use of angles that measure less than 1°. Traditionally 1° is subdivided into 60 minutes, written as 60', and 1' is further subdivided into 60 seconds, written as 60''. An angle of 25 degrees 7 minutes and 49 seconds is written as 25°07'49''. Thus, we have

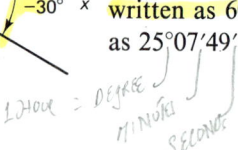

1 right angle = 90°	(90 degrees)
1° = 60'	(60 minutes)
1' = 60''	(60 seconds)

Example 1 One of the acute angles in a right triangle is 35°47'23''. Find the other acute angle.

Solution The two acute angles in a right triangle always add up to 90°. Thus if one acute angle is $\alpha = 35°47'23''$, then the other acute angle is given by

$$\beta = 90° - \alpha = 90° - 35°47'23''.$$

Now we can write 90° as

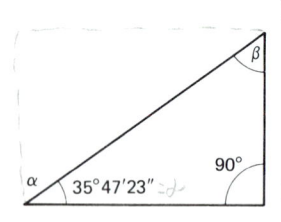

$$90° = 89° + 1° = 89° + 60'$$
$$= 89° + 59' + 1' = 89° + 59' + 60''$$
$$= 89°59'60''.$$

Thus,

$$\beta = 90° - 35°47'23''$$
$$= 89°59'60'' - 35°47'23''$$
$$= (89° - 35°) + (59' - 47') + (60'' - 23'')$$
$$= 54°12'37''.$$

Note With the wide use of electronic calculators, it is becoming common to use decimal degrees—for example 72.25° instead of 72°15'.

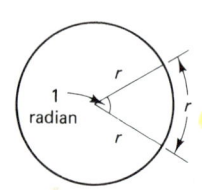

Another unit in which angles can be measured is the ***radian***. One radian is defined as the angle subtended at the center of a circle by an arc whose length is equal to the radius (see adjoining figure). This angle is independent of the choice of radius r.

---

*α and β are the Greek letters *alpha* and *beta*.

Clearly an arc of length $2r$ will subtend an angle of 2 radians at the center; an arc of length $3r$ will subtend an angle of 3 radians at the center and so on. Thus an arc of $2\pi r$ (the whole circumference) will subtend an angle of 2π radians at the center. But the whole circle subtends an angle of $360°$ at the center of a circle, and therefore 2π radians $= 360°$. Thus

$$\pi \text{ radians} = 180°. \tag{1}$$

We can use this relation to convert an angle from radian measure to degree measure or vice versa. When converting an angle from degrees to radians, we multiply by $\pi/180$, and when converting an angle from radians to degrees we multiply by $180/\pi$. Thus, we have

$$\alpha° = \left(\alpha \cdot \frac{\pi}{180}\right) \text{ radians}$$
$$\beta \text{ radians} = \left(\beta \cdot \frac{180}{\pi}\right)°. \tag{2}$$

Example 2 Convert the following to radian measure.
(a) $20°$. (b) $-70°$. (c) $337.5°$.

Solution Using the first formula in (2), we have

(a) $20° = 20\left(\frac{\pi}{180}\right) = \frac{\pi}{9}$ radians.

(b) $-70° = -70\left(\frac{\pi}{180}\right) = -\frac{7\pi}{18}$ radians.

(c) $337.5° = 337.5\left(\frac{\pi}{180}\right) = \frac{15\pi}{8}$ radians.

Example 3 Convert the following to degree measure.
(a) $\frac{\pi}{12}$ radians. (b) $-\frac{17\pi}{120}$ radians. (c) 2.5 radians.

Solution We use the second formula in (2) in each case.

(a) $\frac{\pi}{12}$ radians $= \left(\frac{\pi}{12} \cdot \frac{180}{\pi}\right)° = 15°$.

(b) $-\frac{17\pi}{120}$ radians $= \left(-\frac{17\pi}{120} \cdot \frac{180}{\pi}\right)° = -\frac{51°}{2} = -25.5° = -25°30'$.

(c) 2.5 radians $= \left(2.5 \cdot \frac{180}{\pi}\right)° = 143.24°$ (rounding to two decimal places). Now $0.24° = (0.24)(60') \approx 14'$. Therefore 2.5 radians $\approx 143°14'$.

We give below a table of certain special angles in degrees and radians that occur frequently in trigonometry.

Degrees	0°	30°	45°	60°	90°	120°	135°	150°	180°
Radians	0	$\frac{\pi}{6}$	$\frac{\pi}{4}$	$\frac{\pi}{3}$	$\frac{\pi}{2}$	$\frac{2\pi}{3}$	$\frac{3\pi}{4}$	$\frac{5\pi}{6}$	π

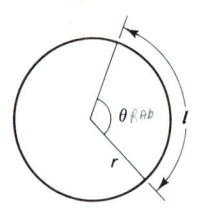

Let us suppose that an arc of length *l* subtends an angle of θ* radians at the center of a circle of radius *r*. Since an angle of 1 radian is subtended by an arc of length *r*, it follows that an angle of θ radians will be subtended by an arc of length $r\theta$. Thus, we must have

$$l = r\theta \qquad (3)$$

When applying Equation (3), one must make sure that *l* and *r* are in the same units (both in feet, or yards, or meters, and so on) and θ *must be in radians.*

Example 4 A horse is tied to a pole with a rope 9 feet long. If the horse moves a distance of 5 yards around the pole, always keeping the rope tight, how large an angle is traced by the rope?

Solution When a horse moves around the pole, keeping the rope tight, he always moves in a circle. Here

$$r = \text{radius of circular path}$$
$$= \text{length of rope} = 9 \text{ feet}$$
$$l = \text{length of arc}$$
$$= \text{distance traveled by horse}$$
$$= 5 \text{ yards} = 15 \text{ feet.}$$

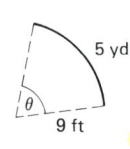

If the angle traced at the center is θ radians, then

$$l = r\theta$$
$$\theta = \frac{l}{r} = \frac{15}{9} = \frac{5}{3} \text{ radians.}$$

Example 5 A circular arc 33 centimeters long subtends an angle of 36° at the center of the circle. What is the radius of the circle? (Approximate π by $\frac{22}{7}$.)

Solution Here *l* = 33 centimeters, and

$$\theta = 36° = 36 \cdot \frac{\pi}{180} = \frac{\pi}{5} \text{ radians.}$$

We know that $l = r\theta$. Hence the radius is

$$r = \frac{l}{\theta} = \frac{33}{\pi/5} = \frac{165}{\pi}$$
$$\approx \frac{165}{\frac{22}{7}} = \frac{(165)(7)}{22}$$
$$= \frac{105}{2} = 52.5 \text{ cm.}$$

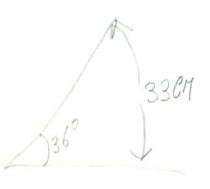

*The Greek letter *theta*.

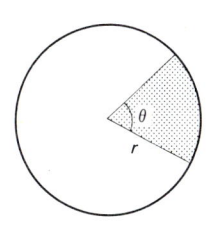

The shaded area in the adjoining figure is called a *sector* of the circle. If a sector of a circle of radius r has an angle of θ radians, then its area A is given by

$$\boxed{A = \tfrac{1}{2} r^2 \theta} \tag{4}$$

(see Exercise 75).

Note that if we put $\theta = 2\pi$ in (4), $= 360°$ we get

$$A = \tfrac{1}{2} r^2 (2\pi) = \pi r^2$$

which is, of course, the correct formula for the area of the complete circle.

Example 6 For a circle of radius 15 feet, find the area of a sector that subtends an angle of 30°.

Solution Here $r = 15$ feet, and

$$\theta = 30° = \frac{\pi}{6} \text{ radians.}$$

Therefore, from (4), the area of the sector is given by

$$A = \frac{1}{2} r^2 \theta$$

$$= \frac{1}{2} \cdot 15^2 \cdot \frac{\pi}{6}$$

$$= \frac{75\pi}{4} \text{ square feet.} \quad = \frac{75 \times 3.14285}{4} = 58.92$$

Two *positive angles* α and β are said to be *complementary* if their sum is 90° or $\pi/2$ radians. Thus 30° and 60° are complementary angles.

Similarly, two positive angles α and β are said to be *supplementary* if their sum is 180° or π radians. For example, 40° and 140° are supplementary angles, as are $\pi/3$ and $2\pi/3$.

Finally, two angles are said to be *coterminal* if in standard position both have the same terminal side. For example, the angles 30° and 390° are coterminal (see figures below), as are $-60°$ and 300°.

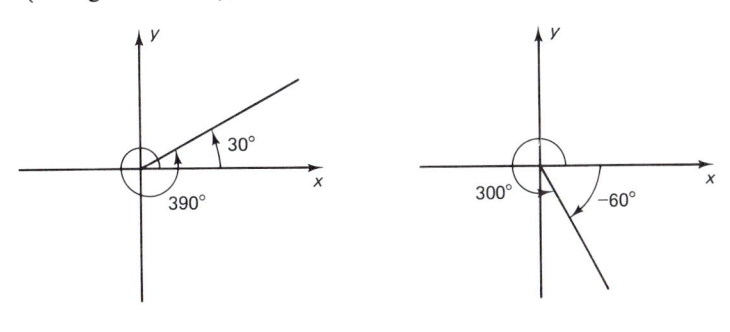

30° and 390° are coterminal angles

$-60°$ and 300° are coterminal angles

Sec. 12.1 Angles

In general, two coterminal angles differ by an integer multiple of 360° or 2π radians. In the above examples, $390° - 30° = 360°$ and $300° - (-60°) = 360°$. If n is any integer, the angles θ (radians) and $2n\pi + \theta$ are coterminal angles. Thus, if an angle is given, other angles with which it is coterminal can be obtained by adding (or subtracting) multiples of 2π or 360° to (or from) the given angle.

Example 7 Find two positive and two negative angles that are coterminal with 160°.

Solution Two positive angles coterminal with 160° are
$$160° + 360° = 520°$$
$$160° + 2(360°) = 880°.$$
Two negative angles coterminal with 160° are
$$160° - 360° = -200°$$
$$160° - 2(360°) = -560°.$$
(Of course there are also others.)

Diagnostic Test 12.1

Fill in the blanks.
1. One right angle = _____ degrees = _____ radians.
2. $\alpha° =$ _____ radians and β radians = _____ degrees.
3. If an arc of length l subtends an angle of θ degrees at the center of a circle of radius r, then $l =$ _____.
4. The area A of the sector of a circle of radius r with an angle of α radians at the center is given by $A =$ _____.
5. If $\alpha + \beta = \pi/2$, then α and β are said to be _____ angles.
6. If $\alpha + \beta = 180°$, then α and β are said to be _____ angles.
7. If two angles differ by an integer multiple of 360°, then they are known as _____ angles.
8. A positive angle α is said to be an acute angle if it is less than _____ radians.
9. If $90° < \theta < 180°$, then θ is an _____ angle.
10. The angle complementary to 40° is _____.
11. The angle supplementary to 60° is _____ (radians).

Exercises 12.1

(1–16) Express the following angles in radian measure.
1. 40°.
2. 130°.
3. 70°.
4. 210°.
5. 325°.
6. 510°.
7. −120°.
8. −225°.
9. −750°.
10. 1140°.
11. 36.5°.
12. −108.25°.

13. 22°30′. 14. 12°30′. 15. 11°15′. 16. 21°20′.

(17–32) Express the following angles in degree measure.

17. $\dfrac{2\pi}{5}$. 18. $\dfrac{3\pi}{2}$. 19. $\dfrac{2\pi}{9}$. 20. $\dfrac{\pi}{10}$. 21. $-\dfrac{5\pi}{6}$.

22. $-\dfrac{\pi}{12}$. 23. $\dfrac{11\pi}{12}$. 24. $\dfrac{13\pi}{6}$. 25. -5π. 26. $\dfrac{4\pi}{3}$.

27. $\dfrac{7\pi}{6}$. 28. $\dfrac{5\pi}{4}$.

29. 1 radian. 30. 4 radians. 31. 0.6 radians. 32. 1.5 radians.

33. One acute angle of a right triangle is 35°27′. Find the other acute angle in degrees.

34. One acute angle of a right triangle is $\pi/9$. Find the other acute angle in degrees.

35. Two angles of a triangle are 27°15′38″ and 53°44′22″. Find the third angle in degrees. (Recall that the three angles in a triangle always add up to 180°.)

36. Two angles of a triangle are $\pi/6$ and $2\pi/5$ radians. Find the third angle in radians.

37. The difference between the two acute angles of a right triangle is $\pi/9$ radians. Express these angles in degrees.

38. The angles of a triangle are to one another in the ratio 2 : 3 : 4. Express them in degrees as well as in radians.

39. One angle of a triangle is 37° and the second is $2\pi/9$ radians. Find the third angle in degrees.

40. If D denotes the degree measure of an angle and θ denotes the radian measure of the same angle, show that
$$\dfrac{D}{9} = \dfrac{20\theta}{\pi}.$$

(41–50) In the following exercises, θ denotes the angle subtended by an arc of length l at the center of a circle of radius r. Find the indicated quantities. (Use $\pi \approx \tfrac{22}{7}$.)

41. $\theta = \pi/6$; $r = 14$ feet; find l.
42. $\theta = 60°$; $r = 63$ centimeters; find l.
43. $\theta = 135°$; $r = 14$ inches; find l.
44. $\theta = 120°$; $l = 88$ centimeters; find r.
45. $\theta = 5\pi/6$; $l = 77$ centimeters; find r.
46. $\theta = 42°$; $l = 27.5$ inches; find r.
47. $l = 121$ inches; $r = 33$ inches; find θ in degrees.
48. $l = 143$ centimeters; $r = 52$ centimeters; find θ in degrees.
49. $l = 11$ feet; $r = 45$ inches; find θ in degrees.
50. $l = 135$ inches; $r = 5$ feet; find θ in degrees.

51. Find the angle in degree measure subtended at the center of a circle of radius 5 feet by an arc of 11 inches. (Use $\pi \approx \tfrac{22}{7}$.)

52. A wire 121 inches long is bent so as to lie along the arc of a circle of 180 inches radius. Find in degrees the angle subtended at the center by the arc. (Use $\pi \approx \tfrac{22}{7}$.)

53. The large hand of a clock is 2 feet 4 inches long. How many inches does its extremity move in 20 minutes? (Use $\pi \approx \tfrac{22}{7}$.)

54. A train is traveling at the rate of 10 miles per hour on a circular curve of a half-mile radius. Through what angle in degrees has it turned in one minute? (Use $\pi \approx \tfrac{22}{7}$.)

55. A horse is tied to a post by a rope. If the horse moves along a circular path always keeping the rope tight and describes a distance of 44 feet when it has traced an angle of 72° at the center, find the length of the rope. (Use $\pi \approx \tfrac{22}{7}$.)

56. A wire of certain length is bent so as to lie along a circular arc of radius 90 centimeters. If the circular wire subtends an angle of 77° at the center, find the length of the wire. (Use $\pi \approx \tfrac{22}{7}$.)

(57–62) Find the angle that lies between 0° and 360° and is coterminal with the given angle.

57. 450°. **58.** 1240°. **59.** −120°. **60.** −750°.
61. 31π/6. **62.** −17π/5.

(63–70) Find the quadrant in which the terminal side of each angle lies.

63. 2π/3. **64.** 6π/5. **65.** −70°. **66.** 215°.
67. 480°. **68.** 1240°. **69.** 31π/6. **70.** −1030°.

(71–72) Find the area of the sector with the given angle and radius.

71. $r = 3$, $\theta = 120°$. **72.** $r = 2$, $\theta = 225°$.

73. A sector of radius 3 has an area of 3. Find its angle in degrees.

74. A sector of angle 135° has an area of 5. Find its radius.

*75. Prove Equation (4) for the area of a sector. [*Hint:* The area equals a fraction $(\theta/2\pi)$ of the area of the whole circle.]

12.2

Right-triangle Trigonometry

In this section we shall define the six trigonometric functions in terms of the sides of a right triangle, and in the following section we shall use them to solve some applied problems. Let ABC be a right triangle with right angle at C (see figure below). Let the acute angle at A be denoted by θ. The side

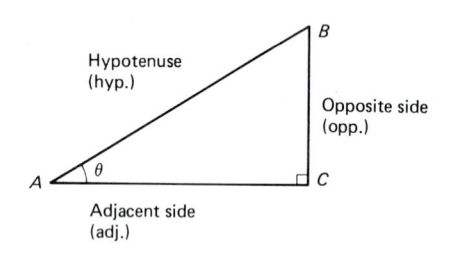

BC opposite to the angle θ is called the **opposite side**, and the side AC is called the **adjacent side**. The side AB is the **hypotenuse**.

We define the trigonometric functions in terms of these three sides as follows:

The ratio $BC/AB = $ opp./hyp. is called the **sine of θ** and is abbreviated as $\sin \theta$.

The ratio $AC/AB = $ adj./hyp. is called the **cosine of θ** and is abbreviated as $\cos \theta$.

The ratio $BC/AC =$ opp./adj. is called the ***tangent of θ*** and is abbreviated as $\tan \theta$.

Thus in terms of the three sides (opposite, adjacent, and hypotenuse relative to the angle θ), we have

$$\sin \theta = \frac{\text{opp.}}{\text{hyp.}}$$
$$\cos \theta = \frac{\text{adj.}}{\text{hyp.}} \qquad (1)$$
$$\tan \theta = \frac{\text{opp.}}{\text{adj.}}$$

The remaining three trigonometric functions are the reciprocals of the above three and are defined as follows:

$$\frac{1}{\tan \theta} = \text{cotangent of } \theta, \text{ abbreviated as } \cot \theta.$$

$$\frac{1}{\cos \theta} = \text{secant of } \theta, \text{ abbreviated as } \sec \theta.$$

$$\frac{1}{\sin \theta} = \text{cosecant of } \theta, \text{ abbreviated as } \csc \theta.$$

Thus, we have

$$\cot \theta = \frac{1}{\tan \theta}, \qquad \sec \theta = \frac{1}{\cos \theta}, \qquad \csc \theta = \frac{1}{\sin \theta}. \qquad (2)$$

From the right triangle ABC, we have

$$\sin \theta = \frac{\text{opp.}}{\text{hyp.}} = \frac{BC}{AB} \quad \text{and} \quad \cos \theta = \frac{\text{adj.}}{\text{hyp.}} = \frac{AC}{AB}.$$

Therefore,

$$\frac{\sin \theta}{\cos \theta} = \frac{BC/AB}{AC/AB} = \frac{BC}{AC} = \frac{\text{opp.}}{\text{adj.}} = \tan \theta.$$

Moreover,

$$\cot \theta = \frac{1}{\tan \theta} = \frac{\cos \theta}{\sin \theta}.$$

Thus, we also have

$$\tan \theta = \frac{\sin \theta}{\cos \theta}, \qquad \cot \theta = \frac{\cos \theta}{\sin \theta}. \qquad (3)$$

We see therefore that all four of the remaining trigonometric functions can be expressed in terms of $\sin \theta$ and $\cos \theta$.

Remark $\sin \theta$ is *not* the product of sin and θ—that is, $\sin \theta \neq (\sin)(\theta)$. It is a single symbol denoting a certain ratio. The same applies to the other trigonometric functions.

[Handwritten note: PYTHAGOREAN THEOREM: $c^2 = a^2 + b^2$, with sketch of right triangle with sides a, b, c]

With regard to the powers of trigonometric functions, the following conventions are followed:

$(\sin \theta)^2$ is written as $\sin^2 \theta$ and read as *sine squared* θ.
$(\cos \theta)^3$ is written as $\cos^3 \theta$ and read as *cosine cubed* θ.
$(\tan \theta)^2$ is written as $\tan^2 \theta$ and read as *tangent squared* θ.

Thus, if n is a positive integer, we write $(\sin \theta)^n = \sin^n \theta$, $(\cos \theta)^n = \cos^n \theta$, and so on. This notation is not generally used for fractional or negative powers.

It is an important fact that any trigonometric function depends only on the angle θ and not on the size of the right triangle. In the two triangles ABC and PQR below, having the same angles,

$$\frac{BC}{AC} = \frac{QR}{PR}.$$

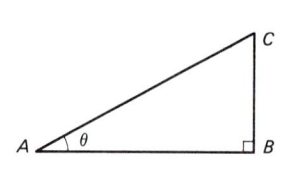

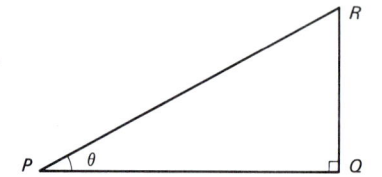

Thus $\sin \theta$ has the same value in both triangles. The same is true for $\cos \theta$, $\tan \theta$, and so on.

Example 1 Let a right triangle be given as shown in the adjoining figure. Write down the six trigonometric functions of θ.

Solution Note that $AC^2 = AB^2 + BC^2$ ($13^2 = 12^2 + 5^2$), so that by Pythagoras's theorem, the triangle is indeed right-angled at B. The side opposite to the angle θ is $AB = 12$, the side adjacent to θ is $BC = 5$, and the hypotenuse is $AC = 13$. Thus,

$$\sin \theta = \frac{\text{opp.}}{\text{hyp.}} = \frac{AB}{AC} = \frac{12}{13}$$

$$\cos \theta = \frac{\text{adj.}}{\text{hyp.}} = \frac{BC}{AC} = \frac{5}{13}$$

$$\tan \theta = \frac{\text{opp.}}{\text{adj.}} = \frac{AB}{BC} = \frac{12}{5}$$

$$\cot \theta = \frac{1}{\tan \theta} = \frac{5}{12}$$

$$\sec \theta = \frac{1}{\cos \theta} = \frac{13}{5}$$

$$\csc \theta = \frac{1}{\sin \theta} = \frac{13}{12}.$$

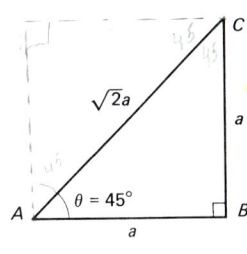

TRIGONOMETRIC FUNCTIONS OF SPECIAL ANGLES

We can determine the trigonometric functions of the special angles 30°, 45°, and 60° by using the Pythagorean theorem. Let us first consider the case $\theta = 45°$. In the right triangle ABC, let $AB = a$. Since angle at C = angle at A (each $= 45°$), the triangle is isosceles and $BC = AB = a$. Therefore using Pythagoras's theorem, we have

$$AC^2 = AB^2 + BC^2 = a^2 + a^2 = 2a^2$$

or

$$AC = \sqrt{2}\, a.$$

Therefore,

$$\sin 45° = \frac{BC}{AC} = \frac{a}{\sqrt{2}\, a} = \frac{1}{\sqrt{2}} = \frac{\sqrt{2}}{2}$$

$$\cos 45° = \frac{AB}{AC} = \frac{a}{\sqrt{2}\, a} = \frac{1}{\sqrt{2}} = \frac{\sqrt{2}}{2}$$

$$\tan 45° = \frac{BC}{AB} = \frac{a}{a} = 1$$

and

$$\cot 45° = \frac{1}{\tan 45°} = 1$$

$$\sec 45° = \frac{1}{\cos 45°} = \sqrt{2}$$

$$\csc 45° = \frac{1}{\sin 45°} = \sqrt{2}.$$

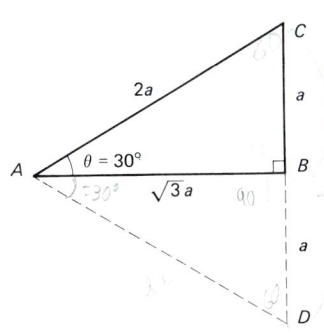

Let us now consider $\theta = 30°$. For this, consider an equilateral triangle ACD, each side of length $2a$. Let AB be the perpendicular from A onto the side CD. Then $BC = \frac{1}{2}CD = a$. Moreover, angle CAB is half of angle $CAD = \frac{1}{2}(60°) = 30°$. Thus, in the right triangle ABC, we have $BC = a$, $AC = 2a$. Now,

$$AB^2 + BC^2 = AC^2$$
$$AB^2 + a^2 = (2a)^2$$
$$AB^2 = 4a^2 - a^2 = 3a^2$$
$$AB = \sqrt{3}\, a.$$

Therefore,

$$\sin 30° = \frac{BC}{AC} = \frac{a}{2a} = \frac{1}{2}$$

$$\cos 30° = \frac{AB}{AC} = \frac{\sqrt{3}\, a}{2a} = \frac{\sqrt{3}}{2}$$

$$\tan 30° = \frac{BC}{AB} = \frac{a}{\sqrt{3}\, a} = \frac{1}{\sqrt{3}} = \frac{\sqrt{3}}{3}.$$

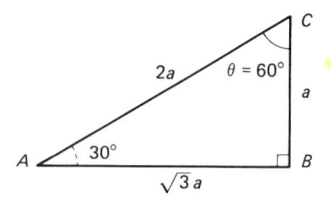

For $\theta = 60°$, we can use the same triangle as for $30°$, because the other acute angle is $90° - 30° = 60°$. Therefore,

$$\sin 60° = \frac{\text{opp.}}{\text{hyp.}} = \frac{\sqrt{3}\,a}{2a} = \frac{\sqrt{3}}{2}$$

$$\cos 60° = \frac{\text{adj.}}{\text{hyp.}} = \frac{a}{2a} = \frac{1}{2}$$

$$\tan 60° = \frac{\text{opp.}}{\text{adj.}} = \frac{\sqrt{3}\,a}{a} = \sqrt{3}.$$

Knowing these, we can calculate their reciprocals.

For $\theta = 0°$ and $\theta = 90°$, we have the following special values:

$$\sin 0° = 0 \quad \text{and} \quad \sin 90° = 1$$
$$\cos 0° = 1 \quad\quad\quad\quad \cos 90° = 0$$
$$\tan 0° = 0 \quad\quad\quad\quad \tan 90° = \text{undefined}$$

These will be explained further in Section 4, but they can also be seen from the tables of trigonometric functions, which we shall describe shortly.

Once we know the values of $\sin \theta$ and $\cos \theta$, we can calculate the remaining four trigonometric functions by using Equations (2) and (3). The values of $\sin \theta$ and $\cos \theta$ for $\theta = 0°, 30°, 45°, 60°,$ and $90°$ are easy to remember from the pattern in the following table.

PROOF??

θ (degrees)	0°	30°	45°	60°	90°
θ (radians)	0	$\pi/6$	$\pi/4$	$\pi/3$	$\pi/2$
$\sin \theta$	$\sqrt{\tfrac{0}{4}}$	$\sqrt{\tfrac{1}{4}}$	$\sqrt{\tfrac{2}{4}}$	$\sqrt{\tfrac{3}{4}}$	$\sqrt{\tfrac{4}{4}}$
$\cos \theta$	$\sqrt{\tfrac{4}{4}}$	$\sqrt{\tfrac{3}{4}}$	$\sqrt{\tfrac{2}{4}}$	$\sqrt{\tfrac{1}{4}}$	$\sqrt{\tfrac{0}{4}}$

Several of the values in the table can, of course, be written more simply. For example,

$$\sqrt{\tfrac{0}{4}} = 0, \quad \sqrt{\tfrac{2}{4}} = \tfrac{\sqrt{2}}{2} = \tfrac{1}{\sqrt{2}}, \quad \text{and} \quad \sqrt{\tfrac{4}{4}} = 1.$$

Example 2 For $\theta = 30°$, show that the following equations are satisfied.
(a) $\sin 2\theta = 2 \sin \theta \cos \theta$.
(b) $\tan 2\theta = \dfrac{2 \tan \theta}{1 - \tan^2 \theta}$.

Solution (a) When $\theta = 30°$, the given equation becomes

$$\sin 60° = 2 \sin 30° \cos 30°. \tag{i}$$

But we have

$$\sin 60° = \frac{\sqrt{3}}{2}, \quad \sin 30° = \frac{1}{2}, \quad \text{and} \quad \cos 30° = \frac{\sqrt{3}}{2}.$$

If we use these values, (i) becomes
$$\frac{\sqrt{3}}{2} = 2 \cdot \frac{1}{2} \cdot \frac{\sqrt{3}}{2}$$
which is true.

(b) When $\theta = 30°$, the given equation becomes
$$\tan 60° = \frac{2 \tan 30°}{1 - \tan^2 30°}. \qquad \text{(ii)}$$

Now $\tan 60° = \sqrt{3}$ and $\tan 30° = 1/\sqrt{3}$. If we use these values, (ii) becomes
$$\sqrt{3} = \frac{2\left(\frac{1}{\sqrt{3}}\right)}{1 - \left(\frac{1}{\sqrt{3}}\right)^2}$$
$$= \frac{2/\sqrt{3}}{1 - \frac{1}{3}} = \frac{2}{\sqrt{3}} \bigg/ \frac{2}{3} = \frac{3}{\sqrt{3}}$$
which is true.

Example 3 Given that α, β, and $\alpha + \beta$ are positive angles less than $\pi/2$ and that $\tan(\alpha + \beta) = \sqrt{3}$ and $\tan(\alpha - \beta) = 1/\sqrt{3}$, find the values of α and β in radians. (Do not use tables.)

Solution We know that $\tan(\alpha + \beta) = \sqrt{3} = \tan(\pi/3)$. Thus
$$\alpha + \beta = \frac{\pi}{3} \qquad \text{(i)}$$

Also, $\tan(\alpha - \beta) = 1/\sqrt{3} = \tan(\pi/6)$. Therefore,
$$\alpha - \beta = \frac{\pi}{6}. \qquad \text{(ii)}$$

Adding (i) and (ii), we obtain
$$2\alpha = \frac{\pi}{3} + \frac{\pi}{6} = \frac{\pi}{2}$$
$$\alpha = \frac{\pi}{4}.$$

From (i), $\beta = \pi/3 - \alpha = \pi/3 - \pi/4 = \pi/12$. Thus,
$$\alpha = \frac{\pi}{4} \quad \text{and} \quad \beta = \frac{\pi}{12}.$$

Example 4 If θ denotes a nonnegative acute angle and $\tan^2 \theta = \tan \theta$, find the value of θ.

Solution The given equation is
$$\tan^2 \theta - \tan \theta = 0$$
or
$$\tan \theta (\tan \theta - 1) = 0.$$

Therefore,

either	$\tan\theta = 0$	or	$\tan\theta = 1$.
i.e.,	$\tan\theta = 0 = \tan 0°$	i.e.,	$\tan\theta = 1 = \tan 45°$
	$\theta = 0°$		$\theta = 45°$

Thus, the two values of θ (nonnegative acute angles) that satisfy the given equation are $\theta = 0°$ and $45°$.

TABLES OF TRIGONOMETRIC FUNCTIONS

If we want to know the value of a trigonometric function for an angle that is not a standard one such as $0°$, $30°$, $45°$, $60°$, or $90°$, then we must use either a table of trigonometric functions or a scientific calculator. The table of trigonometric functions in the Appendix (see Table 4) lists the values of all the trigonometric functions for angles ranging from $0°$ to $90°$ at intervals of 10 minutes.

When looking for the value of a trigonometric function for an angle that is less than $45°$, we read down the left-hand column headed by degrees until we reach the required angle. Then we move horizontally and read the value headed by the required trigonometric function. But when the angle is more than $45°$, we use the right-hand column headed by degrees and the captions given at the bottom (rather than those at the top). Using these tables, you should check the following values:

$$\sin 25°10' = 0.4253, \quad \tan 63°50' = 2.035$$
$$\cos 32°00' = 0.8480, \quad \csc 48°10' = 1.342.$$

It may be noted that the values are given to four decimal places if they are less than 1 and to three decimal places if they are more than 1. Moreover, in the same table, the angles are also listed in radians.

Example 5 Use the table to find the value of θ in degrees and minutes if

(a) $\cos\theta = 0.7509$.   (b) $\cot\theta = 0.6959$.

Solution (a) Looking in the table, we search down the values in the column *headed* by "cos" until we reach 0.7509. Then we read the angle on the left, which is $\theta = 41°20'$.

(b) Looking in the column headed by "cot," we see that it does not contain the value 0.6959 nor, indeed, any value less than 1. Then we look at the column that is captioned "cot" at the bottom and find that the value 0.6959 corresponds to $\theta = 55°10'$ (reading the angle on the *right-hand column* because we are using the *lower caption*).

Sometimes we wish to find the value of a trigonometric function for an angle that is not given in the tables. For example, we might want to find the value of $\sin 35°14'$. In the tables, we can find only the values of $\sin 35°10'$ and $\sin 35°20'$. However, knowing these two values, we can *approximate* the sine of any angle between 35°10′ and 35°20′ by a method known as *linear interpolation*, which we shall explain by working through the following solved examples.

Example 6 Evaluate:
(a) sin 37°16′. (b) cos 53°47′.

Solution (a) Since the angle 37°16′ lies between 37°10′ and 37°20′, sin 37°16′ must also lie between sin 37°10′ and sin 37°20′. Finding these values from the table, we write them in the following three-line arrangement.

$$10'\left[\begin{array}{c} 6'\left[\begin{array}{cc} 37°10' & 0.6041 \\ 37°16' & ? \end{array}\right] \\ 37°20' \quad 0.6065 \end{array}\right] \begin{array}{c} \\ 0.0024 \end{array}$$

$$\begin{array}{cc} \theta & \sin\theta \end{array}$$

Note that as θ increases by 10′ from 37°10′ to 37°20′, the value of sin θ increases by 0.0024. Since 37°16′ is $\frac{6}{10}$ of the distance from 37°10′ to 37°20′, we assume that the value of sin 37°16′ is also $\frac{6}{10}$ of the distance from the value of sin 37°10′ = 0.6041 to sin 37°20′ = 0.6065. Thus, we must *add* $\frac{6}{10}$ of the total increase of 0.0024 to 0.6041. Therefore,

$$\sin 37°16' = 0.6041 + \tfrac{6}{10}(0.0024) \approx 0.6055.$$

(b) From the tables we have

$$\begin{array}{cc} \theta & \cos\theta \end{array}$$

$$10'\left[\begin{array}{c} 7'\left[\begin{array}{cc} 53°40' & 0.5925 \\ 53°47' & ? \end{array}\right] \\ 53°50' \quad 0.5901 \end{array}\right] \begin{array}{c} \\ 0.0024 \end{array}$$

Note that as θ increases from 53°40′ to 53°50′, the value of cos θ *decreases* from 0.5925 to 0.5901. Since 53°47′ is $\frac{7}{10}$ of the distance from 53°40′ to 53°50′, cos 53°47′ is also $\frac{7}{10}$ of the way from 0.5925 to 0.5901. Thus we must *subtract* $\frac{7}{10}$ of the total difference, 0.0024, from 0.5925. Thus

$$\cos 53°47' = 0.5925 - \tfrac{7}{10}(0.0024) \approx 0.5908.$$

Example 7 Use trigonometric tables to find θ in degrees and minutes if
(a) tan θ = 2.75. (b) csc θ = 1.473.

Solution (a) If we look at the column with the "tan" caption at the bottom, we find that our value of 2.75 lies between the values of 2.747 and 2.773, which correspond to θ = 70°00′ and 70°10′. Setting up the three-line interpolation, we have

$$\begin{array}{cc} \text{angle} & \text{tangent of angle} \end{array}$$

$$10'\left[\begin{array}{c} c'\left[\begin{array}{cc} 70°00' & 2.747 \\ \theta & 2.750 \end{array}\right] \\ 70°10' \quad 2.773 \end{array}\right] \begin{array}{c} 0.003 \\ \\ 0.026 \end{array}$$

Setting up a proportion between corresponding differences, we have

$$\frac{c}{10} = \frac{0.003}{0.026} \quad \text{or} \quad c = \frac{30}{26} \approx 1.$$

Thus

$$\theta = 70°00' + c' = 70°01'.$$

(b) In this case we have

$$10' \begin{bmatrix} c' \begin{bmatrix} \text{angle} & \text{csc} \\ 42°40' & 1.476 \\ \theta & 1.473 \end{bmatrix} 0.003 \\ 42°50' \quad 1.471 \end{bmatrix} 0.005.$$

Setting up the proportion, we have

$$\frac{c}{10} = \frac{0.003}{0.005} = \frac{3}{5} \quad \text{or} \quad c = 6'.$$

Thus $\theta = 42°40' + 6' = 42°46'$.

CALCULATORS

You can avoid using tables of trigonometric function if you have a scientific calculator. Most calculators require angles to be expressed in decimal degrees. They provide the values of trigonometric functions to six or eight or sometimes more significant figures. If you have a calculator, check the following values. (Make sure your calculator is in the right mode for degrees or radians.)

$$\cos(77.3°) \approx 0.219846$$
$$\sin(-135.25°) \approx -0.704015$$
$$\tan 4 \approx 1.157821 \quad \text{(this means 4 radians)}$$

Diagnostic Test 12.2

Fill in the blanks.

1. $\tan^2 \theta = ($ _____ $)^2$.

2. $\dfrac{\sin \alpha}{\cos \alpha} = $ _____.

3. $\dfrac{1}{\sec \theta} = $ _____.

4. $\dfrac{1}{\cot \alpha} = $ _____.

5. $\dfrac{1}{\csc \alpha} = $ _____.

6. $(\tan \theta)(\cot \theta) = $ _____.

7. $\tan \pi/4 = $ _____.

8. $\sin \pi/3 = $ _____.

9. $\cos \pi/6 = $ _____.

10. $\sec 0° = $ _____.

11. $\cot 30° = $ _____.

12. $\csc 90° = $ _____.

13. $\sin 0° = $ _____ .   **14.** $\cos 45° = $ _____ .

15. If θ is a positive acute angle and $\sin \theta = \frac{1}{2}$, then $\theta = $ _____ (in degrees).

16. If $0° < \theta < 90°$ and $\cos \theta = \sqrt{2}/2$, then $\theta = $ _____ .

17. If $0° < \theta < 90°$ and $\cos \theta = \sin 60°$, then $\theta = $ _____ (degrees).

18. If α is a positive acute angle and $\sin \alpha = \cos 45°$, then $\alpha = $ _____ (radians).

19. If $0° \leq \beta \leq 90°$ and $\sin \beta = \tan \pi/4$, then $\beta = $ _____ .

20. If $0° < \alpha < \pi/2$ and $\tan \alpha = \cos 0°$, then $\alpha = $ _____ .

Exercises 12.2

(1–6) Find all the trigonometric functions of θ from the data given in the following right triangles.

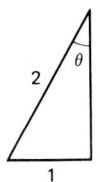

1. (right triangle with legs 4 and 3, angle θ at the vertex with leg 4)

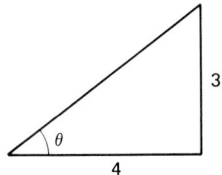

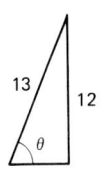

2. (right triangle with legs 1 and 2, angle θ at top)

3. (right triangle with hypotenuse 13 and leg 12, angle θ)

4. (right triangle with legs a and b, angle θ)

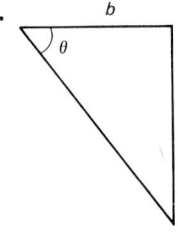

5. (right triangle with legs 2 and 10, angle θ at bottom)

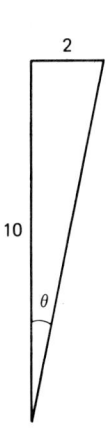

6. (right triangle with sides 10 and 8, angle θ at top)

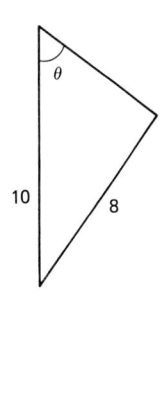

(7–18) Find the values of the following expressions. (Do not use tables.)

7. $\sin^2 30° + \cos^2 30°$.

8. $\tan^2 60° - \sec^2 60°$.

9. $\sin^2 30° + \sin^2 45° + \sin^2 60°$.

10. $\tan^2 30° + \tan^2 45° + \tan^2 60°$.

11. $\cos^2 \dfrac{\pi}{6} + \cos^2 \dfrac{\pi}{3} - \tan^2 \dfrac{\pi}{4}$.

12. $\cos^2 \dfrac{\pi}{4} + \sec^2 \dfrac{\pi}{3} - \csc^2 \dfrac{\pi}{2}$.

13. $\sin 30° \cos 60° + \cos 30° \sin 60°$.

14. $\cos 60° \cos 30° - \sin 60° \sin 30°$.

15. $\cos 60° \cos 45° - \sin 60° \sin 45°$.

16. $\sin 30° \cos 45° + \cos 30° \sin 45°$.

17. $\dfrac{\tan 60° - \tan 30°}{1 + \tan 60° \tan 30°}$.

18. $\dfrac{\tan \dfrac{\pi}{3} + \tan \dfrac{\pi}{4}}{1 - \tan \dfrac{\pi}{4} \tan \dfrac{\pi}{3}}$.

(19–20) Prove the following. (Do not use tables.)

19. $\cos\dfrac{\pi}{4}\cos\dfrac{\pi}{6} + \sin\dfrac{\pi}{4}\sin\dfrac{\pi}{6} = \cos\dfrac{\pi}{3}\cos\dfrac{\pi}{4} + \sin\dfrac{\pi}{3}\sin\dfrac{\pi}{4}$.

20. $\cot^2\dfrac{\pi}{4}\sec^2\dfrac{\pi}{3}\cos\dfrac{\pi}{2} - 15\sin^2\dfrac{\pi}{2}\cos^2\dfrac{\pi}{4} - 4\cos\dfrac{\pi}{6}\cos\dfrac{\pi}{3}\sin^2\dfrac{\pi}{4} = -\dfrac{1}{2}(15 + \sqrt{3})$.

21. Prove that the squares of the sines of the angles 30°, 45°, and 60° are in arithmetic progression.

22. Prove that $\tan^2 \pi/6$, $\tan^2 \pi/4$ and $\tan^2 \pi/3$ are in geometric progression.

(23–32) Show that the following equations are satisfied by the indicated values of angles. (Do not use tables.)

23. $\sin 2\theta = 2\sin\theta\cos\theta$; $\theta = 45°$.

24. $\sin 3A = 3\sin A - 4\sin^3 A$; $A = 30°$.

25. $\cos 2A = \cos^2 A - \sin^2 A$; $A = 45°$.

26. $\cos 3\theta = 4\cos^3\theta - 3\cos\theta$; $\theta = \pi/6$.

27. $\sin 2\alpha = \dfrac{2\tan\alpha}{1+\tan^2\alpha}$; $\alpha = \dfrac{\pi}{6}$.

28. $\cos 2\beta = \dfrac{1-\tan^2\beta}{1+\tan^2\beta}$; $\beta = \dfrac{\pi}{4}$.

29. $\tan 2A = \dfrac{2\tan A}{1-\tan^2 A}$; $A = 30°$.

30. $\sin(A+B) = \sin A\cos B + \cos A\sin B$; $A = 30°$, $B = 60°$.

31. $\cos(A+B) = \cos A\cos B - \sin A\sin B$; $A = 60°$, $B = 30°$.

32. $\tan(A-B) = \dfrac{\tan A - \tan B}{1 + \tan A\tan B}$; $A = \dfrac{\pi}{3}$, $B = \dfrac{\pi}{6}$.

33. If A, B, and $A+B$ are positive acute angles, find A and B in degrees if $\sin(A-B) = \tfrac{1}{2}$ and $\cos(A+B) = \tfrac{1}{2}$. (Do not use tables.)

34. If θ is a nonnegative acute angle, find the values of θ for which the following equation is true. (Do not use tables.)
$$2\cos^2\theta - 3\cos\theta + 1 = 0.$$
[*Hint:* Set $x = \cos\theta$ and solve the quadratic equation for x.]

(35–52) Use trigonometric tables to find the values of the following trigonometric functions.

35. $\sin 36°$.   **36.** $\cos 47°$.   **37.** $\tan 25°$.   **38.** $\sin 74°10'$.   **39.** $\cos 23°30'$.

40. $\tan 54°40'$.   **41.** $\sec 39°20'$.   **42.** $\csc 62°40'$.   **43.** $\cot 32°50'$.   **44.** $\tan\dfrac{\pi}{24}$.

45. $\sec\dfrac{5\pi}{24}$.   **46.** $\sin 43°45'$.   **47.** $\cos 37°27'$.   **48.** $\tan 17°12'$.   **49.** $\sec 32°15'$.

50. $\cot 67°43'$.   **51.** $\csc 55°27'$.   **52.** $\sin 72°18'$.

(53–70) Use trigonometric tables to find the approximate values of θ for which the following equations are true.

53. $\sin\theta = 0.2250$.   **54.** $\cos\theta = 0.8936$.   **55.** $\tan\theta = 0.3346$.   **56.** $\sec\theta = 2.228$.

57. $\cot\theta = 0.4734$.   **58.** $\csc\theta = 2.085$.   **59.** $\sin\theta = 0.6213$.   **60.** $\cos\theta = 0.7512$.

61. $\tan\theta = 1.160$.   **62.** $\sec\theta = 1.261$.   **63.** $\csc\theta = 4.120$.   **64.** $\cot\theta = 0.8750$.

65. $\sin\theta = 0.9390$.   **66.** $\tan\theta = 2.730$.   **67.** $\cos\theta = 0.9625$.   **68.** $\cot\theta = 1.259$.

69. $\sec\theta = 3.125$.   **70.** $\csc\theta = 2.750$.

(71–78) Use a calculator to find the following to six decimal places.

71. $\cos(31.25°)$.   **72.** $\sin(142.6°)$.   **73.** $\tan(-205.6°)$.   **74.** $\cos(-25.8°)$.

75. $\sin(4.35)$.   **76.** $\tan(3.192)$.   **77.** $\cos(-1.95)$.   **78.** $\sin(-20.2)$.

12.3

Applications

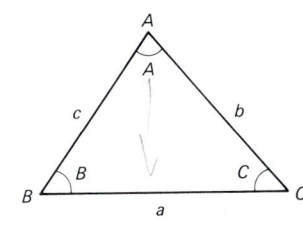

In this section we shall use our knowledge of trigonometric functions to solve some applied problems that involve right triangles. In a triangle ABC, it is customary to denote the angles by the symbols of the corresponding vertices. Thus the angle at A is denoted by A, the angle at B is denoted by B, and so on. The sides opposite to these angles are denoted by the corresponding lower-case letters. Thus the side BC opposite to angle A is denoted by a, the side AC opposite to angle B is denoted by b, and so on. (See the adjoining figure.)

Calculating the values of the remaining sides and angles in a triangle in which some of them are known is called *solving the triangle*. We illustrate the method of solving right triangles by working through the following examples.

Example 1 ABC is a right triangle with right angle at C. Given that $a = 8.50$ and $B = 32°40'$, solve the triangle (that is, find the remaining sides and angles).

Solution In this triangle, we are given that $C = 90°$, $B = 32°40'$, and $a = 8.50$. The sides and angles to be found are b, c, and A.

Since the sum of the two acute angles of a right triangle is 90°, we have

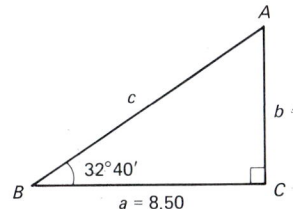

$$A = 90° - B = 90° - 32°40' = 57°20'.$$

Also, from the right triangle ABC we have

$$\frac{b}{a} = \frac{\text{opp.}}{\text{adj.}} = \tan 32°40' = 0.6412 \quad \text{(from tables)}.$$

That is,

$$\frac{b}{8.50} = 0.6412$$

$$b = 8.50(0.6412) = 5.4502 \approx 5.45$$

where we have rounded off to two decimal places, because the given side is also expressed to two decimal places. Furthermore,

$$\frac{c}{a} = \frac{\text{hyp.}}{\text{adj.}} = \sec 32°40' = 1.188$$

$$\frac{c}{8.50} = 1.188$$

$$c = 8.50(1.188) = 10.098 \approx 10.10.$$

Thus the missing parts of the triangle are

$$A = 57°20', \quad b = 5.45, \quad c = 10.10.$$

Note In the above example, after having calculated $b = 5.45$, we can calculate c by using the Pythagorean theorem. This gives

$$c^2 = a^2 + b^2 = (8.50)^2 + (5.45)^2 = 101.9525$$

and so $c = 10.097 \approx 10.10$ to two decimal places. If we made an error in calculating b, then calculating c by this method would also produce an erroneous result. For this reason, when calculating any missing part of the triangle, it is advisable to use only the *given* parts as far as possible.

Example 2 ABC is a right triangle with right angle at C. Given also that $a = 3.2$, $c = 5.7$, solve the triangle.

Solution In this case we are given that

$$C = 90°, \quad a = 3.2, \quad c = 5.7.$$

The parts to be found are b, A, and B. Using Pythagoras's theorem,

$$b^2 = c^2 - a^2 = (5.7)^2 - (3.2)^2$$
$$= 32.49 - 10.24 = 22.25$$

and so

$$b = \sqrt{22.25} = 4.717 \approx 4.7$$

where we round off to one decimal place to match the given sides. Also from the right triangle ABC,

$$\sin A = \frac{\text{opp.}}{\text{hyp.}} = \frac{a}{c} = \frac{3.2}{5.7} = 0.5614. \rightarrow \text{go to table and find corresponding degree.}$$

Therefore,

$$A = 34°09'$$

where we have used the tables and interpolation. Since $A + B = 90°$, we have

$$B = 90° - A = 90° - 34°09' = 55°51'.$$

An important application of trigonometric functions is to problems involving heights and distances. For such problems, we shall need to define two angles, known as the angle of elevation and the angle of depression.

Suppose a person at O looks up at an object at B. Then the positive angle α that the line OB makes with the horizontal line OA is called the **angle of elevation** of B from O. Again, if the observer is at Q and is looking down on an object at R, then the positive angle β that the line QR makes

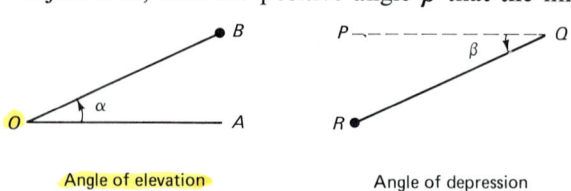

Angle of elevation

Angle of depression

with the horizontal line PQ is called the **angle of depression** of R from Q. (See the above figures.)

Example 3 The angle of elevation of the top of a tree at a point 350 feet from its base is 22°. Find the height of the tree.

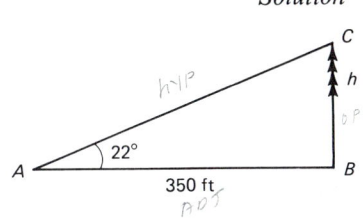

Solution Let BC denote the vertical tree and A the position of the observer who is 350 feet from the base B of the tree—that is, $c = 350$ feet. We are also given the angle of elevation of the top of the tree from A is 22°—that is, $A = 22°$. Now if $h = BC$ denoted the height of the tree, then from the right triangle ABC, we have

$$\frac{h}{350} = \frac{\text{opp.}}{\text{adj.}} = \tan 22° = 0.4040 \quad \text{(from tables)}$$

$$h = 350(0.4040) = 141.4 \text{ ft.}$$

Example 4 The angle of depression of an object on the ground from the top of a tower is 0.71 radians. If the tower is 80 feet high, how far is the object from the base of the tower?

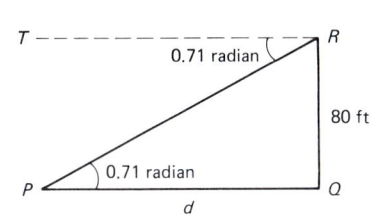

Solution Let QR denote the tower, whose height is 80 feet, and P the object on the ground. The angle of depression of P as observed from the top of the tower R is 0.71 radians—that is, angle $TRP = 0.71$ radians. Then in the right triangle PQR we have

$$\angle RPQ = \angle TRP = 0.71 \text{ rad.} \quad \text{(alternate angles)}$$

Now if $PQ = d$, then from the right triangle PQR we have

$$\frac{d}{80} = \frac{\text{adj.}}{\text{opp.}} = \cot(0.71 \text{ rad.}) = \cot(40°41') = 1.163$$

$$d = 80(1.163) = 93.04 \text{ ft.}$$

Example 5 A person at the point P on the ground observes that the angle of elevation of the top of a tower is 60°. He then moves a distance of 100 feet directly away from the tower to a point Q. At Q, the angle of elevation of the top of the tower is 30°. Find the height of the tower.

Solution Let AB denote the tower and h denote its height. Let x denote the distance AP. Then, from the right triangle ABP, we have

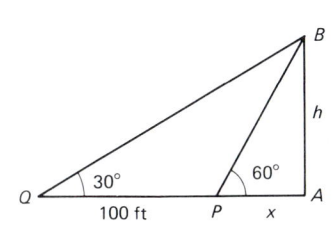

$$\frac{h}{x} = \tan 60° = \sqrt{3}$$

$$h = x\sqrt{3}. \qquad (i)$$

Now from the larger right triangle ABQ, we have

$$\frac{AB}{AQ} = \tan 30°.$$

But $AQ = AP + PQ = x + 100$, and so this equation becomes

$$\frac{h}{x + 100} = \frac{1}{\sqrt{3}}.$$

Sec. 12.3 Applications

Thus,
$$\sqrt{3}\,h = x + 100 \qquad \text{(ii)}$$

We want to determine h. So, we must eliminate x from the two equations (i) and (ii). From (ii) we have $x = \sqrt{3}\,h - 100$, and substituting this in (i) gives

$$h = \sqrt{3}(\sqrt{3}\,h - 100)$$
$$= 3h - 100\sqrt{3}$$
$$2h = 100\sqrt{3}$$
$$h = 50\sqrt{3} = 50(1.732) = 86.6 \text{ ft.}$$

Example 6 A ship is traveling due north. At a certain time the navigator measures the direction of a certain lighthouse L and finds it to be 25° East of North (see the adjacent figure). Half an hour later the ship has traveled 10 miles to B, and the light now lies 55° East of North. How close will the ship pass to L, and how long will it be before the ship reaches the closest point?

Solution Let N be the nearest point to L on the ship's course, and let $NL = d$ and $NB = x$. In the right triangle NBL we have

$$\frac{d}{x} = \tan 55° = 1.4282.$$

Therefore, $d = 1.4282x$. Then in triangle NAL we have

$$\frac{d}{x + 10} = \tan 25° = 0.4663.$$

Therefore,
$$d = 0.4663(x + 10) = 0.4663x + 4.663$$
$$1.4282x = 0.4663x + 4.663.$$

It follows that
$$x = \frac{4.663}{0.9619} \approx 4.85$$

and so
$$d = (1.4282)(4.85) \approx 6.92.$$

The ship will therefore pass 6.92 miles from L. Since it travels 10 miles in 30 minutes, it will cover the distance $x = 4.85$ miles in 14.5 minutes. The nearest point to L will be reached 14.5 minutes after the second measurement.

Exercises 12.3

(1–6) In the following exercises, ABC is a right triangle with right angle at C. Solve the triangle in each case. (Do not use trigonometric tables.)

1. $A = 30°$, $b = 9$.
2. $B = 60°$, $a = 20$.
3. $B = 45°$, $c = 10$.
4. $A = 30°$, $c = 12$.
5. $a = \sqrt{3}$, $b = 1$.
6. $c = 5\sqrt{2}$, $b = 5$.

(7–20) In the following exercises, BAC is a right triangle with right angle at C. Solve the triangle in each case. Evaluate the sides to one decimal place.

7. $B = 37°$, $a = 5$.
8. $A = 25°$, $c = 10$.
9. $A = 43°$, $b = 8$.
10. $B = 63°$, $a = 6$.
11. $B = 52°$, $c = 10$.
12. $A = 47°$, $b = 12$.
13. $A = 25°$, $c = 4$.
14. $B = 27°$, $c = 16$.
15. $a = 5$, $b = 7$.
16. $a = 8$, $c = 12$.
17. $b = 3.7$, $c = 5.2$.
18. $a = 2.5$, $b = 3.1$.
19. $a = 2.1$, $c = 6.3$.
20. $b = 4.3$, $c = 8$.

21. The angle of elevation of the top of a tree from a point A on the ground that is 50 feet from the base of the tree is 30°. Find the height of the tree.

22. The angle of elevation of the top of a building that is 60 feet high from a point P on the ground is 27°. How far is the point P from the base of the building?

23. A kite string is 200 feet long and makes an angle of 37° with the horizontal. How high off the ground is the kite (assuming the string is straight)?

24. A ladder 50 feet long is standing against a wall and makes an angle of 65° with ground. How far is the foot of the ladder from the wall? If the ladder slides 2 feet away from the wall, how far does the top of the ladder move down the wall?

25. The angle of depression of an object A on the ground from the top of the tower is 42°. If the object A is at a distance of 60 feet from the foot of the tower, find the height of the tower.

26. The angle of depression of an object X on the ground from the top of a building that is 100 feet high is 60°. How far is the object from the base of the building?

27. The angle of depression of the top of a building as observed from the top of a tower that is 150 feet high is 60°. If the building is 40 feet from the tower, find the height of the building.

28. The angle of elevation of the top of a building as observed from the top of another building that is 100 feet high is 45°. If the two buildings are 70 feet apart, find the height of the taller building.

29. The angle of elevation of the top of a tower as observed from a point A on the ground is 70° and from a point B on the ground is 35°. If the points A and B are 150 feet apart and the line joining A and B passes through the foot of the tower, find the height of the tower.

30. A person at the point P on the ground observes that the angle of elevation of the top of a tower is 60°. He then moves directly away a distance of 120 feet from the tower to a point Q. At Q, the angle of elevation of the top of the tower is 30°. How far is the point P from the foot of the tower?

31. An observer at a point P measures that the angle of elevation of the top of a tree is 80°. She then moves 20 feet further away from the tree to a point Q and finds that the angle of elevation has decreased to 60°. How tall is the tree?

32. The angles of depression of two points A and B on the ground as observed from the top of a tower 100 feet high are 60° and 30°, respectively. If the line joining the points A and B passes through the foot of the tower, how far apart are the two points A and B?

*33. P and Q are two points on the ground 100 feet apart and in line with the base of a tower. If the angle of elevation of the top of the tower from P is double the angle of elevation from Q, find the angle of elevation of the top of the tower from P, given that the tower is $50\sqrt{3}$ feet high.

*34. A and B are two points on the ground 50 feet apart and in line with the base of a tall tree. If the angle of elevation of the top of the tree at A is double the angle of elevation at B and A is 40 feet away from the base of the tree, find the height of the tree.

35. An airplane is traveling at a constant height of 2000 feet heading directly over the head of an observer on the ground. It takes 30 seconds for the angle of elevation of the plane to increase from 15° to 30°. Find the speed at which the plane is traveling.

36. A car is traveling at a constant speed on a straight road that leads to a tower 550 feet high. An observer at the top of the tower observes that the angle of depression of the car changes from 6° to 12° in one minute. How fast is the car traveling?

37. A flag post is mounted on the top of a building that is 100 feet high. The angles of elevation of the top and bottom of the flag post as observed from a point 60 feet from the building are 60° and 45°, respectively. Find the height of the flag post.

38. The angles of depression of the top and bottom of a building 80 feet high as observed from the top of another building are 30° and 60°, respectively. If the two buildings are 50 feet apart, how high is the taller building?

39. A ladder 25 feet long is placed against a wall and makes an angle of 29° with the wall. Approximately how high is the top of the ladder from the ground? If the foot of the ladder slides away 3 feet from the wall, approximately how far does the top of the ladder slide down?

40. The pilot of an airplane wants to make his approach to an airstrip at an angle of 15° with the horizontal. How far from the strip should he start descending if he is flying at a height of 4000 feet?

41. The driver of a car traveling on a level highway at a constant speed of 40 miles per hour observes that in 15 minutes, the angle of elevation of the top of the mountain ahead of her changes from 5° to 0°. Find the approximate height of the mountain.

42. A hexagon is inscribed in a circle of radius 20 centimeters. Find the perimeter of the hexagon.

43. A circle viewed from an angle must be drawn as an ellipse. If the viewing direction makes an angle of 60° with the perpendicular to the circle, calculate the ratio of major to minor axis of the ellipse.

12.4

Trigonometric Functions

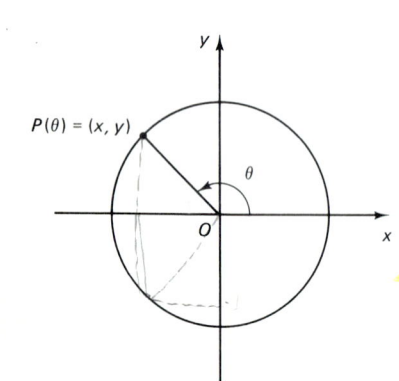

The circle of *unit radius* with center at the origin is called the ***unit circle***. Letting θ be any given real number, we construct an angle of θ radians in the standard position and let $P(\theta)$ denote the point on the unit circle that is also on the terminal side of this angle. In this way for each real number θ we have a unique point $P(\theta)$ on the unit circle. This association of each number θ with a point $P(\theta)$ on the unit circle is often called the ***wrapping function***.

If (x, y) are the coordinates of the point $P(\theta)$, we define the ***sine*** and ***cosine*** functions as follows:

$$\sin \theta = y \quad \text{and} \quad \cos \theta = x. \tag{1}$$

Since θ can be any real number, the domains of the sine and cosine functions defined by (1) are the set of all real numbers. Moreover, for any point (x, y) on the unit circle, x and y can take only values that are between -1 and 1 inclusive—that is, $-1 \leq x \leq 1$ and $-1 \leq y \leq 1$. Thus, the range of each function is the closed interval $[-1, 1]$. We can express this fact as

$$\boxed{\begin{array}{ll} -1 \leq \sin\theta \leq 1 & \text{for all } \theta \\ -1 \leq \cos\theta \leq 1 & \text{for all } \theta. \end{array}} \tag{2}$$

Having defined the sine and cosine functions of θ by (1), we can define the remaining four trigonometric functions as in Section 12.2; that is,

$$\boxed{\begin{array}{ll} \tan\theta = \dfrac{\sin\theta}{\cos\theta} = \dfrac{y}{x}, & \text{provided } x \neq 0 \\[6pt] \cot\theta = \dfrac{1}{\tan\theta} = \dfrac{x}{y}, & \text{provided } y \neq 0 \\[6pt] \sec\theta = \dfrac{1}{\cos\theta} = \dfrac{1}{x}, & \text{provided } x \neq 0 \\[6pt] \csc\theta = \dfrac{1}{\sin\theta} = \dfrac{1}{y}, & \text{provided } y \neq 0. \end{array}} \tag{3}$$

In Section 12.2 we defined the six trigonometric functions of acute angles in terms of the sides of a right triangle. We can easily show that these definitions are consistent with the definitions given in this section. Consider an acute angle of θ radians in the standard position and let $P(\theta) = (x, y)$ be the point on the terminal side of the angle and on the unit circle. Let Q be any other point (different from the origin) on the terminal side of the angle as

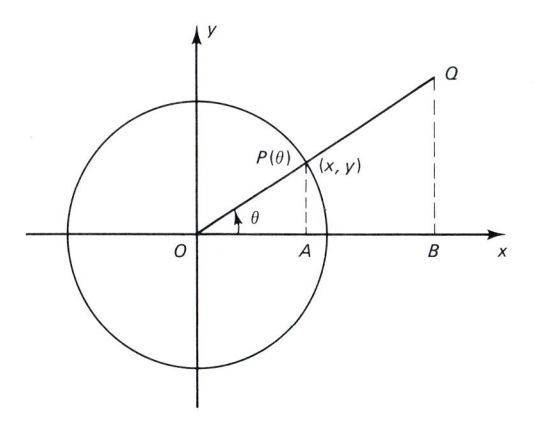

in the above figure. Draw perpendiculars PA and QB from P and Q, respectively, onto the x axis. Then the two right triangles OPA and OQB are similar and therefore

$$\frac{PA}{OP} = \frac{QB}{OQ} \quad \text{and} \quad \frac{OA}{OP} = \frac{OB}{OQ}$$

or

$$\frac{y}{1} = \frac{QB}{OQ} \quad \text{and} \quad \frac{x}{1} = \frac{OB}{OQ}.$$

Hence,

$$\sin\theta = \frac{QB}{OQ} = \frac{\text{opp.}}{\text{hyp.}}$$

and

$$\cos\theta = \frac{OB}{OQ} = \frac{\text{adj.}}{\text{hyp.}}$$

where we have used the definitions (1). Thus we have proved the definitions given in Section 12.2 for the sine and cosine functions. The definitions of the four remaining trigonometric functions are also consistent with those in Section 12.2 because they are defined in terms of the sine and cosine functions in the same way as before.

Let us now return to the unit-circle definitions of the three basic trigonometric functions—sine, cosine, and tangent. If $P(\theta) = (x, y)$ is a point on the unit circle and on the terminal side of an angle of θ radians in standard position, then by definition we have

$$\cos\theta = x, \quad \sin\theta = y, \quad \tan\theta = \frac{y}{x} \quad (x \neq 0). \tag{4}$$

If the angle θ is in the first quadrant—that is, if $0 < \theta < \pi/2$—then x and y are both positive numbers. Thus, from (4), $\sin\theta$, $\cos\theta$, $\tan\theta$, and their reciprocals are all positive. In other words, in the first quadrant all the trigonometric functions are positive. When the point $P(\theta) = (x, y)$ is in the second quadrant, then y is positive and x is negative. Thus in the second quadrant $\sin\theta$ (and $\csc\theta$) is positive whereas $\cos\theta$ (and $\sec\theta$) and $\tan\theta$ (and $\cot\theta$) are negative.

Similarly, when θ is in the third quadrant, x and y are both negative and y/x is positive. Therefore, in view of (4), $\tan\theta$ (and $\cot\theta$) is positive whereas $\sin\theta$ (and $\csc\theta$) and $\cos\theta$ (and $\sec\theta$) are negative. In the fourth quadrant, only $\cos\theta$ (and $\sec\theta$) is positive. For the three basic functions, sine, cosine, and tangent, the function that is positive in each quadrant is shown in the adjacent figure.

The following sentence may help you remember these facts:

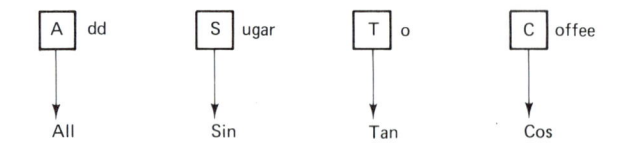

A dd	**S** ugar	**T** o	**C** offee
↓	↓	↓	↓
All	Sin	Tan	Cos

As we go from the first quadrant to the fourth quadrant, the function indicated by each first letter is positive.

From these properties we know, for example, that if $\sin\theta$ is negative and $\tan\theta$ is positive, then θ must be in the third quadrant. Similarly, if $\sec\theta$ is positive and $\cot\theta$ is negative, then θ is in the fourth quadrant.

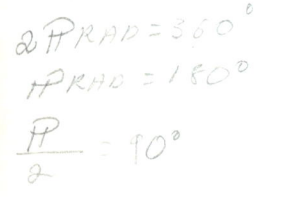

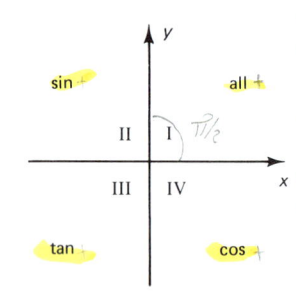

Example 1 Given $\sin\theta = \frac{3}{5}$ and $\pi/2 < \theta < \pi$, find the remaining trigonometric functions of θ.

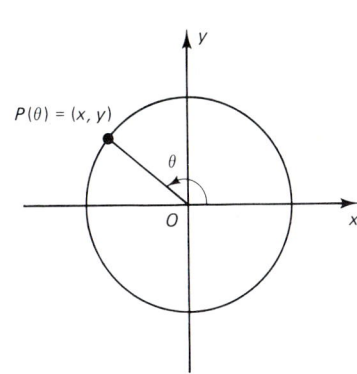

Solution The information that $\pi/2 < \theta < \pi$ implies that θ lies in the second quadrant. Let $P(\theta) = (x, y)$ be the point on the unit circle corresponding to the angle θ. Then by definition we have

$$\sin \theta = y = \tfrac{3}{5} \quad \text{(given)}.$$

That is, $y = \tfrac{3}{5}$. But since the point $P(\theta) = (x, y)$ lies on the unit circle, we have

$$x^2 + y^2 = 1.$$

Thus

$$x^2 + (\tfrac{3}{5})^2 = 1$$
$$x^2 = 1 - \tfrac{9}{25} = \tfrac{16}{25}.$$

Therefore,

$$x = \pm \tfrac{4}{5}.$$

Since the point (x, y) is in the second quadrant, x must be negative and therefore, $x = -\tfrac{4}{5}$. Thus we have

$$y = \tfrac{3}{5}, \quad x = -\tfrac{4}{5}.$$

Then by definition,

$$\cos \theta = x = -\tfrac{4}{5}$$

$$\tan \theta = \frac{y}{x} = \frac{\tfrac{3}{5}}{-\tfrac{4}{5}} = -\frac{3}{4}$$

$$\cot \theta = \frac{1}{\tan \theta} = -\frac{4}{3}$$

$$\sec \theta = \frac{1}{\cos \theta} = -\frac{5}{4}$$

$$\csc \theta = \frac{1}{\sin \theta} = \frac{5}{3}.$$

Example 2 Given that $\tan \theta = \tfrac{5}{12}$ and $\sin \theta$ is negative, find $\sin \theta$ and $\sec \theta$.

Solution Since $\tan \theta$ is positive and $\sin \theta$ is negative, θ must be in the third quadrant. Let $P(\theta) = (x, y)$, as shown in the adjacent figure.

By definition, we have

$$\tan \theta = \frac{y}{x} = \frac{5}{12} \quad \text{(given)}$$

$$y = \frac{5}{12} x. \qquad (i)$$

Since the point (x, y) lies on a unit circle, we also have

$$x^2 + y^2 = 1$$

and, substituting for y from (i), we get

$$x^2 + (\tfrac{5}{12})^2 x^2 = 1.$$

Thus,

$$x^2 = \tfrac{144}{169}$$

Sec. 12.4 Trigonometric Functions   495

which gives $x = \pm\frac{12}{13}$. Since the point (x, y) is in the third quadrant, x and y should both be negative. Therefore, we must take $x = -\frac{12}{13}$. Using this in (i), we have

$$y = \frac{5}{12}\left(-\frac{12}{13}\right) = -\frac{5}{13}.$$

Thus, we have $x = -\frac{12}{13}$, $y = -\frac{5}{13}$. Now, from the definitions,

$$\sin \theta = y = -\frac{5}{13}$$

$$\sec \theta = \frac{1}{\cos \theta} = \frac{1}{x} = -\frac{13}{12}.$$

TRIGONOMETRIC FUNCTIONS OF AXIS ANGLES

We now consider various cases when the point $P(\theta)$ lies on one of the coordinate axes.

When $\theta = 0$ radians $= 0°$, the point $P(0)$ has the coordinates $(1, 0)$. Similarly for $\theta = \pi/2$, π, and $3\pi/2$ radians, the corresponding points on the unit circle have the coordinates $(0, 1)$, $(-1, 0)$, and $(0, -1)$, respectively, as shown in the figure below.

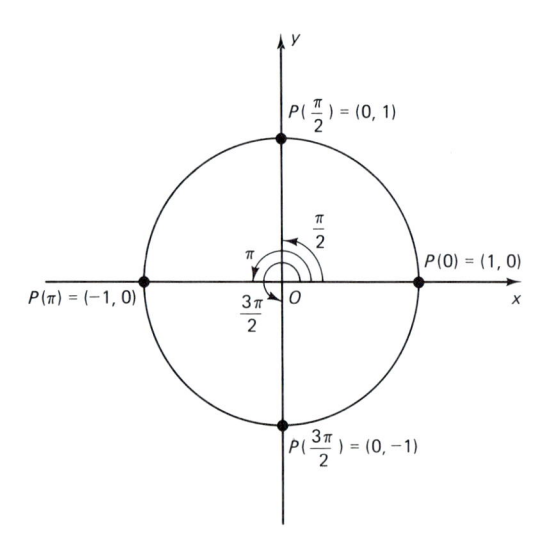

Taking the x coordinates of these points for the cosine function and y coordinates for the sine function, we have

$\sin 0 = 0,$	$\sin \frac{\pi}{2} = 1,$	$\sin \pi = 0,$	$\sin \frac{3\pi}{2} = -1$
$\cos 0 = 1,$	$\cos \frac{\pi}{2} = 0,$	$\cos \pi = -1,$	$\cos \frac{3\pi}{2} = 0.$

(5)

The values of the other four functions can readily be calculated.

For any integer n, the point $P(n\pi)$ lies on the x axis and therefore its y coordinate is zero. Thus,

$$\sin n\pi = 0 \quad \text{for any integer } n. \tag{6}$$

Similarly for each integer n, the point $P(n\pi + \pi/2)$ lies on the y axis and so its x coordinate is zero. Therefore,

$$\cos\left(n\pi + \frac{\pi}{2}\right) = 0 \quad \text{for any integer } n. \tag{7}$$

REFERENCE ANGLE

In order to evaluate any trigonometric function for a nonacute angle, we refer the given angle to an appropriate acute angle. The method is best explained by means of an example.

Example 3 Evaluate:
(a) $\sin 150°$ and $\cos 150°$.
(b) $\sin 240°$ and $\cos 240°$.
(c) $\sin 680°$ and $\cos 680°$.

Solution (a) In the figure below, OP is the terminal side of an angle of $150°$ in standard position. Let OQ be the terminal side of an angle of $30°$ in

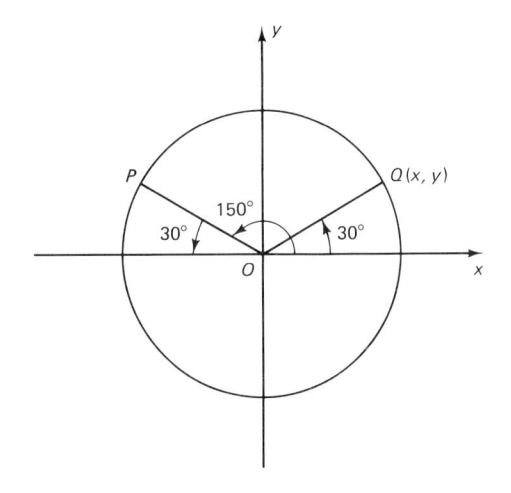

standard position. If Q has coordinates (x, y), then $x = \cos 30°$ and $y = \sin 30°$. But by symmetry, P and Q have the same y coordinate but opposite x coordinates; that is, P is the point $(-x, y)$. These coordinates, by definition, give respectively the cosine and sine of $150°$. Therefore

$$\cos 150° = -x = -\cos 30° = -\frac{\sqrt{3}}{2}$$

$$\sin 150° = y = \sin 30° = \frac{1}{2}.$$

(b) Let OP be the terminal side of an angle of 240° in standard position (see the figure below). If $Q(x, y)$ is on the terminal side of an angle of 60°, then $x = \cos 60°$ and $y = \sin 60°$. By symmetry P and Q are

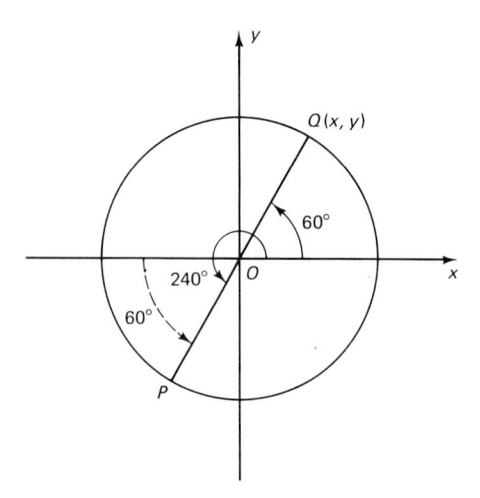

diametrically opposite one another on the unit circle and so P has coordinates $(-x, -y)$. Since these coordinates give the cosine and sine of 240°, we have

$$\cos 240° = -x = -\cos 60° = -\frac{1}{2}$$

$$\sin 240° = -y = -\sin 60° = -\frac{\sqrt{3}}{2}.$$

(c) An angle of 680° consists of one complete revolution (360°) plus 320° and has the terminal side OP as shown in the figure below. Let OQ be

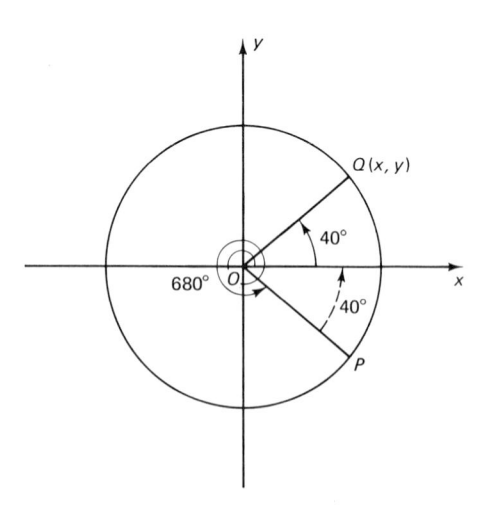

498   Ch. 12   Trigonometry

the terminal side of an angle of 40°. If Q has coordinates (x, y), then $x = \cos 40°$ and $y = \sin 40°$. By symmetry, P must have coordinates $(x, -y)$. Therefore

$$\cos 680° = x = \cos 40° = 0.7660$$
$$\sin 680° = -y = -\sin 40° = -0.6428$$

where the final values of cos 40° and sin 40° have been obtained from the tables.

From these examples we can formulate the following general procedure for determining the trigonometric functions at any given nonacute angle.

(a) Construct the given angle θ in standard position. Let OP be the terminal side.
(b) Determine an acute angle α, called the *reference angle*, as follows:
  (1) If P is in the first or fourth quadrant, then α is the angle between OP and the positive x axis.
  (2) If P is in the second or third quadrant, then α is the angle between OP and the negative x axis.
(c) Construct the angle α in standard position. The sine and cosine of θ can then be related by symmetry to the sine and cosine of α. In each part of Example 3 the terminal side of the angle α was denoted by OQ. We can avoid actually drawing OQ by observing that in every case

$$\sin \theta = \pm \sin \alpha, \quad \cos \theta = \pm \cos \alpha$$

and therefore also $\tan \theta = \pm \tan \alpha$. The signs of $\sin \theta$, $\cos \theta$, and $\tan \theta$ can be determined immediately from the quadrant in which P falls (using the "add sugar to coffee" rule).

Example 4 Evaluate (a) $\sin(-137°20')$, (b) $\tan(1410°)$.

Solution (a) The angle $\theta = -137°20'$ in standard position terminates in the third quadrant (see left figure below). The reference angle α is then 42°40′ (measured from the negative x axis, as shown). Therefore

$$\sin(-137°20') = \pm \sin 42°40'$$
$$= \pm 0.6777.$$

In the third quadrant, sines are negative, so the correct value is -0.6777.

(b) The angle $1410° = 3(360°) + 330°$ thus consists of three complete revolutions plus 330° (right-hand figure). The terminal position P leads to a reference angle $\alpha = 30°$, measured from the positive x axis since P is in the fourth quadrant. Therefore $\tan(1410°) = \pm \tan \alpha = \pm \tan 30° = \pm 1/\sqrt{3}$. Since P is in the fourth quadrant, the tangent is negative, so

$$\tan(1410°) = -1/\sqrt{3}.$$

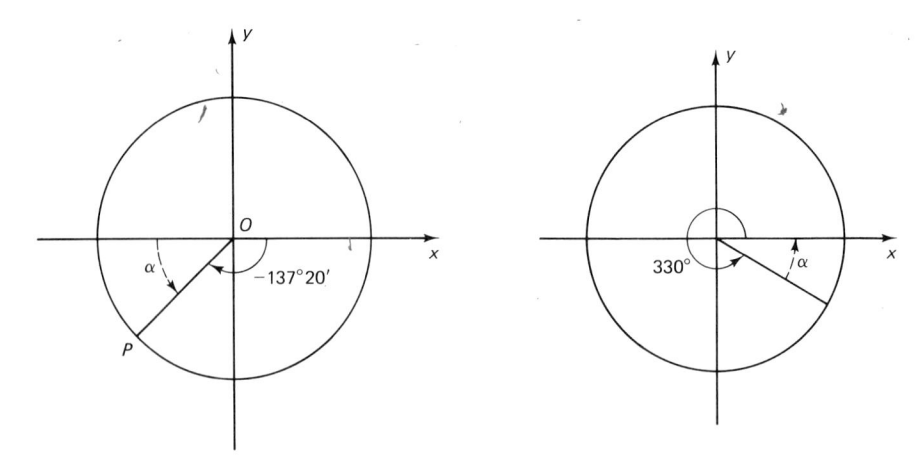

Example 5 Find all the positive values of θ less than $360°$ such that $\sin \theta = -\frac{1}{2}$.

Solution We have
$$\sin \theta = -\tfrac{1}{2} = -\sin 30°. \qquad \text{(i)}$$

Since $\sin \theta$ is negative, θ must be in either the third or the fourth quadrant. From (i) it follows that we must find all the angles θ in the third and the fourth quadrant, whose reference angle is $\alpha = 30°$. If θ is in the third quadrant, then the reference angle $\alpha = 30°$ is measured from the negative x axis. Thus $\theta = 180° + 30° = 210°$. In the fourth quadrant, the reference angle $\alpha = 30°$ is measured from the positive x axis. Thus $\theta = 360° - 30° = 330°$. (See the above figures.) Therefore the two required values of θ are $210°$ and $330°$.

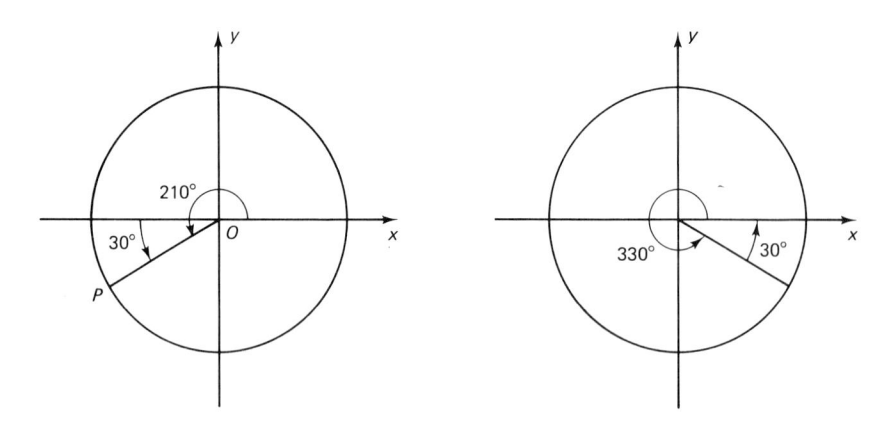

Example 6 Find all the positive values of θ less than 2π that satisfy the equation
$$\cos^2 \theta = \tfrac{1}{2}.$$

Solution The given equation is
$$\cos^2 \theta = \tfrac{1}{2}$$
or
$$\cos \theta = \pm 1/\sqrt{2}.$$
Consider the two possibilities in turn.

(1) $\cos \theta = 1/\sqrt{2} = \cos \pi/4$. Since $\cos \theta$ is positive, θ is in either the first or the fourth quadrant. The solution in the first quadrant is clearly $\theta = \pi/4$. In the fourth quadrant, the reference angle $\pi/4$ corresponds to
$$\theta = 2\pi - \frac{\pi}{4} = \frac{7\pi}{4}$$
(see left figure below). The solutions are therefore
$$\theta = \frac{\pi}{4} \quad \text{and} \quad \theta = \frac{7\pi}{4}.$$

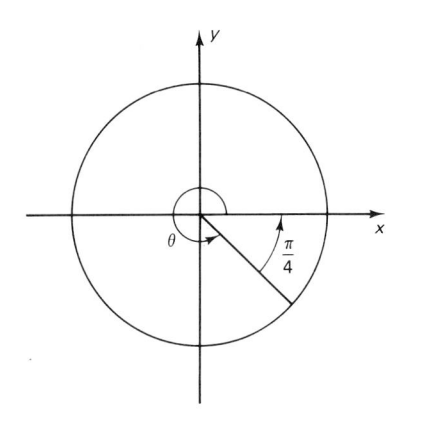

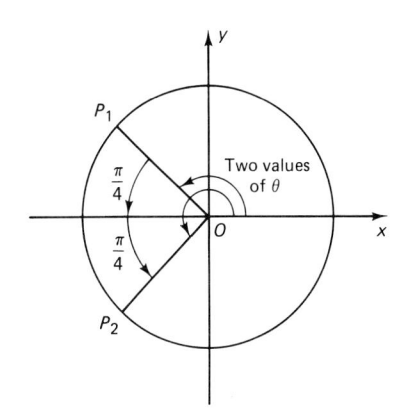

(2) $\cos \theta = -1/\sqrt{2} = -\cos \pi/4$. Since $\cos \theta$ is negative, θ must be in either the second or the third quadrant. The reference angle of $\alpha = \pi/4$ then corresponds to $\theta = \pi - \pi/4 = 3\pi/4$ or $\theta = \pi + \pi/4 = 5\pi/4$. Thus the solutions are
$$\theta = \frac{3\pi}{4} \quad \text{and} \quad \theta = \frac{5\pi}{4}.$$
Hence the solutions of the equation $\cos^2 \theta = \tfrac{1}{2}$ are
$$\theta = \frac{\pi}{4}, \quad \frac{3\pi}{4}, \quad \frac{5\pi}{4}, \quad \text{and} \quad \frac{7\pi}{4}.$$

Diagnostic Test 12.4

Fill in the blanks.
1. _____ $\leq \sin \theta \leq$ _____ for all values of θ.
2. _____ $\geq \cos \theta \geq$ _____ for all values of θ.

3. $|\sec \theta| \geq$ _____ for all values of θ.
4. $|\csc \theta|$ _____ 1 for all values of θ.
5. $\sin \pi =$ _____ .
6. $\cos 3\pi/2 =$ _____ .
7. $\cos \pi =$ _____ .
8. $\sin 3\pi/2 =$ _____ .
9. If $\sin \theta$ is negative and $\cos \theta$ is positive, then θ is in the _____ quadrant.
10. If $\tan \theta$ and $\sin \theta$ are of opposite signs, then θ is in the _____ quadrant.
11. If $\cos \theta$ and $\cot \theta$ are of the same sign, then θ is in the _____ quadrant.
12. If $\sec \theta$ and $\tan \theta$ are of the same sign, then θ is in the _____ quadrant.

Exercises 12.4

(1–14) Find the quadrants in which θ lies so that the following conditions are met.

1. $\sin \theta$ is positive and $\tan \theta$ is negative.
2. $\cos \theta$ is negative and $\cot \theta$ is positive.
3. $\sec \theta$ is positive and $\csc \theta$ is negative.
4. $\sin \theta$ and $\cos \theta$ are both negative.
5. $\sin \theta$ and $\tan \theta$ are both negative.
6. $\sec \theta$ and $\cot \theta$ are both negative.
7. $\sin \theta$ and $\sec \theta$ are both positive.
8. $\sec \theta$ and $\csc \theta$ are both negative.
9. $\cos \theta$ is positive.
10. $\tan \theta$ is negative.
11. $\sin \theta$ is negative.
12. $\sec \theta$ is negative.
13. $\tan \theta$ is positive.
14. $\csc \theta$ is positive.

(15–26) In the following exercises, one trigonometric function of θ and a condition on θ are given. Find the five remaining trigonometric functions of θ.

15. $\sin \theta = \frac{5}{13}$ and θ lies in the second quadrant.
16. $\cos \theta = -\frac{3}{5}$ and θ is in the third quadrant.
17. $\tan \theta = -1/\sqrt{3}$ and θ lies in the fourth quadrant.
18. $\sec \theta = \frac{5}{2}$ and θ lies in the fourth quadrant.
19. $\cot \theta = \frac{3}{4}$ and θ is in third quadrant.
20. $\csc \theta = -2$ and θ is in the fourth quadrant.
21. $\sin \theta = \frac{1}{5}$ and $\cos \theta$ is negative.
22. $\cos \theta = -\frac{1}{3}$ and $\sin \theta$ is negative.
23. $\tan \theta = -\frac{2}{3}$ and $\cos \theta$ is positive.
24. $\sec \theta = -\sqrt{3}$ and $\tan \theta$ is negative.
25. $\cot \theta = \frac{3}{4}$ and $\sin \theta$ is positive.
26. $\csc \theta = -3$ and $\cos \theta$ is negative.

(27–50) Evaluate the following trigonometric functions.

27. $\sin 5\pi/6$.
28. $\cos 2\pi/3$.
29. $\tan 5\pi/3$.
30. $\sec 7\pi/6$.
31. $\csc 3\pi/4$.
32. $\cot 7\pi/4$.
33. $\sin 120°$.
34. $\cos 135°$.
35. $\tan 210°$.
36. $\sec 230°$.
37. $\sin 160°$.
38. $\cos 230°$.
39. $\tan 345°$.
40. $\cos 157°30'$.
41. $\sin 370°20'$.
42. $\sec (-50°)$.
43. $\tan (-30°)$.
44. $\sin (-210°)$.
45. $\cos 390°$.
46. $\sin (-420°)$.
47. $\tan 370°$.
48. $\sec 730°$.
49. $\cos (-460°)$.
50. $\cot (-395°)$.

(51–56) Prove the following. (Do not use tables.)

51. $\sin 420° \cos 390° + \cos (-300°) \sin (-330°) = 1$.
52. $\tan 315° \cot (-405°) + \cot (495°) \tan (-585°) = 2$.

53. $\cot 30° \tan 60° \cos 120° \sin 150° = -\frac{3}{4}$.

54. $\cos 24° + \cos 55° + \cos 125° + \cos 204° + \cos 300° = \frac{1}{2}$.

55. $\dfrac{\sin 135° - \cos 120°}{\sin 135° + \cos 120°} = 3 + 2\sqrt{2}$.

56. $\dfrac{\cos 360° \sin 390° \sec 315°}{\tan 225° \cot 240° \csc(-270°)} = \sqrt{\dfrac{3}{2}}$.

(57–68) Find all the positive values of θ less than 2π that satisfy the following equations.

57. $\cos \theta = \frac{1}{2}$.
58. $\sin \theta = -\sqrt{3}/2$.
59. $\tan \theta = 1$.
60. $\sec \theta = \frac{2}{3}\sqrt{3}$.
61. $\cot \theta = -\sqrt{3}$.
62. $\cos \theta = -\sqrt{2}/2$.
63. $\sin^2 \theta = \frac{1}{4}$.
64. $\cos^2 \theta = \frac{3}{4}$.
65. $\tan^2 \theta = 3$.
66. $\sec^2 \theta = 2$.
67. $\csc^2 \theta = \frac{4}{3}$.
68. $\cot^2 \theta = 1$.

12.5

Graphs of Trigonometric Functions

In this section we construct the graphs of the six trigonometric functions and use them to demonstrate some of the properties of these functions. We start with the graph of the sine function $y = \sin x$, where x denotes the radian measure of the angle. As stated in Section 12.4, the domain of $y = \sin x$ is the set of all real numbers and the range is $-1 \le y \le 1$.*

First we construct the graph when x varies from 0 to 2π by plotting a number of points that lie on it. We list below the values of $y = \sin x$ at intervals of $\pi/6$ radians from 0 to 2π, rounded off to two decimal places. These values can be obtained using appropriate reference angles or by calculator.

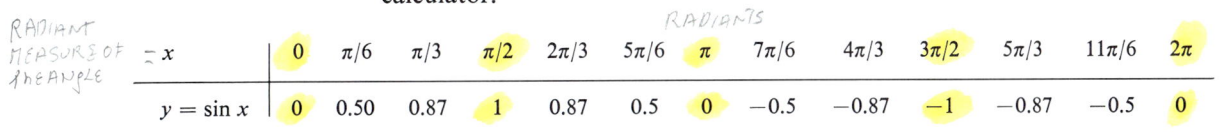

x	0	$\pi/6$	$\pi/3$	$\pi/2$	$2\pi/3$	$5\pi/6$	π	$7\pi/6$	$4\pi/3$	$3\pi/2$	$5\pi/3$	$11\pi/6$	2π
$y = \sin x$	0	0.50	0.87	1	0.87	0.5	0	-0.5	-0.87	-1	-0.87	-0.5	0

Plotting the points given in the above table and joining them by a smooth curve, we obtain the portion of the graph shown in the figure below.

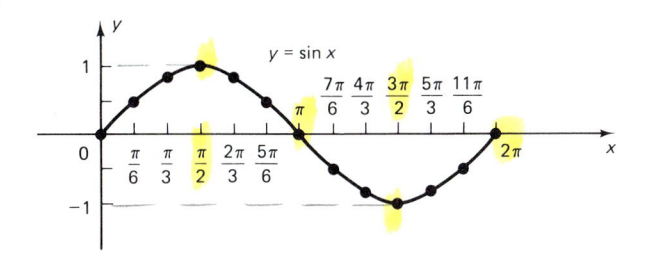

*x and y are no longer used as the coordinates of the point $P(\theta)$. x is the argument and y the value of the sine function.

It is clear that as x increases from 0 to $\pi/2$, $y = \sin x$ increases from 0 to 1. As x further increases from $\pi/2$ to π, y decreases from 1 to 0. As x increases from π to $3\pi/2$, $y = \sin x$ decreases further from 0 to -1, and finally as x increases from $3\pi/2$ to 2π, $y = \sin x$ increases from -1 to 0.

Since any angle x is co-terminal with $(x + 2\pi)$, we have

$$\sin(x + 2\pi) = \sin x.$$

Thus the graph of $\sin x$ repeats itself every 2π radians, as shown in the figure below. The portion of the graph for $2\pi \leq x \leq 4\pi$ is a precise replica of the portion for the basic interval $0 \leq x \leq 2\pi$ as also are the portions for $4\pi \leq x \leq 6\pi$, $-2\pi \leq x \leq 0$, and so on.

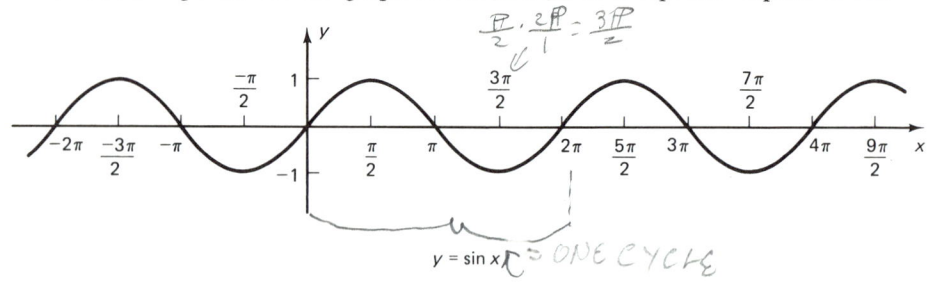

To describe this cyclic behavior of the sine function we say that the sine function is ***periodic*** with ***period 2π***. The graph of $y = \sin x$ over an interval of length 2π, such as $0 \leq x \leq 2\pi$, or $-\pi \leq x \leq \pi$ or $2\pi \leq x \leq 4\pi$, is called ***one cycle*** of the sine curve.

The following properties can easily be observed from the graph of the sine function.

(i) The maximum value of $\sin x$ is 1 and the minimum value is -1. For any periodic function, the quantity

$$\tfrac{1}{2}(\text{maximum value} - \text{minimum value})$$

is called the ***amplitude*** of the function. In this case, the amplitude of the sine function is $\tfrac{1}{2}[1 - (-1)] = 1$.

(ii) $\sin x = 0$, when $x = 0, \pm\pi, \pm 2\pi, \ldots$.

(iii) $\sin x > 0$ in $0 < x < \pi$—that is, when x is in the first or second quadrant, and $\sin x < 0$ in $\pi < x < 2\pi$—that is, when x is in the third or fourth quadrant.

(iv) The graph of the sine function is symmetrical with respect to the origin. Another way of expressing this is the property

$$\boxed{\sin(-x) = -\sin x.}$$

GRAPH OF $y = \cos x$

The cosine function is also ***periodic*** with ***period 2π***, because

$$\cos(x + 2\pi) = \cos x.$$

Thus, we shall plot only the points on the graph in the interval 0 to 2π; then for other values of x we can draw the graph by repeating its form in this basic cycle. We list below the approximate values of $y = \cos x$ at intervals of $\pi/6$ radians from 0 to 2π.

x	0	$\pi/6$	$\pi/3$	$\pi/2$	$2\pi/3$	$5\pi/6$	π	$7\pi/6$	$4\pi/3$	$3\pi/2$	$5\pi/3$	$11\pi/6$	2π
$y = \cos x$	1	0.87	0.5	0	−0.5	−0.87	−1	−0.87	−0.5	0	0.5	0.87	1
$y = \sin x$	0	0.50	0.87	1	0.87	0.50	0	−0.5	−0.87	−1	−0.87	−0.5	0

From the above table of values, it is clear that as x increases from 0 to $\pi/2$, $y = \cos x$ decreases from 1 to 0. As x further increases from $\pi/2$ to π, $\cos x$ further decreases from 0 to −1. As x increases from π to $3\pi/2$, $\cos x$ increases from −1 to 0, and as x increases from $3\pi/2$ to 2π, $\cos x$ further increases from 0 to 1.

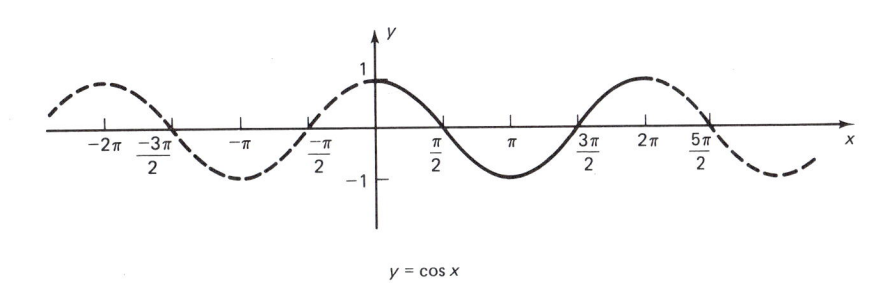

$y = \cos x$

Plotting the points given in the table above and joining them by a smooth curve, we obtain the portion of the graph shown by a solid line in the figure above. We can extend the graph in both directions by using the periodic behavior of $\cos x$ and this is shown in the figure by dashed lines.

The following properties can easily be observed from the graph.

(i) The maximum value of $\cos x$ is 1 and the minimum value is −1. The amplitude of the cosine function is 1.
(ii) $\cos x = 0$ when $x = \pm\pi/2, \pm 3\pi/2, \pm 5\pi/2, \ldots$—that is, odd multiples of $\pi/2$.
(iii) The graph of $y = \cos x$ is symmetrical about the y axis. This is equivalent to the property.

$$\cos(-x) = \cos x.$$

(iv) Comparing the graphs of the cosine and sine functions, we find that if the graph of $\sin x$ is shifted to the left by $\pi/2$ units, we obtain the graph of the cosine function. But shifting the graph of $y = \sin x$ $\pi/2$ units to the left produces the graph of $y = \sin(\pi/2 + x)$ (see section 7.4). Thus we have that

$$\sin\left(\frac{\pi}{2} + x\right) = \cos x$$

for all x. (See Exercise 87 in Section 13.2.)

Sec. 12.5 Graphs of Trigonometric Functions

Example 1 Draw the graph of $y = 3 \sin x$.

Solution For each value of x, the value of $y = 3 \sin x$ is three times as large as the value of $\sin x$. This means that the graphs of $y = 3 \sin x$ and $y = \sin x$ differ only in their vertical scales. The values of $y = 3 \sin x$ vary from -3 to 3, as shown in the figure below.

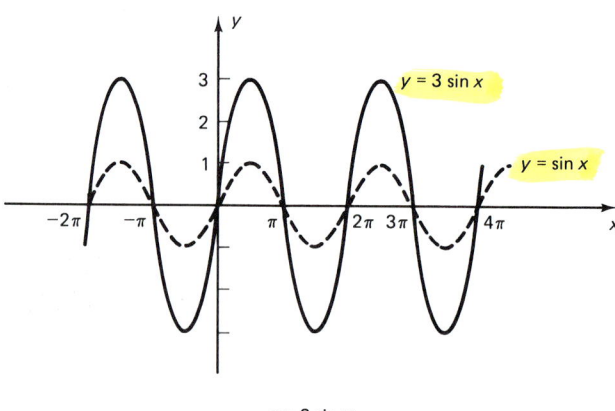

$y = 3 \sin x$

In general, the graph of $y = a \sin x$ (or $a \cos x$) will have the same form as $y = \sin x$ (or $y = \cos x$) with range $-|a| \leq y \leq |a|$ and amplitude $|a|$.

Example 2 Draw the graph of $y = 1 + \cos x$.

Solution The graph of $y = 1 + \cos x$ is displaced one unit upward compared to the graph of $y = \cos x$, as shown in the figure below. The values of y lie between 0 and 2. Thus the amplitude of $y = 1 + \cos x$ is

$$\tfrac{1}{2}(\text{maximum value} - \text{minimum value}) = \tfrac{1}{2}(2 - 0) = 1$$

which is the same as the amplitude of $y = \cos x$. Moreover, the period of $y = 1 + \cos x$ is 2π, the same as the period of $y = \cos x$.

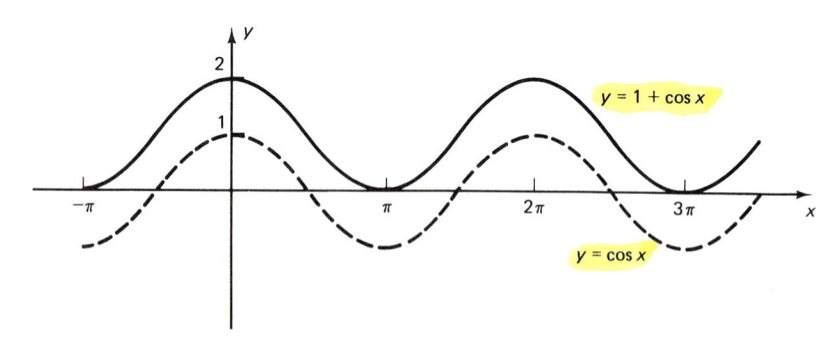

Example 3 Draw the graph of $y = |\sin x|$.

Solution Since $|\sin x| \geq 0$ for each x, and the values of $\sin x$ lie between -1 and 1, the values of $y = |\sin x|$ will lie in the interval $0 \leq y \leq 1$. The graph of $y = |\sin x|$ always lies on or above the x axis, as shown in the following figure. The effect of the absolute value is to reflect the negative portions of the graph of $y = \sin x$ about the x axis. Note that the amplitude of $y = |\sin x|$ is $\frac{1}{2}$ (max. value − min. value) $= \frac{1}{2}(1-0) = \frac{1}{2}$. The period is π (and not 2π) as is evident from the graph.

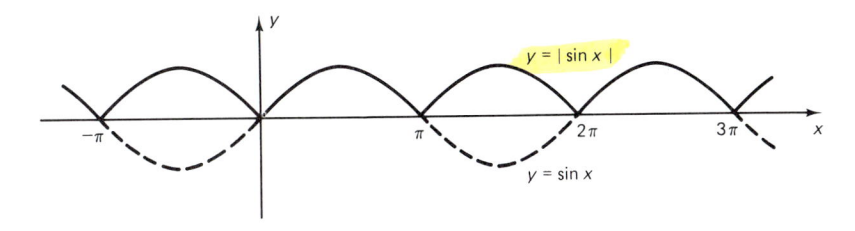

Example 4 Draw the graph of $y = \sin 2x$.

Solution The following table gives the values of $y = \sin 2x$ to two decimal places for different values of x in the interval $0 < x < 2\pi$. These values lead to the graph shown below.

x	0	$\pi/6$	$\pi/4$	$\pi/3$	$\pi/2$	$2\pi/3$	$3\pi/4$	$5\pi/6$	π	$7\pi/6$	$5\pi/4$	$4\pi/3$	$3\pi/2$	$5\pi/3$	$7\pi/4$	$11\pi/6$	2π
$2x$	0	$\pi/3$	$\pi/2$	$2\pi/3$	π	$4\pi/3$	$3\pi/2$	$5\pi/3$	2π	$7\pi/3$	$5\pi/2$	$8\pi/3$	3π	$10\pi/3$	$7\pi/2$	$11\pi/3$	4π
$y = \sin 2x$	0	0.87	1	0.87	0	−0.87	−1	−0.87	0	0.87	1	0.87	0	−0.87	−1	−0.87	0

From the above table it is clear that the set of values of $y = \sin 2x$ for x in the interval $0 \leq x \leq \pi$ are the same as for $\pi \leq x \leq 2\pi$. Thus, the period of $y = \sin 2x$ is π and not 2π. The amplitude of the graph is unchanged and is 1.

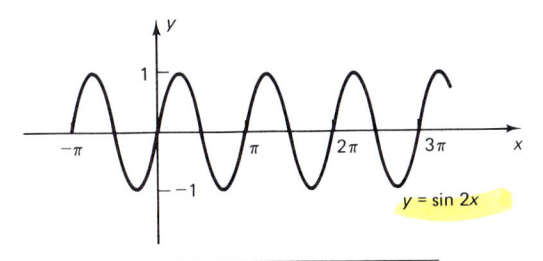

In general, the graph of $y = \sin px$ (or $y = \cos px$) has the same general form as the graph of $y = \sin x$ (or $y = \cos x$) but the period of $\sin px$ (or $\cos px$) is $(2\pi)/|p|$. For example, $y = \sin 4x$ has a period $2\pi/4 = \pi/2$; $y = \cos(x/3)$ has a period $2\pi/\frac{1}{3} = 6\pi$.

Example 5 Draw the graph of $y = \sin(2x + 2\pi/3)$.

Solution We can write

$$y = \sin\left(2x + \frac{2\pi}{3}\right) = \sin 2\left(x + \frac{\pi}{3}\right).$$

The graph can be obtained by displacing the graph of $y = \sin 2x$ obtained in Example 4 by $\pi/3$ units to the left.

Sec. 12.5 Graphs of Trigonometric Functions

$$y = \sin(2x + 2\pi/3)$$

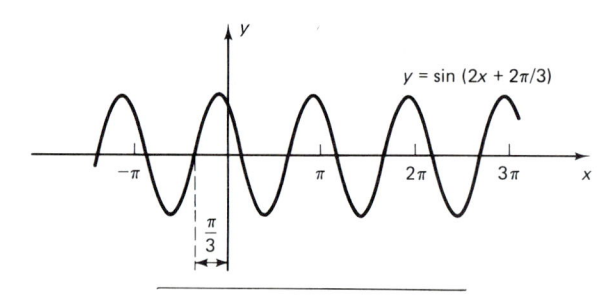

In general, the graph of $y = \sin p(x + c)$ [or $y = \cos p(x + c)$] is the same in form as the graph of $y = \sin px$ (or $y = \cos px$) but is displaced c units to the left if $c > 0$ and $|c|$ units to the right if $c < 0$.

THE GRAPH OF $y = \tan x$

Since we defined the tangent function as

$$\tan x = \frac{\sin x}{\cos x}$$

we must exclude from the domain of $\tan x$ those values of x at which the denominator $\cos x = 0$. These are $x = \pm\pi/2, \pm 3\pi/2, \pm 5\pi/2, \ldots$. Thus the domain of $y = \tan x$ is the set of all real numbers except $x = \pm\pi/2, \pm 3\pi/2$, and so on. As x approaches these values, $y = \tan x$ becomes numerically large without bound.

The values of y for different values of x in the interval $0 \leq x \leq 2\pi$ are given in the following table to two decimal places.

x	0	$\pi/6$	$\pi/3$	$\pi/2$	$2\pi/3$	$5\pi/6$	π
$y = \tan x$	0	0.58	1.73	undefined	-1.73	-0.58	0

x	$7\pi/6$	$4\pi/3$	$3\pi/2$	$5\pi/3$	$11\pi/6$	2π
$y = \tan x$	0.58	1.73	undefined	-1.73	-0.58	0

These lead to the graph shown in the following figure.

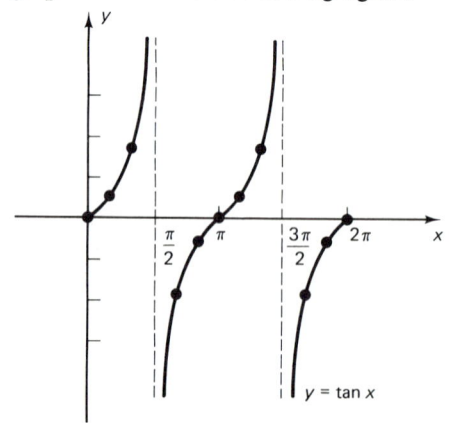

508

Ch. 12   Trigonometry

We see that the graph of $y = \tan x$ repeats itself every π radians. Thus $\tan x$ is **periodic** with **period** π. In this case the graph from $\pi/2$ to $3\pi/2$ represents one cycle of the tangent curve. We can construct the complete graph as shown in the following figure by repeating this cycle in both directions. It

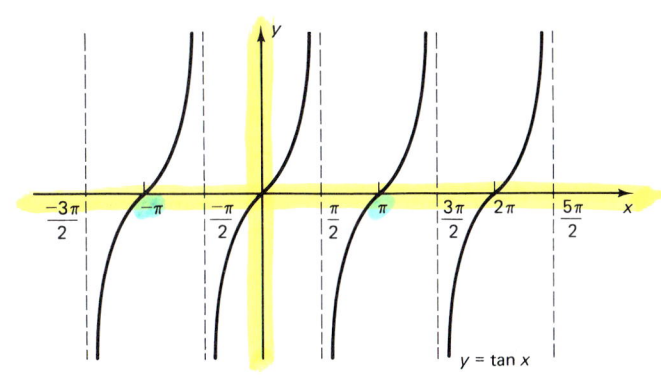

may be observed that the lines $x = \pm\pi/2, \pm 3\pi/2, \ldots$ are **vertical asymptotes** of the tangent curve. These asymptotes are *not* part of the graph of the tangent function.

From the graph, the following properties are readily observed.

(i) The range of $y = \tan x$ is the set of all real numbers.
(ii) Tan $x = 0$, when $x = 0, \pm\pi, \pm 2\pi, \ldots$.
(iii) The graph of $y = \tan x$ is symmetrical about the origin. That is, whenever a point (x, y) lies on the graph, the point $(-x, -y)$ also lies on it. This can also be expressed by the property that

$$\tan(-x) = -\tan x.$$

(iv) The tangent function is periodic with period π.
(v) Since there is no maximum and no minimum value of the tangent function, we cannot assign an amplitude to this function.

THE GRAPHS OF COTANGENT, SECANT, AND
COSECANT FUNCTIONS

We can construct the graphs of the cotangent, secant, and cosecant functions by using the fact that they are the reciprocals of the tangent, cosine, and sine functions, respectively.

To draw the graph of $y = \cot x = 1/\tan x$, we note that $\cot x$ is not defined when $\tan x = 0$—that is, when $x = 0, \pm\pi, \pm 2\pi, \ldots$. Thus the lines $x = 0, \pm\pi, \pm 2\pi, \ldots$ are vertical asymptotes of $y = \cot x$. Moreover at the points where $\tan x$ is undefined—that is, at $x = \pm\pi/2, \pm 3\pi/2, \ldots$—the function $\cot x$ is zero. Note that at $x = \pi/4$, $\tan x$ and $\cot x$ both have the value 1. The graph of $y = \cot x$ is shown by the solid line in the following figure.

Sec. 12.5   Graphs of Trigonometric Functions

$y = \cot x$

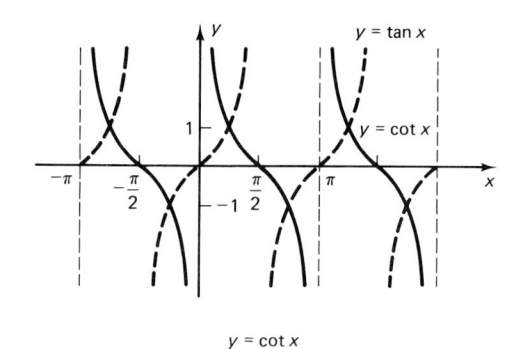

Note that the period of $y = \cot x$ is also π. No amplitude is assigned to a cotangent function.

Proceeding in a similar manner, we find that the graphs of $y = \sec x$ and $y = \csc x$ have the forms shown by solid lines in the following figures. The dotted lines show the graphs of their reciprocal functions $\cos x$ and $\sin x$ for comparison.

Note that $y = \sec x$ and $y = \csc x$ are both periodic with period 2π. Also $y = \sec x$ has vertical asymptotes at $x = \pm\pi/2, \pm 3\pi/2, \ldots$, whereas $y = \csc x$ has vertical asymptotes at $x = 0, \pm\pi, \pm 2\pi, \ldots$. No portion of the graph of $y = \sec x$ or $y = \csc x$ lies in the interval $-1 < y < 1$.

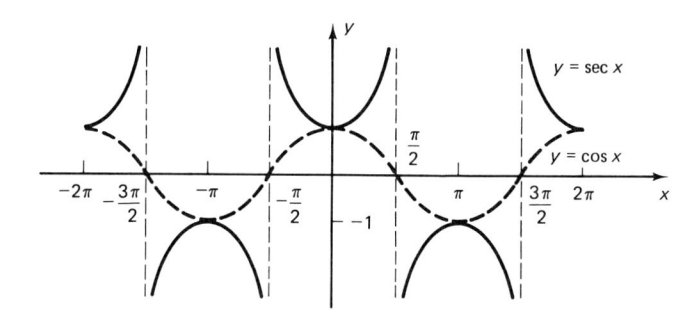

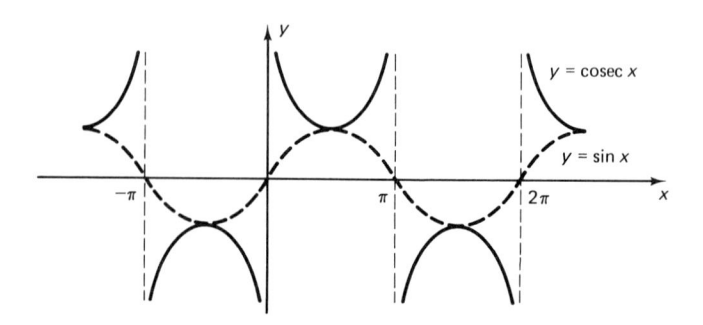

Diagnostic Test 12.5

Fill in the blanks.

1. The graph of $y = 2 \sin x$ has an amplitude of _____ units and a period of _____ units.
2. The graph of $y = \cos 3x$ has a period of _____ units and an amplitude of _____ units.
3. The amplitude of a sine or cosine curve is half the _____ of its maximum and minimum values.
4. The graph of $y = \sin x$ is symmetrical about the _____.
5. The graph of $y = \cos 2x$ is symmetrical about the _____.
6. The graph of $y = \tan x$ is symmetrical about the _____.
7. The vertical asymptotes of the tangent function $y = \tan x$ are given by _____.
8. The function $y = \tan x$ is not defined for those values of x where _____ $= 0$.
9. The values of x for which $\tan x$ becomes zero are given by $x =$ _____.
10. The function $y = \csc x$ is not defined for those values of x where _____ $= 0$.

Exercises 12.5

(1–26) Draw the graphs of the following trigonometric functions on the indicated intervals. In Exercises 1–22, state the amplitude and periods of the function.

1. $y = \sin x$, $-\pi \leq x \leq \pi$.
2. $y = \cos x$, $-\pi \leq x \leq \pi$.
3. $y = 2 \sin x$, $0 \leq x \leq 4\pi$.
4. $y = 3 \cos x$, $-2\pi \leq x \leq 2\pi$.
5. $y = -2 \cos x$, $0 \leq x \leq 2\pi$.
6. $y = -3 \sin x$, $-2\pi \leq x \leq 2\pi$.
7. $y = \sin 3x$, $0 \leq x \leq 2\pi$.
8. $y = \cos 2x$, $-\pi \leq x \leq \pi$.
9. $y = 1 + \sin x$, $-2\pi \leq x \leq 2\pi$.
10. $y = 3 - \cos x$, $0 \leq x \leq 4\pi$.
11. $y = |\cos x|$, $0 \leq x \leq 2\pi$.
12. $y = 2 + \sin x$, $0 \leq x \leq 2\pi$.
13. $y = 2 \sin 3x$, $0 \leq x \leq 2\pi$.
14. $y = -2 \cos(x/2)$, $-2\pi \leq x \leq 2\pi$.
15. $y = 2 \sin(x/3)$, $-3\pi \leq x \leq 3\pi$.
16. $y = 1 + 2 \sin x$, $-\pi \leq x \leq 2\pi$.
17. $y = \sin(x + \pi/3)$, $-4\pi/3 \leq x \leq 8\pi/3$.
18. $y = \cos(x - 2\pi/3)$, $-7\pi/3 \leq x \leq 5\pi/3$.
19. $y = 2 \sin(x + \pi/4)$, $-\pi \leq x \leq \pi$.
20. $y = -\sin(x - \pi/2)$, $0 \leq x \leq 2\pi$.
21. $y = \cos(2x - 120°)$, $0 \leq x \leq 2\pi$.
22. $y = \sin(\frac{1}{2}x + \pi/4)$, $0 \leq x \leq 4\pi$.
23. $y = 2 \tan x$, $-3\pi/2 \leq x \leq 3\pi/2$.
24. $y = \tan 2x$, $-\pi \leq x \leq \pi$.
25. $y = \tan x/2$, $-2\pi \leq x \leq 2\pi$.
26. $y = \tan(x - \pi/2)$, $-3\pi/2 \leq x \leq 3\pi/2$.
27. Sketch the graphs of $y = \sin x$ and $y = \cos x$ in the interval $0 \leq x \leq 2\pi$ on the same graph paper. Read off the values of x in $0 \leq x \leq 2\pi$ for which $\sin x = \cos x$.
28. Sketch the graphs of $y = \sin 2x$ and $y = \sin x$ in the interval $0 \leq x \leq 2\pi$ on the same graph paper. Read from the graph the values of x in $0 \leq x \leq 2\pi$ for which $\sin 2x = \sin x$.

Chapter Review 12

Define and (or) explain the following:

(a) The radian measure of an angle.

(b) The relation between the degree measure and radian measure of an angle.

(c) The relation between the arc of a circle, its radius, and the angle subtended by the arc at the center.

(d) Trigonometric functions as defined by ratios of the sides of a right triangle.

(e) Use of tables of trigonometric functions; linear interpolation.

(f) Trigonometric functions defined on a unit circle.

(g) The reference angle and its use.

(h) The graphs of the sine, cosine, and tangent functions.

(i) Trigonometric functions of axis angles (0°, 90°, 180°, 270°).

(j) Period of a trigonometric function.

(k) Amplitude of a sine or cosine function.

(l) A cycle of the graph of a given function.

Review Exercises on Chapter 12

1. One of the acute angles of a right triangle is 57°25′30″. Find the other acute angle in degrees.

2. Two angles of a triangle are 47°35′ 27″ and 72°24′33″. Find the third angle in radians.

3. The angles of a triangle are to one another in the ratio 1:3:5. Express them in degrees as well as in radians.

4. The angles of a quadrilateral are to one another in the ratio 1:2:3:4. Express them in degrees as well as radians.

5. A circular arc of length 55 centimeters subtends an angle of 30° at the center of a circle. Find the radius of the circle. (Approximate π by $\frac{22}{7}$.)

6. Assuming that the earth's radius is 3960 miles and that it subtends an angle of 57′ at the center of the Moon, find the distance of the center of the Moon from the Earth's center. (Use $\pi \approx \frac{22}{7}$.)

(7–8) Evaluate the following. (Do not use tables.)

7. $4 \cos^2 60° + 4 \tan^2 45° - \csc^2 30°$.

8. $\frac{4}{5} \cot^2 30° + 3 \sin^2 60° - 2 \csc^2 60° - \frac{3}{4} \tan^2 30°$.

(9–10) Prove the following. (Do not use tables.)

9. $\dfrac{\sin 30°}{\cos 45°} + \dfrac{\cot 45°}{\sec 60°} - \dfrac{\sin 60°}{\sin 45°} - \dfrac{\cos 30°}{\sin 90°} = \dfrac{(\sqrt{2}+1)(1-\sqrt{3})}{2}$.

10. $\tan \pi/6 + 2 \tan \pi/4 + \tan \pi/3 = $
$2 \cot \pi/6 = \cot \pi/4 \cot \pi/3 + \cot \pi/6 + \cot \pi/3$.

11. Prove that $\cos^2 \pi/3$, $\cos^2 \pi/4$, and $\cos^2 \pi/6$ are in arithmetic progression.

12. If $A = 60°$ and $B = 30°$, show that
(a) $\sin(A - B) \neq \sin A - \sin B$.
(b) $\cos(A + B) \neq \cos A + \cos B$.

13. If A, B, and $A + B$ denote positive acute angles and $\sin(A + B) = 1$ and $\sin(A - B) = \frac{1}{2}$, find the values of A and B. (Do not use tables.)

14. If $0 \leq \theta \leq 90°$, find the values of θ for which the equation $1 + 2 \sin^2 \theta = 3 \sin \theta$ is true. (Do not use tables.)

15. The angle of elevation of the top of a tower as observed from a point 50 feet from the foot of the tower is 40°. Find the height of the tower.

16. The angle of depression of a boat from the top of a lighthouse is 65°. If the height of the lighthouse is 120 feet above the sea level, find the distance of the boat from the foot of the lighthouse.

17. A and B are two points in line with the foot of a tower. The angles of elevation of the top of the tower at the points A and B are 45° and 30°. If A and B are 70 feet apart, find the height of the tower.

18. The angle of depression of the top of one building as observed from the top of another building 100 feet high is 62°. If the two buildings are 50 feet apart, find the height of the second building.

(19–22) In the following exercises, one trigonometric function of θ and a condition on θ are given. Find the other five trigonometric functions of θ.

19. $\sin \theta = -\frac{2}{3}$ and θ is in the fourth quadrant.

20. $\tan \theta = \sqrt{3}$ and θ is in the third quadrant.

21. $\cos \theta = -\frac{1}{2}$ and $\tan \theta$ is negative.

22. $\sec \theta = 5$ and $\sin \theta$ is negative.

(23–34) Evaluate the following trigonometric functions.

23. $\sin 330°$.
24. $\cos 420°$.
25. $\tan 250°$.
26. $\sec 140°$.
27. $\sin (-315°)$.
28. $\cos (-120°)$.
29. $\tan (-765°)$.
30. $\sec (-690°)$.
31. $\sin 1460°$.
32. $\cos 1310°$.
33. $\tan 1920°$.
34. $\csc 1770°$.

(35–38) Prove the following. (Do not use tables.)

35. $\cos 570° \sin 510° + \sin (-330°) \cos (-390°) = 0$.

36. $\sin 24° + \sin 55° + \sin 204° + \sin 235° + \sin 450° = 1$.

37. $\tan 27° + \tan 63° + \tan 117° + \tan 333° + \tan 225° = 1$.

38. $\sqrt{2} \sin 135° \cos 210° \tan 240° \cot 300° \sec 330° = 1$.

(39–46) Find all the positive values of θ less than 360° that satisfy the following equations.

39. $\sin \theta = \frac{1}{2}$.
40. $\cos \theta = \sqrt{2}/2$.
41. $\tan \theta = -\sqrt{3}$.
42. $\sec \theta = -2$.
43. $\sin^2 \theta = \frac{3}{4}$.
44. $\tan^2 \theta = 1$.
45. $\cos^2 \theta = \frac{1}{2}$.
46. $\sin^2 \theta = 2$.

*47. If A, B, C denote the angles of a triangle ABC, then show that
(a) $\sin (A + B) = \sin C$
(b) $\tan (B + C) + \tan A = 0$.

*48. Show that in a cyclic quadrilateral $ABCD$,
$$\cos A + \cos B + \cos C + \cos D = 0.$$
[Hint: The sum of opposite angles in a cyclic quadrilateral is 180°.]

*(49–50) Reduce the following to its simplest form. (Assume θ is acute.)

49. $\dfrac{\sin (90° - \theta) \cdot \cos (-\theta) \cdot \sec (180° + \theta)}{\cos (180° - \theta) \cdot \sin (180° + \theta) \cdot \csc (90° - \theta)}$.

50. $\dfrac{\cos (90° - \theta)}{\sin (180° + \theta)} + \dfrac{\sin (-\theta)}{\cos (90° - \theta)} + \dfrac{\cot (90° - \theta)}{\tan (180° - \theta)}$.

(51–56) Sketch the graphs of the following on the indicated intervals. State the amplitude and the period in each case.

51. $y = 3 \sin 2x$, $0 \leq x \leq 2\pi$.
52. $y = 2 + \sin x$, $0 \leq x \leq 2\pi$.
53. $y = 3 - 2 \cos x$, $0 \leq x \leq 2\pi$.
54. $y = |\sin 2x|$, $0 \leq x \leq 2\pi$.
55. $y = 1 + |\sin x|$, $0 \leq x \leq 2\pi$.
56. $y = |1 + 2 \sin x|$, $0 \leq x \leq 2\pi$.

13 TRIGONOMETRIC IDENTITIES

13.1

Fundamental Identities

In Sections 12.2 and 12.4 we introduced some basic trigonometric identities, which we used as definitions. We restate these identities here for reference:

$$\csc\theta = \frac{1}{\sin\theta}, \quad \sec\theta = \frac{1}{\cos\theta}, \quad \cot\theta = \frac{1}{\tan\theta}$$
$$\tan\theta = \frac{\sin\theta}{\cos\theta}, \quad \cot\theta = \frac{\cos\theta}{\sin\theta}. \tag{1}$$

The first three of these are sometimes called the *reciprocal identities*.

Consider any point $P(\theta) = (x, y)$ on the unit circle in the following figure. Then by definition we have

$$x = \cos\theta \quad \text{and} \quad y = \sin\theta.$$

The equation of the unit circle is

$$x^2 + y^2 = 1.$$

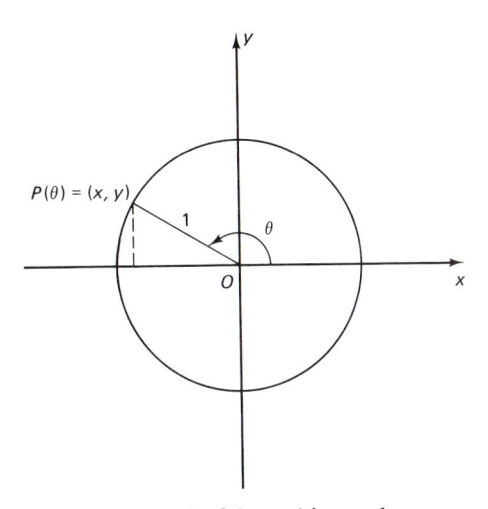

Substituting $x = \cos\theta$ and $y = \sin\theta$ into this, we have

$$(\cos\theta)^2 + (\sin\theta)^2 = 1$$

—that is,

$$\boxed{\cos^2\theta + \sin^2\theta = 1.} \qquad (2)$$

This equation is satisfied by the sine and cosine of any angle θ.

If we divide both sides of (2) by $\cos^2\theta$ and make use of the identities (1), we have

$$\frac{\cos^2\theta}{\cos^2\theta} + \frac{\sin^2\theta}{\cos^2\theta} = \frac{1}{\cos^2\theta}$$

$$1 + \left(\frac{\sin\theta}{\cos\theta}\right)^2 = \left(\frac{1}{\cos\theta}\right)^2$$

$$1 + (\tan\theta)^2 = (\sec\theta)^2$$

$$\boxed{1 + \tan^2\theta = \sec^2\theta.} \qquad (3)$$

This identity again is satisfied by the tangent and secant of any angle θ for which these functions are defined.

If we divide both sides of (2) by $\sin^2\theta$ and make use of (1), we obtain

$$\frac{\cos^2\theta}{\sin^2\theta} + \frac{\sin^2\theta}{\sin^2\theta} = \frac{1}{\sin^2\theta}$$

$$\left(\frac{\cos\theta}{\sin\theta}\right)^2 + 1 = \left(\frac{1}{\sin\theta}\right)^2$$

$$\boxed{\cot^2\theta + 1 = \csc^2\theta.} \qquad (4)$$

Sec. 13.1 Fundamental Identities

Each of the identities (2), (3), (4) can be written in three different forms as follows:

$$\cos^2 \theta + \sin^2 \theta = 1$$
or $\quad\cos^2 \theta = 1 - \sin^2 \theta$ (2′)
or $\quad\sin^2 \theta = 1 - \cos^2 \theta.$

$$\sec^2 \theta - \tan^2 \theta = 1$$
or $\quad\sec^2 \theta = 1 + \tan^2 \theta$ (3′)
or $\quad\tan^2 \theta = \sec^2 \theta - 1.$

$$\csc^2 \theta - \cot^2 \theta = 1$$
or $\quad\csc^2 \theta = 1 + \cot^2 \theta$ (4′)
or $\quad\cot^2 \theta = \csc^2 \theta - 1.$

The identities (2′)–(4′) are known as the ***trigonometric square identities***. You should memorize them in their three different forms. These identities are true for all values of θ for which both sides are defined. For example,

$$\sin^2 70° + \cos^2 70° = 1$$
$$\sec^2 20\pi - \tan^2 20\pi = 1$$
$$\csc^2 3A = 1 + \cot^2 3A$$

and so on.

From the fundamental identities (1)–(4) we can derive many other identities, as illustrated by the following examples.

Example 1 Prove that

$$\cos^4 \theta - \sin^4 \theta = \cos^2 \theta - \sin^2 \theta.$$

Solution Factoring the left side by the difference-of-squares formula, we have

$$\cos^4 \theta - \sin^4 \theta = (\cos^2 \theta)^2 - (\sin^2 \theta)^2$$
$$= (\cos^2 \theta + \sin^2 \theta)(\cos^2 \theta - \sin^2 \theta)$$
$$= (1)(\cos^2 \theta - \sin^2 \theta)$$
$$= \cos^2 \theta - \sin^2 \theta$$

and we have proved the required identity.

Example 2 Prove that

$$3 \sec^2 \theta - \sec^4 \theta + \tan^4 \theta = 2 + \tan^2 \theta.$$

Solution Since the right side involves only one trigonometric function, $\tan \theta$, a useful approach will be to express the left side also in terms of $\tan \theta$. After simplification it should become identical with the right side. Since $\sec^2 \theta = 1 + \tan^2 \theta$, the left side can be written:

$$3 \sec^2 \theta - \sec^4 \theta + \tan^4 \theta$$
$$= 3(1 + \tan^2 \theta) - (1 + \tan^2 \theta)^2 + \tan^4 \theta$$
$$= 3 + 3 \tan^2 \theta - (1 + 2 \tan^2 \theta + \tan^4 \theta) + \tan^4 \theta$$
$$= 2 + \tan^2 \theta$$

as required.

Example 3 Prove that

$$(\csc A - \sin A)(\sec A - \cos A) = \sin A \cos A.$$

Solution Here we start with the left side, which is the more complicated of the two, and express it in terms of sines and cosines. We have

$$(\csc A - \sin A)(\sec A - \cos A) = \left(\frac{1}{\sin A} - \sin A\right)\left(\frac{1}{\cos A} - \cos A\right)$$
$$= \left(\frac{1 - \sin^2 A}{\sin A}\right)\left(\frac{1 - \cos^2 A}{\cos A}\right)$$
$$= \left(\frac{\cos^2 A}{\sin A}\right)\left(\frac{\sin^2 A}{\cos A}\right)$$
$$= \cos A \sin A$$

where in the third line we have used the identities $1 - \cos^2 A = \sin^2 A$ and $1 - \sin^2 A = \cos^2 A$ given in (2').

Example 4 Prove that

$$\frac{1 + \cos \theta}{1 - \cos \theta} - \frac{1 - \cos \theta}{1 + \cos \theta} = 4 \cot \theta \csc \theta.$$

Solution Again we begin with the left side, which is the more complicated:

$$\frac{1 + \cos \theta}{1 - \cos \theta} - \frac{1 - \cos \theta}{1 + \cos \theta} = \frac{(1 + \cos \theta)(1 + \cos \theta)}{(1 - \cos \theta)(1 + \cos \theta)} - \frac{(1 - \cos \theta)(1 - \cos \theta)}{(1 + \cos \theta)(1 - \cos \theta)}$$
$$= \frac{(1 + \cos \theta)^2 - (1 - \cos \theta)^2}{1 - \cos^2 \theta}$$
$$= \frac{(1 + 2 \cos \theta + \cos^2 \theta) - (1 - 2 \cos \theta + \cos^2 \theta)}{\sin^2 \theta}$$

(since $1 - \cos^2 \theta = \sin^2 \theta$)

$$= \frac{4 \cos \theta}{\sin^2 \theta}$$
$$= 4 \frac{\cos \theta}{\sin \theta} \cdot \frac{1}{\sin \theta}$$
$$= 4 \cot \theta \csc \theta$$

as required.

Example 5 Prove that

$$\sqrt{\frac{1+\sin\theta}{1-\sin\theta}} = \frac{1+\sin\theta}{|\cos\theta|} = \frac{|\cos\theta|}{1-\sin\theta}.$$

Solution To prove that the first and second members are equal, we rationalize the first member by multiplying the numerator and denominator by $(1+\sin\theta)$ inside the radical sign. Thus

$$\sqrt{\frac{1+\sin\theta}{1-\sin\theta}} = \sqrt{\frac{(1+\sin\theta)(1+\sin\theta)}{(1-\sin\theta)(1+\sin\theta)}}$$

$$= \sqrt{\frac{(1+\sin\theta)^2}{1-\sin^2\theta}}$$

$$= \sqrt{\frac{(1+\sin\theta)^2}{\cos^2\theta}}$$

$$= \frac{|1+\sin\theta|}{|\cos\theta|}$$

where we have used the property that $\sqrt{a^2} = |a|$ for any real number a. Since $-1 \leq \sin\theta \leq 1$, $1+\sin\theta$ is always nonnegative and so $|1+\sin\theta| = 1+\sin\theta$. Thus,

$$\sqrt{\frac{1+\sin\theta}{1-\sin\theta}} = \frac{1+\sin\theta}{|\cos\theta|}. \tag{i}$$

To prove that the first and third members are equal, we multiply the numerator and denominator by $(1-\sin\theta)$ inside the radical sign on the left and proceed similarly.

While it is not possible to give hard and fast rules for proving identities such as those in the above examples, you may find the following guidelines helpful in providing a somewhat systematic approach to such problems.

1. If one side of the identity involves only one trigonometric function, then express the other side in terms of this function and simplify. (See Example 2.)
2. If one side consists of the sum or difference of two terms and the other side consists of a product of terms, then simplifying the side with the sum or difference will generally be the best approach. (See Example 4.)
3. Often we must express a complicated side of an equation in terms of sines and cosines alone by using identities (1) before we can apply algebraic techniques to simplify it. (See Example 3.)
4. Sometimes a look at the other side of the equation suggests what should be done to the terms on the side with which you are working. (Example 5, for instance.)
5. In some cases you may have to simplify both sides of the identity to a stage where they can be shown to be identical.

The square identities (2)–(4) can also often be used to eliminate θ from two equations involving simple trigonometric functions of θ. This is illustrated in the following solved examples.

Example 6 Eliminate θ from the following equations:
$$x = a + b \sin \theta, \qquad y = c + d \cos \theta.$$

Solution Solving the given equations for $\sin \theta$ and $\cos \theta$, we have
$$\sin \theta = \frac{x-a}{b} \quad \text{and} \quad \cos \theta = \frac{y-c}{d}. \tag{i}$$

But we know that
$$\sin^2 \theta + \cos^2 \theta = 1. \tag{ii}$$

Substituting the values of $\sin \theta$ and $\cos \theta$ from (i) in (ii), we have
$$\left(\frac{x-a}{b}\right)^2 + \left(\frac{y-c}{d}\right)^2 = 1.$$

Thus, we have eliminated θ from the given equations. (The final equation here represents an ellipse—see Section 7.4.)

Example 7 Eliminate θ from the following equations:
$$x = a \sec^3 \theta, \qquad y = b \cot^3 \theta.$$

Solution We have
$$\sec^3 \theta = \frac{x}{a}$$
$$\sec \theta = \left(\frac{x}{a}\right)^{1/3}.$$

From the second of the given equations we have
$$\cot^3 \theta = \frac{y}{b}$$
$$\cot \theta = \left(\frac{y}{b}\right)^{1/3}.$$

This gives
$$\tan \theta = \frac{1}{\cot \theta} = \frac{1}{\left(\frac{y}{b}\right)^{1/3}} = \left(\frac{b}{y}\right)^{1/3}.$$

Now, we know that
$$\sec^2 \theta = 1 + \tan^2 \theta.$$

Thus,
$$\left[\left(\frac{x}{a}\right)^{1/3}\right]^2 = 1 + \left[\left(\frac{b}{y}\right)^{1/3}\right]^2$$
$$\left(\frac{x}{a}\right)^{2/3} = 1 + \left(\frac{b}{y}\right)^{2/3}$$

which is the required result.

TRIGONOMETRIC FUNCTIONS OF $(\pi/2 - \theta)$

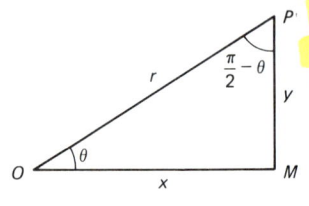

Other useful identities relate the trigonometric functions of $(\pi/2 - \theta)$ to those of θ. Let θ be a positive acute angle. Then from the right triangle OMP we have

$$\sin\left(\frac{\pi}{2} - \theta\right) = \frac{\text{opp.}}{\text{hyp.}} = \frac{x}{r} = \cos\theta.$$

$$\cos\left(\frac{\pi}{2} - \theta\right) = \frac{\text{adj.}}{\text{hyp.}} = \frac{y}{r} = \sin\theta.$$

Therefore,

$$\tan\left(\frac{\pi}{2} - \theta\right) = \frac{\sin\left(\frac{\pi}{2} - \theta\right)}{\cos\left(\frac{\pi}{2} - \theta\right)} = \frac{\cos\theta}{\sin\theta} = \cot\theta.$$

Taking the reciprocals of these functions, we obtain

$$\csc\left(\frac{\pi}{2} - \theta\right) = \frac{1}{\sin\left(\frac{\pi}{2} - \theta\right)} = \frac{1}{\cos\theta} = \sec\theta$$

and so on. Thus, we have

$$\sin\left(\frac{\pi}{2} - \theta\right) = \cos\theta \qquad \csc\left(\frac{\pi}{2} - \theta\right) = \sec\theta$$
$$\cos\left(\frac{\pi}{2} - \theta\right) = \sin\theta \quad \text{and} \quad \sec\left(\frac{\pi}{2} - \theta\right) = \csc\theta$$
$$\tan\left(\frac{\pi}{2} - \theta\right) = \cot\theta \qquad \cot\left(\frac{\pi}{2} - \theta\right) = \tan\theta.$$

Even though we have proved these formulas when θ is an acute angle, they are true for all values of θ.

As an example, we know that $\cos 60° = \frac{1}{2}$. We can also write $\cos 60° = \cos(90° - 30°) = \sin 30°$, which is also $\frac{1}{2}$. Similarly, $\tan 30° = 1/\sqrt{3}$. Also $\tan 30° = \tan(90° - 60°) = \cot 60° = 1/\sqrt{3}$. As a third example we have that $150° = 90° - (-60°)$, and so $\cos 150° = \sin(-60°)$ (both $= -\sqrt{3}/2$) and $\csc(150°) = \sec(-60°)$ (both $= 2$).

Diagnostic Test 13.1

Fill in the blanks.

1. $\dfrac{\sin\theta}{\cos\theta} = $ _TAN θ_ .

2. $\dfrac{1}{\csc\theta} = $ _____ .

3. $1 - \sin^2\theta = $ _cos²θ_ .

4. $1 - \sec^2\theta = $ _____ .

5. $\cos^2 \alpha - 1 = $ _____.
6. $\cot^2 \theta = $ _____ $+ \csc^2 \theta$.
7. $\sin^2 3A + \cos^2 3A = $ _____.
8. $\tan^2 3\theta - \sec^2 3\theta = $ _____.
9. $\dfrac{1}{\sec \theta} = $ _____.
10. $\cos^2 \theta = $ _____ $\sin^2 \theta$.
11. $\sin^2 200° + \cos^2 200° = $ _____.
12. $\tan^2 275° - \sec^2 275° = $ _____.
13. $\cot^2 (-70°) - \csc^2 (-70°) = $ _____.
14. $\tan \theta = \sec \theta$ _____.
15. $\cos \theta$ _____ $= \sin \theta$.
16. $\dfrac{\sec \theta}{\csc \theta} = $ _____.
17. $\sin \theta \sec \theta = $ _____.
18. $\tan \theta \cot \theta = $ _____.
19. $\cos \theta \sec \theta = $ _____.
20. $\csc \theta \sin \theta = $ _____.

Exercises 13.1

(1–62) Prove the following identities.

1. $\sin \theta \sec \theta = \tan \theta$.
2. $\csc \theta \tan \theta = \sec \theta$.
2. $\cos \theta \tan \theta = \sin \theta$.
4. $\cot \theta \sec \theta = \csc \theta$.
5. $\sec^2 \theta (1 - \sin^2 \theta) = 1$.
6. $\csc^2 A - 1 = \csc^2 A \cos^2 A$.
7. $\cos^2 \alpha (\sec^2 \alpha - 1) = \sin^2 \alpha$.
8. $\sin^2 \theta (\csc^2 \theta - 1) = \cos^2 \theta$.
9. $\sin \theta (\csc \theta - \sin \theta) = \cos^2 \theta$.
10. $\cos \alpha (\sec \alpha - \cos \alpha) = \sin^2 \alpha$.
11. $1 - 2 \cos^2 \theta = 2 \sin^2 \theta - 1$.
12. $1 + 2 \tan^2 \theta = 2 \sec^2 \theta - 1$.
13. $\cos^2 \theta - \cos^4 \theta = \sin^2 \theta - \sin^4 \theta$.
14. $\sec^4 A - \sec^2 A = \tan^2 A + \tan^4 A$.
15. $\sec x - \cos x = \tan x \sin x$.
16. $\csc \alpha - \sin \alpha = \cot \alpha \cos \alpha$.
17. $(1 - \sin \theta)(1 + \csc \theta) = \cot \theta \cos \theta$.
18. $(1 - \cos \theta)(1 + \sec \theta) = \sin \theta \tan \theta$.
19. $\sin^4 \theta + \cos^4 \theta = 1 - 2 \sin^2 \theta \cos^2 \theta$.
20. $\sec^4 \theta + \tan^4 \theta = 1 + 2 \sec^2 \theta \tan^2 \theta$.
21. $\sin \theta + \cos \theta \cot \theta = \csc \theta$.
22. $\cos \alpha + \sin \alpha \tan \alpha = \sec \alpha$.
23. $\tan \theta + \cot \theta = \sec \theta \csc \theta$.
24. $\sec^2 A + \csc^2 A = \sec^2 A \csc^2 A$.
25. $(1 + \cos \theta)(\csc \theta - \cot \theta) = \sin \theta$.
26. $(1 - \sin \theta)(\sec \theta + \tan \theta) = \cos \theta$.
27. $\cot A (1 - \tan A) = \cot A - 1$.
28. $\sec \theta (\cos \theta - 1) = 1 - \sec \theta$.
29. $\sin \theta (\tan \theta + \cot \theta) = \sec \theta$.
30. $\cos \alpha (\sin \alpha \tan \alpha + \cos \alpha) = 1$.
31. $\dfrac{1 + \tan^2 \theta}{\csc^2 \theta} = \tan^2 \theta$.
32. $\dfrac{\sec^2 \theta}{1 + \cot^2 \theta} = \tan^2 \theta$.
33. $\dfrac{\cos^2 \alpha}{1 + \sin \alpha} = 1 - \sin \alpha$.
34. $\dfrac{1}{\sec \theta - \tan \theta} = \sec \theta + \tan \theta$.
35. $\dfrac{1 + \sec \theta}{\tan \theta + \sin \theta} = \csc \theta$.
36. $\dfrac{\sin \alpha}{\csc \alpha - \cot \alpha} = 1 + \cos \alpha$.
37. $\dfrac{1 - \sin^2 \theta \sec^2 \theta}{1 - \cos^2 \theta \csc^2 \theta} = -\tan^2 \theta$.
38. $\dfrac{\cot A + \csc A}{\tan A + \sin A} = \cot A \csc A$.
39. $\dfrac{1 - \sin \theta}{1 + \sin \theta} = \dfrac{\csc \theta - 1}{\csc \theta + 1}$.
40. $\dfrac{1 + \cot \theta}{1 - \cot \theta} = \dfrac{\tan \theta + 1}{\tan \theta - 1}$.

41. $\dfrac{1 + \csc \theta}{\cot \theta} = \dfrac{\cot \theta}{\csc \theta - 1}$.

42. $\dfrac{1 + \cos \theta}{\cos \theta} = \dfrac{\tan^2 \theta}{\sec \theta - 1}$.

43. $\dfrac{1 - \sec \theta}{\cos \theta - 1} = \sec \theta$.

44. $\dfrac{\tan \theta - 1}{1 - \cot \theta} = \tan \theta$.

45. $\dfrac{\sec \alpha + \csc \alpha}{\sec \alpha - \csc \alpha} = \dfrac{\sin \alpha + \cos \alpha}{\sin \alpha - \cos \alpha}$.

46. $\dfrac{\tan \alpha - \cot \alpha}{\cos \alpha + \sin \alpha} = \sec \alpha - \csc \alpha$.

47. $\dfrac{\sin \theta + \cos \theta}{1 - \tan^2 \theta} = \dfrac{\cos^2 \theta}{\cos \theta - \sin \theta}$.

48. $\dfrac{\cot \theta - \tan \theta}{\sin \theta \cos \theta} = \csc^2 \theta - \sec^2 \theta$.

49. $\dfrac{1}{1 + \sin \theta} + \dfrac{1}{1 - \sin \theta} = 2 \sec^2 \theta$.

50. $\dfrac{1 + \sin \theta}{1 - \sin \theta} - \dfrac{1 - \sin \theta}{1 + \sin \theta} = 4 \tan \theta \sec \theta$.

51. $\dfrac{\sin \theta}{1 + \cos \theta} + \dfrac{1 + \cos \theta}{\sin \theta} = 2 \csc \theta$.

52. $\dfrac{1}{1 - \cos \theta} + \dfrac{1}{1 + \cos \theta} = 2 \csc^2 \theta$.

53. $\dfrac{\tan \theta}{1 - \cot \theta} + \dfrac{\cot \theta}{1 - \tan \theta} = 1 + \sec \theta \csc \theta$.

54. $\dfrac{\sin \alpha}{1 - \cot \alpha} + \dfrac{\cos \alpha}{1 - \tan \alpha} = \cos \alpha + \sin \alpha$.

55. $\dfrac{\sec \theta}{1 + \sec \theta} - \dfrac{\sec \theta}{1 - \sec \theta} = 2 \csc^2 \theta$.

56. $\dfrac{1}{\tan^2 \theta + 1} - \dfrac{1}{\cot^2 \theta + 1} = 2 \cos^2 \theta - 1$.

57. $\sqrt{\dfrac{1 - \sin \theta}{1 + \sin \theta}} = \dfrac{1 - \sin \theta}{|\cos \theta|} = \dfrac{|\cos \theta|}{1 + \sin \theta}$.

58. $\sqrt{\dfrac{\sec \theta + 1}{\sec \theta - 1}} = \dfrac{1 + \cos \theta}{|\sin \theta|} = \dfrac{|\sin \theta|}{1 - \cos \theta}$.

59. $\sin^6 \theta + \cos^6 \theta = 1 - 3 \sin^2 \theta \cos^2 \theta$.

60. $\sec^6 \theta - \tan^6 \theta = 1 + 3 \sec^2 \theta \tan^2 \theta$.

61. $\log |\sec \theta - \tan \theta| = -\log |\sec \theta + \tan \theta|$.

62. $\log |\csc \theta + \cot \theta| = -\log |\cot \theta - \csc \theta|$.

(63–72) Eliminate θ from the following.

63. $x = a \cos \theta$, $y = a \sin \theta$.

64. $x = a \tan \theta$, $y = a \sec \theta$.

65. $x = a \sec \theta$, $y = a \sin \theta$.

66. $a = \tan \theta$, $b = \cos \theta$.

67. $x = a \sin^3 \theta$, $y = b \cos^3 \theta$.

68. $x = a \sin^4 \theta$, $y = b \cos^4 \theta$.

69. $\sec \theta + \tan \theta = a$, $\sec \theta - \tan \theta = b$.

70. $\sin \theta + \cos \theta = p$, $\cos \theta - \sin \theta = q$.

71. $a \cos \theta + b \sin \theta = p$, $a \sin \theta - b \cos \theta = q$.

[*Hint:* Square and add.]

72. $x = a \sec \theta + b \tan \theta$, $y = a \tan \theta + b \sec \theta$.

[*Hint:* Square and subtract.]

13.2

The Addition Formulas

In this section we derive formulas for the trigonometric functions of the sum or difference of two angles. Such formulas are useful in many applications, especially in the study of calculus. They are known as the *addition formulas*.

We first derive the formula for the cosine of the difference between two angles. Let α and β be two positive angles with $\alpha > \beta$. Let $P_1(\alpha) = (x_1, y_1)$, $P_2(\beta) = (x_2, y_2)$, and $P_3(\alpha - \beta) = (x_3, y_3)$ be the points on the unit circle corresponding to angles α, β, and $\alpha - \beta$, respectively. Then (see the above figures)

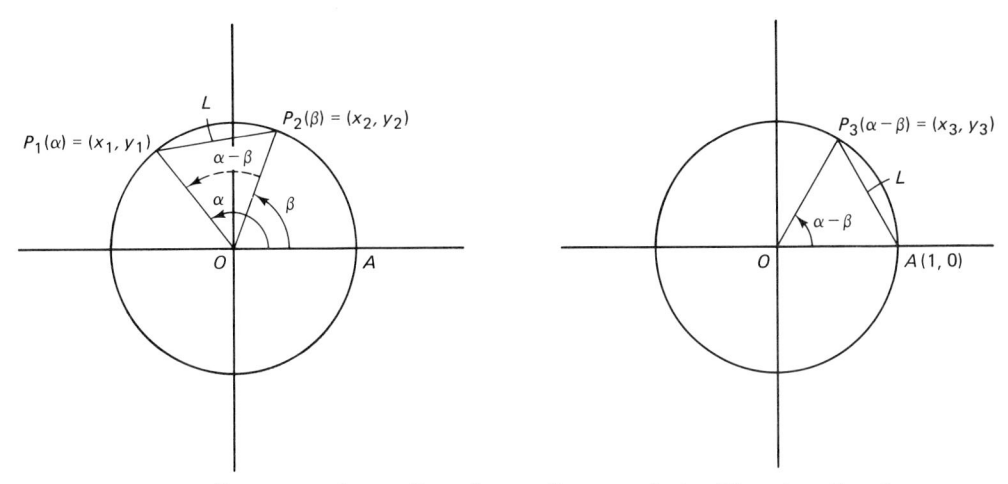

$$\left.\begin{array}{ll} x_1 = \cos \alpha, & y_1 = \sin \alpha \\ x_2 = \cos \beta, & y_2 = \sin \beta \\ x_3 = \cos (\alpha - \beta), & y_3 = \sin (\alpha - \beta). \end{array}\right\} \quad (1)$$

The angle between OA and OP_3 in the second figure is the same as the angle between the radii OP_1 and OP_2 in the first figure (both equal $\alpha - \beta$). Equal angles at the center of a circle subtend chords of equal length, and so the chord AP_3 is equal in length to the chord P_1P_2. We denote each of these lengths by L. Using the distance formula, we have from the first figure above

$$L^2 = P_1P_2^2 = (x_2 - x_1)^2 + (y_2 - y_1)^2$$
$$= (\cos \beta - \cos \alpha)^2 + (\sin \beta - \sin \alpha)^2 \quad \text{[using (1)]}$$
$$= \cos^2 \beta + \cos^2 \alpha - 2 \cos \alpha \cos \beta + \sin^2 \beta - 2 \sin \beta \sin \alpha + \sin^2 \alpha$$
$$= \underbrace{(\cos^2 \beta + \sin^2 \beta)}_{=1} + \underbrace{(\cos^2 \alpha + \sin^2 \alpha)}_{=1} - 2(\cos \alpha \cos \beta + \sin \alpha \sin \beta)$$
$$= 2 - 2 (\cos \alpha \cos \beta + \sin \alpha \sin \beta). \quad (2)$$

Again from the second figure, we have from the distance formula

$$L^2 = AP_3^2 = (x_3 - 1)^2 + (y_3 - 0)^2$$
$$= [\cos (\alpha - \beta) - 1]^2 + [\sin (\alpha - \beta) - 0]^2 \quad \text{[using (1)]}$$
$$= 1 + \cos^2 (\alpha - \beta) - 2 \cos (\alpha - \beta) + \sin^2 (\alpha - \beta)$$
$$= 1 + \underbrace{\cos^2 (\alpha - \beta) + \sin^2 (\alpha - \beta)}_{=1} - 2 \cos (\alpha - \beta)$$
$$= 2 - 2 \cos (\alpha - \beta). \quad (3)$$

Equating the two values of L^2 from (2) and (3), we see that

$$\boxed{\cos (\alpha - \beta) = \cos \alpha \cos \beta + \sin \alpha \sin \beta.} \quad (4)$$

This is the required formula expressing the cosine of the difference between two angles in terms of the sines and cosines of the two angles themselves. In proving this formula, we assumed that α and β are two positive angles with $\alpha > \beta$, but in fact this formula is true for any two angles α and β.

Changing β to $-\beta$ in (4) and noting that $\cos(-\beta) = \cos\beta$ and $\sin(-\beta) = -\sin(\beta)$, we obtain

$$\cos[\alpha - (-\beta)] = \cos\alpha\cos(-\beta) + \sin\alpha\sin(-\beta)$$
$$= \cos\alpha\cos\beta + \sin\alpha(-\sin\beta).$$

Consequently,

$$\boxed{\cos(\alpha + \beta) = \cos\alpha\cos\beta - \sin\alpha\sin\beta.} \qquad (5)$$

This formula gives an expression for the cosine of the sum of two angles.

To derive the formulas for $\sin(\alpha + \beta)$ and $\sin(\alpha - \beta)$, we make use of the fact that $\sin\theta = \cos(\pi/2 - \theta)$ (see Section 13.1). We have

$$\sin(\alpha + \beta) = \cos\left[\frac{\pi}{2} - (\alpha + \beta)\right] = \cos\left[\left(\frac{\pi}{2} - \alpha\right) - \beta\right]$$
$$= \cos\left(\frac{\pi}{2} - \alpha\right)\cos\beta + \sin\left(\frac{\pi}{2} - \alpha\right)\sin\beta \quad \text{[using (4)]}$$
$$= \sin\alpha\cos\beta + \cos\alpha\sin\beta$$

where in the last step we used that $\sin(\pi/2 - \alpha) = \cos\alpha$ and $\cos(\pi/2 - \alpha) = \sin\alpha$. Thus, we have

$$\boxed{\sin(\alpha + \beta) = \sin\alpha\cos\beta + \cos\alpha\sin\beta.} \qquad (6)$$

Replacing β by $-\beta$ in this and making use of the fact that $\cos(-\beta) = \cos\beta$ and $\sin(-\beta) = -\sin\beta$, we obtain that

$$\boxed{\sin(\alpha - \beta) = \sin\alpha\cos\beta - \cos\alpha\sin\beta.} \qquad (7)$$

To derive formulas for $\tan(\alpha + \beta)$ and $\tan(\alpha - \beta)$, we make use of the fact that $\tan\theta = \sin\theta/\cos\theta$. Thus,

$$\tan(\alpha + \beta) = \frac{\sin(\alpha + \beta)}{\cos(\alpha + \beta)} = \frac{\sin\alpha\cos\beta + \cos\alpha\sin\beta}{\cos\alpha\cos\beta - \sin\alpha\sin\beta}.$$

Now divide the numerator and denominator by $\cos\alpha\cos\beta$:

$$\tan(\alpha + \beta) = \frac{\dfrac{\sin\alpha\cos\beta}{\cos\alpha\cos\beta} + \dfrac{\cos\alpha\sin\beta}{\cos\alpha\cos\beta}}{\dfrac{\cos\alpha\cos\beta}{\cos\alpha\cos\beta} - \dfrac{\sin\alpha\sin\beta}{\cos\alpha\cos\beta}}$$

$$= \frac{\dfrac{\sin\alpha}{\cos\alpha} + \dfrac{\sin\beta}{\cos\beta}}{1 - \dfrac{\sin\alpha\sin\beta}{\cos\alpha\cos\beta}}$$

—that is,

$$\tan(\alpha+\beta) = \frac{\tan\alpha + \tan\beta}{1 - \tan\alpha\tan\beta} \tag{8}$$

Changing β to $-\beta$ and noting that $\tan(-\beta) = -\tan\beta$, we obtain

$$\tan(\alpha-\beta) = \frac{\tan\alpha - \tan\beta}{1 + \tan\alpha\tan\beta} \tag{9}$$

All the addition formulas are summarized together with the other most important formulas at the end of the chapter (p. 549).

Example 1 Use the addition formulas to evaluate the following:
(a) $\sin 105°$.
(b) $\tan 75°$.

Solution (a) We can write $105° = 60° + 45°$. Thus, using (6) with $\alpha = 60°$ and $\beta = 45°$, we have

$$\sin 105° = \sin(60° + 45°)$$
$$= \sin 60° \cos 45° + \cos 60° \sin 45°$$
$$= \frac{\sqrt{3}}{2} \cdot \frac{1}{\sqrt{2}} + \frac{1}{2} \cdot \frac{1}{\sqrt{2}} = \frac{\sqrt{3}+1}{2\sqrt{2}}.$$

(b) Using formula (8) for $\tan(\alpha + \beta)$ with $\alpha = 45°$ and $\beta = 30°$, we have

$$\tan 75° = \tan(45° + 30°)$$
$$= \frac{\tan 45° + \tan 30°}{1 - \tan 45° \tan 30°}$$
$$= \frac{1 + 1/\sqrt{3}}{1 - 1 \cdot 1/\sqrt{3}} = \frac{\sqrt{3}+1}{\sqrt{3}-1}$$

after multiplying numerator and denominator by $\sqrt{3}$. Rationalizing the denominator, we obtain

$$\tan 75° = \frac{(\sqrt{3}+1)(\sqrt{3}+1)}{(\sqrt{3}-1)(\sqrt{3}+1)} = \frac{3+1+2\sqrt{3}}{3-1}$$
$$= \frac{4+2\sqrt{3}}{2} = 2 + \sqrt{3}.$$

Example 2 Evaluate the following without using tables or calculators.

$$\sin\frac{2\pi}{9}\sin\frac{\pi}{18} + \cos\frac{2\pi}{9}\cos\frac{\pi}{18}.$$

Solution If we let $\alpha = 2\pi/9$, $\beta = \pi/18$, then we have

$$\sin\frac{2\pi}{9}\sin\frac{\pi}{18} + \cos\frac{2\pi}{9}\cos\frac{\pi}{18} = \sin\alpha\sin\beta + \cos\alpha\cos\beta$$
$$= \cos(\alpha - \beta)$$
$$= \cos\left(\frac{2\pi}{9} - \frac{\pi}{18}\right) = \cos\frac{\pi}{6} = \frac{\sqrt{3}}{2}.$$

Example 3 Given that $\cos \alpha = \frac{1}{3}$ and $0 < \alpha < \pi/2$ and $\sin \beta = -\frac{3}{5}$, where β is in the third quadrant, evaluate $\sin(\alpha + \beta)$ and $\cos(\alpha + \beta)$. Find the quadrant in which the angle $(\alpha + \beta)$ lies.

Solution Since $\cos \alpha = \frac{1}{3}$ and α lies in the first quadrant, we have (since $\sin^2 \alpha = 1 - \cos^2 \alpha$)

$$\sin \alpha = +\sqrt{1 - \cos^2 \alpha} = \sqrt{1 - \frac{1}{9}} = \sqrt{\frac{8}{9}} = \frac{2\sqrt{2}}{3}.$$

Again, because $\sin \beta = -\frac{3}{5}$ and β lies in the third quadrant (where $\cos \beta$ is negative), we have

$$\cos \beta = -\sqrt{1 - \sin^2 \beta} \quad (\text{since } \cos^2 \beta = 1 - \sin^2 \beta)$$
$$= -\sqrt{1 - \frac{9}{25}} = -\sqrt{\frac{16}{25}} = -\sqrt{\frac{4}{5}}.$$

Thus, we have

$$\sin \alpha = \frac{2\sqrt{2}}{3}, \quad \cos \alpha = \frac{1}{3}, \quad \sin \beta = -\frac{3}{5}, \quad \cos \beta = -\frac{4}{5}.$$

Now, from the addition formulas,

$$\sin(\alpha + \beta) = \sin \alpha \cos \beta + \cos \alpha \sin \beta$$
$$= \frac{2\sqrt{2}}{3} \cdot \left(-\frac{4}{5}\right) + \frac{1}{3}\left(-\frac{3}{5}\right)$$
$$= \frac{-3 - 8\sqrt{2}}{15}$$
$$= \frac{-3 + 8\sqrt{2}}{15}$$

and

$$\cos(\alpha + \beta) = \cos \alpha \cos \beta - \sin \alpha \sin \beta$$
$$= \frac{1}{3}\left(-\frac{4}{5}\right) - \left(\frac{2\sqrt{2}}{3}\right)\left(-\frac{3}{5}\right)$$
$$= \frac{6\sqrt{2} - 4}{15}.$$

Note that $\cos(\alpha + \beta) > 0$, because $6\sqrt{2} - 4 > 0$. Since $\sin(\alpha + \beta)$ is negative and $\cos(\alpha + \beta)$ is positive, $\alpha + \beta$ must be in the fourth quadrant.

FORMULAS FOR ALLIED ANGLES

The addition formulas provide many useful identities relating the trigonometric functions of angles such as $(\pi \pm \theta)$, $(\pi/2 \pm \theta)$ or $(3\pi/2 \pm \theta)$ to trigonometric functions of θ. For example, let us take $\alpha = \pi$ and $\beta = \theta$ in formula (7):

$$\sin(\alpha - \beta) = \sin\alpha\cos\beta - \cos\alpha\sin\beta$$
$$\sin(\pi - \theta) = \sin\pi\cos\theta - \cos\pi\sin\theta$$
$$= (0)\cos\theta - (-1)\sin\theta \qquad (\sin\pi = 0, \cos\pi = -1)$$
$$= \sin\theta.$$

Similarly, from (4) we obtain
$$\cos(\alpha - \beta) = \cos\alpha\cos\beta + \sin\alpha\sin\beta$$
$$\cos(\pi - \theta) = \cos\pi\cos\theta + \sin\pi\sin\theta$$
$$= (-1)\cos\theta + (0)\sin\theta$$
$$= -\cos\theta.$$

From (9), since $\tan\pi = 0$, we have
$$\tan(\pi - \theta) = \frac{\tan\pi - \tan\theta}{1 + \tan\pi\tan\theta} = -\tan\theta.$$

These three results together with their reciprocals are summarized as follows.

$$\begin{array}{l} \sin(\pi - \theta) = \sin\theta, \ \cos(\pi - \theta) = -\cos\theta, \ \tan(\pi - \theta) = -\tan\theta \\ \csc(\pi - \theta) = \csc\theta, \ \sec(\pi - \theta) = -\sec\theta, \ \cot(\pi - \theta) = -\cot\theta \end{array} \qquad (10)$$

These identities can be used instead of the reference angle to evaluate trigonometric functions of obtuse angles.

Example 4 Use formulas (10) to evaluate $\sin(2\pi/3)$.

Solution $2\pi/3 = \pi - \pi/3$, so from the first of formulas (10),
$$\sin\frac{2\pi}{3} = \sin\left(\pi - \frac{\pi}{3}\right)$$
$$= \sin\frac{\pi}{3} = \frac{\sqrt{3}}{2}.$$

Example 5 Relate the following to trigonometric functions of θ
(a) $\cos(90° + \theta)$. (b) $\sin(90° + \theta)$. (c) $\tan(90° + \theta)$.

Solution (a) Take $\alpha = 90°$ and $\beta = \theta$ in formula (5):
$$\cos(\alpha + \beta) = \cos\alpha\cos\beta - \sin\alpha\sin\beta$$
$$\cos(90° + \theta) = \cos 90°\cos\theta - \sin 90°\sin\theta$$
$$= (0)\cos\theta - (1)\sin\theta$$
$$= -\sin\theta.$$

(b) Similarly, from formula (6):
$$\sin(90° + \theta) = \sin 90°\cos\theta + \cos 90°\sin\theta$$
$$= (1)\cos\theta + (0)\sin\theta$$
$$= \cos\theta.$$

(c) We cannot evaluate $\tan(90° + \theta)$ by putting $\alpha = 90°$ and $\beta = \theta$ in formula (8) because $\tan \alpha = \tan 90°$ would then be undefined. The simplest way of proceeding is to use the results of parts (a) and (b):

$$\tan(90° + \theta) = \frac{\sin(90° + \theta)}{\cos(90° + \theta)} = \frac{\cos \theta}{-\sin \theta}$$
$$= -\cot \theta.$$

COMBINATIONS OF SINE AND COSINE (OPTIONAL)

Suppose we are interested in sketching the graph of $y = a \cos \theta + b \sin \theta$, where a and b are certain nonzero constants. One way of doing this is to plot directly a set of values of y for different θ and join by a smooth curve. This is not a convenient method, and there is a more efficient alternative. This is to express $a \cos \theta + b \sin \theta$ as the sine or cosine of a single angle and then to construct the graph of this sine or cosine function as we did in Section 12.5.

We can express $a \cos \theta + b \sin \theta$ in terms of the sine or cosine of a single angle as follows. Let us suppose that the constants a and b can be expressed in terms of quantities r and α:

$$a = r \sin \alpha, \qquad b = r \cos \alpha. \tag{11}$$

Then we have

$$a \cos \theta + b \sin \theta = r \cos \theta \sin \alpha + r \sin \theta \cos \alpha$$
$$= r(\sin \theta \cos \alpha + \cos \theta \sin \alpha)$$
$$= r \sin(\theta + \alpha).$$

This has achieved our goal: the only trigonometric function that occurs is a single sine function.

We still need to obtain expressions for r and α. If we square and add the two parts of (11), we obtain

$$a^2 + b^2 = r^2(\sin^2 \alpha + \cos^2 \alpha) = r^2 \cdot 1 = r^2.$$

Thus

$$r = \sqrt{a^2 + b^2}.$$

Also from (11), we have

$$\sin \alpha = \frac{a}{r} = \frac{a}{\sqrt{a^2 + b^2}}, \qquad \cos \alpha = \frac{b}{r} = \frac{b}{\sqrt{a^2 + b^2}}.$$

Summarizing, we have

$$a \cos \theta + b \sin \theta = r \sin(\theta + \alpha)$$

where $r = \sqrt{a^2 + b^2}$, and the angle α is determined from the equations

$$\sin \alpha = \frac{a}{r}, \qquad \cos \alpha = \frac{b}{r}.$$

We can also express $a \cos \theta + b \sin \theta$ as the cosine of a single angle by letting

$$a = r \cos \alpha, \qquad b = r \sin \alpha.$$

We leave it as an exercise to show that
$$a \cos \theta + b \sin \theta = r \cos (\theta - \alpha)$$
where $r = \sqrt{a^2 + b^2}$ and α is given by
$$\cos \alpha = \frac{a}{r}, \quad \sin \alpha = \frac{b}{r}.$$

***Example 6** Express $\sqrt{3} \sin \theta + \cos \theta$ in terms of the sine of a single angle and determine its maximum and minimum values. Hence draw the graph of the function $y = \sqrt{3} \sin \theta + \cos \theta$.

Solution The given function is
$$\sqrt{3} \sin \theta + 1 \cos \theta.$$
Let
$$\sqrt{3} = r \cos \alpha, \; 1 = r \sin \alpha, \tag{i}$$
Then
$$\sqrt{3} \sin \theta + 1 \cos \theta = r \cos \alpha \sin \theta + r \sin \alpha \cos \theta$$
$$= r (\sin \theta \cos \alpha + \cos \theta \sin \alpha)$$
$$= r \sin (\theta + \alpha). \tag{ii}$$

Squaring and adding the two equations in (i), we obtain
$$(\sqrt{3})^2 + 1^2 = r^2(\cos^2 \alpha + \sin^2 \alpha)$$
$$3 + 1 = r^2 \cdot 1$$
$$r = 2.$$

Using this value of $r = 2$ in (i), we obtain
$$\cos \alpha = \frac{\sqrt{3}}{2}, \quad \sin \alpha = \frac{1}{2}.$$

Since $\sin \alpha$ and $\cos \alpha$ are both positive, α must be in the first quadrant. Now $\cos \alpha = \sqrt{3}/2 = \cos \pi/6$ implies $\alpha = \pi/6$. (This value also satisfies $\sin \alpha = \frac{1}{2}$). Substituting these values of $r = 2$ and $\alpha = \pi/6$ in (ii), we obtain
$$\sqrt{3} \sin \theta + \cos \theta = 2 \sin \left(\theta + \frac{\pi}{6}\right).$$

Since the maximum and minimum values of $\sin (\theta + \pi/6)$ are 1 and -1, respectively, the maximum and minimum values of $\sqrt{3} \sin \theta + \cos \theta = 2 \sin (\theta + \pi/6)$ will be 2 and -2, respectively.

The graph of $y = \sqrt{3} \sin \theta + \cos \theta = 2 \sin (\theta + \pi/6)$ is a sine curve moved to the left by $\pi/6$ units and having an amplitude of 2 units. It is shown below.

Diagnostic Test 13.2

Fill in the blanks.

1. $\sin (A + B) = $ _____.
2. $\cos (A - B) = $ _____.
3. $\sin (\theta - \phi) = $ _____.
4. $\cos (\theta + \phi) = $ _____.

5. $\tan(A+B) = $ _____. **6.** $\tan(A-B) = $ _____.

7. $\sin 2A \cos A - \cos 2A \sin A = $ _____.

8. $\cos 3A \cos 2A - \sin 3A \sin 2A = $ _____.

9. $\sin \dfrac{\pi}{9} \cos \dfrac{\pi}{18} + \cos \dfrac{\pi}{9} \sin \dfrac{\pi}{18} = $ _____.

10. $\dfrac{\tan \dfrac{\pi}{10} + \tan \dfrac{\pi}{15}}{1 - \tan \dfrac{\pi}{10} \tan \dfrac{\pi}{15}} = $ _____.

Exercises 13.2

(1–6) Evaluate the following without using tables or calculators.

1. $\sin 75°$. **2.** $\cos 105°$. **3.** $\sin 15°$. **4.** $\cos 75°$. **5.** $\tan 105°$. **6.** $\tan 15°$.

(7–20) Simplify the following. (Do not use tables or calculators.)

7. $\sin 22° \cos 38° + \cos 22° \sin 38°$.

8. $\cos 17° \sin 43° + \sin 17° \cos 43°$.

9. $\sin 70° \cos 40° - \cos 70° \sin 40°$.

10. $\cos \pi/12 \sin \pi/4 - \cos \pi/4 \sin \pi/12$.

11. $\cos \pi/4 \cos \pi/12 - \sin \pi/4 \sin \pi/12$.

12. $\sin \pi/3 \sin \pi/12 + \cos \pi/3 \cos \pi/12$.

13. $\cos 3\theta/2 \cos \theta/2 + \sin 3\theta/2 \sin \theta/2$.

14. $\sin 5\alpha \cos 4\alpha - \cos 5\alpha \sin 4\alpha$.

15. $\sin(A+B)\cos(A-B) - \cos(A+B)\sin(A-B)$.

16. $\sin(A+B)\sin(A-B) + \cos(A-B)\cos(A+B)$.

17. $\dfrac{\tan 38° + \tan 22°}{1 - \tan 38° \tan 22°}$.

18. $\dfrac{\tan \pi/3 - \tan \pi/12}{1 + \tan \pi/3 \tan \pi/12}$.

19. $\dfrac{\tan 70° - \tan 40°}{1 + \tan 70° \tan 40°}$.

20. $\dfrac{\tan \pi/6 + \tan \pi/12}{1 - \tan \pi/6 \tan \pi/12}$.

(21–32) Use addition formulas to prove the following.

21. $\sin(\pi/2 + \theta) = \cos \theta$.

22. $\cos(\pi/2 - \theta) = \sin \theta$.

23. $\sec(\pi/2 + \theta) = -\csc \theta$.

24. $\csc(\pi/2 - \theta) = \sec \theta$.

25. $\sqrt{2}\sin(\theta + \pi/4) = \cos \theta + \sin \theta$.

26. $2\sin(\theta + \pi/6) = \sqrt{3}\sin \theta + \cos \theta$.

27. $2\cos(\theta + \pi/6) = \sqrt{3}\cos \theta - \sin \theta$.

28. $2\sin(\theta + 2\pi/3) = \sqrt{3}\cos \theta - \sin \theta$.

29. $\tan\left(\dfrac{\pi}{4} + \theta\right) = \dfrac{1 + \tan \theta}{1 - \tan \theta}$.

30. $\tan\left(\dfrac{\pi}{4} - \theta\right) = \dfrac{1 - \tan \theta}{1 + \tan \theta}$.

31. $\cot\left(\dfrac{\pi}{4} - \theta\right) = \dfrac{\cot \theta + 1}{\cot \theta - 1}$.

32. $\cot\left(\dfrac{\pi}{4} + \theta\right) = \dfrac{\cot \theta - 1}{\cot \theta + 1}$.

33. Given $\sin \alpha = \tfrac{2}{3}$, $\sin \beta = \tfrac{3}{4}$, where α, β are positive acute angles.
  (a) Find $\sin(\alpha + \beta)$ and $\cos(\alpha + \beta)$. Determine the quadrant in which $(\alpha + \beta)$ lies.
  (b) Find $\sin(\alpha - \beta)$ and $\tan(\alpha - \beta)$. Determine the quadrant in which the angle $(\alpha - \beta)$ lies.

34. Given $\cos \alpha = \tfrac{3}{5}$, $\sin \beta = -\tfrac{5}{13}$ and $0 < \alpha < \pi/2$, $\pi < \beta < 3\pi/2$.
  (a) Find $\cos(\alpha + \beta)$ and $\tan(\alpha + \beta)$. Determine the quadrant in which the angle $(\alpha + \beta)$ lies.
  (b) Find $\sin(\alpha - \beta)$ and $\cos(\alpha - \beta)$. Determine the quadrant in which the angle $(\alpha - \beta)$ lies.

(35–38) Express each of the following functions in terms of the sine of a single angle and hence find its maximum value.

35. $\sqrt{3}\cos\theta + \sin\theta$.
36. $\sqrt{3}\sin\theta - \cos\theta$.
37. $\cos\theta + \sin\theta$.
38. $\cos\theta - \sin\theta$.

(39–42) Express each of the following functions in terms of the cosine of a single angle and hence determine its maximum value.

39. $\sqrt{3}\sin\theta + \cos\theta$.
40. $\sqrt{3}\cos\theta - \sin\theta$.
41. $\sin\theta - \cos\theta$.
42. $3\sin\theta + 4\cos\theta$.

(43–50) Prove the following identities.

43. $(\cos\alpha + \cos\beta)^2 + (\sin\alpha - \sin\beta)^2 = 2 + 2\cos(\alpha + \beta)$.
44. $(\cos\alpha - \cos\beta)^2 + (\sin\alpha - \sin\beta)^2 = 2 - 2\cos(\alpha - \beta)$.
45. $\sin(\alpha + \beta)\sin(\alpha - \beta) = \sin^2\alpha - \sin^2\beta$.
46. $\cos(\alpha + \beta)\cos(\alpha - \beta) = \cos^2\alpha - \sin^2\beta$.
47. $\cos\beta\csc\beta + \cos\alpha\csc\alpha = \dfrac{\sin(\alpha + \beta)}{\sin\alpha\sin\beta}$.
48. $\sin\alpha\sec\alpha - \sin\beta\sec\beta = \dfrac{\sin(\alpha - \beta)}{\cos\alpha\cos\beta}$.
49. $\cot(\alpha + \beta) = \dfrac{\cot\alpha\cot\beta - 1}{\cot\alpha + \cot\beta}$.
50. $\tan(\alpha - \beta) = \dfrac{\cot\beta - \cot\alpha}{1 + \cot\alpha\cot\beta}$.

(51–60) Use the method of Example 5 to find the solutions of the following equations in the interval $0 \leq \theta < 2\pi$.

51. $\sqrt{3}\sin\theta + \cos\theta = 1$.
52. $\sqrt{3}\cos\theta + \sin\theta = 1$.
53. $\sqrt{3}\sin\theta - \cos\theta = \sqrt{3}$.
54. $\sqrt{3}(1 + \cos\theta) = \sin\theta$.
55. $\sin\theta + \cos\theta = 1$.
56. $\cos\theta = 1 + \sin\theta$.
57. $\sqrt{3}\cos\theta + \sin\theta = -1$.
58. $\sqrt{3}\sin\theta + \cos\theta = -1$.
59. $\sqrt{3}(1 - \cos\theta) = \sin\theta$.
60. $\sin\theta = -1 + \cos\theta$.

61. If $f(x) = \cos x$, prove that $\dfrac{f(x+h) - f(x)}{h} = \cos x\left(\dfrac{\cos h - 1}{h}\right) - \sin x\left(\dfrac{\sin h}{h}\right)$.

(62–64) Prove the following formulas.

62. $2\sin\alpha\cos\beta = \sin(\alpha + \beta) + \sin(\alpha - \beta)$.
63. $2\cos\alpha\cos\beta = \cos(\alpha + \beta) + \cos(\alpha - \beta)$.
64. $2\sin\alpha\sin\beta = \cos(\alpha - \beta) - \cos(\alpha + \beta)$.

(Note that the formulas given in Exercises 62–64 express the product of two sines or two cosines or the product of a sine and a cosine in terms of the sum or difference of two sines or cosines.)

(65–74) Use the formulas in Exercises 62–64 to express each of the following products as a sum or difference.

65. $2\sin 2\alpha\cos\alpha$.
66. $2\sin 3\theta\cos\theta$.
67. $2\sin\theta\cos 5\theta$.
68. $2\sin\alpha/2\cos 3\alpha/2$.
69. $2\cos\theta\cos 2\theta$.
70. $\cos 3\theta\cos 4\theta$.
71. $2\sin\theta\sin\theta/2$.
72. $\sin(\alpha + \beta)\sin(\alpha - \beta)$.
73. $\cos(\alpha + \beta)\cos(\alpha - \beta)$.
74. $2\sin(\alpha + \beta)\cos(\alpha - \beta)$.

(75–78) Prove the following formulas.

75. $\sin A + \sin B = 2\sin\dfrac{A+B}{2}\cos\dfrac{A-B}{2}$.
76. $\sin A - \sin B = 2\cos\dfrac{A+B}{2}\sin\dfrac{A-B}{2}$.
77. $\cos A + \cos B = 2\cos\dfrac{A+B}{2}\cos\dfrac{A-B}{2}$.
78. $\cos A - \cos B = 2\sin\dfrac{A+B}{2}\sin\dfrac{B-A}{2}$.

[*Hint:* Express the right sides in Exercises 75–78 in terms of a sum or difference of sines or cosines by using the formulas in Exercises 62–64.]

(The formulas in Exercises 75–78 express the sum or difference of two sines or two cosines as a product.)

*(79–86) Use the formulas in Exercises 75–78 to express the following sums or differences as products.

79. $\sin 3\theta + \sin \theta$.   **80.** $\sin 5\theta + \sin 3\theta$.   **81.** $\sin 5\theta - \sin \theta$.   **82.** $\sin 7\theta - \sin 3\theta$.
83. $\cos 2\theta + \cos \theta$.   **84.** $\cos 4\theta + \cos 2\theta$.   **85.** $\cos 3\theta/2 - \cos \theta/2$.   **86.** $\cos A - \cos 3A$.

(87–90) Use the addition formulas to prove the following identities.

87. $\sin (\pi/2 + \theta) = \cos \theta$, $\cos (\pi/2 + \theta) = -\sin \theta$, $\tan (\pi/2 + \theta) = -\cot \theta$.

88. $\sin (\pi + \theta) = -\sin \theta$, $\cos (\pi + \theta) = -\cos \theta$, $\tan (\pi + \theta) = \tan \theta$.

89. $\sin (270° - \theta) = -\cos \theta$, $\cos (270° - \theta) = -\sin \theta$, $\tan (270° - \theta) = \cot \theta$.

90. $\sin (3\pi/2 + \theta) = -\cos \theta$, $\cos (3\pi/2 + \theta) = \sin \theta$, $\tan (3\pi/2 + \theta) = -\cot \theta$.

(91–94) Using the results of Exercises 87–90, simplify the following.

91. $\dfrac{\cos \theta}{\sin (\pi/2 - \theta)} + \dfrac{\sin (-\theta)}{\sin (\pi + \theta)} - \dfrac{\tan (\pi/2 - \theta)}{\cot \theta}$.

92. $\dfrac{\tan (90° - \theta) \sec (180° - \theta) \sin (-\theta)}{\sin (180° + \theta) \cot (360° - \theta) \csc (90° - \theta)}$.

93. $\dfrac{\sin (-\alpha) \tan (180° + \alpha) \tan (90° - \alpha)}{\cot (90° - \alpha) \cos (360° - \alpha) \sin (180° - \alpha)}$.

94. $\dfrac{\cos (\pi/2 - \theta) \cot (\pi - \theta) \sin (\pi + \theta)}{\sin (\pi - \theta) \tan (\pi/2 - \theta) \sin (2\pi - \theta)}$.

13.3

Multiple-angle Formulas

In this section we derive formulas for $\sin 2\alpha$, $\cos 2\alpha$, and $\tan 2\alpha$ in terms of the trigonometric functions of α. For this we substitute $\beta = \alpha$ in the addition formulas for sine, cosine, and tangent given in the last section. We have

$$\sin (\alpha + \beta) = \sin \alpha \cos \beta + \cos \beta \sin \alpha.$$

When $\beta = \alpha$, this becomes

$$\sin (\alpha + \alpha) = \sin \alpha \cos \alpha + \cos \alpha \sin \alpha$$

$$\sin 2\alpha = 2 \sin \alpha \cos \alpha.$$

Similarly, setting $\beta = \alpha$ in the addition formula for cosine, we have

$$\cos (\alpha + \beta) = \cos \alpha \cos \beta - \sin \alpha \sin \beta$$

$$\cos (\alpha + \alpha) = \cos \alpha \cos \alpha - \sin \alpha \sin \alpha$$

$$\cos 2\alpha = \cos^2 \alpha - \sin^2 \alpha.$$

This can also be written in an alternative form as

$$\cos 2\alpha = \cos^2 \alpha - (1 - \cos^2 \alpha) \quad \text{(since } \sin^2 \alpha = 1 - \cos^2 \alpha\text{)}$$

$$= 2 \cos^2 \alpha - 1$$

or

$$\cos 2\alpha = (1 - \sin^2 \alpha) - \sin^2 \alpha \quad \text{(since } \cos^2 \alpha = 1 - \sin^2 \alpha\text{)}$$

$$= 1 - 2 \sin^2 \alpha.$$

Thus, we have
$$\cos 2\alpha = \cos^2 \alpha - \sin^2 \alpha$$
$$= 2\cos^2 \alpha - 1$$
$$= 1 - 2\sin^2 \alpha.$$

Also, when $\beta = \alpha$, the formula for $\tan(\alpha + \beta)$ becomes
$$\tan(\alpha + \alpha) = \frac{\tan \alpha + \tan \alpha}{1 - \tan \alpha \tan \alpha}$$
$$\tan 2\alpha = \frac{2\tan \alpha}{1 - \tan^2 \alpha}.$$

Summarizing all the formulas we have obtained, we have

$$\sin 2\alpha = 2 \sin \alpha \cos \alpha \tag{1}$$
$$\cos 2\alpha = \cos^2 \alpha - \sin^2 \alpha \tag{2}$$
$$= 2\cos^2 \alpha - 1 \tag{3}$$
$$= 1 - 2\sin^2 \alpha \tag{4}$$
$$\tan 2\alpha = \frac{2\tan \alpha}{1 - \tan^2 \alpha}. \tag{5}$$

These are known as the *double-angle formulas*. They express the sine, cosine, and tangent of an angle (2α) in terms of trigonometric functions of half that angle (α). For example, in view of the above formulas, we can write

$$\sin \theta = \sin\left(2 \cdot \frac{\theta}{2}\right) = 2 \sin \frac{\theta}{2} \cos \frac{\theta}{2}$$
$$\cos 4\theta = \cos(2 \cdot 2\theta) = 1 - 2\sin^2 2\theta = \cos^2 2\theta - \sin^2 2\theta$$
$$\tan 3\theta = \tan\left(2 \cdot \frac{3\theta}{2}\right) = \frac{2 \tan \frac{3\theta}{2}}{1 - \tan^2 \frac{3\theta}{2}}.$$

Example 1 Given that $\sin \alpha = -\frac{3}{5}$ and α is in the fourth quadrant, find the values of:
(a) $\sin 2\alpha$. (b) $\tan 2\alpha$.

Solution (a) We have $\sin \alpha = -\frac{3}{5}$. Now,
$$\cos^2 \alpha = 1 - \sin^2 \alpha = 1 - (-\tfrac{3}{5})^2 = \tfrac{16}{25}$$
and so $\cos \alpha = \pm \frac{4}{5}$. Since α is in the fourth quadrant, $\cos \alpha$ is positive, and we have $\cos \alpha = +\frac{4}{5}$. Thus from (1) we have
$$\sin 2\alpha = 2 \sin \alpha \cos \alpha$$
$$= 2(-\tfrac{3}{5})(\tfrac{4}{5}) = -\tfrac{24}{25}.$$

(b) Since $\sin \alpha = -\frac{3}{5}$ and $\cos \alpha = \frac{4}{5}$,
$$\tan \alpha = \frac{\sin \alpha}{\cos \alpha} = \frac{-\frac{3}{5}}{\frac{4}{5}} = -\frac{3}{4}.$$

Sec. 13.3 *Multiple-angle Formulas*

Therefore
$$\tan 2\alpha = \frac{2 \tan \alpha}{1 - \tan^2 \alpha}$$
$$= \frac{2(-\tfrac{3}{4})}{1 - (-\tfrac{3}{4})^2} = -\frac{24}{7}.$$

Example 2 Prove that
$$\frac{1 + \cos 2\theta}{\sin 2\theta} = \cot \theta.$$

Solution When $\cos 2\theta$ occurs with 1 as we have in the numerator, it is often advantageous to use the formula for $\cos 2\theta$ that cancels the 1. Thus in this case we set $\cos 2\theta = 2\cos^2 \theta - 1$. Hence
$$\frac{1 + \cos 2\theta}{\sin 2\theta} = \frac{1 + (2\cos^2\theta - 1)}{2 \sin \theta \cos \theta}$$
$$= \frac{2 \cos^2 \theta}{2 \sin \theta \cos \theta}$$
$$= \frac{\cos \theta}{\sin \theta}$$
$$= \cot \theta.$$

Example 3 Prove that
$$\frac{\sin 3\theta}{\cos \theta} + \frac{\cos 3\theta}{\sin \theta} = \cot \theta - \tan \theta.$$

Solution Writing the left side as a single fraction with a common denominator, we have
$$\frac{\sin 3\theta}{\cos \theta} + \frac{\cos 3\theta}{\sin \theta} = \frac{\sin 3\theta \sin \theta + \cos 3\theta \cos \theta}{\sin \theta \cos \theta}.$$

Now if we expand $\cos(3\theta - \theta)$ using the formula for $\cos(\alpha - \beta)$, we obtain the expression in the numerator. Thus,
$$\frac{\sin 3\theta}{\cos \theta} + \frac{\cos 3\theta}{\sin \theta} = \frac{\cos(3\theta - \theta)}{\sin \theta \cos \theta}$$
$$= \frac{\cos 2\theta}{\sin \theta \cos \theta}$$
$$= \frac{\cos^2 \theta - \sin^2 \theta}{\sin \theta \cos \theta} \qquad \text{[using (2)]}$$
$$= \frac{\cos^2 \theta}{\sin \theta \cos \theta} - \frac{\sin^2 \theta}{\sin \theta \cos \theta}$$
$$= \frac{\cos \theta}{\sin \theta} - \frac{\sin \theta}{\cos \theta}$$
$$= \cot \theta - \tan \theta$$

which proves the required identity.

Formula (3) gives
$$2\cos^2\alpha - 1 = \cos 2\alpha$$
$$2\cos^2\alpha = 1 + \cos 2\alpha$$
$$\cos^2\alpha = \frac{1 + \cos 2\alpha}{2}.$$

Similarly, solving (4) for $\sin^2\alpha$, we obtain
$$\sin^2\alpha = \frac{1 - \cos 2\alpha}{2}.$$

Thus, we have

$$\boxed{\cos^2\alpha = \frac{1 + \cos 2\alpha}{2}} \qquad (6)$$

$$\boxed{\sin^2\alpha = \frac{1 - \cos 2\alpha}{2}.} \qquad (7)$$

These formulas allow us to express even powers of sines and cosines in terms of the double angles.

Example 4 Express $\sin^4\theta$ in terms of multiple angles without any powers.

Solution Using formula (7), we have
$$\sin^4\theta = (\sin^2\theta)^2 = \left(\frac{1 - \cos 2\theta}{2}\right)^2$$
$$= \frac{1}{4}(1 - 2\cos 2\theta + \cos^2 2\theta)$$
$$= \frac{1}{4}\left[1 - 2\cos 2\theta + \frac{1 + \cos 4\theta}{2}\right]$$

where we have used (6) again with $\alpha = 2\theta$ to reexpress $\cos^2 2\theta$,
$$= \frac{1}{4}\left(\frac{3}{2} - 2\cos 2\theta + \frac{1}{2}\cos 4\theta\right).$$

Replacing α by $\alpha/2$ in formulas (6) and (7) and taking the square root of both sides, we have

$$\boxed{\cos\frac{\alpha}{2} = \pm\sqrt{\frac{1 + \cos\alpha}{2}}} \qquad (8)$$

$$\boxed{\sin\frac{\alpha}{2} = \pm\sqrt{\frac{1 - \cos\alpha}{2}}.} \qquad (9)$$

These are known as the **half-angle formulas**. The sign $+$ or $-$ on the right side must be determined from knowledge of whether $\sin\alpha/2$ or $\cos\alpha/2$ is positive or negative. This in turn would typically come from knowledge of which quadrant $\alpha/2$ lies in.

Example 5 Given $\cos 330° = \sqrt{3}/2$, find the values of $\sin 165°$ and $\cos 165°$.

Solution Using formula (8) with $\alpha = 330°$, we have
$$\cos 165° = \pm\sqrt{\frac{1 + \cos 330°}{2}}.$$

Since an angle of 165° is in the second quadrant, and cosines in the second quadrant are negative, we have

$$\cos 165° = -\sqrt{\frac{1 + \cos 330°}{2}}$$

$$= -\sqrt{\frac{1 + \sqrt{3}/2}{2}}$$

$$= -\sqrt{\frac{2 + \sqrt{3}}{4}}$$

$$= -\sqrt{\frac{4 + 2\sqrt{3}}{8}} \quad \text{(note the trickery in this step)}$$

$$= -\sqrt{\frac{(\sqrt{3})^2 + 1 + 2\sqrt{3}}{8}}$$

$$= -\sqrt{\frac{(\sqrt{3} + 1)^2}{(2\sqrt{2})^2}}$$

$$= -\frac{\sqrt{3} + 1}{2\sqrt{2}}.$$

Using formula (9), and noting that $\sin 165°$ is positive, we have

$$\sin 165° = +\sqrt{\frac{1 - \cos 330°}{2}}$$

$$= \sqrt{\frac{1 - \sqrt{3}/2}{2}}$$

$$= \sqrt{\frac{2 - \sqrt{3}}{4}}$$

$$= \sqrt{\frac{4 - 2\sqrt{3}}{8}}$$

$$= \sqrt{\frac{(\sqrt{3} - 1)^2}{8}}$$

$$= \frac{\sqrt{3} - 1}{2\sqrt{2}}.$$

Example 6 Given $\tan 30° = 1/\sqrt{3}$, find the value of $\tan 15°$.

Solution Using the formula (5) with $\alpha = 15°$, we have
$$\tan 30° = \frac{2 \tan 15°}{1 - \tan^2 15°}.$$

Replacing tan 15° by x for brevity, and using that $\tan 30° = 1/\sqrt{3}$, we have

$$\frac{1}{\sqrt{3}} = \frac{2x}{1-x^2}$$

$$1 - x^2 = 2\sqrt{3}\,x,$$

$$x^2 + 2\sqrt{3}\,x - 1 = 0. \tag{i}$$

This is a quadratic equation for x. From the quadratic formula with $a = 1$, $b = 2\sqrt{3}$, and $c = -1$, the solutions are given by

$$x = \frac{-b \pm \sqrt{b^2 - 4ac}}{2a}$$

$$= \frac{-2\sqrt{3} \pm \sqrt{(2\sqrt{3})^2 - 4 \cdot 1(-1)}}{2(1)}$$

$$= \frac{-2\sqrt{3} \pm \sqrt{12 + 4}}{2}$$

$$= -\sqrt{3} \pm 2.$$

Thus $x = \tan 15° = -\sqrt{3} + 2$ or $-\sqrt{3} - 2$. Since 15° is in the first quadrant, $\tan 15° > 0$ and we must take the positive root:

$$\tan 15° = -\sqrt{3} + 2.$$

We can also express $\sin 2\alpha$ and $\cos 2\alpha$ in terms of $\tan \alpha$ as follows. We have:

$$\sin 2\alpha = 2 \sin \alpha \cos \alpha = \frac{2 \sin \alpha \cos \alpha}{1}$$

$$= \frac{2 \sin \alpha \cos \alpha}{\cos^2 \alpha + \sin^2 \alpha} \quad \text{(since } \cos^2 \alpha + \sin^2 \alpha = 1\text{)}.$$

Now let us divide the numerator and denominator by $\cos^2 \alpha$:

$$\sin 2\alpha = \frac{2 \sin \alpha \cos \alpha / \cos^2 \alpha}{(\cos^2 \alpha + \sin^2 \alpha)/\cos^2 \alpha}$$

$$= \frac{2 \sin \alpha / \cos \alpha}{1 + \sin^2 \alpha / \cos^2 \alpha}$$

$$= \frac{2 \tan \alpha}{1 + \tan^2 \alpha}.$$

Similarly,

$$\cos 2\alpha = \cos^2 \alpha - \sin^2 \alpha$$

$$= \frac{\cos^2 \alpha - \sin^2 \alpha}{\cos^2 \alpha + \sin^2 \alpha} \quad \text{(since } \cos^2 \alpha + \sin^2 \alpha = 1\text{)}.$$

Divide numerator and denominator by $\cos^2 \alpha$:

$$= \frac{1 - \dfrac{\sin^2 \alpha}{\cos^2 \alpha}}{1 + \dfrac{\sin^2 \alpha}{\cos^2 \alpha}}$$

$$= \frac{1 - \tan^2 \alpha}{1 + \tan^2 \alpha}.$$

Thus, we have

$$\sin 2\alpha = \frac{2 \tan \alpha}{1 + \tan^2 \alpha} \qquad (10)$$

$$\cos 2\alpha = \frac{1 - \tan^2 \alpha}{1 + \tan^2 \alpha}. \qquad (11)$$

These two formulas play an important role in the study of calculus. We can use them to determine the values of tan α when either sin 2α or cos 2α is given. The method is similar to that in Example 6, in which tan 2α was given.

We can derive formulas for sin 3α, cos 3α, or tan 3α by making use of the double-angle formulas and the addition formulas of the last section. We illustrate the derivation of the formula for sin 3α in the following solved example.

Example 7 Prove that $\sin 3\alpha = 3 \sin \alpha - 4 \sin^3 \alpha$.

Solution We have

$$\sin 3\alpha = \sin(2\alpha + \alpha)$$
$$= \sin 2\alpha \cos \alpha + \cos 2\alpha \sin \alpha$$
$$= (2 \sin \alpha \cos \alpha) \cos \alpha + (1 - 2 \sin^2 \alpha) \sin \alpha$$
$$= 2 \sin \alpha \cos^2 \alpha + \sin \alpha - 2 \sin^3 \alpha$$
$$= 2 \sin \alpha (1 - \sin^2 \alpha) + \sin \alpha - 2 \sin^3 \alpha$$
$$= 2 \sin \alpha - 2 \sin^3 \alpha + \sin \alpha - 2 \sin^3 \alpha$$
$$= 3 \sin \alpha - 4 \sin^3 \alpha.$$

Diagnostic Test 13.3

Fill in the blanks.

1. $\sin 2A = $ _____ (in terms of sin A, cos A).
2. $\cos 2\theta = $ _____ (in terms of sin θ, cos θ).
   $= $ _____ (in terms of sin θ only).
   $= $ _____ (in terms of cos θ only).
3. $\tan 2\theta = $ _____ (in terms of tan θ).
4. $\dfrac{1 + \cos 2A}{2} = $ _____.
5. $\sin^2 \dfrac{\alpha}{2} = $ _____ (in terms of cos α).
6. $2 \sin \dfrac{\alpha}{2} \cos \dfrac{\alpha}{2} = $ _____.
7. $1 - 2 \cos^2 \dfrac{\theta}{2} = $ _____.
8. $2 \sin^2 \dfrac{\alpha}{2} - 1 = $ _____.

9. $\sin^2 3\theta - \cos^2 3\theta =$ _____.

10. $\dfrac{1 - \tan^2 \theta}{\tan \theta} =$ _____.

11. $1 - 2\sin^2 \dfrac{\pi}{8} =$ _____.

12. $\sin \dfrac{\pi}{12} \cos \dfrac{\pi}{12} =$ _____.

Exercises 13.3

(1–6) In each of the following exercises, find $\sin 2\alpha$, $\cos 2\alpha$, and $\tan 2\alpha$ from the given data.

1. $\sin \alpha = -\frac{3}{5}$ and $\pi < \alpha < 3\pi/2$.
2. $\cos \alpha = \frac{5}{13}$ and $3\pi/2 < \alpha < 2\pi$.
3. $\tan \alpha = -\sqrt{3}$ and $\pi/2 < \alpha < \pi$.
4. $\sec \alpha = -2$ and $\pi < \alpha < 3\pi/2$.
5. $\csc \alpha = 3$ and $\pi/2 < \alpha < \pi$.
6. $\cot \alpha = \frac{5}{12}$ and $\pi < \alpha < 3\pi/2$.

(7–20) Simplify the following without using tables or calculators.

7. $2 \sin \dfrac{\pi}{8} \cos \dfrac{\pi}{8}$.
8. $2 \sin \dfrac{\pi}{12} \cos \dfrac{\pi}{12}$.
9. $2 \cos^2 67\frac{1}{2}° - 1$.
10. $1 - 2 \sin^2 75°$.
11. $2 \sin^2 \dfrac{\pi}{8} - 1$.
12. $1 - 2 \cos^2 105°$.
13. $\dfrac{2 \tan 22\frac{1}{2}°}{1 - \tan^2 22\frac{1}{2}°}$.
14. $\dfrac{2 \tan 67\frac{1}{2}°}{1 + \tan^2 67\frac{1}{2}°}$.
15. $\dfrac{\tan \alpha/2}{1 + \tan^2 \alpha/2}$.
16. $\dfrac{1 - \tan^2 105°}{\tan 105°}$.
17. $\dfrac{1 - \tan^2 75°}{1 + \tan^2 75°}$.
18. $\dfrac{1 + \tan^2 165°}{1 - \tan^2 165°}$.
19. $\dfrac{1 + \tan^2 22\frac{1}{2}°}{\tan 22\frac{1}{2}°}$.
20. $\dfrac{1 + \tan^2 \dfrac{A}{2}}{\tan \dfrac{A}{2}}$.

21. Given $\cos 30° = \sqrt{3}/2$, use the half-angle formulas to evaluate $\sin 15°$, $\cos 15°$, and $\tan 15°$.

22. Given $\cos 150° = -\sqrt{3}/2$, use the half-angle formulas to evaluate $\sin 75°$, $\cos 75°$, and $\tan 75°$.

23. Given $\cos 315° = 1/\sqrt{2}$, evaluate $\sin 157\frac{1}{2}°$ and $\cos 157\frac{1}{2}°$.

24. Given $\cos 225° = -1/\sqrt{2}$, evaluate $\sin 112\frac{1}{2}°$ and $\cos 112\frac{1}{2}°$.

25. Given $\sin 150° = \frac{1}{2}$, find the value of $\tan 75°$.

26. Given $\cos 330° = \sqrt{3}/2$, find the value of $\tan 165°$.

(27–50) Prove the following identities.

27. $\dfrac{\sin 2\theta}{1 + \cos 2\theta} = \tan \theta$.

28. $\dfrac{1 - \cos 2\theta}{\sin 2\theta} = \tan \theta$.

29. $\cos^4 \theta - \sin^4 \theta = \cos 2\theta$.

30. $\tan \theta + \cot \theta = 2 \csc 2\theta$.

31. $\dfrac{1 - \cos 2\theta}{1 + \cos 2\theta} = \tan^2 \theta$.

32. $\dfrac{1 - \cos 2\theta + \sin 2\theta}{1 + \cos 2\theta + \sin 2\theta} = \tan \theta$.

33. $\dfrac{1 + \cos 2A - \sin 2A}{1 - \cos 2A - \sin 2A} = -\cot A$.

34. $\dfrac{\cot \theta - \tan \theta}{\cot \theta + \tan \theta} = \cos 2\theta$.

35. $\cot 2\theta = \dfrac{\cot^2 \theta - 1}{2 \cot \theta}$.

36. $\cot \theta - \cot 2\theta = \csc 2\theta$.

37. $\sec 2\theta + \cot 2\theta = \cot \theta$.

38. $\tan 2\theta - \sec 2\theta = \dfrac{\sin \theta - \cos \theta}{\sin \theta + \cos \theta}$.

39. $\sec^2 A = \dfrac{2 \sec 2A}{1 + \sec 2A}$.

40. $\csc^2 \dfrac{\theta}{2} = \dfrac{2 \sec \theta}{\sec \theta - 1}$.

41. $\dfrac{\tan \theta - \sin \theta}{2 \tan \theta} = \sin^2 \dfrac{\theta}{2}$.

42. $\dfrac{\tan \theta}{1 + \sec \theta} = \tan \dfrac{\theta}{2}$.

43. $\sin^2 \theta = \dfrac{\sec 2\theta - 1}{2 \sec 2\theta}$.

44. $\cot \theta = \dfrac{\sec 2\theta + 1}{\tan 2\theta}$.

45. $\dfrac{\sin 3\theta}{\sin \theta} - \dfrac{\cos 3\theta}{\cos \theta} = 2$.

46. $\dfrac{\sin 2\theta}{\sin \theta} - \dfrac{\cos 2\theta}{\cos \theta} = \sec \theta$.

47. $\cos 3\theta = 4 \cos^3 \theta - 3 \cos \theta$.

48. $\tan 3\theta = \dfrac{3 \tan \theta - \tan^3 \theta}{1 - 3 \tan^2 \theta}$.

49. $\sin 4\theta = 4 \sin \theta \cos \theta - 8 \sin^3 \theta \cos \theta$.

50. $\cos 4\theta = 8 \cos^4 \theta - 8 \cos^2 \theta + 1$
    $= 8 \sin^4 \theta - 8 \sin^2 \theta + 1$.

(51–52) Draw the graphs of the following functions.

51. $f(x) = \sin \dfrac{x}{2} \cos \dfrac{x}{2}$.

52. $g(x) = \cos^2 \dfrac{x}{2} - \sin^2 \dfrac{x}{2}$.

(53–56) Express the following in terms of multiple angles.

53. $\cos^4 \theta$.

54. $\sin^2 \theta \cos^2 \theta$.

55. $\sin \theta \cos^3 \theta$.

56. $\cos \theta \sin^3 \theta$.

13.4

The Inverse Trigonometric Functions

In Section 6.4 we introduced the notion of the inverse of a function. Our aim in the present section is to apply this idea to the three main trigonometric functions, sine, cosine, and tangent. Before doing so, let us briefly review the definition of the inverse of a function.

A function $y = f(x)$ is one-to-one if any horizontal line meets the graph of f in at most one point. In other words, to each value y in the range of f there is exactly one value x in the domain of f. If the function $y = f(x)$ is one-to-one, then a unique inverse of f, denoted by f^{-1}, exists and is given by $x = f^{-1}(y)$. Thus,

> If f is one-to-one and $y = f(x)$, then $x = f^{-1}(y)$.

Sometimes the given function as such is not one-to-one, and so the inverse will not exist. If in such cases we restrict the domain of $f(x)$, so that the new function with the restricted domain is one-to-one, then the inverse will exist for this new function.

Consider the sine function $y = \sin x$. Its domain is the set of all real numbers, and its graph is as shown below. Clearly the function is not one-to-one, because any horizontal line lying between $y = -1$ and $y = +1$ meets the graph in more than one point. Thus the function as such does not have an inverse. For example, the line $y = 0$ meets the graph at $x = 0, \pm\pi, \pm 2\pi$, and so on. So there are infinitely many values of x such that $\sin x = 0$. Similarly for any y in the interval $-1 \leq y \leq 1$ there are infinitely many values of x such that $\sin x = y$.

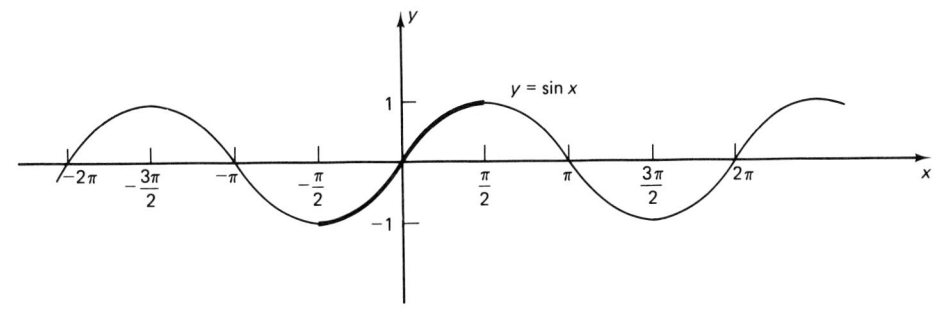

If, however, we restrict the domain of $y = \sin x$ to $-\pi/2 \leq x \leq \pi/2$, then the function becomes one-to-one. The graph of the new (restricted) function is the thick portion of the graph above, and the inverse now exists. The inverse is denoted by $x = \text{Sin}^{-1} y$ (note the capital S) and is called the **principal value of the inverse sine of y**. Thus,

$$\text{If } y = \sin x \text{ and } -\frac{\pi}{2} \leq x \leq \frac{\pi}{2}, \text{ then } x = \text{Sin}^{-1} y. \qquad (1)$$

In words, $\text{Sin}^{-1} y$ *is the angle lying in the interval $[-\pi/2, \pi/2]$ whose sine is equal to y*. Thus,

$$-\frac{\pi}{2} \leq \text{Sin}^{-1} y \leq \frac{\pi}{2}.$$

$\text{Sin}^{-1} y$ is defined only for $-1 \leq y \leq 1$.

Note $\text{Sin}^{-1} y$ is sometimes also denoted by $\text{Arc sin } y$. It may be remarked that in the expression $\text{Sin}^{-1} y$, -1 is *not* an exponent or power; that is,

$$\text{Sin}^{-1} y \neq (\text{Sin } y)^{-1} = \frac{1}{\sin y}.$$

Example 1 Evaluate the following:

(a) $\text{Sin}^{-1} \frac{1}{2}$.

(b) $\text{Sin}^{-1} (-\sqrt{3}/2)$.

Solution (a) Let $\text{Sin}^{-1} \frac{1}{2} = x$. Then by (1) we have

$$\sin x = \tfrac{1}{2}$$

and x lies in the interval $-\pi/2 \leq x \leq \pi/2$. But $\sin x = \tfrac{1}{2}$ if $x = \pi/6$. Thus

$$\text{Sin}^{-1} \frac{1}{2} = \frac{\pi}{6}.$$

(b) Let $\text{Sin}^{-1}(-\sqrt{3}/2) = x$. Then by (1) $\sin x = -\sqrt{3}/2$, and x must be in the interval $-\pi/2 \leq x \leq \pi/2$. Since $\sin x$ is negative, x is in the fourth quadrant. But

$$\text{Sin}\left(-\frac{\pi}{3}\right) = -\sin\frac{\pi}{3} = -\frac{\sqrt{3}}{2} \qquad \text{[because } \sin(-\theta) = -\sin\theta\text{]}.$$

Therefore, $x = -\pi/3$ or $\text{Sin}^{-1}(-\sqrt{3}/2) = -\pi/3$.

Note that also $\sin(5\pi/3) = -\sqrt{3}/2$. However, we cannot take $x = 5\pi/3$, because this does not lie between $-\pi/2$ and $\pi/2$.

From (1), if we substitute $x = \text{Sin}^{-1} y$ in $\sin x = y$, we obtain $\sin(\text{Sin}^{-1} y) = y$. Similarly if we substitute $y = \sin x$ in $x = \text{Sin}^{-1} y$, we obtain $\text{Sin}^{-1}(\sin x) = x$. Thus, we have

$$\sin(\text{Sin}^{-1} y) = y \quad \text{for} \quad -1 \leq y \leq 1 \qquad (2)$$

$$\text{Sin}^{-1}(\sin x) = x \quad \text{for} \quad -\frac{\pi}{2} \leq x \leq \frac{\pi}{2}. \qquad (3)$$

It is particularly important to note the restriction on x in the second of these equations. $\text{Sin}^{-1}(\sin x)$ will not be equal to x if x lies outside the interval $[-\pi/2, \pi/2]$. [See Example 4(b) below.]

If we interchange x and y in (1), we obtain the following equivalent statement:

$$y = \text{Sin}^{-1} x \quad \text{if} \quad x = \sin y \quad \text{and} \quad -\frac{\pi}{2} \leq y \leq \frac{\pi}{2}.$$

The graph of $y = \text{Sin}^{-1} x$ can therefore be obtained by plotting the points (x, y) for which $x = \sin y$ in the interval $-\pi/2 \leq y \leq \pi/2$. It is shown in the adjacent figure.

Next consider the cosine function $y = \cos x$ whose graph is shown below.

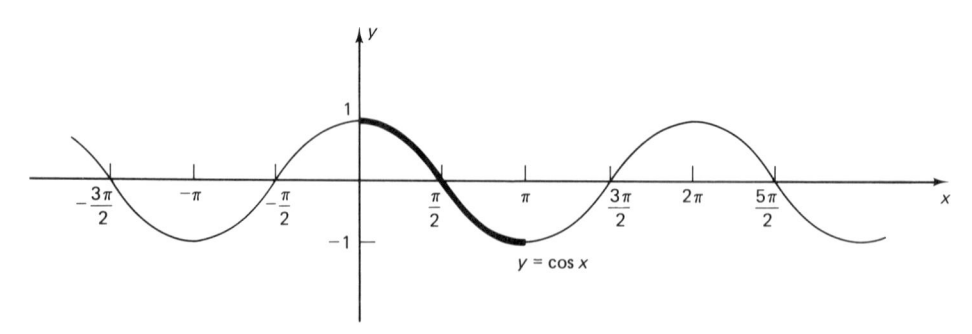

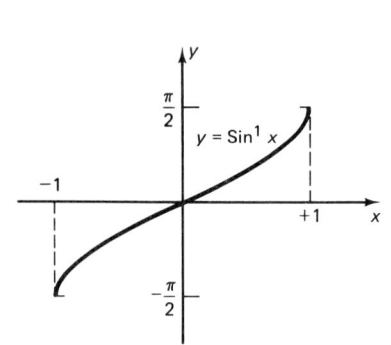

The cosine function is clearly not one-to-one. In this case we restrict the domain to $0 \leq x \leq \pi$, so that the graph of the restricted function is the part of the above figure indicated by a heavy line. The restricted cosine function $y = \cos x$, $0 \leq x \leq \pi$, now becomes one-to-one and has an inverse. The inverse is denoted by $x = \text{Cos}^{-1} y$ (note the capital C) and is called the *principal value of the inverse cosine of y*. Thus,

$$\text{If } y = \cos x \text{ for } 0 \leq x \leq \pi, \text{ then } x = \text{Cos}^{-1} y. \quad (4)$$

In words, $\text{Cos}^{-1} y$ is the angle lying in $[0, \pi]$ whose cosine is equal to y. $\text{Cos}^{-1} y$ is defined only for $-1 \leq y \leq 1$. Further, from (4), it follows that

$$0 \leq \text{Cos}^{-1} y \leq \pi.$$

Upon eliminating x and y in turn from the two equations in (4), we obtain

$$\cos(\text{Cos}^{-1} y) = y \quad \text{for } -1 \leq y \leq 1 \quad (5)$$
$$\text{Cos}^{-1}(\cos x) = x \quad \text{for } 0 \leq x \leq \pi. \quad (6)$$

Again note carefully the restriction on x in the second of these equations.

Example 2 Evaluate the following:

(a) $\text{Cos}^{-1}(\sqrt{2}/2)$. (b) $\text{Cos}^{-1}(-\sqrt{3}/2)$.
(c) $\text{Cos}^{-1}(-0.9272)$. (d) $\text{Cos}^{-1}(2/\sqrt{3})$.

Solution (a) Let $\text{Cos}^{-1}(\sqrt{2}/2) = x$. This gives $\cos x = \sqrt{2}/2$. Thus we must find x in $0 \leq x \leq \pi$ such that $\cos x = \sqrt{2}/2$. But $\cos \pi/4 = \sqrt{2}/2$. Thus

$$x = \frac{\pi}{4} \quad \text{—that is,} \quad \text{Cos}^{-1} \frac{\sqrt{2}}{2} = \frac{\pi}{4}.$$

(b) Let $\text{Cos}^{-1}(-\sqrt{3}/2) = x$. Hence,

$$\cos x = -\frac{\sqrt{3}}{2}.$$

Thus, we must find x in $0 \leq x \leq \pi$ such that $\cos x = -\sqrt{3}/2$. Since $\cos x$ is negative and x is in $0 \leq x \leq \pi$, x must belong to the second quadrant. Since $\cos(5\pi/6) = -\sqrt{3}/2$ and $5\pi/6$ lies in the second quadrant, it must equal x. Thus

$$\text{Cos}^{-1}\left(-\frac{\sqrt{3}}{2}\right) = \frac{5\pi}{6}.$$

(c) Let $\text{Cos}^{-1}(-0.9272) = x$. Then

$$\cos x = -0.9272.$$

From the tables, when $\alpha = 0.3840$ radians, then $\cos \alpha = 0.9272$. As above in (b), the angle x is in the second quadrant and its reference angle α is 0.3840 radians. Therefore

$$x = \pi - 0.3840 \approx 3.1416 - 0.3840 = 2.7576.$$

Thus,

$$\text{Cos}^{-1}(-0.9272) \approx 2.7576 \text{ radians.}$$

(d) If $\cos^{-1}(2/\sqrt{3}) = x$, then $\cos x = 2/\sqrt{3}$. Since $2/\sqrt{3} > 1$, and $\cos x$ cannot be greater than 1, $\cos^{-1}(2/\sqrt{3})$ is not defined.

Interchanging x and y in (4), we obtain the statement

$$y = \cos^{-1} x \quad \text{if} \quad x = \cos y \quad \text{and} \quad 0 \leq y \leq \pi.$$

Thus we obtain the graph of $y = \cos^{-1} x$ by plotting the points (x, y) such that $x = \cos y$ in the interval $0 \leq y \leq \pi$. The graph is shown in the adjoining figure.

Finally let us consider the tangent function $y = \tan x$. If we restrict the domain of the tangent function to $-\pi/2 < x < \pi/2$, then the function becomes one-to-one. The graph of the restricted function is shown in the figure below. The restricted tangent function has an inverse, which we denote by $x = \text{Tan}^{-1} y$ (note the capital T) and which is called the *princpial value of the inverse tangent of y*. Thus,

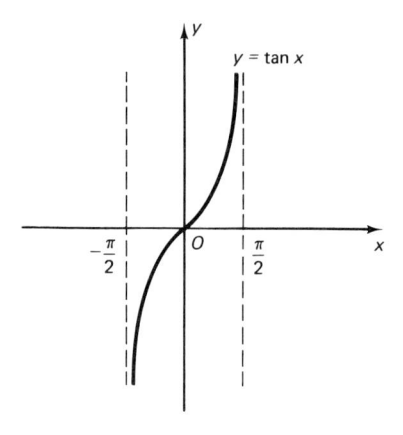

$$\text{If} \quad y = \tan x \quad \text{for} \quad -\frac{\pi}{2} < x < \frac{\pi}{2}, \quad \text{then} \quad x = \text{Tan}^{-1} y. \quad (7)$$

$\text{Tan}^{-1} y$ is defined for all y in $-\infty < y < \infty$. By definition, we have

$$-\frac{\pi}{2} < \text{Tan}^{-1} y < \frac{\pi}{2}.$$

By eliminating x and y in turn from the two equations in (7), we obtain

$$\tan(\text{Tan}^{-1} y) = y \quad \text{for all } y \qquad (8)$$

$$\text{Tan}^{-1}(\tan x) = x \quad \text{for} \quad -\frac{\pi}{2} < x < \frac{\pi}{2}. \qquad (9)$$

Example 3 Evaluate the following:
(a) $\text{Tan}^{-1}(\sqrt{3})$.
(b) $\text{Tan}^{-1}(-1)$.

Solution (a) If $\text{Tan}^{-1}\sqrt{3} = x$, then $\tan x = \sqrt{3}$. Thus we are looking for x in $-\pi/2 < x < \pi/2$ such that $\tan x = \sqrt{3}$. But $\tan \pi/3 = \sqrt{3}$. Thus $x = \pi/3$; that is,
$$\text{Tan}^{-1}\sqrt{3} = \pi/3.$$

(b) If $\text{Tan}^{-1}(-1) = x$, then $\tan x = -1$. So, we want to find x in $-\pi/2 < x < \pi/2$ such that $\tan x = -1$. Now,
$$\tan\left(-\frac{\pi}{4}\right) = -\tan\frac{\pi}{4} = -1$$
[since $\tan(-\theta) = -\tan\theta$]. Therefore, $x = -\pi/4$; that is, $\text{Tan}^{-1}(-1) = -\pi/4$.

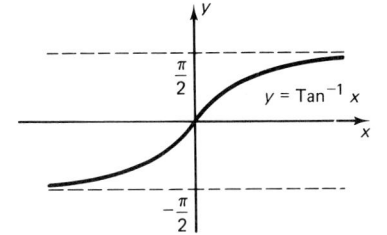

Interchanging x and y in (7), we obtain

$$y = \text{Tan}^{-1} x \quad \text{if} \quad x = \tan y \quad \text{and} \quad -\frac{\pi}{2} < y < \frac{\pi}{2}.$$

The graph of $y = \text{Tan}^{-1} x$ is shown in the adjoining figure.

Example 4 Evaluate the following:
(a) $\text{Tan}^{-1}(\cos 4\pi/3)$.
(b) $\text{Sin}^{-1}(\sin 2\pi/3)$.

Solution (a) $\text{Tan}^{-1}(\cos 4\pi/3)$: Now $\cos 4\pi/3 = -\frac{1}{2}$. Thus,
$$\text{Tan}^{-1}\left(\cos\frac{4\pi}{3}\right) = \text{Tan}^{-1}\left(-\frac{1}{2}\right) = x \quad (\text{say}).$$

Then, $\tan x = -\frac{1}{2}$. Therefore, we look for x in $-\pi/2 < x < \pi/2$ such that $\tan x = -\frac{1}{2} = -0.5$. From the trigonometric tables we have
$$\tan(0.4625) = 0.4986, \quad \tan(0.4654) = 0.5022.$$

By interpolation, therefore,
$$\tan(0.4636) \approx 0.5000.$$

Thus, we can write
$$\tan x = -\tfrac{1}{2} = -\tan(0.4636)$$
$$= \tan(-0.4636)$$
[because $\tan(-\theta) = -\tan\theta$]. Thus we have $x = -0.4636$; that is,
$$\text{Tan}^{-1}\left(\cos\frac{4\pi}{3}\right) = -0.4636.$$

(b) $\text{Sin}^{-1}(\sin 2\pi/3)$: One might be tempted to use formula (3) and say that
$$\text{Sin}^{-1}\left(\sin\frac{2\pi}{3}\right) = \frac{2\pi}{3}.$$

This would be incorrect, however, because $\text{Sin}^{-1} y$ must always lie between $-\pi/2$ and $\pi/2$, both inclusive. Instead, we use that

$$\sin \frac{2\pi}{3} = \frac{\sqrt{3}}{2}.$$

Thus, $\text{Sin}^{-1} (\sin 2\pi/3) = \text{Sin}^{-1} \sqrt{3}/2$, which can be shown to be $\pi/3$.

Example 5 Without using tables or calculator, evaluate $\sec(\text{Sin}^{-1} \frac{1}{3})$.

Solution Let $\text{Sin}^{-1} \frac{1}{3} = x$, so that $\sin x = \frac{1}{3}$. Since x has to satisfy $-\pi/2 \leq x \leq \pi/2$ and $\sin x$ is positive, x must be in the first quadrant. We must evaluate $\sec(\text{Sin}^{-1} \frac{1}{3}) = \sec x$. Thus we know that $\sin x = \frac{1}{3}$ and we require the value of $\sec x$. Since x is a positive acute angle, we can construct a right triangle OMP, as in the figure, in which x is the angle at O. We let $MP = 1$ unit. Then

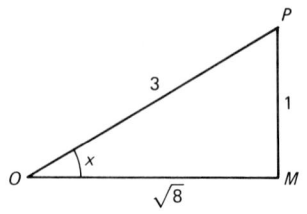

$$\sin x = \frac{MP}{OP} = \frac{1}{3} \quad \text{(given)}.$$

Since $MP = 1$, $OP = 3$. Thus, from Pythagoras's theorem,

$$OM^2 = OP^2 - PM^2 = 9 - 1 = 8$$
$$OM = \sqrt{8}.$$

Consequently,

$$\sec x = \frac{\text{hyp.}}{\text{adj.}} = \frac{3}{\sqrt{8}}.$$

Thus,

$$\sec\left(\text{Sin}^{-1} \frac{1}{3}\right) = \frac{3}{\sqrt{8}}.$$

Example 6 Prove that if $0 < x < 1$, $\text{Sin}^{-1} x = \text{Tan}^{-1} (x/\sqrt{1-x^2})$.

Solution Let $\text{Sin}^{-1} x = \theta$. Then $x = \sin\theta$, and θ is a positive acute angle. The identity to be proved can then be written

$$\theta = \text{Tan}^{-1} \frac{x}{\sqrt{1-x^2}}.$$

Using (8) above, this becomes

$$\tan\theta = \frac{x}{\sqrt{1-x^2}}.$$

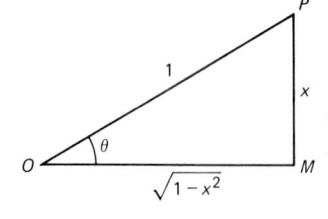

Thus, the given problem can be rephrased as follows: If $x = \sin\theta$, prove that $\tan\theta = x/\sqrt{1-x^2}$. We construct a right triangle OMP in which θ is the angle at O (see the adjacent figure). Then we have

$$\sin\theta = \frac{\text{opp.}}{\text{hyp.}} = \frac{MP}{OP} = x \quad \text{(given)}$$

Therefore if we choose $OP = 1$, we will have $MP = x$. By the Pythagorean theorem we have

$$OM^2 = OP^2 - MP^2$$
$$= 1 - x^2$$
$$OM = \sqrt{1-x^2}.$$

Thus,
$$\tan\theta = \frac{\text{opp.}}{\text{adj.}} = \frac{MP}{OP} = \frac{x}{\sqrt{1-x^2}}$$
which proves the result.

Note The result in this example can also be proved when $-1 < x < 0$. In this case we let $\text{Sin}^{-1} x = -\theta$. In the triangle OMP we then have $MP = -x = |x|$, and the rest follows.

Example 7 Evaluate $\sin(\text{Sin}^{-1}\frac{1}{3} + \text{Tan}^{-1} 2)$.

Solution We let
$$\text{Sin}^{-1}\tfrac{1}{3} = \alpha \quad \text{and} \quad \text{Tan}^{-1} 2 = \beta$$
so that $\sin\alpha = \frac{1}{3}$ and $\tan\beta = 2$. Thus the task becomes, given that $\sin\alpha = \frac{1}{3}$ and $\tan\beta = 2$, find the value of $\sin(\text{Sin}^{-1}\frac{1}{3} + \text{Tan}^{-1} 2) = \sin(\alpha + \beta)$. Now, we have
$$\sin(\alpha + \beta) = \sin\alpha\cos\beta + \cos\alpha\sin\beta. \tag{i}$$
We know the value of $\sin\alpha$ and need $\cos\alpha$ to use in (i). We construct a right triangle, as in the figure, with α as one of the angles. Taking the hypotenuse as 3 and the opposite side as 1 (since $\sin\alpha = \frac{1}{3}$), we obtain the adjacent side equal to $\sqrt{8}$ (from Pythagoras's theorem). Hence $\cos\alpha = \sqrt{8}/3$.

For β we again construct a right triangle in which $\tan\beta = 2$ (as in the second figure). From this it follows that $\cos\beta = 1/\sqrt{5}$ and $\sin\beta = 2/\sqrt{5}$.

Substituting these values in (i), we obtain
$$\sin(\alpha + \beta) = \sin\alpha\cos\beta + \cos\alpha\sin\beta$$
$$= \frac{1}{3} \cdot \frac{1}{\sqrt{5}} + \frac{\sqrt{8}}{3} \cdot \frac{2}{\sqrt{5}}$$
$$= \frac{1 + 4\sqrt{2}}{3\sqrt{5}}.$$
Thus,
$$\sin\left(\text{Sin}^{-1}\frac{1}{3} + \text{Tan}^{-1} 2\right) = \frac{1 + 4\sqrt{2}}{3\sqrt{5}}.$$

Diagnostic Test 13.4

Fill in the blanks.

1. $\text{Sin}^{-1}\frac{1}{2} = $ _____.
2. $\text{Cos}^{-1}\sqrt{3}/2 = $ _____.
3. _____ $\leq \text{Sin}^{-1}\theta \leq$ _____.
4. _____ $\leq \text{Cos}^{-1} x \leq$ _____.
5. $\sin(\text{Sin}^{-1} x) = $ _____ for all x in _____.
6. $\text{Sin}^{-1}(\sin x) = $ _____ for all x in _____.
7. $\cos(\text{Cos}^{-1} x) = $ _____ for all x in _____.
8. $\text{Cos}^{-1}(\cos x) = $ _____ for all x in _____.

Sec. 13.4 *The Inverse Trigonometric Functions*

9. $\tan(\text{Tan}^{-1} x) = $ _____ for all x in _____.

10. $\text{Tan}^{-1}(\tan x) = $ _____ for all x in _____.

Exercises 13.4

(1–18) Evaluate the following. (Do not use tables or calculator.)

1. $\text{Sin}^{-1}(\sqrt{3}/2)$.
2. $\text{Sin}^{-1}(\sqrt{2}/2)$.
3. $\text{Cos}^{-1} \frac{1}{2}$.
4. $\text{Cos}^{-1} 0$.
5. $\text{Tan}^{-1}(\sqrt{3}/3)$.
6. $\text{Tan}^{-1} 0$.
7. $\text{Sin}^{-1} 1$.
8. $\text{Tan}^{-1} 1$.
9. $\text{Sin}^{-1} 0$.
10. $\text{Cos}^{-1} 1$.
11. $\text{Sin}^{-1}(-\frac{1}{2})$.
12. $\text{Sin}^{-1}(-1)$.
13. $\text{Cos}^{-1}(-\frac{1}{2})$.
14. $\text{Cos}^{-1}(-1)$.
15. $\text{Sin}^{-1}(-1/\sqrt{2})$.
16. $\text{Cos}^{-1}(-\sqrt{2}/2)$.
17. $\text{Tan}^{-1}(-\sqrt{3})$.
18. $\text{Tan}^{-1}(-1/\sqrt{3})$.

(19–24) Use trigonometric tables to find the approximate values of the following.

19. $\text{Sin}^{-1}(0.2250)$.
20. $\text{Cos}^{-1}(0.9848)$.
21. $\text{Tan}^{-1}(0.4841)$.
22. $\text{Sin}^{-1}(-0.4514)$.
23. $\text{Cos}^{-1}(-0.8923)$.
24. $\text{Tan}^{-1}(-2.605)$.

(25–56) Evaluate the following without using tables or calculator.

25. $\cos(\text{Sin}^{-1} \frac{1}{2})$.
26. $\sin(\text{Cos}^{-1} \frac{1}{2})$.
27. $\tan(\text{Sin}^{-1} 0)$.
28. $\sec(\text{Sin}^{-1} \sqrt{3}/2)$.
29. $\cot(\text{Sin}^{-1} 1)$.
30. $\csc(\text{Tan}^{-1} 1)$.
31. $\sin(\text{Cos}^{-1} \frac{3}{5})$.
32. $\cos(\text{Tan}^{-1} 2)$.
33. $\sec(\text{Sin}^{-1} \frac{5}{13})$.
34. $\tan[\text{Sin}^{-1}(-\frac{1}{2})]$.
35. $\sin[\text{Cos}^{-1}(-\frac{4}{5})]$.
36. $\cot[\text{Tan}^{-1}(-\frac{1}{2})]$.
37. $\text{Sin}^{-1}(\cos 3\pi/4)$.
38. $\text{Cos}^{-1}(\sin 30°)$.
39. $\text{Tan}^{-1}(\cos 180°)$.
40. $\text{Sin}^{-1}(\tan 5\pi/4)$.
41. $\text{Tan}^{-1}(\tan 30°)$.
42. $\text{Tan}^{-1}(\tan 2\pi/3)$.
43. $\text{Sin}^{-1}(\sin 30°)$.
44. $\text{Sin}^{-1}(\sin 150°)$.
45. $\text{Cos}^{-1}(\cos 60°)$.
46. $\text{Cos}^{-1}(\cos 210°)$.
47. $\text{Sin}^{-1}(2 \sin \pi)$.
48. $\text{Sin}^{-1}(2 \cos \pi/3)$.
49. $\text{Sin}^{-1}(3 \tan 0)$.
50. $\text{Sin}^{-1}(2 \sin \pi/6)$.
51. $\text{Cos}^{-1}(2 \sin \pi/6)$.
52. $\text{Cos}^{-1}(5 \cos \pi/2)$.
53. $\text{Cos}^{-1}(\frac{1}{2} \tan \pi/4)$.
54. $\text{Cos}^{-1}(\frac{1}{2} \tan 3\pi/4)$.
55. $\text{Tan}^{-1}(2 \sin 2\pi/3)$.
56. $\text{Tan}^{-1}(3 \cos 3\pi/2)$.
57. $\sin(\text{Sin}^{-1} \frac{1}{2} + \text{Cos}^{-1} \frac{1}{2})$.
58. $\cos(\text{Tan}^{-1} \sqrt{3} + \text{Cos}^{-1} 1)$.
59. $\tan(\text{Sin}^{-1} 1 - \text{Tan}^{-1} \sqrt{3})$.
60. $\sin(\text{Cos}^{-1} \frac{1}{2} + \text{Tan}^{-1} \sqrt{3})$.

(61–72) Express the following in terms of x.

61. $\tan(\text{Sin}^{-1} x)$.
62. $\sin(\text{Tan}^{-1} x)$.
63. $\cos(\text{Tan}^{-1} x)$.
64. $\sec(\text{Sin}^{-1} x)$.
65. $\sin(\text{Cos}^{-1} x)$.
66. $\cot(\text{Cos}^{-1} x)$.
67. $\csc(\text{Sin}^{-1} x)$.
68. $\cot(\text{Tan}^{-1} x)$.
69. $\sec(\text{Cos}^{-1} x)$.
70. $\tan(\pi - \text{Cos}^{-1} x)$.
71. $\sin(\pi - \text{Sin}^{-1} x)$.
72. $\cos(\pi + \text{Tan}^{-1} x)$.

(73–78) Prove the following identities.

73. $\text{Cos}^{-1} x = \text{Sin}^{-1} \sqrt{1 - x^2}$.
74. $\text{Tan}^{-1} x = \text{Sin}^{-1} \left(\dfrac{x}{\sqrt{1 + x^2}} \right)$.
75. $\text{Tan}^{-1} x = \text{Cos}^{-1} \dfrac{1}{\sqrt{1 + x^2}}$.
76. $\text{Cos}^{-1} x = \text{Tan}^{-1} \left(\dfrac{\sqrt{1 - x^2}}{x} \right)$.
*77. $\text{Tan}^{-1} x + \text{Tan}^{-1} y = \text{Tan}^{-1} \dfrac{x + y}{1 - xy}$.
*78. $\text{Tan}^{-1} \dfrac{2x}{1 - x^2} = 2 \text{Tan}^{-1} x$.

Chapter Review 13

1. Describe and or explain the following:
   (a) The trigonometric-square identities.
   (b) Addition formulas for sine, cosine, and tangent.
   (c) Double-angle formulas for sine, cosine, and tangent.
   (d) Half-angle formulas for sine and cosine.
   *(e) Method of expressing $a \cos \theta + b \sin \theta$ in terms of the sine or cosine of a single angle.
   *(f) Method of solving an equation of the form $a \cos \theta + b \sin \theta = c$ for $0 \leq \theta < 2\pi$.
   (g) Definition of inverse sine, inverse cosine, and inverse tangent functions.
   (h) The domains and ranges of these functions and their graphs.
   (i) Method of evaluating the inverse trigonometric functions.

2. *Summary of the most useful trigonometric identities*

$$\sec \theta = \frac{1}{\cos \theta}, \quad \csc \theta = \frac{1}{\sin \theta}, \quad \cot \theta = \frac{1}{\tan \theta}$$

$$\tan \theta = \frac{\sin \theta}{\cos \theta}, \quad \cot \theta = \frac{\cos \theta}{\sin \theta}$$

$$\cos^2 \theta + \sin^2 \theta = 1$$
$$\cos^2 \theta = 1 - \sin^2 \theta$$
$$\sin^2 \theta = 1 - \cos^2 \theta$$

$$\sec^2 \theta - \tan^2 \theta = 1$$
$$\sec^2 \theta = 1 + \tan^2 \theta$$
$$\tan^2 \theta = \sec^2 \theta - 1$$

$$\csc^2 \theta - \cot^2 \theta = 1$$
$$\csc^2 \theta = 1 + \cot^2 \theta$$
$$\cot^2 \theta = \csc^2 \theta - 1$$

$$\cos(\alpha + \beta) = \cos \alpha \cos \beta - \sin \alpha \sin \beta$$
$$\cos(\alpha - \beta) = \cos \alpha \cos \beta + \sin \alpha \sin \beta$$

$$\sin(\alpha + \beta) = \sin \alpha \cos \beta + \cos \alpha \sin \beta$$
$$\sin(\alpha - \beta) = \sin \alpha \cos \beta - \cos \alpha \sin \beta$$

$$\tan(\alpha + \beta) = \frac{\tan \alpha + \tan \beta}{1 - \tan \alpha \tan \beta}$$

$$\tan(\alpha - \beta) = \frac{\tan \alpha - \tan \beta}{1 - \tan \alpha \tan \beta}$$

$$\sin 2\alpha = 2 \sin \alpha \cos \alpha$$
$$\cos 2\alpha = \cos^2 \alpha - \sin^2 \alpha = 2 \cos^2 \alpha - 1$$
$$= 1 - 2 \sin^2 \alpha$$

$$\tan 2\alpha = \frac{2 \tan \alpha}{1 - \tan^2 \alpha}$$

$$\sin^2 \alpha = \frac{1 - \cos 2\alpha}{2}$$

$$\cos^2 \alpha = \frac{1 + \cos 2\alpha}{2}.$$

Review Exercises on Chapter 13

(1–14) Prove the following identities.

1. $(\sec \theta + \tan \theta)(\csc \theta - 1) = \cot \theta$.
2. $(\sin \theta + \cos \theta)(\tan \theta + \cot \theta) = \sec \theta + \csc \theta$.
3. $(\cos \theta + p \sin \theta)^2 + (\sin \theta - p \cos \theta)^2 = 1 + p^2$.
4. $(\sec \theta - \cos \theta)(\csc \theta - \sin \theta)(\tan \theta + \cot \theta) = 1$.
5. $(1 + \sin \theta + \cos \theta)^2 = 2(1 + \sin \theta)(1 + \cos \theta)$.
6. $(\sec \theta + \csc \theta)(\cos \theta - \sin \theta) = \cot \theta - \tan \theta$.
7. $\dfrac{\sin \alpha + \cos \beta}{\cos \alpha + \sin \beta} = \dfrac{\cos \alpha - \sin \beta}{\cos \beta - \sin \alpha}$.
8. $\dfrac{1 + \cos \theta}{1 - \cos \theta} = (\csc \theta + \cot \theta)^2$.

9. $\dfrac{\sin^2 \theta}{\sec \theta - 1} = \cos^2 \theta + \cos \theta$.

10. $\tan^2 \theta - \sin^2 \theta = \tan^2 \theta \sin^2 \theta$.

11. $\dfrac{1}{\tan \theta + \sin \theta} = \dfrac{\cos \theta \csc \theta}{1 + \cos \theta}$.

12. $\dfrac{1 + \cos \theta}{\sin \theta} + \dfrac{\sin \theta}{1 + \cos \theta} = 2 \csc \theta$.

13. $\dfrac{\sin \theta}{1 + \cos \theta} + \dfrac{\sin \theta \cos \theta}{1 - \cos \theta} = \csc \theta + \cot \theta \cos \theta$.

14. $\dfrac{\tan \alpha}{1 - \cos \alpha} - \dfrac{\sin \alpha}{1 + \cos \alpha} = \cot \alpha + \sec \alpha \csc \alpha$.

(15–18) Eliminate θ from the following equations.

15. $x = a \sec^3 \theta$, $y = b \tan^3 \theta$.

16. $x = a \sin^4 \theta$, $y = a \cos^4 \theta$.

17. $x \cos \theta = p$, $y \tan \theta = q$.

18. $x \sec \theta = a$, $y \csc \theta = b$.

19. Given $\sin \alpha = \tfrac{3}{5}$, $\cos \beta = \tfrac{5}{13}$, and $0 < \alpha, \beta < \pi/2$, find $\cos(\alpha + \beta)$ and $\sin(\alpha - \beta)$.

20. Given $\sin \theta = -\tfrac{2}{3}$, $\cos \phi = \tfrac{1}{3}$, and $\pi < \theta < 3\pi/2$, $3\pi/2 < \phi < 2\pi$, find the values of $\sin(\theta + \phi)$, $\cos(\theta + \phi)$, and $\tan(\theta - \phi)$.

(21–24) Evaluate the following without using tables or calculators.

21. $\sin 31° \sin(-14°) + \cos 31° \cos(-14°)$.

22. $\sin 95° \cos(-40°) - \cos 95° \sin(-40°)$.

23. $\sin 99° \cos 9° - \cos 99° \sin 9°$.

24. $\sin 55° \sin 35° - \cos 55° \cos 35°$.

(25–36) Prove the following.

25. $\sin(\pi/6 + \theta) + \cos(\pi/3 + \theta) = \cos \theta$.

26. $\sin(120° + \alpha) + \cos(210° + \alpha) = 0$.

27. $\tan \alpha + \tan \beta = \dfrac{\sin(\alpha + \beta)}{\cos \alpha \cos \beta}$.

28. $\tan \theta + \cot \phi = \cos(\theta - \phi) \sec \theta \csc \phi$.

29. $\csc \alpha - \cot \alpha = \tan \alpha/2$.

30. $\tan \alpha - \sin \alpha = 2 \tan \alpha \sin^2 \alpha/2$.

31. $\cos 2\theta = 1 - 8 \sin^2 \theta/2 \cos^2 \theta/2$.

32. $(\tan 2\theta - \sec 2\theta)(1 + \tan \theta) = \tan \theta - 1$.

33. $\sin^2(45° + \theta) - \sin^2(45° - \theta) = \sin 2\theta$.

34. $\tan(45° + A) - \tan(45° - A) = 2 \tan 2A$.

35. $\tan(45° + \theta) + \tan(45° - \theta) = 2 \sec 2\theta$.

36. $\cos^2(\pi/4 - \theta/2) - \sin^2(\pi/4 - \theta/2) = \sin \theta$.

37. If $A + B + C = \pi$, then show that
$\tan A + \tan B + \tan C = \tan A \tan B \tan C$
[Hint: $\tan(A + B) = \tan(\pi - C) = -\tan C$.]

38. If $A + B + C = \pi/2$, then show that
$\tan A \tan B + \tan B \tan C + \tan C \tan A = 1$.

39. Given $\sin \theta = -\tfrac{2}{3}$ and $-\pi < \theta < -\pi/2$, find $\sin(\pi/6 - \theta)$.

40. Given $\cos \theta = \tfrac{1}{3}$, evaluate $\cos 2\theta$.

41. Given $\tan \alpha = \tfrac{2}{3}$ and $\sin \beta = \tfrac{1}{5}$, $\pi/2 < \beta < \pi$, find $\tan(\alpha + \beta)$.

42. Given $\sin \alpha = \tfrac{1}{3}$, $\cos \beta = -\tfrac{3}{4}$, and $\pi/2 < \alpha < \pi$, $\pi < \beta < 3\pi/2$, find the value of $\tan(\alpha - \beta)$.

43. Given $\sin 330° = -\tfrac{1}{2}$, find $\tan 165°$.

44. Given $\tan 315° = -1$, find $\tan 157\tfrac{1}{2}°$.

45. Sketch the graph of $f(x) = \sqrt{3} \cos x - \sin x$.

46. Sketch the graph of $g(x) = \sin 2x \cos x - \cos 2x \sin x$.

*(47–50) Find the positive values of x that are less than $360°$ and satisfy the following equations.

47. $\sin x + \cos x = -1$.

48. $\sqrt{3} \sin x - \cos x = 1$.

49. $3 \sin x - 4 \cos x = 2.5$.

50. $12 \sin x + 5 \cos x = 6.5$.

(51–60) Evaluate the following without using tables or calculator.

51. $\sin(\text{Sin}^{-1} \tfrac{1}{3})$.

52. $\cos(\text{Cos}^{-1} \tfrac{2}{3})$.

53. $\tan[\text{Tan}^{-1}(-5)]$.

54. $\sin (\text{Cos}^{-1} \frac{3}{4})$.
55. $\sec (\text{Tan}^{-1} 7)$.
56. $\cot [\text{Sin}^{-1} (-\frac{2}{3})]$.
57. $\sin (\text{Sin}^{-1} 1 + \text{Cos}^{-1} 1)$.
58. $\cos (\text{Sin}^{-1} \frac{1}{2} + \text{Cos}^{-1} \frac{1}{2})$.
59. $\text{Sin}^{-1} (\cos 330°)$.
60. $\text{Cos}^{-1} (\tan 225°)$.

(61–62) Prove the following.

*61. $\text{Sin}^{-1} x + \text{Cos}^{-1} x = \pi/2 \quad (-1 \leq x \leq 1)$.

*62. $\text{Sin}^{-1} \dfrac{2x}{1+x^2} = 2 \text{Tan}^{-1} x \quad (-\infty < x < \infty)$.

14 TRIGONOMETRIC APPLICATIONS

14.1

The Law of Sines

In Section 12.3 we used the trigonometric functions to calculate the sides and angles in right triangles. These functions are also useful for similar calculations in triangles that do not contain a right angle. Such triangles are often called *oblique triangles*.

Recall that in the triangle ABC the three angles at the respective vertices are denoted by A, B, and C and the three sides are denoted by a, b, and c as shown in the adjoining figure. The three angles and the three sides are called the *parts* of the triangle. If the values of certain of these parts are known then the remaining parts can be calculated. Four cases can arise.

(a) *Three sides (SSS)*. In this case the values of a, b, and c are known and we need to calculate the three angles A, B, and C.

(b) *Two sides and the included angle (SAS)*. We might, for example, know the lengths of the sides a and b and the angle C included between them and need to calculate the third side c and the remaining two angles A and B.

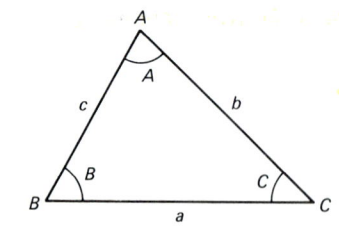

(c) *One side and two angles (ASA)*. We might, for example, know the side a and the angles A and B and need to calculate b and c and angle C. (Of course, if two angles are known the third can be found immediately since $A + B + C = 180°$.)

(d) *Two sides and an opposite angle* (*SSA*). We might, for example, know the sides a and b and the angle A and need to calculate c and the angles B and C.

The process of carrying out the above types of calculation is called *solving the triangle*. Generally this is most easily done by using one or both of two formulas known as the *law of sines* and the *law of cosines*. In this section we shall discuss the first of these formulas, which will enable us to solve problems of the ASA and SSA types. Problems of the SSS and SAS types are solved using the law of cosines.

In the triangle ABC, let AD be the perpendicular from A onto BC. The figures below illustrate respectively the cases when the angles B and C are both acute and when B is obtuse. Let AD have length h. Then in the right triangle ACD we have

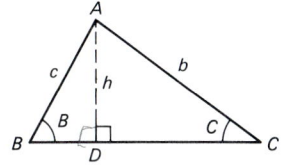

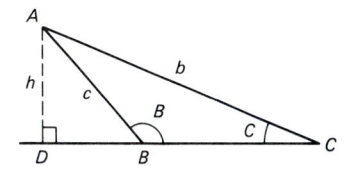

$$h = b \sin C.$$

Similarly in the right triangle ABD we have $h = c \sin B$ in the top figure and $h = c \sin (180° - B)$ in the bottom figure. Since* $\sin (180° - \theta) = \sin \theta$ for any θ, we have in both cases

$$h = c \sin B.$$

Thus

$$b \sin C = c \sin B$$

or

$$\frac{b}{\sin B} = \frac{c}{\sin C}.$$

In an exactly similar way we can drop a perpendicular from B onto AC. That would lead to the equation

$$\frac{a}{\sin A} = \frac{c}{\sin C}.$$

Combining these two equations, we have

$$\boxed{\frac{a}{\sin A} = \frac{b}{\sin B} = \frac{c}{\sin C}}$$

which is known as the *law of sines*.

---

*See Section 13.2. This identity is used frequently in solving triangles.

Sec. 14.1   The Law of Sines

Example 1 Solve the triangle in which $B = 30°$, $C = 80°$, and $b = 2$ centimeters.

Solution The triangle is drawn in the adjacent figure. The third angle is found immediately:
$$A = 180° - B - C$$
$$= 180° - 30° - 80° = 70°.$$

To find c we use the bc portion of the law of sines:
$$\frac{c}{\sin C} = \frac{b}{\sin B}$$
$$\frac{c}{\sin 80°} = \frac{2}{\sin 30°}$$
$$c = \frac{2(\sin 80°)}{\sin 30°}$$
$$= \frac{2(0.9848)}{(0.5)}$$
$$= 3.939.$$

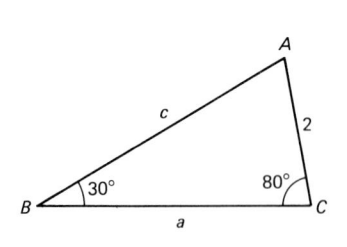

Similarly, to find a we use the ab portion of the law of sines:
$$\frac{a}{\sin A} = \frac{b}{\sin B}$$
$$\frac{a}{\sin 70°} = \frac{2}{\sin 30°}$$
$$a = \frac{2(\sin 70°)}{\sin 30°}$$
$$= \frac{2(0.9397)}{0.5}$$
$$= 3.759.$$

Thus the solution is $a = 3.759$ centimeters, $c = 3.939$ centimeters, and $A = 70°$.

Example 1 is typical of problems of ASA type—that is, those that involve solving a triangle when one side and two angles are given. Such problems arise often in surveying. The method of triangulation used in surveying consists of fixing a base line whose length and bearing (that is, direction with respect to north) are both measured. Then in order to locate the position of any other point, one measures the bearings of this point from the ends of the base line.

Example 2 A surveyor's base line AB is 55 meters long and the bearing of A from B is $21°40'$ east of north (see left figure on next page). A third point C is measured to have bearings of $134°10'$ east of north from A and $72°50'$ east of north from B. Calculate the distances AC and BC.

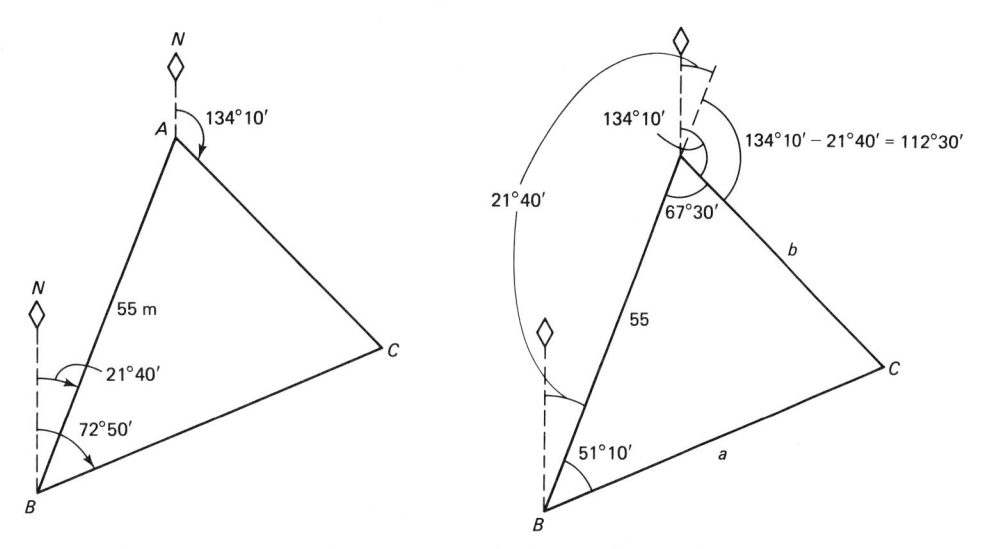

Solution We first calculate the angles A and B in triangle ABC (see right figure above):

$$B = 72°50' - 21°40' = 51°10'$$
$$A = 180° - 112°30' = 67°30'.$$

Then

$$C = 180° - A - B$$
$$= 180° - 67°30' - 51°10'$$
$$= 61°20'.$$

Using the law of sines, we have

$$\frac{a}{\sin A} = \frac{b}{\sin B} = \frac{c}{\sin C}$$

$$\frac{a}{\sin 67°30'} = \frac{b}{\sin 51°10'} = \frac{55}{\sin 61°20'}$$

$$\frac{a}{0.9239} = \frac{b}{0.7790} = \frac{55}{0.8774}.$$

Therefore,

$$a = \frac{55(0.9239)}{0.8774} = 57.91$$

$$b = \frac{55(0.7790)}{0.8774} = 48.83.$$

Thus C is 48.83 meters from A and 57.91 meters from B.

Problems of SSA type (those in which two sides and an opposite angle are specified) may have one solution, two solutions, or no solutions at all. For this reason, this type of problem is sometimes called the ***ambiguous case***.

Example 3 Solve the triangle in which $b = 2$, $c = 3$, and $B = 30°$.

Solution From the law of sines, we have

$$\frac{a}{\sin A} = \frac{b}{\sin B} = \frac{c}{\sin C}$$

$$\frac{a}{\sin A} = \frac{2}{\sin 30°} = \frac{3}{\sin C}. \qquad (i)$$

From the bc part of this formula we thus obtain

$$\sin C = \frac{3 \sin 30°}{2} = \frac{3(0.5)}{2} = 0.75.$$

The acute angle whose sine is 0.75 is $48°35'$. But the angle $180 - 48°35' = 131°25'$ also has a sine of 0.75 [since $\sin(180° - \theta) = \sin \theta$ for any angle θ]. Therefore there are two possibilities for the angle C: $C_1 = 48°35'$ and $C_2 = 131°25'$. As can be seen from the figures below, each of these two possibilities represents an acceptable solution for the triangle.

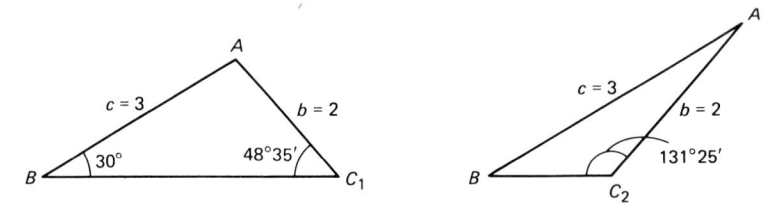

The values of A in the two cases are

(a) $A_1 = 180° - B - C_1 = 180° - 30° - 48°35' = 101°25'$.
(b) $A_2 = 180° - B - C_2 = 180° - 30° - 131°25' = 18°35'$.

Finally we obtain the third side a from the ab part of the law of sines. From (i):

$$a = \frac{2 \sin A}{\sin 30°} = \frac{2 \sin A}{0.5} = 4 \sin A.$$

For the two solutions we thus have

(a) $a_1 = 4 \sin 101°25' = 4(0.9802) = 3.921$.
(b) $a_2 = 4 \sin 18°35' = 4(0.3187) = 1.275$.

Thus the two triangles are given by

(a) $a_1 = 3.921$, $A_1 = 101°25'$, and $C_1 = 48°35'$.
(b) $a_2 = 1.275$, $A_2 = 18°35'$, and $C_2 = 131°25'$.

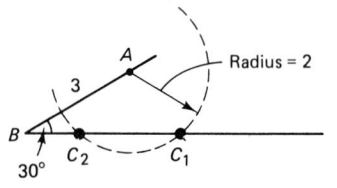

It is instructive to consider how we construct the triangle in Example 3 using protractor, ruler, and compasses. First we draw two lines at an angle of $30°$, their intersection being the vertex B. Along one of these lines we mark off a distance of 3 units, obtaining the vertex A (see the adjacent figure). To find the third vertex C, we must locate a point that lies on the second of the lines and is 2 units from A. So with compasses set at a radius of 2 units we draw a circle centered at A, and the point where this circle cuts the second line must be C. But there are obviously two such points of intersection, C_1 and C_2. These correspond to the two solutions we found for the triangle.

Sometimes the ambiguous case has no solution.

Example 4 Solve the triangle in which $b = 1$, $c = 3$, and $B = 30°$.

Solution This example is the same as Example 3 except for the value of b. Proceeding as in that example, we use the bc part of the law of sines to calculate C:

$$\frac{b}{\sin B} = \frac{c}{\sin C}$$

$$\frac{1}{\sin 30°} = \frac{3}{\sin C}$$

$$\sin C = 3 \sin 30°$$
$$= 3(0.5)$$
$$= 1.5.$$

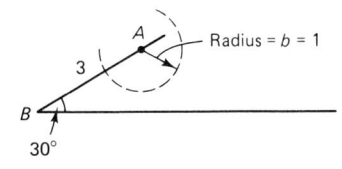

Since no angle can have a sine greater than 1, we conclude that there is *no solution* in this case.

Suppose we try to construct the triangle in Example 4. This is illustrated in the adjacent figure. The vertices A and B are constructed as before. The third vertex C lies at the intersection of the horizontal line and the circle of radius 1 unit centered at A. But in this case the circle does not intersect the horizontal line, so there can be no solution.

Sometimes only one solution exists.

Example 5 Solve the triangle in which $b = 4$, $c = 3$, and $B = 30°$.

Solution The bc part of the law of sines gives

$$\sin C = \frac{c \sin B}{b} = \frac{3 \sin 30°}{4} = \frac{3(0.5)}{4} = 0.375.$$

The acute angle C whose sine is 0.375 is 22°01′. The second solution for C is then $180° - 22°01' = 157°59'$. The values of A in these two cases are then as follows:

(a) $C_1 = 22°01'$
$A_1 = 180° - B - C_1 = 180° - 30° - 22°01' = 127°59'.$
(b) $C_2 = 157°59'$
$A_2 = 180° - B - C_2 = 180° - 30° - 157°59' = -7°59'.$

Since in case (b) A has turned out to be negative, this case does not provide a valid solution. The only solution is case (a):

$$C = 22°01', \quad A = 127°59'.$$

Sec. 14.1 The Law of Sines

We complete the solution in this case by using the *ab* part of the law of sines:

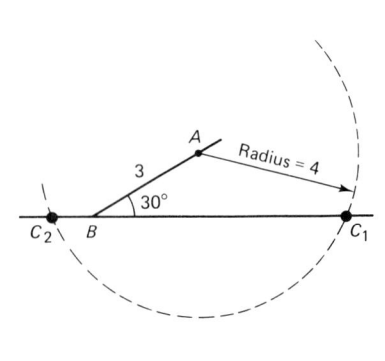

$$a = \frac{b \sin A}{\sin B} = \frac{4 \sin 127°59'}{\sin 30°} = \frac{4(0.7882)}{0.5}$$
$$= 6.306.$$

The solution is therefore $a = 6.306$, $A = 127°59'$, $C = 22°01'$.

The geometrical construction in Example 5 is shown in the adjacent figure. In this case the radius of the the circle centered at A is 4 units. Although this circle does intersect the horizontal line at two points, one of these points, C_2, lies to the left of B. Thus in the triangle ABC_2 the angle of 30° at B is not an interior angle, so this triangle is not a valid solution. The only solution is the triangle ABC_1, and this is the one whose parts were calculated.

Example 6 Solve the triangle in which $b = 4$, $a = 3$, and $B = 120°$.

Solution Here the given angle is obtuse. We have

$$\sin A = \frac{a \sin B}{b} = \frac{3 \sin 120°}{4} = \frac{3 \sin 60°}{4}$$
$$= \tfrac{3}{4}(0.8660) = 0.6495.$$

Therefore the acute angle A is 40°30′. In this case the second solution for A cannot be valid, since the triangle already contains one obtuse angle, B, and cannot contain a second. Then

$$C = 180° - A - B = 180° - 40°30' - 120° = 19°30'.$$

Finally, we have

$$c = \frac{b \sin C}{\sin B} = \frac{4 \sin 19°30'}{\sin 120°} = \frac{4(0.3338)}{0.8660} = 1.542.$$

The solution is therefore $c = 1.542$, $A = 40°30'$, and $C = 19°30'$.

The precise conditions which govern the number of solutions in the ambiguous case are given in Exercise 34.

Diagnostic Test 14.1

Fill in the blanks.

1. $\dfrac{\sin A}{a} = \dfrac{(\underline{\quad\quad})}{b}$.

2. $\dfrac{\sin C}{\sin B} = \underline{\quad\quad}$.

3. The law of sines can be used to solve triangles in the _____ and _____ cases.

4. When $C = 90°$, the law of sines reduces to $\sin A = $ _____ and $\sin B = $ _____.

Exercises 14.1

(1–18) Calculate the remaining parts in the triangle ABC.

1. $A = 45°, B = 60°, b = 3$.
2. $A = 30°, C = 45°, b = 5$.
3. $B = 45°, b = 3, c = 4$.
4. $B = 60°, b = 7, c = 8$.
5. $A = 60°, a = 8, c = 7$.
6. $C = 35°, a = 2, c = 4$.
7. $B = 110°, C = 30°, a = 2$.
8. $A = 20°, C = 150°, c = 8$.
9. $A = 60°, c = 3, a = 2$.
10. $B = 80°, a = 2, b = 4$.
11. $B = 120°, a = 1, b = 3$.
12. $A = 160°, a = 2, c = 3$.
13. $B = 110°, a = 3, b = 4$.
14. $C = 50°, a = 5, c = 3$.
15. $A = 40°, a = 2, b = 4$.
16. $B = 20°, b = 4, c = 5$.
17. $A = 35°, a = 3, b = 5$.
18. $A = 100°, a = 5, b = 1$.

19. Point A lies 400 yards due north of point B. A surveyor determines that a third point C has a direction 45°E of N from A and 30°E of N from B. Find the distances of C from A and B.

20. A lies 2 miles due west of B. From A, the point C lies in the direction 50°E of N while from B, C lies 330°E of N (i.e., 30°W of N). Find the distances of C from A and B.

21. A ship is traveling in a direction 20°E of N. At a certain time a lighthouse is measured to have a direction of 40°E of N. When the ship has traveled 2 miles further along its course, the same lighthouse has a direction of 70°E of N. Calculate the distance between the ship and the lighthouse at the second observation.

22. Two boats leave the same point, one traveling in a direction 20°E of N and the other in a direction 100°E of N. After the first boat has traveled 4 miles, it observes that the second boat lies due south. How far has the second boat traveled and what is the distance between it and the first boat?

23. A building stands at the top of a hill that slopes upward at a constant angle of 25° to the horizontal. From a certain point on the hillside the top of the building has an angle of elevation of 60° above the horizontal. From a point 100 feet farther down the hillside the angle of elevation is 45°. How tall is the building?

24. How long a shadow does the building in Exercise 23 cast down the hillside when the altitude of the sun is 70° above the horizontal?

25. Two ships start from the same point, ship A traveling at 10 miles per hour and ship B at 8 miles per hour. After half an hour ship A observes that B lies in a direction due west. If A had been traveling in a direction 40°E of N, in what direction must B have been traveling?

26. A is 120 yards from B and 200 yards from C. From B, A lies in the direction 120°E of N while C lies in the direction 70°E of N. Calculate the distance from B to C and the direction of C from A.

27. The altitude of the sun is 45° above the horizontal. A pole 30 feet long is tilted directly away from the sun and casts a shadow that is 40 feet long. At what angle is the pole tilted to the vertical?

28. The altitude of the sun is 40° above the horizontal. A man 6 feet tall stands on a roadway that slopes directly away from the sun. If the length of his shadow is 12 feet, find the angle at which the roadway slopes. Find also the distance between the tip of his head and the tip of his shadow.

(29–32) Use the law of sines to show that in the triangle ABC the following relations hold.

29. $\dfrac{b+c}{a} = \dfrac{\sin B + \sin C}{\sin (B+C)}$.

30. $\dfrac{c-a}{b} = \dfrac{\sin C - \sin A}{\sin (A+C)}$.

31. $\dfrac{a+b-c}{a} = (1 - \cos B) + \dfrac{b}{a}(1 - \cos A)$.

32. $\dfrac{a+b-c}{a+b+c}$
$= \dfrac{\sin A\,(1 - \cos B) + \sin B\,(1 - \cos A)}{\sin A\,(1 + \cos B) + \sin B\,(1 + \cos A)}$.

33. The area of the triangle ABC is given by the expression $\tfrac{1}{2}bc \sin A$ (see Section 14.2, p. 565) Use the law of sines to express the area in terms of a, c, and B and in terms of a, b, and C.

*34. By analyzing the geometrical construction in the SSA case when b, c and B are given, prove that: If $B < 90°$, no solution exists for $b < c \sin B$, two exist for $c \sin B < b < c$, and one solution exists if either $b = c \sin B$ or $b \geq c$. If $B \geq 90°$, no solution exists if $b \leq c$ and one exists if $b > c$.

14.2

The Law of Cosines

When the three sides of a triangle are given or when two sides and the angle included between them are given, the triangle cannot be solved by means of the law of sines. In such cases a group of formulas called the law of cosines must be used. They are derived as follows.

In the triangle ABC, drop a perpendicular AD from A onto BC. The two figures below illustrate the two cases when the angle B is acute or obtuse, respectively. Let $h = AD$ and $d = BD$. Then when B is acute, the distance CD equals $a - d$, and in the right triangle ABD, $d = c \cos B$. Therefore $CD = a - c \cos B$.

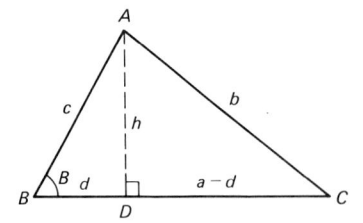

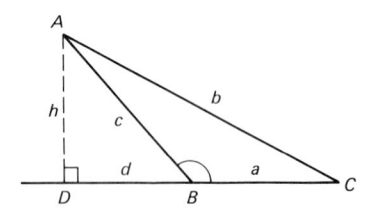

When B is obtuse, as in the second figure, the distance CD equals $a + d$. But now in the right triangle ABD the angle $ABD = 180° - B$, and so $d = c \cos (180° - B) = -c \cos B$. Therefore
$$CD = a + d = a + (-c \cos B) = a - c \cos B.$$

Consequently in both cases $CD = a - c \cos B$.

Now we apply Pythagoras's theorem to the two triangles ABD and ACD. We obtain
$$AB^2 = AD^2 + BD^2$$
or
$$\begin{aligned} c^2 &= h^2 + d^2 \\ &= h^2 + (\pm c \cos B)^2 \\ &= h^2 + c^2 \cos^2 B \end{aligned} \tag{i}$$

and
$$AC^2 = AD^2 + CD^2$$
or
$$\begin{aligned} b^2 &= h^2 + (a - c \cos B)^2 \\ &= h^2 + a^2 - 2ac \cos B + c^2 \cos^2 B. \end{aligned} \tag{ii}$$

Subtracting (ii) from (i), we obtain
$$\begin{aligned} c^2 - b^2 &= h^2 + c^2 \cos^2 B - (h^2 + a^2 - 2ac \cos B + c^2 \cos^2 B) \\ &= -a^2 + 2ac \cos B. \end{aligned}$$

It follows by rearranging the terms that
$$b^2 = a^2 + c^2 - 2ac \cos B.$$

Two other formulas can be derived similarly. We thus have the following *law of cosines*:

$$\boxed{\begin{aligned} a^2 &= b^2 + c^2 - 2bc \cos A \\ b^2 &= a^2 + c^2 - 2ac \cos B \\ c^2 &= a^2 + b^2 - 2ab \cos C. \end{aligned}} \tag{1}$$

These formulas state that *the square of any side of a triangle is equal to the sum of the squares of the other two sides minus twice their product times the cosine of the angle between them.*

These formulas can also be rewritten in a form that expresses the cosines of the angles in terms of the sides:

$$\boxed{\cos A = \frac{b^2 + c^2 - a^2}{2bc}, \quad \cos B = \frac{a^2 + c^2 - b^2}{2ac}, \quad \cos C = \frac{a^2 + b^2 - c^2}{2ab}.}$$
$$\tag{2}$$

These formulas state that *the cosine of any angle in a triangle is equal to the sum of the squares of the two adjacent sides minus the square of the opposite side all divided by twice the product of the adjacent sides.*

Example 1 Find the angles in the triangle whose three sides are $a = 6, b = 3, c = 4$.

Solution Since we wish to calculate the angles, it is quicker to take the law of cosines in the form (2). We have

$$\cos A = \frac{b^2 + c^2 - a^2}{2bc}$$

$$= \frac{3^2 + 4^2 - 6^2}{2(3)(4)}$$

$$= \frac{-11}{24}$$

$$= -0.4583.$$

Since $\cos A$ is negative, A must be obtuse. The reference angle A_r is the acute angle such that $\cos A_r = 0.4583$, and from the table we find that $A_r = 62°43'$. Then $A = 180° - A_r = 117°17'$.

Using the law of cosines again, we get

$$\cos B = \frac{a^2 + c^2 - b^2}{2ac}$$

$$= \frac{6^2 + 4^2 - 3^2}{2(6)(4)}$$

$$= \frac{43}{48}$$

$$= 0.8958.$$

Therefore $B = 26°23'$.

Finally,

$$C = 180° - A - B = 180° - 117°17' - 26°23' = 36°20'.$$

The angles are therefore $A = 117°17'$, $B = 26°23'$, and $C = 36°20'$.

We could, of course, also compute C by using the law of cosines a third time, but this would involve more work. However, let us do it as a check:

$$\cos C = \frac{a^2 + b^2 - c^2}{2ab}$$

$$= \frac{6^2 + 3^2 - 4^2}{2(6)(3)}$$

$$= \frac{29}{36}$$

$$= 0.8056$$

and from the tables we again find $C = 36°20'$.

In calculations such as Example 1, we can if we wish use the law of sines to calculate the second angle once the first has been found using the law of cosines. For instance, having calculated $A = 117°17'$ in Example 1, we could calculate B as follows:

$$\sin B = \frac{b \sin A}{a} = \frac{3 \sin 117°17'}{6}$$

$$= \frac{\sin 62°43'}{2} = \frac{0.8888}{2}$$

$$= 0.4444$$

from which $B = 26°23'$. However the angle $180° - 26°23' = 153°37'$ also has a sine equal to 0.4444, and this must be considered as a possible solution for B. In the present example we can discard this second possibility immediately, since the triangle already has one obtuse angle. In some cases, however, we can discard the second solution only by computing the third angle in the triangle. For this reason it is usually better, when possible, to use the cosine formula to compute angles, because the value of its cosine determines an angle uniquely for angles between $0°$ and $180°$ whereas the value of its sine does not.

Example 2 Solve the triangle for which $a = 3$, $b = 5$, and $C = 50°$.

Solution We are given two sides and the included angle, as illustrated in the adjacent figure. We can compute c by using the law of cosines in the form (1):

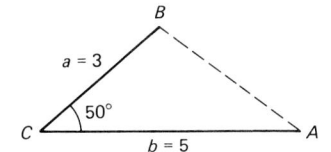

$$c^2 = a^2 + b^2 - 2ab \cos C$$
$$= 3^2 + 5^2 - 2(3)(5) \cos 50°$$
$$= 9 + 25 - 30(0.6428)$$
$$= 14.72.$$

Therefore $c = 3.836$.

Again we have a choice of using either the law of sines or the law of cosines to compute the remaining angles. For the reason mentioned earlier we use the law of cosines. In the form (2) it gives

$$\cos A = \frac{b^2 + c^2 - a^2}{2bc}$$

$$= \frac{25 + 14.72 - 9}{2(5)(3.836)}$$

$$= \frac{30.72}{38.36}$$

$$= 0.8008.$$

Therefore $A = 36°48'$.

Finally, $B = 180° - A - C = 180° - 50° - 36°48' = 93°12'$. The solution is therefore $A = 36°48'$, $B = 93°12'$, and $c = 3.836$.

Example 3 A flagpole 100 feet high stands on top of a mound whose sides slope at $25°$ to the horizontal. The pole casts a shadow 200 feet long down the side of the mound. What is the sun's altitude?

Solution In the figure below, AC represents the flagpole and BC its shadow down the side of the mound. In the triangle ABC we are given that $a = 200$ and $b = 100$. Also the angle $C = 90° + 25° = 115°$. Thus we know two sides and the included angle. First we use the law of cosines to calculate the third side:

$$c^2 = a^2 + b^2 - 2ab \cos C$$
$$= 200^2 + 100^2 - 2(200)(100) \cos 115°$$
$$= 100^2[4 + 1 - 4(-\cos 65°)].$$

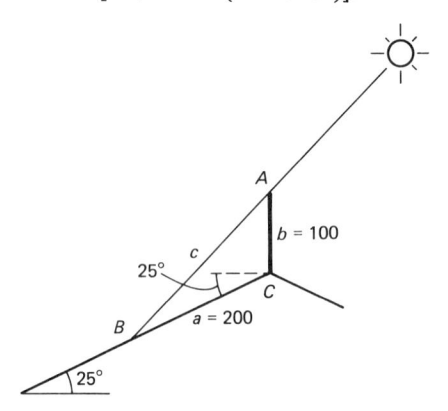

Here we have factored out 100^2 to keep the numbers small, and we have used the fact that $\cos 115° = \cos(180° - 65°) = -\cos 65°$. Then

$$c^2 = 10,000[5 + 4(0.4226)] = 6.691 \times 10^4.$$

Therefore $c = 258.7$. Now we use the law of cosines again to find the angle B:

$$\cos B = \frac{a^2 + c^2 - b^2}{2ac}$$
$$= \frac{200^2 + 6.691 \times 10^4 - 100^2}{2(200)(258.7)}$$
$$= \frac{4 + 6.691 - 1}{4(2.587)}$$
$$= 0.9365.$$

Therefore $B = 20°32'$.

Finally the altitude of the sun is equal to $25° + B$, or $45°32'$ (see the figure below).

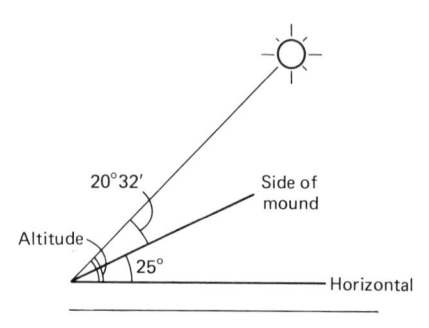

Now that we have examined all four cases of the solution of triangles, let us collect the general results in a single table.

Type of problem	Given values	Method of solution	Uniqueness of solution
SSS	a, b, c	Law of cosines	Unique solution exists provided that the longest side does not exceed the sum of the two shorter sides
SAS	a, b, C or b, c, A or a, c, B	Law of cosines	Unique solution always exists
ASA	One of $a, b,$ or c and two of $A, B,$ or C	Third angle from $A + B + C = 180°$, then law of sines	Unique solution always exists provided the given angles do not exceed 180°
SSA	a, b, A or a, b, B or b, c, B or b, c, C or $a, c\,A$ or a, c, C	Law of sines	May be no solution, one solution or two solutions

AREA OF A TRIANGLE

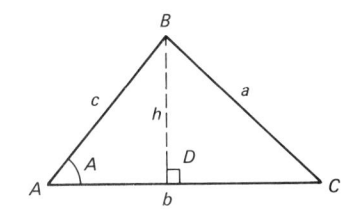

The area of a triangle can be expressed in various ways in terms of the sides and angles. Consider the adjoining figure, in which BD is the perpendicular from B onto AC, and has length h. Then

$$\text{area} = \tfrac{1}{2}(\text{base})(\text{height})$$
$$= \tfrac{1}{2}bh.$$

But from the right triangle BAD we have $h = c \sin A$. Therefore

$$\text{area} = \tfrac{1}{2}bc \sin A.$$

In a similar way we can derive the two alternative expressions for the area of the triangle, area $= \tfrac{1}{2}ac \sin B = \tfrac{1}{2}ab \sin C$. (See also Exercise 33 of Section 14.1) Thus:

$$\boxed{\text{area} = \tfrac{1}{2}bc \sin A = \tfrac{1}{2}ac \sin B = \tfrac{1}{2}ab \sin C.}$$

This formula states that *the area of a triangle is equal to half the product of any two sides times the sine of the angle between them.*

Example 4 Calculate the area of the triangle given in Example 2.

Solution In Example 2 we were given $a = 3$, $b = 5$, and $C = 50°$. Therefore we use the formula

$$\text{area} = \tfrac{1}{2}ab \sin C$$
$$= \tfrac{1}{2}(3)(5) \sin 50°$$
$$= (7.5)(0.7660)$$
$$= 5.745 \text{ square units.}$$

The above formulas express the area of a triangle in terms of two sides and the included angle. An alternative formula expresses the area in terms of the lengths of the three sides. Let $s = \tfrac{1}{2}(a + b + c)$ denote half the sum of the sides of the triangle (i.e., half the perimeter). Then

$$\text{area} = \sqrt{s(s-a)(s-b)(s-c)}.$$

This is known as *Hero's formula*. (See Exercise 36.)

Example 5 Calculate the area of the triangle given in Example 1.

Solution In Example 1 we were given $a = 6$, $b = 3$, and $c = 4$. Then

$$s = \tfrac{1}{2}(a+b+c) = \tfrac{1}{2}(6+3+4) = 6.5.$$

So from Hero's formula

$$\text{area} = \sqrt{(6.5)(6.5-6)(6.5-3)(6.5-4)}$$
$$= \sqrt{\tfrac{455}{16}}$$
$$= 5.333 \text{ square units.}$$

Diagnostic Test 14.2

Fill in the blanks.

1. $a^2 = $ _____. (law of cosines)
2. $\cos C = $ _____. (law of cosines)
3. In the ASA case we can solve the triangle using the law of _____.
4. In the SSA case we can solve the triangle using the law of _____.
5. In the SSS case we can solve the triangle using the law of _____.
6. In the SAS case we can solve the triangle using the law of _____.
7. The ambiguous case is the _____ case.
8. In the SSA case there are _____ solutions. (How many?)
9. When $C = 90°$, the law of cosines $c^2 = a^2 + b^2 - 2ab \cos C$ reduces to _____.
10. In terms of a, b, and C, the area of the triangle is _____.

11. Hero's formula expresses the area of the triangle in terms of _____.

12. In terms of a, b, and c, the area is _____ where $s = $ _____.

Exercises 14.2

(1–8) Calculate the remaining parts in the triangle ABC.

1. $a = 4, b = 5, c = 6$.
2. $a = 6, b = 4, c = 7$.
3. $a = 4, b = 3, c = 2$.
4. $a = 10, b = 5, c = 6$.
5. $b = 3, c = 4, A = 30°$.
6. $a = 8, b = 2, C = 60°$.
7. $a = 2, c = 10, B = 110°$.
8. $a = 10, c = 15, B = 140°$.

9–16. Calculate the area of each of the triangles in Exercises 1 through 8, using only the parts of the triangles that are given.

17. Three pieces of wood of lengths 8 feet, 10 feet, and 15 feet are joined at their ends to form a triangular frame. Calculate the angles in the triangle and the area that it encloses.

18. A loop of wire 30 centimeters long is passed around three pegs P, Q, and R to form a triangle. If $PQ = 12$ centimeters and $QR = 7$ centimeters, calculate the angles in the triangle.

19. Two boats leave the same point traveling at 10 and 12 miles per hour in different directions. After one hour they are 5 miles apart. What is the angle between their two directions?

20. The three sides of a triangular plot of land are measured to be 150 feet, 240 feet, and 300 feet. Calculate the angle between the two longest sides and the area of the plot.

21. A flagpole 80 feet high is erected on a hillside that slopes at 15° to the horizontal. The pole is supported by guy wires that are attached to the top of the pole and are anchored to the ground at a distance of 40 feet from the base of the pole. What are the lengths of the guys that are located directly up and directly down the hill?

22. At a certain time the flagpole in Exercise 21 casts a shadow 100 feet long directly up the hillside. What is the altitude of the sun?

23. An orienteer walks 360 meters from A to B, then turns to the left through 75° and walks 400 meters from B to C. Through what angle must she now turn so as to head directly back to A and how far must she walk to get there?

24. Two ships leave the same point, one traveling at 15 miles per hour in the direction 120°E on N and the second at 8 miles per hour in the direction 230°E of N. How far apart are the ships after one hour? In what direction does the first ship lie from the second?

25. An airplane travels at a speed of 140 miles per hour in the direction 30°E of N relative to the air. There is a wind of 30 miles per hour blowing *from* the direction 80°E of N. At what speed and in what direction does the airplane travel relative to the ground?

26. A surveyor S knows that he is 300 yds from a tree T and 240 yds from a rock R. The angle TSR is measured to be 78°. Calculate the distance between T and R.

*27. Using the law of sines only, show that in the triangle ABC,

$$\frac{a-b}{a+b} = \frac{\tan\frac{1}{2}(A-B)}{\tan\frac{1}{2}(A+B)}.$$

Show how this formula can be used to solve the triangle when a, b, and C are given (SAS case).

28. Use Exercise 27 to solve Exercises 5 through 8.

(29–34) Show that in any triangle ABC the following equations are satisfied.

29. $a^2 + b^2 + c^2 = 2(bc \cos A + ca \cos B + ab \cos C)$.

30. $\dfrac{\cos A}{a} + \dfrac{\cos B}{b} + \dfrac{\cos C}{c} = \dfrac{a^2 + b^2 + c^2}{2abc}$.

31. $(a + b - c)(a + b + c) = 4ab \cos^2 \dfrac{C}{2}$.

$\left[\text{Hint: } \cos^2 \dfrac{C}{2} = \dfrac{1 + \cos C}{2}.\right]$

32. $(a + b - c)(c + a - b) = 4bc \sin^2 \dfrac{A}{2}$.

33. $(a + b + c)^2 = 4\left(bc \cos^2 \dfrac{A}{2} + ca \cos^2 \dfrac{B}{2} + ab \cos^2 \dfrac{C}{2}\right)$.

34. $(a + b - c)^2 = 4\left(ab \cos^2 \dfrac{C}{2} - bc \sin^2 \dfrac{A}{2} - ca \sin^2 \dfrac{B}{2}\right)$.

*35. Use the cosine formula to derive the sine formula.

$\left[\text{Hint: } \dfrac{\sin A^2}{a^2} = \dfrac{1 - \cos^2 A}{a^2}, \text{ and so on.}\right]$

*36. Prove Hero's formula. [*Hint:* Write (area)^2 = $\frac{1}{4}b^2c^2(1 - \cos^2 A)$; then use the cosine formula to express $(1 - \cos A)(1 + \cos A)$ in terms of $a, b,$ and c.]

14.3

Polar Coordinates and the Complex Plane

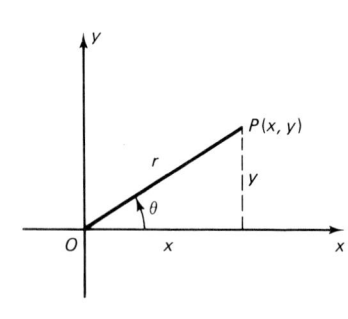

As we know, any point P lying in a given plane can be represented by its rectangular Cartesian coordinates (x, y). There are alternative ways of representing points in a plane by pairs of numbers. The most important of these is the system of polar coordinates.

Let $P(x, y)$ be any point in the plane except the origin, as in the adjacent figure. Let r be the distance OP from the origin to P and let θ be the angle between OP and the positive x axis, measured in the usual positive sense (i.e., counterclockwise). Then r and θ are called the ***polar coordinates*** of the point P; r is called the ***radial coordinate*** and θ the ***angular coordinate*** or the ***polar angle***.

From the above figure we see that

$$x = r \cos \theta, \qquad y = r \sin \theta. \tag{1}$$

These equations enable us to calculate the Cartesian coordinates of P if we are given the polar coordinates. They remain true whatever quadrant P lies in. For example, the following figure illustrates the case of a point in the third quadrant. In this case x, y, $\cos \theta$, and $\sin \theta$ are all negative, and relations (1) still hold.

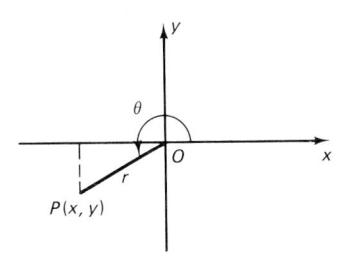

Example 1 Find the Cartesian coordinates of the points whose polar coordinates are as follows:

(a) $r = 4, \theta = \dfrac{3\pi}{4}$.

(b) $r = 6, \theta = \dfrac{-5\pi}{6}$.

Solution The two points are denoted by P and Q on the adjacent figure.

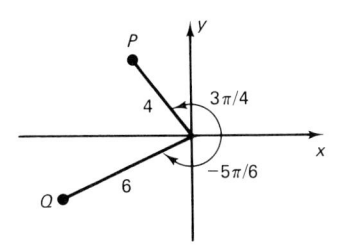

(a) For P:

$$x = r \cos \theta = 4 \cos \frac{3\pi}{4}$$

$$= -4 \cos \frac{\pi}{4} = -4\left(\frac{1}{\sqrt{2}}\right) = -2\sqrt{2} \approx -2.83$$

$$y = r \sin \theta = 4 \sin \frac{3\pi}{4}$$

$$= 4 \sin \frac{\pi}{4} = 4\left(\frac{1}{\sqrt{2}}\right) = 2\sqrt{2} \approx 2.83.$$

(b) For Q:

$$x = r \cos \theta = 6 \cos \left(-\frac{5\pi}{6}\right) = 6 \cos \frac{5\pi}{6}$$

$$= -6 \cos \frac{\pi}{6} = -6\frac{\sqrt{3}}{2} = -3\sqrt{3} \approx -5.20$$

$$y = r \sin \theta = 6 \sin \left(-\frac{5\pi}{6}\right) = -6 \sin \frac{5\pi}{6}$$

$$= -6 \sin \frac{\pi}{6} = -6\left(\frac{1}{2}\right) = -3.$$

As in the above example, it is more usual to use radians as units for polar angles, although occasionally degrees are used. We can also see that the polar angle is not uniquely determined. For example, the point Q in Example 1 could also be represented by $\theta = 7\pi/6$ rather than $-5\pi/6$, or P could be represented by $\theta = 11\pi/4$ rather than $3\pi/4$. In general, *we can add or subtract any integral multiple of 2π to a polar angle and it will still represent the same point.*

Given the Cartesian coordinates of a point, we can calculate the polar coordinates as follows. First, we note that

$$x^2 + y^2 = r^2 \cos^2 \theta + r^2 \sin^2 \theta = r^2(\cos^2 \theta + \sin^2 \theta) = r^2 \cdot 1 = r^2.$$

Therefore $r = \sqrt{x^2 + y^2}$. Second,

$$\frac{y}{x} = \frac{r \sin \theta}{r \cos \theta} = \frac{\sin \theta}{\cos \theta} = \tan \theta.$$

Thus we have

$$r = \sqrt{x^2 + y^2}, \qquad \tan \theta = \frac{y}{x}.$$

These equations determine r uniquely but they are not sufficient to determine θ, since the angles θ and $\theta + \pi$ have the same tangent. This ambiguity can be resolved most easily by making use of the quadrant in which the given point lies.

Example 2 Find the polar coordinates of the following points (x, y).
(a) $(1, \sqrt{3})$. (b) $(4, -3)$. (c) $(-3, 0)$. (d) $(0, 2)$.

Solution (a) $x = 1, y = \sqrt{3}$. Therefore

$$r = \sqrt{x^2 + y^2} = \sqrt{1 + 3} = 2$$

$$\tan \theta = \frac{y}{x} = \frac{\sqrt{3}}{1} = \sqrt{3}.$$

There are two angles whose tangent is $\sqrt{3}$, namely $\pi/3$ and $4\pi/3$ (or $-2\pi/3$), as illustrated in the adjoining figure. Since the given point lies in the first quadrant, the correct value is $\theta = \pi/3$. Therefore $r = 2$, $\theta = \pi/3$.

(b) $x = 4, y = -3$. Therefore

$$r = \sqrt{x^2 + y^2} = \sqrt{16 + 9} = 5$$

$$\tan \theta = \frac{y}{x} = \frac{-3}{4} = -0.75.$$

From the tables we find that the reference angle whose tangent is 0.75 is 36°52′, or 0.6435 radians. Therefore the two possible values of θ are $\pi - 0.6435 = 2.4981$ or -0.6435. Since $x > 0$ and $y < 0$ the point is in the fourth quadrant and the second of these values is the correct one. Therefore $r = 5, \theta = -0.6435$.

(c) $x = -3, y = 0$.

$$r = \sqrt{x^2 + y^2} = \sqrt{9 + 0} = 3.$$

Points on the negative x axis all have polar angle equal to π. Therefore $r = 3, \theta = \pi$.

(d) $x = 0, y = 2$.

$$r = \sqrt{x^2 + y^2} = \sqrt{0 + 4} = 2.$$

Points on the positive y axis all have polar angle equal to $\pi/2$. Therefore $r = 2, \theta = \pi/2$.

Note Points on the positive x axis have $\theta = 0$, those on the positive y axis have $\theta = \pi/2$, those on the negative x axis have $\theta = \pi$, and those on the negative y axis have $\theta = 3\pi/2$. The polar angle is not defined at the origin, which has radial coordinate $r = 0$. The formula $\tan \theta = y/x$ cannot be used on the y axis, since there the denominator x is zero.

THE COMPLEX PLANE

The set of all real numbers can be represented geometrically by the real-number line, each point on the line corresponding to one and only one real number. In a similar way we can represent complex numbers by the points in a plane. The complex number $a + bi$ (see Section 9.2) is represented by the point whose Cartesian coordinates are (a, b). That is, the x coordinate is taken as the real part of the complex number and the y coordinate as the coefficient of i in its imaginary part.

The figure below illustrates the geometrical representation of several complex numbers. Each point is labeled with the corresponding complex number.

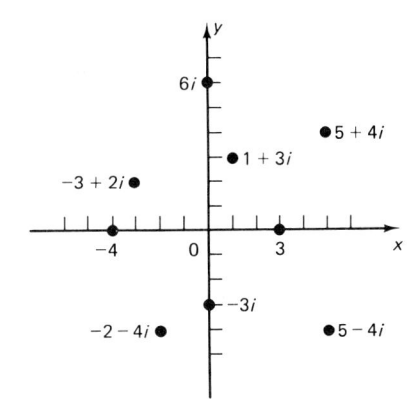

The x axis contains purely real numbers (i.e., numbers with zero imaginary parts). It is called the *real axis*. The y axis is called the *imaginary axis*, since it contains only purely imaginary numbers. This geometrical representation of complex numbers is referred to as the *complex plane* (or the *Argand diagram*).

Note that the points representing $a + bi$ and its complex conjugate $a - bi$ are reflections of one another in the x axis (for example, the points $5 + 4i$ and $5 - 4i$ in the above figure).

The use of polar coordinates in the complex plane leads to an extremely important way of writing complex numbers. In the adjacent figure, the complex number $a + bi$ is represented by the point $P(a, b)$ in the complex plane. If P has polar coordinates r and θ, then as usual $a = r \cos \theta$ and $b = r \sin \theta$. Therefore

$$a + bi = r \cos \theta + (r \sin \theta)i$$
$$= r(\cos \theta + i \sin \theta).$$

(The i is written before the $\sin \theta$ for convenience.)

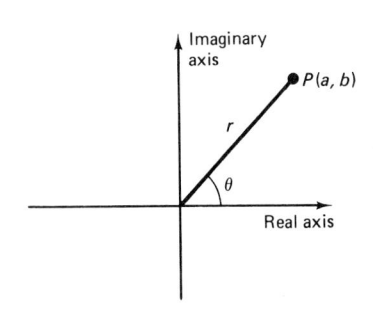

The form $r(\cos\theta + i\sin\theta)$ is called the **polar form** of the complex number $a + bi$. The radial coordinate r is called the **modulus** of $a + bi$, and the polar angle θ is called the **argument** of $a + bi$. The modulus and argument are given by the usual expressions for polar coordinates:

$$r = \sqrt{a^2 + b^2}, \qquad \tan\theta = \frac{b}{a}.$$

Example 3 Express each of the following complex numbers in polar form.
(a) $3 - 3\sqrt{3}\,i$. (b) $-2 + 2i$. (c) $-4i$.

Solution (a) For $3 - 3\sqrt{3}\,i$ we have

$$r = \sqrt{3^2 + (-3\sqrt{3})^2} = \sqrt{9 + 27} = 6$$

$$\tan\theta = \left(-\frac{3\sqrt{3}}{3}\right) = -\sqrt{3}.$$

Therefore $\theta = -\pi/3$ (since the point lies in the fourth quadrant—see the adjoining figure). Alternatively we could, for example, take $\theta = 5\pi/3$. Thus we have the polar form

$$3 - 3\sqrt{3}\,i = 6\left[\cos\left(-\frac{\pi}{3}\right) + i\sin\left(-\frac{\pi}{3}\right)\right].$$

(b) For $-2 + 2i$,

$$r = \sqrt{(-2)^2 + 2^2} = 2\sqrt{2}$$

$$\tan\theta = \frac{2}{-2} = -1$$

Therefore $\theta = 3\pi/4$ (see the adjacent figure) and so

$$-2 + 2i = 2\sqrt{2}\left[\cos\left(\frac{3\pi}{4}\right) + i\sin\left(\frac{3\pi}{4}\right)\right].$$

(c) For $-4i$, $r = 4$ and $\theta = -\pi/2$ or $3\pi/2$ (see the above figure). Therefore

$$-4i = 4\left[\cos\left(\frac{3\pi}{2}\right) + i\sin\left(\frac{3\pi}{2}\right)\right].$$

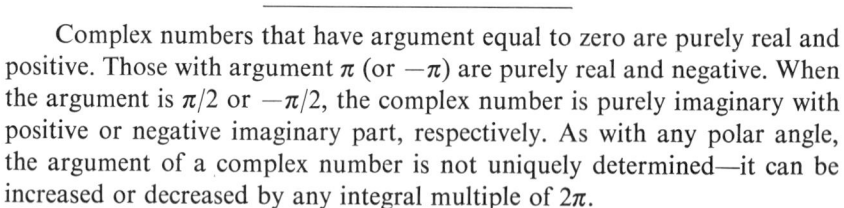

Complex numbers that have argument equal to zero are purely real and positive. Those with argument π (or $-\pi$) are purely real and negative. When the argument is $\pi/2$ or $-\pi/2$, the complex number is purely imaginary with positive or negative imaginary part, respectively. As with any polar angle, the argument of a complex number is not uniquely determined—it can be increased or decreased by any integral multiple of 2π.

If $a + bi = r(\cos\theta + i\sin\theta)$, then its **complex conjugate** is obtained by reversing the sign of the imaginary part:

$$\overline{a + bi} = a - bi = r(\cos\theta - i\sin\theta)$$
$$= r[\cos(-\theta) + i\sin(-\theta)].$$

Therefore $\overline{a + bi}$ has the same modulus as $a + bi$ but the negative of its argument.

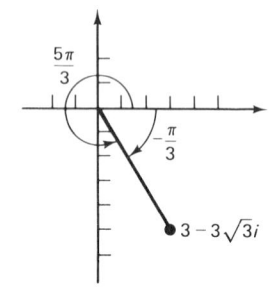

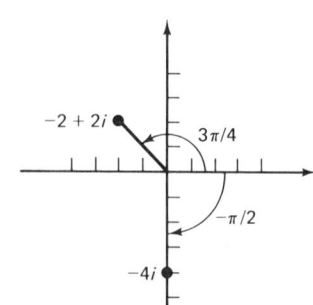

For example, from Example 3(b), $-2 + 2i$ has modulus $2\sqrt{2}$ and argument $3\pi/4$. Its conjugate, $-2 - 2i$, has modulus $2\sqrt{2}$ and argument $-3\pi/4$.

MULTIPLICATION AND DIVISION

Multiplication and division of complex numbers becomes very easy if the numbers are written in polar form. (In contrast, addition and subtraction become complicated.) The method is based on the following theorem, which we prove at the end of the section.

THEOREM 1

$$(\cos\theta + i\sin\theta)(\cos\phi + i\sin\phi) = \cos(\theta + \phi) + i\sin(\theta + \phi)$$

$$\frac{\cos\theta + i\sin\theta}{\cos\phi + i\sin\phi} = \cos(\theta - \phi) + i\sin(\theta - \phi).$$

It follows directly from this theorem that if $a + bi = r(\cos\theta + i\sin\theta)$ and $c + di = s(\cos\phi + i\sin\phi)$ are two complex numbers in polar form then their product and quotient are given by

$$(a + bi)(c + di) = rs[\cos(\theta + \phi) + i\sin(\theta + \phi)] \qquad (2)$$

$$\frac{a + bi}{c + di} = \frac{r}{s}[\cos(\theta - \phi) + i\sin(\theta - \phi)]. \qquad (3)$$

Thus to multiply two complex numbers we simply multiply their moduli and add their arguments. To divide one complex number by another we divide their respective moduli and subtract the argument of the denominator from the argument of the numerator.

Example 4 Use polar forms to evaluate the following.

(a) $(3 - 3\sqrt{3}i)^2$.

(b) $\dfrac{-2 + 2i}{-2 - 2i}$.

Solution (a) From Example 3(a),

$$3 - 3\sqrt{3}i = 6\left[\cos\left(-\frac{\pi}{3}\right) + i\sin\left(-\frac{\pi}{3}\right)\right].$$

The two complex numbers in Equation (2) must each be taken equal to $3 - \sqrt{3}i$ in this example. Therefore $r = s = 6$ and $\theta = \phi = -\pi/3$. Then

$$(3 - 3\sqrt{3}i)^2 = rs[\cos(\theta + \phi) + i\sin(\theta + \phi)]$$

$$= 6^2\left[\cos\left(-\frac{\pi}{3} - \frac{\pi}{3}\right) + i\sin\left(-\frac{\pi}{3} - \frac{\pi}{3}\right)\right]$$

$$= 36\left[\cos\left(-\frac{2\pi}{3}\right) + i\sin\left(-\frac{2\pi}{3}\right)\right]$$

$$= 36\left(-\frac{1}{2} - i\frac{\sqrt{3}}{2}\right)$$

$$= -18 - 18\sqrt{3}i.$$

(b) From Example 3(b),
$$-2 + 2i = 2\sqrt{2}\left(\cos\frac{3\pi}{4} + i\sin\frac{3\pi}{4}\right)$$

and its conjugate
$$-2 - 2i = 2\sqrt{2}\left[\cos\left(\frac{-3\pi}{4}\right) + i\sin\left(-\frac{3\pi}{4}\right)\right].$$

In Equation (3) therefore we must take $r = s = 2\sqrt{2}$, $\theta = 3\pi/4$ and $\phi = -3\pi/4$. Then

$$\frac{-2 + 2i}{-2 - 2i} = \frac{r}{s}[\cos(\theta - \phi) + i\sin(\theta - \phi)]$$

$$= \frac{2\sqrt{2}}{2\sqrt{2}}\left\{\cos\left[\frac{3\pi}{4} - \left(-\frac{3\pi}{4}\right)\right] + i\sin\left[\frac{3\pi}{4} - \left(-\frac{3\pi}{4}\right)\right]\right\}$$

$$= \cos\frac{3\pi}{2} + i\sin\frac{3\pi}{2}$$

$$= 0 + i(-1)$$

$$= -i.$$

You should check these results using the standard methods of multiplication and division (see Section 9.2).

We conclude this section by proving Theorem 1. We have, using the multiplication rule for complex numbers,

$$(\cos\theta + i\sin\theta)(\cos\phi + i\sin\phi) = (\cos\theta)(\cos\phi) - (\sin\theta)(\sin\phi)$$
$$+ [(\cos\theta)(\sin\phi) + (\sin\theta)(\cos\phi)]i$$
$$= (\cos\theta\cos\phi - \sin\theta\sin\phi)$$
$$+ (\cos\theta\sin\phi + \sin\theta\cos\phi)i$$
$$= \cos(\theta + \phi) + i\sin(\theta + \phi)$$

where in the last step we have used the addition formulas for sine and cosine (see Section 13.2).

In order to prove the second part of Theorem 1, we multiply numerator and denominator by $\cos\phi - i\sin\phi$, the complex conjugate of the denominator:

$$\frac{\cos\theta + i\sin\theta}{\cos\phi + i\sin\phi} = \frac{(\cos\theta + i\sin\theta)(\cos\phi - i\sin\phi)}{(\cos\phi + i\sin\phi)(\cos\phi - i\sin\phi)}$$

$$= \frac{\cos\theta\cos\phi - \sin\theta(-\sin\phi) + [\cos\theta(-\sin\phi) + \sin\theta\cos\phi]i}{\cos^2\phi + \sin^2\phi + 0i}$$

$$= \frac{(\cos\theta\cos\phi + \sin\theta\sin\phi) + (\sin\theta\cos\phi - \cos\theta\sin\phi)i}{1 + 0i}$$

$$= \cos(\theta - \phi) + i\sin(\theta - \phi)$$

as required.

Diagnostic Test 14.3

Fill in the blanks.

1. The transformation from polar to Cartesian coordinates is

$$x = \underline{}, \qquad y = \underline{}.$$

2. The transformation from Cartesian to polar coordinates is

$$r = \underline{}, \qquad \tan \theta = \underline{}.$$

3. The point $r = 2$, $\theta = \pi/3$ has $(x, y) = \underline{}$.
4. The point $(x, y) = (1, -1)$ has $r = \underline{}$ and $\theta = \underline{}$.
5. The point $(x, y) = (0, -2)$ has $r = \underline{}$ and $\theta = \underline{}$.
6. The polar coordinates $r = 4$, $\theta = -\pi$ correspond to $(x, y) = \underline{}$.
7. In polar form, $a + bi = r(\underline{})$. r is called the $\underline{}$ and θ the $\underline{}$ of the complex number.
8. $2i = \underline{}$ in polar form.
9. $-3 = \underline{}$ in polar form.
10. $1 + i = \underline{}$ in polar form.
11. $\overline{1 + i} = \underline{}$ in polar form.
12. If $a + bi$ has modulus s and argument α, then $\overline{a + bi}$ has modulus $\underline{}$ and argument $\underline{}$.
13. $(\cos \alpha + i \sin \alpha)(\cos \beta + i \sin \beta) = \underline{}$.

(14–20) Suppose that $3 - 4i$ has modulus r and argument α and $(-1 + 2i)$ has modulus s and argument β. Express in terms of r, s, α, and β.

14. The modulus and argument of $(3 - 4i)(-1 + 2i)$ are $\underline{}$ and $\underline{}$.
15. The modulus and argument of $(-1 + 2i) \div (3 - 4i)$ are $\underline{}$ and $\underline{}$.
16. The modulus and argument of $(-1 + 2i)^4$ are $\underline{}$ and $\underline{}$.
17. The modulus and argument of $(3 + 4i) \div (-1 + 2i)$ are $\underline{}$ and $\underline{}$.
18. The modulus and argument of $(3 - 4i)^2$ are $\underline{}$ and $\underline{}$.
19. The modulus and argument of $(-1 + 2i)^{-3}$ are $\underline{}$ and $\underline{}$.
20. The modulus and argument of $(3 - 4i)^2(-1 + 2i)$ are $\underline{}$ and $\underline{}$.

Exercises 14.3

(1–8) Find the Cartesian coordinates of the points whose polar coordinates are as follows.

1. $r = 1$, $\theta = \dfrac{5\pi}{2}$.
2. $r = 3$, $\theta = -\dfrac{\pi}{6}$.
3. $r = 2$, $\theta = \dfrac{5\pi}{6}$.
4. $r = 5$, $\theta = -\dfrac{3\pi}{2}$.
5. $r = 3\sqrt{2}$, $\theta = \dfrac{5\pi}{4}$.
6. $r = 2$, $\theta = -\dfrac{\pi}{4}$.
7. $r = 4$, $\theta = 9\pi$.
8. $r = 0$, $\theta = \pi$.

(9–20) Find the polar coordinates of the points whose Cartesian coordinates are as follows. (Give the polar angle in the range $-\pi < \theta \leq \pi$ in each case.)

9. (2, 2).   **10.** (0, 7).   **11.** (0, −5).   **12.** (−3, 0).

13. $(-\sqrt{3}, 1)$.   **14.** $\left(\frac{1}{\sqrt{3}}, -1\right)$.   **15.** (−6, −8).   **16.** (−12, 5).

17. (−10, 0).   **18.** $(-\frac{1}{3}, -\frac{1}{4})$.   **19.** $(\frac{1}{4}, -1)$.   **20.** (2, −6).

(21–22) Plot the following points in the complex plane.

21. $3 + i, -4i, -2, -4 + 2i$.

22. $-3 - 2i, 4 - 4i, 2i, -5 + i$.

(23–36) Express each complex number in polar form.

23. 5.   **24.** $3i$.   **25.** $-6i$.   **26.** −8.

27. $-1 + i$.   **28.** $1 - \sqrt{3}i$.   **29.** $-3 - 3i$.   **30.** $-2 + \frac{2}{\sqrt{3}}i$.

31. $-4\sqrt{3} - 4i$.   **32.** $-6 + 6i$.   **33.** −6.   **34.** $-5i$.

35. $3 - 4i$.   **36.** $-4 + 3i$.

(37–44) Form the product and quotient (first number divided by second) of the two given complex numbers by changing to polar form and using Equations (2) and (3) of this section.

37. $1 - \sqrt{3}i, -2 + \frac{2}{\sqrt{3}}i$.   **38.** $1 - i, -6 + 6i$.   **39.** $4 + 4i, -3 + 3i$.

40. $\sqrt{3} + i, 2 - 2\sqrt{3}i$.   **41.** $5 - 5i, -2i$.   **42.** $2i, -3 + 3\sqrt{3}i$.

43. $-4, 2 - 2\sqrt{3}i$.   **44.** $-2, -4 - 4i$.

14.4

De Moivre's Theorem

In Section 3 we established a formula for the product of two complex numbers in polar form. The first part of Theorem 1 states that

$$(\cos \theta + i \sin \theta)(\cos \phi + i \sin \phi) = \cos(\theta + \phi) + i \sin(\theta + \phi). \quad (1)$$

Put $\phi = \theta$ in this equation. We obtain that

$$(\cos \theta + i \sin \theta)^2 = \cos 2\theta + i \sin 2\theta. \quad (2)$$

Next, we have

$$(\cos \theta + i \sin \theta)^3 = (\cos \theta + i \sin \theta)(\cos \theta + i \sin \theta)^2$$
$$= (\cos \theta + i \sin \theta)(\cos 2\theta + i \sin 2\theta) \quad \text{[from (2)]}$$
$$= \cos(\theta + 2\theta) + i \sin(\theta + 2\theta)$$

where in the last step we have used (1) with $\phi = 2\theta$. Thus we have

$$(\cos \theta + i \sin \theta)^3 = \cos 3\theta + i \sin 3\theta.$$

If we continue in this way, we obtain the result that for any positive integer n

$$(\cos\theta + i\sin\theta)^n = \cos n\theta + i\sin n\theta. \quad (3)$$

This is known as *de Moivre's theorem.**

Example 1 Evaluate $(-1+i)^{10}$.

Solution The modulus of $(-1+i)$ is $r = \sqrt{(-1)^2 + 1^2} = \sqrt{2}$ and the argument is given by $\theta = 3\pi/4$. Thus

$$-1 + i = \sqrt{2}\left(-\frac{1}{\sqrt{2}} + \frac{1}{\sqrt{2}}i\right) = 2^{1/2}\left(\cos\frac{3\pi}{4} + i\sin\frac{3\pi}{4}\right).$$

Therefore,

$$(-1+i)^{10} = (2^{1/2})^{10}\left(\cos\frac{3\pi}{4} + i\sin\frac{3\pi}{4}\right)^{10}$$

$$= 2^5\left[\cos\left(10\cdot\frac{3\pi}{4}\right) + i\sin\left(10\cdot\frac{3\pi}{4}\right)\right]$$

$$= 32\left(\cos\frac{15\pi}{2} + i\sin\frac{15\pi}{2}\right)$$

$$= 32[0 + (-1)i]$$

$$= -32i.$$

Example 2 Find three complex numbers with moduli all equal to unity each of whose cubes is equal to -1.

Solution Suppose one of the numbers we are looking for is $(\cos\theta + i\sin\theta)$ in polar form (modulus = 1 as given). Then we must have

$$(\cos\theta + i\sin\theta)^3 = -1$$
$$\cos 3\theta + i\sin 3\theta = -1.$$

Thus
$$\cos 3\theta = -1 \quad\text{and}\quad \sin 3\theta = 0.$$

It follows that 3θ must be an odd multiple of π. That is, the possible solutions are $3\theta = \pi$, $3\theta = 3\pi$, $3\theta = 5\pi$, $3\theta = 7\pi$, and so on. Thus $\theta = \pi/3$ or π or $5\pi/3$ or $7\pi/3$ and so on.

However, $\theta = 7\pi/3 = 2\pi + \pi/3$ gives the same complex number $(\cos\theta + i\sin\theta)$ as does $\theta = \pi/3$; $\theta = 9\pi/3$ gives the same as $\theta = \pi$, and so on. We can in fact obtain only three different complex numbers:

$$\theta = \frac{\pi}{3}: \quad \cos\theta + i\sin\theta = \frac{1}{2} + \frac{\sqrt{3}}{2}i$$

$$\theta = \pi: \quad \cos\theta + i\sin\theta = -1$$

$$\theta = \frac{5\pi}{3}: \quad \cos\theta + i\sin\theta = \frac{1}{2} - \frac{\sqrt{3}}{2}i.$$

---

*A rigorous proof of de Moivre's theorem can be completed using the method of mathematical induction. It also holds for negative integers n.

The same type of method as used in Example 2 can be used to solve the polynomial equation

$$x^n - a = 0$$

where n is any positive integer and a is any number, real or complex. We know from Section 9.3 that there are n solutions of this equation, some of which could be coincident although in fact they are all distinct. Since the equation can be written $x^n = a$, the solutions are called the **nth roots** of a. They are given by the following list where $a = r(\cos\theta + i\sin\theta)$.

$$\sqrt[n]{r}\left(\cos\frac{\theta}{n} + i\sin\frac{\theta}{n}\right), \quad \sqrt[n]{r}\left(\cos\frac{\theta + 2\pi}{n} + i\sin\frac{\theta + 2\pi}{n}\right),$$
$$\sqrt[n]{r}\left(\cos\frac{\theta + 4\pi}{n} + i\sin\frac{\theta + 4\pi}{n}\right), \quad \ldots, \tag{4}$$
$$\sqrt[n]{r}\left[\cos\frac{\theta + 2(n-1)\pi}{n} + i\sin\frac{\theta + 2(n-1)\pi}{n}\right].$$

To summarize:

There are n distinct nth roots of any nonzero complex number, $r(\cos\theta + i\sin\theta)$, and they are given by

$$\sqrt[n]{r}\left(\cos\frac{\theta + 2\pi k}{n} + i\sin\frac{\theta + 2\pi k}{n}\right), \quad k = 0, 1, 2, \ldots, n-1. \tag{5}$$

Example 3 Find the three cube roots of 1.

Solution In polar form we write $1 = \cos 0 + i\sin 0$. That is, $r = 1$ and $\theta = 0$. From the list (4), we then obtain the following roots (with $n = 3$):

$$\sqrt[n]{r}\left(\cos\frac{\theta}{n} + i\sin\frac{\theta}{n}\right) = \cos\frac{0}{3} + i\sin\frac{0}{3} = 1$$

$$\sqrt[n]{r}\left(\cos\frac{\theta + 2\pi}{n} + i\sin\frac{\theta + 2\pi}{n}\right) = \cos\frac{0 + 2\pi}{3} + i\sin\frac{0 + 2\pi}{3}$$
$$= -\frac{1}{2} + \frac{\sqrt{3}}{2}i$$

$$\sqrt[n]{r}\left(\cos\frac{\theta + 4\pi}{n} + i\sin\frac{\theta + 4\pi}{n}\right) = \cos\frac{0 + 4\pi}{3} + i\sin\frac{0 + 4\pi}{3}$$
$$= -\frac{1}{2} - \frac{\sqrt{3}}{2}i.$$

The three cube roots of 1 are therefore 1, $-\frac{1}{2} + \sqrt{3}i/2$, and $-\frac{1}{2} - \sqrt{3}i/2$.

It is interesting to plot these three roots in the complex plane (see the adjacent figure). All three lie on the circle centered at the origin with radius 1 (called the **unit circle**). Furthermore their positions are **equally spaced** around the circumference of the unit circle.

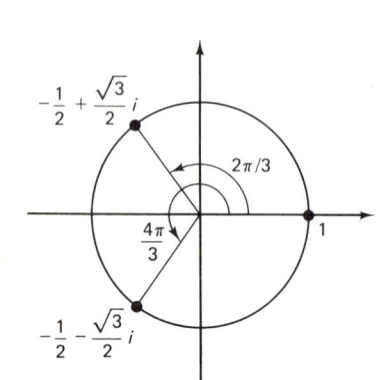

Example 4 Find the four fourth roots of $-8 + 8\sqrt{3}\,i$.

Solution In polar form

$$-8 + 8\sqrt{3}\,i = 16\left(-\frac{1}{2} + \frac{\sqrt{3}}{2}i\right) = 16\left(\cos\frac{2\pi}{3} + i\sin\frac{2\pi}{3}\right).$$

Thus $r = 16$ and $\theta = 2\pi/3$. From the general equation (5) with $n = 4$, the roots are given by

$$\sqrt[n]{r}\left(\cos\frac{\theta + 2\pi k}{n} + i\sin\frac{\theta + 2\pi k}{n}\right)$$

$$= \sqrt[4]{16}\left[\cos\frac{1}{4}\left(\frac{2\pi}{3} + 2\pi k\right) + i\sin\frac{1}{4}\left(\frac{2\pi}{3} + 2\pi k\right)\right]$$

$$= 2\left[\cos\left(\frac{\pi}{6} + k\frac{\pi}{2}\right) + i\sin\left(\frac{\pi}{6} + k\frac{\pi}{2}\right)\right]$$

with $k = 0, 1, 2, 3$. These four values of k give, respectively,

$k = 0$:

$$2\left(\cos\frac{\pi}{6} + i\sin\frac{\pi}{6}\right) = 2\left(\frac{\sqrt{3}}{2} + \frac{1}{2}i\right) = \sqrt{3} + i$$

$k = 1$:

$$2\left(\cos\frac{2\pi}{3} + i\sin\frac{2\pi}{3}\right) = 2\left(-\frac{1}{2} + \frac{\sqrt{3}}{2}i\right) = -1 + \sqrt{3}\,i$$

$k = 2$:

$$2\left(\cos\frac{7\pi}{6} + i\sin\frac{7\pi}{6}\right) = 2\left(-\frac{\sqrt{3}}{2} - \frac{1}{2}i\right) = -\sqrt{3} - i$$

$k = 3$:

$$2\left(\cos\frac{5\pi}{3} + i\sin\frac{5\pi}{3}\right) = 2\left(\frac{1}{2} - \frac{\sqrt{3}}{2}i\right) = 1 - \sqrt{3}\,i.$$

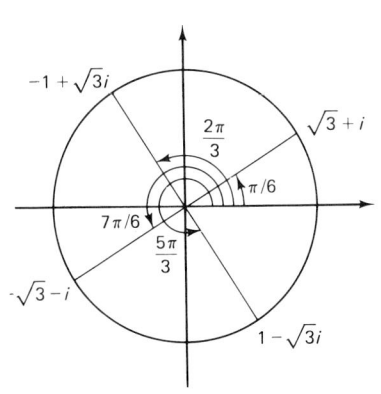

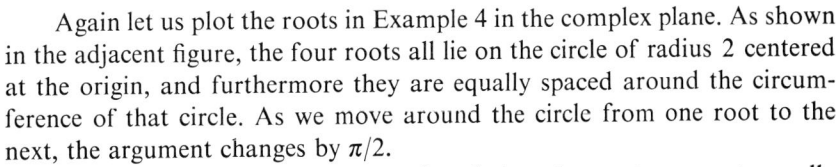

Again let us plot the roots in Example 4 in the complex plane. As shown in the adjacent figure, the four roots all lie on the circle of radius 2 centered at the origin, and furthermore they are equally spaced around the circumference of that circle. As we move around the circle from one root to the next, the argument changes by $\pi/2$.

In both Examples 3 and 4 we found that the roots were at equally spaced points around the circumference of a certain circle in the complex plane. This result generalizes. The nth roots of $r(\cos\theta + i\sin\theta)$, as given in the list (4), all have modulus $\sqrt[n]{r}$. This means that they all lie on the circle of radius $\sqrt[n]{r}$ centered at the origin. Furthermore, as we move from one root to the next the argument increases by $2\pi/n$. Thus the n roots are equally spaced around the circle, the polar angles of neighboring roots differing by $2\pi/n$.

Exercises 14.4

(1–10) Use de Moivre's theorem to evaluate the following (in the form $a + bi$).

1. $(-i)^7$.
2. i^{15}.
3. $(1 + i)^5$.
4. $(2 - 2i)^4$.
5. $(-3 + 3i)^4$.
6. $(-\frac{1}{2} - \frac{1}{2}i)^5$.
7. $(\sqrt{3} - i)^6$.
8. $(\sqrt{3} + i)^7$.
9. $(-2 + 2\sqrt{3}i)^5$.
10. $(\sqrt{3} - 3i)^6$.

(11–14) De Moivre's theorem also holds for negative integers n. Use this to evaluate the following.

11. $(1 - i)^{-4}$.
12. $(\frac{1}{2} + \frac{1}{2}i)^{-5}$.
13. $(-\sqrt{3} + 3i)^{-5}$.
14. $(2\sqrt{3} - 2i)^{-3}$.

(15–18) Find the two square roots of each.

15. i.
16. $-i$.
17. $-1 + \sqrt{3}i$.
18. $-1 - \sqrt{3}i$.

(19–20) Find the three cube roots of each.

19. i.
20. $-27i$.

(21–22) Find the four fourth roots of each.

21. $-8 + 8\sqrt{3}i$.
22. -256.

(23–24) Find the six sixth roots of each.

23. -1.
24. 1.

25. Let ω denote either of the complex cube roots of 1. Show that the other complex cube root is ω^2. Show that $1 + \omega + \omega^2 = 0$.

26. With the notation in Exercise 25 show that the three cube roots of -1 are $-\omega$, $-\omega^2$, and $-\omega^3$.

Chapter Review 14

Review the following topics.
(a) The law of sines.
(b) The law of cosines.
(c) The procedure for solving a triangle in the cases SSS, SAS, SSA, and ASA.
(d) The geometrical construction in the SSA case and the conditions for two, one, or no solutions.
(e) Expressions for the area of a triangle, Hero's formula.
(f) The transformation from polar coordinates to Cartesian coordinates.
(g) The transformation from Cartesian coordinates to polar coordinates.
(h) The complex plane, real axis, and imaginary axis.
(i) Polar form of complex numbers, modulus, and argument.
(j) Multiplication and division of complex numbers in polar form.
(k) De Moivre's theorem.
(l) Formula for the nth roots of a complex number.
(m) Geometrical configuration of the nth roots in the complex plane.

Review Exercises on Chapter 14

(1–10) Calculate the remaining parts in the triangle *ABC*.

1. $a = 20, A = 20°, B = 30°$.
2. $b = 5, A = 35°, B = 125°$.
3. $a = 5, b = 7, B = 50°$.
4. $a = 3, c = 2.5, C = 115°$.
5. $b = 3, c = 4, B = 25°$.
6. $a = 10, b = 9, A = 70°$.
7. $a = 2, b = 2.5, c = 4$.
8. $a = 11, b = 8, c = 4$.
9. $b = 5, c = 1, A = 60°$.
10. $a = 3, c = 6, B = 55°$.

11–14. Calculate the areas of the triangles in Exercises 7 through 10, using only the parts of the triangles that are given.

15. Show that the area of a triangle is given by $\frac{1}{2}a^2 \sin B \sin C / \sin A$. Use this formula to calculate the area of the triangle in Exercise 1.

*16. Show that if b, c, and B are given, we can calculate the area of the triangle from the formula $\frac{1}{2}c \sin B(c \cos B \pm \sqrt{b^2 - c^2 \sin^2 B})$. [*Hint:* Area $= \frac{1}{2}bc \sin(B + C)$ and $\sin C = (c/b) \sin B$.]

17. A ship leaves a certain point traveling due north at 8 miles per hour. Fifteen minutes later a second ship leaves the same point traveling at 12 miles per hour. Forty-five minutes later the second ship observes that the first ship lies in the direction 20°E of N. In what direction was the second ship traveling?

18. Point *A* lies in a direction 330°E of N from *B* at a distance of 500 meters. From *B*, point *C* lies in the direction 15°E of N while from *A* it lies in the direction 110°E of N. Calculate the distances of *C* from *A* and *B*.

19. A triangular plot of land has three sides equal to 180 feet, 240 feet, and 360 feet. Calculate the three angles at the corners and the area.

20. From a point *C*, *A* lies 500 yards in the direction 50°E of N and *B* lies 350 yards in the direction 120°E of N. How far apart are *A* and *B*?

21. A speedboat is traveling due west at 40 miles per hour. At one time it is in the direction 85°E of N from a point *A* and one minute later it is in the direction 65°E of N. How far from *A* is the boat at each of these two times?

22. A TV aerial is erected on a hillside that slopes at 20° to the horizontal. Guy wires 100 feet long are attached to the aerial at a point 90 feet above its base. Calculate the distances directly up and down the slope from the base of the aerial at which the guy wires will be anchored to the ground.

*(23–24) Solve the following triangles approximately without using trigonometric tables by using the approximation that when θ is small $\sin \theta \approx \theta$ in radians.

*23. $A = 2°, B = 2°40', c = 12$.

*24. $A = 3°30', a = 14, b = 10$.

*25. An airplane flies at an altitude of 2000 feet. How far away along the surface of the earth can its pilot see? What is the angle of depression of her horizon (called the angle of dip)? (Radius of the earth $= 2.09 \times 10^7$ feet.)

26. Mount Baker is 10,800 feet high. From a distance of 20 miles at approximately sea level, what is the angle of elevation of its peak?

(27–28) Find the Cartesian coordinates of the points whose polar coordinates are given.

27. $r = 4, \theta = 7\pi/6$.

28. $r = 6, \theta = 15\pi/4$.

(29–32) Find the polar coordinates of the points whose Cartesian coordinates are given. (Take $-\pi < \theta \leq \pi$.)

29. $(-6, 6)$. **30.** $(-3, 0)$. **31.** $(1, -4)$. **32.** $(-4, -3)$.

(33–36) Express each complex number in polar form.

33. $\sqrt{3} - i$. **34.** $-5 + 5i$. **35.** $-5 + 2i$. **36.** $-1 - 2i$.

(37–38) Evaluate the product and quotient of the following pairs of complex numbers by expressing them in polar form.

37. $\sqrt{3} - i$, $4 + 4\sqrt{3}\,i$. **38.** $-3 + 3i$, $\frac{1}{2} + \frac{1}{2}i$.

(39–40) Evaluate, using de Moivre's theorem.

39. $(-3 + \sqrt{3}\,i)^5$. **40.** $\left(\dfrac{\sqrt{3}}{2} - \dfrac{\sqrt{3}}{2}i\right)^6$.

41. Find the three cube roots of $-i$.

42. Find the three cube roots of $1 + i$. [$\cos 15° = (1 + \sqrt{3})/2\sqrt{2}$, $\sin 15° = (\sqrt{3} - 1)/2\sqrt{2}$.]

APPENDIX TABLES

TABLE 1. FOUR-PLACE COMMON LOGARITHMS

BASE OF 10

N	0	1	2	3	4	5	6	7	8	9
1.0	.0000	.0043	.0086	.0128	.0170	.0212	.0253	.0294	.0334	.0374
1.1	.0414	.0453	.0492	.0531	.0569	.0607	.0645	.0682	.0719	.0755
1.2	.0792	.0828	.0864	.0899	.0934	.0969	.1004	.1038	.1072	.1106
1.3	.1139	.1173	.1206	.1239	.1271	.1303	.1335	.1367	.1399	.1430
1.4	.1461	.1492	.1523	.1553	.1584	.1614	.1644	.1673	.1703	.1732
1.5	.1761	.1790	.1818	.1847	.1875	.1903	.1931	.1959	.1987	.2014
1.6	.2041	.2068	.2095	.2122	.2148	.2175	.2201	.2227	.2253	.2279
1.7	.2304	.2330	.2355	.2380	.2405	.2430	.2455	.2480	.2504	.2529
1.8	.2553	.2577	.2601	.2625	.2648	.2672	.2695	.2718	.2742	.2765
1.9	.2788	.2810	.2833	.2856	.2878	.2900	.2923	.2945	.2967	.2989
2.0	.3010	.3032	.3054	.3075	.3096	.3118	.3139	.3160	.3181	.3201
2.1	.3222	.3243	.3263	.3284	.3304	.3324	.3345	.3365	.3385	.3404
2.2	.3424	.3444	.3464	.3483	.3502	.3522	.3541	.3560	.3579	.3598
2.3	.3617	.3636	.3655	.3674	.3692	.3711	.3729	.3747	.3766	.3784
2.4	.3802	.3820	.3838	.3856	.3874	.3892	.3909	.3927	.3945	.3962
2.5	.3979	.3997	.4014	.4031	.4048	.4065	.4082	.4099	.4116	.4133
2.6	.4150	.4166	.4183	.4200	.4216	.4232	.4249	.4265	.4281	.4298
2.7	.4314	.4330	.4346	.4362	.4378	.4393	.4409	.4425	.4440	.4456
2.8	.4472	.4487	.4502	.4518	.4533	.4548	.4564	.4579	.4594	.4609
2.9	.4624	.4639	.4654	.4669	.4683	.4698	.4713	.4728	.4742	.4757
3.0	.4771	.4786	.4800	.4814	.4829	.4843	.4857	.4871	.4886	.4900
3.1	.4914	.4928	.4942	.4955	.4969	.4983	.4997	.5011	.5024	.5038
3.2	.5051	.5065	.5079	.5092	.5105	.5119	.5132	.5145	.5159	.5172
3.3	.5185	.5198	.5211	.5224	.5237	.5250	.5263	.5276	.5289	.5302
3.4	.5315	.5328	.5340	.5353	.5366	.5378	.5391	.5403	.5416	.5428
3.5	.5441	.5453	.5465	.5478	.5490	.5502	.5514	.5527	.5539	.5551
3.6	.5563	.5575	.5587	.5599	.5611	.5623	.5635	.5647	.5658	.5670
3.7	.5682	.5694	.5705	.5717	.5729	.5740	.5752	.5763	.5775	.5786
3.8	.5798	.5809	.5821	.5832	.5843	.5855	.5866	.5877	.5888	.5899
3.9	.5911	.5922	.5933	.5944	.5955	.5966	.5977	.5988	.5999	.6010
4.0	.6021	.6031	.6042	.6053	.6064	.6075	.6085	.6096	.6107	.6117
4.1	.6128	.6138	.6149	.6160	.6170	.6180	.6191	.6201	.6212	.6222
4.2	.6232	.6243	.6253	.6263	.6274	.6284	.6294	.6304	.6314	.6325
4.3	.6335	.6345	.6355	.6365	.6375	.6385	.6395	.6405	.6415	.6425
4.4	.6435	.6444	.6454	.6464	.6474	.6484	.6493	.6503	.6513	.6522
4.5	.6532	.6542	.6551	.6561	.6571	.6580	.6590	.6599	.6609	.6618
4.6	.6628	.6637	.6646	.6656	.6665	.6675	.6684	.6693	.6702	.6712
4.7	.6721	.6730	.6739	.6749	.6758	.6767	.6776	.6785	.6794	.6803
4.8	.6812	.6821	.6830	.6839	.6848	.6857	.6866	.6875	.6884	.6893
4.9	.6902	.6911	.6920	.6928	.6937	.6946	.6955	.6964	.6972	.6981
5.0	.6990	.6998	.7007	.7016	.7024	.7033	.7042	.7050	.7059	.7067
5.1	.7076	.7084	.7093	.7101	.7110	.7118	.7126	.7135	.7143	.7152
5.2	.7160	.7168	.7177	.7185	.7193	.7202	.7210	.7218	.7226	.7235
5.3	.7243	.7251	.7259	.7267	.7275	.7284	.7292	.7300	.7308	.7316
5.4	.7324	.7332	.7340	.7348	.7356	.7364	.7372	.7380	.7388	.7396
N	0	1	2	3	4	5	6	7	8	9

TABLE 1. FOUR-PLACE COMMON LOGARITHMS (*concluded*)

N	0	1	2	3	4	5	6	7	8	9
5.5	.7404	.7412	.7419	.7427	.7435	.7443	.7451	.7459	.7466	.7474
5.6	.7482	.7490	.7497	.7505	.7513	.7520	.7528	.7536	.7543	.7551
5.7	.7559	.7566	.7574	.7582	.7589	.7597	.7604	.7612	.7619	.7627
5.8	.7634	.7642	.7649	.7657	.7664	.7672	.7679	.7686	.7694	.7701
5.9	.7709	.7716	.7723	.7731	.7738	.7745	.7752	.7760	.7767	.7774
6.0	.7782	.7789	.7796	.7803	.7810	.7818	.7825	.7832	.7839	.7846
6.1	.7853	.7860	.7868	.7875	.7882	.7889	.7896	.7903	.7910	.7917
6.2	.7924	.7931	.7938	.7945	.7952	.7959	.7966	.7973	.7980	.7987
6.3	.7993	.8000	.8007	.8014	.8021	.8028	.8035	.8041	.8048	.8055
6.4	.8062	.8069	.8075	.8082	.8089	.8096	.8102	.8109	.8116	.8122
6.5	.8129	.8136	.8142	.8149	.8156	.8162	.8169	.8176	.8182	.8189
6.6	.8195	.8202	.8209	.8215	.8222	.8228	.8235	.8241	.8248	.8254
6.7	.8261	.8267	.8274	.8280	.8287	.8293	.8299	.8306	.8312	.8319
6.8	.8325	.8331	.8338	.8344	.8351	.8357	.8363	.8370	.8376	.8382
6.9	.8388	.8395	.8401	.8407	.8414	.8420	.8426	.8432	.8439	.8445
7.0	.8451	.8457	.8463	.8470	.8476	.8482	.8488	.8494	.8500	.8506
7.1	.8513	.8519	.8525	.8531	.8537	.8543	.8549	.8555	.8561	.8567
7.2	.8573	.8579	.8585	.8591	.8597	.8603	.8609	.8615	.8621	.8627
7.3	.8633	.8639	.8645	.8651	.8657	.8663	.8669	.8675	.8681	.8686
7.4	.8692	.8698	.8704	.8710	.8716	.8722	.8727	.8733	.8739	.8745
7.5	.8751	.8756	.8762	.8768	.8774	.8779	.8785	.8791	.8797	.8802
7.6	.8808	.8814	.8820	.8825	.8831	.8837	.8842	.8848	.8854	.8859
7.7	.8865	.8871	.8876	.8882	.8887	.8893	.8899	.8904	.8910	.8915
7.8	.8921	.8927	.8932	.8938	.8943	.8949	.8954	.8960	.8965	.8971
7.9	.8976	.8982	.8987	.8993	.8998	.9004	.9009	.9015	.9020	.9025
8.0	.9031	.9036	.9042	.9047	.9053	.9058	.9063	.9069	.9074	.9079
8.1	.9085	.9090	.9096	.9101	.9106	.9112	.9117	.9122	.9128	.9133
8.2	.9138	.9143	.9149	.9154	.9159	.9165	.9170	.9175	.9180	.9186
8.3	.9191	.9196	.9201	.9206	.9212	.9217	.9222	.9227	.9232	.9238
8.4	.9243	.9248	.9253	.9258	.9263	.9269	.9274	.9279	.9284	.9289
8.5	.9294	.9299	.9304	.9309	.9315	.9320	.9325	.9330	.9335	.9340
8.6	.9345	.9350	.9355	.9360	.9365	.9370	.9375	.9380	.9385	.9390
8.7	.9395	.9400	.9405	.9410	.9415	.9420	.9425	.9430	.9435	.9440
8.8	.9445	.9450	.9455	.9460	.9465	.9469	.9474	.9479	.9484	.9489
8.9	.9494	.9499	.9504	.9509	.9513	.9518	.9523	.9528	.9533	.9538
9.0	.9542	.9547	.9552	.9557	.9562	.9566	.9571	.9576	.9581	.9586
9.1	.9590	.9595	.9600	.9605	.9609	.9614	.9619	.9624	.9628	.9633
9.2	.9638	.9643	.9647	.9652	.9657	.9661	.9666	.9671	.9675	.9680
9.3	.9685	.9689	.9694	.9699	.9703	.9708	.9713	.9717	.9722	.9727
9.4	.9731	.9736	.9741	.9745	.9750	.9754	.9759	.9763	.9768	.9773
9.5	.9777	.9782	.9786	.9791	.9795	.9800	.9805	.9809	.9814	.9818
9.6	.9823	.9827	.9832	.9836	.9841	.9845	.9850	.9854	.9859	.9863
9.7	.9868	.9872	.9877	.9881	.9886	.9890	.9894	.9899	.9903	.9908
9.8	.9912	.9917	.9921	.9926	.9930	.9934	.9939	.9943	.9948	.9952
9.9	.9956	.9961	.9965	.9969	.9974	.9978	.9983	.9987	.9991	.9996
N	0	1	2	3	4	5	6	7	8	9

TABLE 2. NATURAL LOGARITHMS

	0.00	0.01	0.02	0.03	0.04	0.05	0.06	0.07	0.08	0.09
1.0	0.0000	0.0100	0.0198	0.0296	0.0392	0.0488	0.0583	0.0677	0.0770	0.0862
1.1	0.0953	0.1044	0.1133	0.1222	0.1310	0.1398	0.1484	0.1570	0.1655	0.1740
1.2	0.1823	0.1906	0.1989	0.2070	0.2151	0.2231	0.2311	0.2390	0.2469	0.2546
1.3	0.2624	0.2700	0.2776	0.2852	0.2927	0.3001	0.3075	0.3148	0.3221	0.3293
1.4	0.3365	0.3436	0.3507	0.3577	0.3646	0.3716	0.3784	0.3853	0.3920	0.3988
1.5	0.4055	0.4121	0.4187	0.4253	0.4318	0.4383	0.4447	0.4511	0.4574	0.4637
1.6	0.4700	0.4762	0.4824	0.4886	0.4947	0.5008	0.5068	0.5128	0.5188	0.5247
1.7	0.5306	0.5365	0.5423	0.5481	0.5539	0.5596	0.5653	0.5710	0.5766	0.5822
1.8	0.5878	0.5933	0.5988	0.6043	0.6098	0.6152	0.6206	0.6259	0.6313	0.6366
1.9	0.6419	0.6471	0.6523	0.6575	0.6627	0.6678	0.6729	0.6780	0.6831	0.6881
2.0	0.6931	0.6981	0.7031	0.7080	0.7130	0.7178	0.7227	0.7275	0.7324	0.7372
2.1	0.7419	0.7467	0.7514	0.7561	0.7608	0.7655	0.7701	0.7747	0.7793	0.7839
2.2	0.7885	0.7930	0.7975	0.8020	0.8065	0.8109	0.8154	0.8198	0.8242	0.8286
2.3	0.8329	0.8372	0.8416	0.8459	0.8502	0.8544	0.8587	0.8629	0.8671	0.8713
2.4	0.8755	0.8796	0.8838	0.8879	0.8920	0.8961	0.9002	0.9042	0.9083	0.9123
2.5	0.9163	0.9203	0.9243	0.9282	0.9322	0.9361	0.9400	0.9439	0.9478	0.9517
2.6	0.9555	0.9594	0.9632	0.9670	0.9708	0.9746	0.9783	0.9821	0.9858	0.9895
2.7	0.9933	0.9969	1.0006	1.0043	1.0080	1.0116	1.0152	1.0188	1.0225	1.0260
2.8	1.0296	1.0332	1.0367	1.0403	1.0438	1.0473	1.0508	1.0543	1.0578	1.0613
2.9	1.0647	1.0682	1.0716	1.0750	1.0784	1.0818	1.0852	1.0886	1.0919	1.0953
3.0	1.0986	1.1019	1.1053	1.1086	1.1119	1.1151	1.1184	1.1217	1.1249	1.1282
3.1	1.1314	1.1346	1.1378	1.1410	1.1442	1.1474	1.1506	1.1537	1.1569	1.1600
3.2	1.1632	1.1663	1.1694	1.1725	1.1756	1.1787	1.1817	1.1848	1.1878	1.1909
3.3	1.1939	1.1970	1.2000	1.2030	1.2060	1.2090	1.2119	1.2149	1.2179	1.2208
3.4	1.2238	1.2267	1.2296	1.2326	1.2355	1.2384	1.2413	1.2442	1.2470	1.2499
3.5	1.2528	1.2556	1.2585	1.2613	1.2641	1.2669	1.2698	1.2726	1.2754	1.2782
3.6	1.2809	1.2837	1.2865	1.2892	1.2920	1.2947	1.2975	1.3002	1.3029	1.3056
3.7	1.3083	1.3110	1.3137	1.3164	1.3191	1.3218	1.3244	1.3271	1.3297	1.3324
3.8	1.3350	1.3376	1.3403	1.3429	1.3455	1.3481	1.3507	1.3533	1.3558	1.3584
3.9	1.3610	1.3635	1.3661	1.3686	1.3712	1.3737	1.3762	1.3788	1.3813	1.3838
4.0	1.3863	1.3888	1.3913	1.3938	1.3962	1.3987	1.4012	1.4036	1.4061	1.4085
4.1	1.4110	1.4134	1.4159	1.4183	1.4207	1.4231	1.4255	1.4279	1.4303	1.4327
4.2	1.4351	1.4375	1.4398	1.4422	1.4446	1.4469	1.4493	1.4516	1.4540	1.4563
4.3	1.4586	1.4609	1.4633	1.4656	1.4679	1.4702	1.4725	1.4748	1.4770	1.4793
4.4	1.4816	1.4839	1.4861	1.4884	1.4907	1.4929	1.4952	1.4974	1.4996	1.5019
4.5	1.5041	1.5063	1.5085	1.5107	1.5129	1.5151	1.5173	1.5195	1.5217	1.5239
4.6	1.5261	1.5282	1.5304	1.5326	1.5347	1.5369	1.5390	1.5412	1.5433	1.5454
4.7	1.5476	1.5497	1.5518	1.5539	1.5560	1.5581	1.5602	1.5623	1.5644	1.5665
4.8	1.5686	1.5707	1.5728	1.5748	1.5769	1.5790	1.5810	1.5831	1.5851	1.5872
4.9	1.5892	1.5913	1.5933	1.5953	1.5974	1.5994	1.6014	1.6034	1.6054	1.6074
5.0	1.6094	1.6114	1.6134	1.6154	1.6174	1.6194	1.6214	1.6233	1.6253	1.6273
5.1	1.6292	1.6312	1.6332	1.6351	1.6371	1.6390	1.6409	1.6429	1.6448	1.6467
5.2	1.6487	1.6506	1.6525	1.6544	1.6563	1.6582	1.6601	1.6620	1.6639	1.6658
5.3	1.6677	1.6696	1.6715	1.6734	1.6752	1.6771	1.6790	1.6808	1.6827	1.6845
5.4	1.6864	1.6882	1.6901	1.6919	1.6938	1.6956	1.6974	1.6993	1.7011	1.7029

$\ln(N \cdot 10^m) = \ln N + m \ln 10$, $\quad \ln 10 = 2.3026$

TABLE 2. NATURAL LOGARITHMS (*concluded*)

	0.00	0.01	0.02	0.03	0.04	0.05	0.06	0.07	0.08	0.09
5.5	1.7047	1.7066	1.7084	1.7102	1.7120	1.7138	1.7156	1.7174	1.7192	1.7210
5.6	1.7228	1.7246	1.7263	1.7281	1.7299	1.7317	1.7334	1.7352	1.7370	1.7387
5.7	1.7405	1.7422	1.7440	1.7457	1.7475	1.7492	1.7509	1.7527	1.7544	1.7561
5.8	1.7579	1.7596	1.7613	1.7630	1.7647	1.7664	1.7682	1.7699	1.7716	1.7733
5.9	1.7750	1.7766	1.7783	1.7800	1.7817	1.7834	1.7851	1.7867	1.7884	1.7901
6.0	1.7918	1.7934	1.7951	1.7967	1.7984	1.8001	1.8017	1.8034	1.8050	1.8066
6.1	1.8083	1.8099	1.8116	1.8132	1.8148	1.8165	1.8181	1.8197	1.8213	1.8229
6.2	1.8245	1.8262	1.8278	1.8294	1.8310	1.8326	1.8342	1.8358	1.8374	1.8390
6.3	1.8406	1.8421	1.8437	1.8453	1.8469	1.8485	1.8500	1.8516	1.8532	1.8547
6.4	1.8563	1.8579	1.8594	1.8610	1.8625	1.8641	1.8656	1.8672	1.8687	1.8703
6.5	1.8718	1.8733	1.8749	1.8764	1.8779	1.8795	1.8810	1.8825	1.8840	1.8856
6.6	1.8871	1.8886	1.8901	1.8916	1.8931	1.8946	1.8961	1.8976	1.8991	1.9006
6.7	1.9021	1.9036	1.9051	1.9066	1.9081	1.9095	1.9110	1.9125	1.9140	1.9155
6.8	1.9169	1.9184	1.9199	1.9213	1.9228	1.9242	1.9257	1.9272	1.9286	1.9301
6.9	1.9315	1.9330	1.9344	1.9359	1.9373	1.9387	1.9402	1.9416	1.9430	1.9445
7.0	1.9459	1.9473	1.9488	1.9502	1.9516	1.9530	1.9544	1.9559	1.9573	1.9587
7.1	1.9601	1.9615	1.9629	1.9643	1.9657	1.9671	1.9685	1.9699	1.9713	1.9727
7.2	1.9741	1.9755	1.9769	1.9782	1.9796	1.9810	1.9824	1.9838	1.9851	1.9865
7.3	1.9879	1.9892	1.9906	1.9920	1.9933	1.9947	1.9961	1.9974	1.9988	2.0001
7.4	2.0015	2.0028	2.0042	2.0055	2.0069	2.0082	2.0096	2.0109	2.0122	2.0136
7.5	2.0149	2.0162	2.0176	2.0189	2.0202	2.0215	2.0229	2.0242	2.0255	2.0268
7.6	2.0282	2.0295	2.0308	2.0321	2.0334	2.0347	2.0360	2.0373	2.0386	2.0399
7.7	2.0412	2.0425	2.0438	2.0451	2.0464	2.0477	2.0490	2.0503	2.0516	2.0528
7.8	2.0541	2.0554	2.0567	2.0580	2.0592	2.0605	2.0618	2.0631	2.0643	2.0656
7.9	2.0669	2.0681	2.0694	2.0707	2.0719	2.0732	2.0744	2.0757	2.0769	2.0782
8.0	2.0794	2.0807	2.0819	2.0832	2.0844	2.0857	2.0869	2.0882	2.0894	2.0906
8.1	2.0919	2.0931	2.0943	2.0956	2.0968	2.0980	2.0992	2.1005	2.1017	2.1029
8.2	2.1041	2.1054	2.1066	2.1078	2.1090	2.1102	2.1114	2.1126	2.1138	2.1150
8.3	2.1163	2.1175	2.1187	2.1190	2.1211	2.1223	2.1235	2.1247	2.1258	2.1270
8.4	2.1282	2.1294	2.1306	2.1318	2.1330	2.1342	2.1353	2.1365	2.1377	2.1389
8.5	2.1401	2.1412	2.1424	2.1436	2.1448	2.1459	2.1471	2.1483	2.1494	2.1506
8.6	2.1518	2.1529	2.1541	2.1552	2.1564	2.1576	2.1587	2.1599	2.1610	2.1622
8.7	2.1633	2.1645	2.1656	2.1668	2.1679	2.1691	2.1702	2.1713	2.1725	2.1736
8.8	2.1748	2.1759	2.1770	2.1782	2.1793	2.1804	2.1815	2.1827	2.1838	2.1849
8.9	2.1861	2.1872	2.1883	2.1894	2.1905	2.1917	2.1928	2.1939	2.1950	2.1961
9.0	2.1972	2.1983	2.1994	2.2006	2.2017	2.2028	2.2039	2.2050	2.2061	2.2072
9.1	2.2083	2.2094	2.2105	2.2116	2.2127	2.2138	2.2148	2.2159	2.2170	2.2181
9.2	2.2192	2.2203	2.2214	2.2225	2.2235	2.2246	2.2257	2.2268	2.2279	2.2289
9.3	2.2300	2.2311	2.2322	2.2332	2.2343	2.2354	2.2364	2.2375	2.2386	2.2396
9.4	2.2407	2.2418	2.2428	2.2439	2.2450	2.2460	2.2471	2.2481	2.2492	2.2502
9.5	2.2513	2.2523	2.2534	2.2544	2.2555	2.2565	2.2576	2.2586	2.2597	2.2607
9.6	2.2618	2.2628	2.2638	2.2649	2.2659	2.2670	2.2680	2.2690	2.2701	2.2711
9.7	2.2721	2.2732	2.2742	2.2752	2.2762	2.2773	2.2783	2.2793	2.2803	2.2814
9.8	2.2824	2.2834	2.2844	2.2854	2.2865	2.2875	2.2885	2.2895	2.2905	2.2915
9.9	2.2925	2.2935	2.2946	2.2956	2.2966	2.2976	2.2986	2.2996	2.3006	2.3016

TABLE 3. EXPONENTIAL FUNCTIONS

x	e^x	e^{-x}	x	e^x	e^{-x}
0.00	1.0000	1.0000	0.45	1.5683	0.6376
0.01	1.0101	0.9900	0.46	1.5841	0.6313
0.02	1.0202	0.9802	0.47	1.6000	0.6250
0.03	1.0305	0.9704	0.48	1.6161	0.6188
0.04	1.0408	0.9608	0.49	1.6323	0.6126
0.05	1.0513	0.9512	0.50	1.6487	0.6065
0.06	1.0618	0.9418	0.51	1.6653	0.6005
0.07	1.0725	0.9324	0.52	1.6820	0.5945
0.08	1.0833	0.9231	0.53	1.6989	0.5886
0.09	1.0942	0.9139	0.54	1.7160	0.5827
0.10	1.1052	0.9048	0.55	1.7333	0.5769
0.11	1.1163	0.8958	0.56	1.7507	0.5712
0.12	1.1275	0.8869	0.57	1.7683	0.5655
0.13	1.1388	0.8781	0.58	1.7860	0.5599
0.14	1.1503	0.9694	0.59	1.8040	0.5543
0.15	1.1618	0.8607	0.60	1.8221	0.5488
0.16	1.1735	0.8521	0.61	1.8044	0.5434
0.17	1.1853	0.8437	0.62	1.8589	0.5379
0.18	1.1972	0.8353	0.63	1.8776	0.5326
0.19	1.2092	0.8270	0.64	1.8965	0.5273
0.20	1.2214	0.8187	0.65	1.9155	0.5220
0.21	1.2337	0.8106	0.66	1.9348	0.5169
0.22	1.2461	0.8025	0.67	1.9542	0.5117
0.23	1.2586	0.7945	0.68	1.9739	0.5066
0.24	1.2712	0.7866	0.69	1.9937	0.5016
0.25	1.2840	0.7788	0.70	2.0138	0.4966
0.26	1.2969	0.7711	0.71	2.0340	0.4916
0.27	1.3100	0.7634	0.72	2.0544	0.4868
0.28	1.3231	0.7558	0.73	2.0751	0.4819
0.29	1.3364	0.7483	0.74	2.0959	0.4771
0.30	1.3499	0.7408	0.75	2.1170	0.4724
0.31	1.3634	0.7334	0.76	2.1383	0.4677
0.32	1.3771	0.7261	0.77	2.1598	0.4630
0.33	1.3910	0.7189	0.78	2.1815	0.4584
0.34	1.4049	0.7118	0.79	2.2034	0.4538
0.35	1.4191	0.7047	0.80	2.2255	0.4493
0.36	1.4333	0.6977	0.81	2.2479	0.4449
0.37	1.4477	0.6907	0.82	2.2705	0.4404
0.38	1.4623	0.6839	0.83	2.2933	0.4360
0.39	1.4770	0.6771	0.84	2.3164	0.4317
0.40	1.4918	0.6703	0.85	2.3396	0.4274
0.41	1.5068	0.6637	0.86	2.3632	0.4232
0.42	1.5220	0.6570	0.87	2.3869	0.4190
0.43	1.5373	0.6505	0.88	2.4109	0.4148
0.44	1.5527	0.6440	0.89	2.4351	0.4107

TABLE 3. EXPONENTIAL FUNCTIONS (*concluded*)

x	e^x	e^{-x}	x	e^x	e^{-x}
0.90	2.4596	0.4066	2.75	15.643	0.0639
0.91	2.4843	0.4025	2.80	16.445	0.0608
0.92	2.5093	0.3985	2.85	17.288	0.0578
0.93	2.5345	0.3946	2.90	18.174	0.0550
0.94	2.5600	0.3906	2.95	19.106	0.0523
0.95	2.5857	0.3867	3.00	20.086	0.0498
0.96	2.6117	0.3829	3.05	21.115	0.0474
0.97	2.6379	0.3791	3.10	22.198	0.0450
0.98	2.6645	0.3753	3.15	23.336	0.0429
0.99	2.6912	0.3716	3.20	24.533	0.0408
1.00	2.7183	0.3679	3.25	25.790	0.0388
1.05	2.8577	0.3499	3.30	27.113	0.0369
1.10	3.0042	0.3329	3.35	28.503	0.0351
1.15	3.1582	0.3166	3.40	29.964	0.0334
1.20	3.3201	0.3012	3.45	31.500	0.0317
1.25	3.4903	0.2865	3.50	33.115	0.0302
1.30	3.6693	0.2725	3.55	34.813	0.0287
1.35	3.8574	0.2592	3.60	36.598	0.0273
1.40	4.0552	0.2466	3.65	38.475	0.0260
1.45	4.2631	0.2346	3.70	40.447	0.0247
1.50	4.4817	0.2231	3.75	42.521	0.0235
1.55	4.7115	0.2122	3.80	44.701	0.0224
1.60	4.9530	0.2019	3.85	46.993	0.0213
1.65	5.2070	0.1920	3.90	49.402	0.0202
1.70	5.4739	0.1827	3.95	51.935	0.0193
1.75	5.7546	0.1738	4.00	54.598	0.0183
1.80	6.0496	0.1653	4.10	60.340	0.0166
1.85	6.3598	0.1572	4.20	66.686	0.0150
1.90	6.6859	0.1496	4.30	73.700	0.0136
1.95	7.0287	0.1423	4.40	81.451	0.0123
2.00	7.3891	0.1353	4.50	90.017	0.0111
2.05	7.7679	0.1287	4.60	99.484	0.0101
2.10	8.1662	0.1225	4.70	109.95	0.0091
2.15	8.5849	0.1165	4.80	121.51	0.0082
2.20	9.0250	0.1108	4.90	134.29	0.0074
2.25	9.4877	0.1054	5.00	148.41	0.0067
2.30	9.9742	0.1003	5.20	181.27	0.0055
2.35	10.486	0.0954	5.40	221.41	0.0045
2.40	11.023	0.0907	5.60	270.43	0.0037
2.45	11.588	0.0863	5.80	330.30	0.0030
2.50	12.182	0.0821	6.00	403.43	0.0025
2.55	12.807	0.0781	7.00	1096.6	0.0009
2.60	13.464	0.0743	8.00	2981.0	0.0003
2.65	14.154	0.0707	9.00	8103.1	0.0001
2.70	14.880	0.0672	10.00	22026.	0.00005

TABLE 4. TRIGONOMETRIC FUNCTIONS

Degrees	Radians	sin	cos	tan	cot	sec	csc		
0° 00′	.0000	.0000	1.0000	.0000	—	1.000	—	1.5708	90° 00′
10	029	029	000	029	343.8	000	343.8	679	50
20	058	058	000	058	171.9	000	171.9	650	40
30	.0087	.0087	1.0000	.0087	114.6	1.000	114.6	1.5621	30
40	116	116	.9999	116	85.94	000	85.95	592	20
50	145	145	999	145	68.75	000	68.76	563	10
1° 00′	.0175	.0175	.9998	.0175	57.29	1.000	57.30	1.5533	89° 00′
10	204	204	998	204	49.10	000	49.11	504	50
20	233	233	997	233	42.96	000	42.98	475	40
30	.0262	.0262	.9997	.0262	38.19	1.000	38.20	1.5446	30
40	291	291	996	291	34.37	000	34.38	417	20
50	320	320	995	320	31.24	001	31.26	388	10
2° 00′	.0349	.0349	.9994	.0349	28.64	1.001	28.65	1.5359	88° 00′
10	378	378	993	378	26.43	001	26.45	330	50
20	407	407	992	407	24.54	001	24.56	301	40
30	.0436	.0436	.9990	.0437	22.90	1.001	22.93	1.5272	30
40	465	465	989	466	21.47	001	21.49	243	20
50	495	494	988	495	20.21	001	20.23	213	10
3° 00′	.0524	.0523	.9986	.0524	19.08	1.001	19.11	1.5184	87° 00′
10	553	552	985	553	18.07	002	18.10	155	50
20	582	581	983	582	17.17	002	17.20	126	40
30	.0611	.0610	.9981	.0612	16.35	1.002	16.38	1.5097	30
40	640	640	980	641	15.60	002	15.64	068	20
50	669	669	978	670	14.92	002	14.96	039	10
4° 00′	.0698	.0698	.9976	.0699	14.30	1.002	14.34	1.5010	86° 00′
10	727	727	974	729	13.73	003	13.76	981	50
20	756	756	971	758	13.20	003	13.23	952	40
30	.0785	.0785	.9969	.0787	12.71	1.003	12.75	1.4923	30
40	814	814	967	816	12.25	003	12.29	893	20
50	844	843	964	846	11.83	004	11.87	864	10
5° 00′	.0873	.0872	.9962	.0875	11.43	1.004	11.47	1.4835	85° 00′
10	902	901	959	904	11.06	004	11.10	806	50
20	931	929	957	934	10.71	004	10.76	777	40
30	.0960	.0958	.9954	.0963	10.39	1.005	10.43	1.4748	30
40	989	987	951	992	10.08	005	10.13	719	20
50	.1018	.1016	948	.1022	9.788	005	9.839	690	10
6° 00′	.1047	.1045	.9945	.1051	9.514	1.006	9.567	1.4661	84° 00′
10	076	074	942	080	9.255	006	9.309	632	50
20	105	103	939	110	9.010	006	9.065	603	40
30	.1134	.1132	.9936	.1139	8.777	1.006	8.834	1.4573	30
40	164	161	932	169	8.556	007	8.614	544	20
50	193	190	929	198	8.345	007	8.405	515	10
7° 00′	.1222	.1219	.9925	.1228	3.144	1.008	8.206	1.4486	83° 00′
10	251	248	922	257	7.953	008	8.016	457	50
20	280	276	918	287	7.770	008	7.834	428	40
30	.1309	.1305	.9914	.1317	7.596	1.009	7.661	1.4399	30
40	338	334	911	346	7.429	009	7.496	370	20
50	367	363	907	376	7.269	009	7.337	341	10
8° 00′	.1396	.1392	.9903	.1405	7.115	1.010	7.185	1.4312	82° 00′
10	425	421	899	435	6.968	010	7.040	283	50
20	454	449	894	465	6.827	011	6.900	254	40
30	.1484	.1478	.9890	.1495	6.691	1.011	6.765	1.4224	30
40	513	507	886	524	6.561	012	6.636	195	20
50	542	536	881	554	6.435	012	6.512	166	10
9° 00′	.1571	.1564	.9877	.1584	6.314	1.012	6.392	1.4137	81° 00′
		cos	sin	cot	tan	csc	sec	Radians	Degrees

Appendix Tables

TABLE 4. TRIGONOMETRIC FUNCTIONS (*continued*)

Degrees	Radians	sin	cos	tan	cot	sec	csc		
9° 00′	.1571	.1564	.9877	.1584	6.314	1.012	6.392	1.4137	**81° 00′**
10	600	593	872	614	197	013	277	108	50
20	629	622	868	644	084	013	166	079	40
30	.1658	.1650	.9863	.1673	5.976	1.014	6.059	1.4050	30
40	687	679	858	703	871	014	5.955	1.4021	20
50	716	708	853	733	769	015	855	992	10
10° 00′	.1745	.1736	.9848	.1763	5.671	1.015	5.759	1.3963	**80° 00′**
10	774	765	843	793	576	016	665	934	50
20	804	794	838	823	485	016	575	904	40
30	.1833	.1822	.9833	.1853	5.396	1.017	5.487	1.3875	30
40	862	851	827	883	309	018	403	846	20
50	891	880	822	914	226	018	320	817	10
11° 00′	.1920	.1908	.9816	.1944	5.145	1.019	5.241	1.3788	**79° 00′**
10	949	937	811	974	066	019	164	759	50
20	978	965	805	.2004	4.989	020	089	730	40
30	.2007	.1994	.9799	.2035	4.915	1.020	5.016	1.3701	30
40	036	.2022	793	065	843	021	4.945	672	20
50	065	051	787	095	773	022	876	643	10
12° 00′	.2094	.2079	.9781	.2126	4.705	1.022	4.810	1.3614	**78° 00′**
10	123	108	775	156	638	023	745	584	50
20	153	136	769	186	574	024	682	555	40
30	.2182	.2164	.9763	.2217	4.511	1.024	4.620	1.3526	30
40	211	193	757	247	449	025	560	497	20
50	240	221	750	278	390	026	502	468	10
13° 00′	.2269	.2250	.9744	.2309	4.331	1.026	4.445	1.3439	**77° 00′**
10	298	278	737	339	275	027	390	410	50
20	327	306	730	370	219	028	336	381	40
30	.2356	.2334	.9724	.2401	4.165	1.028	4.284	1.3352	30
40	385	363	717	432	113	029	232	323	20
50	414	391	710	462	061	030	182	294	10
14° 00′	.2443	.2419	.9703	.2493	4.011	1.031	4.134	1.3265	**76° 00′**
10	473	447	696	524	3.962	031	086	235	50
20	502	476	689	555	914	032	039	206	40
30	.2531	.2504	.9681	.2586	3.867	1.033	3.994	1.3177	30
40	560	532	674	617	821	034	950	148	20
50	589	560	667	648	776	034	906	119	10
15° 00′	.2618	.2588	.9659	.2679	3.732	1.035	3.864	1.3090	**75° 00′**
10	647	616	652	711	689	036	822	061	50
20	676	644	644	742	647	037	782	032	40
30	.2705	.2672	.9636	.2773	3.606	1.038	3.742	1.3003	30
40	734	700	628	805	566	039	703	974	20
50	763	728	621	836	526	039	665	945	10
16° 00′	.2793	.2756	.9613	.2867	3.487	1.040	3.628	1.2915	**74° 00′**
10	822	784	605	899	450	041	592	886	50
20	851	812	596	931	412	042	556	857	40
30	.2880	.2840	.9588	.2962	3.376	1.043	3.521	1.2828	30
40	909	868	580	994	340	044	487	799	20
50	938	896	572	.3026	305	045	453	770	10
17° 00′	.2967	.2924	.9563	.3057	3.271	1.046	3.420	1.2741	**73° 00′**
10	996	952	555	089	237	047	388	712	50
20	.3025	979	546	121	204	048	356	683	40
30	.3054	.3007	.9537	.3153	3.172	1.049	3.326	1.2654	30
40	083	035	528	185	140	049	295	625	20
50	113	062	520	217	108	050	265	595	10
18° 00′	.3142	.3090	.9511	.3249	3.078	1.051	3.236	1.2566	**72° 00′**
		cos	sin	cot	tan	csc	sec	Radians	Degrees

Appendix Tables

TABLE 4. TRIGONOMETRIC FUNCTIONS (*continued*)

Degrees	Radians	sin	cos	tan	cot	sec	csc		
18° 00′	.3142	.3090	.9511	.3249	3.078	1.051	3.236	1.2566	**72° 00′**
10	171	118	502	281	047	052	207	537	50
20	200	145	492	314	018	053	179	508	40
30	.3229	.3173	.9483	.3346	2.989	1.054	3.152	1.2479	30
40	258	201	474	378	960	056	124	450	20
50	287	228	465	411	932	057	098	421	10
19° 00′	.3316	.3256	.9455	.3443	2.904	1.058	3.072	1.2392	**71° 00′**
10	345	283	446	476	877	059	046	363	50
20	374	311	436	508	850	060	021	334	40
30	.3403	.3338	.9426	.3541	2.824	1.061	2.996	1.2305	30
40	432	365	417	574	798	062	971	275	20
50	462	393	407	607	773	063	947	246	10
20° 00′	.3491	.3420	.9397	.3640	2.747	1.064	2.924	1.2217	**70° 00′**
10	520	449	387	673	723	065	901	188	50
20	549	475	377	706	699	066	878	159	40
30	.3578	.3502	.9367	.3739	2.675	1.068	2.855	1.2130	30
40	607	529	356	772	651	069	833	101	20
50	636	557	346	805	628	070	812	072	10
21° 00′	.3665	.3584	.9336	.3839	2.605	1.071	2.790	1.2043	**69° 00′**
10	694	611	325	872	583	072	769	1.2014	50
20	723	638	315	906	560	074	749	985	40
30	.3752	.3665	.9304	.3939	2.539	1.075	2.729	1.1956	30
40	782	692	293	973	517	076	709	926	20
50	811	719	283	.4006	496	077	689	897	10
22° 00′	.3840	.3746	.9272	.4040	2.475	1.079	2.669	1.1868	**68° 00′**
10	869	773	261	074	455	080	650	839	50
20	898	800	250	108	434	081	632	810	40
30	.3927	.3827	.9239	.4142	2.414	1.082	2.613	1.1781	30
40	956	854	228	176	394	084	595	752	20
50	985	881	216	210	375	085	577	723	10
23° 00′	.4014	.3907	.9205	.4245	2.356	1.086	2.559	1.1694	**67° 00′**
10	043	934	194	279	337	088	542	665	50
20	072	961	182	314	318	089	525	636	40
30	.4102	.3987	.9171	.4348	2.300	1.090	2.508	1.1606	30
40	131	.4014	159	383	282	092	491	577	20
50	160	041	147	417	264	093	475	548	10
24° 00′	.4189	.4067	.9135	.4452	2.246	1.095	2.459	1.1519	**66° 00′**
10	218	094	124	487	229	096	443	490	50
20	247	120	112	522	211	097	427	461	40
30	.4276	.4147	.9100	.4557	2.194	1.099	2.411	1.1432	30
40	305	173	088	592	177	100	396	403	20
50	334	200	075	628	161	102	381	374	10
25° 00′	.4363	.4226	.9063	.4663	2.145	1.103	2.366	1.1345	**65° 00′**
10	392	253	051	699	128	105	352	316	50
20	422	279	038	734	112	106	337	286	40
30	.4451	.4305	.9026	.4770	2.097	1.108	2.323	1.1257	30
40	480	331	013	806	081	109	309	228	20
50	509	358	001	841	066	111	295	199	10
26° 00′	.4538	.4384	.8988	.4877	2.050	1.113	2.281	1.1170	**64° 00′**
10	567	410	975	913	035	114	268	141	50
20	596	436	962	950	020	116	254	112	40
30	.4625	.4462	.8949	.4986	2.006	1.117	2.241	1.1083	30
40	654	488	936	.5022	1.991	119	228	054	20
50	683	514	923	059	977	121	215	1.1025	10
27° 00′	.4712	.4540	.8910	.5095	1.963	1.122	2.203	1.0996	**63° 00′**
		cos	sin	cot	tan	csc	sec	Radians	Degrees

592   Appendix Tables

TABLE 4. TRIGONOMETRIC FUNCTIONS (*continued*)

Degrees	Radians	sin	cos	tan	cot	sec	csc		
27° 00′	.4712	.4540	.8910	.5095	1.963	1.122	2.203	1.0996	**63° 00′**
10	741	566	897	132	949	124	190	966	50
20	771	592	884	169	935	126	178	937	40
30	.4800	.4617	.8870	.5206	1.921	1.127	2.166	1.0908	30
40	829	643	857	243	907	129	154	879	20
50	858	669	843	280	894	131	142	850	10
28° 00′	.4887	.4695	.8829	.5317	1.881	1.133	2.130	1.0821	**62° 00′**
10	916	720	816	354	868	134	118	792	50
20	945	746	802	392	855	136	107	763	40
30	.4974	.4772	.8788	.5430	1.842	1.138	2.096	1.0734	30
40	.5003	797	774	467	829	140	085	705	20
50	032	823	760	505	816	142	074	676	10
29° 00′	.5061	.4848	.8746	.5543	1.804	1.143	2.063	1.0647	**61° 00′**
10	091	874	732	581	792	145	052	617	50
20	120	899	718	619	780	147	041	588	40
30	.5149	.4924	.8704	.5658	1.767	1.149	2.031	1.0559	30
40	178	950	689	696	756	151	020	530	20
50	207	975	675	735	744	153	010	501	10
30° 00′	.5236	.5000	.8660	.5774	1.732	1.155	2.000	1.0472	**60° 00′**
10	265	025	646	812	720	157	1.990	443	50
20	294	050	631	851	709	159	980	414	40
30	.5323	.5075	.8616	.5890	1.698	1.161	1.970	1.0385	30
40	352	100	601	930	686	163	961	356	20
50	381	125	587	969	675	165	951	327	10
31° 00′	.5411	.5150	.8572	.6009	1.664	1.167	1.942	1.0297	**59° 00′**
10	440	175	557	048	653	169	932	268	50
20	469	200	542	088	643	171	923	239	40
30	.5498	.5225	.8526	.6128	1.632	1.173	1.914	1.0210	30
40	527	250	511	168	621	175	905	181	20
50	556	275	496	208	611	177	896	152	10
32° 00′	.5585	.5299	.8480	.6249	1.600	1.179	1.887	1.0123	**58° 00′**
10	614	324	465	289	590	181	878	094	50
20	643	348	450	330	580	184	870	065	40
30	.5672	.5373	.8434	.6371	.1570	1.186	1.861	1.0036	30
40	701	398	418	412	560	188	853	1.0007	20
50	730	422	403	453	550	190	844	977	10
33° 00′	.5760	.5446	.8387	.6494	1.540	1.192	1.836	.9948	**57° 00′**
10	789	471	371	536	530	195	828	919	50
20	818	495	355	577	520	197	820	890	40
30	.5847	.5519	.8339	.6619	1.511	1.199	1.812	.9861	30
40	876	544	323	661	501	202	804	832	20
50	905	568	307	703	1.492	204	796	803	10
34° 00′	.5934	.5592	.8290	.6745	1.483	1.206	1.788	.9774	**56° 00′**
10	963	616	274	787	473	209	731	745	50
20	992	640	258	830	464	211	773	716	40
30	.6021	.5664	.8241	.6873	1.455	1.213	1.766	.9687	30
40	050	688	225	916	446	216	758	657	20
50	080	712	208	959	437	218	751	628	10
35° 00′	.6109	.5736	.8192	.7002	1.428	1.221	1.743	.9599	**55° 00′**
10	138	760	175	046	419	223	736	570	50
20	167	783	158	089	411	226	729	541	40
30	.6196	.5807	.8141	.7133	1.402	1.228	1.722	.9512	30
40	225	831	124	177	393	231	715	483	20
50	254	854	107	221	385	233	708	454	10
36° 00′	.6283	.5878	.8090	.7265	1.376	1.236	1.701	.9425	**54° 00′**
		cos	sin	cot	tan	csc	sec	Radians	Degrees

Appendix Tables

TABLE 4. TRIGONOMETRIC FUNCTIONS (*concluded*)

Degrees	Radians	sin	cos	tan	cot	sec	csc		
36° 00′	.6283	.5878	.8090	.7265	1.376	1.236	1.701	.9425	54° 00′
10	312	901	073	310	368	239	695	396	50
20	341	925	056	355	360	241	688	367	40
30	.6370	.5948	.8039	.7400	1.351	1.244	1.681	.9338	30
40	400	972	021	445	343	247	675	308	20
50	429	995	004	490	335	249	668	279	10
37° 00′	.6458	.6018	.7986	.7536	1.327	1.252	1.662	.9250	53° 00′
10	487	041	969	581	319	255	655	221	50
20	516	065	951	627	311	258	649	192	40
30	.6545	.6088	.7934	.7673	1.303	1.260	1.643	.9163	30
40	574	111	916	720	295	263	636	134	20
50	603	134	898	766	288	266	630	105	10
38° 00′	.5632	.6157	.7880	.7813	1.280	1.269	1.624	.9076	52° 00′
10	661	180	862	860	272	272	618	047	50
20	690	202	844	907	265	275	612	.9018	40
30	.6720	.6225	.7826	.7954	1.257	1.278	1.606	.8988	30
40	749	248	808	.8002	250	281	601	959	20
50	778	271	790	050	242	284	595	930	10
39° 00′	.6807	.6293	.7771	.8098	1.235	1.287	1.589	.8901	51° 00′
10	836	316	753	146	228	290	583	872	50
20	865	338	735	195	220	293	578	843	40
30	.6894	.6361	.7716	.8243	1.213	1.296	1.572	.8814	30
40	923	383	698	292	206	299	567	785	20
50	952	406	679	342	199	302	561	756	10
40° 00′	.6981	.6428	.7660	.8391	1.192	1.305	1.556	.8727	50° 00′
10	.7010	450	642	441	185	309	550	698	50
20	039	472	623	491	178	312	545	668	40
30	.7069	.6494	.7604	.8541	1.171	1.315	1.540	.8639	30
40	098	517	585	591	164	318	535	610	20
50	127	539	566	642	157	322	529	581	10
41° 00′	.7156	.6561	.7547	.8693	1.150	1.325	1.524	.8552	49° 00′
10	185	583	528	744	144	328	519	523	50
20	214	604	509	796	137	332	514	494	40
30	.7243	.6626	.7490	.8847	1.130	1.335	1.509	.8465	30
40	272	648	470	899	124	339	504	436	20
50	301	670	451	952	117	342	499	407	10
42° 00′	.7330	.6691	.7431	.9004	1.111	1.346	1.494	.8378	48° 00′
10	359	713	412	057	104	349	490	348	50
20	389	734	392	110	098	353	485	319	40
30	.7418	.6756	.7373	.9163	1.091	1.356	1.480	.8290	30
40	447	777	353	217	085	360	476	261	20
50	476	799	333	271	079	364	471	232	10
43° 00′	.7505	.6820	.7314	.9325	1.072	1.367	1.466	.8203	47° 00′
10	534	841	294	380	066	371	462	174	50
20	563	862	274	435	060	375	457	145	40
30	.7592	.6884	.7254	.9490	1.054	1.379	1.453	.8116	30
40	621	905	234	545	048	382	448	087	20
50	650	926	214	601	042	386	444	058	10
44° 00′	.7679	.6947	.7193	.9657	1.036	1.390	1.440	.8029	46° 00′
10	709	967	173	713	030	394	435	999	50
20	738	988	153	770	024	398	431	970	40
30	.7767	.7009	.7133	.9827	1.018	1.402	1.427	.7941	30
40	796	030	112	884	012	406	423	912	20
50	825	050	092	942	006	410	418	883	10
45° 00′	.7854	.7071	.7071	1.000	1.000	1.414	1.414	.7854	45° 00′
		cos	sin	cot	tan	csc	sec	Radians	Degrees

ANSWERS TO TESTS AND ODD-NUMBERED EXERCISES

CHAPTER 1

Test 1.1: **1.** Rational. **2.** Irrational. **3.** Rational. **4.** Irrational. **5.** Real. **6.** Rational. **7.** Irrational. **8.** 1; $\neq 0$. **9.** $\neq 0$. **10.** ≤ 0. **11.** 2, 3. **12.** 2, 2, 2.

Exercises 1.1

1. -3. **3.** -3. **5.** -7. **7.** 7. **9.** 1. **11.** 0. **13.** 4. **15.** -17. **17.** -10. **19.** 14. **21.** 4. **23.** -14. **25.** Prime. **27.** Not prime; $20 = 2 \cdot 2 \cdot 5$. **29.** Prime. **31.** Not prime; $62 = 2 \cdot 31$. **33.** Not prime; $60 = 2 \cdot 2 \cdot 3 \cdot 5$. **35.** Not prime; $144 = 2 \cdot 2 \cdot 2 \cdot 2 \cdot 3 \cdot 3$.

Test 1.2: **1.** Negative. **2.** $>$. **3.** $<$; $=$. **4.** $=$. **5.** $-x$. **6.** $-x$. **7.** Negative. **8.** All. **9.** -1. **10.** $x + 3$.

Exercises 1.2

1. $>$. **3.** $>$. **5.** $<$. **7.** $=$. **9.** $<$. **11.** $>$. **13.** $=$. **15.** $=$. **17.** $=$. **19.** False. **21.** True. **23.** True. **25.** True. $-1 < +1$. **29.** $x > -5$. **31.** $4 < z < 6$. **33.** $-2 < y < -1$. **35.** $p \leq 2$. **37.** $|x| < 2.5$. **39.** $|x + 2| \leq 3$. **41.** $|x + 6| = \frac{1}{2}|x - 6|$. **43.** $-1, -\sqrt{2}, -\frac{3}{2}, 3, 5, -6$. **45.** $-2, -4.5, -1, -5.5, 3$. **47.** -3. **49.** -1. **51.** 0.

Test 1.3: **1.** $7x$. **2.** $5x$. **3.** $-4x$. **4.** $8x/3$. **5.** $10x - 5y$. **6.** $-x - 2y$. **7.** $-2x + 6y$. **8.** $-6x - 3y$. **9.** $b - a$. **10.** b. **11.** ab. **12.** 6.

Exercises 1.3

1. $3x + 6y$. **3.** $4x - 2y$. **5.** $-x + 6$. **7.** $3x - 12$. **9.** $2x + 4$. **11.** $-xy + 6x$. **13.** $6x - 2y$. **15.** $-3x + 4z$. **17.** $5x - 3y$. **19.** $43x - 22y$. **21.** xyz. **23.** $2x^2 + 6x$. **25.** $-6a + 2a^2$.

A-1

27. $2x^2 + 8x$. **29.** $-x^2 + 4x - 2$. **31.** 1. **33.** $x - 2y$. **35.** $8x - 4$. **37.** 0. **39.** $1 + (2/x)$. **41.** $-3/2 + 1/(2x)$. **43.** $(1/y) + (1/x)$.

Test 1.4: **1.** $a/(a + b)$; no simplification possible. **2.** a/c; b/c. **3.** ac/bd. **4.** ad/bc. **5.** ab/c. **6.** ac/b. **7.** a/bc. **8.** b/a. **9.** b/a; $\neq 0$. **10.** $a/2$.

Exercises 1.4

1. $\frac{5}{27}$. **3.** $\frac{4}{25}$. **5.** $\frac{9}{80}$. **7.** $2x^2/15$. **9.** $3x^2y/20$. **11.** $8x/9y$. **13.** $4x^2y/3$. **15.** $2x/45y$. **17.** $\frac{3}{4}$. **19.** $\frac{3}{4}$. **21.** $\frac{1}{4}$. **23.** $7y/3$. **25.** $2x/3y$. **27.** $y/4x$. **29.** $\frac{4}{15}$. **31.** $\frac{8}{15}$. **33.** $\frac{1}{3}$. **35.** $2xy$. **37.** $\frac{27}{4}$. **39.** $\frac{1}{12}$. **41.** $\frac{1}{6}$. **43.** $9x/2$. **45.** $5x/2$. **47.** $30y^2/x$.

Test 1.5: **1.** $7/x$. **2.** $-2/x$. **3.** $7x/12$. **4.** $-2x/35$. **5.** $(a + b)/c$. **6.** $(ac + ab)/bc$. **7.** $(ad + bc)/bd$. **8.** $(xb - xa)/ab$. **9.** 24. **10.** $2b$.

Exercises 1.5

1. $-\frac{1}{3}$. **3.** $\frac{9}{8}$. **5.** $\frac{1}{6}$. **7.** $\frac{19}{24}$. **9.** $23x/30$. **11.** $(15t + 22x)/36$. **13.** $3/2x$. **15.** $7/(24x)$. **17.** $-a/(3b)$. **19.** $37x/(24y)$. **21.** $(14x + 9)/(12x^2)$. **23.** $(x^2 + 1)/x$. **25.** $(ad + bc)/bd$. **27.** $(2y + 3x)/x^2y$. **29.** $(xq + py)/p^2q$. **31.** $(x^2 - y^2)/xy$. **33.** $\frac{50}{27}$. **35.** $\frac{119}{135}$. **37.** $19x/44y$. **39.** $23a/31b$.

Test 1.6: **1.** a^{m+n}. **2.** a^{m-n}. **3.** x^{ab}. **4.** 1. **5.** a^pa^q. **6.** x^a/x^b. **7.** $a^m - a^n$. **8.** $p^x + p^y$. **9.** $(ab)^p$. **10.** $(x/y)^a$. **11.** $16a^4$. **12.** $1/a^m$. **13.** $a^3/2$. **14.** xy^{-6}. **15.** -1. **16.** 1. **17.** a^pb^q. **18.** x^m/y^n. **19.** $m = n$. **20.** 1; $\neq 0$. **21.** $ab/(a + b)$. **22.** xy. **23.** $a^n + b^n$. **24.** 0. **25.** -1.

Exercises 1.6

1. $2^{10} = 1024$. **3.** a^{21}. **5.** $-x^{10}$. **7.** y^7. **9.** $1/a^2$. **11.** $9/x^5$. **13.** $32/x$. **15.** $x^{10}y^7z^3$. **17.** x^4/y^2. **19.** x^2y. **21.** 16. **23.** $3^6 = 729$. **25.** x^7. **27.** x^2. **29.** 1. **31.** $-x^9$. **33.** $1/x^8y^5$. **35.** $-8y^2$. **37.** -3. **39.** $4b/a^{11}$. **41.** $x^6 - 2x^3$. **43.** $2x^6 + 6$. **45.** $2x^6 - x^5 - 3x^2$. **47.** $2x/(x + 2)$. **49.** $1/(x + y)$. **51.** $15/4x^2$. **53.** $(9y^2 + 4x^2)/30x^3y$. **55.** $5x^2/6$.

Test 1.7: **1.** b^2; ≥ 0. **2.** b^n. **3.** b^n; ≥ 0. **4.** $a \geq 0$. **5.** $-a$. **6.** 3; 2. **7.** -2; not defined.

Exercises 1.7

1. 9. **3.** $\frac{5}{4}$. **5.** -2. **7.** 3. **9.** Not defined. **11.** $\frac{5}{6}$. **13.** $8x^2$. **15.** $2a/3b$. **17.** $2\sqrt{2x}$. **19.** $3x^2/\sqrt{3x}$. **21.** $-2\sqrt[3]{x^2}$. **23.** $(-4/x)\sqrt[5]{x/2}$. **25.** $2x^2y\sqrt{2y}$. **27.** $(2/y)\sqrt{x/y}$. **29.** $9x^2/16$. **31.** $8xy^3\sqrt{x}$. **33.** $4b\sqrt[3]{4a^2b}$. **35.** $2xy^2\sqrt{2x}$. **37.** $x\sqrt[3]{2y}$. **39.** $(2x/y)\sqrt{2}$. **41.** $(p/12q^3)\sqrt[3]{2}$. **43.** $11\sqrt{5}$. **45.** $2\sqrt{2}$. **47.** $14\sqrt{7}$. **49.** $2\sqrt{5}$. **51.** $3\sqrt{2}$. **53.** $\frac{2}{15}\sqrt{35}$. **55.** $\frac{1}{3}\sqrt[3]{9}$. **57.** $(x/2y)\sqrt[3]{y^2}$. **59.** $\sqrt{2pq}/2q^2$. **61.** $\sqrt[3]{2xy^2z}/2y$.

Test 1.8: **1.** $\frac{1}{2}$. **2.** $1/n$. **3.** m/n. **4.** $-\frac{7}{3}$. **5.** $a^{1/n}$. **6.** $a = b$. **7.** $|a| = |b|$.

Exercises 1.8

1. 9. **3.** $\frac{5}{4}$. **5.** -2. **7.** 27. **9.** 9. **11.** 4. **13.** $\frac{1}{27}$. **15.** 2.5. **17.** $\frac{1}{18}$. **19.** 3. **21.** $5^{4/3}$. **23.** $8x^3$. **25.** $2x^{3/4}$. **27.** $\frac{1}{2}x^{-1/3}$. **29.** $x^{13/12}$. **31.** $2xy^{-2}$. **33.** $x^{5/6}$. **35.** 1. **37.** $x^{4/7}y^{1/5}$. **39.** p^4q^8. **41.** $a^{-5/6}$. **43.** 1. **45.** 1. **47.** 1. **49.** 3^{3n}. **51.** $x + 1$. **53.** $1 - 2/x$. **55.** $3^{1/3}$. **57.** $3^{1/2}$.

Review Exercises on Chapter 1

1. $-\frac{1}{2} < x < \frac{3}{2}$. **3.** -2. **5.** $10a - 25b$. **7.** $4a^2 - 2a$. **9.** $2y + (3/x)$. **11.** $p^2/20$. **13.** $3y^2/4$.
15. $7/12x$. **17.** $\frac{34}{5}$. **19.** $19x^2/12$. **21.** $x/2$. **23.** $-40p^{10}q^9$. **25.** $2x^{7/2} - x^{1/2}$. **27.** $-(4x/y^2)\sqrt[3]{x}$.
29. $-x^2\sqrt{2x}$. **31.** $1/x$. **33.** $r^{-1/30}$. **35.** $z^{1/10} - 2z^{-1/15}$. **37.** $3^{11n/4} \cdot 2^{7n/2}$. **39.** $2\sqrt{3xy/x^2y}$.

CHAPTER 2

Test 2.1. **1.** bi. **2.** xy^2; -5. **3.** 7. **4.** Literal parts. **5.** $5xy^2z$. **6.** $2x^2 + x - 10$. **7.** $(3x^2/2) - y^2$.
8. $3x^2 - 7$.

Exercises 2.1

1. Monomial; 2. **3.** Monomial; 2. **5.** Trinomial; 2. **7.** 5. **9.** 24. **11.** 6. **13.** 21.5. **15.** -15.
17. $\frac{200}{3}$ km/hr. **19.** $4x + 2y$. **21.** $3a + 10b + 6$. **23.** $5\sqrt{a} + 3\sqrt{b}$. **25.** $t^3 + 12t^2 + 3t - 5$.
27. $\sqrt{x} + 3\sqrt{2y}$. **29.** $8x + y$. **31.** $9x + 3y$. **33.** $8x^2 + 2xy + 11y^2$. **35.** $2x^3 - 2x^2y + 3xy^2 - y^3$.
37. $-3x^4y + 4x^3y^2 + 10xy^4$. **39.** $2x^2 - (3x/2)$. **41.** $x + 2y$. **43.** $x + 7 - (5/x) + (4/x^2)$.
45. $t^{3/2} - 2t^{1/2} + 7/\sqrt{t}$. **47.** $4x - 2y$. **49.** $(p^2/2q) - 3p + q + (2q^2/p)$. **51.** $3x^2 + 135x - 90$.
53. $4a^2 - 54a + 24$.

Test 2.2 **1.** $a^2 + 2ab + b^2$. **2.** $x^2 - 2xa + a^2$. **3.** $4x^2 + 12xy + 9y^2$. **4.** $9x^2 - 30xy + 25y^2$.
5. $6x^2 - 17x + 12$. **6.** $4x^2 + x - 14$. **7.** $a^2 - 16$. **8.** $9x^2 - 49$. **9.** $2x^2y + 8x^2 - 6xy - 24x$.
10. $6xy - 15x + 24y - 60$. **11.** $46x - 20$. **12.** $\sqrt{3} + \sqrt{2}$.

Exercises 2.2

1. $xy + 3x + y + 3$. **3.** $ab + 4a - 2b - 8$. **5.** $3xy - 15x - 2y + 10$. **7.** $4x^2 - 12x + 9$.
9. $x^3 + 4x^2 + 7x + 6$. **11.** $xy + 2x - 3y - 6$. **13.** $6xy - 8x + 3y - 4$. **15.** $3a^2 + 2a - 8$.
17. $2x^3 + x^2 - 8x + 21$. **19.** $a - b$. **21.** $x^2 + y^2 - z^2 + 2xy$. **23.** $x^5 + 2x^3 - x^2 - 2$.
25. $x^2 + 14x + 48$. **27.** $y^2 - 11 + (24/y^2)$. **29.** $x^2 + 20x + 100$. **31.** $x^2 - 10x + 25$. **33.** $x^2 - 16$.
35. $9t^2 + 24xt + 16x^2$. **37.** $4t^2 - 25x^2$. **39.** $8x^2 + 18y^2$. **41.** $2x^2 - 2x\sqrt{6y} + 3y$.
43. $x^4y^2 + 2xy^2 + (y^2/x^2)$. **45.** $12xy$. **47.** $x^{2a} + 3x^a + 2$. **49.** $12p^2q - 3pq^2$. **51.** $2a^3 + 8a$.
53. $\frac{1}{2}(3 - \sqrt{7})$. **55.** $\frac{1}{2}(\sqrt{5} - \sqrt{3} + \sqrt{10} - \sqrt{6})$. **57.** $\frac{1}{2}(3 - \sqrt{3})$. **59.** $(\sqrt{x} + \sqrt{y})/(x - y)$.
61. $\sqrt{x+2} + \sqrt{2}$. **63.** $-\frac{2}{3}(\sqrt{x+3} + 2\sqrt{x})$.

Test 2.3: **1.** $(x)(x - 5)$. **2.** $(5xy)(x - 4y)$. **3.** $(2x)(3x^2 - 2xy + 4y^2)$. **4.** $(ab)(a - b)$. **5.** $(x - y)(x + y)$.
6. $(2a - 3b)(2a + 3b)$. **7.** $(5u - 6v)(5u + 6v)$. **8.** $(7ab - 2u/v)(7ab + 2u/v)$. **9.** $(\sqrt{2}x - \sqrt{3}y)(\sqrt{2}x + \sqrt{3}y)$.
10. $(2a - \sqrt{5}b)(2a + \sqrt{5}b)$. **11.** $(y + a)(y^2 - ay + a^2)$. **12.** $(2y - 3x)(4y^2 + 6xy + 9x^2)$. **13.** $(x + 1)(x^2 + 1)$.
14. $(x + y)(1 + a)$.

Exercises 2.3

1. $3(a + 2b)$. **3.** $2y(2x - 3z)$. **5.** $5xy(2x^2z + 3yz^2 - y^3)$. **7.** $4x^2y^2z(2x^3y + xz^3 - 3z^4)$. **9.** $(x - 4)(x + 4)$.
11. $3(t - 6a)(t + 6a)$. **13.** $xy(x - 5y)(x + 5y)$. **15.** $x(x - 1)(x + 1)(x^2 + 1)$. **17.** $(x - 3)(x^2 + 3x + 9)$.
19. $(3u + 2v)(9u^2 - 6uv + 4v^2)$. **21.** $x^2(x - 1)(x^2 + x + 1)$. **23.** $(x - 2)(y + 4)$. **25.** $(a - 2)(v - u)$.
27. $2(3x + 2y)(z - 4)$. **29.** $(x - 3)(x + 3)(x - 5)$. **31.** $(x - 3)(x + 3)(y - 2)(y + 2)$.
33. $(x - 2)(x + 2)(x^2 + z^2)$. **35.** $(a + b + 2)(a^2 + b^2 - ab - 2a + 4b + 4)$.
37. $(x - 2y)(x + 2y)(x^4 + 4x^2y^2 + 16y^4) = (x - 2y)(x + 2y)(x^2 - 2xy + 4y^2)(x^2 + 2xy + 4y^2)$.
39. $(x + a)(a + 1)$. **41.** $(x^2 - 2xy + 2y^2)(x^2 + 2xy + 2y^2)$. **43.** $(x - 2y)(x + 2y)(x^2 + 4y^2)$.
45. $(y - 1)(x + 2)(x^2 - 2x + 4)$. **47.** $(x + 3)(x - 1)(x + 1)(x^2 + 1)$. **49.** $(x + 3y)(2x - z + 2)$. **51.** $4xy$.
53. $(x^n - 5)(x^n + 5)$.

Test 2.4: **1.** $(x-3)(x-4)$. **2.** $(y+2)(y+3)$. **3.** $(a+4)(a+2)$. **4.** $(t-6)(t+1)$. **5.** $(x-7)(x+3)$. **6.** $(t+7)(t-3)$. **7.** $(x+1)(x+1)$. **8.** $(x-3)(x-3)$. **9.** $(t+2)(t+2)$. **10.** $(y-1)(y-1)$.

Exercises 2.4

1. $(x+1)(x+2)$. **3.** $(x-2)(x+1)$. **5.** $(x-3)(x-4)$. **7.** $(x-3)(x+3)$. **9.** $x(x-4)$. **11.** $(x-11)(x-1)$. **13.** $(x-9)(x-6)$. **15.** $(x-10)(x+8)$. **17.** $2(x+3)(x-2)$. **19.** $5y^2(y+7)(y-2)$. **21.** $(2x+3)(x+1)$. **23.** $t(2t-5)(3t+4)$. **25.** $(3t-2)^2$. **27.** $(2x-3)(5x+2)$. **29.** $(t+2)(2t-7)$. **31.** $2xy(3x+5)(x-1)$. **33.** $(3q+8)(q+4)$. **35.** $2pq(2p+9)(p-2)$. **37.** $(x+y)(x+5y)$. **39.** $(p-5q)(p+4q)$. **41.** $(2t-3u)(t+2u)$. **43.** $(2a-3b)(3a+5b)$. **45.** $(x+y+1)(x+y+2)$. **47.** $(x^n+2)(3x^n+1)$. **49.** $(x-1)(2x+3)$. **51.** $(\sqrt{3}x-\sqrt{2})(\sqrt{3}x+\sqrt{2})$. **53.** $(x+2a-2)(x+2a+2)$. **55.** $(x+y+z)(x+y-z)$.

Test 2.5: **1.** $3x/(x+2)$. **2.** $(x+2)/2$. **3.** $(x+2)/(x+1)$. **4.** $(x+1)/(x-1)$. **5.** $(x+2)/(x(x+1))$. **6.** $(2x-3)/(2x+1)$. **7.** $1/2x$. **8.** $x/2$.

Exercises 2.5

1. $(x+2y)/(2x+3y)$. **3.** $(1+5x^2)/(x(1+5x^3))$. **5.** $t/2(t-1)$. **7.** $2(t-2)/(t+1)$. **9.** $2(x^3-1)/(x^3+x^2+1)$. **11.** $(x-3)/(x+1)$. **13.** 2. **15.** $x-2$. **17.** $(5x+7)/(x+2)$. **19.** $2(x^2+x+3)/(x+2)(2x-1)$. **21.** $(3+4x-3x^2)/(x^2-1)$. **23.** $(2-x)/(2x-1)(x+1)$. **25.** $2/(x-1)(x-2)(x-3)$. **27.** $(x+1)(x-3)/(x+3)(x-1)^2$. **29.** $(10+4x-2x^2)/(x^2-1)(x+3)$. **31.** $(x-1)(x+2)$. **33.** $(x+2)(2x-1)(x-2)(2x+1)$. **35.** $-\frac{2}{3}(x+1)$. **37.** 3. **39.** $2/(x-1)$. **41.** $(x-1)(2x-1)/(x+1)^2$. **43.** $(x-1)(x+1)/(x-2)$. **45.** $(x-1)(x+2)/(x-2)(x-5)$. **47.** $(x+y)^2/xy$. **49.** $-(x^2+y^2)/(x+y)^2$. **51.** $-1/x(x+h)$.

Review Exercises on Chapter 2

1. 2. **3.** -6. **5.** $9x-9y$. **7.** $12-x^4$. **9.** $(t/2)-(5/2t)$. **11.** $(2-x^2)/12x$. **13.** $xy-4(x+y)+16$. **15.** $6x+\sqrt{x}-2$. **17.** $2x^2-5x+10$. **19.** $\frac{1}{11}(10+3\sqrt{5})$. **25.** 3.146. **27.** $(x+2)(x+5)$. **29.** $(2p-7)(p+4)$. **31.** $x^2(3x-4)$. **33.** $2(3x-2y)(9x^2+6xy+4y^2)$. **35.** $2(x-1)(x^2+x+1)$. **37.** $2x(x-5y+2x^2)$. **39.** $(x+5)(x-1)$. **41.** $(x+2)(x-1)$. **43.** $(y-5)(y+2)$. **45.** $(2t+u)(5t-u)$. **47.** $(4x-3)(2x-3)$. **49.** $(p+q+4)(p+q-1)$. **51.** $(3x+1)/(x+2)(x-3)$. **53.** $(x^2-3x+3)/(x-2)(x-1)^2$. **55.** $1/(x-y)(p-q)$. **57.** $(a+b)(a-2)/2(a-2b)$. **59.** x^2-1. **61.** $(6x^3-33x^2+55x-17)/(x-2)(2x-1)$.

CHAPTER 3

Test 3.1: **1.** $\frac{7}{2}$. **2.** $-\frac{5}{3}$. **3.** 12. **4.** 20. **5.** 5. **6.** -2. **7.** $-b/a$. **8.** -6. **9.** $-p/q$. **10.** 0.

Exercises 3.1

1. Yes. **3.** No. **5.** $-\frac{1}{2}$ yes, $\frac{1}{3}$ no. **7.** 4. **9.** $\frac{17}{5}$. **11.** No solution. **13.** $-\frac{19}{7}$. **15.** $-\frac{1}{3}$. **17.** 1. **19.** -2. **21.** 6. **23.** 3. **25.** $\frac{4}{3}$. **27.** 2. **29.** No solution. **31.** 1. **33.** $\frac{8}{3}$. **35.** No solution. **37.** (i) $(cz-by)/a$, (ii) $(cz-ax)/y$. **39.** (i) $(a-S)/(l-S)$, (ii) $(a-S+rS)/r$. **45.** $\frac{8}{3}$. **47.** No. **49.** Yes.

Exercises 3.2

1. (a) $x+4$, (b) $2x-3$, (c) $(x/2)+2$. **3.** 21, 22. **5.** 176, 178. **7.** 15. **9.** 25 yr. **11.** 18 yr. **13.** 7 nickels, 5 dimes, and 2 quarters. **15.** 17. **17.** 40 lb and 30 lb, respectively. **19.** 13.5 lb. **21.** 15 oz. **23.** 10 oz. **25.** 30 oz. **27.** 60 oz. **29.** $3\frac{1}{3}$ hr. **31.** 1 pm. **33.** 25 min. **35.** $\frac{20}{23}$ hr. **37.** 60 min.

39. $8.00.   **41.** $2200 and $700.   **43.** 210 hr.   **45.** $5000 at 8% and $3000 at 10%.
47. $12,000 at 9% and $6000 at 6%.

Test 3.3:   **1.** ± 3.   **2.** 0, 4.   **3.** $\pm a$.   **4.** 0, ± 5.   **5.** 2, 3.   **6.** $-5, -3$.   **7.** 0, 5.   **8.** 0, 3.

Exercises 3.3

1. $-2, -3$.   **3.** 2, 7.   **5.** 3.   **7.** 3, 4.   **9.** $\pm\frac{5}{3}$.   **11.** 0, 8.   **13.** $\frac{3}{2}$.   **15.** 0, 7.   **17.** 2, -1.   **19.** 2, $\frac{1}{2}$.
21. 2, -1.   **23.** 4, $-\frac{4}{3}$.   **25.** $-3, 1$.   **27.** $-6, -\frac{2}{3}$.   **29.** 6, $-\frac{1}{2}$.   **31.** $\pm 1, \pm 2$.   **33.** $\pm\sqrt{2}$.   **35.** 1, 8.
37. 1, -32.   **39.** 1, -3.   **41.** 4, 11.   **43.** 11, 13.   **45.** 5 cm, 12 cm.   **47.** 4 in., 6 in.
49. (a) 1 sec, 4 sec; (b) 5 sec; (c) 100 ft.   **51.** $2 or $4.   **53.** $4 or $5.

Test 3.4:   **1.** $(-b \pm \sqrt{b^2 - 4ac})/2a$.   **2.** $(2q \pm \sqrt{4q^2 - 12pr})/2p$.   **3.** $b^2 - 4ac$.   **4.** Real.   **5.** $b^2 - 4ac = 0$.
6. $b^2 - 4ac < 0$.

Exercises 3.4

1. $-2, -4$.   **3.** $3 \pm 2\sqrt{2}$.   **5.** 2.   **7.** $\frac{1}{2}(3 \pm \sqrt{13})$.   **9.** $\frac{1}{2}(2 \pm \sqrt{7})$.   **11.** $-1 \pm \sqrt{5}$.   **13.** $\frac{3}{2}, -\frac{2}{3}$.
15. $-2\sqrt{3}, \sqrt{3}/2$.   **17.** $\frac{3}{4}$.   **19.** $\frac{1}{5}(-5 \pm \sqrt{10})$.   **21.** $4 \pm \sqrt{21}$.   **23.** $-1 \pm \sqrt{5}$.   **25.** $\pm\sqrt{\frac{11}{6}}$.
27. 0, $\frac{7}{4}$.   **29.** $\frac{1}{4}(3 \pm \sqrt{17})$.   **31.** No real solution.   **33.** $-1, -3$.   **35.** No real solution.   **37.** 2, -2.
39. $-1, \frac{1}{8}$.   **41.** $(1/g)(-u \pm \sqrt{u^2 + 2gs})$.   **43.** $(1/s)(1 \pm \sqrt{1 - s^2})$.   **45.** 1.   **47.** $k = 0$.
49. (a) $x = y \pm \sqrt{4y^2 - 1}$, (b) $y = \frac{1}{3}(-x \pm \sqrt{4x^2 + 3})$.   **53.** 3 in.   **55.** $180 per month.
57. $(30 \pm 5\sqrt{2}) \simeq $37.07 or 22.93.

Test 3.5:   **1.** No solution.   **2.** $x = 4$.   **3.** No.   **4.** No.   **5.** $x = \frac{7}{3}$.   **6.** No.   **7.** No.   **8.** No.

Exercises 3.5

1. 3, $-\frac{1}{2}$.   **3.** 3, $\frac{1}{3}$.   **5.** $\frac{3}{2}, \frac{5}{3}$.   **7.** ± 6.   **9.** 4, $-\frac{3}{7}$.   **11.** -4.   **13.** $\frac{7}{3}$.   **15.** No solution.   **17.** 3, $-\frac{1}{2}$.
19. 6, $\frac{40}{13}$.   **21.** $\frac{3}{2}$.   **23.** $\frac{1}{5}$.   **25.** 4.   **27.** No solution.   **29.** 5.   **31.** 2.   **33.** 0.   **35.** No solution.
37. No solution.   **39.** 0.   **41.** 5.   **43.** 2.   **45.** 2.   **47.** No solution.   **49.** 30 ft and 20 ft.
51. 40 mph and 30 mph.   **53.** $50.   **55.** $\frac{5}{4}$.   **57.** 5 mph.   **59.** 16 ft or 9 ft.
61. (a) $\frac{7}{2}$ miles or $\frac{25}{6}$ miles; (b) no solution: the distance cannot be covered in 2 hr; (c) no solution in $0 \leq x \leq 5$.

Review Exercises on Chapter 3

1. $\frac{3}{4}$.   **3.** $\frac{8}{13}$.   **5.** No solution.   **7.** -1.   **9.** 5.   **11.** $ab/(a - b)$.   **13.** pqr.   **15.** $-5, -8$.   **17.** $\frac{1}{2}, -\frac{1}{3}$.
19. 1.   **21.** 3, -2.   **23.** $\frac{7}{2}$.   **25.** $-1 \pm \sqrt{2}$.   **27.** 1.   **29.** No solution.   **31.** 2.   **33.** 1, -2.   **35.** 1.
37. $\frac{77}{4}$.   **39.** $\frac{9}{5}$.   **41.** 1, -4.   **43.** 2, -2.   **45.** 3.   **47.** $RR_1/(R_1 - R)$.   **49.** $abd/(bd - ad - ab)$.
51. $25,000 at 10% and $75,000 at 8%.   **53.** 100 min after the start of slow runner.   **55.** 24 mph.   **57.** 8 ft.
59. 4% and 8%.   **61.** $\frac{1}{4}$.   **63.** 3 ft.

CHAPTER 4

Test 4.1:   **1.** $\in$.   **2.** $\notin$.   **3.** $\subset$.   **4.** $\subset$.   **5.** $\in$.   **6.** $\notin$.   **7.** $\subset$.   **8.** $\subset$.   **9.** $\subset$.   **10.** $\notin$.
11. $\subset$.   **12.** $\notin$.   **13.** $\subset$.   **14.** Is a subset of.   **15.** $\neq$.   **16.** $=$.   **17.** $\subset$.

Exercises 4.1

1. $\{-1, 0, 1, 2, 3, 4\}$.   **3.** $\{-19, -18, -17, \ldots, -1\}$.   **5.** $\{2, 3, 5, 7, 11, 13, 17, 19\}$.   **7.** $\{2, \frac{3}{2}, \frac{4}{3}, \frac{5}{4}, \ldots\}$.
9. $\{1, \frac{1}{2}, \frac{1}{4}, \frac{1}{6}, \frac{1}{10}, \frac{1}{12}, \frac{1}{16}, \frac{1}{18}\}$.   **11.** $\{2x \mid x$ is an integer and $x < 50\}$.   **13.** $\{x \mid x$ is a prime number less than 30$\}$.

15. {$3n \mid n$ is a natural number.}   **17.** {$n^2 \mid n$ is a natural number}.   **19.** $\left\{\dfrac{n}{n+1} \mid n \text{ is a natural number}\right\}$.   **21.** [3, 8].
23. (5, 8].   **25.** (−7, −3).   **27.** [5, ∞).   **29.** (−∞, −3).   **31.** $2 \leq x \leq 5$.   **33.** $-3 < x < 2$.   **35.** $x \leq 3$.
37. (a) True; (b) false; (c) false; (d) true; (e) false; (f) false; (g) true; (h) false; (i) false; (j) true; (k) false; (l) false; (m) false; (n) true; (o) true; (p) true; (q) true.   **39.** $A \subseteq B$ and $D \subseteq C$.

Test 4.2:   **1.** <.   **2.** >.   **3.** > 0.   **4.** =.   **5.** >.   **6.** < 0.   **7.** >.   **8.** > 0.   **9.** <.   **10.** <.
11. ≤.

Exercises 4.2

1. (−∞, 2).   **3.** [−5, ∞).   **5.** (2, ∞).   **7.** [7, ∞).   **9.** $(-\frac{13}{6}, \infty)$.   **11.** $(\frac{1}{8}, \infty)$.   **13.** $(-\infty, -\frac{16}{7}]$.
15. $(-\frac{1}{2}, \infty)$.   **17.** $(-\infty, \frac{1}{18}]$.   **19.** No solution.   **21.** (−∞, ∞).   **23.** (−3, 1).   **25.** (2, ∞).
27. No solution.   **29.** $x = -2$.   **31.** $5000.   **33.** More than 1500.   **35.** At least 112,000 copies.
37. 45 sq km.

Test 4.3:   **1.** 1; 3.   **2.** −3; 2.   **3.** $x < 1$ or $x > 3$.   **4.** $x < -1$ or $x > 5$.   **5.** $a < x < b$.
6. $x \leq a$ or $x \geq b$.   **7.** All x.   **8.** All x.   **9.** $-3 < x < 3$.

Exercises 4.3

1. $2 < x < 5$.   **3.** $-3 \leq x \leq \frac{5}{2}$.   **5.** $x < -\frac{1}{3}$ or $x > 2$.   **7.** $-3 < x < 3$.   **9.** $3 < x < 4$.
11. $y < -2$ or $y > 5$.   **13.** $y < -2$ or $y > \frac{3}{2}$.   **15.** No solution.   **17.** No solution.   **19.** All x.   **21.** $x = 3$.
23. All x.   **25.** No solution.   **27.** All x.   **29.** $-\frac{1}{3}(4 + \sqrt{37}) \leq x \leq \frac{1}{3}(\sqrt{37} - 4)$.   **31.** $-3 < x < 0$ or $x > 3$.
33. (a) $k < -4$ or $k > 4$; (b) $-4 < k < 4$.   **35.** (a) $0 < p < 4$; (b) $p < 0$ or $p > 4$.   **37.** 70 yd.   **39.** 2 in.
41. $1 < t < 4$.   **43.** (a) $p < 11$; (b) at least 20.   **45.** $5 \leq p \leq 7$.   **47.** $225 \leq $ rent ≤ 255.
49. $x < -\frac{3}{2}$ or $x > 0$.   **51.** $1 < x < 4$.   **53.** $x > 2$.   **55.** $\frac{3}{2} < x < \frac{8}{3}$.   **57.** $x < 0$ or $x \geq 8$.
59. $x < -1$ or $x > \frac{3}{2}$.   **61.** $x < -2$.   **63.** $x < 2$.   **65.** No solution.   **67.** All $x \neq 2$.
69. $0 < $ denominator < 3.   **71.** $-3 < k < -1$.

Test 4.4:   **1.** $|x|$.   **2.** <.   **3.** >.   **4.** >.   **5.** $-a < x < a$.   **6.** $x < -a$ or $x > a$.
7. Empty set (no solution).   **8.** All x.   **9.** $-2 < x < 2$.   **10.** $x < -3$ or $x > 3$.   **11.** $x = \pm 3$.
12. $x = -1, 5$.   **13.** No.   **14.** No.   **15.** $|x| < 3$.   **16.** $|x| > 5$.

Exercises 4.4

1. $1, -\frac{1}{2}$.   **3.** $\frac{3}{2}, -1$.   **5.** $\frac{1}{2}$.   **7.** No solution.   **9.** No solution.   **11.** $\frac{27}{17}, \frac{33}{19}$.   **13.** $(-\frac{11}{3}, -1)$.
15. $(-\infty, -\frac{1}{3}]$, and $[1, \infty)$.   **17.** $[\frac{1}{4}, \frac{13}{4}]$.   **19.** All x except $\frac{7}{2}$.   **21.** $x = -\frac{4}{3}$.   **23.** No solution.   **25.** (−∞, ∞).
27. (−∞, ∞).   **29.** No solution.   **31.** (−3, −1) and (1, 3).   **33.** No solution.
35. (−2, 2), $(-\infty, -\sqrt{10})$, and $(\sqrt{10}, \infty)$.   **37.** (−2, 1).   **39.** [2, 3], (−∞, −1], and [6, ∞).

Review Exercises on Chapter 4

1. $x > -8$.   **3.** $x > \frac{37}{8}$.   **5.** $x > -\frac{5}{24}$.   **7.** No solution.   **9.** $x < \frac{1}{2}$ or $x > 2$.   **11.** $\frac{1}{3} \leq x \leq 3$.
13. $-\frac{3}{2} < x < 1$.   **15.** All x.   **17.** No solution.   **19.** $0 < x < 3$ or $x > 4$.   **21.** $x < -\frac{3}{2}$ or $x > 1$.
23. $x < 2$.   **25.** $\frac{1}{5} < x < 1$.   **27.** $x \leq 1$ or $x \geq \frac{5}{2}$.   **29.** $x < -4$ or $x > 10$.   **31.** No solution.
33. No solution.   **35.** $1 < t < 2$.   **37.** Denominator ≤ -2 or denominator > 0.   **39.** $200.

CHAPTER 5

Test 5.1:   **1.** 5; 2.   **2.** $y > 0$.   **3.** $y < 0$.   **4.** $x > 0$.   **5.** $x < 0$.   **6.** $y = 0$.   **7.** $x = 0$.   **8.** Origin.
9. $> 0; > 0$.   **10.** Second.   **11.** $> 0; < 0$.   **12.** First or third.   **13.** Second or fourth.   **14.** Negative.

15. First or third. **16.** y-axis. **17.** Undefined. **18.** $\sqrt{(x_2 - x_1)^2 + (y_2 - y_1)^2}$. **19.** $\sqrt{x_1^2 + y_1^2}$.
20. $|x_2 - x_1|$. **21.** $|y_1 - y_2|$. **22.** $\left(\dfrac{a+c}{2}, \dfrac{b+d}{2}\right)$.

Exercises 5.1

1.

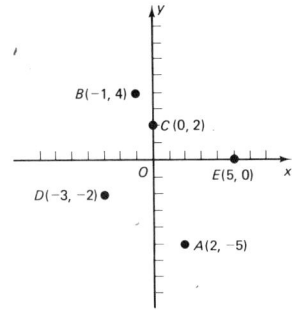

A: IV, B: II, D: III.
C and E: no quadrant

3. $\sqrt{5}$; $(3, -\tfrac{1}{2})$. **5.** $\tfrac{1}{2}\sqrt{29}$, $(-\tfrac{3}{4}, \tfrac{3}{2})$. **7.** $\sqrt{a^2 + b^2}$; $(a/2, b/2)$. **9.** C. **15.** 0 or -6. **17.** 19 or -5.
19. -3 or 1. **21.** $(4, 0)$ and $(-2, 0)$. **23.** $Q(-5, 2)$. **25.** $x^2 + y^2 - 4x + 6y - 12 = 0$. **27.** $y = 4$.
29. $3x^2 + 3y^2 + 12x - 22y + 35 = 0$. **31.** $(3, 3)$ and $(5, 5)$.

Test 5.2: **1.** y axis. **2.** x. **3.** $-x$; $-y$. **4.** y axis. **5.** Origin. **6.** y axis. **7.** x axis, y axis, and the origin.
8. x axis, y axis, and the origin. **9.** x axis.

Exercises 5.2

1. y axis. **3.** x axis. **5.** Origin. **7.** y axis. **9.** Origin. **11.** All. **13.** Origin. **15.** y axis.

17.

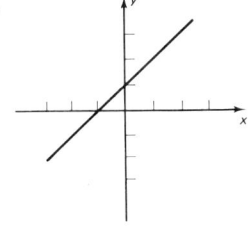

19.

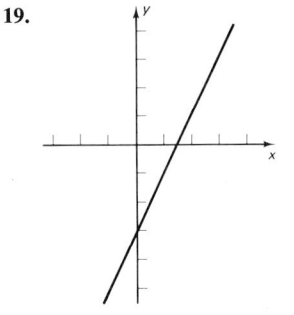

21.

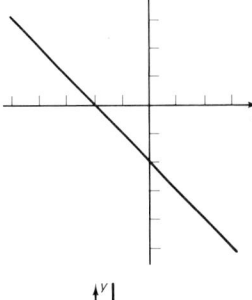

23.

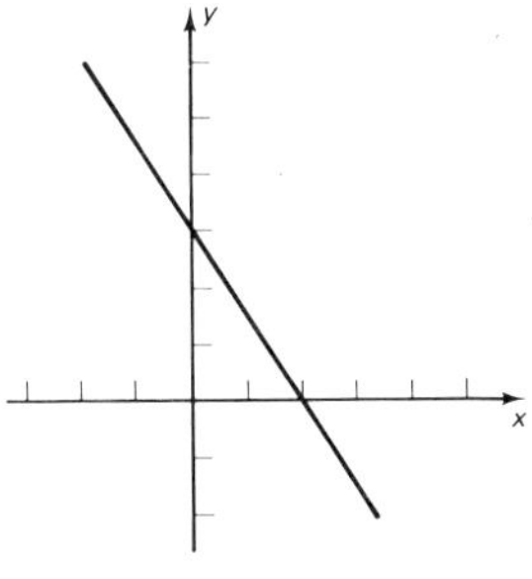

25.

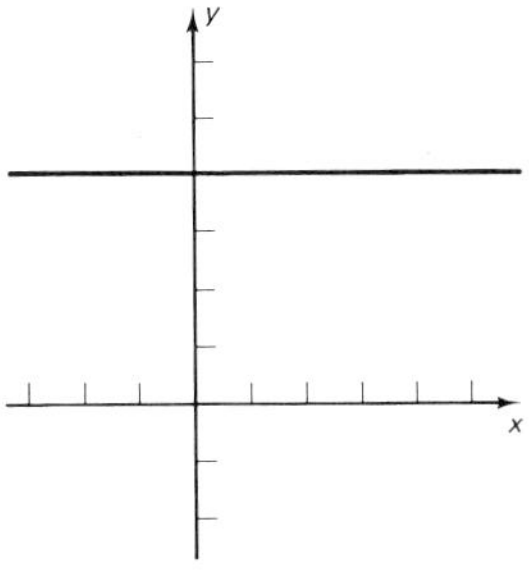

27.

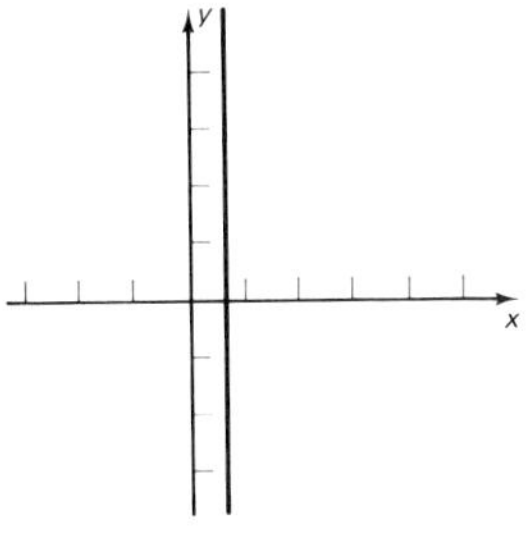

Answers to Tests and Odd-numbered Exercises

A-7

29.

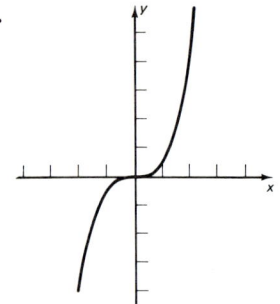

31.

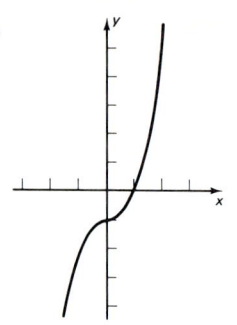

33.

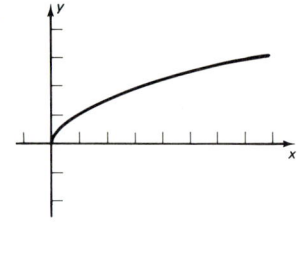

35.

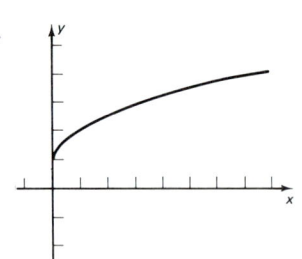

37.

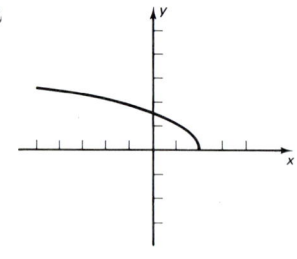

39.

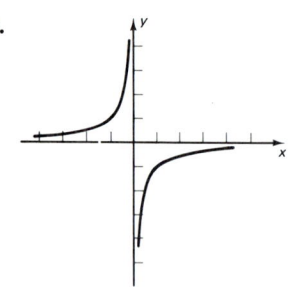

Test 5.3: **1.** Slope. **2.** Zero. **3.** No. **4.** $(y_2 - y_1)/(x_2 - x_1)$. **5.** Positive. **6.** Negative. **7.** >.
8. Falls. **9.** Rises. **10.** $y_1 = y_2$. **11.** $a = c$. **12.** Vertical. **13.** Horizontal. **14.** $y = 0$. **15.** $y = q$.
16. $x = a$. **17.** $y - y_1 = m(x - x_1)$. **18.** $y = mx + b$. **19.** $y = -3$. **20.** $x = -2$. **21.** $y -$.
22. $y -$. **23.** $-2; 3$. **24.** $1/m$. **25.** A straight line.

Exercises 5.3

1. 2. **3.** 0. **5.** Undefined. **7.** $y = 5x - 9$. **9.** $y = 4$. **11.** $y = 6x - 19$. **13.** $x - 3 = 0$.
15. $y = -2x + 5$. **17.** $y = 3x - 9$. **19.** $4x - 3y + 12 = 0$. **21.** $y = 1$. **23.** $y = x$. **25.** $\frac{3}{2}; -3$.
27. $2; -3$. **29.** $0; \frac{3}{2}$. **35.** $-5 < k < 3$. **39.** $y = 5x - 13; \frac{13}{5}$ and -13. **41.** $5x + 3y = 7$.
43.

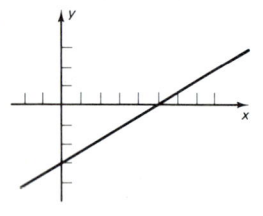

45.

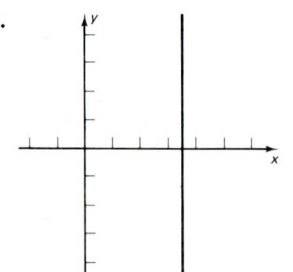

47.

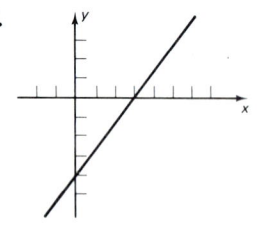

Test 5.4: **1.** $m_1 = m_2$. **2.** $-1/m_1$. **3.** Parallel. **4.** Parallel. **5.** Perpendicular. **6.** Parallel.

Exercises 5.4

1. Parallel. **3.** Perpendicular. **5.** Perpendicular. **7.** Parallel. **9.** Perpendicular. **11.** Neither.
13. Neither. **15.** $3x - 4y - 8 = 0$. **17.** $x = 4$. **19.** $3x - 2y - 5 = 0$. **21.** $y + 2 = 0$.
23. $3x + y - 10 = 0$. **33.** $y = 2x - 3$. **35.** $y = 2x + 5$. **39.** (a) $2x + 5y = 280$; (b) 40.
41. $5x + 8y = 100$. **43.** $p = 1.70 - 0.00005x$. **45.** $p = 0.00025x + 0.50$.

A-8

Answers to Tests and Odd-numbered Exercises

Exercises 5.5

1. $x = 2, y = 1$.   3. $x = 1, y = 6$.   5. $p = 2, q = 1$.   7. No solution.
9. Coordinates of any point on the line $2x + 3y = 6$.   11. $x = 1, y = -1$.   13. $x = -1, y = -1$.
15. $x = y = 6$.   17. No solution.   19. Coordinates of any point on the line $3x - 5y = 15$.   21. $x = y = 6$.
23. $x = 18, y = 6$.   25. No solution.   27. $x = 2, y = 3$.
29. Coordinates of any point on the line $5x - y = 6$.
31. $x = 30, p = 5$.   33. $x = 10, p = 10$.   35. $a = 2, b = -3$.   37. (a) $k \neq 18$; (b) no real number; (c) $k = 18$.
39. 15 oz of 10 carat and 20 oz of 24 carat gold.   41. 10 lb and 40 lb.   43. 2 to 1.
45. $x = 5, y = -2$ or $x = -2, y = 5$.   47. $x = 3, y = -1$ or $x = -\frac{7}{3}, y = \frac{5}{3}$.   49. $x = 2, y = 1$ or $x = \frac{1}{2}, y = 4$.
51. $x = 1, y = 1$.   53. $x = 2, y = 1$ or $x = 1, y = 2$.   55. $x = 3, p = 4$.   57. $x = 5, p = 12$.
59. $x = 2, y = 1, z = 2$.   61. $x = 3, y = -2, z = 1$.   63. $x = 3, y = -2, z = 1$.

Review Exercises on Chapter 5

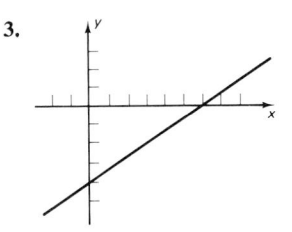

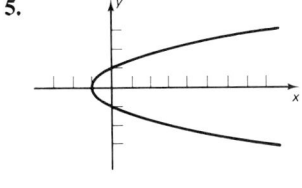

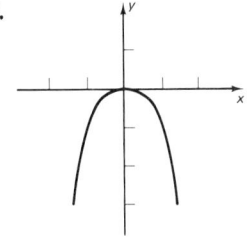

1. $\sqrt{82}$.   3.   5.   7.

9. $y = 2x - 1$.   11. $x = 2$.   13. $x = 2$.   15. $x = 2$.   17. $2x - 3y - 11 = 0$.   19. $4x - 6y = 3$.
21. $(3, 0)$ and $(0, -2)$.   23. $(-\frac{1}{3}, 0)$ and $(0, \frac{1}{4})$.   27. $x + y = 3$.   29. $x + 4y = 6$.   31. $x = y = 1$.
33. $x = 6, y = 12$.   35. $x = y = \frac{1}{6}$.   37. No solution.   39. $x = y = 1$ or $x = -\frac{1}{2}, y = -2$.
41. $x = 2, y = \frac{1}{2}$ or $x = \frac{1}{2}, y = 2$.   43. No real solution.   45. $x = 2, y = -1$.
47. No solution ($2^x = -1$ has no solution for x).   49. $x = 1, y = 1$ or $x = -1/3, y = \frac{5}{3}$.   51. $x = \frac{1}{2}, y = \frac{1}{3}, z = 1$.
53. No solution.   55. $a = \frac{1}{3}, b = -\frac{4}{3}, c = 3$.
57. (a) $C = 5x + 200$, where C is the cost of producing x units; (b) $200, $5.   59. 60 units of A and 80 units of B.
61. (a) $p = 42 - 0.06x$; (b) $x = 150, p = 33$.

CHAPTER 6

Test 6.1:   1. 3.   2. -28.   3. $-7; -7$.   4. $1/(x + a)$.   5. $x \geq -2$.   6. $x > 1$.   7. $x \geq 1$.
8. $\{x \mid x \neq 1\}$.   9. $\{x \mid x \geq 1$ and $x \neq 2\}$.   10. $\{y \mid y \geq 0\}$.   11. Vertical; at most.

Exercises 6.1

1. $2, -4, -10$.   3. $0, -6, -6$.   5. $-1, -1, -1$.   7. $1, 5,$ undefined.   9. $\sqrt{2}$, undefined, $\sqrt{5}$.
11. (a) 7; (b) 9; (c) 49; (d) $\frac{21}{4}$; (e) 19; (f) $3c^2 - 5c + 7$; (g) $3(c + h)^2 - 5(c + h) + 7$; (h) $6c + 3h - 5$.
13. (a) 6; (b) 11; (c) 12; (d) $7 + 2h$ and $3h - 9$.   15. (a) 2; (b) 7; (c) -1; (d) 1; (e) undefined; (f) 7 and $5 - 4h + h^2$.
17. All real numbers.   19. All real numbers.   21. All real numbers except 2.   23. All real numbers *not* less than 2.
25. All real numbers greater than $\frac{3}{2}$.   27. All non-negative real numbers.   29. All positive real numbers except 1.
31. $\{x \mid -1 \leq x \leq 1\}$.   33. $A(L) = (1/4\pi)L^2$.   35. $D(t) = 20t$.   37. $D(t) = 5t$.
39. $C(x) = 15x + 3000; R(x) = 25x; P(x) = 10x - 3000$.   41. $A(x) = 100x - x^2$.
43. $C(x) = 5.5x^2 + 1800/x$.

45. $C(x) = \begin{cases} 25x & \text{if } 0 \leq x \leq 50 \\ 250 + 20x & \text{if } x > 50 \end{cases}$

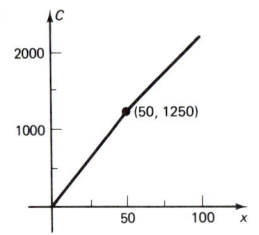

47. No. **49.** No. **51.** Yes. **53.** 15. **55.** $\frac{1}{9}$. **57.** $S = 4/p^2$; 16. **59.** 562.5 ft. **61.** $V = 10^7 a^2$.
63. $\lambda = 400/f$; 6.

Test 6.2: **1.** Linear. **2.** Straight line. **3.** Horizontal. **4.** Parabola. **5.** Rectangular hyperbola.
6. Absolute value function. **7.** Of all non-negative numbers. **8.** $\{y \mid y \geq -2\}$. **9.** -1. **10.** Polynomial; 7.

Exercises 6.2

1. Algebraic. **3.** Rational. **5.** Algebraic (note that $|x^2 + 3x + 2| = \sqrt{(x^2 + 3x + 2)^2}$. **7.** Polynomial; 9.
9. Rational.

11.

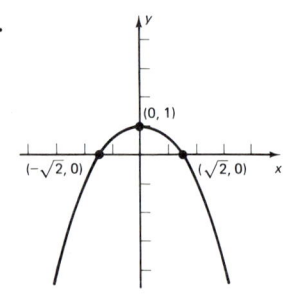

13.

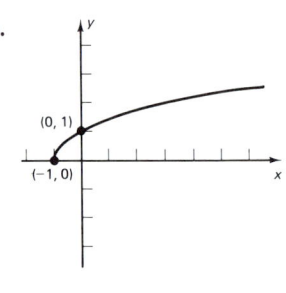

15.

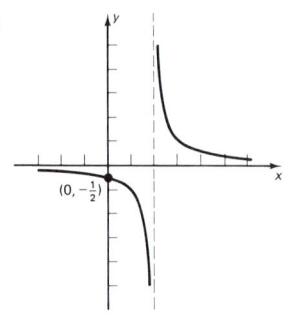

17. 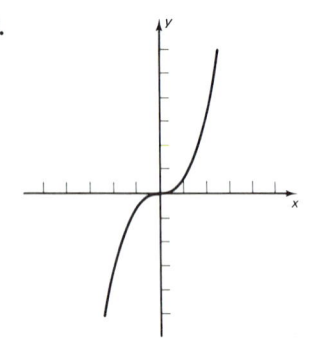 Domain = all real numbers

19. Domain = $\{x \mid x > 0\}$

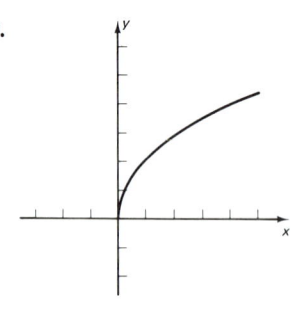

21. Domain = $\{x \mid x < 3\}$

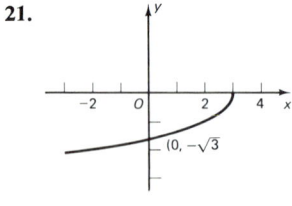

23. Domain = $\{t \mid t \neq -2\}$

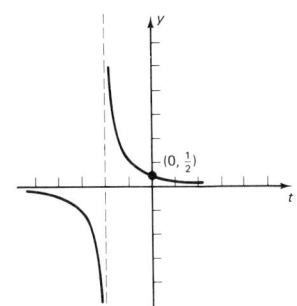

25. Domain = $\{y \mid y \neq 2\}$

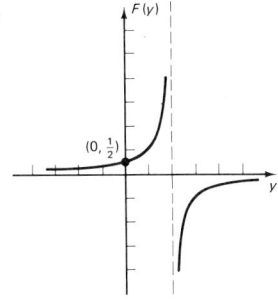

27. Domain = All real numbers

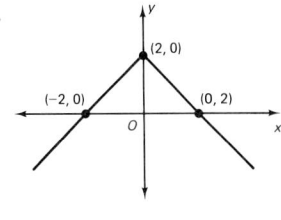

29. Domain = All real numbers

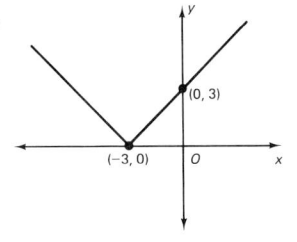

31. Domain = All real numbers

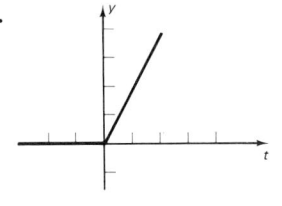

33. Domain = $\{x \mid x \neq 3\}$

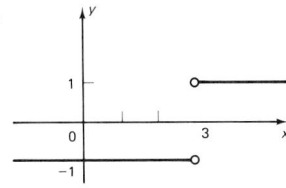

35. 36 ft, 64 ft, 84 ft, and 96 ft. At $t = \frac{5}{2}$, the object is at the highest point. At $t = 5$, the object returns back to the ground.

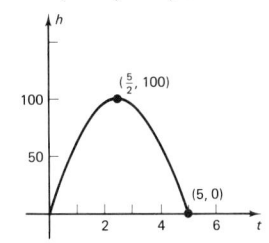

37. $P = 4\sqrt{A}$.

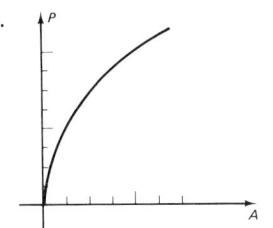

39. $f(x) = 2x + |x - 2|$.

Test 6.3: **1.** $f(x) + g(x); f(x) \cdot g(x); g(x)/f(x); f(x) \neq 0$. **2.** Equal to. **3.** Of all real numbers. **4.** Of all real numbers except 2. **5.** Of all $x \neq 3$. **6.** $f(g(x))$. **7.** 3; 7. **8.** 2; 4.

Exercises 6.3

1. -2. 3. -8. 5. Undefined. 7. -2. 9. Undefined. 11. Undefined. 13. 1. 15. $\frac{3}{4}$.
17. Undefined.
19. $((f \pm g)(x) = x^2 \pm 1/(x-1); (fg)(x) = x^2/(x-1); (f/g)(x) = x^2(x-1); (g/f)(x) = 1/x^2(x-1);$
    $D_{f \pm g} = D_{fg} = D_{f/g} = \{x \mid x \neq 1\}, D_{g/f} = \{x \mid x \neq 0, 1\}$.
21. $(f \pm g)(x) = \sqrt{x-1} \pm 1/(x+2); (fg)(x) = \sqrt{x-1}/(x+2); (f/g)(x) = (x+2)\sqrt{x-1};$
    $(g/f)(x) = 1/(x+2)\sqrt{x-1}; D_{f \pm g} = D_{fg} = D_{f/g} = \{x \mid x \geq 1\}; D_{g/f} = \{x \mid x > 1\}$.
23. $(f \pm g)(x) = (x+1)^2 \pm 1/(x^2-1); (fg)(x) = (x+1)/(x-1)(f/g)(x) = (x+1)^3(x-1);$
    $(g/f)(x) = 1/(x+1)^3(x-1); D_{f/g} = D_{fg} = D_{f/g} = D_{g/f} = \{x \mid x \neq \pm 1\}$. 25. 4. 27. $\frac{1}{4}$. 29. Undefined.
31. 1. 33. 1. 35. 1. 37. Undefined. 39. 3. 41. $(f \circ g)(x) = (1+x)^2; (g \circ f)(x) = 1 + x^2$.
43. $(f \circ g)(x) = 1/(\sqrt{x}+2); (g \circ f)(x) = 1 + 1/\sqrt{x+1}$. 45. $(f \circ g)(x) = (x-3)^2 + 2; (g \circ f)(x) = x^2 - 1$.
47. $(f \circ g)(x) = 5; (g \circ f)(x) = 4$. 51. $R = 243(t+1)^5 + 3\sqrt{3t+3}; 9 + 3^{10}$. 53. $R = x(2000-x)/15$.
55. $3a + 4b = 7$.

Test 6.4: 1. $f(x_1) \neq f(x_2)$. 2. $g^{-1}(v)$. 3. $f(x)$. 4. $1/y$. 5. $x+2$. 6. Not defined. 7. $\sqrt{y-1}$.
8. Horizontal; at most. 9. $y = x$. 10. x; range. 11. x; domain.

Exercises 6.4

1. -3. 3. 1. 5. Not defined. 7. 1. 9. 2.

11. $x = f^{-1}(y) = -(y+4)/3$

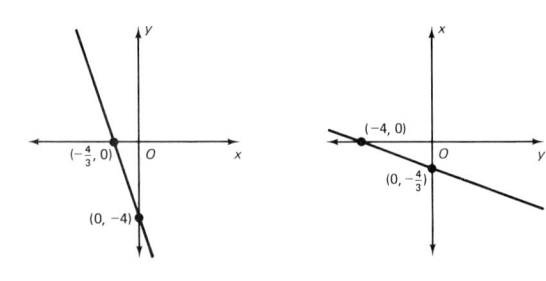

$y = -3x - 4$ $x = -\frac{1}{3}(y+4)$

13. $x = f^{-1}(p) = 10 - 5p/2$:

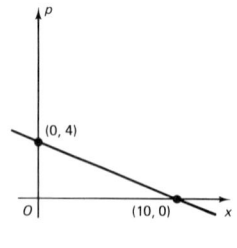

$p = 4 - \frac{2}{5}x$ 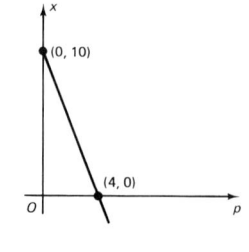 $x = 10 - \frac{5}{2}p$

15. $x = f^{-1}(y) = (y^2 + 4)/3$, $(y > 0)$:

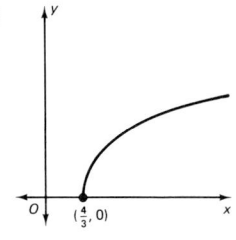

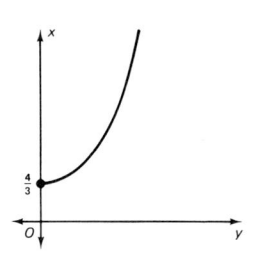

$y = \sqrt{3x-4}$; $x = \frac{1}{3}(y^2+4)$

17. $x = f^{-1}(y) = y^{1/5}$:

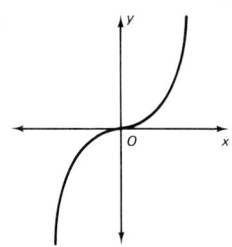

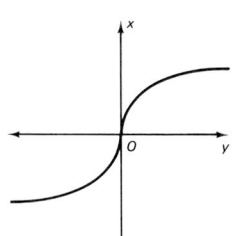

$y = x^5$; $x = y^{1/5}$

19. $x = f^{-1}(y) = 4 - y^2$, $(y > 0)$:

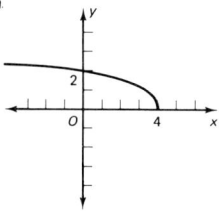

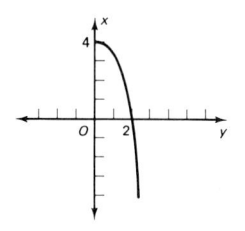

$y = \sqrt{4-x}$; $x = 4 - y^2$, $(y > 0)$

21. $x = f^{-1}(y) = 2 + 1/y$:

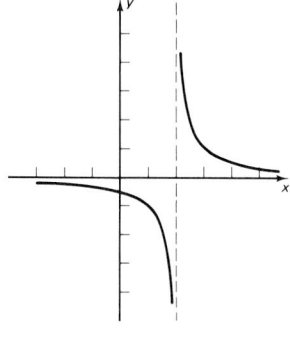

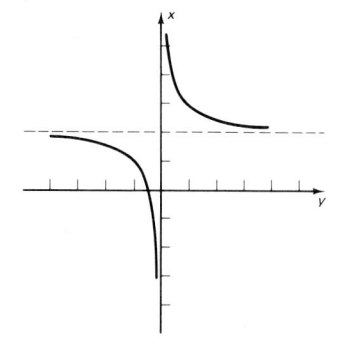

$y = \dfrac{1}{x-2}$; $x = 2 + \dfrac{1}{y}$

23. $x = f^{-1}(g) = 3(1-y)/(1+2y)$

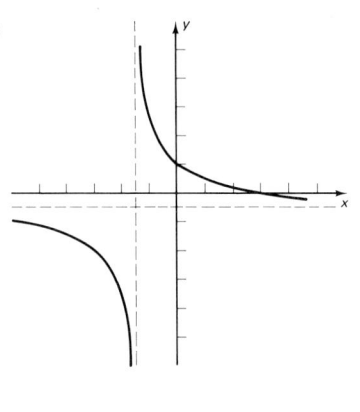

$y = \frac{3-x}{2x+3}$

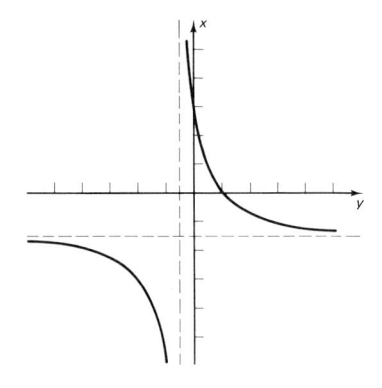

$x = \frac{3(1-y)}{1+2y}$

25. $f^{-1}(x) = (x-2)/3$. **27.** $f^{-1}(x) = (5-3x)/(5x-3)$.
29. $x = f^{-1}(y) = -1 + \sqrt{y}$ if $x \geq -1$; $x = f^{-1}(y) = -1 - \sqrt{y}$ if $x \leq -1$.
31. $x = f^{-1}(y) = y^{3/2}$ if $x \geq 0$; $x = f^{-1}(y) = -y^{3/2}$ if $x \leq 0$.
33. $x = \frac{1}{2}(-3 + \sqrt{4y+1})$ if $x \geq -\frac{3}{2}$; $x = -\frac{1}{2}(3 + \sqrt{4y+1})$ if $x \leq -\frac{3}{2}$. **35.** No. **39.** $t = \frac{1}{4}\sqrt{s}$.
41. $t = (3-4v)/(v-2)$.
43. $x = \begin{cases} c/10 & \text{if } c > 500 \\ (c-350)/3 & \text{if } c \leq 500. \end{cases}$

Test 6.5: **1.** 2. **2.** 2. **3.** $+\infty$. **4.** $-\infty$. **5.** $\frac{1}{2}$; $-\frac{1}{2}$. **6.** $x = 1$; $x = -1$. **7.** $x = 0$ and $x = -1$.

Exercises 6.5

1. $y \to 0$ as $x \to \pm\infty$; $y \to -\infty$ as $x \to 4^-$; $y \to +\infty$ as $x \to 4^+$.
3. $y \to 0$ as $x \to \pm\infty$; $y \to +\infty$ as $x \to 3^-$; $y \to -\infty$ as $x \to 3^+$.
5. $y \to 1$ as $x \to \pm\infty$; $y \to -\infty$ as $x \to -2^+$; $y \to +\infty$ as $x \to -2^-$.
7. $y \to 1$ as $x \to \pm\infty$; $y \to +\infty$ as $x \to -1^+$; $y \to +\infty$ as $x \to -1^-$.
9. $y \to 1$ as $x \to \pm\infty$; $y \to +\infty$ as $x \to -1^+$; $y \to -\infty$ as $x \to -1^-$. $y \to -\infty$ as $x \to -2^+$; $y \to +\infty$ as $x \to -2^-$.
11. $y \to 1$ as $x \to \pm\infty$; $y \to +\infty$ as $x \to 1^-$; $y \to -\infty$ as $x \to 1^+$. $y \to +\infty$ as $x \to -1^+$; $y \to -\infty$ as $x \to -1^-$.

13. **15.** 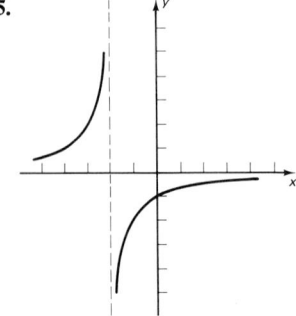 **17.**

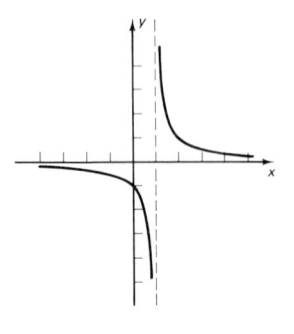

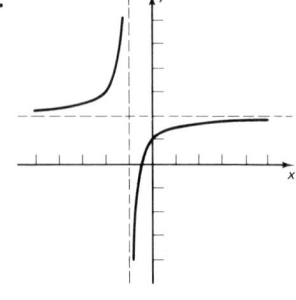

19. **21.** 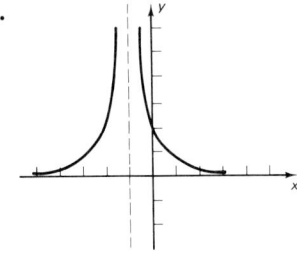 **23.**

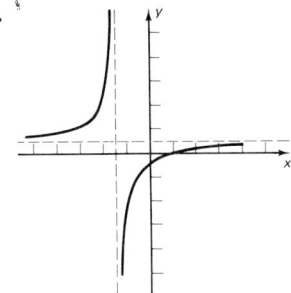

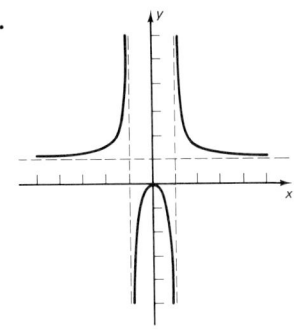

25. 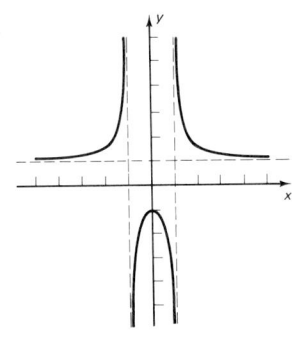 **27.** 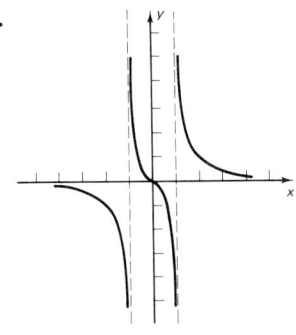 **29.**

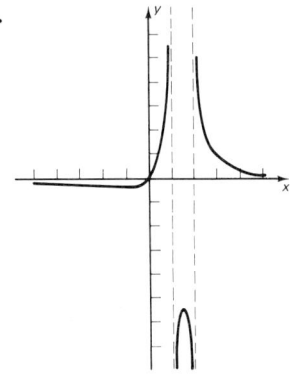

Review Exercises on Chapter 6

1. $\frac{13}{2}$ and 2.   **3.** (a) 5; (b) 1; (c) 19; (d) 4.   **5.** All real numbers except $-\frac{3}{2}$.   **7.** All real numbers except zero.
9. $V = (4\pi/3)(S/4\pi)^{3/2}$.
11. $d = \begin{cases} 5t & \text{if } 0 \le t \le 1 \\ \sqrt{16 + (7t - 4)^2} & \text{if } 1 < t \le 2 \\ \sqrt{16 + (t - 12)^2} & \text{if } 2 < t \le 3 \\ \sqrt{9t^2 + (16 - 4t)^2} & \text{if } 3 < t \le 4. \end{cases}$

13. (a) $f(x) = x^2$ is the simplest example; (b) $f(x) = x^3$ is the simplest example (apart from $f(x) = x$); (c) $f(x) = kx$, (k is any constant).

15.

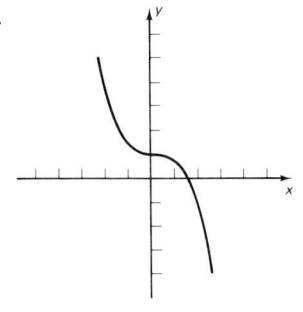

Domain = All real numbers

17.

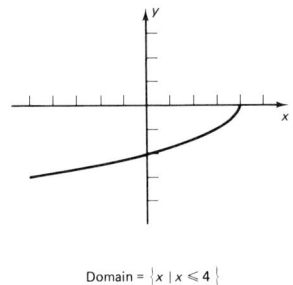

Domain = $\{x \mid x \le 4\}$

19.

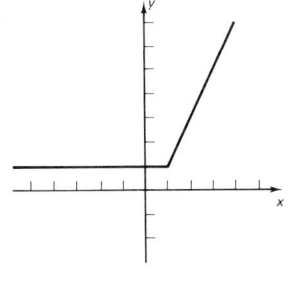

Domain = All real numbers

Answers to Tests and Odd-numbered Exercises

21. Undefined.   **23.** 16.   **25.** Undefined.   **27.** x.
29. $(f \circ g)(x) = (x+1)^2 - |x+1|$; $(g \circ f)(x) = (x+1-\sqrt{x})^2$.   **31.** $x = f^{-1}(y) = (y-2)/2$.
33. $x = f^{-1}(y) = (3+y)/2y$.   **35.** $x = f^{-1}(y) = \frac{1}{2}(3+y^{1/4})$.
37. $x = \begin{cases} y-2 & \text{if } y \geq 3 \\ \frac{1}{2}(y-1) & \text{if } y < 3. \end{cases}$

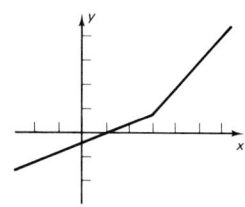

39. $y \longrightarrow 3$ as $x \longrightarrow \pm\infty$; $y \longrightarrow -\infty$ as $x \longrightarrow 3^-$; $y \longrightarrow +\infty$ as $x \longrightarrow 3^+$.
41. $y \longrightarrow 1$ as $x \longrightarrow \pm\infty$; $y \longrightarrow -\infty$ as $x \longrightarrow -1^+$ or as $x \longrightarrow -1^-$.

43.

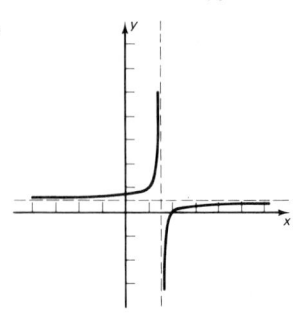

45.

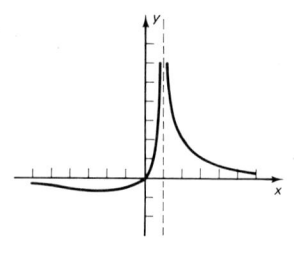

47.

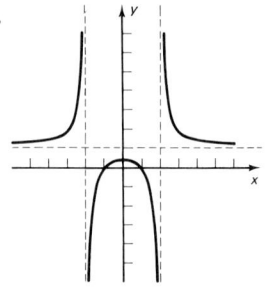

49. $ps + rq = s + q$.

CHAPTER 7

Test 7.1:   **1.** Circle.   **2.** An empty set.   **3.** An empty set.   **4.** The point $(0, 0)$.   **5.** The point $(2, -5)$.
6. $A = B$ and $AC < 0$.   **7.** $(x-h)^2 + (y-k)^2 = r^2$.   **8.** Circle; 4; $(2, -3)$.

Exercises 7.1

1. $x^2 + y^2 - 4y - 21 = 0$.   **3.** $x^2 + y^2 + 6x - 7 = 0$.   **5.** $x^2 + y^2 + 4x + 10y + 28 = 0$.
7. $x^2 + y^2 - 12x + 12y + 36 = 0$.   **9.** $x^2 + y^2 + 4x - 16y + 59 = 0$.   **11.** $x^2 + y^2 - 6y + 7 = 0$.
13. $x^2 + y^2 + 4x - 8y - 60 = 0$.   **15.** Yes; $(-1, -1)$, 1.   **17.** Yes; $(2, 4)$, 4.   **19.** Yes; $(-\frac{3}{2}, \frac{5}{2})$, $\sqrt{\frac{15}{2}}$.
21. Yes; $(\frac{5}{4}, -1)$, $\frac{7}{4}$.   **23.** Yes; $(\frac{1}{3}, -\frac{2}{3})$, $\sqrt{\frac{26}{27}}$.   **25.** Yes; $(-1, -3)$, 0.   **27.** $(-1, 0)$, $(0, -1)$.
29. $(2, 0)$, $(0, 4 \pm 2\sqrt{3})$.   **31.** $((-3 \pm \sqrt{5})/2, 0)$, $(0, (5 \pm \sqrt{21})/2)$.   **33.** $((5 \pm \sqrt{33})/4, 0)$, $(0, (-2 \pm \sqrt{6})/2)$.

35. $((3 \pm \sqrt{42})/9, 0), (0, (-6 \pm \sqrt{69})/9)$.   **37.** $x^2 + y^2 - 4x + 6y - 37 = 0$.
39. $x^2 + y^2 - 4x - 6y + 4 = 0; (0, 3 \pm \sqrt{5})$.   **41.** $x^2 + y^2 - 3x - y = 0$ **43.** $x^2 + y^2 - 2y = 0$.
45. $x^2 + y^2 + 5x - 5y + 10 = 0$.

47.

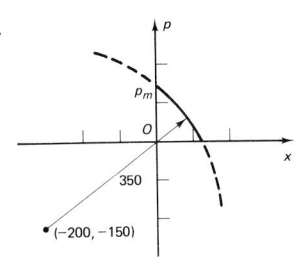

$p_m = \$137.23$

49.

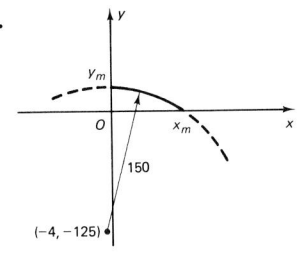

Max. apples (x_m) = 78.92 kg
Max. apple cider (y_m) = 24.95 liters

51. Yes; $y = \sqrt{9 - x^2}$.   **53.** No.

55.   **57.**

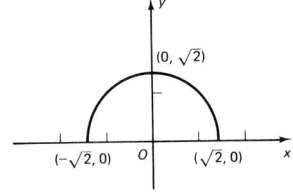

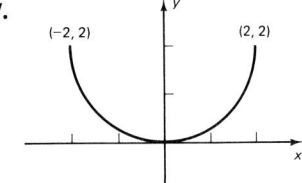

Test 7.2:   **1.** Parabola; $(0, 0)$; upward; downward.   **2.** Parabola; (k, h); upward; downward.
3. Parabola; $(2, 2)$; downward.   **4.** Parabola; $(0, -3)$; to the left.   **5.** Parabola; $(0, 0)$; to the right; to the left.
6. Parabola; upward; downward.   **7.** Half-parabola; $(0, 0)$; to the right; $a > 0$; $a < 0$.
8. Half-parabola; $(0, 0)$; to the left; $a > 0$; $a < 0$.   **9.** Half-parabola; (h, k).   **10.** Perpendicular.   **11.** Axis.
12. Vertex.

Exercises 7.2

1. $(0, 0)$.   **3.** $(0, 2)$.   **5.** $(0, -2)$.   **7.** $(2, -3)$.   **9.** $(-\frac{1}{3}, -\frac{7}{3})$.   **11.** $(\frac{1}{3}, \frac{2}{3})$.   **13.** $(\pm 2, 0), (0, -4)$.
15. $(0, 1)$.   **17.** $(-1, 0), (-2, 0), (0, 2)$.   **19.** $(0, 0), (3, 0)$.   **21.** $(0, 4)$.   **23.** $(1 \pm \sqrt{3}, 0), (0, 2)$.

25.   **27.** 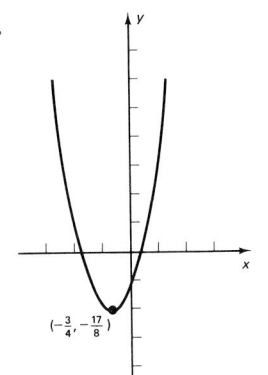   **29.**

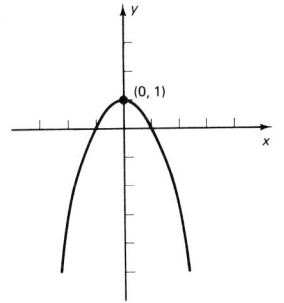

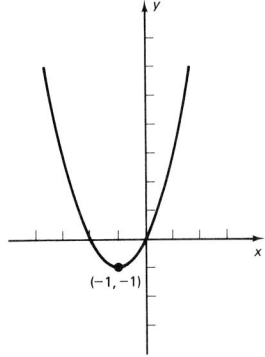

31.

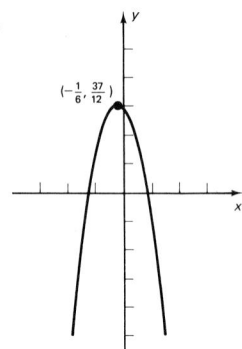

[Graph of downward parabola with vertex $(-\frac{1}{6}, \frac{37}{12})$]

33.

[Graph of downward parabola with vertex $(2, 0)$]

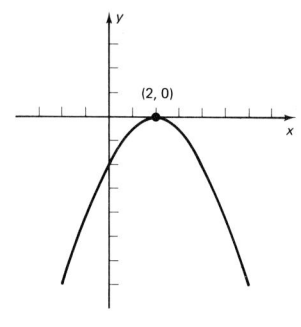

35. 8 and 8. **37.** $-\frac{5}{2}$ and $\frac{5}{2}$. **39.** $(\frac{12}{13}, \frac{18}{13})$. **41.** 15,625 sq yd. **43.** $x = 10$. **45.** $22.50; $202,500.
47. $175; $6125. **49.** 600 units; $3600. **51.** (a) 3000 units; $90,000; (b) 1750 units; $28,625.
53. $(-b/2a, (4ac - b^2)/4a)$. **55.** $b^2 = 4ac$.

Test 7.3: **1.** An ellipse; $(\pm\sqrt{2}, 0); (0, \pm\sqrt{3})$. **2.** $x^2/4 + y^2/25 = 1$. **3.** $x^2/9 + y^2/1 = 1; x^2/1 + y^2/9 = 1$.
4. $4; 2\sqrt{3}$. **5.** Hyperbola; $(\pm\sqrt{2}, 0)$. **6.** Asymptotes. **7.** Asymptotes. **8.** (Rectangular) hyperbola.
9. $y = \pm(\sqrt{\frac{3}{2}})x$. **10.** $y = \pm(\sqrt{\frac{2}{3}})x$. **11.** Sum. **12.** Hyperbola.

Exercises 7.3

1. $(\pm 4, 0), (0, \pm 3)$, 8 and 6. **3.** $(\pm 1/\sqrt{5}, 0), (0, \pm 1/\sqrt{20}), 2/\sqrt{5}$ and $1/\sqrt{5}$. **5.** $(\pm 1, 0), (0, \pm\sqrt{3}), 2\sqrt{3}$ and 2.

7.

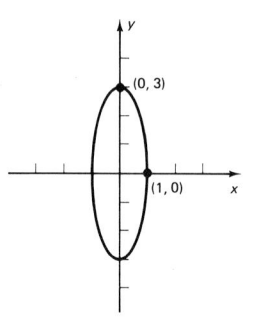

[Ellipse with vertices $(0, 3)$ and $(1, 0)$]

9.

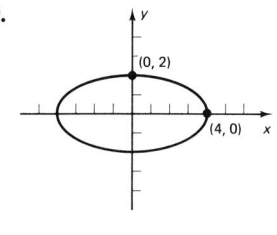

[Ellipse with vertices $(0, 2)$ and $(4, 0)$]

11.

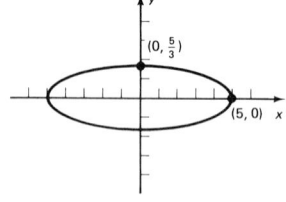

[Ellipse with vertices $(0, \frac{5}{3})$ and $(5, 0)$]

13.

[Ellipse with vertices $(0, 2)$ and $(\sqrt{3}, 0)$]

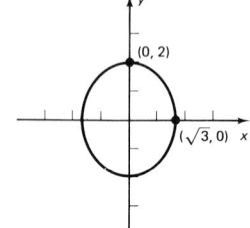

15. $x^2/4 + y^2/16 = 1$.   **17.** $4x^2 + 9y^2 = 1$.   **19.** $(\pm\sqrt{5}, 0)$, $x = \pm\sqrt{5}\, y$.   **21.** $(\pm\sqrt{2}/2, 0)$, $x = \pm\sqrt{2}\, y$.
23. $(\pm\sqrt{8/3}, 0)$, $\sqrt{3}\, x = \pm\sqrt{2}\, y$.   **25.** $(0, \pm 2)$, $2x = \pm\sqrt{3}\, y$.   **27.** $(0, \pm\sqrt{3})$, $\sqrt{5}\, y = \pm\sqrt{6}\, x$.

29.

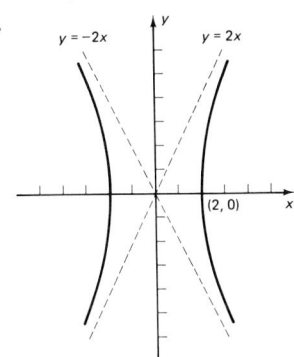

31.

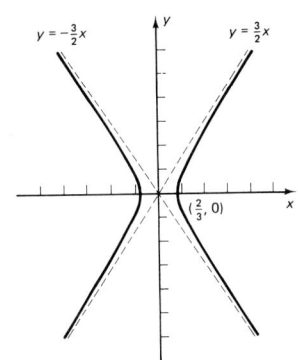

33.

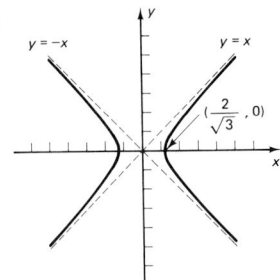

35.

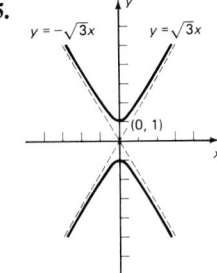

37. $x^2 - y^2 = 4$.   **39.** $4x^2 - 9y^2 = 4$.   **41.** $y^2 - 9x^2 = 4$.   **43.** $x^2 - y^2 = 4$.   **45.** $9y^2 - 9x^2 = 2$.
49. $x^2/625 + y^2/324 = 1$.   **51.** $20\sqrt{1+m^2}$ cm and 20 cm.

Test 7.4:   **1.** Vertical parabola.   **2.** Horizontal parabola.   **3.** Vertical parabola.   **4.** Horizontal parabola.
5. Circle.   **6.** An ellipse.   **7.** A hyperbola.   **8.** A or C (not both) zero.   **9.** $AB < 0$.   **10.** $A = B$.
11. $A \neq B$ and $AB > 0$.

Exercises 7.4

1. $(x-3)^2 + 4(y+2)^2 = 16$.   **3.** $2(x+2)^2 - 3(y-2)^2 = 1$.   **5.** $y - 1 = 3(x+2)^2$.
7. $(y+2)^2 + 4(x+4) = 0$.   **9.** $2(x+2)^2 + 2(x+2) + 3(y-2) = 0$.
11. $(x+2)^2/1 - (y+3)^2/2 = -1$; hyperbola.   **13.** $y - \frac{1}{8} = -2(x + \frac{3}{4})^2$; vertical parabola.
15. $(x+4)^2/25 + (y+2)^2/(\frac{25}{2}) = 1$; ellipse.   **17.** $(x - \frac{1}{8})^2 + (y + \frac{1}{8})^2 = \frac{9}{32}$; circle.
19. $(x - \frac{3}{2})^2/\frac{3}{4} + (y - 3)^2/\frac{3}{2} = 1$; ellipse.

Review Exercises on Chapter 7

1. $x^2 + y^2 + 4x - 8y + 4 = 0$.   **3.** $x^2 + y^2 - 2x + 4y - 15 = 0$.   **5.** $4x^2 + 4y^2 - 4x - 12y + 1 = 0$.
7. $(-2, -2)$, 3; $(-2 \pm \sqrt{5}, 0)$, $(0, -2 \pm \sqrt{5})$.   **9.** Not a circle.   **11.** $(-5, 2)$, 5; $(-5 \pm \sqrt{21}, 0)$, $(0, 2)$.
13. $p = 5$ or -3.

Answers to Tests and Odd-numbered Exercises

15.

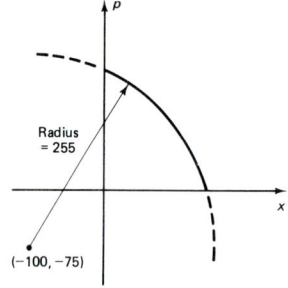

Radius = 255
(−100, −75)
Highest price = $159.57

17. $(0, -2)$; $(\pm 2, 0)$, $(0, -2)$. **19.** $(4, -4)$; $(2, 0)$, $(6, 0)$, $(0, 12)$. **21.** $(\frac{1}{4}, \frac{121}{8})$; $(3, 0)$, $(-\frac{5}{2}, 0)$, $(0, 15)$.

23.

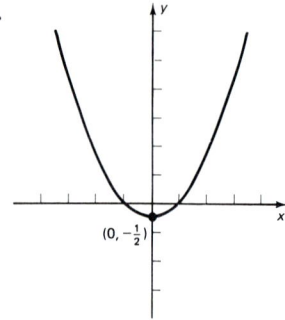

$(0, -\frac{1}{2})$

25.

$(-1, \frac{9}{4})$

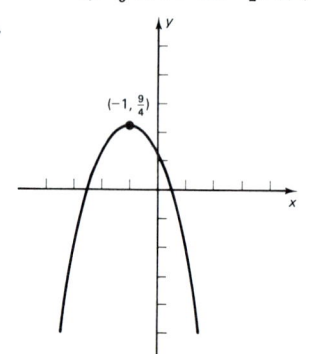

27. (a) $R = 108p - 0.4p^2$; (b) $R = 270y - 2.5y^2$; (c) $135 per month. **29.** $(\frac{64}{25}, \frac{27}{25})$. **31.** 72.5 ft.
33. Ellipse; $(\pm\sqrt{\frac{2}{3}}, 0), (0, \pm\sqrt{\frac{2}{3}})$. **35.** Hyperbola; $(0, \pm\sqrt{3}/3)$. **37.** Hyperbola; $(0, \pm\sqrt{2})$.
39. $y = \pm(\sqrt{2}/3)x$. **41.** $64x^2 + 9y^2 = 144$. **43.** $x^2 - 2y^2 = 9$.
45. $(x + \frac{1}{6})^2/(\frac{29}{72}) + (y + \frac{1}{4})^2/(\frac{29}{48}) = 1$; ellipse. **47.** $x - \frac{1}{2} = -\frac{1}{2}(y + 1)^2$; horizontal parabola.

CHAPTER 8

Test 8.1: **1.** Exponential; base. **2.** $0 < a < 1$. **3.** Natural exponential. **4.** 2.718. **5.** Of all x; of all $y > 0$.
6. Above; below. **7.** $<$. **8.** $>$. **9.** $>$. **10.** $y = 0$. **11.** Falls. **12.** Rises. **13.** Falls.

Exercises 8.1

1.

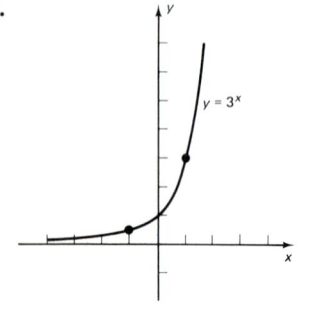

$y = 3^x$

3.

$y = (\frac{1}{3})^x$

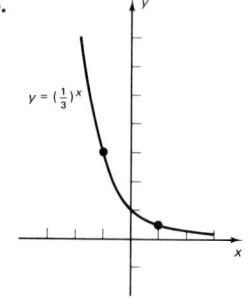

$5y^3x^2 - 7xy^2 + 3x - 7$
$8y^3x^2 - 3y^2x + x + 5$

$8x^2 + 11x + 6$

$\log_a u = v$

$a^v = u$

$\log_2 P = Q$

$2^Q = P$

$\log_7 7 =$

$7^7 = 7$

$\log_3 1 =$

$3^x = 1$

$\log_2(-1) =$

$2^x = -1$

```
6300
6400            GDP85@WE   =   GDP85@EU                                         00007000
6500              C85@WE   =    C85$@EU  *  RXECU%US$@EC                        00007100
6600              CPI@WE   =   CPI85@EU                                         00007200
6700           JQIND@WE   = JQIND85@EU                                          00007300
6800                                                                            00007400
6900           RXFF%ECU@FR  =  RXFF%US$@FR  /  RXECU%US$@EC                     00007500
7000           RXDM%ECU@GY  =  RXDM%US$@GY  /  RXECU%US$@EC                     00007600
7100           RXIL%ECU@IT  =  RXIL%US$@IT  /  RXECU%US$@EC                     00007700
7200           RXUP%ECU@UK  =  RXUP%US$@UK  /  RXECU%US$@EC                     00007800
7300           RXFM%ECU@FN  =  RXFM%US$@FN  /  RXECU%US$@EC                     00007900
7400           RXSK%ECU@SW  =  RXSK%US$@SW  /  RXECU%US$@EC                     00008000
7500                                                                            00008100
7600      ... ############################################## TEMP ############# 00008200
7700                                                                            00008300
7800      LOOP R BY REGION2NL BEGIN                                             00008400
7900         CR<OVER>GDP85!'@'IR ,  GDP85@IT                                    00008500
8000         CR<OVER>JQIND!'@'IR ,  JQIND@IT                                    00008600
8100      END                                                                   00008700
8200                                                                            00008800
8300      ... ############################################# TEMP -- ADV85NL --- 00008900
8400                                                                            00009000
8500      ADV85NL=EMPTY(NL)                                                     00009100
8600      LOOP M BY MARKET2NL BEGIN                                             00009200
8700        LOOP R BY REGION1NL BEGIN                                           00009300
8800          ADV85NL= ADV85NL CONCAT NL(ADV85!'@'IM!'@'IR)                     00009400
8900        END                                                                 00009500
9000      END                                                                   00009600
9100                                                                            00009700
9200      DIST<OVER,0,84:1 TO 93:4,SRC=CL/EPFSA>FROM A,WITH LINEAR,$ADV85NL     00009800
9300                                                                            00009900
9400      CR<OVER> ADV85@DM@WE,  0 FOR*                                         00010000
9500                                                                            00010100
9600      LOOP R BY REGION4NL BEGIN                                             00010200
9700        CR<OVER> ADV85@DM@WE, ADV85@DM!'@'IR                                00010300
9800      END                                                                   00010400
9900                                                                            00010500
10000     ADV85@DM@WE = ADV85@DM@WE / 4                                         00010600
10100                                                                           00010700
10200     LOOP R BY REGION2NL BEGIN                                             00010800
10300       CR<OVER> ADV85@DM!'@'IR , ADV85@DM@WE                               00010900
10400     END                                                                   00011000
10500                                                                           00011100
10600     LOOP R BY REGION1NL BEGIN                                             00011200
10700       CR<OVER>  ADV85@PR!'@'IR,   ADV85@NP!'@'IR + ADV85@MG!'@'IR +&&     00011300
10800                 ADV85@DM!'@'IR                                            00011400
10900       CR<OVER>  ADV85@TO!'@'IR,   ADV85@NP!'@'IR + ADV85@MG!'@'IR +&&     00011500
11000                 ADV85@TV!'@'IR + ADV85@RA!'@'IR + ADV85@CI!'@'IR +&&     00011600
11100                 ADV85@OU!'@'IR + ADV85@DM!'@'IR                           00011700
11200     END
11300
11400     ...                                                             QNL ---
11500
11600     QEPINL=EMPTY(NL)
11700
```

5.

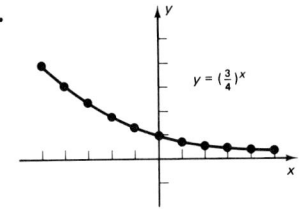

[Graph: $y = (\frac{3}{4})^x$]

7.

[Graph: $y = (\frac{1}{2})^{-x} = 2^x$]

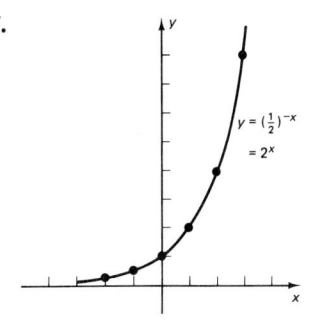

9. 20.0855. **11.** 0.1353. **13.** 1.5068. **15.** 0.3499. **17.** 0.0055.

19.

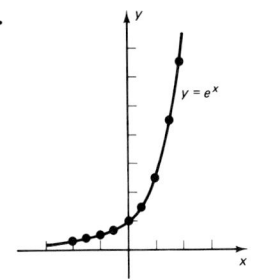

[Graph: $y = e^x$]

21. $p = 30$ million, 34.6 years after 1960.

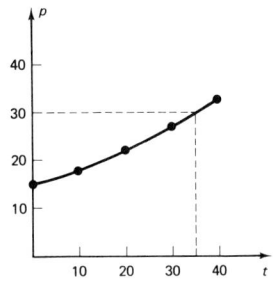

[Graph of p vs t]

23. 409,400; 274,400. **25.** 1.98%. **27.** (a) $2019; (b) 18.1%. **29.** (a) 80,000; (b) 1,280,000; (c) $P = 5000(2^{4t/3})$.
31. $2000(1.06)^2 = $2247.20. **33.** $2000(1.06)^8 = $3187.70. **35.** $100(1.08)^2 = $116.64.
37. $100(1.08)^{10} = $215.89. **39.** 7.18%. **41.** $2000(1.03)^4 = $2251.02. **43.** $2000(1.06)^8 = $3187.70.
45. (a) $1000(1 + \frac{2}{300})^{60} = $1489.85; (b) $1000(1.0825)^4 = $1486.41. The monthly compounding is better.
47. 8.76%. **49.** 10.766 billions. **51.** $9.164 millions.

Test 8.2: **1.** $\log_a y$. **2.** $\log_2 8$. **3.** a^v. **4.** 2^q. **5.** 1. **6.** 0. **7.** Not defined. **8.** 1. **9.** x. **10.** 4.
11. 1; $\neq 1$ and > 0. **12.** Power; a; y. **13.** Of all $x > 0$. **14.** Of all real numbers. **15.** (1, 0).
16. Of all $x \neq 0$; of all real numbers. **17.** Of all $t > -3$; of all real numbers. **18.** $\log_{10} x$. **19.** a^x.

Exercises 8.2

1. $\log_2 8 = 3$. **3.** $\log_{16} 4 = \frac{1}{2}$. **5.** $\log_8 \frac{1}{2} = -\frac{1}{3}$. **7.** $\log_{27} \frac{1}{81} = -\frac{4}{3}$. **9.** $\log_{125} 25 = \frac{2}{3}$.
11. $\log_{(8/27)} \frac{3}{2} = -\frac{1}{3}$. **13.** $3^3 = 27$. **15.** $(\frac{1}{2})^{-3} = 8$. **17.** $(\frac{1}{9})^{5/2} = \frac{1}{243}$. **19.** $4^{-1/2} = \frac{1}{2}$. **21.** 9. **23.** $\frac{5}{3}$.
25. 8. **27.** $\frac{7}{3}$. **29.** 100. **31.** $p/2$. **33.** x. **35.** $\frac{1}{3}$. **37.** $\sqrt{2}$. **39.** $\frac{5}{2}$.

41.

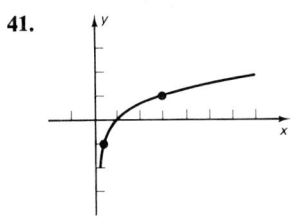

[Graph]

43.

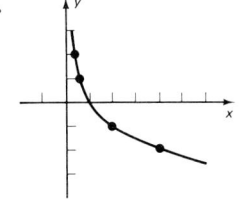

[Graph]

45.

[Graph]

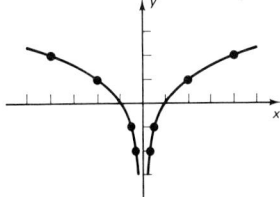

47. 13. **49.** -3. **51.** 1, -5. **53.** 3. **55.** No real solution. **57.** -1. **59.** 2. **61.** 5, $\frac{1}{3}$. **63.** 2, $\frac{1}{2}$.

Test 8.3: **1.** 10^x. **2.** e^t. **3.** e^x. **4.** 10^x. **5.** $\log_2 x$. **6.** $\ln x$. **7.** x. **8.** u. **9.** 1. **10.** 0.
11. 0. **12.** 1. **13.** $\ln 3$. **14.** xy. **15.** x/y. **16.** 1. **17.** 3. **18.** -2. **19.** $-$. **20.** 3.

Exercises 8.3

1. $\log x + \log y - \log z$. **3.** $2 \log x + \log z$. **5.** $\frac{1}{2} \log x + \frac{1}{2} \log z - \log y$. **7.** $\frac{1}{2}(3 \log x - \log y - \log z)$.
9. $\frac{1}{2} \log x + \frac{1}{3} \log y + \frac{1}{4} \log z$. **11.** $\log (x/y^2)$. **13.** $\log (\sqrt{y}/z^2)$. **15.** $\log (10z\sqrt{y})$. **17.** $\log_2 (y^2/16z^{1/3})$.
19. $\log_3 (x\sqrt{z}/27)$. **21.** $1 - x$. **23.** $1 + x$. **25.** $\frac{3}{2} - x$. **27.** $1 - a$. **29.** $2 - 2a + b/2$. **31.** $\frac{1}{4}(b + 2)$.
33. $1 - a - 3b/2$. **35.** 1.2267. **37.** 0.2231. **39.** 4.4332. **41.** 6.3852. **43.** -1.0759. **45.** -4.6918.
49. $-\frac{1}{4}$. **51.** No solution. **53.** $\frac{3}{2}$. **55.** 3, 1.

Test 8.4: **1.** $0.3478; \bar{2}$. **2.** $0.2573; -3$. **3.** $0.2200; -4$. **4.** $\bar{2}.16$. **5.** 1.17. **6.** 1.92. **7.** 10.

Exercises 8.4

1. 0.8751. **3.** 0.2330. **5.** 1.3304. **7.** 3.4099. **9.** $\bar{2}.4330$. **11.** $\bar{3}.1761$. **13.** 5.46. **15.** 17.2
17. 0.179. **19.** 0.00170. **21.** 19.7 **23.** 7.25. **25.** 356. **27.** 133. **29.** 0.00913. **31.** 1.51. **33.** 137.
35. 0.0207. **37.** 0.0000580. **39.** 0.00000669. **41.** 35.1. **43.** 0.0116. **45.** 0.0270. **47.** 332.
49. 0.00387. **51.** 0.785. **53.** 52.3. **55.** 0.773. **57.** 0.619. **59.** 2.19. **61.** 0.528. **63.** 1.20.

Test 8.5: **1.** $\log 2/\log 1.08$ (or $\log_{1.08} 2$). **2.** $\log b/\log a$ (or $\log_a b$). **3.** $(1/k) \ln y$. **4.** e. **5.** $P(1 + i/k)^{nk}$.
6. Pe^{in}. **7.** $-(1/k) \ln 2$.

Exercises 8.5

1. 1.398. **3.** 4.64. **5.** 3.17. **7.** 0.436. **9.** 0.774. **11.** 0.312. **13.** ± 2.155. **15.** $\log c/(\log a - \log b)$.
17. 46.3 yr after 1976. **19.** 38.39 yr after 1976. **21.** After 41.56 yr. **23.** 23.22 months. **25.** 6.12 yr.
27. 5.27 yr. **29.** 5.86 yr. **31.** 40.75 months. **33.** 7.85%. **35.** 3.23 g. **37.** 2101.58 yr. **39.** 1.244×10^{-4}.
41. 7366 yr. **43.** 8.66 yr. **45.** 6.93. **47.** 8.2% quarterly compounding is better. **49.** 60 Db. **51.** $\frac{1}{100}$.
53. 31.62. **55.** 8. **57.** 48.

Test 8.6: **1.** $\ln a$. **2.** $200; -\ln 0.6$. **3.** $\log_y x$. **4.** 1. **5.** 1. **6.** $\ln 10$. **7.** 2. **8.** 3. **9.** $1/\ln a$.
10. $\log_a x$.

Exercises 8.6

1. $y = (7.389)^t$. **3.** $y = (0.819)^t$. **5.** $y = e^{(0.916)x}$. **7.** $y = e^{-0.223x}$. **9.** $y = 3e^{0.693t}$. **11.** $y = 5e^{0.039t}$.
13. $P = 5000e^{0.924t}$. **15.** $V = 10,000e^{-0.223t}$. **17.** 1.4898. **19.** 2.5845. **21.** 1.6609. **23.** $x/(x - 1)$.
25. $x/(x + 1)$. **27.** $z/(1 - z)$. **29.** $2z/(2z - 1)$. **31.** $(x + 2y)/(2x + y)$.

Review Exercises on Chapter 8

1.

3. 0.1738. **5.** 0.1496. **7.** $802.89. **9.** $576.19. **11.** 0.66%. **13.** $\log_{(1/9)} 27 = -\frac{3}{2}$. **15.** $\frac{8}{3}$. **19.** $\frac{1}{4}$.

21. -4.3741. **23.** $1 - 2x$. **25.** $1/(3 - 6x)$. **27.** 4.83. **29.** 7.68. **31.** $n = 0.568$. **33.** $x = 1$.
35. $x = 0.5204$. **37.** 83 min. **39.** (a) \$176.23; (b) \$180.61; (c) \$182.21. **41.** 28.49 yr.
43. (a) 4.14%; (b) $e^{0.04055t}$; (c) 18.15 yr from 1970. **45.** 6.18%; 6.76 yr. **47.** 0.5, 0.8, 1.143, 1.684.
49. $y = 4000e^{0.0862t}$. **51.** $2/(x - 1)$.

CHAPTER 9

Test 9.1: **1.** Quotient; remainder. **2.** -2. **3.** Divisor. **4.** Constant.
5. $3x^3 - 10x^2 + 6$; $x - 3$; $3x^2 - x - 3$; -3.
6. $\underline{-1 \ | \ 1 \quad 0 \quad 1}$ **7.** 0.
$ \dfrac{ \ -1 \ 1}{1 \ -1 \ 2}$; $x - 1$; 2.

Exercises 9.1

1. $x - 3 + 1/(x - 2)$. **3.** $t + 1 + 2/(t - 1)$. **5.** $x + 1$. **7.** $2 + (2x - 3)/(x^2 - x + 2)$.
9. $x^2 + x + 2 + 2/(x + 1)$. **11.** $x^3 + 2x^2 + 4x + 8 + 16/(x - 2)$. **13.** $3x^2 + x + \frac{9}{2} + (x + \frac{27}{2})/(2x^2 - 3)$.
15. $x^2 - x/2 - \frac{7}{4} + (13x/4 + \frac{27}{4})/(2x^2 + x + 1)$. **17.** $2 - 1/(x - 2)$. **19.** $x + 9 + 26/(x - 3)$.
21. $2x - 3 + 13/(x + 3)$. **23.** $2x^2 + 4x + 9 + 11/(x - 2)$. **25.** $x^2 + 3x - 11 + 45/(x + 4)$.
27. $x^3 - 2x^2 + 4x - 8 + 16/(x + 2)$. **29.** $2x^3 + 2x^2 - 2x + 3$. **31.** $x^4 + 2x^3 + 5x^2 + 10x + 21 + 42/(x - 2)$.

Test 9.2: **1.** Discriminant; non-negative. **2.** $\frac{1}{4}$. **3.** Imaginary unit. **4.** 2; $-5i$. **5.** $a = 0$.
6. -1; $-i$; 1; i. **7.** 1; i. **8.** $-1 - i$. **9.** 3; -2. **10.** $ac - bd$; $(ad + bc)$. **11.** $3 + 2i$. **12.** $7 - 3i$.
13. 5. **14.** Zero. **15.** $-3 - 7i$. **16.** $3i$. **17.** $2bi$. **18.** $a^2 + b^2$. **19.** $\frac{1}{2}$; $(-\frac{1}{2})$. **20.** $c - id$.

Exercises 9.2

1. $\pm\sqrt{3}\,i$. **3.** $1 \pm i$. **5.** $(-3 \pm i)/2$. **7.** $-2, i - 2$.
9. $x = \pm\sqrt{-c}$. Parabola $y = x^2 + c$ meets the x axis at two points if $c < 0$, at one point if $c = 0$, and not at all if $c > 0$.
11. $x = -2 \pm \sqrt{c + 4}$. Parabola $y = c - 4x - x^2$ meets the x axis at two points if $c > -4$, at one point if $c = -4$, and not at all if $c < -4$. **13.** $8 - 2i$. **15.** $4 - 3i$. **17.** $2 - 3i$. **19.** $5 + 2i$. **21.** $1 + 3i$. **23.** $0 + 2i$.
25. $17 + 7i$. **27.** $7 + 4i$. **29.** $2 - 3i$. **31.** $\frac{1}{2} - \frac{1}{2}i$. **33.** $-\frac{5}{13} - \frac{12}{13}i$. **35.** $\frac{19}{13} - \frac{4}{13}i$. **37.** $-2 - 2i$.
39. $4 + 2i$. **41.** $-\frac{1}{2} + \frac{5}{2}i$. **43.** $-2 + 4i$. **45.** $-16i$. **49.** $a = \frac{1}{2}, b = 1$. **51.** $a = 1, b = -1$.
53. $3 + i$. **55.** $-\frac{1}{25}(3 + 4i)$. **57.** $5 + i$.

Test 9.3: **1.** $f(c)$. **2.** 1. **3.** -1. **4.** $x + 2$. **5.** $f(c) = 0$. **6.** Is not. **7.** $a(x - \alpha)(x - \beta)$.
8. $a(x - \alpha)(x - \beta)(x - \gamma)$. **9.** At most n. **10.** $2(x - \frac{1}{2})(x - 2)$. **11.** 4; 2; 1. **12.** $0, i, -i; 1, 2, 2$.
13. $12x^2(x - i/\sqrt{2})^2(x + i/\sqrt{2})^2$.

Exercises 9.3

7. 28. **9.** -2. **11.** 27. **13.** -9. **15.** $(x - 1)$ and $(x + 1)$ both are factors. **17.** $(x - 1)$ is a factor only.
19. $(x + 1)$ is a factor only. **21.** Yes. **23.** Yes. **25.** $k = -5$. **27.** $k = 2$. **29.** $k = 4$.
31. $\frac{1}{2}(3 \pm \sqrt{17})$; $(x - \frac{3}{2} - \sqrt{17}/2)(x - \frac{3}{2} + \sqrt{17}/2)$. **33.** $0, 2, -2$; $x(x - 2)(x + 2)$.
35. $\pm\sqrt{2}, \pm\sqrt{2}\,i$; $(x - \sqrt{2})(x + \sqrt{2})(x - \sqrt{2}\,i)(x + \sqrt{2}\,i)$.
37. $1, \frac{1}{2}(-1 \pm \sqrt{3}\,i)$; $(x - 1)[x + \frac{1}{2} - (\sqrt{3}/2)i][x + \frac{1}{2} + (\sqrt{3}/2)i]$. **39.** $1(m = 2)$, $-1(m = 1)$; l.c. $= 1$.
41. $-\frac{1}{2}(m = 2), -2(m = 1), \sqrt{3}\,i(m = 3), -\sqrt{3}\,i(m = 3)$, l.c. $= 4$.
43. $0(m = 3), \sqrt{3}/2(m = 2), -\sqrt{3}/2(m = 2)$, l.c. $= 32$. **45.** $f(x) = (x - 2)(x + 1)^2$.
47. $f(x) = 4(x - 2)(x + \frac{1}{2})^2$. **49.** $f(x) = (x - 1)(x - \frac{1}{2} - \sqrt{3}\,i/2)(x - \frac{1}{2} + \sqrt{3}\,i/2)$.
51. $f(x) = 2(x - \frac{1}{2})(x + 2 - i)(x + 2 + i)$.

Test 9.4: **1.** $2 - 3i$. **2.** $-3 - 4i$. **3.** 6; 15. **4.** 2 or 0. **5.** 3 or 1. **6.** 2.6.

Exercises 9.4

1. $x^2 - 4x + 5$. **3.** $x^3 - 2x^2 + 16$. **5.** $x^4 + 3x^2 - 6x + 10$. **7.** $x^4 - 3x^3 + 9x^2 + 13x$. **9.** 3. **11.** 4.
13. 6. **15.** 4. **17.** $x^4 - 2x^3 + 3x^2 - 2x + 2$. **19.** $x^6 - 2x^5 + 5x^4 - 10x^3 + 4x^2 - 8x$. **21.** $2, \pm i$.
23. $1, 2, \tfrac{1}{2}$. **25.** $-\tfrac{1}{2}, (-1 \pm \sqrt{7}\,i)/2$. **27.** $\tfrac{4}{3}, -1 \pm \sqrt{3}\,i$. **29.** $1, 1, 3, \tfrac{1}{2}$. **31.** $1, \tfrac{3}{2}, 1 \pm \sqrt{5}$.
33. $-\tfrac{1}{2}, \tfrac{1}{4}, \pm \sqrt{2}\,i$. **35.** 2.73. **37.** -2.87. **39.** 1.65. **41.** -2.96. **43.** $(2, 3), (-1, 0)$.
45. $(0, 1), (2, 3),$ and $(-3, -2)$. **47.** $(0, 1), (3, 4), (-2, -1)$. **49.** $(-1, 0)$.

Review Exercises on Chapter 9

1. $x - 3 + (3x - 7)/(x^2 - 3)$. **3.** $3x^2 - 3 + (2x + 2)/(2x^2 + x + 2)$. **5.** $x^2 + 4x + 17 + 63/(x - 4)$.
7. $2x^5 - 6x^4 + 23x^3 - 68x^2 + 204x - 612 + 1834/(x + 3)$.
9.

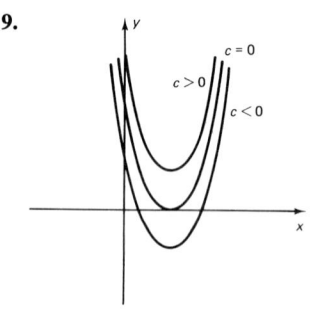

[Graph showing three parabolas labeled $c = 0$, $c > 0$, $c < 0$ on xy-axes]

11. $26 + 7i$. **13.** $0 + 10i$. **15.** $\tfrac{19}{13} - \tfrac{1}{13}i$. **17.** $\tfrac{18}{13} - \tfrac{4}{13}i$. **19.** $a = -\tfrac{7}{2}, b = -\tfrac{1}{2}$. **21.** $a = 3$ or -1 and $b = 0$.
23. 2. **25.** Yes. **27.** $k = -\tfrac{25}{9}$. **29.** $x^2(x + 1 + 2i)(x + 1 - 2i)$.
31. $2(x - \tfrac{3}{2})(x + \tfrac{1}{4} - \sqrt{7}\,i/4)(x + \tfrac{1}{4} + \sqrt{7}\,i/4)$. **33.** $3(x + \tfrac{1}{3})(x + \tfrac{3}{4} - \sqrt{7}\,i/4)(x + \tfrac{3}{4} + \sqrt{7}\,i/4)$.
35. $2(x - 2)(x - \tfrac{1}{2})(x + \tfrac{1}{6} + \sqrt{23}\,i/6)(x + \tfrac{1}{6} - \sqrt{23}\,i/6)$. **37.** $x^4 - 8x^3 + 42x^2 - 104x + 169$.
39. $x^4 - 3x^3 + 4x^2 - 12x$. **41.** $-3, -\tfrac{2}{3}, \tfrac{1}{2}$. **43.** 3.64. **45.** $-1.53, -0.35, 1.88$.

CHAPTER 10

Test 10.1: **1.** $T_2 - T_1 = T_3 - T_2 = \ldots$. **2.** $x, x + y, x + 2y$. **3.** $a + (n - 1)d$. **4.** $a - (p - 1)d$.
5. $(n/2)[2d + (n - 1)a]$. **6.** $3p + 7$. **7.** $5 - 2(3) = -1$. **8.** T_{n-5}. **9.** T_n.

Exercises 10.1

1. 39; 59. **3.** $-7, 7 - 2n$. **5.** $(n + 11)/3$. **7.** $(29 - r)/4$. **9.** 54. **11.** $5.3 - 0.3n$. **13.** 27th. **15.** 13.
17. 1335. **19.** $n(n + 1)/2$. **21.** $\tfrac{1}{2}n(43 - 3n)$. **23.** 414. **25.** 60.5. **27.** 12. **29.** 21. **33.** 3, 5, 7.
35. $-1, 1, 3, 5$ or $2 - \sqrt{31}, 2 - \sqrt{31}/3, 2 + \sqrt{31}/3, 2 + \sqrt{31}$. **37.** $120. **39.** $165. **41.** (a) 30; (b) $125.
43. $1275. **45.** 9.

Test 10.2: **1.** $a_2/a_1 = a_3/a_2 = a_4/a_3 = \ldots$. **2.** G.P. **3.** A.P. **4.** G.P. **5.** A.P.
6. Neither A.P. nor G.P. **7.** A.P. **8.** G.P. **9.** ar^{n-1}. **10.** $a(1 - r^n)/(1 - r)$. **11.** $a/(1 - r); |r| < 1$.
12. 2. **13.** Not defined. **14.** $0.7 + 0.07 + 0.007 + \cdots$. **15.** $n + 1 + 2^{1-n}$.

Exercises 10.2

1. 768. **3.** 81. **5.** $3^{n-3}/2^{n-2}$. **7.** $(-2)^{n-3}/3^{n-2}$. **9.** 10th. **11.** 8th. **13.** 16, 24, 36; $3^9/2^5$.
17. $3^{12} - 1$. **19.** $121(\sqrt{3} - 3)$. **21.** $2^n - 1$. **23.** $4(3^n 2^{-n} - 1)/9$. **25.** $9[1 - (\tfrac{2}{3})^7]/4$. **27.** 2. **29.** $\tfrac{9}{20}$.
31. $(a + br)/(1 - r^2)$. **39.** $\tfrac{4}{9}$. **41.** $\tfrac{23}{99}$. **43.** $\tfrac{91}{110}$. **45.** $3(3^p - 1)/2$. **49.** $2938.66. **51.** $r = 16, T_1 = 80$.
53. $256(2^{1/3}), 256(2^{2/3})$.

Test 10.4: **1.** $n(n-1)$; $8 \cdot 7 \cdot 6 = 336$. **2.** $n(n-1)/2$; 56. **3.** $n-2$; 3. **4.** 1; 1. **5.** $n(n-1)\ldots(n-r+1) = n!/(n-r)!$ **6.** $n!/r!(n-r)!$. **7.** $r!$ **8.** $n(n-1)$. **9.** 1. **10.** $_nC_r$; $_nP_r$. **11.** $_5P_3 = 60$. **12.** $_5P_3$. **13.** 5! **14.** $_8C_5$.

Exercises 10.4

1. (a) 380; (b) 9; (c) $\frac{28}{3}$. **3.** (a) 90; (b) 360; (c) 120. **5.** (a) 45; (b) 56; (c) 190. **7.** 19. **9.** 8. **11.** 3. **13.** 7. **15.** 20. **17.** 4. **19.** 7. **25.** 24. **27.** 6. **29.** 8!; 7! **31.** (2!)(4!)(5!) = 5760. **37.** $(7!)_8P_5$. **35.** 56; 21. **37.** (a) 120; (b) 560. **39.** 1200. **41.** 451. **43.** $_{19}C_5 = 11{,}628$. **45.** $4(_{48}C_3) = 69{,}184$. **47.** (a) 21; (b) 35; (c) 31. **49.** 1584. **51.** 5.

Test 10.5: **1.** $_nC_0x^n + {_nC_1}x^{n-1}a + {_nC_2}x^{n-1}a^2 + \cdots + {_nC_n}a^n$. **2.** $_nC_{p-1}x^{n-p+1}a^{p-1}$. **3.** $_9C_6x^3(-a)^6$. **4.** $_{10}C_6x^4y^6$. **5.** $(n-r+2)$. **6.** Equal. **7.** $n+1$. **8.** n. **9.** Two. **10.** One.

Exercises 10.5

1. $a^7 + 7a^6b + 21a^5b^2 + 35a^4b^3 + 35a^3b^4 + 21a^2b^5 + 7ab^6 + b^7$.
3. $32x^5 + 40x^4y + 20x^3y^2 + 5x^2y^3 + (\frac{5}{8})xy^4 + y^5/32$.
5. $32x^5/243 + 40x^3/27 + 20x/3 + 15x^{-1} + 135x^{-3}/8 + 243x^{-5}/32$.
7. $a^6 - 12a^5b + 60a^4b^2 - 160a^3b^3 + 240a^2b^4 - 192ab^5 + 64b^6$.
9. $32p^5 - 240p^4q + 720p^3q^2 - 1080p^2q^3 + 810pq^4 - 243q^5$. **11.** $p^{10} - 5p^7q + 10p^4q^2 - 10pq^3 + 5q^4/p^2 - p^5/q^5$.
13. $567x^3y^6/16$. **15.** $(-1)^{n-1}(_{3n}C_{n-1})x^3$. **17.** $252x^2$. **19.** -252. **21.** $189x^{17}/8$, $-21x^{19}/16$. **23.** 252.
25. 112. **27.** $\frac{495}{256}$. **29.** 2268. **31.** No term independent of x. **35.** $a = \frac{9}{2}$. **37.** 724.

Review Exercises on Chapter 10

1. $3n - 18$. **3.** p, $(2p+q)/3$, $(p+2q)/3$, q. **5.** $21n - n^2$. **7.** $n^2 + 2n$. **13.** 10th. **15.** $na + b^n$. **17.** 81.
21. $9(1 - 3^{-p/2})(3 + \sqrt{3})$. **31.** 10. **33.** $r = n - 1$. **37.** 1680. **39.** 100; 36. **41.** 14,400.
43. (a) 315; (b) 560. **45.** (a) 91; (b) 435. **47.** $(x-y)^6 = x^6 - 6x^5y + 15x^4y^2 - 20x^3y^3 + 15x^2y^4 - 6xy^5 + y^6$.
49. $-(_{15}C_7)a^8b^7$, $(_{15}C_8)a^7b^8$. **51.** 8064. **53.** $-10{,}206x^4y^5$.

CHAPTER 11

Exercises 11.1

1. $x = 2, y = 1$. **3.** $u = 4, v = -1$. **5.** No solution. **7.** $x = 1, y = 2, z = 3$. **9.** $x = 1, y = 2, z = 3$.
11. $x_1 = 1, x_2 = 2, x_3 = -1$. **13.** $p = -1, q = 2, r = 2$. **15.** $x = 2 + z, y = 3 - 2z$, z arbitrary.
17. No solution. **19.** No solution. **21.** $x = 6z - 4, y = 10 - 11z$, z arbitrary.
23. $x = -1, y = 3, z = -3, t = 2$. **25.** $x = 1, y = 1, z = -1, w = 2$. **27.** No solution.
29. $u = 1, v = 2, w = -1$. **31.** No solution. **33.** No solution.
35. $x = (3-z)/5, y = (7+6z)/5$, z arbitrary. **37.** $a = \frac{7}{24}, b = -\frac{3}{4}, c = \frac{1}{3}$.
39. 100, 150, and 200 units of A, B, and C. **41.** \$6000 at 6%, \$6800 at 8%, and \$7200 at 10%.
43. $i_1 = -\frac{16}{13}, i_2 = \frac{28}{13}, i_3 = \frac{12}{13}$.

Test 11.2: **1.** $x = 1, y = 5, u = 2, v = 3$. **2.** $\begin{bmatrix} 9 & 12 \\ 15 & -21 \end{bmatrix}$. **3.** $m = p, n = q$. **4.** Undefined.
5. $x = 1, y = 3, t = 5$. **6.** 0. **7.** C.

Exercises 11.2

1. A: 2×2; B: 2×3; C: -3×1; D: 3×3; E: 2×3; F: 2×2; G: 1×3; H: 1×1. **3.** $A = \begin{bmatrix} 0 & 1 \\ 1 & 2 \end{bmatrix}$.
5. Any matrix of the form $\begin{bmatrix} 0 & x & y \\ -x & 0 & z \\ -y & -z & 0 \end{bmatrix}$. **7.** $\begin{bmatrix} 6 & 12 \\ 3 & 9 \end{bmatrix}$. **9.** $\begin{bmatrix} -2 & 4 & -6 \\ 4 & -2 & 8 \\ -6 & 0 & -4 \end{bmatrix}$. **11.** $\begin{bmatrix} 2 & 0 & 5 \\ 0 & 6 & -1 \end{bmatrix}$.

13. $\begin{bmatrix} 2 & 3 \\ -4 & 6 \\ 3 & -3 \end{bmatrix}$. 15. $\begin{bmatrix} -4 & 13 \\ 1 & 6 \end{bmatrix}$. 17. $\begin{bmatrix} 2 & 1 & 12 \\ 13 & 4 & -12 \\ 5 & 10 & 21 \end{bmatrix}$. 19. $x = 1, y = 4$. 21. $x = 1, y = 5, z = -2$.

23. $x = 6, y = 5, z = 2, u = 2, v = 3, t = -2$. 25. $x = 1, y = 2, z = 3, t = 4, u = -1, v = 2, w = 5$.
27. $x = 0, y = 1, z = 2, t = -1, u = -2, v = 3, w = 0$. 29. (a) $\begin{bmatrix} 11 & 13 & 16 \\ 14 & 11 & 13 \\ 9 & 16 & 7 \\ 17 & 10 & 11 \end{bmatrix}$. (b) $\begin{bmatrix} 12 & 14.4 & 18 \\ 15.6 & 12 & 14.4 \\ 9.6 & 18 & 7.2 \\ 19.2 & 10.8 & 12 \end{bmatrix}$.

Test 11.3: 1. $n = p; q = m$. 2. $m = n$. 3. $m \times q$. 4. $q \times p$. 5. $n \times n$. 6. 0. 7. Is not necessarily.
8. n. 9. $A; n \times n$. 10. $\begin{bmatrix} 2 & 3 \\ -4 & 5 \end{bmatrix}; \begin{bmatrix} x \\ y \end{bmatrix}; \begin{bmatrix} 7 \\ 3 \end{bmatrix}$. 11. $AB = BA$.

Exercises 11.3

1. 3×3. 3. 2×4. 5. 2×5. 7. [23]. 9. [5 -3 4]. 11. $\begin{bmatrix} 18 \\ 28 \end{bmatrix}$. 13. $\begin{bmatrix} 6 & -1 \\ -5 & -12 \end{bmatrix}$.

15. $\begin{bmatrix} 11 \\ -6 \\ 32 \end{bmatrix}$. 17. $\begin{bmatrix} 5 & 7 & 8 & 21 \\ 14 & 9 & 16 & 41 \end{bmatrix}$. 19. $\begin{bmatrix} -25 & 14 \\ -58 & 32 \end{bmatrix}$. 21. $\begin{bmatrix} 8 & 33 \\ 5 & -22 \end{bmatrix}$. 23. $\begin{bmatrix} 4 & 12 \\ 12 & 16 \end{bmatrix}$.

25. (a) $\begin{bmatrix} 9 & 5 \\ 0 & 4 \end{bmatrix}$; (b) $\begin{bmatrix} 6 & 0 \\ 3 & 7 \end{bmatrix}$; (c) No. 27. $A = [3 \ -1]$. 29. $A = \begin{bmatrix} 1 \\ -1 \end{bmatrix}$. 31. $A = \begin{bmatrix} 3 & 0 \\ 0 & 1 \end{bmatrix}$.

33. $\begin{bmatrix} 2 & 3 \\ 1 & 4 \end{bmatrix} \begin{bmatrix} x \\ y \end{bmatrix} = \begin{bmatrix} 7 \\ 5 \end{bmatrix}$. 35. $\begin{bmatrix} 1 & 2 & 3 \\ 2 & -1 & 4 \\ 0 & 3 & -2 \end{bmatrix} \begin{bmatrix} x \\ y \\ z \end{bmatrix} = \begin{bmatrix} 8 \\ 13 \\ 5 \end{bmatrix}$. 37. $\begin{bmatrix} 2 & 1 & 0 & -1 \\ 0 & 3 & 2 & 4 \\ 1 & -2 & 4 & 1 \end{bmatrix} \begin{bmatrix} x \\ y \\ z \\ u \end{bmatrix} = \begin{bmatrix} 0 \\ 5 \\ 12 \end{bmatrix}$.

39. $\begin{bmatrix} -2a & -2b \\ a & b \end{bmatrix}$, $(a, b$ arbitrary$)$. 41. $A^n = \begin{bmatrix} 1 & 0 \\ 0 & 1 \end{bmatrix}$. 43. $\begin{bmatrix} 1 & 0 \\ 1 - 2^{-n} & 2^{-n} \end{bmatrix}$.

Test 11.4: 1. $PQ = QP = I$. 2. $m = n$. 3. $n \times n$. 4. The same identity matrix. 5. Nonexistent.
6. $A^{-1}B$. 7. QP^{-1}. 8. P^{-1}. 9. A. 10. A^{-1} exists. 11. C.

Exercises 11.4

1. $\begin{bmatrix} \frac{1}{2} & 0 \\ 0 & \frac{1}{3} \end{bmatrix}$. 3. $\begin{bmatrix} 0 & \frac{1}{3} \\ \frac{1}{5} & -\frac{2}{15} \end{bmatrix}$. 5. $\frac{1}{7}\begin{bmatrix} -4 & 5 \\ 3 & -2 \end{bmatrix}$. 7. $\begin{bmatrix} -2 & -1 \\ -\frac{3}{2} & -\frac{1}{2} \end{bmatrix}$. 9. No inverse. 11. No inverse.

13. $\begin{bmatrix} 1 & 0 & 0 \\ 0 & \frac{1}{2} & 0 \\ 0 & 0 & \frac{1}{3} \end{bmatrix}$. 15. $\frac{1}{11}\begin{bmatrix} -1 & 2 & 6 \\ -2 & 4 & 1 \\ 6 & -1 & -3 \end{bmatrix}$. 17. $\begin{bmatrix} -2 & -\frac{1}{3} & \frac{4}{3} \\ 1 & \frac{2}{3} & -\frac{2}{3} \\ \frac{1}{2} & -\frac{1}{3} & -\frac{1}{6} \end{bmatrix}$. 19. No inverse. 21. $\frac{1}{14}\begin{bmatrix} 3 & 7 & 1 \\ 1 & -7 & 5 \\ -5 & -7 & 3 \end{bmatrix}$.

23. No inverse. 25. $\begin{bmatrix} -6 & \frac{7}{2} & \frac{9}{2} & -\frac{3}{2} \\ 3 & -2 & -2 & 1 \\ -4 & \frac{5}{2} & \frac{7}{2} & -\frac{3}{2} \\ 7 & -4 & -5 & 2 \end{bmatrix}$. 27. $x = 2, y = 1$. 29. $u = 1, v = 2$. 31. $x = 1, y = 2, z = -1$.

33. $u = 1, v = 0, w = 3$. 43. $(3L - 4K)/7$ units on first line and $(5K - 2L)/7$ units on the second line.

Test 11.5: 1. $ad - bc$. 2. (i) d; (ii) c. 3. $(-1)^{3+2}\begin{bmatrix} 1 & 5 \\ 7 & 1 \end{bmatrix} = 34$. 4. $(-1)^{i+j}$. 5. Equal to. 6. $\Delta \neq 0$.

Exercises 11.5

1. $(an - cl)$. 3. $-(am - bl)$. 5. 25. 7. -38. 9. $3a + 2b$. 11. 32. 13. -192. 15. $(ac - b^2)$.
17. -21. 19. 35. 21. 0. 23. 6. 25. adf. 27. 5. 29. $x = 3$. 31. $x = 6; -1$.

33. $x = 1, y = -1$. **35.** $x = 1, y = 2$. **37.** $x = 6, y = 10$. **39.** No solution. **41.** $x = 1, y = -1, z = -1$.
43. $x = -1, y = -1, z = 3$. **45.** $x = 3, y = 1, z = 2$. **47.** No solution. **49.** $x = -1, y = 2, z = 1$.
51. $x = (6a + 9b - c)/4, y = (c - 2a - b)/4, z = (c - 2a - 3b)/2$.
53. (a) $i_1 = (3E_1 - 2E_2)/13, i_2 = (7E_2 - 4E_1)/26, i_3 = (2E_1 + 3E_2)/26$;
(b) $i_1 = [E_1(R_2 + R_3) - E_2R_3/R], i_2 = [E_2(R_1 + R_3) - E_1R_3]/R$
$i_3 = (E_1R_2 + E_2R_1)/R$, where $R = R_1R_2 + R_2R_3 + R_3R_1$.

Review Exercises on Chapter 11

1. $\begin{bmatrix} 2 & 1 & 0 \\ 3 & 2 & 1 \end{bmatrix}$. **3.** $a = 5, b = -2, c = 2, d = -1$. **5.** $x = 4, y = 3, z = 2, u = 2, v = 3, w = -4$.
7. $\begin{bmatrix} 1 & -11 \\ 20 & 10 \end{bmatrix}$. **9.** $\begin{bmatrix} 7 & 1 \\ 11 & 9 \end{bmatrix}$. **11.** $\begin{bmatrix} 9 & 1 \\ -5 & -9 \end{bmatrix}$. **13.** $x = 2, y = 1$. **15.** $x = 1, y = -1, z = 2$.
17. $x = 1, y = 1, z = -1$. **19.** $x = -1, y = 2$. **21.** $x = 1, y = -1$. **23.** $\mathbf{x} = \begin{bmatrix} 1 & -1 \\ 0 & 1 \end{bmatrix}$. **25.** $\frac{1}{11}\begin{bmatrix} 5 & 3 \\ -2 & 1 \end{bmatrix}$.
27. $\frac{1}{(a^2 + b^2)}\begin{bmatrix} a & -b \\ b & a \end{bmatrix}$. **29.** No inverse. **31.** $\frac{1}{3}\begin{bmatrix} -2 & 1 & 1 \\ -8 & 7 & -2 \\ 7 & -5 & 1 \end{bmatrix}$. **33.** No inverse. **35.** $x = 1, y = 1$.
37. $x = 1, y = 1, z = -1$. **39.** $ad - bc$. **41.** -26. **43.** 0. **45.** $x = 2, y = 0$. **47.** $x = \frac{4}{5}, y = \frac{9}{5}, z = \frac{14}{5}$.

CHAPTER 12

Test 12.1: **1.** 90; $\pi/2$. **2.** $\alpha\pi/180; 180\beta/\pi$. **3.** $\pi r\theta/180$. **4.** $\frac{1}{2}r^2\alpha$. **5.** Complementary. **6.** Supplementary.
7. Coterminal. **8.** $\pi/2$. **9.** Obtuse. **10.** 50°. **11.** $2\pi/3$.

Exercises 12.1

1. $2\pi/9$. **3.** $7\pi/18$. **5.** $65\pi/36$. **7.** $-2\pi/3$. **9.** $-25\pi/6$. **11.** $73\pi/360$. **13.** $\pi/8$. **15.** $\pi/16$. **17.** 72°.
19. 40°. **21.** $-150°$. **23.** 165°. **25.** $-900°$. **27.** 210°. **29.** 57.3°. **31.** 34.4°.
33. 54°33′. **35.** 99°. **37.** 35° and 55°. **39.** 103°. **41.** $\frac{22}{3}$ ft. **43.** 33 in. **45.** 29.4 cm. **47.** 210°.
49. 168°. **51.** 10.5°. **53.** $\frac{17.6}{3}$ in. **55.** 35 ft. **57.** 90°. **59.** 240°. **61.** $7\pi/6$. **63.** Second.
65. Fourth. **67.** Second. **69.** Third. **71.** 3π square, units. **73.** 38.2°.

Test 12.2: **1.** $\tan\theta$. **2.** $\tan\alpha$. **3.** $\cos\theta$. **4.** $\tan\alpha$. **5.** $\sin\alpha$. **6.** 1. **7.** 1. **8.** $\sqrt{3}/2$. **9.** $\sqrt{3}/2$.
10. 1. **11.** $\sqrt{3}$. **12.** 1. **13.** 0. **14.** $\sqrt{2}/2$. **15.** 30°. **16.** 45°. **17.** 30°. **18.** $\pi/4$. **19.** 90°.
20. $\pi/4$.

Exercises 12.2

1. $\sin\theta = \frac{3}{5}, \cos\theta = \frac{4}{5}, \tan\theta = \frac{3}{4}, \cot\theta = \frac{4}{3}, \sec\theta = \frac{5}{4}, \csc\theta = \frac{5}{3}$.
3. $\sin\theta = \frac{12}{13}, \cos\theta = \frac{5}{13}, \tan\theta = \frac{12}{5}, \cot\theta = \frac{5}{12}, \sec\theta = \frac{13}{5}, \csc\theta = \frac{13}{12}$.
5. $\sin\theta = 1/\sqrt{26}, \cos\theta = 5/\sqrt{26}, \tan\theta = \frac{1}{5}, \cot\theta = 5, \sec\theta = \sqrt{26}/5, \csc\theta = \sqrt{26}$.
7. 1. **9.** $\frac{3}{2}$. **11.** 0. **13.** 1. **15.** $(\sqrt{2} - \sqrt{6})/4$. **17.** $\sqrt{3}/3$. **33.** $A = 45°, B = 15°$. **35.** 0.5878.
37. 0.4663. **39.** 0.9171. **41.** 1.293. **43.** 1.550. **45.** 1.260. **47.** 0.7939. **49.** 1.182. **51.** 1.214.
53. 13°. **55.** 18°30′. **57.** 64°40′. **59.** 38°25′. **61.** 49°14′. **63.** 14°03′. **65.** 69°53′. **67.** 15°44′.
69. 71°20′. **71.** 0.854912. **73.** -0.479120. **75.** -0.935053. **77.** -0.370181.

Exercises 12.3

1. $B = 60°, a = 3\sqrt{3}, c = 6\sqrt{3}$. **3.** $A = 45°, a = b = 5\sqrt{2}$. **5.** $A = 60°, B = 30°, c = 2$.
7. $A = 53°, b = 3.8, c = 6.3$. **9.** $B = 47°, a = 7.5, c = 10.9$. **11.** $A = 38°, a = 6.2, b = 7.9$.
13. $B = 65°, a = 1.7, b = 3.6$. **15.** $A = 35°32′, B = 54°28′, c = 8.6$. **17.** $A = 44°38′, B = 45°22′, a = 3.7$.

19. $A = 19°28'$, $B = 70°32'$, $b = 5.9$.   **21.** 28.9 ft.   **23.** 120 ft.   **25.** 54.0 ft.   **27.** 80.7 ft.   **29.** 141 ft.
31. 49.9 ft.   **33.** 60°.   **35.** 90.9 mph.   **37.** 73.2 ft.   **39.** 21.9 ft, 2.0 ft.   **41.** 1.736 miles.   **34.** 2:1.

Test 12.4:   **1.** -1; 1.   **2.** 1; -1.   **3.** 1.   **4.** $\geq$.   **5.** 0.   **6.** 0.   **7.** -1.   **8.** -1.   **9.** Fourth.
10. Second or third.   **11.** First or second.   **12.** First or second.

Exercises 12.4

1. Second.   **3.** Fourth.   **5.** Fourth.   **7.** First.   **9.** First or fourth.   **11.** Third or fourth.
13. First or third.
15. $\cos\theta = -\frac{12}{13}$, $\tan\theta = -\frac{5}{12}$, $\cot\theta = -\frac{12}{5}$, $\sec\theta = -\frac{13}{12}$, $\csc\theta = \frac{13}{5}$.
17. $\sin\theta = -\frac{1}{2}$, $\cos\theta = \sqrt{3}/2$, $\cot\theta = -\sqrt{3}$, $\sec\theta = 2\sqrt{3}/3$, $\csc\theta = -2$.
19. $\sin\theta = -\frac{4}{5}$, $\cos\theta = -\frac{3}{5}$, $\tan\theta = \frac{4}{3}$, $\sec\theta = -\frac{5}{3}$, $\csc\theta = -\frac{5}{4}$.
21. $\cos\theta = -2\sqrt{6}/5$, $\tan\theta = -\sqrt{6}/12$, $\cot\theta = -2\sqrt{6}$, $\sec\theta = -5\sqrt{6}/12$, $\csc\theta = 5$.
23. $\sin\theta = -2\sqrt{13}/13$, $\cos\theta = 3\sqrt{13}/13$, $\cot\theta = -\frac{3}{2}$, $\sec\theta = \sqrt{13}/3$, $\csc\theta = -\sqrt{13}/2$.
25. $\sin\theta = \frac{4}{5}$, $\cos\theta = \frac{3}{5}$, $\tan\theta = \frac{4}{3}$, $\sec\theta = \frac{5}{3}$, $\csc\theta = \frac{5}{4}$.
27. $\frac{1}{2}$.   **29.** $-\sqrt{3}$.   **31.** $\sqrt{2}$.   **33.** $\sqrt{3}/2$.   **35.** $\sqrt{3}/3$.   **37.** 0.3420.   **39.** -0.2679.   **41.** 0.1794.
43. -0.5774.   **45.** 0.8660.   **47.** 0.1763.   **49.** -0.1736.   **57.** $\pi/3, 5\pi/3$.   **59.** $\pi/4, 5\pi/4$.   **61.** $5\pi/6, 11\pi/6$.
63. $\pi/6, 5\pi/6, 7\pi/6, 11\pi/6$.   **65.** $\pi/3, 2\pi/3, 4\pi/3, 5\pi/3$.   **67.** $\pi/3, 2\pi/3, 4\pi/3, 5\pi/3$.

Test 12.5:   **1.** 2; 2π.   **2.** $2\pi/3$; 1.   **3.** Difference.   **4.** Origin.   **5.** y axis.   **6.** Origin.
7. $x = (n + \frac{1}{2})\pi$, n any integer.   **8.** $\cos x$.   **9.** $n\pi$, n any integer.   **10.** $\sin x$.

Exercises 12.5

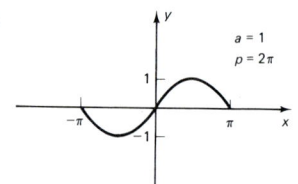

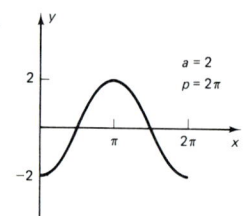

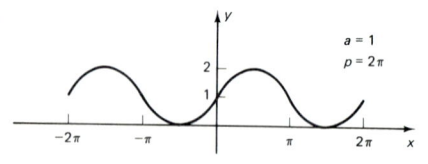

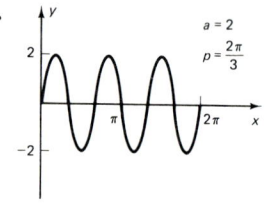

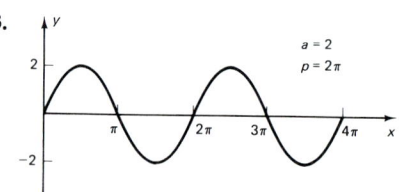

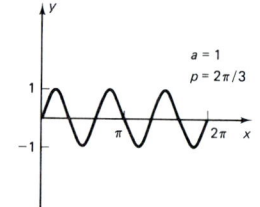

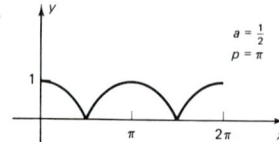

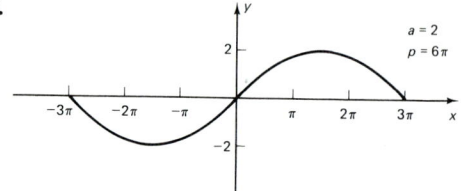

17. [graph: a = 1, p = 2π]

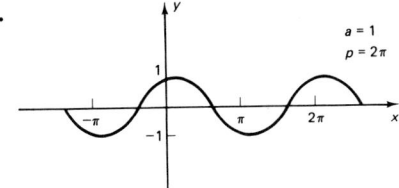

19. [graph: a = 2, p = 2π]

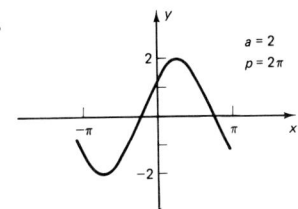

21. [graph: a = 1, p = 2π]

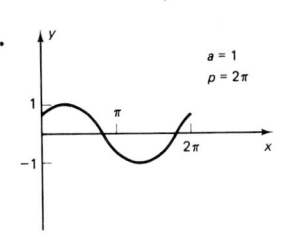

23. [graph]

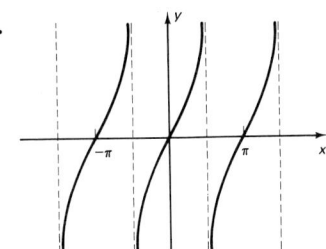

25. [graph]

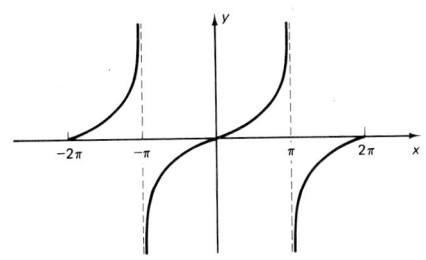

27. [graph showing y = cos x and y = sin x, with x = π/4 and x = 5π/4]

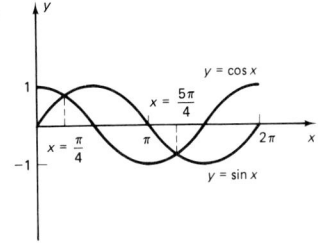

Review Exercises on Chapter 12

1. 32°34′30″. **3.** 20°, 60°, 100°; π/9, π/3, 5π/9. **5.** 105 cm. **7.** 1. **13.** $A = 60°, B = 30°$. **15.** 42.0 ft.
17. 95.6 ft. **19.** $\cos\theta = \sqrt{5}/3$, $\tan\theta = -2\sqrt{5}/5$, $\cot\theta = -\sqrt{5}/2$, $\sec\theta = 3\sqrt{5}/5$, $\csc\theta = -\frac{3}{2}$.
21. $\sin\theta = \sqrt{3}/2$, $\tan\theta = -\sqrt{3}$, $\cot\theta = -\sqrt{3}/3$, $\sec\theta = -2$, $\csc\theta = 2\sqrt{3}/3$. **23.** −0.5. **25.** 2.747.
27. 0.7071. **29.** −1. **31.** 0.3420. **33.** −1.732. **39.** 30°, 150°. **41.** 120°, 300°. **43.** 60°, 120°, 240°, 300°.
45. 45°, 135°, 225°, 315°. **49.** $-\cot\theta$.

51. [graph: Amplitude = 3, Period = π]

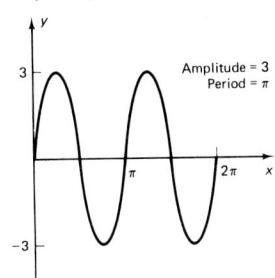

53. [graph: Amplitude = 2, Period = 2π]

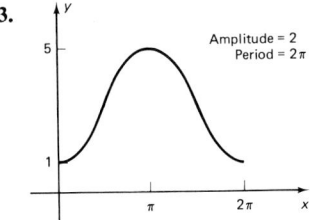

55. [graph: Amplitude = $\frac{1}{2}$, Period = π]

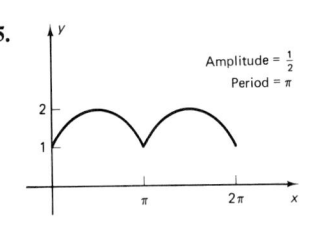

CHAPTER 13

Test 13.1: **1.** $\tan\theta$. **2.** $\sin\theta$. **3.** $\cos^2\theta$. **4.** $-\tan^2\theta$. **5.** $-\sin^2\alpha$. **6.** −1. **7.** 1. **8.** −1.
9. $\cos\theta$. **10.** 1 −. **11.** 1. **12.** −1. **13.** −1. **14.** $\sin\theta$. **15.** $\tan\theta$. **16.** $\tan\theta$. **17.** $\tan\theta$.
18. 1. **19.** 1. **20.** 1.

Exercises 13.1

63. $x^2 + y^2 = a^2$. **65.** $a^2/x^2 + y^2/a^2 = 1$. **67.** $(x/a)^{2/3} + (y/b)^{2/3} = 1$. **69.** $ab = 1$.
71. $a^2 + b^2 = p^2 + q^2$.

Test 13.2: **1.** $\sin A \cos B + \cos A \sin B$. **2.** $\cos A \cos B + \sin A \sin B$. **3.** $\sin \theta \cos \phi - \cos \theta \sin \phi$.
4. $\cos \theta \cos \phi - \sin \theta \sin \phi$. **5.** $(\tan A + \tan B)/(1 - \tan A \tan B)$. **6.** $(\tan A - \tan B)/(1 + \tan A \tan B)$.
7. $\sin (2A - A) = \sin A$. **8.** $\cos (3A + 2A) = \cos 5A$. **9.** $\sin (\pi/9 + \pi/18) = \sin \pi/6 = \frac{1}{2}$.
10. $\tan (\pi/15 + \pi/10) = \tan \pi/6 = 1/\sqrt{3}$.

Exercises 13.2

1. $(\sqrt{6} + \sqrt{2})/4$. **3.** $(\sqrt{6} - \sqrt{2})/4$. **5.** $-(2 + \sqrt{3})$. **7.** $\sqrt{3}/2$. **9.** $\frac{1}{2}$. **11.** $\frac{1}{2}$. **13.** $\cos \theta$.
15. $\sin 2B$. **17.** $\sqrt{3}$. **19.** $\sqrt{3}/3$.
33. (a) $(2\sqrt{7} + 3\sqrt{5})/12$; $(\sqrt{35} - 6)/12$; second. (b) $(2\sqrt{7} - 3\sqrt{5})/12$; $27\sqrt{7} - 32\sqrt{5}$; fourth.
35. $2 \sin (\theta + \pi/3)$; 2. **37.** $\sqrt{2} \sin (\theta + \pi/4)$; $\sqrt{2}$. **39.** $2 \cos (\theta - \pi/3)$; 2. **41.** $-\sqrt{2} \cos (\theta + \pi/4)$; $\sqrt{2}$.
51. $0, 2\pi/3$. **53.** $\pi/2, 5\pi/6$. **55.** $0, \pi/2$. **57.** $5\pi/6, 3\pi/2$. **59.** $0, \pi/3$. **65.** $\sin 3\alpha + \sin \alpha$.
67. $\sin 6\theta - \sin 4\theta$. **69.** $\cos 3\theta + \cos \theta$. **71.** $\cos (\theta/2) - \cos (3\theta/2)$. **73.** $\frac{1}{2}(\cos 2\alpha + \cos 2\beta)$.
79. $2 \sin 2\theta \cos \theta$. **81.** $2 \cos 3\theta \sin 2\theta$. **83.** $2 \cos (3\theta/2) \cos (\theta/2)$. **85.** $-2 \sin \theta \sin (\theta/2)$. **91.** 1. **93.** $-\csc \alpha$.

Test 13.3: **1.** $2 \sin A \cos A$. **2.** $\cos^2 \theta - \sin^2 \theta; 1 - 2\sin^2 \theta; 2\cos^2 \theta - 1$. **3.** $2 \tan \theta/(1 - \tan^2 \theta)$.
4. $\cos^2 A$. **5.** $(1 - \cos \alpha)/2$. **6.** $\sin \alpha$. **7.** $-\cos \theta$. **8.** $-\cos \alpha$. **9.** $-\cos 6\theta$. **10.** $2 \cot 2\theta$.
11. $\sqrt{2}/2$. **12.** $\frac{1}{4}$.

Exercises 13.3

1. $\frac{24}{25}; \frac{7}{25}; \frac{24}{7}$. **3.** $-\sqrt{3}/2, -\frac{1}{2}, \sqrt{3}$. **5.** $-4\sqrt{2}/9, \frac{7}{9}, -4\sqrt{2}/7$. **7.** $\sqrt{2}/2$. **9.** $-\sqrt{2}/2$. **11.** $-\sqrt{2}/2$.
13. 1. **15.** $\frac{1}{2} \sin \alpha$. **17.** $-\sqrt{3}/2$. **19.** $2\sqrt{2}$. **21.** $(\sqrt{6} - \sqrt{2})/4, (\sqrt{6} + \sqrt{2})/4, 2 - \sqrt{3}$.
23. $\frac{1}{2}\sqrt{2 - \sqrt{2}}, -\frac{1}{2}\sqrt{2 + \sqrt{2}}$. **25.** $2 + \sqrt{3}$.

51.

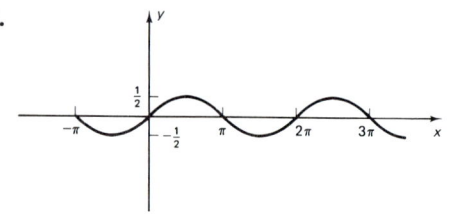

53. $\frac{1}{8}(3 + 4 \cos 2\theta + \cos 4\theta)$. **55.** $\frac{1}{8}(\sin 4\theta + 2 \sin 2\theta)$.

Test 13.4: **1.** $\pi/6$. **2.** $\pi/6$. **3.** $-\pi/2; \pi/2$. **4.** $0; \pi$. **5.** $x; -1 \leq x \leq 1$. **6.** $x; -\pi/2 \leq x \leq \pi/2$.
7. $x; -1 \leq x \leq 1$. **8.** $x; 0 \leq x \leq \pi$. **9.** $x; -\infty < x < \infty$. **10.** $x; -\pi/2 < x < \pi/2$.

Exercises 13.4

1. $\pi/3$. **3.** $\pi/3$. **5.** $\pi/6$. **7.** $\pi/2$. **9.** 0. **11.** $-\pi/6$. **13.** $2\pi/3$. **15.** $-\pi/4$. **17.** $-\pi/3$.
19. 0.2269. **21.** 0.4508. **23.** 2.6732. **25.** $\sqrt{3}/2$. **27.** 0. **29.** 0. **31.** $\frac{4}{5}$. **33.** $\frac{13}{12}$. **35.** $\frac{3}{5}$.
37. $-\pi/4$. **39.** $-\pi/4$. **41.** $\pi/6$. **43.** $\pi/6$. **45.** $\pi/3$. **47.** 0. **49.** 0. **51.** 0. **53.** $\pi/3$. **55.** $\pi/3$.
57. 1. **59.** $\sqrt{3}/3$. **61.** $x/\sqrt{1 - x^2}$. **63.** $1/\sqrt{1 + x^2}$. **65.** $\sqrt{1 - x^2}$. **67.** $1/x$. **69.** $1/x$. **71.** x.

Review Exercises on Chapter 13

15. $(x/a)^{2/3} - (y/b)^{2/3} = 1.$ **17.** $p^2(y^2 + q^2) = x^2y^2.$ **19.** $-\frac{16}{65}, -\frac{33}{65}.$ **21.** $\sqrt{2}/2.$ **23.** 1.
39. $(2\sqrt{3} - \sqrt{5})/6.$ **41.** $(75 - 13\sqrt{6})/106.$ **43.** $\sqrt{3} - 2.$
45.

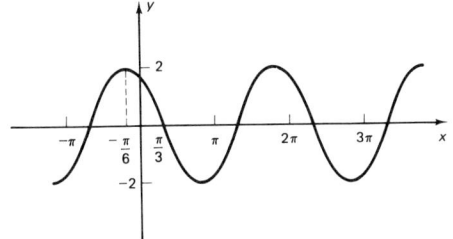

47. 180°, 270°. **49.** $30° + \alpha, 150° + \alpha$, where $\alpha = 53°08'.$ **51.** $\frac{1}{3}.$ **53.** $-5.$ **55.** $5\sqrt{2}.$ **57.** 1. **59.** $\pi/3.$

CHAPTER 14

Test 14.1: **1.** $\sin B.$ **2.** $c/b.$ **3.** ASA; SSA. **4.** $a/c; b/c.$

Exercises 14.1

1. $C = 75°, a = 2.45, c = 3.35.$ **3.** $C_1 = 70°32', A_1 = 64°28', a_1 = 3.83; C_2 = 109°28', A_2 = 25°32', a_2 = 1.83.$
5. $B = 70°44', C = 49°16', b = 8.72.$ **7.** $A = 40°, b = 2.92, c = 1.56.$ **9.** No solution.
11. $A = 16°47', C = 43°13', c = 2.37.$ **13.** $A = 44°49', C = 25°11', c = 1.81.$ **15.** No solution.
17. $B_1 = 72°56', C_1 = 72°04', c_2 = 4.98; B_2 = 107°04', C_2 = 37°56', c_2 = 3.22.$ **19.** 773 yd and 1093 yd.
21. 1.37 miles. **23.** 83.6 ft. **25.** 16°45' either E of N or W of N. **27.** 25°32' or 64°28'.
33. Area $= \frac{1}{2} ac \sin B = \frac{1}{2} ab \sin C.$

Test 14.2: **1.** $b^2 + c^2 - 2bc \cos A.$ **2.** $(a^2 + b^2 - c^2)/(2ab).$ **3.** Sines. **4.** Sines. **5.** Cosines.
6. Cosines. **7.** SSA. **8.** 0, 1 or 2. **9.** $c^2 = a^2 + b^2.$ **10.** $\frac{1}{2}ab \sin C.$ **11.** The three sides.
12. $\sqrt{s(s-a)(s-b)(s-c)}; (a + b + c)/2.$

Exercises 14.2

1. $A = 41°24', B = 55°46', C = 82°49'.$ **3.** $A = 104°29', B = 46°34', C = 28°57'.$
5. $a = 2.05, B = 46°56', C = 103°04'.$ **7.** $b = 10.85, A = 9°59', C = 60°01'.$ **9.** 9.92. **11.** 2.90. **13.** 3.
15. 9.40. **17.** 29°32', 38°03', 112°25'; 36.98 sq ft. **19.** 24°09'. **21.** 79.6 ft and 98.3 ft. **23.** 144°49'; 603.4 m.
25. 122.9 mph; 19°13' E of N.

Test 14.3: **1.** $r \cos \theta; r \sin \theta.$ **2.** $\sqrt{x^2 + y^2}; y/x.$ **3.** $(1, \sqrt{3}).$ **4.** $\sqrt{2}; -\pi/4.$ **5.** $2; 3\pi/2.$ **6.** $(-4, 0).$
7. $\cos \theta + i \sin \theta$; modulus; argument. **8.** $2(\cos \pi/2 + i \sin \pi/2).$ **9.** $3(\cos \pi + i \sin \pi).$
10. $\sqrt{2}[\cos (\pi/4) + i \sin (\pi/4)].$ **11.** $\sqrt{2}[\cos (-\pi/4) + i \sin (-\pi/4)].$ **12.** $s; -\alpha.$
13. $\cos (\alpha + \beta) + i \sin (\alpha + \beta).$ **14.** $rs; \alpha + \beta.$ **15.** $s/r; \beta - \alpha.$ **16.** $s^4; 4\beta.$ **17.** $r/s; -\alpha - \beta.$
18. $r^2; 2\alpha.$ **19.** $s^{-3}; -3\beta.$ **20.** $r^2s; 2\alpha + \beta.$

Exercises 14.3

1. $(0, 1).$ **3.** $(-\sqrt{3}, 1).$ **5.** $(-3, -3).$ **7.** $(-4, 0).$ **9.** $(2\sqrt{2}, \pi/4).$ **11.** $(5, -\pi/2).$ **13.** $(2, 5\pi/6).$
15. $(10, -2.214).$ **17.** $(10, \pi).$ **19.** $(\sqrt{17}/4, -1.326).$

21.

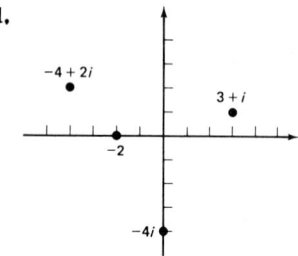

[Graph showing points −4 + 2i, 3 + i, −2, −4i on complex plane]

23. $5(\cos 0 + i \sin 0)$. **25.** $6(\cos 3\pi/2 + i \sin 3\pi/2)$. **27.** $\sqrt{2}(\cos 3\pi/4 + i \sin 3\pi/4)$.
29. $3\sqrt{2}(\cos 5\pi/4 + i \sin 5\pi/4)$. **31.** $8(\cos 7\pi/6 + i \sin 7\pi/6)$.
33. $6(\cos \pi + i \sin \pi)$. **35.** $5 \cos(5.36) + i \sin(5.36)$.
37. $(8/\sqrt{3})(\cos \pi/2 + i \sin \pi/2) = (8/\sqrt{3})i$; $(\sqrt{3}/2)(\cos 5\pi/6 + i \sin 5\pi/6) = -\frac{3}{4} + (\sqrt{3}/4)i$.
39. $24(\cos \pi + i \sin \pi) = -24$; $\frac{4}{3}(\cos 3\pi/2 + i \sin 3\pi/2) = -(\frac{4}{3})i$.
41. $10\sqrt{2}(\cos 5\pi/4 + i \sin 5\pi/4) = -10 - 10i$, $(5\sqrt{2}/2)(\cos \pi/4 + i \sin \pi/4) = \frac{5}{2} + \frac{5}{2}i$.
43. $16(\cos 2\pi/3 + i \sin 2\pi/3) = -8 + 8\sqrt{3} i$; $\cos 4\pi/3 + i \sin 4\pi/3 = -\frac{1}{2} - (\sqrt{3}/2)i$.

Exercises 14.4

1. i. **3.** $-4 - 4i$. **5.** -324. **7.** -64. **9.** $-512 - 512\sqrt{3} i$. **11.** $-\frac{1}{4}$. **13.** $-\sqrt{3}/1728 + (1/576)i$.
15. $\pm[1/\sqrt{2} + (1/\sqrt{2})i]$. **17.** $\pm(1/\sqrt{2})(1 + \sqrt{3} i)$. **19.** $-i, \pm\sqrt{3}/2 + (\frac{1}{2})i$.
21. $\pm(\sqrt{3} + i), \pm(-1 + \sqrt{3} i)$. **23.** $\pm i, \pm(\sqrt{3}/2 + (\frac{1}{2})i), \pm(\sqrt{3}/2 - (\frac{1}{2})i)$.

Review Exercises on Chapter 14

1. $b = 29.2$, $c = 44.8$, $C = 130°$. **3.** $A = 33°10'$, $C = 96°50'$, $c = 9.07$.
5. $C_1 = 34°18'$, $A_1 = 120°42'$, $a_1 = 6.10$; $C_2 = 145°42'$, $A_2 = 9°18'$, $a_2 = 1.15$.
7. $A = 24°09'$, $B = 30°45'$, $C = 125°06'$. **9.** $a = 4.58$, $B = 109°06'$, $C = 10°53'$. **11.** 2.05. **13.** 2.17.
15. 223.98 sq. units. **17.** $217°42'$ E of N. **19.** $26°23', 36°20', 117°16'$; 19,200 sq ft. **21.** 0.82 mile and 0.17 mile.
23. $a \simeq \frac{36}{7}, b \simeq \frac{48}{7}, C = 175°20'$. **25.** 54.8 miles (stat.); 33.6'. **27.** $(-2\sqrt{3}, -2)$. **29.** $(6\sqrt{2}, 3\pi/4)$.
31. $(\sqrt{17}, -1.33)$. **33.** $2[\cos (11\pi/6) + i \sin (11\pi/6)]$. **35.** $\sqrt{29}[\cos (2.76) + i \sin (2.76)]$. **37.** $8\sqrt{3} + 8i$; $-(\frac{1}{4})i$.
39. $144(3 + \sqrt{3} i)$. **41.** $i, \pm\sqrt{3}/2 - (\frac{1}{2})i$.

INDEX

A

Abscissa, 154
Absolute value:
    definition, 8
    equations, 146
    function, 222
    inequalities, 147
Addition and subtraction:
    of complex numbers, 349
    of expressions, 46
    of matrices, 434
Addition formulas, 522
Addition principle of equations, 81
Additive inverse, 14
Adjacent side, 476
Algebraic expressions, 45
Ambiguous case, 555
Amplitude, 504
Angle:
    acute, 470
    complementary, 473
    coterminal, 473

Angle (*cont.*):
    obtuse, 470
    of depression, 489
    of evaluation, 488
    polar, 568
    reference, 497, 499
    right, 470
    round, 470
    straight, 470
    supplementary, 473
    vertex of an, 468
Angular coordinate, 568
Antilogarithm, 314
Area of triangle, 560, 565
    Hero's formula for, 566
Argand diagram, 571
Argument:
    of a complex number, 572
    of a function, 206
Arithmetic progression (A.P.), 378
    common difference of, 379
    nth term of, 379
    sum of n terms of, 381

ASA, 552, 565
Associative property, 11-12
Asymptotes, 221, 239, 275, 509
Augmented matrix, 420
Axes:
   coordinate, 153
   imaginary, 571
   real, 571
   x and y, 153
Axioms for complex numbers, 351

B

Base-change formula, 330-332
   of a logarithm, 297
   of an exponential function, 289
Binomial, 45
Binomial coefficients, 411
Binomial expansion, 411
Binomial expression, 409
Binomial-square formula, 54
Binomial theorem, 411
Brightness of stars, 327

C

Cancellation of common factors, 19
Cartesian coordinate system, 154
Cartesian coordinates, 154
Cartesian plane, 153
Characteristic of a logarithm, 313
Circle, 251
   unit, 492
Closed interval, 129
Coefficient, 45
   leading, 361
   matrix, 444
Cofactor, 457
Combinations, 403
Common difference of an A.P., 379
Common factor, 19
Common logarithm, 305, 311
Common ratio of a G.P., 386

Commutative property, 11
Complete factorization, 359
Completing the square, 107
Complex number(s), 349
   argument of, 572
   conjugate of, 353, 572
   imaginary part of, 349
   modulus of, 572
   nth roots of, 578
   polar form of, 572
   real part of, 349
Complex plane, 568, 571
Composition of functions, 228
Compound interest, 291, 323
Conic sections, 250-286
Constant function, 218
Constant of proportionality, 213
Constant term, 45
Coordinates, 154
Coordinate axes, 153
Cosine function, 476, 492
Cosines, law of, 560
Cramer's rule, 460-62
Cube root, 33
Cycle, 504

D

Decay:
   constant, 321
   radioactive, 321
Decibel, 325
Decimal, 3
Degree:
   measure of angles, 469
   of a polynomial, 338
DeMoivre's theorem, 576
Denominator, 18
   least common (LCD), 23
   rationalizing the, 36, 56
Dependent, system, 195
Dependent variable, 206
Descarte's rule of signs, 370

Determinants, 456
Diagonal elements, 433
Direct variation, 213
Directrix, 257
Discriminant, 110
Distance formula, 155
Distributive properties, 13
Division:
    by zero, 3
    of complex numbers, 354
    of fractions, 18, 74
    synthetic, 341
Divisor, 339
Domain of a function, 206

E

e, 280-281
Earthquake, magnitude of, 326
Ellipse, 269
    axes of, 272
    center of, 272
    foci of, 269
    major axis of, 272
    minor axis of, 272
    standard equation of, 270
Empty set, 128
Equations, 80-125, 190-200
    equivalent, 82
    fractional, 116
    general linear, 177
    graph of, 163
    irrational, 119
    linear, 84
    quadratic, 99-116
    root of, 87
    solution of, 87
    system of, 190-200
    trigonometric, 500
Exponents, 27-42
    fractional, 38
    laws of, 40
Exponential functions, 287-295
    decaying, 280

Exponential functions (*cont.*):
    growing, 280
    natural, 280

F

Factor theorem, 358
Factorial, 401
Factoring, 59-69
    by formulas, 60, 62
    by grouping, 61
    quadratic expressions, 65-69
    summary of, 63
Factorization theory, 356-363
Factors, 59
Finite sequence, 378
Finite set, 128
Focus (foci), 257, 269
    of a hyperbola, 273
    of a parabola, 257
    of an ellipse, 269
Formula:
    addition, 522
    base change, 330
    distance, 155
    double-angle, 533
    for allied angles, 526
    half-angle, 535
    Hero's, 566
    mid-point, 158
    multiple angle, 532
    point-slope, 174
    quadratic, 106
    slope-intercept, 175
Function(s), 205-39
    absolute-value, 222
    algebraic, 224
    amplitude of, 504
    composition of two, 228
    constant, 218
    cosine, 476, 492
    cubic, 224
    domain of, 206

Function(s) (*cont.*):
   equality of, 248
   even, 248
   exponential, 287-295
   graphs of, 208
   inverse, 232-238
   inverse trigonometric, 540
   linear, 218
   logarithmic, 299
   odd, 248
   one-to-one, 233
   periodic, 504
   polynomial, 223
   power, 219
   quadratic, 224
   range of, 206
   rational, 224
   sine, 476, 492
   tangent, 477
   transcendental, 224
   trigonometric, 492
   wrapping, 492
Fundamental theorem of algebra, 360

G

General equation of a circle, 252
General equation of conics, 280-282
General term, 379, 411
Geometric progression (G.P.), 386-393
   common ratio of, 386
   nth term of, 386
   sum of an infinite, 390
   sum to n terms, 388
Graphing by slope, 176
Graphs:
   of equations, 163
   of functions, 208

H

Half-life, 322
Hero's formula, 566
Horizontal line, 177
Horizontal asymptotes, 239

Hyperbola, 273
   asymptotes of, 275
   foci of, 273
   rectangular, 222, 278
   standard equation of, 273

I

i, 348-355
Identity:
   elements, 14
   matrix, 433
   trigonometric, 514-20
Imaginary numbers, 347
Imaginary part, 349
Imaginary unit, 348
Inconsistent system, 195
Independent system, 195
Independent variable, 206
Induction, mathematical, 393-397
Inductive step, 394
Inequalities, 132
   absolute-value, 146
   linear, 132
   quadratic, 139
   solution of, 133
Infinite G.P., 390
Infinite sequence, 378
Infinity (∞), 240
Integers, 2
Intensity of earthquake, 326
Intercept, 175
Interest:
   compound, 291, 329
   nominal rate of, 293
Interpolation, linear, 271
Intervals, 129-130
   bounded, 130
   closed, 129
   end-points of, 130
   open, 129
   semiclosed, 130
   unbounded, 130
Inverse:
   additive, 14
   functions, 232

Inverse (*cont.*):
    matrix, 448-455
    multiplicative, 14
    proportion, 214
    trigonometric functions, 540
Irrational equations, 119
Irrational numbers, 3

L

Law(s):
    of cosines, 553, 560
    of exponents, 40
    of sines, 552
Leading coefficient, 361
Least common denominator (LCD), 23
Less than, 6
Like terms, 47
Linear equations, 84
Linear functions, 218
Linear inequalities, 132
Linear interpolation, 371, 483
Linear system of equations, 190
Line(s):
    general equation of, 179
    horizontal, 177
    intercept form of, 181
    parallel, 181
    perpendicular, 183
    point-slope formula of, 174
    slope of, 172
    slope-intercept formula of, 175
    straight, 171-189
    two-point form of, 181
    vertical, 177
Logarithmic form, 298
Logarithm(s), 297-333
    base change formula of, 330
    characteristic of, 313
    common, 305, 311
    elementary properties of, 301
    mantissa of, 313
    natural, 308
    properties of, 304
Long division, 339
Loudness, 325

M

Major axis, 272
Mantissa, 313
Mathematical induction, 393-397
Matrix (matrices), 431
    addition of, 434
    augmented, 420
    coefficient, 444
    column, 431
    elements of, 431
    equality of, 433
    identity, 433
    inverse of, 449
    invertible, 449
    nonsingular, 449
    products of, 438
    reduced, 422
    row, 431
    scalar multiplication of, 433
    singular, 449
    square, 433
    unit, 433
    zero, 432
Method of:
    addition, 191
    arcs, 53
    row reduction, 422
    substitution, 195
Midpoint formula, 158
Minor axis, 272
Minor of a determinant, 457
Modulus, 572
Monomial, 45
Multiplication principle, 82
Multiplicative inverse, 14

N

Napierian logarithms, 308
Natural exponential functions, 280
Natural logarithms, 308
Natural numbers, 2
Null set, 128
Number(s):
    complex, 349

Number(s) (*cont.*):
  line, 3
  imaginary, 347
  irrational, 3
  natural, 2
  prime, 5
  properties of real, 10
  rational, 2
  real, 3
Numerator, 18, 71, 339

O

Oblique triangle, 552
Odd function, 248
One-to-one function, 233
Open interval, 129
Opposite side, 476
Ordinate, 154
Origin, 153

P

Parabola, 165, 257-267
  axis of, 257
  directrix of, 257
  focus of, 257
  horizontal, 266
  vertex of, 257
Parallel lines, 181
Pascal's triangle, 410
Periodic function, 504
Period of a function, 504
Permutations, 398
Perpendicular lines, 183
Point-slope formula, 174
Polar angle, 568
Polar coordinates, 568
Polar form of complex numbers, 572
Polynomial(s), 223, 338
  complete factorization of, 359
  degree of, 338
  function, 223
  with integer coefficients, 367
  with real coefficients, 364

Polynomial(s) (*cont.*):
  zeros of, 361
Power, 27
Prime numbers, 5
Principal value, 541, 543-544
Proportionality constant, 213

Q

Quadrants, 155
Quadratic equations, 99-116
Quadratic formula, 106
Quadratic function, 224
Quadratic inequalities, 139
Quotient, 339

R

Radial coordinate, 568
Radian measure, 470
Radicals, 33, 34
Range of a function, 206
Rational functions, 224
Rational numbers, 2
Rationalizing the denominator, 36, 56
Real-number line, 4
Real numbers, 3
Real part of a complex number, 349
Reciprocal, 14
Reference angle, 497, 499
Remainder, 339
Remainder theorem, 357
Repeating decimals, 3, 390
Richter scale, 326
Right triangle, 476
Rise, 172
Run, 172

S

SAS, 552, 565
Scalar multiple, 433
Scientific notation, 312

Sequence, 378
Side:
    initial, 468
    terminal, 468
Sine(s) function, 492
    law of, 552
Slope-intercept formula, 175
Slope of a line, 171
Solution:
    of a system, 190
    of a triangle, 487, 553
    of an equation, 81
    of an inequality, 133
Square matrix, 433
Square root, 33
SSA, 553, 565
SSS, 552, 565
Standard position of an angle, 468
Stellar magnitude, 327
Substitution method, 195
Symmetry, 165-168
    about origin, 166
    about x axis, 166
    about y axis, 165
Synthetic division, 341
System of linear equations, 190-200

T

Table(s):
    of common logarithms, 584-585
    of exponential functions, 588-589
    of natural logarithms, 586-587
    of trigonometric functions, 590-594
Tangent function, 477
Terminal side of an angle, 468
Term(s):
    of a sequence, 378
    of an expression, 45
Triangle:
    area of, 560, 565
    oblique, 552
    Pascal's, 410
    right, 476

Triangle (*cont.*):
    solution of, 487, 553
Trigonometric function, 492
Trigonometric identities, 514-520
Trinomials, 45

U

Unit:
    circle, 492
    imaginary, 348

V

Variable, 46
    dependent, 206
    independent, 206
Vertex:
    of a parabola, 257
    of an angle, 468
Vertical line test, 209

W

Wrapping function, 492

X

x axis, 153
x coordinate, 154

Y

y axis, 153
y coordinate, 154
y intercept, 175

Z

Zero of a polynomial, 361

Special Products

1. $(a + b)^2 = a^2 + 2ab + b^2$
2. $(a - b)^2 = a^2 - 2ab + b^2$
3. $(a + b)^3 = a^3 + 3a^2 b + 3ab^2 + b^3$
4. $(a - b)^3 = a^3 - 3a^2 b + 3ab^2 - b^3$

Factoring

1. $a^2 - b^2 = (a + b)(a - b)$
2. $a^2 \pm 2ab + b^2 = (a \pm b)^2$
3. $a^3 + b^3 = (a + b)(a^2 - ab + b^2)$
4. $a^3 - b^3 = (a - b)(a^2 + ab + b^2)$

Laws of Exponents

1. $a^m a^n = a^{m+n}$
2. $(a^m)^n = a^{mn}$
3. $\dfrac{a^m}{a^n} = a^{m-n}$
4. $(ab)^n = a^n b^n$
5. $\left(\dfrac{a}{b}\right)^n = \dfrac{a^n}{b^n}$
6. $a^{-x} = \dfrac{1}{a^x}$
7. $a^0 = 1$

Radical Properties

1. $(\sqrt[n]{a})^n = a$
2. $\sqrt[n]{a^n} = a$, if $a \geq 0$
3. $\sqrt[n]{ab} = \sqrt[n]{a}\sqrt[n]{b}$
4. $\sqrt[n]{\dfrac{a}{b}} = \dfrac{\sqrt[n]{a}}{\sqrt[n]{b}}$

Quadratic Formula

The solutions of $ax^2 + bx + c = 0$ are given by $x = \dfrac{-b \pm \sqrt{b^2 - 4ac}}{2a}$

Properties of Inequalities

1. If $a < b$, then $a + c < b + c$.
2. If $a < b$ and $c > 0$, then $ac < bc$.
3. If $a < b$ and $c < 0$, then $ac > bc$.

Inequalities Involving Absolute Values

1. $|u| < a$, for $a > 0$ if and only if $-a < u < a$.
2. $|u| > a$, for $a > 0$ if nd only if $a < -a$ or $u > a$.

Lines

1. Slope: $m = \dfrac{y_2 - y_1}{x_2 - x_1}$
2. Point-slope equation: $y - y_1 = m(x - x_1)$
3. Slope-intercept equation: $y = mx + b$

Laws of Logarithms

1. $\log_a (xy) = \log_a x + \log_a y$
2. $\log_a \dfrac{1}{x} = -\log_a x$
3. $\log_a \dfrac{x}{y} = \log_a x - \log_a y$
4. $\log_a (x^c) = c \log_a x$
5. $\log_b x = \dfrac{\log_a x}{\log_a b}$
6. $\log_a 1 = 0$
7. $\log_a a = 1$
8. $a^{\log_a x} = x$
9. $\log_a (a^x) = x$

Permutations and Combinations

1. $n! = n(n - 1)(n - 2) \cdots 3 \cdot 2 \cdot 1$
2. $_nP_k = \dfrac{n!}{(n - k)!}$
3. $\dbinom{n}{k} = {_nC_k} = \dfrac{n!}{k!(n - k)!}$

Binomial Theorem

$(x + y)^n = {_nC_0} x^n y^0 + {_nC_1} x^{n-1} y^1 + \cdots + {_nC_{n-1}} x^1 y^{n-1} + {_nC_n} x^0 y^n$

Sequences

A.P. $T_n = a + (n - 1)d$

$S_n = \dfrac{n}{2}[a + l] = \dfrac{n}{2}[2a + (n - 1)d]$

G.P. $T_n = ar^{n-1}$

$S_n = \dfrac{a(1 - r^n)}{1 - r}$, $S_\infty = \dfrac{a}{1 - r}$ ($|r| < 1$)